Areas of High Fern and Lycophyte Diversity

This map shows 3 major areas of native pteridophyte diversity in Texas. Many species in East Texas (in various shades of green) are from wet habitats in the Pineywoods (area in darkest green). In the Trans-Pecos most are found in the "sky islands," high elevation areas such as the Chisos, Franklin, Guadalupe, Davis, or Rosillos mountains (see page 43). In the Edwards Plateau, ferns and lycophytes are concentrated in special microhabitats such as moist canyons, cave entrances, and rock outcrops. A few species from marginal counties are considered to be in two areas. A number of species are widespread, occurring in two or even all three of the areas.

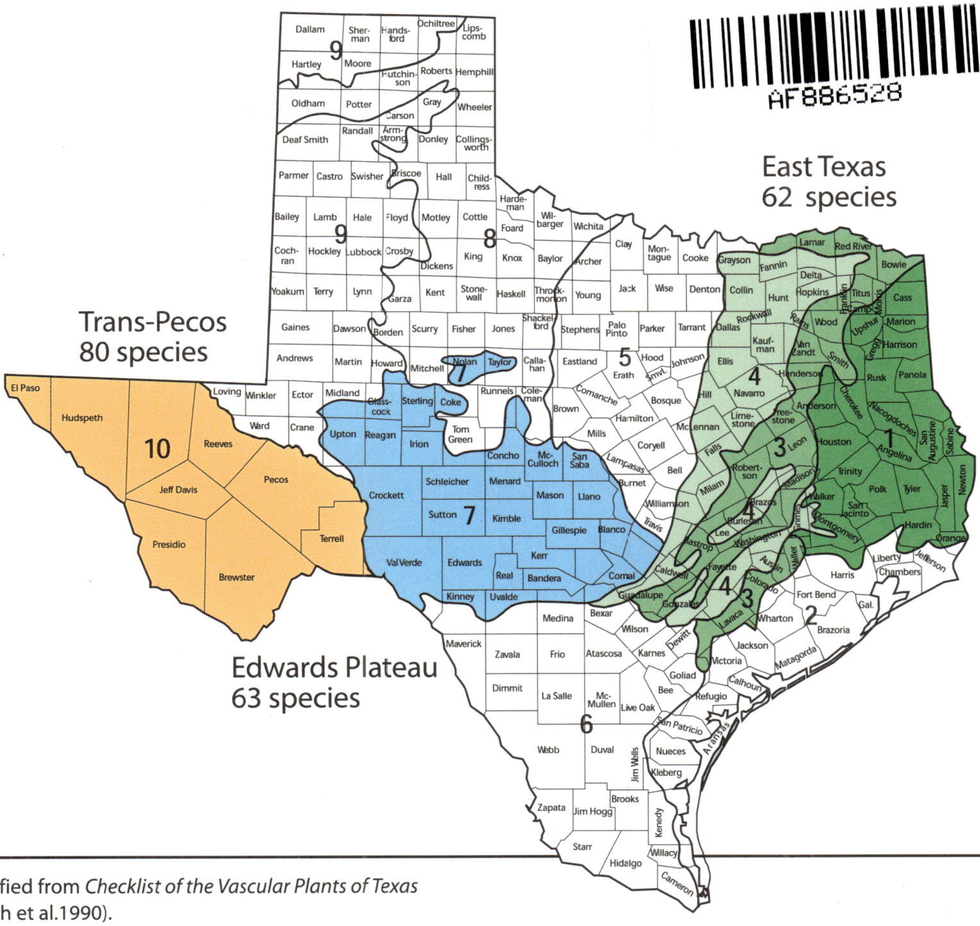

Modified from *Checklist of the Vascular Plants of Texas* (Hatch et al.1990).
Nearly identical maps have been used in numerous works on Texas including Gould (1962) and Correll and Johnston (1970).

1 Pineywoods
2 Gulf Prairies and Marshes
3 Post Oak Savannah
4 Blackland Prairies
5 Cross Timbers and Prairies
6 South Texas Plains
7 Edwards Plateau
8 Rolling Plains
9 High Plains
10 Trans-Pecos, Mountains and Basins

THE FERNS AND LYCOPHYTES OF TEXAS

FRONTISPIECE

Adiantum Capillus-Veneris L.: fig. 1, frond and rhizome, ×⅔; fig. 2, fertile pinnule, ×3. *Adiantum tricholepis* Fée: fig. 3, pinna, ×⅔; fig. 4, fertile segment, ×3; fig. 5, erect-ascending rhizome, ×⅔.

CORRELL, D.S. 1956. Ferns and fern allies of Texas. Texas Research Foundation, Renner.

THE FERNS AND LYCOPHYTES OF TEXAS

George M. Diggs, Jr. &
Barney L. Lipscomb

Project Coordinator
Robert George

Center for Environmental Studies and
Department of Biology, Austin College
Sherman, Texas

AUSTIN COLLEGE

Botanical Research Institute of Texas (BRIT)
Fort Worth, Texas

BRIT

The Ferns and Lycophytes of Texas
Botanical Research Institute of Texas Press, Fort Worth 76107

First edition 2014
Printed in Korea

ISBN: 13-978-1889878-37-9

Production: Rebecca Horn, rhorngraphics, rlrchorn@verizon.net
Cover Design: Rebecca Horn, rhorngraphics
Cover: Fern habitat in the Pineywoods of East Texas. Original watercolor painting by Pam Chamberlain (BRIT).
Front cover: Cinnamon fern (*Osmundastrum cinnamomeum*), Bald-cypress (*Taxodium distichum*), Southern leopard frog (*Rana sphenocephala*), Texas leaf-cutting ant (*Atta texana*).
Back cover: Cinnamon fern (*Osmundastrum cinnamomeum*), Bald-cypress (*Taxodium distichum*), Eastern Phoebe (*Sayornis phoebe*).

The Ferns and Lycophytes of Texas
By George M. Diggs, Jr. and Barney L. Lipscomb

Distribution of copies by:
Botanical Research Institute of Texas Press
1700 University Dr.
Fort Worth, Texas 76107-3400, USA
Telephone: 1-817-332-4441
Fax: 1-817-332-4112
Website: http://www.brit.org/brit-press/
Email: jbrit@brit.org

BRIT
PRESS
2014
Botanical Miscellany, No. 38
ISSN: 0833-1475
28 Feb 2014

DEDICATION

This book is dedicated to Donovan Stuart Correll, 1908–1983, prolific Texas botanist and author of numerous articles and books.[1] Correll was educated at Duke University (A.B., A.M.) and Harvard University (Ph.D.). He served as a gunnery officer in the U.S. Navy from 1944 to 1946. Subsequently, until 1956, he worked at the United States Department of Agriculture on a variety of research projects.

He had numerous connections with Texas. In 1946 he began to collect plants under the auspices of the Southern Methodist University Herbarium (SMU) in Dallas. In 1956 he moved to Texas and became Head of the Research Department at the Texas Research Foundation, Renner, near Dallas. His groundbreaking book, *Manual of the Vascular Plants of Texas* (Correll & Johnston 1970), is still the only statewide flora available. From 1971 to 1973, he was Director of the SMU Herbarium and also served as Program Director for Systematic Biology at the National Science Foundation.

In 1973 he joined the staff of the Fairchild Tropical Garden, Miami, Florida, where he worked until his death in 1983. He was a valued personal friend of one of the authors (BLL) of this book.

Bernice Schubert (1984), who wrote an obituary of Correll, described him personally as:

> "...a talented, versatile man of charm, good humor, enormous energy, and great good will. He was a generous colleague and a good friend to many of us. He had a wide range of interests, had traveled extensively, and maintained his enthusiasm and joyful spirit throughout."

His personality can also be seen in "The Little Runt: A Botanical Christmas Story" (Correll 1978), originally written as a letter to his children on the occasion of a visit to a Christmas tree lot.

Correll's work has inspired generations of Texas botanists.

[1] e.g., Ferns and Fern Allies of Louisiana (1942, with C.A. Brown); Native Orchids of North America, North of Mexico (1950); Orchids of Guatemala (1952–1953, with O. Ames); Ferns and Fern Allies of Texas (1956); Ferns and Fern Allies of Chihuahua, Mexico (1962, with I.W. Knobloch); Manual of the Vascular Plants of Texas (1970, with M.C. Johnston); Aquatic and Wetland Plants of Southwestern United States (1972, with his wife, Helen B. Correll); and Flora of the Bahama Archipelago, Including the Turks and Caicos (1982, with H.B. Correll).

AUTHORS' NOTE

In a work such as this, it is inevitable that omissions and errors, both large and small, escape attention. Because of the possibility of future editions, we would appreciate corrections, suggestions, or additions from individuals using the book. Such information can be sent to:

George M. Diggs, Jr.
gdiggs@austincollege.edu

Barney L. Lipscomb
barney@brit.org

Also, we hope that this book will spur additional interest in Texas pteridophytes. Plant specimens, particularly county, regional, or state records would be much appreciated and can be deposited at a number of herbaria in the state including:

Botanical Research Institute of Texas Herbarium (BRIT)
1700 University Drive, Fort Worth, Texas 76107-3400, U.S.A.
Phone: 817-332-4441

Such specimens will be important scientific contributions, will have permanent protection, and will be important resources for the future (Prather et al. 2004a, 2004b[2]).

Online material on Texas ferns can be found at: fernsoftexas.org

[2] Prather, L.A., O. Alvarez-Fuentes, M.H. Mayfield, and C.J. Ferguson. 2004a. The decline of plant collecting in the United States: A threat to the infrastructure of biodiversity studies. Syst. Bot. 29:15–28.

Prather, L.A., O. Alvarez-Fuentes, M.H. Mayfield, and C.J. Ferguson. 2004b. Implications of the decline in plant collecting for systematic and floristic research. Syst. Bot. 29:216–220.

THE FERNS AND LYCOPHYTES OF TEXAS

TABLE OF CONTENTS — PAGE

PREFACE VIII
ACKNOWLEDGMENTS IX
INTRODUCTION 1
"RECORD-HOLDERS" AND "PRIZE-WINNERS" 1
THE PLANTS INCLUDED 1
RECENT DISCOVERIES 3
FERN CLASSIFICATION 5
DIVERSITY—WHY SO MANY FERNS AND LYCOPHYTES IN TEXAS? 9
THE LIFE OF FERNS—STRUCTURE AND REPRODUCTION 16
STRUCTURE 16
THE SEXUAL LIFE CYCLE 16
MODIFICATIONS OF THE FERN LIFE CYCLE 21
HETEROSPORY 22
ASEXUAL REPRODUCTION 22
POLYPLOIDY 23
ADAPTATIONS TO DRYNESS AND OTHER SPECIAL CONDITIONS 24
AVOIDING DROUGHT 25
DESICCATION TOLERANCE 25
PREVENTING WATER LOSS 25
PHOTOPROTECTION—FROM EXCESSIVE LIGHT 28
RAPID REHYDRATION 29
REPRODUCTIVE ADAPTATIONS 29
AQUATIC ADAPTATIONS 29
PROTECTIVE ADAPTATIONS 30
WHERE TEXAS FERNS AND LYCOPHYTES OCCUR—BIOGEOGRAPHY 30
LONG-DISTANCE DISPERSAL 30
VICARIANCE 32
THE ASIAN CONNECTION 32
CONNECTIONS ACROSS TEXAS 36
THE MEXICAN CONNECTION 37
OTHER DISTRIBUTIONAL PHENOMENA 38
ECONOMIC USES AND ETHNOBOTANY 41
CONSERVATION CONCERNS 43
THE FUTURE 44
KEYS TO FAMILIES OF FERNS AND LYCOPHYTES 47
TAXONOMIC TREATMENTS 53
LYCOPODIOPHYTA (CLUB-MOSSES) 54
POLYPODIOPHYTA (FERNS) 95
APPENDICES 315
APPENDIX 1 (ILLUSTRATION SOURCES) 315
APPENDIX 2 (A FEW PUBLICLY ACCESSIBLE LOCATIONS TO SEE FERNS AND LYCOPHYTES IN TEXAS) 317
APPENDIX 3 (SOURCES OF ADDITIONAL INFORMATION ON FERNS AND LYCOPHYTES) 320
GLOSSARY 322
LITERATURE CITED 335
INDEX 361
COLOPHON 375
AUTHORS AND INSTITUTIONS 376

PREFACE

For some people all ferns and similar plants seem much the same and are rarely noticed. They lack the striking cones of conifers or the brightly colored flowers and obvious fruits of flowering plants. They are, however, among the most interesting of all Texas plants. Some of them are extremely rare, occurring in only one Texas county and nowhere else in the entire U.S.A. Others thrive in the most extreme conditions—in crevices on huge exposed granite outcrops, underwater on the bottoms of ponds or lakes, or on the vertical walls of the well-like, 140-foot deep Devil's Sinkhole, a collapsed limestone cavern. Still others are not rooted in anything solid and live as wanderers, endlessly floating on the surfaces of lakes or ponds. In contrast some cling to the exposed branches of trees high above the ground and survive only on the moisture they catch from passing rain or fog. A few are even tropical species that have almost miraculously made their way to Texas from much farther south—their rare, lonesome Texas populations occurring many hundreds of miles from others of their kind.

Ferns and similar plants also vary tremendously in size and form. One Texas species has leaves reaching nearly 4 m (13 ft) in length. Another is a tiny floating aquatic plant often less than a cm (less than ½ in) in total size—a person can hold hundreds of such plants in a cupped hand. Another even lacks both leaves and roots, its naked stems fulfilling all its needs. Some have leaves that look very "fern-like" while others more closely resemble grasses or even mosses. They do share some things in common—they all have vascular or conducting tissue and they do not produce seeds, instead reproducing by microscopic spores. Also, in a sense they are all living fossils—members of lineages that have survived for hundreds of millions of years, since long before the dinosaurs.

The sheer number of ferns and related species (known collectively as "pteridophytes") in Texas is also fascinating. The state has a surprising 138 species, including those native to Texas and some that have naturalized. However, more interesting than the total number of species is the number of natives—127 species are indigenous to Texas, the most of any state in the continental U.S.A. The large number found in Texas is particularly surprising given that most people associate ferns and related plants with humid, even tropical conditions, just the opposite of much of Texas. However, many species adapted to the dry western part of the state have amazing adaptations that allow survival even under desperately dry conditions. For example, some have very small leaves with dense coverings of hairs, strangely shaped scales, or waterproof or light-reflective, wax-like materials. Some called resurrection plants even have the ability to essentially dry out, curl up, and go into a state of suspended animation until moisture is available.

Texas ferns and similar plants definitely offer many surprises—even for those of us who have long studied them. Our goal in this book is to make these fascinating plants of Texas more understandable and appreciated by the general public as well as botanists.

ACKNOWLEDGMENTS

Our deepest thanks to Susan Tracy, whose love and enthusiasm for ferns was the stimulus for us to write this book, and for her encouragement and support throughout the project.

Several individuals were indispensable members of the team that produced this work. Special thanks go to Project Coordinator Robert George for locating and scanning line drawings, obtaining color photographs, producing diagrams, laying out the illustration pages, coordinating copyright issues, managing the project databases, and doing innumerable other things—he is the glue that held everything together; to James Peck for providing invaluable advice on ferns and lycophytes; to Billie Turner for allowing us to use and modify county distribution maps that he produced; to Jim Blassingame for donating his extensive collection of Texas ferns and lycophytes to BRIT and providing advice and insight on Texas ferns; to Robert J. O'Kennon for allowing access to his plant collections (at BRIT) and providing advice and assistance in many ways; and to Rebecca Horn (rhorngraphics) for her professional design and layout skills.

We could not have done the project without help from a number of individuals with expertise in pteridology. These include James Blassingame, Sandy Hancock, James Peck, Jack Stanford, Carl Rothfels, and George Yatskievych.

A number of other colleagues provided help in a variety of ways including Walter Holmes, Jeffrey Mink, Jackie Poole, Tiana Rehman, Jill Reid, Jason Singhurst, and Susan Yarborough.

We are also indebted to a number of individuals who proofed extensive portions of the manuscript: Darrell Brandon, Kerry Brock, Brooke Byerley, Robert George, Gus Hall, Sonnia Hill, Marian McCarley, Melissa Skees, and Sue Tracy.

Curators and other individuals at various Texas herbaria are owed a debt of gratitude for their hospitality. These include Llewellyn Densmore, Arthur Elliott, Walter Holmes, Stephan Hatch, Richard Kazmaier, Michael Powell, Monique Reed, Jack Stanford, Beryl Simpson, Tom Wendt, and Lindsay Woodruff.

Recognition goes also to the numerous individuals who have collected pteridophytes specimens in Texas for more than one hundred and fifty years. Early collections are housed in herbaria such as ASTC, BRIT (including SMU and VDB), HBC, SBSC, SRSC, TAES, TAMU, and TEX (including LL and RUNYON), and represent irreplaceable windows to a time before the vegetation of the state was radically altered. While not all collectors can be recognized here, specimens collected by the following have been of particular importance in allowing the completion of this work: Jim Blassingame, Edwin Bridges, Larry Brown, William Carr, Donovan Correll, Victor Cory, Stephan Hatch, L.C. Hinckley, Walter Holmes, Marshall Johnston, Stanley Jones, William McCart, Barbara McRoberts, Michael McRoberts, Rogers McVaugh, Elray Nixon, Robert O'Kennon, Steve Orzell, Ernest J. Palmer, Harris B. Parks, Michael Powell, Jeff Quayle, Julien Reverchon, Chester Rowell, Lloyd Shinners, Jason Singhurst, G.M. Soxman, Omar Sperry, Jack Stanford, Thomas Starbuck, Julian Steyermark, Connie Taylor, John Taylor, Benjamin Tharp, R. Dale Thomas, Billie Turner, Barton Warnock, Geraldine Watson, Eula Whitehouse, Charles Wright, and Mary Sophie Young.

We would like to thank the following individuals for the use of their color photographs: Steve Baskhauf; Brooke Byerley; Keith Bradley; Tom Chester; Jim Conrad; J.R. Crellin/floralimages.co.uk; Alan Cressler; George Diggs; John Eichler/Viridans Images; Robert George; Petei Guth; Kwan Han@www.natureloveyou.sg; Tanya Harvey; Sonnia Hill; John Hilty, Illinois Wildflowers: www.illinoiswildflowers.info; Bruce Horst; Chad Husby; Ann Larson; Daniel Layotn; Max Licher; Joseph Marcus, Courtesy LBJ Wildflower Center; Jeff McMillian; Jordan Metzgar; Mississippi Museum of Natural Science (MS-DWFP); Missouri Botanical Garden; Lorena Babcock Moore; Robin Moran; Roy Morey; Robert J. O'Kennon; Jim Peck; P.B. Pelser; Tiana Rehman; Carl Rothfels; Eleanor Saulys; Al Schneider, www.swcoloradowildflowers.com; Stan Shebs; Jason Singhurst; Jack Stanford; Forest & Kim Starr; Richie Steffen, Elizabeth C. Miller Botanical Garden; Tom Stuart; Rebecca K. Swadek; Dan Tenaglia-missouriplants.com; Sue Tracy; Carlos Velasco; Sally and Andy Wasowski, Courtesy LBJ Wildflower Center; W. Cole Weatherby; Carol W. Witham; Sharon Yarborough; George Yatskievych; and Dale A. Zimmerman Herbarium, Western N. Mex. Univ.

Since this book is an outgrowth of work on *Shinners & Mahler's Illustrated Flora of North Central Texas* (Diggs et al. 1999) and the *Illustrated Flora of East Texas* (Vol. 1) (Diggs et al. 2006), thanks again to all the individuals who contributed to those works.

In addition to original observations and measurements, material for the keys and descriptions has been obtained from a variety of sources listed in the literature cited. Of particular assistance were the following works: *Ferns and Fern Allies of Texas* (1956), *Flora of North America North of Mexico, Vol. 2, Pteridophytes and Gymnosperms* (Flora of North America Editorial Committee 1993), *Steyermark's Flora of Missouri* (Yatskievych 1999), *Ferns and fern allies of the Trans-Pecos and adjacent areas* (Yarborough & Powell 2002), and the *Illustrated Flora of East Texas* (Diggs et al. 2006).

This project is part of the ongoing collaboration between the Austin College Center for Environmental Studies and the Botanical Research Institute of Texas. Without the support of both institutions it would not have been possible.

Finally, thanks to George Diggs' partner, Kerry G. Brock.

THE FERNS AND LYCOPHYTES OF TEXAS

INTRODUCTION

The beauty, variety, and fascinating biology of ferns and similar plants catch our imaginations and almost demand our attention. More than a half-century ago they captured the interest of the Texas botanist Donovan Stewart Correll. In 1956 he published the landmark *Ferns and Fern Allies of Texas*, his first major work on Texas botany. No book since that time has focused on these unique plants throughout Texas despite the fact that a great deal has changed. Since the mid-1950s, the number of ferns and similar plants known for Texas has increased by nearly 29%, including some new to science. Also, the scientific naming (nomenclature) of these plants has changed dramatically, and perhaps most importantly, our understanding of their biology and evolutionary relationships has greatly improved. Indeed, some changes have been so dramatic that Haufler (2007) recently noted, "a revolution began in evolutionary studies" of ferns and similar plants "about 25 years ago." As a result, now seems to be an appropriate time for a new volume that follows in Correll's tradition and focuses on these often rare but always interesting plants.

"Record-Holders" and "Prize-Winners"

One of the reasons we are attracted to ferns and similar plants is that they are endlessly fascinating and full of surprises. Before beginning to discuss their biology, it is worthwhile to look at a few of the unofficial record-holders and prize-winners among the species that occur in Texas. Table 1 gives an overview of a number of our favorites; more information about these plants can be found in the species treatments. Although some of our choices are subjective, reading about these "champion" species will give you an intuitive sense of how special some of the ferns and similar plants that occur in Texas really are.

The Plants Included: "Pteridophytes"

As indicated by Correll's title, his book dealt with more than ferns—in addition, it included the "fern allies," those plants that in some ways resemble ferns and that are alluded to in such phrases as "ferns and similar plants." Indeed, almost all fern books cover these similar and equally interesting plants and this book is no exception. Botanists have traditionally referred to ferns and fern allies collectively as "pteridophytes." The word is from the Greek, *pteron*, feather or wing, due to the form of the leaves, and -phyte, from the Greek, *phuton*, plant. Scientists who study pteridophytes are referred to as pteridologists and their subject as pteridology. The pteridophytes all share some things in common:

1. they have vascular or conducting tissue;
2. their life cycles include two very different alternating phases or generations (explained in the section titled "The Sexual Life Cycle" on page 16);
3. their fertilization involves swimming sperm;
4. they are the surviving descendants of evolutionary lineages reaching back hundreds of millions of years to shortly after plants colonized land; and
5. perhaps most significantly, they are "seed-free"—unlike gymnosperms and flowering plants, they lack seeds and instead reproduce by microscopic, single-celled structures called spores. (In fact, in Shakespeare's time they were considered mystical plants because it was not understood how they could reproduce without flowers, fruits, or seeds [Moran 2004].)

Seedless reproduction is the ancestral condition of the earliest land plants, and lacking these structures can be a serious drawback. Seeds, basically an embryo (baby plant) packaged with stored food and surrounded by a protective waterproof coating, have several important advantages. They have freed seed plants from the necessity of a moist environment for reproduction, they have a supply of food that allows the embryo to grow rapidly, they are capable of remaining dormant for long periods of time, and they can be modified to allow effective dispersal by animals. Indeed, such advantages have allowed seed plants to be incredibly successful—more than 95% of all plants alive today reproduce by seeds.

TABLE 1. Texas "Record-Holders" and "Prize-Winners."

Category	Species	Description
Smallest Size	*Azolla caroliniana*, CAROLINA MOSQUITO FERN/p. 112	Whole plant is about the size of a dime
Largest Size	*Dennstaedtia globulifera*, BEADED CUPLET FERN[1] /p. 130	Leaves are up to about 4 m/13 ft long and 2 m/6.5 ft or more wide
Strangest Leaf	*Lygodium japonicum*, JAPANESE CLIMBING FERN /p. 164	The tip of the leaf midrib continues to grow indefinitely, with the leaf becoming twining, climbing, and vine-like (true stem/rhizome below ground). Leaves reach at least 6 m/20 ft long (reports to 30 m)
Strangest Habitat	Tie: *Pilularia americana*, AMERICAN PILLWORT or *Tectaria heracleifolia*, BROAD HALBERD FERN/p. 173, 293	Lives underwater on the bottom of lakes or 24m/80 ft below ground level on the vertical walls of Devil's Sinkhole, a collapsed limestone cavern
Rarest Species	*Asplenium septentrionale*, FORKED SPLEENWORT /p. 106	Known in Texas from only 1 collection at Boot Springs in the Chisos Mountains
Most Unique	*Isoëtes lithophila*, ROCK QUILLWORT/p. 58	Known in the world only from 4 counties in the Llano Uplift of central Texas
Best Traveler	*Asplenium platyneuron*, EBONY SPLEENWORT/p. 102	Occurs in eastern North America with very distant populations in South Africa
Best Survivor	*Selaginella lepidophylla*, RESURRECTION PLANT/p. 79	Can survive long periods of time, possibly years, in a dried up, "ball-like" condition and then rapidly revive upon wetting
Strangest Reproducer	*Asplenium palmeri*, PALMER'S SPLEENWORT/p. 102	The tip of the leaf elongates into a whip-like extension ending in a bud capable of rooting and giving rise to a new plant
Fastest Reproducer	*Salvinia molesta*, GIANT SALVINIA/p. 291	Some authorities indicate that populations can double in size in a little > 2 days
Strangest Structure	*Psilotum nudum*, WHISK FERN/p. 206	Looks like a whisk broom since it lacks roots and true leaves and consists only of naked photosynthetic stems
Worst Weed	*Salvinia molesta*, GIANT SALVINIA/p. 291	This invasive exotic is considered by some to be the world's worst weed; it can rapidly cover lakes forming mats 1 m/3 ft thick
Most Poisonous	*Astrolepis cochisensis*, JIMMY FERN/p. 219	An unidentified poison causes arched back, rapid pulse, gasping, trembling (known as "the jimmies"), and collapse in livestock; death results from respiratory paralysis
Biggest Surprise	*Dennstaedtia globulifera*, BEADED CUPLET FERN /p. 130	Known in the U.S. only from inside the entrance of one cave in western Texas; it is primarily a species of wet tropical forests in the mountains of Latin America
Oldest Species	*Onoclea sensibilis*, SENSITIVE FERN/p. 175	57 million year old Paleocene Epoch fossils are virtually identical to modern members of this species
Most Widespread Worldwide	*Pteridium aquilinum*, BRACKEN FERN/p. 132	This or a very closely related species occurs in multiple habitats on every continent except Antarctica

[1]The winner of this prize is somewhat subjective. The leaves of *Lygodium japonicum* are longer, but the leaves of *Dennstaedtia globulifera* are many times wider, more massive, and still quite long.

On the other hand, microscopic spores do have some advantages. They are so small that ferns can make tremendous numbers of them, and they are so lightweight that they can easily be blown or otherwise carried for hundreds or even thousands of miles and reach distant locales. In fact, atmospheric sampling has "... found fern spores at high altitudes, including in the jet stream" (Moran 2008). Long-distance dispersal by wind is thought to be "... far more common in pteridophytes than in seed plants" (Wagner & Smith 1993). The high percentage of ferns on remote oceanic islands in comparison with mainland areas demonstrates this advantage (Smith 1972, 1993; Kramer 1993; Moran 2004, 2008). Likewise, the isolated occurrences of a number of tropical ferns far removed from others of their kind, in tiny pockets of appropriate habitat in Texas, is testament to the dispersal ability of spores. Thus the fern life cycle itself can help these plants to colonize remote places. However, although some ferns are remarkably well-suited for long-distance dispersal, it should be noted that probably less than 10% of ferns can accomplish self-fertilization (most ferns have mechanisms to prevent this and thus ensure cross-fertilization). We now know there is great variation in the probability that different fern species will successfully colonize over long distances, with some being excellent candidates for long-distance dispersal and others not so (Peck et al. 1990; Wild & Gagnon 2005; Ranker & Geiger 2008). Details of the fern life cycle will be discussed in a later section.

These pteridophytes, seedless vascular plants reproducing by spores, include more than 10,000 species, with some authorities suggesting as many as 13,000 or more. This is a sizable number, but these relatively unusual plants still represent less than five percent of all known living vascular plants. Such modern pteridophytes are representatives of evolutionary lineages that have astonishingly long histories. From fossil evidence we know that pteridophytes were once extremely abundant and dominated many of the planet's ecosystems, some becoming tree-like in size. For example, lycophytes, horsetail relatives, and other ferns were among the dominant plants in the extensive swamps of the Carboniferous Period (359–299 million years ago). Over geologic time, the compressed ancient remains of these swamp plants became coal (Hoshizaki & Moran 2001); in fact, the word "carboniferous," means "coal-bearing." In this form, the stored energy of sunlight captured in swamp forests hundreds of millions of years ago provides the modern world with much of its energy.

Because of their similarities, earlier taxonomists often formally grouped ferns and similar plants together as Division Pteridophyta (e.g., Correll 1956). Later, they were often divided into four distinct groups: Division Lycopodiophyta (the lycophytes, including club-mosses, spike-mosses, and quillworts), Division Polypodiophyta (the true ferns), Division Psilophyta (the whisk ferns or psilophytes), and Division Equisetophyta (the horsetails). Even though they have only a few species each, the visually distinctive whisk ferns and horsetails have often been segregated into their own groups because they were thought to represent separate evolutionary lineages.

Recent Discoveries

In the past several decades, molecular studies (including DNA sequence data) have revolutionized our understanding of plant evolutionary relationships. No longer do we have to rely solely on comparing the form and structure of living organisms or interpreting the fossil remains of long dead plants, as DNA now provides us a very different and much more detailed way to learn about the evolutionary history of organisms. In a real sense it gives us an exciting new window into the evolutionary history of the species with which we share our planet.

Based on all this new information, it is now clear that ferns and similar plants actually represent only two evolutionarily distinct lineages, which are recognized in this book. These are the **lycophytes** (club-mosses, spike-mosses, and quillworts), technically referred to as the Division Lycopodiophyta, and the **ferns,** known as the Polypodiophyta (including not only "typical" ferns, but also both the horsetails, a group of 15 species, and the psilophytes, comprised of 4 to 8 species). Genetic and fossil evidence now shows that about 400 million years ago a major evolutionary split occurred. This split gave rise on one hand to the lycophytes, and on the other hand the remaining vascular plants. The latter lineage—the remaining vascular plants, subsequently split into a fern branch and a branch including all seed plants: the gymnosperms (conifers and their relatives) and angiosperms (flowering plants). So, although lycophytes and ferns do share a number of characteristics (e.g., reproduction by spores), it can be noted from Figure 1 that ferns are actually more closely related to seed plants (i.e., they share a more recent common ancestor) than they are to lycophytes—therefore, lycophytes and ferns need to be treated as separate taxonomic groups.

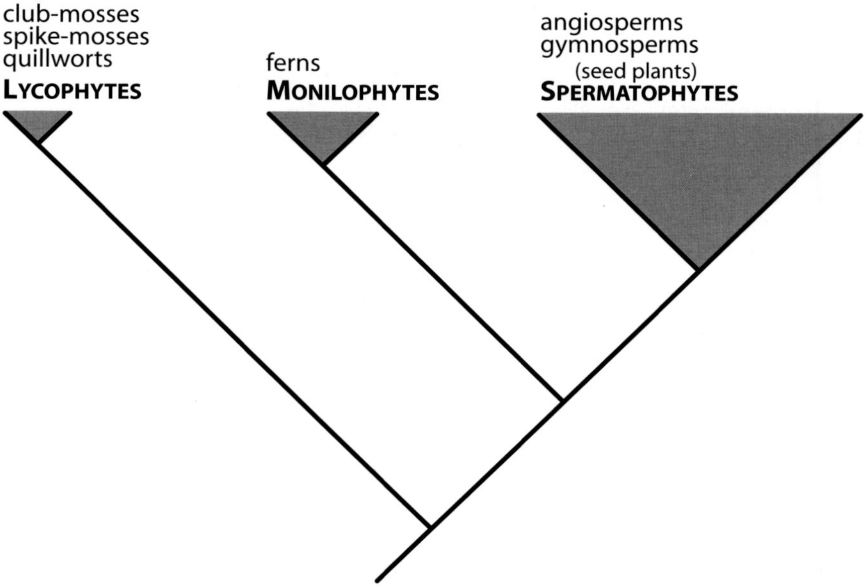

FIG. 1. Phylogeny showing evolutionary relationships of major groups of vascular plants. Modified from Smith et al. (2006). The size of the triangles is only a rough approximation of the number of species in the groups.

It is thus now known that among living vascular plants the lycophytes are the basal lineage—they are on a separate evolutionary branch from all other vascular plants (Fig. 1). As such they are appropriately recognized taxonomically as being extremely distinct—as the Division Lycopodiophyta, a different taxonomic division from the ferns. There are approximately 1,200 to 1,400 species of lycophytes living worldwide today, less than 1% of vascular plants. These club-mosses, spike-mosses, and quillworts are living representatives of the world's oldest surviving group of vascular plants. They are an ancient line that separated from all other vascular plants hundreds of millions of years ago and have been evolving independently ever since. Because they are such a distinct lineage that has survived from the early days of land plant evolution, they are of particular interest to botanists. The lycophytes differ from ferns not just in their DNA—they also look different. First, they produce a single sporangium (spore case that produces spores) on the upper side of a leaf or in the leaf axil, where the leaf joins the stem. Secondly, their leaves, called lycophylls (also known as microphylls) are unlike those of any other plant—these leaves are typically small, but more significantly they have only a single unbranched vein. Of the three families in this ancient group presently living, all three occur in Texas: the Isoëtaceae (QUILLWORT FAMILY, Fig. 2), the Lycopodiaceae (CLUB-MOSS FAMILY, Fig. 3), and the Selaginellaceae (SPIKE-MOSS FAMILY, Fig. 4).

The ferns are currently referred to by a variety of technical names including Division Polypodiophyta (as done in this volume), Class Polypodiopsida, Monilophytes (Smith et al. 2006), or Division Monilophyta (Weakley 2010). Whatever it is called, the fern lineage includes the vast majority of seedless vascular plants (between 9,000 and 11,500 species) and represents about 3.6% of living vascular plant species. Because of all the new insights into plant relationships brought about by molecular information, our formal system of nomenclature is undergoing significant change—hence the varied and still unsettled use of these names. Despite such problems, the evolutionary relationships are now becoming clearer, significantly improving our system of classification. For example, it was only a few years ago that Wagner and Smith (1993) noted that "... whisk ferns (*Psilotum*) and horsetails (*Equisetum*) are so distinctive that we are still in a quandary as to their relationships." Instinctively, their unique morphologies made their initial classification as

FIG. 2. ***Isoëtes butleri*****, BUTLER'S QUILLWORT, a Texas native Isoëtaceae.**

FIG. 3. ***Lycopodiella appressa*****, CHAPMAN'S CLUB-MOSS, a Texas native Lycopodiaceae.**

something quite apart from ferns unquestionable. However, today the group Psilophyta is no longer formally recognized as a separate group, nor is the Equisetophyta. DNA has provided evidence that these two morphologically unusual groups are just types of ferns—very strange looking, but ferns nonetheless. As a result they are now considered to be ferns and recognized as Class Psilotopsida and Class Equisetopsida respectively, within the Division Polypodiophyta (but see Rothwell & Nixon 2006 and Rothwell & Stockey 2008 for dissenting views).

Fern Classification

Let us turn briefly to the classification of the ferns, the Division Polypodiophyta. The goal of classification by taxonomists/systematists is to accurately reflect evolutionary relationships. In recent years, with an increasing number of new discoveries, our understanding of these relationships has been completely revolutionized. As Moran (2004) noted, "… there has never been a more exciting time to be a systematist."

We now know that there are 11 orders or major groups of ferns (Fig. 5 and Table 2; for more detail see the phylogenetic tree in "end paper" pages adjacent to back cover of this book) (following Moran 2004, Smith et al. 2006, and Rothfels et al. 2012b). While the state lacks four primarily tropical groups (Fig. 5), the rest of these orders occur in Texas. The fern orders are grouped into four classes. One of these classes, the mainly tropical Marattiopsida, has no known members in Texas, but the other three are all represented (Table 2).

The seven orders in Texas are relatively easy to learn and most Texas ferns are in one order, the giant Polypodiales. The Ophioglossales and Psilotales are more closely related to each other than to any other groups and share the lowest or basal branch on the family tree of ferns (Fig. 5). These two orders are relicts of ancient lineages that have survived from the early days of fern evolution. The **Ophioglossales** is made up of only one family, Ophioglossaceae, including *Ophioglossum*, the ADDER'S-TONGUES, and *Botrychium*, the GRAPE FERNS, which all have a unique leaf structure: the leaves are divided into a sterile photosynthetic portion and a fertile, spore-bearing portion. Texas examples of the order, *Ophioglossum petiolatum* and *Botrychium biternatum* can be seen in Figure 6. The extremely distinctive **Psilotales** also has a single family, Psilotaceae (e.g., *Psilotum*, the WHISK FERNS), including just 2 genera, and only 4 to 8 species. Their reduced, very simple structure superficially resembles fossils of some of the earliest land plants. They were finally accepted as ferns based on molecular data. *Psilotum nudum*, the only representative in Texas (Fig. 7), lacks both roots and well-defined leaves. The dichotomously branching (forking regularly into two equal branches), green, photosynthetic stems serve multiple functions.

The **Equisetales** (apparently distantly related to the very different looking tropical Marattiales) likewise has only one living family, Equisetaceae, and just one genus, *Equisetum*, with only about 15 species, three of which grow in Texas (Fig. 8). These HORSETAILS or SCOURING-RUSHES are so distinctive that they also were only recently realized to be ferns. The plant body consists primarily of hollow, jointed, green stems; the leaves are whorled, tiny, and very inconspicuous; and the sporangia are borne in small cone-like structures at the tips of the stems.

© GEORGE DIGGS

FIG. 4. ***Selaginella apoda*, MEADOW SPIKE-MOSS, a Texas native Selaginellaceae.**

The **Osmundales** is the next branch on the fern family tree, and it differs from the orders mentioned previously in an important way: it has sporangia with thin walls only a single cell thick (in contrast to walls several cells thick). Because the Osmundales and the remaining ferns (more than 95% of living fern species excluding only the small Ophioglossales, Psilotales, Marattiales, and Equisetales) all share this unique characteristic of the sporangia, they are referred to as leptosporangiate (Greek: *lepto-*, thin, referring to the thin-walled sporangia). Various molecular markers also unite the leptosporangiate ferns. We are fortunate in Texas to have two members of the Osmundaceae or CINNAMON FERN family, *Osmunda regalis* and *Osmundastrum cinnamomeum* (Fig. 9). The Osmundaceae is a family of strikingly beautiful ferns, but it is also interesting for other reasons. For example, from the fossil record we know some species have hardly changed for 200 million years—they are truly living fossils.

The **Schizaeales** has two families in Texas, the Anemiaceae (FLOWERING FERN family, genus *Anemia*—Fig. 10) and the naturalized Lygodiaceae (CLIMBING FERN family, genus *Lygodium*). Although superficially very different (one diminutive and growing on rocks, the other a twining vine in forests), these and other members of the Schizaeales share many DNA similarities and a characteristic of the sporangium: the annulus completely encircles the sporangium near its tip. (The annulus

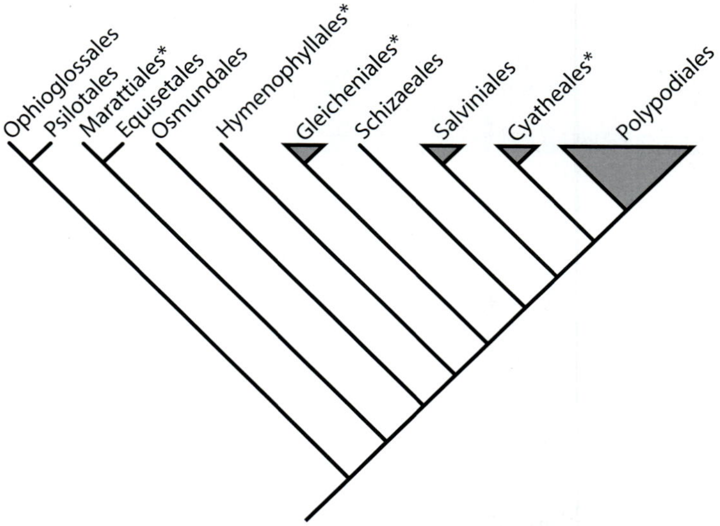

FIG. 5. Fern phylogeny showing the evolutionary relationships of the eleven orders. The four primarily tropical groups with no representatives in Texas are indicated by *. The size of the triangles is only a rough approximation of the number of species in the various orders. Modified from Moran (2004).

TABLE 2. Classification of Texas pteridophytes followed in this book.

- **Division Lycopodiophyta** (LYCOPHYTES)
  - **Class Lycopodiopsida** (LYCOPHYTES)
    - Order Isoëtales (QUILLWORTS)
      - Isoëtaceae (QUILLWORT FAMILY)
    - Order Lycopodiales (CLUB-MOSSES)
      - Lycopodiaceae (CLUB-MOSS FAMILY)
    - Order Selaginellales (SPIKE-MOSSES)
      - Selaginellaceae (SPIKE-MOSS FAMILY)
      - Marsileaceae (WATER-CLOVER FAMILY)
- **Division Polypodiophyta** (FERNS)
  - **Class Psilotopsida** (WHISK FERNS & ADDER'S-TONGUES)
    - Order Ophioglossales (ADDER'S-TONGUES, etc.)
      - Ophioglossaceae (ADDER'S-TONGUE FAMILY)
    - Order Psilotales (WHISK FERNS)
      - Psilotaceae (WHISK FERN FAMILY)
  - **Class Equisetopsida** (HORSETAILS)
    - Order Equisetales (HORSETAILS)
      - Equisetaceae (HORSETAIL FAMILY)
  - **Class Polypodiopsida (LEPTOSPORANGIATE FERNS)**
    - Order Osmundales (CINNAMON FERNS)
      - Osmundaceae (ROYAL FERN FAMILY)
    - Order Schizaeales (CLIMBING FERNS)
      - Anemiaceae (FLOWERING FERN FAMILY)
      - Lygodiaceae (CLIMBING FERN FAMILY)
    - Order Salviniales (FLOATING FERNS, WATER-CLOVERS)
      - Azollaceae (MOSQUITO FERN FAMILY)
      - Salviniaceae (FLOATING FERN FAMILY)
    - Order Polypodiales (POLYPODS)
      - Aspleniaceae (SPLEENWORT FAMILY)
      - Athyriaceae (LADY FERN FAMILY)
      - Blechnaceae (CHAIN FERN FAMILY)
      - Cystopteridaceae (BLADDER FERN FAMILY)
      - Dennstaedtiaceae (BRACKEN or CUPLET FERN FAMILY)
      - Dryopteridaceae (WOOD FERN FAMILY)
      - Lomariopsidaceae (FRINGED FERN FAMILY)
      - Onocleaceae (SENSITIVE FERN FAMILY)
      - Polypodiaceae (POLYPODY FERN FAMILY)
      - Pteridaceae (MAIDENHAIR FERN FAMILY)
      - Tectariaceae (HALBERD FERN FAMILY)
      - Thelypteridaceae (MARSH FERN FAMILY)
      - Woodsiaceae (CLIFF FERN FAMILY)

LEFT: © W. COLE WEATHERBY; RIGHT: © ALAN CRESSLER

FIG. 6. Native Texas examples of the Ophioglossales: *Botrychium biternatum*, SOUTHERN GRAPE FERN (left), and *Ophioglossum vulgatum*, SOUTHERN ADDER'S-TONGUE (right).

LEFT: © KIM AND FOREST STARR; RIGHT: © ALAN CRESSLER

FIG. 7. *Psilotum nudum*, WHISK FERN, a Texas native member of the Psilotales.

FIG. 8. *Equisetum hyemale*, SCOURING-RUSH HORSE-TAIL, a Texas native member of the Equisitales.

is the ring of thick-walled cells around the sporangium that opens the sporangium and allows the release of spores.)

The **Salviniales** is represented in the state by three families, the Azollaceae (MOSQUITO FERN family, genus *Azolla*), the Marsileaceae (WATER-CLOVER family, genera *Marsilea* and *Pilularia*), and the introduced Salviniaceae (FLOATING FERN family, genus *Salvinia*). Two Texas members of the order, *Azolla microphylla* and *Marsilea macropoda* can be seen in Figure 11. Most members of this order are aquatic and are very un-fern-like in appearance—the leaves of *Marsilea*, for example, resemble a four-leaf clover, while *Azolla* includes the world's smallest ferns, with whole plants roughly the size of a dime and leaves only about 1/32 of an inch long (about 1 mm). However, Salviniales share a characteristic different from all other ferns—they make two different types of spores and are thus referred to as heterosporous, rather than homosporous.

The final order found in Texas is the **Polypodiales,** which includes the rest of the state's families and species—indeed, the great majority. This is not surprising since worldwide it is the most diverse of all fern groups, encompassing nearly 80% of all ferns and occurring virtually all over the globe. Figure 12 shows several Texas representatives of the Polypodiales. As can be seen, many species in this order would appear typically fern-like to most people. The Polypodiales is the most evolutionarily derived of the ferns—meaning it is most different from the ancestral ferns. In terms of structure, all members of the order are characterized by a particular type of annulus (Fig. 13). The polypods have a vertically oriented annulus that only partially encircles the sporangium or spore case, the annulus being interrupted by the stalk of the sporangium (the annulus is thus said to be "incomplete") (Moran 2004). Even this most derived order has a distant evolutionary origin. Moran (2007) noted that new "... fossils have pushed back the first appearance of polypodiaceous ferns ... to a minimum age of 120 million years ago." Recent molecular data (e.g., Schuettpelz & Pryer

Fig. 9. ***Osmundastrum cinnamomeum*****, CINNAMON FERN, a Texas native member of the Osmundales.**

2008; Rothfels et al. 2012a) have greatly improved our understanding of relationships in the Polypodiales and have resulted in significant changes in family concepts within the order. For example, in this book we are recognizing six families, totaling six genera (Athyriaceae—*Athyrium*, Cystopteridaceae—*Cystopteris*, Lomariopsidaceae—*Nephrolepis*, Onocleaceae—*Onoclea*, Tectariaceae—*Tectaria*, and Woodsiaceae—*Woodsia*) that in a recent Texas flora were all placed in the Dryopteridaceae (Diggs et al. 2006). These families had been erroneously classified in that family based on superficial similarities resulting from convergent evolution—in other words, similar selection pressures made them appear more closely related than they really were. While our understanding of fern relationships is much improved, it is far from complete and additional changes in classification will almost certainly occur as more studies are done.

Diversity—Why so Many Ferns and Lycophytes in Texas?

At the time Correll published his book in 1956, 107 species of ferns and lycophytes were known for Texas, 104 of these native to the state. That was a substantial increase over the number listed in 1883 (Davenport noted 35 ferns) and in 1903 (Bush cited 59 ferns; Reverchon included 51 ferns and 15 "fern allies"). Currently, we know of 146 taxa (taxonomic kinds including varieties and subspecies) of ferns and lycophytes representing 138 species, including both those native to the state and those that have naturalized. Naturalized species are those that were intentionally or accidentally introduced to Texas from elsewhere in the world and now grow wild without human help. These 138 species make Texas the second most diverse state in the continental U.S. in terms of total pteridophyte diversity. Only Florida, which has a total of 152 species (including 39 that have naturalized), has more total species than Texas.

The most interesting statistic, however, is that Texas has 127 native species. Thus, there are 127 species growing in the state without the intervention of humans (the number increases to 129 if two well established hybrids, *Selaginella* ×*neomexicana* and *Equisetum* ×*ferrissii* are included). Texas, therefore, has more natives than any other state in the continental U.S. Florida, with its rich tropical flora, is a close second, with 113 native species (Wunderlin & Hansen 2000), and Virginia third with 100 (Weakley et al. 2012). Comparative numbers for a few other states recently studied include about 100 for California (Hickman 1993), 81 for Arkansas (J.H. Peck, pers. comm.; Peck 2011b), 76 for Kentucky (Jones 2005), 70 for Oklahoma (Taylor & Taylor 1994), 68 for Missouri (Yatskievych 1999), and 31 for Nebraska (Kaul et al. 2006; S. Rolfsmeier, pers. comm.).

The large number of ferns and lycophytes in Texas is particularly surprising to people who assume these plants are found only in humid, even wet tropical conditions, something not occurring in most of Texas. In North America as a whole, ferns and lycophytes do reach their greatest abundance and diversity in rich, moist, forest habitats (Wagner & Smith 1993). Within the state, East Texas, the wettest region (Fig. 14), is reasonably rich in ferns and lycophytes—62[2] native spe-

[2]Diggs et al. (2006) treated 65 native species in the *Illustrated Flora of East Texas*, but three of these species were in counties straddling the extreme western border of East Texas in habitats that were technically Edwards Plateau.

©CARLOS VELASCO

FIG. 10. ***Anemia mexicana*****, MEXICAN FERN, a Texas native member of the Schizaeales.**

cies, a number comparable to many whole states. But counter-intuitively, the dry basin and mountain area known as the Trans-Pecos, in the far western part of Texas, is even richer (Fig. 15). That region, technically part of the Chihuahuan Desert, has 80 different natives (Yarborough & Powell 2002; Windham & Yatskievych 2003a; O'Kennon & Diggs 2011). The Edwards Plateau[3], another seemingly inhospitable area, has 63 native species. Some of these species are found widely, occurring in two or even all three of these areas, but the numbers are deceptive. While the Trans-Pecos has many more species than either East Texas or the Edwards Plateau, biomass is a completely different story. If all the tissue of ferns and lycophytes in the Trans-Pecos was somehow magically weighed, it wouldn't total as much as that found in a several mile stretch of the Neches River bottomland in the Big Thicket area of East Texas. There one can hike waist or even shoulder deep through long stretches of fern thicket in an East Texas river bottom. In contrast, most of the species in the Trans-Pecos are either small or rare. The diversity of Texas pteridophytes is also remarkable when compared to the continental U.S. and Canada. Of the 424 native species found in this vast area, Texas' 127 represents 30% of the total in less than 4% of the geographic area (Wagner & Smith 1993; Zech et al. 1998; Windham & Yatskievych 2003a; O'Kennon & Diggs 2011). However, Texas is rich only in comparison with some temperate areas and cannot compete with the tropics. For example, the tiny, tropical country of Costa Rica, one-thirteenth the size of Texas, has more than 1,100 pteridophyte species, and a single tree in a wet tropical forest can host as many as 45 species (Schuettpelz & Trapnell 2006; Moran 2008).

[3]For the purposes of this book, the Edwards Plateau is defined as that area west of the Cross Timbers and Prairies and the Blackland Prairie, south of the Rolling Plains and the High Plains, north of the South Texas Plains, and east of the Pecos River.

FIG. 11. Texas examples of the Salviniales: ***Marsilea macropoda*****, LARGE-FOOT PEPPERWORT (top), and** ***Azolla microphylla*****, MEXICAN MOSQUITO FERN.**

The question then becomes, in comparison with the rest of the U.S. and Canada, why is Texas so diverse in ferns and lycophytes, or technically, why does it have high species richness? A number of interrelated factors contribute to this diversity:

1. **Size.—**Although the large size of Texas is a factor, it certainly is not the main one. Alaska, for example, which is nearly 2.2 times larger than Texas, has only 70 native species of ferns and lycophytes (Flora of North America Editorial Committee 1993; Stensvold 2010). In reality, sheer size is probably of relatively little importance in terms of diversity.
2. **Southern latitude.—**A general rule for many groups of plants is that the number of species within a particular habitat is higher in areas with warmer, wetter climates, and more generally that species richness tends to increase from the poles to the tropics (e.g., Marks & Harcombe 1975; Hillebrand 2004). Costa Rica, mentioned above, is an example of this phenomenon. Indeed, Texas' southern location in the U.S. is an important reason for its high diversity. Florida also is rich in species due to its southern location, but the lack of a land connection to the south probably limits diversity in that state.
3. **Geographic, geologic, and edaphic variation.—**There is a great deal of geographic, geologic, and edaphic (pertaining to soil) variation across Texas. Elevations vary widely, from sea level along the coast, to 2,667 m/8,749 ft at the highest point, Guadalupe Peak. Soils, which reflect underlying rock layers and a variety of other factors, also vary a great deal—from waterlogged to excessively drained and from extremely fertile to highly leached and mineral poor. Soils are also variable in acidity, ranging from highly acidic in parts of East Texas to highly alkaline in the desert basins of the Trans-Pecos. Further, there is significant variation in terms of exposed rocks, including the striking feature known as Enchanted Rock (Fig. 16) (a huge, pink granite dome rising 130 m/425 ft above the surrounding area in the Central Mineral Region of central Texas); extensive limestone on the Edwards Plateau; and the tremendous variety of rocky substrates exposed in the Trans-Pecos. A great deal of this variation occurs on a very local scale and provides numerous special microhabitats that allow a large number of species to survive in small areas. Enchanted Rock State Natural Area alone, for example, has at least 29 species of ferns and lycophytes, while the entire state of Nebraska has only 31 species (Butterwick 1979; Walters & Wyatt 1982; Kaul et al. 2006; Plant Resources Center 2009; Bob O'Kennon collections at BRIT).

Such local variation is well known in Texas. For example, quoting Correll (1956), "... along some of the streams on the Edwards Plateau masses of the delicately herbaceous *Adiantum Capillus-Veneris* cascade from dripping limestone walls above extensive colonies of the mesophytic *Dryopteris normalis* [now treated as *Thelypteris kunthii*] while a few feet away

©ALAN CRESSLER

FIG. 12. Texas representatives of the Polypodiales: *Phegopteris hexagonoptera*, BROAD BEECH FERN (upper left), *Athyrium filix-femina*, SOUTHERN LADY FERN (upper right), *Dryopteris ludoviciana*, SOUTHERN WOOD FERN (lower left), and *Thelypteris kunthii*, WIDESPREAD MAIDEN FERN (lower right).

xerophytic species of *Cheilanthes*, *Notholaena*, *Pellaea*, and *Selaginella* grow on sun-baked rocks under the most adverse conditions of exposure and drought."

4. **Climate variation.—**Climate varies dramatically across the state. Average annual rainfall ranges from nearly 152 cm/60 in at the southeastern portion of the state on the border with Louisiana to about 20 cm/8 in at El Paso near the state's western margin (Fig. 17). Likewise, temperatures vary greatly, with the subtropical climate of far southern Texas being strikingly different from the sometimes frigid areas of the Panhandle. Brownsville and McAllen in the far south average only three freezes per year, while Amarillo in the Panhandle has 111 (Bomar 1995).

5. **Proximity to source floras and biogeographic history.—**Another reason for the high diversity of Texas pteridophytes is the occurrence in the state of species typical of other regions: the eastern deciduous forests, southeastern swamps, deserts of the southwestern U.S. and Mexico, and even the tropics. The result is that the Texas pteridophyte flora is a complex and diverse mixture derived from several major and quite different floristic provinces (Thorne 1993). In particular the forests of far East Texas are essentially a continuation of the eastern deciduous forests that stretch to the Atlantic and contain a significant Arcto-Tertiary or Boreotropical element, a belt of similar, mainly temperate vegetation but with tropical components that extended around much of the northern hemisphere during portions of the Tertiary geologic period (65–1.8 million years ago). At times during the Tertiary, conditions such as sea level and temperature were very different than at present; plants and animals were able to migrate broadly amongst the northern continents via land connections that are no longer present (see further discussion on page 32). During some

particularly warm spells of the Tertiary such forests extended north to extremely high latitudes (55° to even 80°, near the North Pole). A large number of descendants of this flora survive in the eastern U.S. including many Texas pteridophytes, such as *Onoclea sensibilis*, *Osmunda regalis*, *Osmundastrum cinnamomeum*, *Thelypteris palustris*, and *Equisetum hyemale*. Other Texas genera including *Azolla* (MOSQUITO FERNS) and *Woodwardia* (CHAIN FERNS) are also known from very high latitudes during the Tertiary (Collinson 2001).

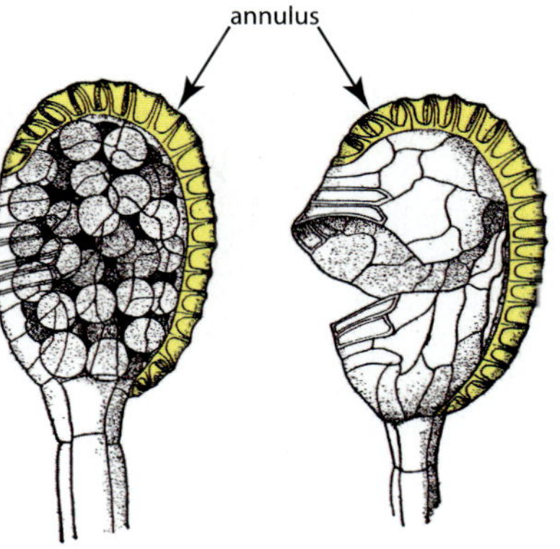

FIG. 13. Annulus (ring of cells functioning in opening of sporangium) characteristic of the order Polypodiales. The polypods have a vertically oriented annulus that only partially encircles the sporangium or spore case, the annulus being interrupted by the stalk of the sporangium. From Robbins and Rickett (1939).

On the other hand, the flora of the Trans-Pecos has a strong link to the dry-adapted area flora of Mexico (McLaughlin 2007), particularly that component known as the Madro-Tertiary. That flora, named after the Mexican Sierra Madre Mountains, is thought to have gradually evolved during the Tertiary Period in the southern part of western North America in a relatively dry zone between areas of deciduous and tropical vegetation (Axelrod 1958). Such plants ". . . seem to have originated as the arid areas of North America expanded through the Tertiary—for the past 65 million years, and especially in the last 15 million years" (Thorne 1993). The evolution and expansion of this flora was in response to a variety of changes including the uplift of the mountains of western North America and changes in rainfall patterns. Many of our desert-adapted fern and lycophyte species trace their origin to this Madro-Tertiary flora; examples probably include *Astrolepis* species, the STAR-SCALED CLOAK FERNS, and *Cheilanthes* species, the LIP FERNS. A significant number of our dry-adapted species are mainly Mexican in distribution and reach the extreme northern limit of their ranges in Texas (Fig. 18). In fact, of the 127 Texas native pteridophytes, 80 (63%) also occur in Mexico (Mickel & Smith 2004).

6. **High habitat diversity.**—Texas has a broad range of distinct habitats, each supporting a unique assemblage of species—there are 10 major vegetational areas (see map inside front cover), each with multiple varied habitats. Habitat diversity is obviously a combined reflection of many of the other factors previously mentioned (e.g., geography, geology, climate). For example, the diverse habitats of far East Texas share many characteristics with those of the southeastern U.S. and they are remarkably different from those of the Trans-Pecos, with its basins and mountain ranges and habitats similar to the rest of the Chihuahuan Desert (Fig. 19). The many different habitats are certainly an important contributor to the diversity of ferns and lycophytes in the state. The distinct habitats of East Texas, the Edwards Plateau, and the Trans-Pecos all individually have relatively rich pteridophyte floras. Together they make the state the most diverse in the nation.

Conclusions about diversity.—The diversity of Texas pteridophytes has not been studied enough to draw many definitive conclusions. However, several things do seem clear. 1) The evolutionary relationships of Texas ferns and lycophytes show clear affinities to those of two source areas—the eastern U.S. and Mexico. 2) Habitat diversity and thus the appropriate conditions for a variety of different pteridophyte species are concentrated in three areas: East Texas, the Edwards Plateau, and the Trans-Pecos. Extensive parts of the state (e.g., Blackland Prairies, High Plains, Rolling Plains, South Texas Plains) host few species. 3) In comparison with most other states, the southern latitude of Texas, similar to that of Florida, definitely contributes to its high diversity. On the basis of these general observations, it seems that proximity to source floras, high habitat diversity, and

Fig. 14. (Left) A typical fern habitat in a moist forest in East Texas.

Fig. 15. (Below) Map showing 3 major areas of native pteridophyte diversity in Texas. Many species in East Texas (in various shades of green) are from wet habitats in the Pineywoods (area in darkest green). In the Trans-Pecos most are found in the "sky islands," high elevation areas such as the Chisos, Franklin, Guadalupe, Davis, or Rosillos mountains (see page 43). In the Edwards Plateau, ferns and lycophytes are concentrated in special microhabitats such as moist canyons, cave entrances, and rock outcrops. A few species from marginal counties are considered to be in two areas. A number of species are widespread, occurring in two or even all three of the areas.

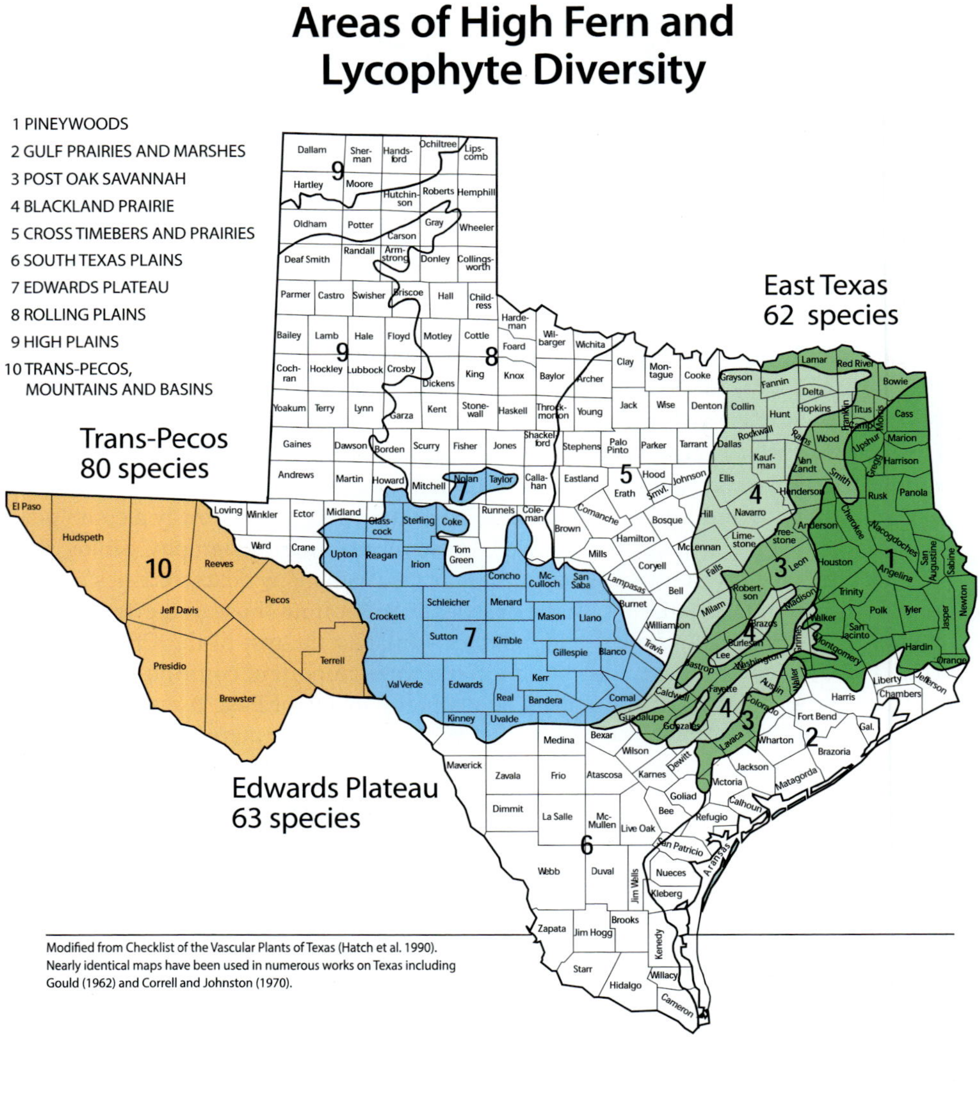

Modified from Checklist of the Vascular Plants of Texas (Hatch et al. 1990). Nearly identical maps have been used in numerous works on Texas including Gould (1962) and Correll and Johnston (1970).

© TIANA REHMAN

FIG. 16. Photograph of Enchanted Rock, a rich pteridophyte habitat in the Central Mineral Region of Texas.

FIG. 17. Color map of annual precipitation (in inches) averaged over the period 1971–2000. Used with permission of Spatial Climate Analysis Service, Oregon State Univ. © 2006.

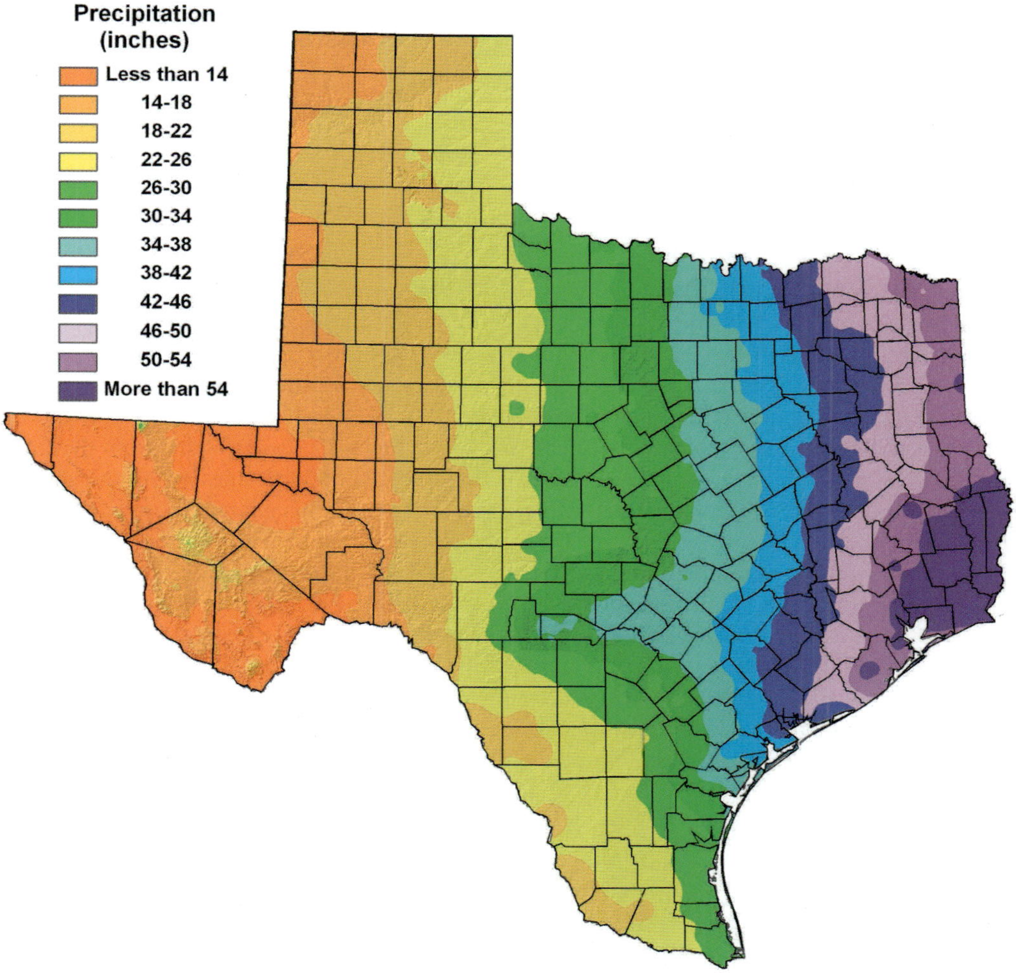

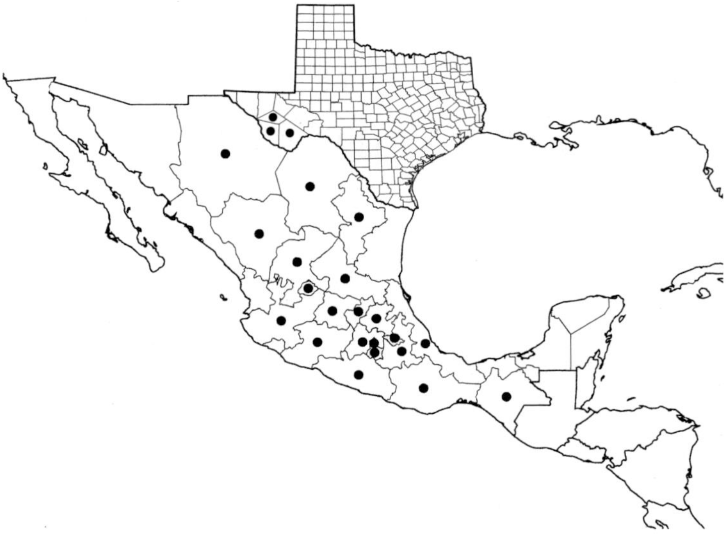

FIG. 18. Distribution of *Pellaea cordifolia*, HEART-LEAF CLIFF-BRAKE, a species widespread in Mexico and known in U.S. only from the Trans-Pecos of Texas. This species is among the 63% of Texas native pteridophytes also known from Mexico, most of these being dry-adapted species. Map adapted from Turner et al. (2003) and Mickel and Smith (2004).

southern location are among the most important factors resulting in the diversity seen in Texas ferns and lycophytes.

THE LIFE OF FERNS—STRUCTURE AND REPRODUCTION

Structure

Ferns and lycophytes are quite variable in structure and form. Living species range from tropical tree ferns that reach 24 m (nearly 80 ft) in height to tiny free-floating aquatics less than a cm (approximately 0.4 in) in total size in the genus *Azolla*. Leaves (also called fronds) also vary widely both in form and in size—from many meters long to only about 1 mm (approximately 1⁄32 of an in) long. The varied appearance of Texas pteridophytes is illustrated in Figure 20.

With relatively few exceptions, most Texas species are perennial herbs with a stem that is typically creeping and is either underground (called a rhizome), in rock crevices, or on the surface of rocks or tree limbs. Often, a few relatively large leaves arise from this stem and provide the entire height of the plant. In most ferns the young leaves develop in a characteristic pattern—these immature leaves are referred to as croziers or fiddleheads. The word fiddlehead, referring to the tight coil-like scrollwork at the top of a violin, also aptly describes the developing fern leaf that uncoils as it matures. The immature leaves are alternatively known as croziers, after the stylized coil at the top of the staff of high-ranking officials of some Christian religions. This is thought to be in reference to the curved staffs traditionally carried by shepherds.

In order to understand ferns, a few basic terms are helpful. Figure 21 shows basic terminology associated with a typical fern, while Figure 22 illustrates some frequently used terms associated with fern leaves.

The Sexual Life Cycle

Ferns and lycophytes have a sexual life cycle that is more complicated than that of most animals—they have two very different free-living phases or plant bodies that alternate during the life cycle. One is much larger and obvious and called the sporophyte, while the other is much smaller (often smaller than a thumbnail) and known as the gametophyte. These separate independent plant bodies are connected in the life cycle as shown in Figure 23 and as explained below.

Fig. 19. A fern habitat in the Trans-Pecos.

The type of fern body that is familiar to most people, because it is larger, is the sporophyte. It is usually diploid (2*n*), with 2 sets of chromosomes, similar in this respect to ourselves and a variety of other vertebrates. The sporophyte is so named because it produces microscopic reproductive structures called spores. As noted by Simpson and Dyer (1999), a spore "...is a single cell so small that up to about 100 could be placed on a newsprint full-stop [a period]." Even the largest of the usually dust-like spores are only about the size of a grain of sand. The spores are typically produced on the leaves within tiny structures called sporangia. These structures are grouped together into clusters known as sori (singular: sorus—from the Greek word meaning heap). The sori are frequently noticeable as dark spots on the backs of fern leaves. In many species each sorus is protected by a very delicate specialized covering known as an indusium or by the curved edge of a leaf, a false indusium. In many ferns the tiny globe-like sporangia are stalked and under drying conditions actually "throw" or catapult the microscopic spores away from the leaf surface. Because their reproductive structures are not obvious (like seeds or flowers) and because ferns and lycophytes have vascular tissue, in the past they were often called by the now nearly obsolete term vascular cryptogams (Greek: *cryptos*, "hidden" and *gamos*, "to marry").

As noted in an earlier section, all ferns and lycophytes are seedless. Reproduction occurs via their microscopic spores. These spores are created by a special type of cell division called meiosis or reduction division that results in spores with one-half the number of chromosomes (*n*) found in the cells of the sporophyte (2*n*). This reduction in chromosome number is crucial, because in sexually reproducing species each parent contributes chromosomes to the offspring. Without a mechanism to produce cells with half as many chromosomes, two organisms cannot have their reproductive cells combine in such a way that the resulting offspring have the same number of chromosomes as the parents. If this halving of chromosome number did not occur in sexually reproducing species, each successive generation would have twice as many chromosomes.

These spores subsequently undergo mitosis (regular cell division where chromosome number remains constant) and develop into the other plant body, the gametophyte. This body is a tiny, inconspicuous, separate, free-living, multicellular plant also called a prothallus. It is typically haploid (*n*), with one-half the chromosomes/genetic material of the sporophytes. The fully developed gametophyte is typically only about the size of a thumbnail or less, and is usually quite dependent on constant moisture. These tiny plants, which look almost like mosses or liverworts, are referred to as gametophytes, because they ultimately produce the gametes or sex cells—eggs and sperm, each with only half the chromosomes (*n*) typical of the species. Eventually, if conditions are appropriate, a swimming sperm fertilizes a stationary egg resulting in the production of a zygote, which reestablishes the correct diploid (2*n*) number of chromosomes. This zygote is thus the first cell of

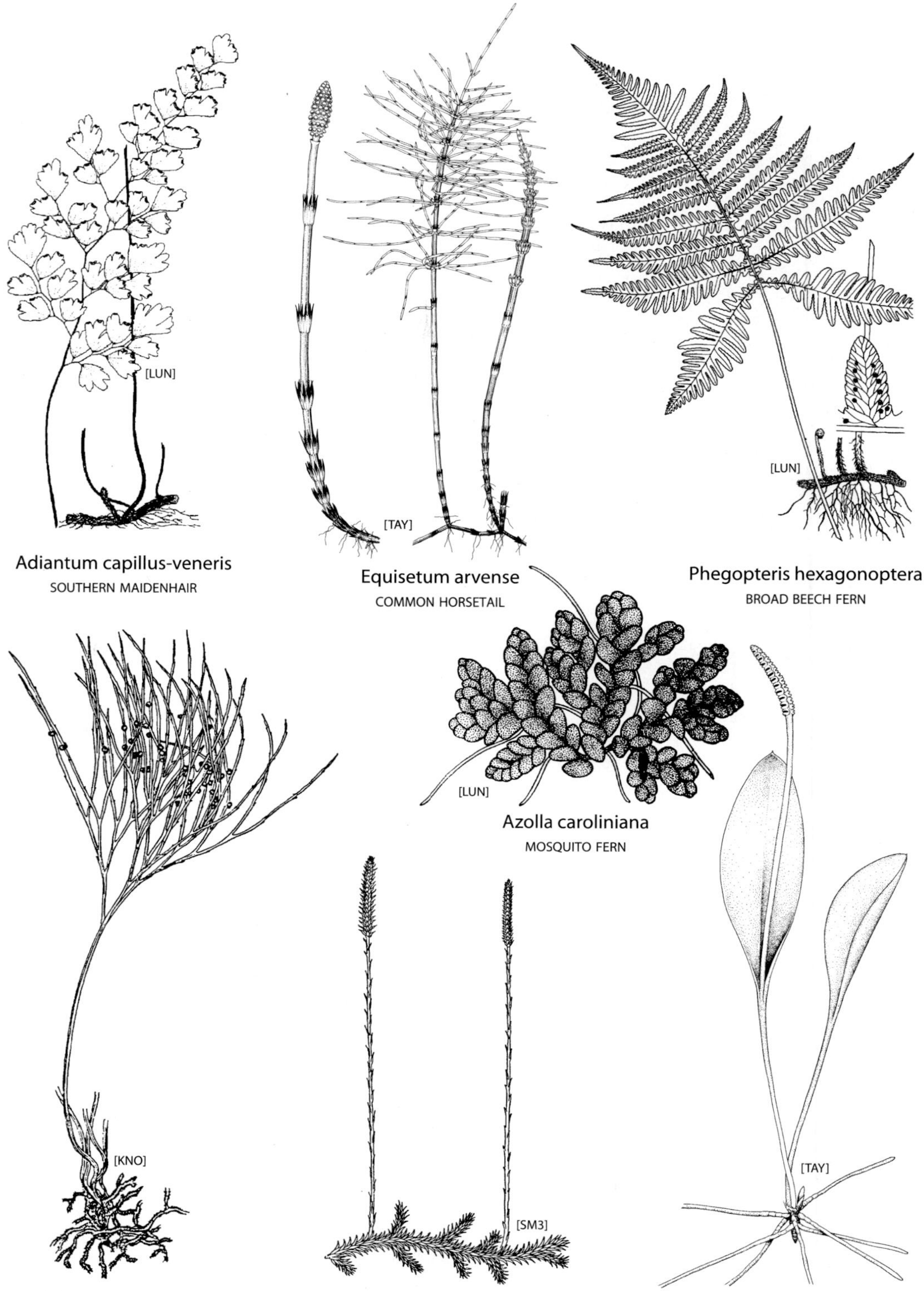

Fig. 20. Varied appearances of Texas pteridophytes.

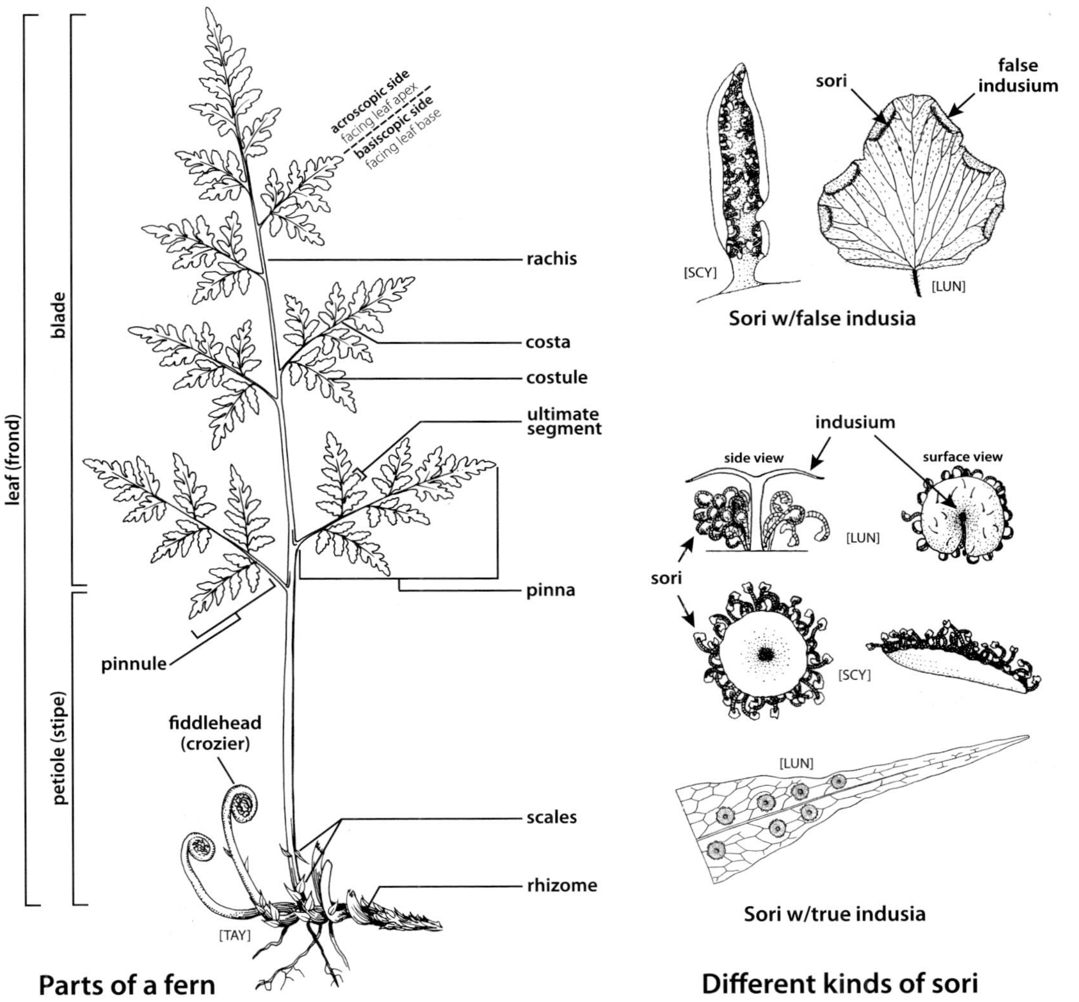

FIG. 21. Basic fern terminology.

the new sporophyte generation. The baby sporophyte—an embryo—then grows into the type of familiar sporophyte plant body we began with. Over time ferns and lycophytes go back and forth between a 2*n* sporophyte phase or generation and a 1*n* gametophyte generation—hence this type of life cycle is often referred to as an alternation of generations. It should be noted that although some gametophytes can produce both eggs and sperm, typically a sperm from one gametophyte swims to and fertilizes an egg produced by a different gametophyte (intergametophytic crossing). This cross-fertilization or outcrossing increases genetic variation, the driving force behind sexual reproduction. In fact, a variety of mechanisms have evolved in ferns and lycophytes to promote outcrossing and prevent inbreeding and its adverse genetic consequences, collectively known as inbreeding depression.

One might ask, why do fern and lycophyte species bother with this bizarre life cycle with two strikingly different free-living forms? The answer is that these different forms can serve very different functions or exploit different environmental conditions far more efficiently than either could do alone (Hughes & Otto 1999; Sheffield 2008). Sporophytes are typically perennial and can persist indefinitely, often producing incredible numbers (100s of millions) of spores over many years.

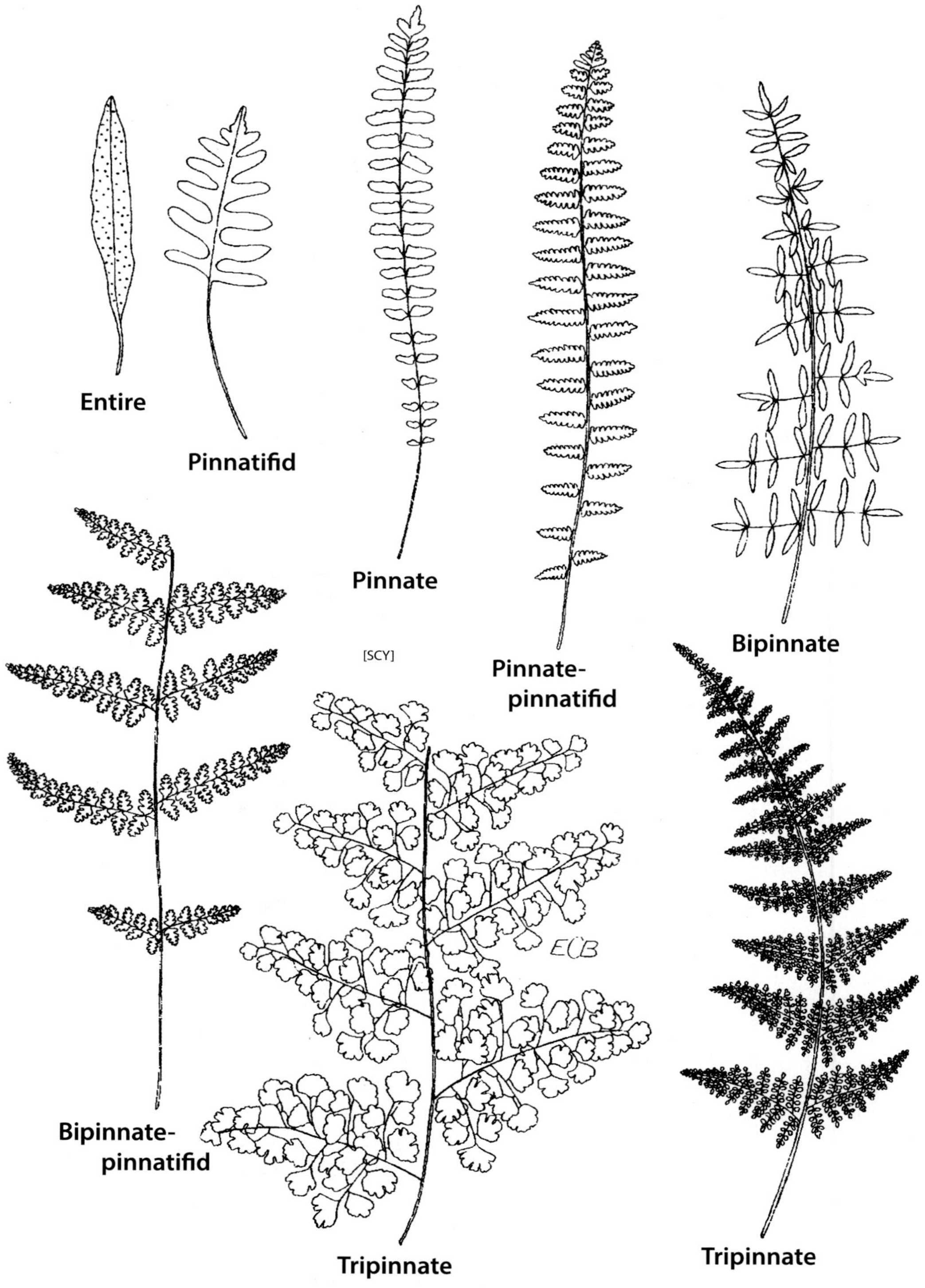

FIG. 22. Frequently used terms for fern leaves.

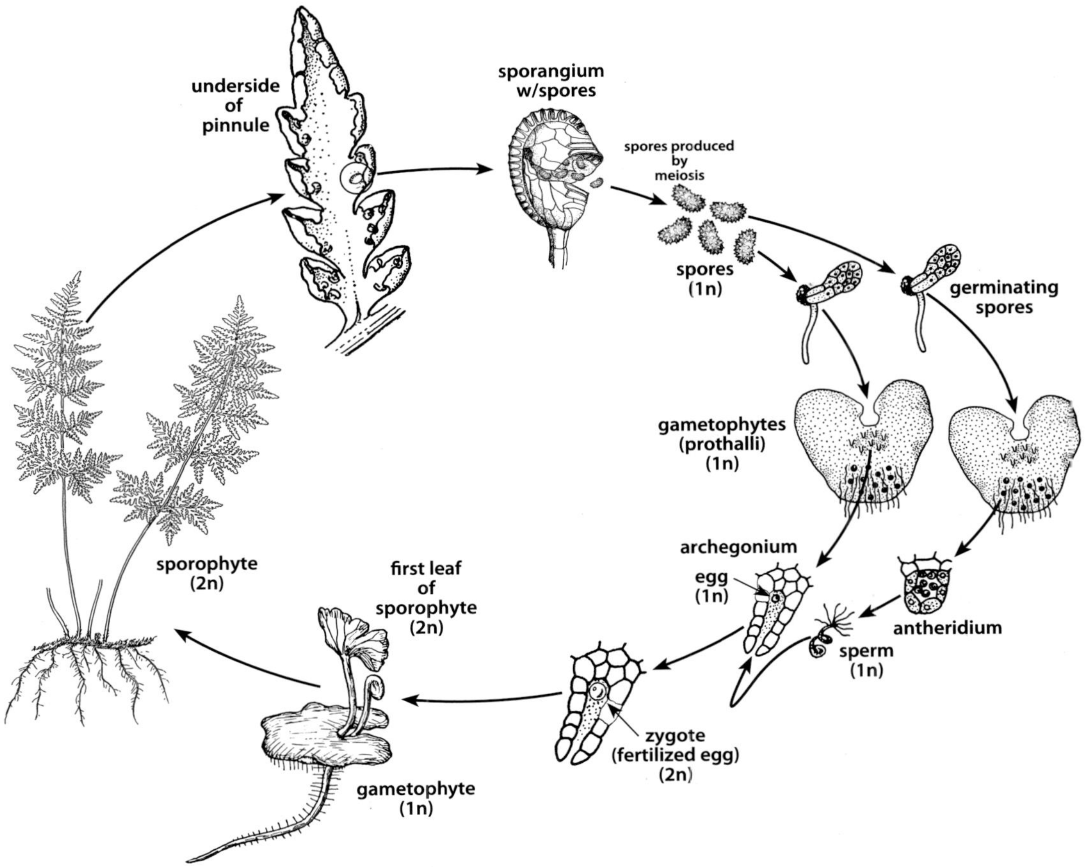

Fig. 23. Typical fern life cycle based on *Cheilanthes wrightii*, Wright's lip fern. Various components from and modified from Correll (1956), Dittmer et al. (1954), Taylor (1984), Robbins and Rickett (1939).

In contrast, spores, the first cells of the gametophyte generation, can travel very long distances, and then grow into gametophytes that can potentially produce both sperm and eggs. These gametes can then produce new sporophytes via fertilization. The two generations are thus specialized and very efficient in quite different ways, ultimately increasing the potential for survivability of the species.

Modifications of the Fern Life Cycle

As described above and illustrated in Figure 23, ferns and lycophytes, like the majority of other plants and animals, normally reproduce sexually. This process by definition involves the fusion of gametes, an egg and a sperm, typically from different gametophytes. Although there is a great deal of research into the significance of sexual reproduction, the enormous variation generated via the process seems to be critically important for organisms living in a world where conditions constantly change. Sexual reproduction ensures that offspring are unique combinations of genetic material from both parents—this variation in the offspring increases the probability that at least some of them will survive despite a changing environment.

However, there are numerous deviations from the basic life cycle described above, many of which are related to sex. For example, in a minority of fern and lycophyte species (probably less than 10%) a single bisexual gametophyte can produce both sperm and eggs that can successfully join (known technically as intragametophytic selfing) (Ranker & Geiger 2008; Schneller 2008). In these species it only takes a single spore, developing into a bisexual gametophyte, for a new sporophyte to be produced. The disadvantage is that since this is "self-fertilization," no genetic variation is generated. However, there are a variety of ecological factors that favor such gametophyes that can "self-fertilize." For example, pioneer or colonizer species that inhabit transitory environments or species that inhabit rare environments (e.g., rock crevices) need to be particularly good at dispersing and have thus been selected to be able to produce a new sporophyte following the dispersal of a single spore. This is critical when "the chance of more than one spore arriving at a site is highly unlikely" (Schneller 2008; also Peck et al. 1990).

Heterospory

Another variation in life cycle concerns the type of spore produced. Most ferns make only one kind of spore (they are referred to as homosporous), leading to a gametophye which can potentially make either eggs or sperm or both. However, a few species produce two different kinds of spores (they are termed heterosporous). One type of spore develops into a gametophyte only capable of making eggs, the other into a gametophyte that makes only sperm. Worldwide, only seven known genera of ferns and lycophytes have two spore types. Six of the seven known heterosporous genera occur in Texas: *Azolla*, *Isoëtes*, *Marsilea*, *Pilularia*, *Salvinia*, and *Selaginella*; the exception, *Regnellidium,* is native to South America.

Asexual Reproduction

Yet another deviation from the basic life cycle is asexual reproduction. Many ferns and lycophytes can reproduce both sexually and asexually. In some of these cases asexual reproduction serves as the primary but not the sole way of reproducing, while in other instances it serves only as a backup. However, it has been estimated that 10% of ferns and lycophytes can only reproduce asexually—for these species it is their only mode of reproduction (Walker 1984; Gastony & Windham 1989; Haufler 2008). Some species, such as *Pteridium aquilinum*, while capable of reproducing sexually, can spread into large clonal colonies by the formation of additional individuals from long, creeping rhizomes. The tiny floating aquatic *Azolla caroliniana* reproduces primarily via fragmentation of existing plants—using this mechanism the populations can grow extremely rapidly. This species can become so abundant that at certain times of the summer it can turn the surface of a pond from green to a striking red color. The invasive aquatic *Salvinia molesta* only reproduces asexually and is so effective that populations are reported to double in little more than two days (Moran 2004). Other species such as *Cystopteris bulbifera* (with the appropriate common name BULBLET BLADDER FERN) and *Asplenium palmeri* have evolved unusual ways of reproducing asexually. In the BULBLET BLADDER FERN the leaves produce bulblets that can grow into new plants, while in *Asplenium palmeri* the tip of the leaves can end in a bud capable of giving rise to a new plant and rooting (Fig. 24). The new plants are thus genetically identical copies of the parent—in other words, clones.

Apogamy.—One particularly interesting kind of asexual reproduction seen in a number of Texas species (including *Pellaea atropurpurea*) is known as apogamy (literally, "without" "marriage"; from the Greek: *apo*, away from, and *gamos*, marriage), meaning the production of a sporophyte without fusion of sex cells. This process has gone by a variety of names, including apomixis and agamospory (Sheffield & Bell 1987; Gastony & Windham 1989; Sheffield). Whatever the name, gametophytes are formed and functional antheridia (which can produce sperm) can even be present, but not functional archegonia (and thus no eggs)—therefore there will be no fertilization. Rather the gametophyte tissue directly gives rise to the new sporophyte through regular cell division (mitosis). In apogamous species spore formation does not result in the usual halving of chromosome number; rather the spores (and thus the gametophytes that grow from them) have the same number of chromosomes as the sporophyte, the main plant body that produced them. Each spore grows into a tiny gametophyte that divides by mitosis to directly give rise to a new sporophtye—without any gametes involved (Fig. 25). In other words, both the halving of chromosome number by meiosis and the doubling by fertilization have been bypassed. In essence the new sporophyte is simply a genetically identical clone of the parent sporophyte, no different than if a cutting had been made, a rhizome had sprouted a new stem, or a leaf tip had taken root and formed a new plant.

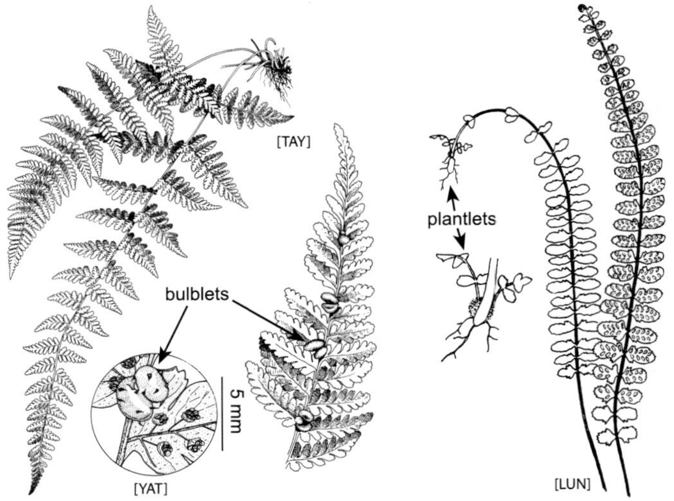

Fig. 24. Examples of vegetative reproduction in Texas ferns showing bulblets in *Cystopteris bulbifera*, BULBLET BLADDER FERN (left), and plantlets in *Asplenium palmeri*, PALMER'S SPLEENWORT (right).

This method of asexual reproduction is quite common in species that live in dry environments and has two major advantages (Sheffield 2008). First, the life cycle step from gametophyte to sporophyte does not require water, because there will be no fertilization. Second, it results in a much shorter life cycle. Since gamete formation and fertilization are skipped, the gametophyte simply develops a new sporophyte directly from gametophyte tissue by budding. This reduces the risk of delicate tissues drying out, and increases the chance of survival. Moran (2004) estimated that 5 to 10% of ferns worldwide are apogamous, many of these in dry areas.

Since ferns have evolved numerous mechanisms of asexual reproduction (e.g., creeping rhizomes, fragmentation, bulblets, rooting of leaf tips, etc.) one then might ask, what is the advantage of making spores that must develop into a delicate gametophyte before giving rise to another sporophyte? The answer, as Sheffield (2008) noted, is "... that spores have enormously greater dispersal potential than the more cumbersome bulbils, plantlets, offshoots, and stolons..." that non-apogamous species use to reproduce asexually. In other words, spores let apogamous species spread efficiently over much greater distances. However, there is a trade-off—recent research (Beck et al. 2011) suggests that in the genus *Astrolepis* (STAR-SCALED CLOAK FERNS; Pteridaceae) the apogamous asexual species are on average much younger evolutionarily than the sexual parental species they evolved from. In other words, apogamous species appear to suffer from a higher rate of extinction. Thus it seems that "...asexuality limits evolutionary potential in *Astrolepis*."

Polyploidy

Yet another reproductive complication is that many ferns and lycophytes have more than two sets of chromosomes—a condition known as polyploidy, which often results from an error in meiosis (the type of cell division normally resulting in the halving of chromosome number). Plants are much more flexible than animals in being able to survive with different numbers of chromosomes. Why this is the case is not completely clear. The standard situation in most organisms, including humans, is to have two sets of chromosomes, known as the diploid condition. Many plants, however, have more than two sets. For example, a number of ferns have three sets and are known as triploids. Some Texas examples include *Asplenium resiliens*, certain individuals of *Astrolepis cochisensis*, *Notholaena aliena*, and *Pellaea atropurpurea*. These three sets cannot be divided into two equal groups as is necessary in order for sexual reproduction to occur—as a result such plants

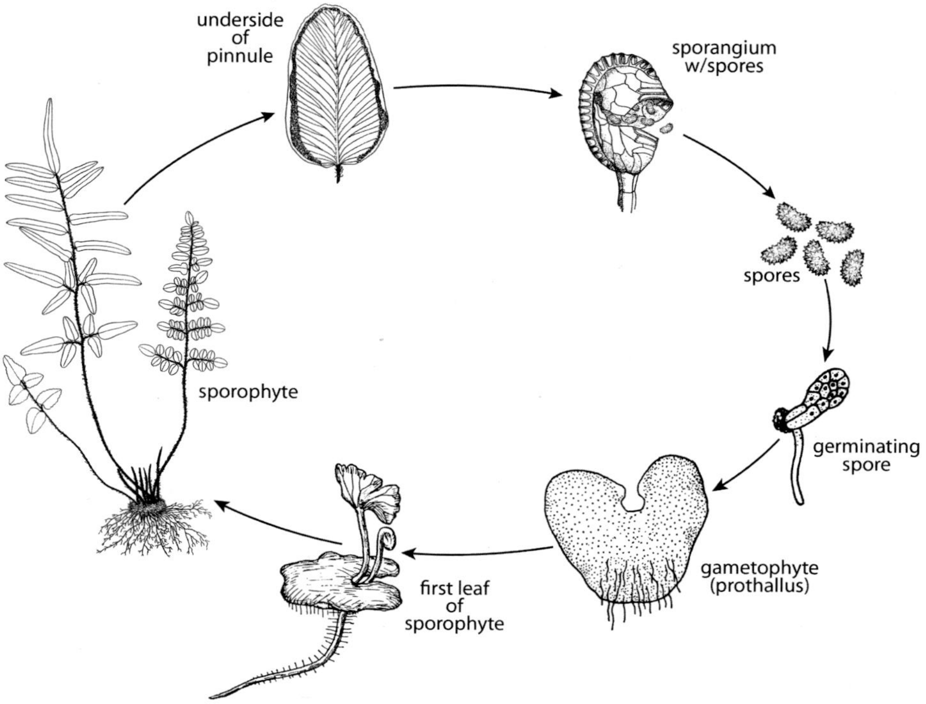

Fig. 25. Life cycle of *Pellaea atropurpurea*, PURPLE CLIFF-BRAKE, a triploid species that reproduces via apogamy. Such asexually reproducing species do not undergo fertilization and thus have a rapid life cycle less dependent on water. Various components from and modified from Correll (1956); Taylor (1984); Robbins and Rickett (1939).

can only reproduce asexually. Although this might seem to be a serious disadvantage, ferns have evolved remarkably effective mechanisms of asexual reproduction (such as apogamy discussed earlier) and these species are often widespread and common. Other species have four sets of chromosomes (tetraploids) or some other even number. These plants are generally fertile because the even number of sets can be divided during meiosis into two equal groups, and the resulting spores are fertile. Such polyploids can result from chromosome doubling within a species (autopolyploidy) or from doubling following hybridization as a result of the union of gametes from two different species (allopolyploidy). As it turns out, polyploidy is an extremely important mechanism for the formation of new species of ferns—most ferns living today have evolved via polyploidy, including many that occur in Texas. In fact, roughly one-third to one-half of fern species that have been studied in detail are polyploids of recent origin, and polyploidy is thought to have been involved in the evolution of at least 95% of all ferns at some point in their history (Haufler 1987, 1996; Schuettpelz et al. 2008; Wood et al. 2009; Barker & Wolf 2010).

Adaptations to Dryness and Other Special Conditions

Texas pteridophytes have a variety of unusual adaptations that allow them to thrive in their respective environments. Because more than half of the state's species live in seasonally dry or even desert areas, many are remarkably well-adapted to dry (referred to as xeric) conditions—they have evolved to not just survive, but to thrive in the face of limited water, high temperatures, high light intensities, and harsh soil conditions (Hevly 1963). There is not one best solution to these environmental challenges, and dry area species have evolved an impressive variety of ways to either avoid or tolerate the extreme conditions that would make life impossible for most ferns (see summary in

Table 3). In fact, most species probably use a combination of these strategies in the face of extreme conditions.

Avoiding Drought

Some species simply **avoid** dry conditions by losing their leaves—they die back to their rhizomes and become dormant during periods of drought. They may appear dead or under severe stress, but are simply waiting for rain to spring back to activity. Others avoid drought altogether by effectively becoming annuals. For example, *Marsilea mollis* grows in mud or shallow water in desert washes, but if drought occurs, the plants die, and after heavy rains their life cycle begins again with spore germination. The spores have survived the dry period by being protected in very desiccation-resistant, modified, bean-shaped pinnae called sporocarps (Hietz 2010). Other dry area species occupy rare microhabitats with adequate year-round moisture such as around seeps or springs or in sheltered rock crevices (Andersen 1996).

Desiccation Tolerance

Other desert-adapted species are amazingly **desiccation-tolerant**—their tissues are able to lose more than 90% of their water and still survive (Alpert 2000). Despite being able to withstand repeated cycles of drying and wetting, when dry the brown leaves appear dead, are quite delicate, and basically crumble into powder if treated roughly. Examples of such desiccation-tolerant plants include species of *Cheilanthes, Notholaena, Pellaea, Pleopeltis*, and *Selaginella*. In fact, some authorities now estimate that 5–10% of all fern and lycophyte species can tolerate desiccation (Hietz 2010). Although some plants are able to withstand desperately dry conditions for only limited periods of time, others, including a number of species of *Cheilanthes*, *Pellaea*, and *Selaginella*, can survive in an "air-dry state" for several years. For example, the lycophyte *Selaginella lepidophylla* essentially dries out: its flat main branches curl inward, the entire plant becomes ball-like and brownish, and it goes into a virtual state of suspended animation until moisture is again available. Within hours of wetting, the plant uncurls and quickly becomes green and active, possibly surviving years of being dry—the name RESURRECTION PLANT seems quite appropriate (Fig. 26). Even the gametophytes of some fern species have evolved to be remarkably desiccation-tolerant. Experiments show that some can survive five years of desiccation (Pickett 1931).

Preventing Water Loss

Many other species tolerate dry conditions by having evolved a multitude of ways to **prevent the loss of water.** Some species have thick cuticles and leathery leaf textures that minimize water loss, or special scales, hairs, or waxes that protect the leaf surface. Others have leaves that curl or roll up under drought conditions, thereby reducing the amount of exposed surface and thus moisture loss. Still others have very small leaves with little surface area through which to lose water. Texas ferns and lycophytes display all of these adaptations. For example, *Pellaea* species, such as *P. atropurpurea*, have thickened, very leathery leaf textures that both prevent water loss and maintain leaf shape when they do dry out. Other members of the genus, such as *P. truncata*, have a thick cuticle that gives the leaf blades a striking bluish-green or grayish green color.

Probably the most noticeable of these adaptations are the variously colored and shaped scales and hairs. For example, *Astrolepis* (STAR-SCALED CLOAK FERNS) have intricately shaped, often overlapping scales that can partially or in some cases even completely cover leaf surfaces. These are particularly important on the lower leaf surfaces where they protect the stomates, the gas exchange pores through which water loss can also occur. The scales obviously shade the leaf and reflect light, thereby reducing the amount of light (and thus heat) reaching the leaf surface. In other words, they act as "... insulators against heat and reflectors of incoming solar radiation" (Hevly 1963). In addition, they are also thought to decrease water loss by creating air spaces next to the leaf surface, where the air remains at a relative humidity approaching that inside the leaf (Hevly 1963), thus greatly reducing the movement of dry desiccating air across the leaf surface. The scales also are able to trap dew and rainwater, thus at least temporarily increasing moisture adjacent to the leaf surface. For example, the lower leaf surfaces of *Astrolepis windhamii* are completely covered with visually obvious lanceo-

TABLE 3. Adaptations to avoid and tolerate dry conditions. Many desert species utilize a variety of different methods, and some mechanisms serve more than one purpose.

General Category	Mechanism	Explanation
Avoiding/Evading Drought	Drop leaves during drought	Reduces surface area & thus water loss
	Drought-induced dormancy	Plants become inactive, reducing need for water
	Annual or ephemeral habit	Plants die with drought and spores start new generation upon rains
	Occupy microhabitats with moisture	Adequate water supply maintained
Preventing Water Loss	Thick cuticle	Reduces water loss through leaf surface & in some cases reflects light
	Leathery leaf texture	Reduces water loss through leaf surface & reduces damage caused by water stress-induced tissue collapse
	Scales & hairs	Reduces heat & light reaching leaf; creates humid air spaces next to the leaf surface; trap dew and rainwater
	Whitish or yellowish, wax-like farina	Reflects light (and thus heat) and acts as a barrier preventing water loss
	Leaf curling	Reduces effective surface area and thus water loss & light exposure
	Small leaf size (microphylly)	Reduces surface area & thus water loss
	Leaves divided into many tiny segments	Allows better heat dissipation through convection—reduces heat load
	Leaf orientation	Reduces direct sun exposure and thus excess light and heat
Desiccation Tolerance	Physiological & biochemical mechanisms	Reduces damage associated with water loss (e.g., production of special molecules to prevent membrane damage)
	Decreased cell size	Less chance of cell damage/change in cell shape upon desiccation
	Photoprotection	Prevents light-induced damage see below and text
	Desiccation-tolerant gametophytes	Allows gametophyte survival & thus reproduction
	Osmotic protection	High sugar content in cells prevents water loss-related damage
Photoprotection—from excessive light	Leaf orientation	Minimizes direct light exposure & thus protects from free radical damage caused by photoinhibition (reduced photosynthesis)
	Leaf curling	Minimizes light exposure
	Wax-like farina	Minimizes light exposure
	Scales & hairs	Minimizes light exposure
	Biochemical mechanisms	Special pigments and antioxidants reduce light-induced free radical damage
Rapid Rehydration	Scales with dead disk-like portion and a stalk of living cells	Absorbs water & channels it into the leaf allowing rehydration in as little as 15 minutes
Reproductive	Apogamy (asexual reproduction with sporophyte formed directly from gametophyte without gametes)	Allows reproduction under unfavorable moisture conditions

late scales (Fig. 27). The scales of some *Astrolepis* species can have very intricate shapes. These stellate or star-shaped, multi-branched scales are useful not only in providing desiccation resistance but facilitate species identification by botanists (Fig. 28). In a similar manner, some *Cheilanthes*-species (LIP FERNS) have such dense coatings of hairs that the actual leaf tissue is obscured. Like scales, such hairs both reflect light and prevent loss of water.

Another easily observed adaptation is the obvious dense undercoating of a wax-like material. The genus *Notholaena* (CLOAK FERNS) is an excellent example. *Notholaena* species typically have the lower leaf surfaces completely covered with a striking whitish or yellowish farina (a wax-like flavonoid produced by short glandular hairs) (Fig. 29) (Wollenweber & Schneider 2000). This material reflects light and also provides a barrier preventing water loss, allowing *Notholaena* species to survive under extremely hot and dry conditions. A specific example is *Notholaena standleyi*, whose leaf segments under dry conditions "...are rolled inward to form a tight yellow whitish ball" (Correll 1956). Surprisingly, when stressed by lack of water, their leaves curl so that the lower (farina-covered) surfaces are exposed to the sun. This might seem counterintuitive, since the stomates, located on the underside become more exposed, a fact that may signify the importance that farina has both on preventing water loss and for its light-reflective properties.

TOP: © ALAN CRESSLAR; BOTTOM: © PETE I GUTH

Fig. 26. Photograph of RESURRECTION PLANT, *Selaginella lepidophylla*. Top: desiccated; bottom: hydrated.

Numerous species in a variety of genera have very small leaves (microphylly) that are finely divided into tiny segments. In fact, almost all Texas ferns of seasonally dry or desert habitats exhibit microphylly, the most common adaptation for xerophytic ferns (Hevly 1963). *Argyrochosma dealbata* is an excellent example—its tiny, 3 to 5-pinnate leaves reach at most 15 cm long (less than 6 in) and are divided into dozens of tiny separate segments. Having small leaves with relatively little surface area reduces water loss, while the reduced segment size helps dissipate heat and thus prevents overheating and tissue death (Hevly 1963). By contrast, a large continuous leaf surface has a deep boundary layer of stagnant air, inhibiting heat dissipation through convection (= transfer of heat by air moving across the surface). When a leaf consists of many tiny segments, the boundary layer over a small surface is much shallower, and heat dissipation through convection is more efficient. For example, a "leaf [or separate leaf segment] up to ½ inch (10 mm) across can stay below the lethal tissue temperature of about 115°F (46°C) on a calm day with its stomates closed. A larger leaf requires transpiration through open stomates for evaporative cooling" (Dimmitt 2009). Therefore, having leaves divided into tiny segments allows a fern to survive higher temperatures while preventing the loss of its most critical resource: water. The bottom line is that large leaves are just too difficult or costly (in terms of water loss) to keep cool.

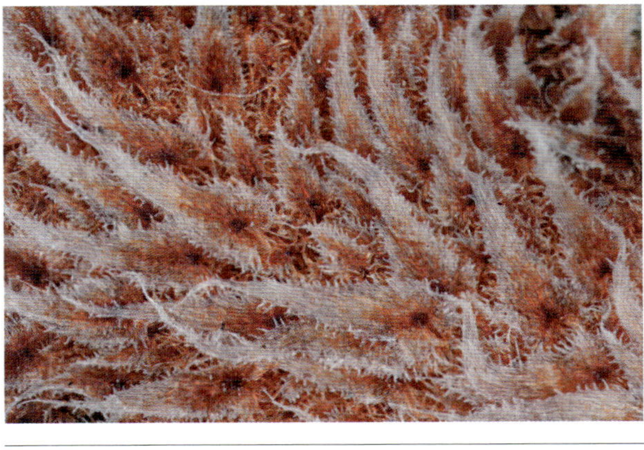

© DALE A. ZIMMERMAN HERBARIUM, WESTERN N. MEX. UNIV.

Fig. 27. Example of scales completely covering the lower leaf surface of *Astrolepis windhamii*, WINDHAM'S SCALY CLOAK FERN.

Other desert ferns have evolved more dramatic adaptations to survive their harsh climates. During dry weather, the ultimate leaf segments and sometimes the entire leaves of these plants curl up in a manner that protects the lower leaf surface. In fact, this is the norm in species whose lower surfaces lack a dense layer of white or yellow farina or a thick coating of hairs or scales to protect the stomates. This type of curling minimizes the total amount of leaf surface exposed, and most importantly, shields the stomates. Water loss can thus be significantly reduced. The exact way in which leaves curl is species-specific—many species curl so as to protect the underside of the leaves. However, some species (e.g., those with farina or dense undercoatings of scale or hairs) curl in the opposite direction, purposely exposing the highly light-reflective, farina-covered or otherwise well-protected underside as we have seen in the case of *Notholaena standleyi*, on page 27. In these species, the unprotected upper surface apparently most needs the benefits provided by curling. In addition to reducing the amount of exposed surface subject to water loss, curling can also dramatically decrease the leaf's total exposure to light, which, as explained below, may be of great importance.

© DALE A. ZIMMERMAN HERBARIUM, WESTERN N. MEX. UNIV.

Fig. 28. Example of stellate (star-shaped) scales on leaves of *Astrolepis sinuata*, WAVY CLOAK FERN.

The variation in curling and the dramatically different appearances of the upper and lower leaf surfaces in numerous species (e.g., *Bommeria hispida*, *Cheilanthes lindheimeri*, *Notholaena copelandii*) show the complexity of adaptation to dry environments. The often nearly scaleless and hairless upper leaf surfaces may in general allow relatively unobstructed air movement and thus better dissipation of heat through convection, while the lower surfaces are generally better protected. However, this generalization has many exceptions—there are dry-adapted species with virtually no scales or hairs on either leaf surface, and others with scales, hairs, or farina on both.

© DALE A. ZIMMERMAN HERBARIUM, WESTERN N. MEX. UNIV.

Fig. 29. Example of wax-like farina on leaves of *Notholaena standleyi*, STAR CLOAK FERN.

Photoprotection—From Excessive Light

While many adaptations are focused on reducing water loss, mechanisms to cope with too much light are also important in desiccation-tolerant ferns and lycophytes. When leaves lose much of their water, the process of photosynthesis is inhibited, a phenomenon known as photoinhibition. However, light is still absorbed by chloroplasts, producing free electrons that would normally be passed along the biochemical pathway of photosynthesis to make food for the plant. But when photosynthesis is inoperative, the electrons cannot be used normally and instead are passed to other highly reactive oxidizing molecules. These wreak havoc in the plant's tissues—irreversible damage and death of the leaves can result from such photo-oxidation. Dry area plants have evolved a variety of ways of coping with this. Some species simply orient their leaves so as to minimize direct sun on the leaf surfaces. Leaf or stem curling is another of the obvious physical ways—in the curled condition very little leaf tissue is exposed to light. The light-reflective, wax-like farina on leaf surfaces of *Notholaena* species is probably another mechanism to reduce the amount of light actually reaching the photosynthetic tissues. Scales and hairs do the same thing. Biochemical mechanisms are also important in providing photoprotection—special pigments and antioxidant molecules have evolved that reduce the damage associated with excessive light. Such biochemical protection is present in numerous Texas species including *Selaginella lepidophylla* and *Pleopeltis polypodioides* (Hietz 2010). There is a variety of other biochemical and physiological adaptations to dry conditions that, although not visibly obvious, are nonetheless essential—heat-tolerant enzymes are just one more example. In fact, many species that can tolerate extreme desiccation will survive only if they dry out slowly and have time to make gradual physiological adjustments—for example, some *Selaginella* species must dehydrate over a period of five to seven days; if they dry out more quickly, they die (Dimmitt 2010).

Rapid Rehydration

A rather different type of adaptation to xeric conditions can be seen in *Pleopeltis polypodioides*, which grows as an epiphyte. As a result of its soil-less habitat, it is often under conditions of extreme moisture stress, and uses epidermal scales in a somewhat different way. The common name, RESURRECTION FERN, refers to the leaves "... which become brown and appear dead during dry periods, but 'resurrect' themselves after rains and rapidly become green and lush" (Nelson 2000). This occurs because special scales on the lower surface of the leaf are able to absorb water and quickly channel it into the leaf. Each peltate scale is made up of a disk-like portion composed of numerous dead cells and a stalk of living cells. When water is present, it is drawn into the dead cells by capillary action, taken up by the living stalk cells, and subsequently transported into the leaf tissue. The uptake can occur in as little as 15 minutes (Moran 1998, 2004) and the plant can quickly resume metabolic activity. This can easily and quickly be demonstrated by placing the dried plant in water with a little soap to reduce surface tension (Jack Stanford, pers. comm.). Such plants are able to go through numerous desiccation and rehydration cycles.

Reproductive Adaptations

A final category of adaptations to dry environments are those involving reproduction. As mentioned earlier, apogamy results in a shorter life cycle and also skips gamete formation and fertilization—thereby avoiding the most water-sensitive part of the life cycle. A different reproductive approach is seen in two species in the Trans-Pecos, *Cystopteris bulbifera* and *Asplenium palmeri*. Both of these ferns reproduce vegetatively, completely bypassing the gametophyte generation. In *Cystopteris bulbifera* the leaves produce bulblets that can grow into new plants, while in *Asplenium palmeri* the tip of the leaves can end in a bud capable of rooting and giving rise to a new plant (Fig. 24). *Marsilea mollis* also avoids drought using a reproductive adaptation. When its transitory wet environment dries up, the spores survive in sporocarps, which are highly modified, tightly folded pinnae. In fact, Marsileaceae apparently only produce sporocarps when their habitat dries out. The sporocarps then lie dormant (potentially for years) and only release their spores when water again becomes available. Further, in contrast to most ferns which take months or even years to produce mature leaves capable of making spores, Marsileaceae can produce spores quite quickly, an extremely valuable characteristic given the seasonal, transitory, or highly unpredictable environments such as those seen in deserts.

Aquatic Adaptations

In contrast to the ferns of arid regions, the tiny floating aquatics known as MOSQUITO FERNS (genus *Azolla*) face very different environmental problems, and have evolved special adaptations for life in a water environment. Their extremely small size ensures maximum contact with the water and allows for easy dispersal. The upper leaf lobes are hollow and inhabited by a symbiotic nitrogen-fixing cyanobacterium, *Anabaena azollae*, which combines or "fixes" atmospheric nitrogen with hydrogen to make ammonia, which can then be utilized by the plants (Moore 1969; Shi & Hall 1988; Hoshizaki & Moran 2001). This phenomenon is possibly an adaptation for a free-floating lifestyle, since obtaining nitrogen may be more difficult for a plant that is not rooted in a solid substrate.

Different aquatic adaptations are seen in the introduced free-floating species, *Salvinia molesta*.

For example, the hairs on the upper leaf surfaces are 4-branched, the branches joined at their tips—resembling a "cage" or an "egg beater" (Fig. 30). This layer of elaborate hairs traps air, causing the plants to right themselves if turned over in the water. The plants are virtually unsinkable, both because of air trapped between these special waxy hairs, and also because of the buoyancy provided by spongy tissue with large air spaces (aerenchyma) inside the leaf lobes. *Salvinia minima*, another introduced species, has hairs that function in a similar fashion, but in this case the branches of the 4-branched hairs are not joined at their tips.

© BARRY A. RICE

FIG. 30. Four-branched, "egg beater" hairs of *Salvinia molesta*, GIANT SALVINIA.

Protective Adaptations

Finally, other ferns have interesting adaptations not only to their physical environment, but also to their biological surroundings. As described in more detail in the various species accounts, many protect themselves from herbivores using elaborate and sometimes deadly chemical defense mechanisms that deter animals from eating them. For example, *Astrolepis cochisensis* produces an unidentified poison that in livestock causes incoordination, arched back, rapid pulse, gasping, trembling (known as "the jimmies"), and prostration; death can result from respiratory paralysis. Likewise, *Pteridium aquilinum*, the widely known BRACKEN FERN, makes a variety of toxic and carcinogenic compounds. If its tissues are damaged they can even release cyanide, and it can poison insects by making chemicals that mimic the insects' molting hormones. BRACKEN FERN is also famous as an example of allelopathy (the chemical inhibition of one species by another). Essentially, it is able to outcompete other plants by releasing inhibitory chemicals (called allelochemicals) into the environment that prevent the growth and development of nearby plants—thus it poisons its potential competitors.

WHERE TEXAS FERNS AND LYCOPHYTES OCCUR—BIOGEOGRAPHY

Where Texas ferns and lycophytes occur—their distribution patterns—turns out to be quite interesting. Several Texas pteridophytes have **disjunct** distributions, which means that some populations are widely isolated geographically from other populations. In some cases they are separated by deserts or mountain ranges, while sometimes they are divided merely by great distance. How can these things be explained? Biogeographers (scientists who study the distribution of organisms) have two basic explanations for this type of pattern. The disjunct population could have been established by **long-distance dispersal** (e.g., a spore was blown a great distance by the wind and happened to fall in a favorable spot[4]) or by **vicariance** (the species was once much more widespread but has become fragmented [vicariated] with survivors in only parts of its former range). Both long-distance dispersal and vicariance are important factors in pteridophyte distributions. Long-distance dispersal is sometimes jokingly referred to as the "constipated bird" explanation, since in the case of flowering plants birds sometimes spread seeds long distances by carrying them internally. Likewise, vicariance is sometimes referred to as the "survivor" explanation because widely separated populations are survivors of once much more widespread distributions.

Long-Distance Dispersal

Evidence suggests that the distribution patterns of many Texas pteridophytes are the result of long-distance dispersal. For example, four species are tropical in origin. *Adiantum tricholepis*, *Blechnum appendiculatum*, *Dennstaedtia globulifera*, and *Tectaria heracleifolia* are all rare in Texas and disjunct hundreds of miles north of others of their kind in one or a few special and unusual locations. How

[4]An even stranger possibility is that disjunct species found at or near cave entrances could have had their spores carried long distances on the fur of migrating, cave-dwelling bats.

FIG. 31. Disjunct distribution of *Pellaea wrightiana*, WRIGHT'S CLIFF-BRAKE. From Windham (1993c).

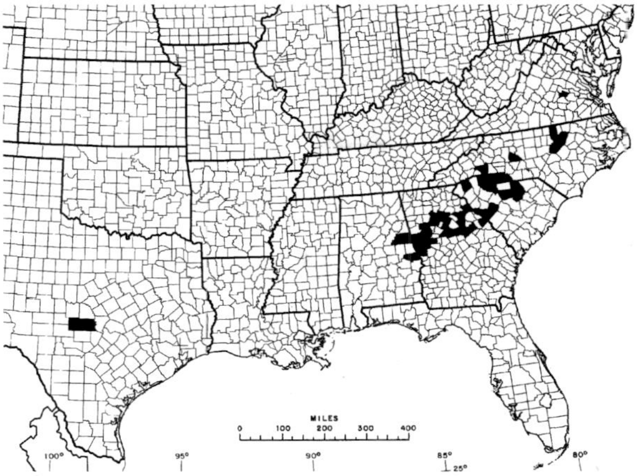

FIG. 32. North American distribution of *Isoëtes piedmontana*, PIEDMONT QUILWOFT. Modified from Heafner and Bray (2005).

they got to Texas is an excellent question. Fern spores are dust-like, lightweight, and incredibly numerous and can be blown hundreds or even thousands of miles, sometimes finding the perfect conditions for growth in tiny pockets of appropriate habitat (microhabitats). Moran (2004) calculated that a **single leaf** of a WOOD FERN (genus *Dryopteris*) only 25 inches long could produce 7,305,216 spores! Further, the spores of most fern species can remain viable at room temperature for 2–4 years, with a number of them capable of surviving for decades (Page 1976). Some can even survive passage through the gut of earthworms (Dyer & Lindsay 1992), storage in liquid nitrogen, or exposure to temperatures up to 70° C (Simpson & Dyer 1999). And for some ferns, particularly homosporous species adapted to transitory or rare environments, it only takes one spore to begin a new population. This is because in these dispersal-adapted species the gametophyte can make both eggs and sperm and is thus self-fertile. It therefore may give rise to a new sporophyte or, in the case of apogamous species, the gametophyte can directly give rise vegetatively to a new sporophyte.

At least seven Texas ferns display a similar, dramatic disjunction—they are found primarily in the southwestern or western U.S., with remarkably isolated disjunct populations in the southern Appalachian Mountains. A good example is *Astrolepis sinuata*. It has a broad distribution in the southwestern U.S., but an isolated population in Georgia represents a disjunction of about 1,530 km/950 mi. Likewise, *Pellaea wrightiana* is widespread in southwestern U.S. and then strikingly disjunct about 1,450 km/900 mi to the mountains of North Carolina (Fig. 31). Other examples are summarized in Table 4, and more information can be found in the species treatments. Long-distance dispersal seems the most likely explanation for most of these examples, though it is possible they are the "result of the regional expansion of entire xeric communities during past periods of elevated temperature" (Beck et al. 2012). In a sense, the opposite type of disjunction is seen in *Isoëtes piedmontana*. This species is widespread in the southeastern U.S. from Virginia to Alabama, but disjunct populations were recently described from two counties in the Llano Uplift area of central Texas, a separation of many hundreds of miles/kilometers (Fig. 32).

Another distributional anomaly is that of two species only known in Texas from caves in Val Verde County—*Dryopteris cinnamomea* and *Dennstaedtia globulifera*. Apparently the moist shady conditions near cave openings are pefect for these unusual members of the Texas flora. Such moist microhabitats are literally habitat islands in a sea of otherwise extremely inhospitable terrain, and are well known in other parts of the U.S. (Walick et al. 1996; Farrar 1998). These rare Texas plants are separated by long distances from larger populations of their species in Mexico. How their spores arrived in these unusual habitats is an interesting question: possibly by wind, or on the fur of migrating bats?

There are two even stranger disjunct distributions involving Texas ferns, the first shown by

TABLE 4. Disjunct distributions between the western or southwestern U.S. and the southern Appalachian Mountains.

Asplenium septentrionale, FORKED SPLEENWORT	Scattered in the mountains in the western U.S., then disjunct from Black Mesa in far western Oklahoma and the Chisos Mts. in the Trans-Pecos of Texas to one location in Arkansas and then disjunct again about 1,127 km/700 mi east to West Virginia
Astrolepis integerrima, SOUTHWESTERN CLOAK FERN	In western U.S. from Arizona and Nevada east to central Texas, then disjunct about 1,000 km/620 mi to rock outcrops in Alabama
Astrolepis sinuata, WAVY SCALE CLOAK FERN	Widespread in southwestern U.S. and disjunct about 1,530 km/950 mi/ to Georgia
Cheilanthes eatonii, EATON'S LIP FERN	Widespread in southwestern U.S. east to Arkansas, then disjunct about 1,046 km/650 mi to isolated stations in Virginia and West Virginia
Cheilanthes feei, SLENDER LIP FERN	Widespread in the western U.S. east to Arkansas and Missouri, then disjunct to one location in Tennessee and then disjunct again 450 km/289 mi to Virginia
Pellaea ternifolia subsp. *arizonica*, ARIZONA CLIFF-BRAKE	Populations in Arizona and the Texas Trans-Pecos, disjunct nearly 1,930 km/1,200 mi to South Carolina
Pellaea wrightiana, WRIGHT'S CLIFF-BRAKE	Widespread in southwestern U.S. and disjunct about 1,450 km/900 mi to North Carolina

Pellaea ternifolia. Populations occur widely in the New World from Texas south through Mexico to South America and then in Hawaii—a separation of at least 4,000 km/2,485 mi. By comparing molecular markers of Hawaiian ferns with those of ferns from other parts of the world, Geiger et al. (2007) were able to explain other similar New World–Hawaii fern distributions. They suggested four climate- or weather-based spore dispersal mechanisms to explain how ferns and lycophytes came to inhabit the very isolated Hawaiian Islands (Fig. 33). Of these four, trade winds and westward moving tropical storms are the most likely ways spores from the New World arrived in Hawaii. A second and perhaps even more surprising disjunct distribution is seen in *Asplenium platyneuron*. This species is widespread and common in eastern North America, including eastern Texas, but is remarkably disjunct across the Atlantic Ocean to southern Africa—a distance of approximately 12,000 km/roughly 7,500 mi. Wagner et al. (1993) noted that "no other North American fern has this distribution." The remote disjunct occurrence in Africa is almost certainly the result of the long-distance dispersal by spores (Moran & Smith 2001; Moran 2008).

Vicariance

As in the case with long-distance dispersal, there is evidence that vicariance can explain the modern distributions of a number of Texas ferns and lycophytes. Plants that today are separated by vast distances, sometimes continents apart, can be shown to have been continuously connected in the distant past. Thus their widely separated modern populations are lucky survivors or relicts of what were once much more widespread distributions.

The Asian Connection

Many Texas plants, including some pteridophytes, are only found in two widely separated parts of the world: eastern North America and eastern Asia. An excellent example is *Onoclea sensibilis*, widely known as SENSITIVE FERN, whose striking disjunction is best explained by vicariance (Fig. 34). In the distant geologic past, the position of the continents, climate, and sea level were sometimes quite different than at present. As a result, during a number of particularly warm spells plants could migrate between the Eurasian and North American continents across land bridges connecting North America to both Europe and Asia—between western North America and eastern Asia across the Bering Land Bridge (also called Beringia) and between eastern North America and western Europe across the North Atlantic Land Bridge. At some periods of the past, these land connections are known to have been extensive, providing excellent migration corridors. Even today, the Bering Strait between Alaska and Siberia is only about 85 km/53 mi wide and 30–50 m/98–160 ft deep. Compared to the dramatic changes known to have occurred in the geologic past, it would take a relatively modest change in sea level to once again connect these land masses.

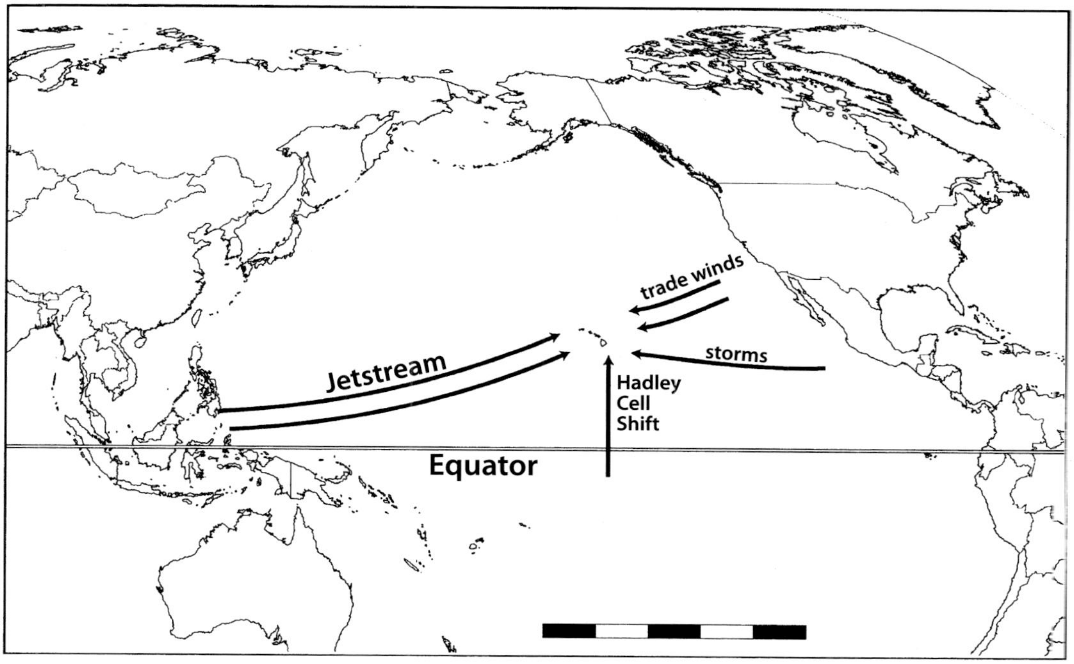

Fig. 33. Map of the Pacific Basin illustrating the possible climate- or weather-based phenomena that might lead to common paths of airborne dispersal of fern spores to Hawaii. From Geiger et al. (2007). Spores of *Pellaea ternifolia*, TRANS-PECOS CLIFF-BRAKE, (native from Texas south through Mexico to South America) were perhaps blown westward to Hawaii by trade winds or storms.

An example of the remarkable temperature difference in the far north between today and portions of the Tertiary geologic period can be seen by looking at the Paleocene/Eocene thermal maximum (PETM) dating to around 55 million years ago. That was a time of widespread, extreme climatic warming, during which the Arctic Ocean can only be described as subtropical, with surface temperatures over 23° C/73.4° F (Moran et al. 2006). Climate varied widely during the Tertiary, with plants able to move across high latitudes during a number of different warm periods.

The fossil record shows that many plants had extensive distributions across this well-connected and much warmer northern hemisphere. Good examples of ferns with fossils showing such a pattern include *Azolla, Onoclea, Osmunda, Osmundastrum, Thelypteris*, and *Woodwardia*. Temperate forests often with some tropical plants thus occurred very broadly, and reached their maximum extent in the mid-Tertiary during a period of unusually high global temperatures. Such forests extended much farther north than would be possible today, occurring in Alaska, Greenland, and northern Canada. During the middle Eocene (approximately 45 million years ago), lush forests, dominated by deciduous conifers but also with various angiosperms (including hickories and oaks) and ferns (e.g., *Osmunda*), occurred at latitudes of 75° to 80°, well north of the Arctic Circle. A particularly well-preserved example of such a forest is known from Axel Heiberg Island in the Canadian High Arctic (paleolatitude 80° N). *Onoclea* and *Osmunda* were abundant, *Equisetum* was common, and both angiosperms (e.g., *Betula*, BIRCH) and gymnosperms (e.g., *Metasequoia*, DAWN REDWOOD) were present (Greenwood & Basinger 1993, 1994; Jahren & Sternberg 1999; McIver & Basinger 1999; Collinson 2002). This widespread northern hemisphere flora has been variously referred to as the Arcto-Tertiary flora (since it occurred in the north during the Tertiary), and in more recent years, as the Boreotropical flora (since it was northern and included some tropical plants). Whatever the name, the resulting flora was characterized by a diverse combination of temperate deciduous and tropical plants, and developed into a widespread forest that "… was once more or less continuously distributed throughout the Northern Hemisphere" (Xiang et al. 1998).

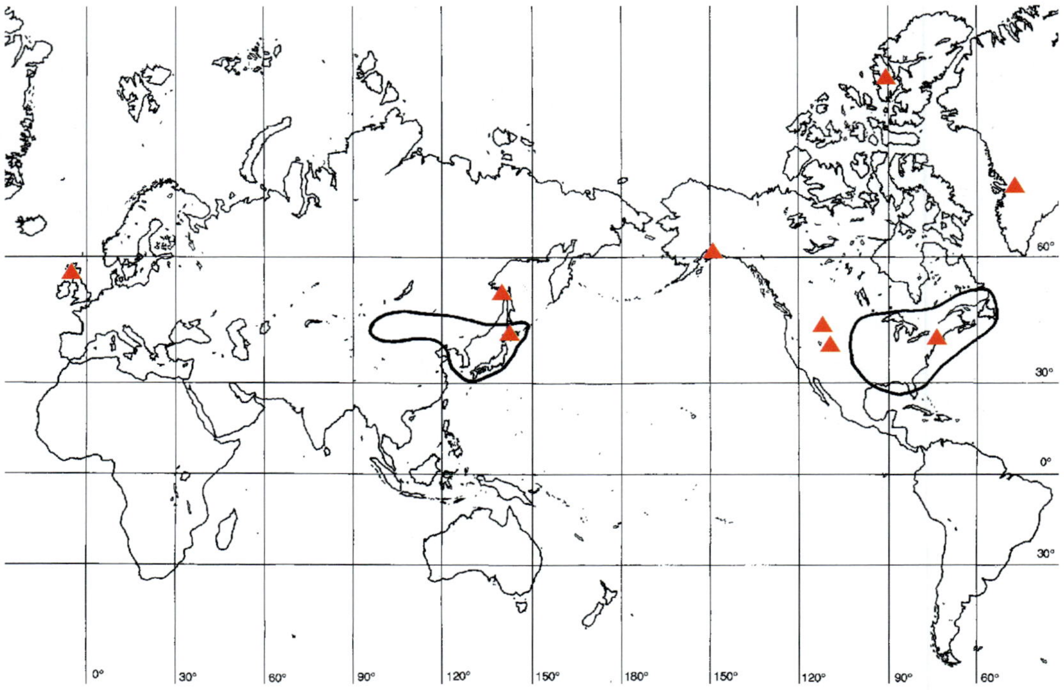

FIG. 34. Present distribution of *Onoclea sensibilis* in eastern North America and eastern Asia (represented by areas enclosed by solid lines) and known fossil occurrences (indicated by red triangles). Modified from Kato (1993b) with additional data from Collinson (2001).

The present widespread northern hemisphere distributions of many plants still reflect this ancient pattern—for example, *Thelypteris palustris* now occurs broadly across the northern hemisphere (Fig. 35). The modern distribution of *Equisetum hyemale* (Fig. 36) may give additional insight. In this species, variety *affine* occurs in North America and easternmost Asia, while variety *hyemale* occurs in Europe and western Asia. This is possibly a reflection of the fact that the land connection between North America and Asia persisted longer than the connection between North America and Europe (Sauer 1988; Graham 1999a).

The widely separated modern populations of many plants is thus a reflection of the numerous changes that have occurred in the tens of millions of years since the mid-Tertiary. These include alterations in the shapes of the northern landmasses, fluctuations in sea levels, plate tectonic movements, mountain building, glaciation, and other changes in the climate. As a result, the ranges of many once widely distributed plants have been greatly restricted, and many have become extinct in part or most of their previous ranges. A significant number of plants that were once much more widespread now survive in only one or a few areas—in "Tertiary refugia." The eastern North America and eastern Asia pattern is just one such example. Though this disjunct distribution pattern has complex and multiple origins, the consensus is that the eastern North America–eastern Asia pattern is in general a relict of the maximum development of northern hemisphere temperate forests (with tropical elements) in the Tertiary, with greater survival in eastern North America and eastern Asia, and higher rates of extinction in Europe and western North America, through glaciation and mountain building (e.g., Li 1952; Boufford & Spongberg 1983; Kato & Iwatsuki 1983; Wu 1983; Tiffney 1985a, 1985b; Basinger 1991; Graham 1993a, 1999a; Frederiksen 1995; Xiang et al. 1998; Wen 1999, 2001; Dilcher 2000; Donoghue et al. 2001; Xiang & Soltis 2001; Qian 2002; Williams et al. 2003a, 2003b; Moran 2004; Xiang et al. 2004; Richter & LePage 2005; Jahren 2007).

According to Wen (1999), about 65 genera of modern seed plants have this eastern North American-eastern Asian disjunct distribution. As mentioned earlier, SENSITIVE FERN is a pteridophyte exam-

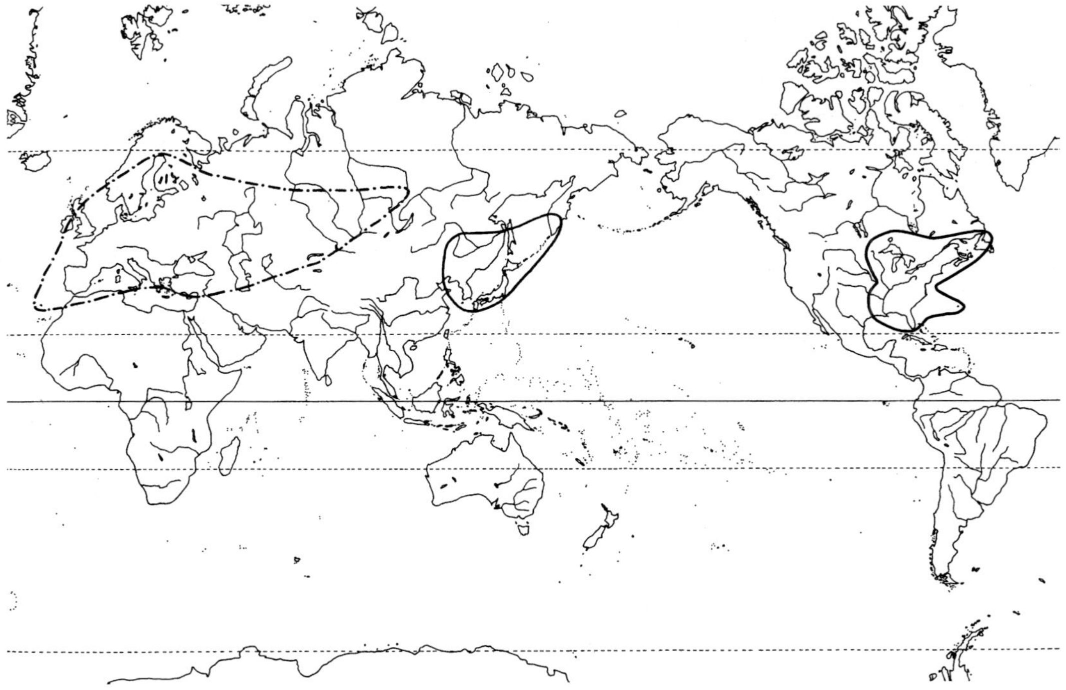

FIG. 35. Modern distribution of *Thelypteris palustris*, SOUTHERN MARSH FERN, reflecting a widespread, northern hemisphere, Arcto-Tertiary (also called Boreotropical) distribution during the Tertiary Period. The plants in North America and eastern Asia are var. *pubescens*, while those in Europe and western Asia are var. *palustris*. Modified from Kato and Iwatsuki (1983, fig. 3, p. 727). Used with permission of Missouri Botanical Garden Press, St. Louis.

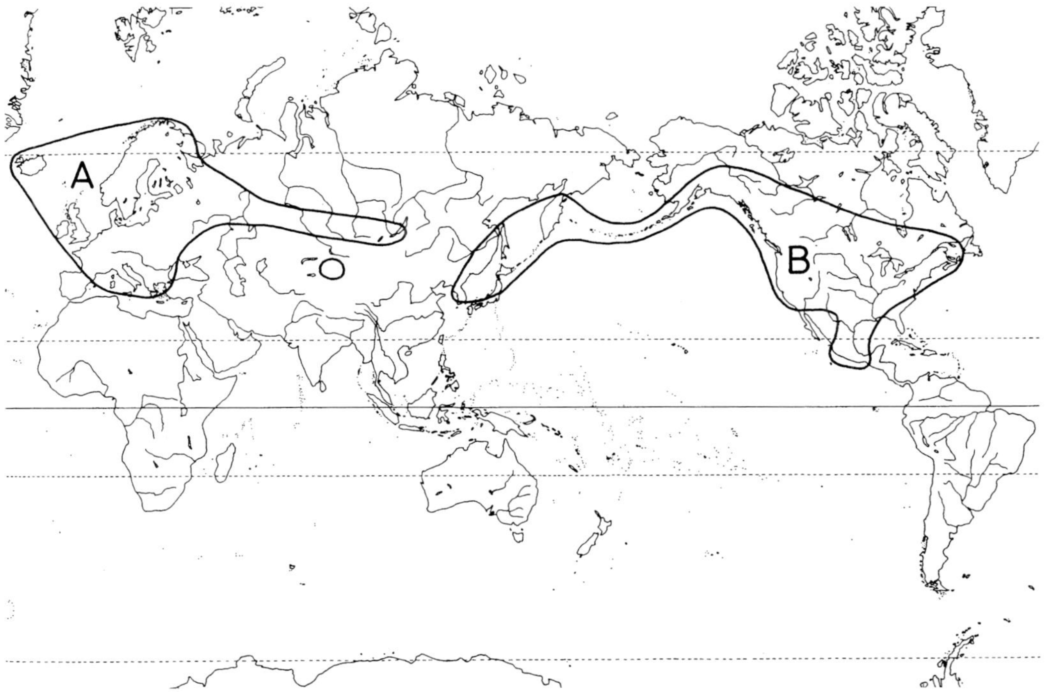

FIG. 36. Modern distribution of *Equisetum hyemale* (A: subsp. *hyemale*, B: subsp. *affine*), SCOURING-RUSH HORSETAIL. Note the North America-East Asia connection of *E. hyemale* subsp. *affine*. From Kato and Iwatsuki (1983, fig. 2, p. 726). Used with permission of Missouri Botanical Garden Press, St. Louis.

ple; the distribution of SENSITIVE FERN fossils from the Tertiary in western Canada, Alaska, interior North America, Greenland, the British Isles, and Japan (Fig. 34) lends further support. Surprisingly, 40% of the fern species of the northernmost island of Japan (Hokkaido) and northeastern North America are common to both regions—a remarkable holdover from the Tertiary Period (Kato & Iwatsuki 1983). This phenomenon occurs not only in ferns—many Texas flowering plant genera also share this same eastern North America-eastern Asia disjunction. These include *Carya* (the HICKORIES and PECAN), *Sassafras* (SASSAFRAS), and the well-known ornamental vine *Wisteria* (WISTERIA). Differing amounts of extinction in the various parts of the Northern Hemisphere resulting from climate and geological changes since the Tertiary are the most likely explanation for this fascinating modern disjunct distribution pattern (Barrington 1993).

Connections Across Texas

Another example of vicariance on a more local scale is seen in the disjunct Texas distribution of *Osmundastrum cinnamomeum*, the well-known CINNAMON FERN. Within Texas (Fig. 37), CINNAMON FERN is widespread in East Texas in the Pineywoods and adjacent Post Oak Savannah. Isolated populations are also disjunct southwest to Gonzales County in south central Texas and even farther to the west to Uvalde County in the western Edwards Plateau. In Gonzales County, the plants in specific microhabitats provide a window into what Texas was like thousands of years ago. CINNAMON FERN occurs here in the Ottine wetlands—an isolated, spring-fed, relict wetland ecosystem on the floodplain of the San Marcos River (Fleenor & Taber 2009). The area is also home to a disjunct remnant population of *Sabal minor* (DWARF PALMETTO) and many other flowering plants as well as populations of *Cystopteris protrusa* (Fig. 38), *Woodsia obtusa*, *Woodwardia areolata*, and *Woodwardia virginica*. These ferns all reach the western or southern limit of their distributions in the Ottine wetlands. A number of these species are found in and seem much more at home in the bog, swamp, and bottomland forests of the Big Thicket of far East Texas. The Ottine populations are relicts of Ice Age times when such favorable habitats were much more widespread in the state. The reason these plants still occur in the Ottine wetlands today is the permanence of its springs, acidic seeps, bogs, and other wetlands which exist because of the unique geological and soil conditions that have been present in this location since glacial times. Rainwater percolates through the very porous, acidic, sandy soil derived from Eocene sandstone layers (roughly 50 million years old; e.g., Carrizo Formation), reaches an impenetrable rock or clay layer beneath and moves horizontally, eventually seeping out to the surface as springs and seeps (MacRoberts & MacRoberts 2001; MacRoberts et al. 2002b; Taber & Fleenor 2005). Over thousands of years the Texas climate has become drier, and moisture-loving vegetation has become restricted to the far eastern part of the state. However, wet conditions have persisted further west in the Ottine wetlands, providing a tiny island of favorable habitat surrounded by a sea of extensive dry areas where water-dependent species cannot possibly survive. This type of phenomenon is sometimes referred to as "biological inertia" or "vegetational inertia" (Gorham 1957; Summerfield 1972; Cole 1985; Anderson 1987; Peck 2011a). As noted by Gorham (1957) under certain conditions "… once having colonized an area, [some plants] continue to exist there long after the conditions suitable for their establishment have disappeared."

This interpretation makes even more sense when studies of pollen and spores from Texas bogs, as well as larger scale studies and analyses, are considered. This research indicates that near the end of the last glacial maximum about 15,000 years ago, species currently present only much farther north (e.g., white spruce now in the northern U.S. and Canada) were present in Texas, suggesting a very different, cooler, moister climate (Potzger & Tharp 1943, 1947, 1954; Graham & Heimsch 1960; Bryant 1977; Holloway & Bryant 1984; Bryant & Holloway 1985; Delcourt & Delcourt 1993; Stahle & Cleaveland 1995). Western populations of some plants, such as the ferns mentioned above, have been able to survive in isolation for thousands of years due to the unique conditions of the Ottine wetlands and similar microhabitats.

In Uvalde County on the Edwards Plateau, CINNAMON FERN is disjunct to a different type of isolated microhabitat, again reflecting survival in a special area following long-term climate change at the end of the last Ice Age. In the case of the Edwards Plateau, for CINNAMON FERN and other

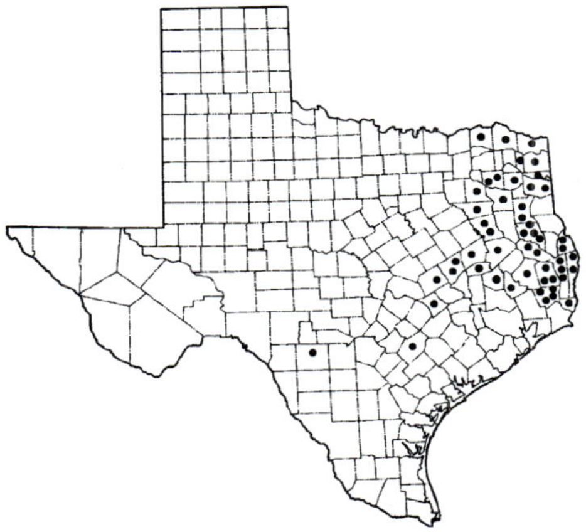

FIG. 37. Texas distribution of *Osmundastrum cinnamomeum*, CINNAMON FERN. Modified from Turner et al. (2003).

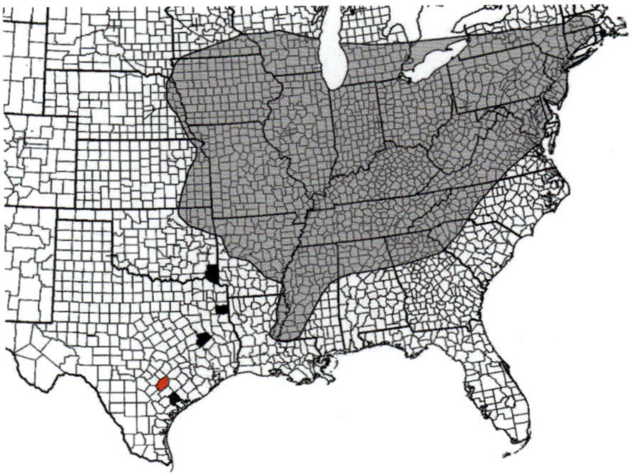

FIG. 38. Distribution map of *Cystopteris protrusa* (SOUTHERN BLADDER FERN) in the U.S. One of the disjunct Texas locations is from the Ottine wetlands in Gonzales County (indicated in red). Modified from Haufler et al. (1993) and Turner et al. (2003).

disjuncts the microhabitats include moist, steep-sided, shady canyons along streams (e.g., Frio River and Sabinal River) associated with the Edwards Escarpment. Such canyons are similar to "rockhouses" or rockhouse shelters in the eastern U.S. that harbor rare species, including unusual gametophyte-only ferns. As discussed by Farrar (1998), "rockhouses" are defined as deep recesses in bedrock cliffs that maintain microhabitats highly buffered from the extremes in temperature and moisture typical of the surrounding macroenvironment. Like "rockhouses," in comparison with the outside environment, narrow canyons on the Edwards Plateau tend to be more shaded, have higher temperatures during the winter, lower temperatures during the summer, and have lower rates of evaporation and higher relative humidities (Walck et al. 1996). Numerous other Texas plants have similar distributions, including the lycophyte *Selaginella apoda*, *Apios americana* (GROUNDNUT), *Bignonia capreolata* (CROSS-VINE), *Hamamelis virginiana* (WITCH-HAZEL), *Lindera benzoin* (SPICEBUSH), and *Tilia americana* (AMERICAN BASSWOOD). They have widespread populations in East Texas and isolated disjunct populations in tiny pockets of favorable habitat scattered in the generally dry environment of the Edwards Plateau (Fig. 39). Only because of the special microclimate provided by these habitats have such species been able to survive the drying and warming that has occurred since the end of the last glacial period.

The Mexican Connection

A final example of vicariance can be seen in *Selaginella apoda* (Figs. 4, 40). This is one of a number of plant species that occur broadly across the eastern U.S. as far west as Texas and then reappear disjunctly many hundreds of miles away in the highlands of Mexico and, in some cases, Guatemala. Well-known trees that show this pattern include *Fagus grandifolia* (AMERICAN BEECH), *Nyssa sylvatica* (BLACK-GUM), and *Liquidambar styraciflua* (SWEETGUM). Numerous deciduous forest tree genera display it as well including *Acer*, *Carpinus*, *Cercis*, *Fraxinus*, *Juglans*, *Magnolia*, *Prunus*, and *Ulmus*. A variety of shrubs and herbaceous plants also show this disjunction—a few examples include *Mitchella repens* (PARTRIDGE BERRY), *Epifagus virginiana* (BEECHDROPS), the lycophyte *Lycopodiella alopecuroides*, and the ferns *Ophioglossum vulgatum*, *Osmunda regalis*, and *Osmundastrum cinnamomeum*. This surprising distribution pattern is thought to be the result of a period of widespread cooling (e.g., Antarctic glaciation) that occurred during the middle Miocene (15–14 million years ago). During that extremely cool period, deciduous forests from the eastern U.S. were able to spread south into Mexico and Guatemala. In the words of Graham (1999a), a "major decline

Fig. 39. Fern habitat on the Edwards Plateau.

in temperature occurred in the Middle Miocene," and "global temperatures reached new lows, allowing the deciduous forest to reach its southernmost extent." Subsequently, during the Pliocene (5.3–1.8 million years ago) and later, the climate warmed and dried, and other types of vegetation (e.g., prairie and dry area shrub communities) spread and replaced the deciduous forests in southern Texas and most of Mexico. As a result, the once continuous deciduous forests became fragmented, and the remnants of that flora survived south of Texas only in very limited isolated pockets of appropriate microclimate in the highlands of Mexico and Guatemala. In other words, the plants survived in tiny isolated islands of suitable mountain habitat usually above 1,000 m (3,281 ft) where conditions remained favorable for temperate plants (Miranda & Sharp 1950; Carlson 1954; Martin & Harrell 1957; Thorne 1993; Graham 1993a, 1993b, 1999a, 1999b; Panero & Dávila 1998; Burnham & Graham 1999).

Other Distributional Phenomena

Several other biogeographical examples involving Texas pteridophytes are worth mentioning. One lycophyte, *Isoëtes lithophila*, is known only from Texas—it is found nowhere else in the world; in other words, it is **endemic** to Texas. It has been found in only four counties in the geologically fascinating area referred to as the Llano Uplift or Central Mineral Region (the location of Enchanted Rock). This is an area over 80 km (50 mi) wide in parts of 10 counties that is an island of Precambrian igneous rocks (granites, gneisses, and schists) surrounded by escarpments of Paleozoic and Mesozoic sedimentary limestones and sandstones (Walters & Wyatt 1982). The evolutionary history of *Isoëtes lithophila,* commonly known as ROCK QUILLWORT, is not clear, but this isolated, rock-loving species may well have evolved as a specialist on the relatively unusual rocks of the Central Mineral Region.

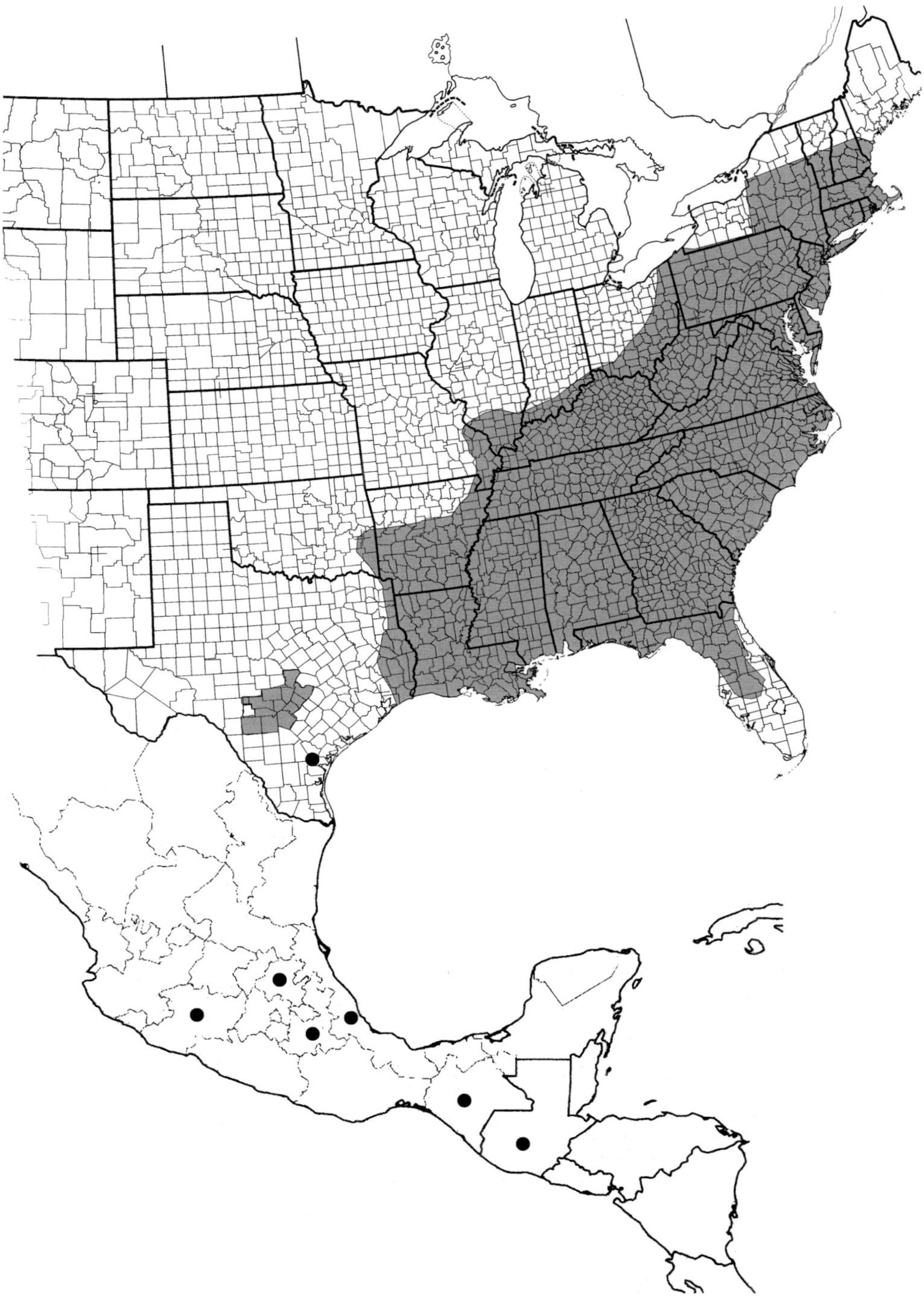

FIG. 40. Modern distribution of *Selaginella apoda*, MEADOW SPIKE-MOSS, possibly reflecting a mid-Miocene extension of the eastern deciduous forest into Mexico during a period of widespread cooling. Modified from Valdespino (1993) and Mickel and Smith (2004).

Another distributional phenomenon emphasizes the uniqueness of Texas' pteridophyte flora. A significant number of species occurring in the state, although not endemic, are found nowhere else in the U.S. There are 17 such species (19 when the two endemic *Isoëtes* species are counted), as well as one variety and two subspecies, that are known in the U.S. only in the state of Texas (Table 5). The 19 species make up 15% of Texas' native pteridophytes. They are mostly dry-adapted species that extend north from Mexico in appropriate habitats of the Chihuahuan Desert. Numerous other species are only known in the U.S. from Texas and Arizona and/or New Mexico.

Many fern and lycophyte species are very limited in their occurrence in Texas—21 are known from only one county. Some extreme examples include *Cheilanthes yavapensis*, known only from extreme western Texas in El Paso County, and *Dryopteris celsa*, known only from Bowie County in the northeastern corner of the state. In the far northern part of Texas, *Pellaea glabella* has been found only in Potter County in the High Plains. Perhaps most striking, *Llavea cordifolia* was only reported from Texas in 2011 (based on a 1992 collection) (O'Kennon & Diggs 2011). Its sole known U.S. location is in Presidio County only 13 km (6.9 miles) from the border with Mexico.

Numerous species—a total of 40 mostly dry-adapted pteridophytes—are known in the state only from the Trans-Pecos. These include some biogeographically interesting examples: *Ophioglossum polyphyllum*, only recently reported (1998) as present in the U.S.; *Notholaena greggii*, with all known U.S. populations within 25 km of the Mexican border; *Pleopeltis polylepis*, in the U.S. known only from one canyon (Limpia Canyon) in the Davis Mountains of Jeff Davis County; and the bulblet-producing, clonal *Cystopteris bulbifera*, known in Texas only from the Guadalupe Mountains of Culberson County.

At the other end of the state, a number of eastern deciduous forest species barely extend into Texas, in the very mesic area of extreme East Texas near the Louisiana border. These include *Dryopteris ludoviciana*, *Phegopteris hexagonoptera*, and *Thelypteris hispidula*. The only confirmed report of *Palhinhaea cernua* in the state is from two plants that were discovered and photographed but not collected in Jasper County. *Psilotum nudum* is known from only two Texas collections—its very rare occurrence in eastern Texas is slightly beyond its main range in the eastern U.S. The presence of both of these latter species in Texas is possibly due to long-distance dispersal of spores by major storms or hurricanes.

Nine species (Table 6) are likewise known in the state only from the third center of Texas pteridophyte diversity, the Edwards Plateau, which includes the Central Mineral Region. Three striking examples include *Adiantum tricholepis*, in the U.S. known only from Bandera and Medina counties; the mostly tropical *Tectaria heracleifolia*, found in a few Edwards Plateau counties in unusual habitats such as sink-holes, cave entrances, or railroad tunnels; and *Cheilanthes leucopoda*, known in the U.S. only from limestone areas in four counties on the Edwards Plateau. A number of other species are known in the state primarily but not exclusively from the Edwards Plateau. In fact, 63 native species are found in the Edwards Plateau (Fig. 15). As has long been recognized (Seigler & Lockwood 1975), the Edwards Plateau, with its numerous microhabitats (caves, moist canyons in a dry landscape, extensive limestone, huge granite outcrops, etc.), is "well known for its unusual ferns" including a number of species that here reach the "extreme northern limit of their ranges." Many of these species are probably in Texas as the result of long-distance dispersal.

A discussion of the distribution of Texas ferns and lycophytes would be incomplete without mentioning naturalized species—11 such species are known for the state. Two particularly problematic introduced ferns are worthy of special focus: *Lygodium japonicum*, native to eastern Asia, and *Salvinia molesta*, native to South America. *Lygodium japonicum* is one of the few ferns in the world that is vine-like and capable of twining around and climbing up other plants. In some areas (e.g., Jack Gore Baygall Unit of the Big Thicket National Preserve in Hardin County), it is so abundant and integrated into the vegetation that it appears native, but its dense foliage can prevent the growth of underlying native vegetation. *Salvinia molesta* is even worse—this small floating aquatic has been referred to by some as the world's worst weed. It has been widely introduced around the world, including in eastern Texas, and it often causes serious problems. The

TABLE 5. Pteridophyte taxa in the U.S. known only from Texas.

Adiantum tricholepis, HAIRY MAIDENHAIR
Anemia mexicana, MEXICAN FERN
Astrolepis sinuata subsp. *mexicana*, MEXICAN SCALY CLOAK FERN (subsp. *sinuata* is known elsewhere in U.S.)
Cheilanthes aemula, TEXAS LIP FERN
Cheilanthes kaulfussii, KAULFUSS' LIP FERN
Cheilanthes leucopoda, WHITE-FOOT LIP FERN
Dennstaedtia globulifera, BEADED CUPLET FERN
Dryopteris cinnamomea, CINNAMON WOOD FERN
Isoëtes lithophila, ROCK QUILLWORT (endemic to Texas)
Isoëtes texana, TEJAS QUILLWORT (endemic to Texas)
Llavea cordifolia, LLAVE'S FERN
Notholaena aliena, MEXICAN CLOAK FERN
Notholaena copelandii, COPELAND'S CLOAK FERN
Notholaena greggii, GREGG'S CLOAK FERN
Notholaena nealleyi, NEALLEY'S CLOAK FERN
Pellaea cordifolia, HEART-LEAF CLIFF-BRAKE
Pellaea ovata, ZIGZAG CLIFF-BRAKE
Pellaea ternifolia subsp. *ternifolia*, TRANS-PECOS CLIFF-BRAKE (subsp. *arizonica* is known from Arizona)
Phanerophlebia umbonata, CHISOS HOLLY FERN
Pleopeltis polylepis var. *erythrolepis*, RED-SCALE POLYPODY (the only variety of this species known for the U.S.)
Selaginella viridissima, GREEN SPIKE-MOSS
Thelypteris ovata var. *lindheimeri*, LINDHEIMER'S MAIDEN FERN (var. *ovata* is known from the southeastern U.S.)

plants grow very rapidly and cover the surface of lakes and streams, resulting in floating mats that shade and crowd out native plants. Additionally, the mats (sometimes up to a meter thick) reduce oxygen content, degrade water quality, and can hinder boats, clog irrigation and drainage canals, and block water intakes. The mats can become so dense that they are able to temporarily support the weight of a cinder block. As a result, this species is federally listed as a noxious weed, and its further distribution is prohibited in the U.S. by federal law. Biological control using the salvinia weevil, a tiny insect that feeds on the plant, is being tried in Texas, and initial indications are that it is effective in helping to control GIANT SALVINIA in the state (Tipping et al. 2008).

ECONOMIC USES AND ETHNOBOTANY

Texas ferns and lycophytes are generally considered to be of little economic importance. At present, their most important positive economic contributions in the state are as ornamentals in landscapes, but these plants have had profound impacts based on their geological past. As noted earlier, ferns and lycophytes were among the dominant plants in the extensive swamps of the Carboniferous Period (359–299 million years ago), and their compressed remains became preserved as coal (Hoshizaki & Moran 2001); in fact, readily recognizable pteridophyte fossils are sometimes clearly visible in coal.

Historically, ferns and lycophytes have also had many uses—they have provided dyes, medicines, flavorings, beverages, and have been used as food (May 1978). For example, the leaves of *Ophioglossum petiolatum*, a species naturalized in Texas, are used in parts of Asia (e.g., Taiwan) in traditional Chinese medicine. They are dried, ground into a powder, and made into a tea (Moran 2004). Of special interest as a food is *Pteridium aquilinum*, native to Texas but also widespread in many parts of the world. In Asia, particularly Japan and Korea, the fiddleheads (also known as croziers) of this species are eaten as a cooked vegetable and as a salty snack food (Hodge 1973; Moran 2004). However, since they have high concentrations of poisonous plant defense chemicals and are potentially mutagenic and carcinogenic, they should not be eaten. While in the U.S. they are blanched for only a few minutes, Asians steep the croziers for one to two hours; it is possible that the Asian method of preparation might remove toxins and thus be safe, while ours would not be so (J. Peck, pers. comm.). The plant defense chemicals of a number of Texas species also make them potentially dangerous to animals. For example, *Dryopteris filix-mas*, known in Texas only from the Trans-Pecos but widespread in various parts of the world, is one of the oldest known vermifuges (a

TABLE 6. Pteridophyte species known in Texas only from the Edwards Plateau.

Adiantum tricholepis, HAIRY MAIDENHAIR	*Blechnum appendiculatum*, HAMMOCK FERN
Cheilanthes leucopoda, WHITE-FOOT LIP FERN	*Dennstaedtia globulifera*, BEADED CUPLET FERN
Dryopteris cinnamomea, CINNAMON WOOD FERN	*Isoëtes lithophila*, ROCK QUILLWORT (endemic to Texas)
Isoëtes piedmontana, PIEDMONT QUILLWORT	*Pteris vittata*, LADDER BRAKE (introduced)
Tectaria heracleifolia, BROAD HALBERD FERN	

medicine that expels intestinal worms). Its use dates to ancient Greek times. The drug aspidium, derived from the rhizomes, paralyzes tapeworms (which can then be flushed from the bowel) but is dangerous because it also paralyzes voluntary muscles of patients, and overdoses are potentially fatal. Livestock can also be affected upon eating the apical buds of the rhizomes—severe digestive disturbances and neurological problems including blindness can result (Correll 1956; Mabberley 1997; Yarborough & Powell 2002). Several species, including *Onoclea sensibilis*, *Pteridium aquilinum*, and *Equisetum* species (HORSETAILS) are thought to contain thiaminase, an enzyme that destroys thiamine and causes Vitamin B1 deficiency. Toxicity symptoms in animals include walking in circles, difficulty in chewing (with food sometimes falling from the mouth), loss of muscular control, trembling, staggering, seizures, and death. *Astrolepis cochisensis*, commonly called JIMMY FERN, is also known to be poisonous to livestock. Symptoms range from incoordination to arched back, rapid pulse, gasping, trembling (known as "the jimmies"), prostration, and even death from respiratory paralysis (Weathers 1998; Burrows & Tyrl 2001).

Some ferns and lycophytes have also had unusual historical uses. *Lycopodium* species (CLUB-MOSSES, in the broad sense) have long had their trailing stems woven into Christmas wreaths and decorations (Correll 1956), while the fibrous root masses of *Osmunda regalis* and *Osmundastrum cinnamomeum* have been used as a substrate for the cultivation of orchids. Because of the high silica content of their stems, *Equisetum* (SCOURING-RUSHES) species were sometimes used in the past to scour cast iron pots and pans and as an abrasive in polishing. Perhaps an even more unusual use was found historically for the very flammable (due to the presence of volatile oils) spores of some Lycopodiaceae (CLUB-MOSSES). Because the powder-like spores burn with a quick flash, they were formerly used in fireworks, signal fires, for stage-lighting, and in photography as flash powder (Whitebread 1941; Jones & Luchsinger 1986). In the past, when pills were hand-made, the powdery spores were used by pharmacists to coat pills and prevent them from sticking together. They were also once used as a dusting agent on rubber surgical gloves, condoms, and suppositories, but today they are considered dangerous for such uses (and have been replaced) because they can cause inflammation, adhesions, granulomas, and allergic reactions including severe asthma attacks. It is surprising that use of the spores (called "vegetable sulfur") on some condoms continued until the late 1980s (Whitebread 1941; Balick & Beitel 1988, 1989; Hoshizaki & Moran 2001).

Currently, probably the most important roles of Texas pteridophytes include their ecological and symbolic importance. In some parts of East Texas, large stands of fern species such as *Pteridium aquilinum* and *Osmundastrum cinnamomeum* are a significant component of the natural vegetation. *Azolla* (MOSQUITO FERN) species in Texas almost certainly add to the nitrogen balance of the ecosystems in which they are found, due to the nitrogen-fixing, symbiotic blue-green bacteria living inside the hollow leaves. However, their importance here is insignificant when compared to other parts of the world. In Asia, *Azolla* has long been used in rice fields as a biofertilizer, and the genus is considered by some to be the world's most economically important fern. Its use is thought to have begun in Vietnam by the 11th century. In one technique, the tiny, floating *Azolla* plants were grown between rows of rice where they could die, sink to the bottom, and add nitrogen. In another method, the plants were scattered on flooded rice paddies, and after about a month of rapid growth they would form a dense velvety mat on the water's surface. The paddies were then drained, stranding the *Azolla* on the mud where it could be worked into the soil, decay, and release its nitrogen—all in preparation for a new crop of rice to be planted (Moran 1997, 2005). Lastly, as previously noted, two species, *Lygodium japonicum* and *Salvinia molesta* have ecological significance as invasive weeds. In parts of southeastern Texas (e.g., the Big Thicket) *Lygodium japonicum* can invade native forests, become one of the dominants, and hinder the growth of native forest floor vegetation. *Salvinia molesta* is possibly even more of a threat. Sometimes considered one of the world's worst weeds, it can cover the surface of lakes or ponds and cause a host of problems for native plants, animals, and humans.

Although these are important or interesting uses, the symbolic importance of most Texas fern and lycophyte species as ecological indicators exceeds any direct or even indirect economic benefit. Many species are rare, and their continued existence as part of the Texas flora is threatened—they thus serve as an "early warning system" giving important information about Texas ecosystems. If these and other rare components of Texas' native vegetation are eliminated, then the long-term stability of our ecosystems begins to come into question.

CONSERVATION CONCERNS

Although Texas is rich in ferns and lycophytes in comparison with much of the U.S., many of the species occur in extremely small numbers or in very limited areas of the state—in one or a few counties or in rare or isolated pockets of appropriate microhabitat. The combination of being rare and having very specific habitat needs makes such plants particularly vulnerable, and thus many Texas species are susceptible to being lost from the state's flora. Two species of particular concern are *Isoëtes lithophila* and *Isoëtes texana*, the state's only endemic pteridophytes. *Isoëtes lithophila* is known only from temporary pools on granite or gneiss outcrops in the Llano Uplift area of central Texas in Burnet, Mason, Gillespie, and Llano counties (Poole et al. 2007a, 2007b, 2010) and nowhere else in the world. Likewise, *Isoëtes texana* is confirmed globally only from Aransas and Calhoun cos. in the Texas Coastal Bend. In all, we consider 60 of the 127 Texas natives (approx. 47%) to be of conservation concern at the state level. By this, we mean that they are so limited in occurrence or so rare in the state as to be in danger of becoming extinct locally (i.e., in Texas), or, in the cases *of Isoëtes lithophila* or *Isoëtes texana,* of becoming extinct globally. These species are marked with a conservation concern symbol ⚠ following their individual write-ups. As discussed earlier, many of these species occur nowhere else in the U.S. other than Texas and many are found in only one county, sometimes in only one location or in populations consisting of only a handful of plants. In actual fact, however, probably even a higher percentage of the Texas species are actually of conservation concern. For example, the vast majority of the ferns and lycophytes in the Trans-Pecos occur in the "Madrean sky islands" (isolated high elevation areas [e.g., Chisos Mts., Davis Mts., Guadalupe Peak] similar to habitats found in the Sierra Madre Mountains of Mexico [Windham & Yatskievych 2003a]). Although these "sky islands" are widely scattered in the Trans-Pecos, they occupy extremely tiny areas/islands (e.g., mountain peaks) separated by vast "oceans" of inhospitable desert habitat where most pteridophytes could not possibly grow. Therefore most of these pteridophytes are limited to extremely small areas, have small population numbers, and are thus vulnerable to a variety of threats.

Despite such considerations, the number of Texas ferns and lycophytes officially recognized as being rare or of conservation concern is shockingly low. Poole et al. (2007a, 2007b, 2010) for example, listed only two species, *Isoëtes lithophila* and *Selaginella viridissima*, as rare for Texas. Why ferns and lycophytes are largely ignored in this regard is unclear, but Yatskievych (2003) noted the same trend at the national level—of all the species listed under the federal Endangered Species Act, there are only six ferns and lycophytes from the entire continental U.S. Ironically, it was a fern, *Lygodium palmatum* (native to the eastern U.S.), that was the inspiration for the first plant conservation law passed in the U.S. It had been collected so extensively during the 1800s for Christmas decorations that in 1869 the Connecticut legislature enacted a law protecting the plant (Hoshizaki & Moran 2001). Because of their rarity, some such species should not be collected at all even for scientific purposes, while others should only be collected in a way that does not kill the plant (removing only one or a few leaves, etc.). A good example of sensitivity to a plants' survival was displayed by MacRoberts and MacRoberts (1995a) who made the first discovery of *Palhinhaea cernua* in Texas—the entire population consisted of only two plants, which were photographed but not collected. Likewise, in a 1939 article about collecting the rare *Tectaria heracleifolia* from a sink-hole in west Texas, Parks noted, "After having secured a number of fern specimens, without molesting the roots of the plants, the return trip was started." Since even relatively common species can be threatened by over-collecting for either scientific or commercial purposes, collecting ferns and lycophytes from the wild should be done only with full awareness of the effects on the population (Norton et al. 1994).

There are several other particularly serious threats to Texas plants, including pteridophytes. First and most important is habitat transformation or destruction—in East Texas this can range from the conversion of rich mixed deciduous forests to pine monocultures, to inundation of watersheds for reservoirs, or to fragmentation or even the total destruction of large areas due to development. Further west, abusive overgrazing can result in serious habitat degradation. The introduction of exotic plant species can also be a threat to native Texas plants. These nonindigenous species can aggressively invade native ecosystems, reproduce extensively, and occupy the habitat of indigenous species. In some cases, they even dominate communities and occur in nearly pure monocultures, completely changing nearly all aspects of an ecosystem.

A longer-term conservation concern is the effect that rising atmospheric levels of CO_2 and other heat-trapping gases might have on Texas ecosystems. The consensus of the world's leading climate scientists is that global temperatures will continue to rise. Further, precipitation patterns are also predicted to change. Both of these factors are of concern when considering Texas ferns and lycophytes since many of them survive only in very limited microhabitats or at the extreme periphery of their ranges. They are thus likely to be especially susceptible to changes in temperature and precipitation levels and resulting changes in water loss rates. Ironically, it may be plants in East Texas, the wetter part of the state, that are most susceptible to such problems. Hundreds of species of plants (including many ferns and lycophytes) reach the southwestern limit of their ranges in East Texas and are rare or very limited in occurrence in the state. Many of these species are probably Ice Age relicts, surviving only in a few areas of favorable microclimate or geology (Kral 1966c; MacRoberts & MacRoberts 1997a). Such species are particularly vulnerable to changes of various sorts including habitat transformation and climate change (Given 1993). Although the future consequences of climate change are extremely difficult to predict, some Texas pteridophytes will likely be at risk.

It is particularly difficult to garner support to protect ferns and lycophytes—as noted earlier, most have no easily observed, direct economic benefit. However, biologists and other people aware of the bigger picture understand the importance of the world's biodiversity and that it is our responsibility to protect the amazing diversity of species with which we share the planet for future generations.

The Future

On a more positive note, it is exciting to contemplate what other new and possibly rare ferns are yet to be discovered in the state. Of the 138 total species now known in Texas, 31 were not known in the state when Correll published his *Ferns and Fern Allies of Texas* in 1956—a significant number have thus been discovered in a few decades. *Dryopteris celsa* was only recognized in the state in 1988 (the specimen was collected in 1925 but misidentified—Peck & Peck 1988). *Palhinhaea cernua* was reported as new to Texas in 1995. The problematic introduced weed, *Salvinia molesta*, was first observed in the state in 1997. *Ophioglossum polyphyllum* was first reported as occurring in the U.S. and Texas in 1998. The first Texas report of the introduced species *Pteris vittata* was in 1998 (collected in 1987). *Pellaea villosa* was only realized to be a distinct species in 2003. The strikingly disjunct *Isoëtes piedmontana* was only recognized as occurring in Texas in 2005. *Dryopteris marginalis* was first discovered in Texas in 2009 and reported in 2010. The first report of *Llavea cordifolia* for the U.S. and Texas was in 2011 (collected but misidentified in 1992). Finally, the Texas endemic *Isoëtes texana* was described as new to science in mid-2011. Based on these recent finds, it is obvious that important discoveries are still being made in the state and that there is much yet to be learned about Texas plants. We hope you will be inspired by this book to look for and find yet more interesting ferns or lycophytes, maybe even an undescribed species!

KEY TO FAMILIES OF FERNS AND LYCOPHYTES

KEY TO FAMILIES OF FERNS AND LYCOPHYTES

1. Plants small floating aquatics (sometimes stranded on mud); leaf blades < 4 cm long (usually much less).
   2. Leaf blades 7–30(–38) mm long, the upper (= adaxial) surface with conspicuous 4-branched hairs; roots absent but submerged leaves dissected into numerous root-like segments ____ (*Salvinia*) **Salviniaceae,** p. 289
   2. Leaf blades extremely small, usually < 1 mm long, the upper surface without branched hairs; roots unbranched ____ (*Azolla*) **Azollaceae,** p. 110
1. Plants not small floating aquatics, either terrestrial (rooted in soil or mud) OR growing on rocks or tree trunks OR if aquatic, the plants either rooted on bottom or much larger in size; leaf blades often, but not always, much > than 4 cm long.
   3. Stems nearly naked (without leaves), green, photosynthetic, erect, repeatedly branched dichotomously (= into 2 equal parts); leaves absent but minute scale-like stem appendages present, these 0.7–2.5 mm long; (2–)3 sporangia fused into a (2–)3-lobed cluster (= synangium); species extremely rare in TX, known from only two historical collections in East TX ____ (*Psilotum*) **Psilotaceae,** p. 206
   3. Stems various, not naked, usually not green (except in *Equisetum*), erect OR not so, often underground, not repeatedly branched dichotomously; leaves present, these varying from scale-like to large and compound; sporangia not fused into a (2–)3-lobed cluster; species widespread and common in TX.
      4. Stems conspicuously jointed, green and hollow, the segments separating easily at the joints (= nodes), erect and very obvious; leaves reduced to small, non-obvious, essentially non-photosynthetic (non-green) scales in whorls at the nodes; sporangia (= spore cases) in terminal strobili (= cone-like structures) on stems without green leaves ____ (*Equisetum*) **Equisetaceae,** p. 153
      4. Stems not jointed, not green and hollow, typically horizontal at or below the surface of the ground (called a rhizome) and usually not at all obvious; leaves usually green, obvious; sporangia either on the surface of leaves or leaf-like structures OR in terminal strobili on leafy stems OR in short-stalked sporocarps at leaf bases.
         5. Leaves simple, linear, grass-like or thread-like, the blades not expanded; spore-bearing structures embedded in leaf bases or on very short stalks (1–2 mm long) at leaf bases; plants often rooted in mud or in temporary pools or even submerged on lake bottoms.
            6. Leaves tightly clustered together, arising from a corm-like rootstock, quill-like (larger at base); sporangia embedded in the leaf bases, borne one per leaf ____ (*Isoëtes*) **Isoetaceae,** p. 55
            6. Leaves scattered along a creeping rhizome, separate, not clustered, not quill-like; sporangia in stalked (stalks 1–2 mm long), globose, hairy sporocarps (= nut-like or hard bean- or pea-like sporangia-bearing cases) arising at the base of the leaves, the sporangia numerous per sporocarp ____ (*Pilularia*) **Marsileaceae,** p. 172
         5. Leaves compound or simple, with expanded blades OR leaves needle-like or scale-like, neither grass-like nor thread-like; spore-bearing structures neither embedded in leaf bases nor on very short stalks at leaf bases; plants rooted in various substrates including mud, soil, rocks, and tree trunks.
            7. Plants distinctly vine-like (CLIMBING FERNS); leaves with indeterminate growth, climbing, vine-like, to 3(+) m long, the rachis elongate, twining, flexuous, stem-like (true stem below ground); sporangia borne on finger-like lobes of the fertile pinnules (= subdivisions of pinnae); species known in TX only from the s part of the Pineywoods ____ (*Lygodium*) **Lygodiaceae,** p. 164
            7. Plants not vine-like: leaves various but neither climbing, vine-like, nor with a twining rachis; sporangia variously borne, but not on finger-like lobes of the fertile pinnules; including species widespread in TX.
               8. Leaf blades deeply 4-parted (resembling a 4-leaf clover), on petioles usually much longer than the blades; sporangia in sporocarps borne near the base of the plant ____ (*Marsilea*) **Marsileaceae,** p. 167
               8. Leaf blades variously compound or simple but not 4-parted; petioles usually much shorter than leaf blades to absent; sporangia not in sporocarps near the base of the plant.
                  9. Leaves 8 mm or less long; plants with numerous, small, usually overlapping or divergent, scale-like or needle-like leaves, each with a single unbranched vein (this type of leaf is a microphyll); stems well-exposed above the ground surface, covered with the numerous small leaves; sporangia in the axils of the microphylls, these microphylls aggregated into cone-like strobili.
                     10. Sporangia in cylindrical strobili; leaves 4–8 mm long; plants homosporous (= with 1 spore type of a single size); species of moist habitats of e ½ of TX ____ **Lycopodiaceae,** p. 63
                     10. Sporangia in ± 4-angled strobili; leaves 1–3 mm long; plants heterosporous (= with 2 spore types which are of different sizes); species typically of dry habitats (with 1 exception), mainly (but not exclusively) in w ½ of TX ____ (*Selaginella*) **Selaginellaceae,** p. 74

9. Leaves usually much more than 10 mm long; plants with relatively few large leaves with numerous branched veins (this type of leaf is a megaphyll); stems underground rhizomes or short crowns or caudices, not well-exposed above the ground surface and relatively inconspicuous; sporangia in clusters (= sori) on the surface of the leaf blades (the blades can sometimes be considerably modified).
  11. Plants with 1(–2) leaves per stem; leaves with 2 distinct parts, the sterile portion either entire or ternately (= in 3 parts) to pinnately compound to dissected, the fertile portion being an elongate stalk with a spike-like or panicle-like sporangia-bearing terminal part ______ **Ophioglossaceae,** p. 177
  11. Plants usually with numerous leaves per stem; leaves not as described above.
    12. Lowermost 2 pinnae (= primary divisions of a leaf) of the fertile leaf greatly elongated and bearing the sporangia near their tips ______ (*Anemia*) **Anemiaceae,** p. 97
    12. Lowermost 2 pinnae of the fertile leaf neither greatly elongated nor bearing the sporangia near their tips.
      13. Fertile and sterile leaves either completely different OR fertile portion of fertile leaves dramatically different from sterile portion (the fertile portion with little or only a very reduced amount of blade tissue).
        14. Pinnae auricled basally (= with ear-like lobe), the auricle on the acroscopic side (= side toward the leaf apex) of the pinna, the pinnae margins bristle-toothed; sori round, the indusia round, peltate ______ (*Polystichum*) **Dryopteridaceae,** p. 149
        14. Pinnae not auricled basally, the margins not bristle-toothed; sori various but neither round nor with peltate indusia.
          15. Sori linear-oblong, in chain-like rows, one on each side of and immediately adjacent to midveins of the very reduced, narrow pinnae or pinnules ______ (*Woodwardia*) **Blechnaceae,** p. 119
          15. Sori various, but not linear-oblong in chain-like rows adja cent to midveins of the very reduced, narrow pinnae or pinnules.
            16. Plants aquatic or semiaquatic, usually rooted in mud in quiet or moving water of springs and adjacent rivers and lakes; leaf blades with adventitious buds or small plantlets in notches along margins; in TX known only along the Balcones Fault at the edge of the Blackland Prairie and Edwards Plateau ______ (*Ceratopteris*) **Pteridaceae,** p. 226
            16. Plants terrestrial, sometimes in wet habitats, but usually not actually in water; leaf blades without adventitious buds or plantlets; including plants widespread in TX.
              17. Fertile leaves completely different from sterile leaves, with little or almost no photosynthetic tissue on any pinnae, solely sporangia-bearing.
                18. Sterile leaves 1-pinnatifid (= deeply divided but not completely pinnate), the rachis (= central axis of a fern frond/leaf) with a conspicuous flange or wing of photosynthetic tissue; fertile leaves glabrous ______ (*Onoclea*) **Onocleaceae,** p. 175
                18. Sterile leaves 1-pinnate-pinnatifid, the rachis without a flange or wing of tissue; fertile leaves conspicuously pubescent (*Osmundastrum*) **Osmundaceae,** p. 197
              17. Fertile leaves with numerous ± normal photosynthetic pinnae essentially identical to those of sterile leaves, as well as dramatically different sporangia-bearing pinnae at blade tip.
                19. Fertile pinnae without inrolled margins, the sori not at all hidden, visually obvious on the extremely reduced pinnae; leaf blades 2-pinnate; including species rather widespread in the e ½ of TX ______ (*Osmunda*) **Osmundaceae,** p. 195
                19. Fertile pinnae with inrolled margins partially covering sori, the sori at least partially hidden from view; leaf blades 3-pinnate at least basally; extremely rare species of the Trans-Pecos, known

in TX only from Presidio Co. near the U.S.-Mexico border_________________________(*Llavea*) **Pteridaceae,** p. 256

13. Fertile and sterile leaves or portions of leaves similar or somewhat modified, the fertile portion never so different as to have a dramatically reduced amount of blade tissue (leaves not extremely dimorphic).
    20. Sori marginal or submarginal (= located at or near the edges of the leaves) with leaf margins recurved over the sori, protecting them and forming a false indusium (= thin scale-like outgrowth covering the clusters of sporangia) (except not recurved in *Astrolepis*—which has stellate or coarsely ciliate scales on the adaxial leaf surfaces, and in *Bommeria*—which has small pentagonal leaf blades); leaf blades with lower surfaces sometimes glabrous, but often with conspicuous hairs, scales, or wax-like farina.
        21. Plants stout, to 1 m or more tall; leaf blades usually triangular in shape or nearly so, with 3 main divisions (each division usually bipinnate), the lower surfaces glabrous or with hairs; petioles greenish or ± straw-colored; stems and petiole bases with slender hairs but without scales______ (*Pteridium*) **Dennstaedtiaceae,** p. 130
        21. Plants slender, usually 0.1–0.5 m tall; leaf blades usually neither triangular nor with 3 main divisions, the lower surfaces sometimes glabrous but typically with conspicuous hairs, scales, or wax-like farina; petioles often (but not always) dark brown or black; stems and petiole bases often (but not always) with scales_____________________________ **Pteridaceae,** p. 208
    20. Leaf margins not recurved to form a false indusium; sori variously located on the abaxial (= beneath) leaf surfaces, often near veins, occasionally near the margins of the leaves; adaxial leaf surfaces without stellate or coarsely ciliate scales, varying from glabrous to with pubescence, but the pubescence usually not conspicuous.
        22. Leaf blades simple; species extremely rare in TX and known only from the Trans-Pecos.
            23. Leaf blades linear; sori elongate along the veins ___________ (*Asplenium*) **Aspleniaceae,** p. 99
            23. Leaf blades lanceolate to oblanceolate; sori round to oval _________________________(*Pleopeltis*) **Polypodiaceae,** p. 200
        22. Leaf blades not simple, varying from 1-pinnate or 1-pinnatifid to more divided; including common and widespread species occurring broadly across TX.
            24. Sori EITHER linear-oblong, in chain-like rows, one on each side of and immediately adjacent to midveins of the pinnae or pinnules OR continuous as long lines along the length of the midveins of the pinnae or pinnules __________**Blechnaceae,** p. 117
            24. Sori various, but neither in chain-like rows nor continuous as long lines along midveins of pinnae or pinnules.
                25. Sori elongate along the veins (but neither in rows nor continuous); leaf blades 1-pinnate______ (*Asplenium*) **Aspleniaceae,** p. 99
                25. Sori round or variously shaped, not elongate (in our species, except elongate in *Athyrium* with leaf blades 2-pinnate-pinnatifid); leaf blades in our species 1-pinnatifid, 1-pinnate, or more than 1-pinnate or pinnatifid.
                    26. Leaf blades in our species only 1-pinnate or 1-pinnatifid, the pinnae themselves neither pinnate nor pinnatifid (but large basal lobes or auricles sometimes present on pinnae).
                        27. Leaf blades only 1-pinnatifid (in our species), divided nearly but not all the way to the rachis (= midrib); leaf segments essentially as wide at base as towards their tips; margins of leaf segments (in our species) without any teeth or basal auricle; indusia lacking __________ (*Pleopeltis*) **Polypodiaceae,** p. 200

27. Leaf blades 1-pinnate, divided all the way to the rachis; pinnae narrowed basally to a very narrow petiole-like attachment to the rachis; pinnae with teeth on the margins, sometimes with basal auricles; indusia present (but sometimes shriveled at maturity).
    28. Pinnae 1–2(–3) pairs in addition to a single terminal pinna (pinnae sometimes with large basal lobes or auricles); plants primarily of the Edwards Plateau, in TX known e only to Comal Co. near w margin of Blackland Prairie ________ *(Tectaria)* **Tectariaceae,** p. 293
    28. Pinnae four or more pairs (usually many more) in addition to a single terminal pinna; plants widespread in East TX and also in the Trans-Pecos.
        29. Leaf blades usually with 5–18 pairs of lateral pinnae (in addition to the terminal pinna) ________ **Dryopteridaceae,** p. 135
        29. Leaf blades with numerous pinnae (> than 18 pairs of lateral pinnae).
            30. Sori only on the uppermost, reduced, fertile pinnae (limited to distal ⅓ to ½ of leaf); indusia orbicular, not at all kidney-shaped; pinnae margins with bristly teeth; stolons absent ________ *(Polystichum)* **Dryopteridaceae,** p. 149
            30. Sori not restricted to the uppermost pinnae, the fertile pinnae not reduced; indusia orbicular-kidney-shaped; pinnae margins without bristly teeth (small non-bristly teeth may be present); stolons present, wiry and widely creeping ________ *(Nephrolepis)* **Lomariopsidaceae,** p. 163

26. Leaf blades more than 1-pinnate, the pinnae themselves either pinnate or pinnatifid.
    31. Lower leaf surfaces with at least some transparent needle-like hairs (particularly on the costae and veins, sometimes even on the intervening tissue or on the indusia—use hand lens or scope); ultimate leaf segments often (but not always) entire; petioles with 2 crescent-shaped vascular bundles basally, distally these united into a U-shape ________ **Thelypteridaceae,** p. 295
    31. Lower leaf surfaces without transparent needle-like hairs; ultimate leaf segments variously incised, serrate, dentate, crenate, or lobulate, not entire; petioles with 2 or more round or oblong vascular bundles.
        32. Sori elongate, straight to hooked or curved; indusia with elongate attachment along one side, the attachment as long as the elongate sori ________ *(Athyrium)* **Athyriaceae,** p. 107
        32. Sori round or nearly so; indusia variously attached, but attachment not elongate.
            33. Sori marginal in sinuses at vein tips, in cup-like indusia; leaf blades extremely large, 1–3 m long, 3(–4) pinnate; species known in TX only froma single cave in Val Verde Co. ________ *(Dennstaedtia)* **Dennstaedtiaceae,** p. 130
            33. Sori not marginal (except in one rare species of *Dryopteris*), the indusia various; leaf blades less than 1 m long (often much less), variously 1 or more pinnate or pinnate-pinnatifid; species widespread in TX.
                34. Leaf blades relatively large, 10–30 cm wide, 35–90 cm long; pinnules (= subdivisions of pinnae) of middle pinnae typically 5 mm or more wide at their bases, only shallowly toothed and thus not appearing subdivided; indusia attached at a narrow sinus, the round-kidney-shaped indusia usually obvious with the naked eye or at least with 10x magnification; petiole bases with 3 or more vascular bundles; all species occurring in TX rare ________ *(Dryopteris)* **Dryopteridaceae,** p. 139
                34. Leaf blades usually smaller, 3.5–10 cm wide, 8–40(–60) cm long; pinnules of middle pinnae typically < 5 mm wide at their bases, at least some so deeply toothed or incised as to appear subdivided; indusia not attached at a narrow sinus (rather attached below sori and at first cup-like and completely enclosing sori, but later splitting into several irregular lobes or flaps spreading around sori OR attached along one side of sori, hood- or pocket-like, arching over sori) (NOTE: indusia are often inconspicuous in mature sori); petiole bases with 2 vascular bundles; including species widespread and common in East TX and widely scattered but rare elsewhere in the state.
                    35. Petioles with scattered light brown scales and glandular pubescence, glabrate with age; indusia attached below sori and at first cup-like and completely enclosing sori, but later splitting into several irregular lobes or flaps spreading around sori (sometimes inconspicuous in mature sori); leaf blades often broadest above the middle; veins of ultimate leaf segments (= smallest subdivisions of leaf) not reaching segment margins (use magnification; important if sori not available); plants widespread and common in e ½ of TX, and occasional in the Trans-Pecos ________ *(Woodsia)* **Woodsiaceae,** p. 308
                    35. Petioles glabrous except at very base; indusia attached by one side, hood- or pocket-like, arching over sori (sometimes inconspicuous in mature sori) towards segment margin; leaf blades broadest at base or at or just below middle; veins of ultimate leaf segments reaching segment margins; plants widely scattered (East TX to Trans-Pecos), but very rare in TX ________ *(Cystopteris)* **Cystopteridaceae,** p. 123

Ferns and Lycophytes

Ferns and Lycophytes (Polypodiophyta and Lycopodiophyta)

Ferns and lycophytes are seedless vascular plants reproducing by single-celled spores—they lack seeds. Collectively, they conservatively include ca. 10,300–11,900 species in 238–248 genera, with some authorities indicating as many as 13,000 or even more species in 350 genera. About 80% of the diversity is in the tropics (Roos 1996; Moran 2004; Mabberley 2008). This is a sizable number of species, but these relatively unusual plants still represent less than five percent of all known living vascular plants, the vast majority of which are flowering plants. Ferns and similar plants were much more ecologically important in the geologic past. They were among the dominant plants in the extensive swamps of the Carboniferous Period (359–299 mya), forming large forests. Over geologic time, the compressed remains of these ancient plants became coal (Hoshizaki & Moran 2001), and in this form they now supply much of the energy upon which modern society is based.

In the past, ferns and those similar plants sometimes called fern allies were grouped together as Division Pteridophyta; they were commonly referred to as the pteridophytes. Later they were often segregated into four separate divisions (Equisetophyta, Lycopodiophyta, Psilophyta, and Polypodiophyta) because they were thought to represent separate evolutionary lineages and to reflect the great morphological diversity among them. In the recent decades DNA sequence data have revolutionized our understanding of evolutionary relationships—no longer do we have to rely solely on comparing the form and structure of plants or on interpreting fossils. It is now apparent that ferns and similar plants should be divided into only two evolutionarily distinct lineages—the lycophytes (CLUB-MOSSES, SPIKE-MOSSES, and QUILLWORTS), technically referred to as the Lycopodiophyta or Lycophyta, and the ferns or Polypodiophyta (including the HORSETAILS and the PSILOPHYTES). It is now clear that among living vascular plants the lycophytes are the basal lineage—they are on a separate evolutionary branch from all other vascular plants (see Fig. 1 on page 4). They are an ancient group surviving from the early days of land plant evolution. Based on their evolutionary isolation, they are appropriately recognized taxonomically as being extremely distinct—in the Lycopodiophyta, a separate taxonomic division from the ferns. The ferns, which include the vast majority of seedless vascular plants, are referred to by a variety of scientific or formal names, including Division Polypodiophyta (as done here), Division Monilophyta (Weakley 2010), or even without taxonomic rank (following cladistic principles) as Moniliformopses (Kendrick & Crane 1997) or Monilophytes[5] (Smith et al. 2006). In the past they were sometimes referred to as the Filicophyta or Pterophyta (Bell & Woodcock 1983; Raven et al. 1986). Because of all the new insights into plant relationships brought about by molecular information, the formal system of botanical nomenclature is undergoing significant change. Therefore we are in a period where our naming system is in flux. Despite some issues with technical names, the evolutionary relationships are now becoming clearer which has significantly improved our system of classification. For example, the Psilophyta and Equisetophyta, considered until recently to be "fern allies," are no longer formally recognized as divisions. DNA has provided convincing evidence that these two morphologically unusual groups are ferns—strange looking, but ferns nonetheless. As a result they are now recognized as Order Psilotales and Order Equisetales respectively within the Division Polypodiophyta (e.g., Pryer et al. 2001; Schuettpelz et al. 2006; but see Rothwell & Nixon 2006 and Rothwell & Stockey 2008 for dissenting views).

REFERENCES: Thieret 1980; Tryon & Tryon 1982; Lellinger 1985; Snyder & Bruce 1986; Kramer & Green 1990; Wagner & Smith 1993; Hasebe et al. 1995; Manhart 1995; Kenrick & Crane 1997; Doyle 1998; Nauman et al. 2000; Nelson 2000; Hoshizaki & Moran 2001; Pryer et al. 2001; Yarborough & Powell 2002; Mickel & Smith 2004; Moran 2004; Ranker & Haufler 2008.

[5]The derivation of the word Monilophyte (from Latin: monilo, necklace or string of beads and Greek: phyt, plant) is in reference to an obscure technical feature of stem anatomy—"the position and ontogeny of protoxylem in the lobed primary xylem of early fossil groups"—Crane & Kendrick 1997

DIVISION LYCOPODIOPHYTA
CLUB-MOSSES, SPIKE-MOSSES, QUILLWORTS

A group of 1,200–1,400 species in 12–17 genera arranged in three families. Each of the three families represents ancient relictual lineages extending back well into the Paleozoic; they are now recognized in three different orders (Flora of North America Editorial Committee 1993; Bateman 1996; Kessler 2010). Altogether they represent less than 1% of living vascular plants worldwide. Members of all three families, Isoëtaceae, Lycopodiaceae, and Selaginellaceae, occur in TX. Extinct members of this ancient division (e.g., Lepidodendrales—scale trees to 55 m or 165 ft tall and 2 m wide at base) were dominants of the Carboniferous forests that formed present-day coal and oil deposits. They were so abundant that one type of coal, cannel coal (from "candle" coal, an old English name for coal that lights easily and burns with a steady, bright flame), is made up almost entirely of the spores of scale trees. The Lycopodiophyta, sometimes referred to collectively as the lycophytes, are the oldest surviving vascular plant group, dating to about 400 million years ago in the Early Devonian Period (416–359 mya). The lycophyte lineage is thus thought to have split from the lineage containing all other living vascular plants shortly after the appearance of vascular plants ca. 410 millon years ago (Banks et al. 2011). The group was much more abundant during the Late Paleozoic, its ecological dominance ending by the end of the Carboniferous about 299 million years ago (Morrell & Faragher 1929; Benson 1979; Bell & Woodcock 1983; Jones & Luchsinger 1986; Raven et al. 1986; Bateman 1996; Moran 2004). As would be expected given their ancient origin, lycophytes have been present on every landmass throughout the history of land plants, and today they are found on every "continent except Antarctica and almost every major island group in the world" (DiMichele & Skog 1992). Lycophytes are characterized by a unique type of leaves referred to as lycophylls or microphylls (= leaves with an intercalary meristem, a single vein, and that arise without a gap in the stem vasculature), and the group is sometimes referred to as the Microphyllophyta (Woodland 1997). Chloroplast DNA data (Raubeson & Jansen 1992) show that the living Lycopodiophyta (*Lycopodium* sensu lato, *Selaginella*, and *Isoëtes*) share with the bryophytes (mosses, liverworts, hornworts) a particular mutation in contrast to all other vascular plants. Also as in bryophytes, most lycophytes also have biflagellated sperm, in contrast to the multiflagellated sperm of other pteridophytes (Renzaglia & Garbary 2001). These data, other molecular studies (e.g., Qiu et al. 2006; Banks et al. 2011), and morphological evidence all indicate that the lycophytes are only distantly related to other vascular plants, and that among living vascular plants the lycophytes are the basal lineage. They are thought to have been evolving separately from other vascular plants for ca. 400 million years (Raubeson & Stein 1995; Kenrick & Crane 1997; Doyle 1998; Duff 2000; Pryer et al. 2001; Pryer et al. 2004a, 2004b; Qiu et al. 2006).

REFERENCES: Benson 1979; Bell & Woodcock 1983; Jones & Luchsinger 1986; Raven et al. 1986; Snyder & Bruce 1986; Bold et al. 1987; DiMichele & Skog 1992; Gensel 1992; Wagner & Smith 1993; Raubeson & Jansen 1992; Raubeson & Stein 1995; Woodland 1997; Gensel & Berry 2001; Duff 2000; Pryer et al. 2004a, 2004b; Ranker & Haufler 2008; Banks et al. 2011.

ISOËTACEAE Dumort.
QUILLWORT FAMILY

An ancient, nearly cosmopolitan family of only one genus with ca. 200 (Schuettpelz & Hoot 2006) to perhaps as many as 350 species. Several other genera are known from fossils (Lellinger 1985; Holmes et al. 2005). Hickey et al. (2003) noted that South America is the modern center for both morphological and taxonomic diversity. QUILLWORTS are small, tufted plants with a cormose buried stem, and hollow, elongate, quill-like, superficially grass-like leaves. They range from perennial evergreen aquatics in glacial lakes to ephemeral terrestrials; some even invade old fields and roadside ditches (Hoot et al. 2006). As noted earlier, along with other Lycopodiophyta, they are part of an ancient evolutionary lineage that has been separate from all other vascular plants for about 400 million years (Pryer et al. 2004a). These morphologically distinctive plants are superficially very unlike other Lycopodiophyta, in part because of extreme reduction. But as in other members of the division, the leaves (lycophylls) have a single vein and other similarities, and all lycophytes are linked by molecular evidence. Ligules (= minute, tongue-like, basal protuberances on leaves) are present as in the Selaginellaceae, and the spores are differentiated into microspores and megaspores (also as in the Selaginellaceae, which are the closest living relatives of QUILLWORTS). The Isoëtaceae are an ancient phylogenetically isolated group; "fossils assigned to the *Isoëtes* lineage have been reported from the Late Devonian" with some of these species being tree-like in habit (Schuettpelz & Hoot 2006). Wikström (2001) noted that fossil evidence suggests the minimum age for the split between the lineages leading to Isoëtaceae and Selaginellaceae is 377 million years ago. Fossils similar to living species are present by the Jurassic (Schuettpelz & Hoot 2006; Jansson et al. 2008) or possibly even the Triassic (Retallack 1997), and there has been remarkably little morphological change during the subsequent more than 150 million years. QUILLWORTS are the only living plants that have rhizomorphs (= root-like stems that have nutrient absorption and anchoring functions usually carried out by roots), and are also the only extant plants exhibiting rhizotaxy (= a regular and precise spiral pattern of "rootlets" around the main "root," unlike the irregular arrangement seen in all other living vascular plants). Having rhizomorphs and exhibiting rhizotaxy are some of the reasons why the Isoëtaceae are thought to be the closest living relatives of the Lepidodentrales (giant rhizomorphic and rhizotaxic scale trees that have been extinct since about 225 million years ago) (Bateman 1996; Pigg 2001; Moran 2004). Though tree-sized and capable of branching, these scale trees bore superficially grass-like leaves in some ways resembling those of *Isoëtes*. QUILLWORTS also are unique among living plants in that their leaves have four air chambers (visible in cross-section) running lengthwise—such structures are generally thought to be aquatic adaptations to provide buoyancy and facilitate gas exchange; Taylor and Hickey (1992) thus interpreted the air chambers to mean that all living *Isoëtes* had aquatic ancestors. They apparently get CO_2 through the "roots" which absorb the gas from the sediment where the plant is anchored—the "roots" have a large central air space for transporting CO_2 to the chambered leaves. This is yet another similarity shared with the extinct Lepidodentrales (Moran 2004). Because of this different method of CO_2 capture, the leaves of some *Isoëtes* species lack stomates (the pores through which plants typically obtain gases including CO_2). The long linear leaves have a resemblance to the quills of feathers formerly used as writing implements. This in combination with the Old English, *wort*, (from Anglo-Saxon: *wyrt*), an herb, root, or plant, gives rise to the common name. (Order Isoëtales)

FAMILY RECOGNITION IN THE FIELD: the TX species are small, tufted, wet area plants with a cormose stem and hollow, elongate, quill-like, superficially grass-like leaves, giving them the appearance of garden chives, tiny green onions, or small sterile spike-rushes or grasses; sporangia are in the leaf bases at or below ground level.

REFERENCES: Pfeiffer 1922; Correll 1949, 1956, 1966; Jermy 1990a; Pigg 1992, 2001; Taylor et al. 1993.

ISOËTES L. QUILLWORT

Plants tufted perennials, on seasonally saturated soils, shallow or temporary pools, shallow depressions on rock outcrops, or ponds and swales; stem buried, cormose, often globose, usually 2-lobed, persisting after the leaves die back during hot dry weather (in terrestrial species); leaves quill-like, linear, hollow, tightly clustered together, in cross-section having four air chambers; sporangia solitary, embedded in broadened basal cavity of leaf (sporophyll) with ligule inserted above, typically partly covered by a velum (= membranous flap of tissue extending downward over sporangium); spores of 2 types (plant heterosporous), the megaspores with an equatorial ridge (which, equator-like, encircles the spore) and 3 other converging ridges, the intervening surfaces varying microscopically from smooth to tuberculate or echinate to cobwebby, the microspores much smaller.

Within the genus there is little morphological variation among living species despite its ancient origins, worldwide distribution, and diverse ecological adaptations (Hoot et al. 2006; Schuettpelz & Hoot 2006). Because of the highly conserved (= unchanged from ancestral forms) and similar morphology, species are often extremely difficult to distinguish; in fact, they are among the most difficult pteridophytes to identify in TX. Knowledge about habitat, velum morphology, and usually microscopic examination of the mature megaspores (for surface texture and size) are often necessary for identification of species. In fact, detailed microscopy of the megaspores (or in some cases use of a hand lens) is essential for definitive identification of the five species known in TX (the ranges of four overlap in c TX, e.g., Llano Co.) (Holmes et al. 2005). Botanist John Thieret, author of *Louisiana Ferns and Fern Allies* (1980), was driven to say, "The certain identification of many species of *Isoëtes* is a matter for specialists—or even God." The vegetative characters given in the key to species, although useful, should be considered as "trends" rather than dependable ways of separating the species. Interspecific hybrids are known in the genus. The spores are reported to be dispersed in the excreta of earthworms. Megaspores in the genus "may be up to 750 µm in diameter, among the largest megaspores in the plant kingdom" (Musselman 2002). The key to species is adapted in part from Holmes et al. (2005). The name of the genus is written as *Isoetes* by some and as *Isoëtes* by others (with a diacritical sign, the diaeresis, i.e., ë). According to the International Code of Botanical Nomenclature (ICBN; McNeill et al. 2006), "Diacritical signs are not used in Latin names. ... [however,] The diaeresis, indicating that a vowel is to be pronounced separately from the preceding vowel (as in *Cephaëlis, Isoëtes*), is permissible. ..." We are therefore following Taylor et al. (1993, in *Flora of North America*) and many other recent authors (e.g., Hoot & Taylor 2001; Hoot et al. 2006; Schuettpelz & Hoot 2006) in using the diaeresis. (Greek: *isos*, equal, and *etos*, year, referring to the evergreen habit of some species)

REFERENCES: Pfeiffer 1939; Landry & Thieret 1973; Taylor et al. 1975, 1985; Taylor & Taylor 1981; Boom 1982; Lott et al. 1982; Hickey 1986a, 1986b; Taylor & Hickey 1992; Heafner 1997; Retallack 1997; Duff 2000; Hoot & Taylor 2001; Musselman 2002; Rydin & Wikström 2002; Hickey et al. 2003; Budke et al. 2005; Heafner & Bray 2005; Holmes et al. 2005; Turner et al. 2005; Brunton & Britton 2006; Hoot et al. 2004, 2006; Schuettpelz & Hoot 2006; Poole et al. 2007a; Singhurst et al. 2011; Norton Taylor et al. 2012.

1. Plants aquatic with leaves lying flat on surface of water, in permanent ponds; leaves to 62 cm long; velum covering entire sporangium; megaspores smooth to obscurely rugulose; endemic species known only from Aransas and Calhoun cos. in TX Coastal Bend ____ **I. texana**
1. Plants typically ± upright, the leaves usually rising out of the water, in seasonally saturated soils or temporary or shallow pools, on various substrates including granite, gneiss, or sandstone rocks and calcareous soils; leaves various, from 3–40 cm long; velum covering from < ¼ to < ¾ of sporangium (except in *I. lithophila* which is endemic to rock outcrops in c TX); megaspores various; including species from the Central Mineral Region to far East TX.
   2. Megaspores light gray to gray-brown or bronze, the surfaces microscopically smooth to obscurely rugulose (= with slight ridges/wrinkles) or obscurely tuberculate (= with minute warts/tubercles), glabrous; velum covering entire sporangium; leaves typically 3–12(–20+) cm long; species endemic to 4 counties in the Central Mineral Region ____ **I. lithophila**
   2. Megaspores white, the surfaces microscopically obscurely rugulose to tuberculate or echinate (= with minute prickles), sometimes cobwebby; velum covering from < ¼ to < ¾ of sporangium; leaves various, from 7–40 cm long; including species widespread in the e ½ of TX including the Central Mineral Region.
      3. Megaspores (0.36–)0.48–0.65 mm in diam., the surfaces appearing densely cobwebby (visible using a microscope); leaves at base with outer surface pale or sometimes somewhat brownish, not shiny; plants dioecious; species predominantly on calcareous substrates, but also on sandstone or granite ____ **I. butleri**
      3. Megaspores 0.25–0.45 mm in diam., the surfaces not appearing cobwebby; leaves at base with outer surface blackish and shiny OR pale to brownish and not shiny; plants monoecious; species on granite or non-calcareous substrates.
         4. Megaspores with surfaces sparsely rugulose with low, fused ridges; leaves usually 15–40 cm long, the outer surface of leaf bases usually black (sometimes brownish or rarely pale) and shiny; species of various substrates, often on sandstone or non-calcareous substrates, but also on granite, widespread in e ½ of TX ____ **I. melanopoda**
         4. Megaspores with surfaces tuberculate or echinate; leaves usually 7–15(–20) cm long, brownish and not shiny at base; species of granite outcrops known in TX only from Burnet, Llano, and Mason cos. in the Central Mineral Region ____ **I. piedmontiana**

Isoëtes butleri Engelm., (for George Dexter Butler, 1850–1910, lawyer, teacher, botanist, and correspondent of George Engelmann, who discovered the species in what is now OK), BUTLER'S QUILLWORT. Leaves to 15(–30) cm long, dull green to gray-green or yellow-green, often lax and twisted, the outer surfaces white to tan or brown at base, not shiny; velum covering < ¼ of sporangium; megaspores white, the surfaces densely cobwebby, microscopically obscurely tuberculate; $2n = 22$ (Taylor et al. 1993). Seasonally saturated soils, temporary or shallow pools, usually on calcareous soils, but also on sandstone or granite (in these cases there is probably a source of calcium nearby influencing pH—Lott et al. 1982); Mason (*R. McVaugh 8322*, BRIT, annotated K.D. Heafner 2005) and Llano (*Correll & Johnston 17313*, TEX-LL, annotated by W.C. Holmes & J. Singhurst 2004, SEM at TEX-LL; Lott et al. 1982), cos. on Llano Uplift of c TX and nearby Comal Co. (Turner et al. 2003), Sabine (*Singhurst 11749*, BAYLU) and San Augustine (*Singhurst 11750*, BAYLU) cos. on glauconite deposits

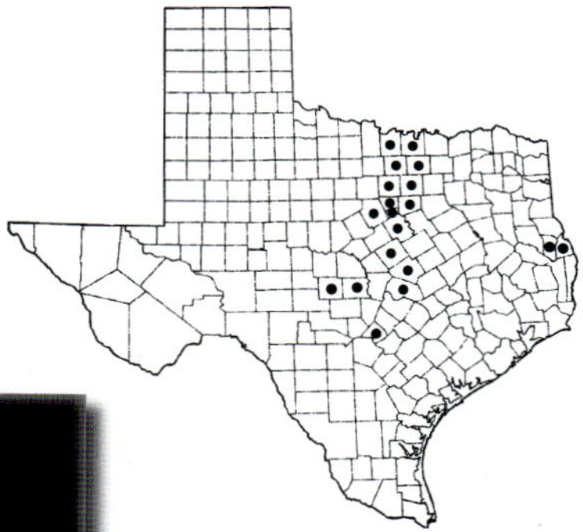

note cobwebby
megaspore surface

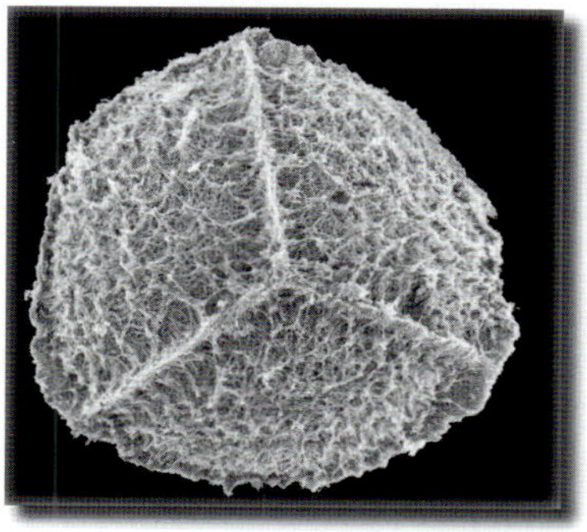

megaspore (SEM)
[PRC]

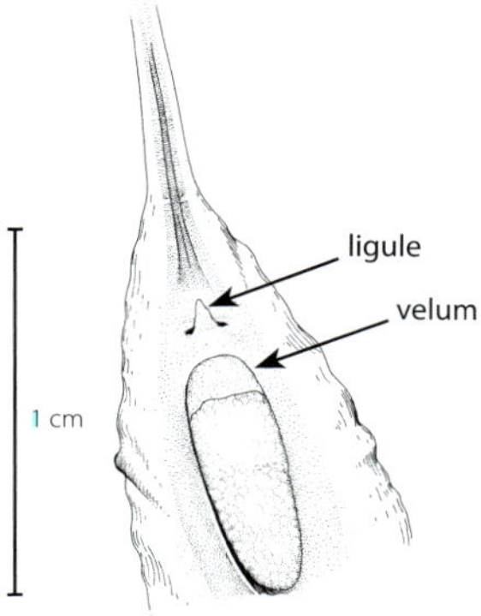

adaxial (inner) side of base of
sporophyll (fertile leaf)
w/megasporangium
[TAY]

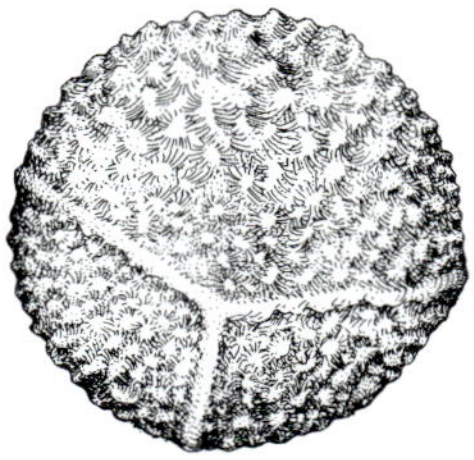

megaspore
[TAY]

Alan Cressler

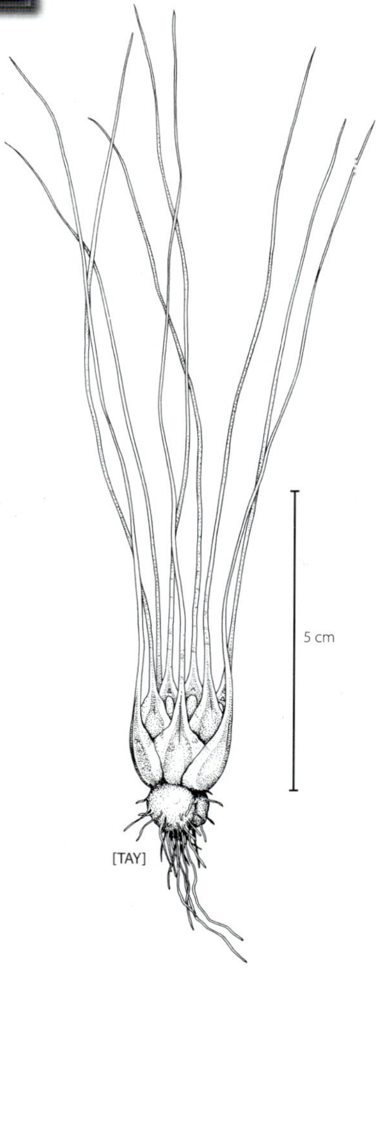

[TAY]

Isoëtes butleri

BUTLER'S QUILLWORT

(Weches Formation) in far East TX, and recently discovered in a number of counties in the Cross Timbers and Prairies (e.g., Tarrant Co, *O'Kennon et al. 22367*, BRIT; Bell and Coreyll cos., Hansen 2010). Norton Taylor et al. (2012) have recently greatly expanded the distribution of the species in TX by making collections in Bosque, Cooke, Denton, Erath, Hood, Johnson, Montague, Parker, Somervell, Williamson, and Wise cos. (specimens at BRIT). Collections are also known from a number of counties in Oklahoma—e.g., Choctaw, Johnston, and McCurtain cos. (Taylor & Taylor 1981) including some just across the Red River from TX; IL s to GA w to KS and TX. Spores mature in late spring. This species was reported as new to Texas by Lott et al. (1982). It was for many years known in TX only from the Llano Uplift of c TX but was later found to be disjunct to far East TX on the Weches Formation (Holmes et al. 2005). Very recently (2010–2012) it has been discovered as disjunct to a third region of TX, the Cross Timbers and Prairies in another geologically unusual setting—calcareous seeps on a number of limestone strata. For example, it was initially discovered on a "muhly seep" (= seeping calcareous slope dominated by the grass *Muhlenbergia reverchonii*, SEEP MUHLY) on the Walnut limestone; the identification has been confirmed based on SEM examination of megaspore ornamentation patterns (Tiana Franklin Rehman, pers. comm.). Taylor et al. (1993) noted that the "leaves yellow, wither, and disappear by late spring." Unlike what is typical in *Isoëtes* (monoecious—developing megasporophylls early in the growing season and microsporophylls later), *I. butleri* is the only species known to be dioecious (Turner et al. 2005). This is one of several pteridophytes present in East TX with disjunct populations in the Central Mineral Region of the Edwards Plateau. Other examples are *Ophioglossum crotalophoroides*, BULBOUS ADDER'S-TONGUE, *Ophioglossum nudicaule*, SLENDER ADDER'S-TONGUE, and *Onoclea sensibilis*, SENSITIVE FERN. The inconspicuousness, visibility primarily in the spring, superficial similarity to a little tuft of some non-reproductively active grass or sedge, and habitat specificity of this species are probably the reasons explaining why there were relatively few collections in TX until recently.

Isoëtes lithophila N. Pfeiff., (Greek: lithos, *rock*, and *philos*, love), ROCK QUILLWORT. Leaves typically 3–12 (–20+) cm long, erect, not or only slightly twisted, bright green, the outer surfaces pale toward base; velum covering the entire sporangium; megaspores gray to gray-brown or bronze, 0.25–0.40 mm in diam., the surfaces not cobwebby, microscopically smooth to obscurely tuberculate; 2*n* = 22 (Taylor et al. 1993). Temporary vernal pools on granite or gneiss outcrops in the Llano Uplift area of c TX in Burnet (*V.L. Cory 49507*, *S.D. Jones et al. 10051*, BRIT, det. W.C. Holmes, *G.M. Soxman 448*, TEX-LL, det. Holmes & Singhurst, SEM at TEX-LL), Mason (*S. & S. Pence 67*, BRIT, det. W.C. Holmes), Llano (*W. Krodel et al. 17t278*, BRIT, det. W.C. Holmes, *H. Bold s.n.*, 1963, TEX-LL, det. Holmes & Singhurst), and Gillespie (Poole et al. 2002, 2010) cos.; endemic to c TX (Kartesz 1999; Carr 2002b, 2002c). Sporulating late winter–Jun. Poole et al. (2007a) discussed it in detail and noted that the epithet is often erroneously spelled "lithophylla." This rare species is one of only two pteridophytes endemic to TX (the other is *I. texana*). It is also one of only two pteridophytes included in *Rare Plants of Texas* (Poole et al. 2007a). (Conservation Concern—TOES 1993: V; Poole et al. 2007a, 2007b, 2010: G2S2)

Isoëtes melanopoda J. Gay & Durieu ex Durieu, (Greek: *melas, melanos*, black, and *pous* or *podium*, foot), BLACK-FOOT QUILLWORT. Leaves usually 15–40 cm long, erect, not or little twisted, often grass-green, the outer surface usually blackish and shiny toward very base (sometimes brownish or rarely pale); velum covering < ¾ of sporangium; megaspores white, 0.25–0.45 mm in diam., the surfaces not cobwebby, microscopically with obscure sparse ridges; 2*n* = 22 (Taylor et al. 1993). Seasonally saturated soils, temporary or shallow pools, usually on non-calcareous substrates (Taylor et al. 1993); widely scattered in e ½ of TX (e.g., Madison Co., *A.K. Neill 2545*, BRIT) (including far East TX (e.g., Hardin, Sabine, and Shelby cos., BAYLU; Holmes et al. 2005) where it has often been "overlooked" (Holmes et al. 2005); Pineywoods and n Gulf Prairies and Marshes w to Cross Timbers and Prairies and e Edwards Plateau; e U.S. from NJ s to GA w to NE and TX, also UT. Sporulating Mar–Oct. This species commonly has black and shiny phyllopodia (= hardened, sclerified leaf bases) persisting from the previous year (Heafner & Bray 2005). Brunton and Britton (2006) recently named a new subspecies of *I. melanopoda* (subsp. *silvatica*) that occurs in the e U.S. from NJ s to AL and w to MS and TN; TX populations are of subsp. *melanopoda*. Subspecies *silvatica* typically has wider leaves (average 1.5–2.18 mm wide vs. 0.51–1.1 mm), pale leaf bases (vs. black), larger megaspores with bolder ornamentation, and habitat preference for deciduous woodland swamp (vs. open grassland/prairie).

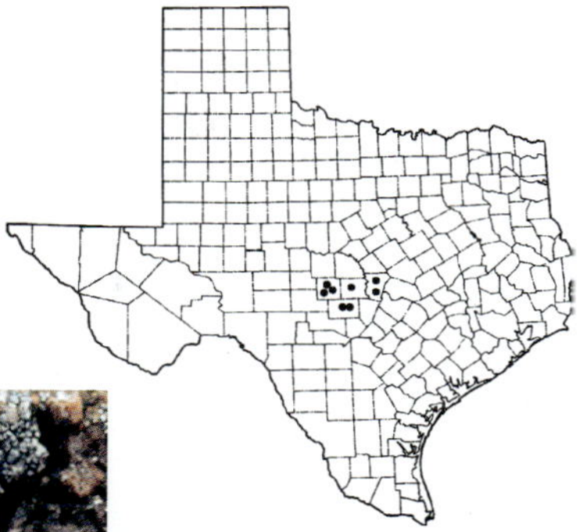

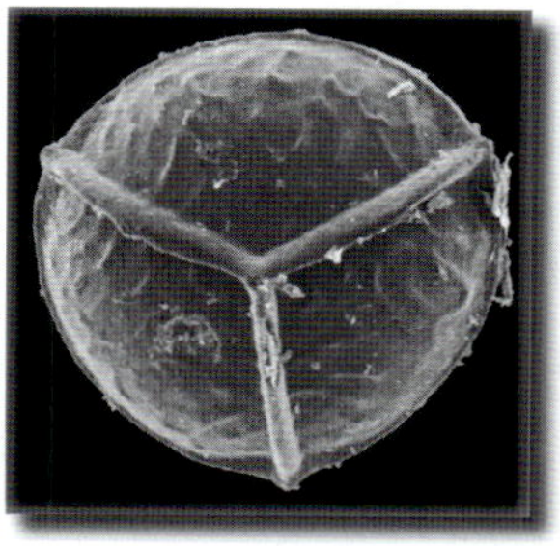

megaspore (SEM)
[PRC]

Jack Stanford

Chad Husby

Chad Husby

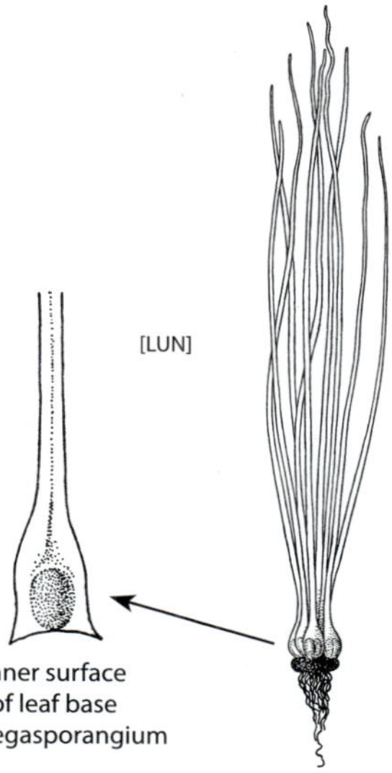

[LUN]

inner surface
of leaf base
w/megasporangium

Isoëtes lithophila

ROCK QUILLWORT

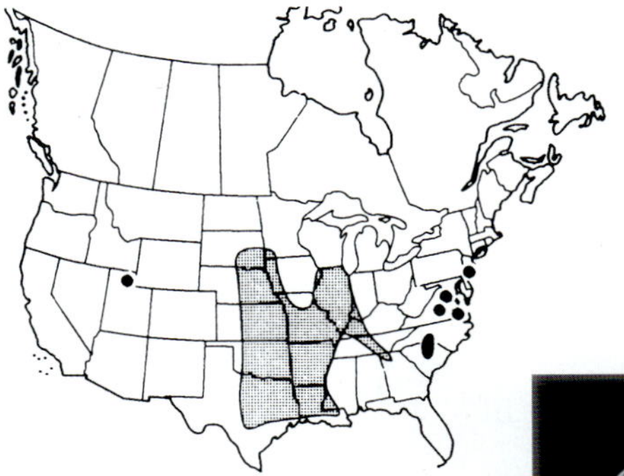

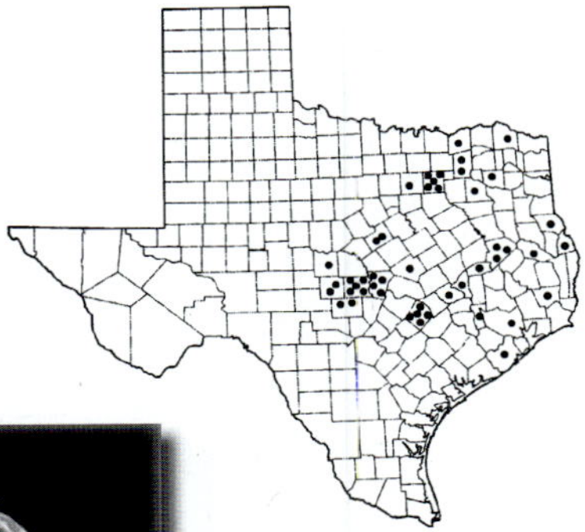

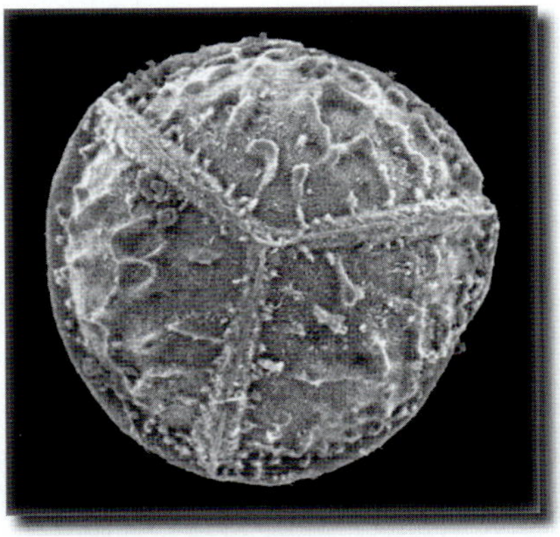

megaspore (SEM)
[PRC]

Jeff McMillian

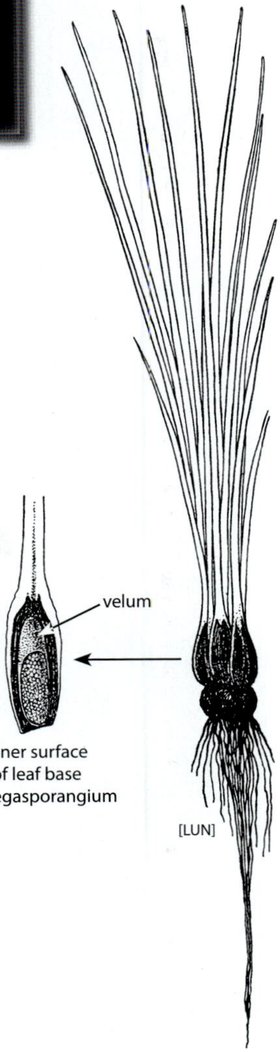

Isoëtes melanopoda

BLACK-FOOT QUILLWORT

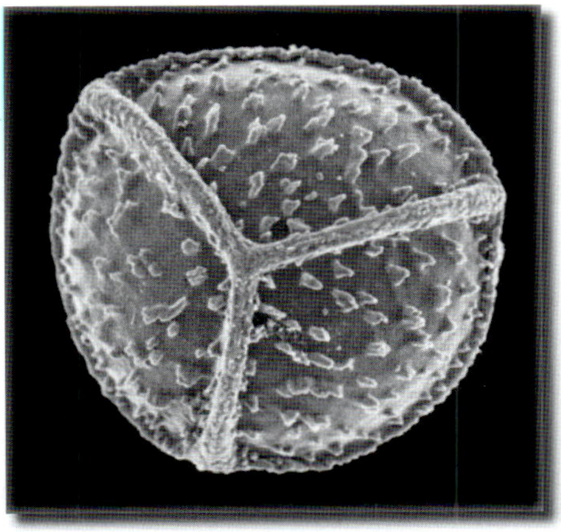

megaspore (SEM)
[PRC]

Alan Cressler

Alan Cressler

Isoëtes piedmontana

PIEDMONT QUILLWORT

Jason Singhurst

note fairly smooth megaspore surface

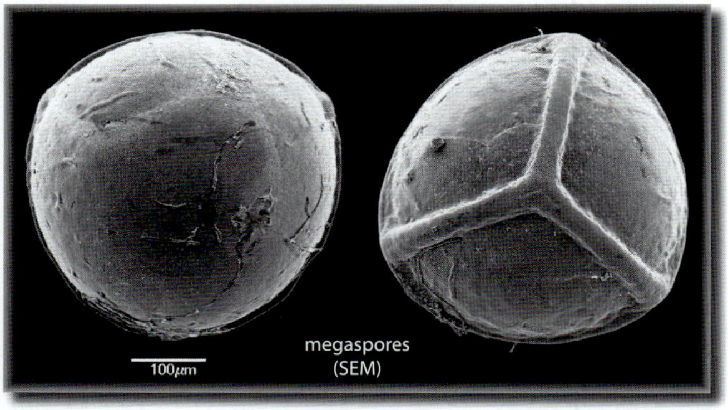

[PHY]

Jason Singhurst

Isoëtes texana

TEJAS QUILLWORT

Isoëtes piedmontana (N. Pfeiff.) C.F. Reed, (Latin: *pes*, *pedis*, a foot, and *montanus*, pertaining to the mountains, = between the mountains and the coastal plain, the naming author saying, "It is apparently of fairly wide distribution in the piedmont"), PIEDMONT QUILLWORT. Leaves usually 7–15(–20) cm long, brownish at base, not shiny, rarely with phyllopods; velum covering <⅓ of sporangium; megaspores white, not cobwebby, the surfaces microscopically tuberculate to echinate; $2n$ = 22, 33, 44 (Heafner & Bray 2005). Granite outcrops in temporary pools or shallow depressions; Mason (*L.L. Sanchez & J.D. Stanford 2364*, BRIT, BAYLU; Holmes et al. 2005), Burnet (*Correll & Johnston 17332*, TEX-LL, annotated Holmes & Singhurst, SEM at TEX-LL), and Llano (*Carr 10569, 11625*, TEX-LL, annotated Holmes & Singhurst; Holmes et al. 2005) cos. in Llano Uplift area (Central Mineral Region) of c TX; se U.S., VA to AL, disjunct w to TX. [*I. virginica* var. *piedmontana* N. Pfeiff.] Sporulating at least in Apr–May. This disjunct species was first recognized for TX by Holmes et al. (2005); see Fig. 32 and page 31 for a discussion of the disjunction. They noted that it is the only *Isoëtes* species in the TX flora with megaspores having "a tuberculate to echinate ornamentation pattern not obscured by perine projections." They also pointed out that it was possible that the species was a hybrid between *I. lithophila* and *I. melanopoda*, but stated that their detailed study did not support that conclusion. Heafner and Bray (2005) suggested that based on their evidence, "*I. piedmontana* is best recognized as a complex of cryptic species that [has] evolutionarily distinct origins in their respective geographic ranges" and that ongoing studies will be needed to determine the relatedness of the TX populations to those in the e U.S. It is thus quite possible that the TX populations represent a separate species that will eventually be recognized as distinct. Because of its rareness and limited distribution in the state, we consider this species to be of conservation concern in TX. ⚠

Isoëtes texana Singhurst, A.E. Rushing, & W.C. Holmes, (of Texas, the species being endemic to the state), TEJAS QUILLWORT. Plant aquatic; leaves evergreen, green-yellow, 24–62 cm long, pliant, gradually thinning from base to pointed apex, laying flat on water surface; velum covering whole sporangium; megaspores white, 0.35–0.40 mm, the surfaces not cobwebby, microscopically smooth to obscurely rugulose. The species is known only from "freshwater ponds and interdunal swales in neutral to moderately alkaline sands of Pleistocene and to Holocene barrier islands" (Singhurst et al. 2011); Aransas (*Singhurst 18338*, BAYLU) and Calhoun (*Singhurst 18336* [holotype: BAYLU; isotypes: BRIT, TEX]; *Singhurst, Hanks, and Hensen 18339, 18340, 18341, 18342*, and *18343* [all BAYLU]) cos. in the TX Coastal Bend; endemic to coastal TX (however, there are reported but unconfirmed occurrences in LA and MS—J. Singhurst, pers. comm.). Sporulating Jun and Oct (collections known only from these months). Singhurst et al. (2011) noted that "The new species resembles *Isoëtes flaccida* A. Braun of Florida and Georgia and keys near its vicinity in both Lellinger (1985) and Taylor et al. (1993). The resemblance is mainly expressed in length of the leaves and aquatic habitat." However, *Isoëtes flaccida* has microspores with tuberculate to prominently rugulate surfaces and bright green leaves and is separated geographically from *I. texana* by about 1200 km (746 mi). This species is one of only two pteridophytes endemic to TX (the other is *I. lithophila*). *Isoëtes texana* is the most recently described species of the TX flora—the publication date is 5 May 2011 (Singhurst et al. 2011). The species is of conservation concern; Singhurst et al. (2011) consider it globally rare with a ranking of G1S1. They note that "Immediate threats to this species include development and expansion of ponds for livestock use and human recreation (fishing). The population in Aransas National Wildlife Refuge in Aransas County is well protected. The Calhoun County occurrences are on private land and appear to be reasonably secure at this time. Rising sea level may present a long-term threat to the species as they appear susceptible to salt water intrusion." (Singhurst et al. 2011: G1S1) ⚠

LYCOPODIACEAE P. Beauv. ex Mirb.

CLUB-MOSS FAMILY

Perennials with horizontal and upright stems; roots produced along the horizontal stems; leaves (microphylls) numerous, small, with a single unbranched vein; upright stems simple or branched, serving as peduncle-like supporting structures for the terminal strobili (= cones); sporangia solitary per sporophyll (= spore-bearing leaf); spores all of 1 kind (plants homosporous).

A diverse ancient family with a long fossil history dating to the late Paleozoic Era (Thomas 1992). CLUB-MOSSES are homosporous and lack ligules (= minute, tongue-like, basal protuberances on leaves), differing in these ways from both Isoetaceae and Selaginellaceae, the other two lycophyte families. This and other evidence suggests that of the three, the Lycopodiaceae represent the basal branch among the living members of the Lycopodiophyta. As noted earlier, along with other Lycopodiophyta, they are part of an ancient, isolated evolutionary lineage that has been separate from all other vascular plants for around 400 million years since the early Devonian (Wikström 2001; Pryer et al. 2004a). The family "is widely regarded as a relict group, because living species bear a striking resemblance to early fossils and modern species diversity is low" (Wikström & Kenrick 2000). The Lycopodiaceae are a cosmopolitan family and contain 10–15 genera and ca. 350–400+ species (Wagner & Beitel 1993). They are terrestrial or epiphytic, evergreen, coarsely moss-like, spore-bearing vascular plants with scale- or needle-like leaves containing a single vein; ligules are absent, spores are all of one type, and the sporangia are often in "club-like" strobili (= cones). All North American species were previously recognized in the large and very diverse genus *Lycopodium*, which more recently has been divided into three subfamilies (of Lycopodiaceae) and seven segregate genera (Wagner & Beitel 1993). It should be noted that the evidence for dividing *Lycopodium* is very strong—the segregate genera differ by numerous morphological and molecular characters and by chromosome number, and are thought to represent ancient evolutionary lineages that have been separate for many tens to perhaps hundreds of million years. In fact, based on fossil evidence Wikström (2001) puts the minimum age for the split between *Lycopodium* and *Lycopodiella* at 208 mya. Three of the segregates, *Lycopodiella*, *Palhinhaea*, and *Pseudolycopodiella*, are treated here. Although virtually all modern authors subdivide *Lycopodium*, exact taxonomic opinion on the family differs, with various authors recognizing fewer genera (Øllgaard (1987, 1992) or in some cases (e.g., Haines 2003) even more than the seven of Wagner and Beitel (1993). Certain species were in the past gathered for making Christmas wreaths, and in some areas (e.g., Appalachian Mts.) this practice resulted in populations being nearly eliminated. The dust-like spores are very rich in oils, quite flammable, and burn with a quick flash—as a result they were formerly used in fireworks, signal fires, by magicians, for stage-lighting, and in photography as flash powder (Whitebread 1941; Jones & Luchsinger 1986). The spores also repel water and resist wetting. Thus, when pills were hand-made, the spores were used by pharmacists to coat pills and prevent them from sticking together. Further, they were once used as a dusting powder on rubber surgical gloves, condoms, and suppositories, but they are considered dangerous for such uses (and have been replaced) because they can cause inflammation, adhesions, granulomas, and allergic reactions including severe asthma attacks (Whitebread 1941; Balick & Beitel 1988, 1989; Hoshizaki & Moran 2001; Runneson 2010). The spores were formerly also used medicinally. These various uses and unusual properties (e.g., ignite explosively when thrown into flames) gave rise to interesting names including vegetable sulfur, druid's flour, vegetable brimstone, and witch meal (Fernie 2008). They have even been used by chemists in experiments to visually estimate Avogadro's number (King & Neilsen 1958; King 1961). Temperate species are reported to be difficult to cultivate, often requiring specific mycorrhizal fungi (Hoshizaki & Moran 2001). Lycopodiaceae are unique among living Lycopodiophyta in being homosporous (having only one type of spore, versus heterosporous—having microspores and megaspores). Kramer (1993) suggested this is why some Lycopodiaceae species have such wide distributions (e.g., *Pseudolycopodiella caroliniana*—North to South America, Africa, e Asia). As Kramer (1993) pointed out, "That the small size and weight of the spores of homosporous pteridophytes facilitates long-range dispersal is indicated by the fact that the species of the homosporous Lycopodiaceae with their tiny, light spores tend to be more widely distributed than the species of the heterosporous Selaginellaceae with their very much larger megaspores and usually also larger microspores." Family name from *Lycopodium*, CLUB-MOSS, now recognized as a mainly temperate and subarctic genus of 15–25 species (Wagner & Beitel 1993). (Greek: *lykos*, wolf, and *pous* or *podium*, foot, in reference to the resemblance of the branch tips of the type species, *L. clavatum* L., to a wolf's paw, the numerous bristle-tipped leaves imparting a somewhat fur-like appearance—Moran 2004) (Order Lycopodiales)

FAMILY RECOGNITION IN THE FIELD: evergreen, superficially somewhat moss-like herbs with stems covered by numerous, small, linear to linear-lanceolate or lanceolate, 1-veined leaves; horizontal stems lying flat on the ground with upright stems terminating in cylindrical, spore-producing cones.

REFERENCES: Correll 1949, 1956, 1966; Cranfill 1981; Holub 1983; Øllgaard 1987, 1990, 1992; Thomas 1992; Wagner & Beitel 1992, 1993; Wikström & Kenrick 1997, 2000; Wikström 2001; Haines 2003; Weakley 2010.

1. Upright stems (bearing strobili) with many branches; strobili (= cones) many per upright stem, nodding or pendant at the tips of the many branches ______ **Palhinhaea**
1. Upright stems unbranched; strobili solitary per upright stem, erect.
   2. Upright, unbranched stems with crowded leaves; sporophylls shaped similarly to leaves of upright stems, linear to lanceolate; leaves of horizontal stems usually but not always with teeth, all the same size, not appearing 2-ranked (not all the same size in *L. prostrata*, which is rare in TX); horizontal stems lying flat on ground OR strongly arching ______ **Lycopodiella**
   2. Upright, unbranched stems with only scattered, scale-like, subulate (= awl-shaped) leaves, the surface of the upright stems easily seen between the scattered leaves; sporophylls dramatically differently shaped than narrower leaves of upright stems, broadly ovate to deltate basally narrowing to acuminate apically; leaves of horizontal stems without teeth, not all the same size, the lateral leaves larger than medial leaves and appearing 2-ranked; horizontal stems lying flat on ground ______ **Pseudolycopodiella**

LYCOPODIELLA Holub BOG CLUB-MOSS

Plants perennial; horizontal stems evergreen at least at apex, prostrate or arching, densely leafy; upright, unbranched stems scattered along horizontal stems, densely leafy; leaves linear to linear-lanceolate or lanceolate, entire to conspicuously toothed; strobili solitary, terminating upright stems; sporophylls appressed to wide-spreading, similar to but generally slightly longer than other leaves; sporangia globose or subglobose, solitary at base of upper side of sporophylls.

Lycopodiella has traditionally been treated as part of a more broadly defined *Lycopodium*. Molecular evidence, tentatively time-calibrated using fossils, places the split between *Lycopodium* and *Lycopodiella* during the early Jurassic Period (208 mya) (Wikström & Kenrick 2000), providing support for its recognition at the generic level. *Lycopodiella* is a n temperate and tropical American genus variously considered to have as few as 8–10 species (Wagner & Beitel 1993) or as many as 40 (Mabberley 2008). Species of *Lycopodiella* hybridize readily, and all possible hybrids (Thieret 1980; Wagner & Beitel 1993) between the three TX species are known from the state: *L. alopecuroides* × *L. prostrata* (Jefferson Co.—TEX-LL), *L. alopecuroides* × *L. appressa* (Jasper and Newton cos.—TEX-LL), and *L. appressa* × *L. prostrata* (Jasper Co.—TEX-LL; Snyder & Bruce 1986). The situation is the same in LA, and Thieret (1980) pointed out, "Further compounding the difficulties [in identification] is that all three species and all three hybrids may grow together in the same ditch. To make positive identification of these plants, both vegetative and reproductive parts are needed. Even with these, identification m[a]y be troublesome except for the experts. And, on occasion, even the experts may be puzzled as to the identity of a particular specimen." The genus has a base chromosome number (*x*) of 78 (Wagner & Beitel 1993). (Name derived from the genus *Lycopodium*, from Greek: *lykos*, wolf, and *pous* or *podium*, foot, in reference to the resemblance of the bristly, almost furry-appearing branch tips to a wolf's paw, plus the Latin diminutive suffix, *-ella*, signifying tiny or small)

REFERENCES: Cranfill 1981; Øllgaard 1987; Wikström & Kenrick 2000.

1. Strobili 3–6(–9) mm wide (including sporophylls), only slightly wider than supporting upright stem including its leaves; sporophylls appressed; leaves of horizontal stems (and upright stems) sparsely or not toothed; horizontal stems prostrate ______ **L. appressa**
1. Strobili 11–20 mm wide (including sporophylls), usually 1.5–3 times wider than supporting upright stem; sporophylls usually ± spreading or wide-spreading; leaves of horizontal stems (and also typically upright stems) usually conspicuously toothed; horizontal stems prostrate OR strongly arching.
   2. Horizontal stems arching, rooting at intervals; leaves on horizontal stems all ± the same size, usually 0.5–0.7 mm wide; horizontal stems (excluding leaves) 2–4 mm in diam., ca. 8–11 mm wide with leaves ______ **L. alopecuroides**
   2. Horizontal stems prostrate, rooting throughout; leaves on horizontal stems of two sizes, the largest (lateral) leaves usually 0.8–1.8 mm wide, visibly wider than medial leaves; horizontal stems (excluding leaves) 1–1.5 mm in diam., ca. 12–19 mm wide with leaves (due to lateral leaves being nearly perpendicular to stem) ______ **L. prostrata**

Lycopodiella alopecuroides (L.) Cranfill, (from *Alopecurus*, foxtail grass, and Greek: *-eides*, -oides, like or having the form of, apparently in reference to the resemblance of the strobili to the seed heads of that grass, which superficially resemble a fox's tail), FOXTAIL BOG CLUB-MOSS, FOXTAIL CLUB-MOSS. Horizontal stems long-creeping, arching, rooting at intervals, 2–4 mm in diam. (excluding leaves); leaves monomorphic, linear to linear-lanceolate, ca. 5–7 mm long, ca. 0.5–0.7 mm wide, with conspicuous marginal teeth, spreading to ascending; upright stems unbranched, (6–)10–30(–45) cm tall, densely covered with leaves; strobili 2–6(–11)

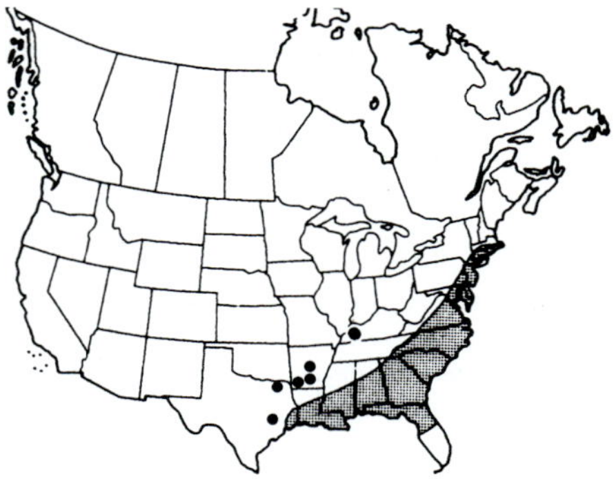

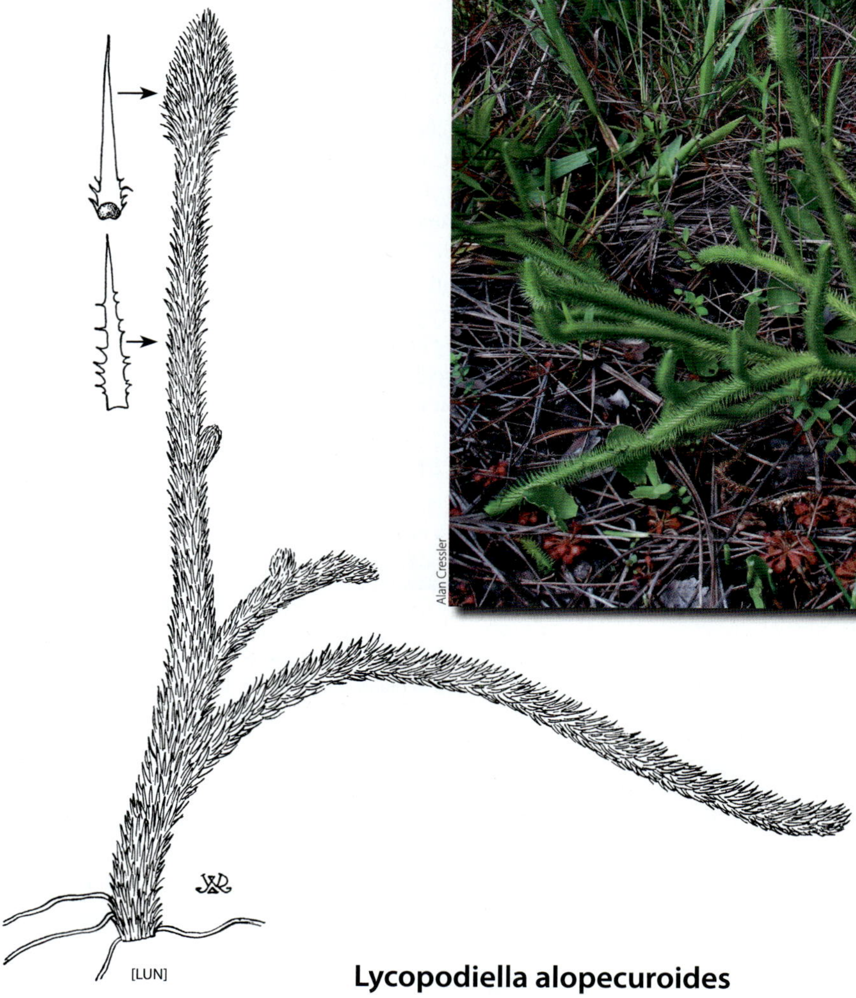

Lycopodiella alopecuroides

FOXTAIL BOG CLUB-MOSS

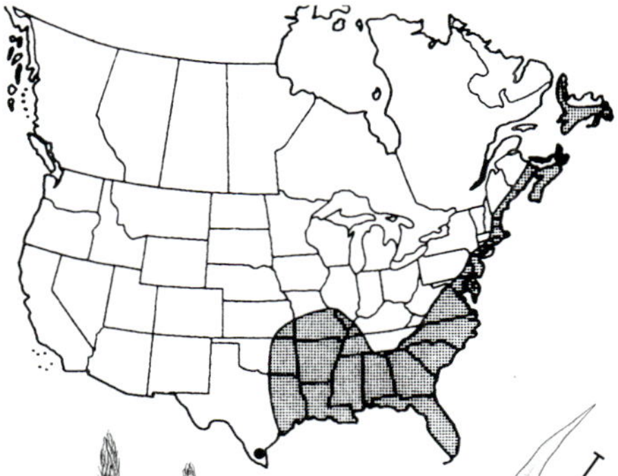

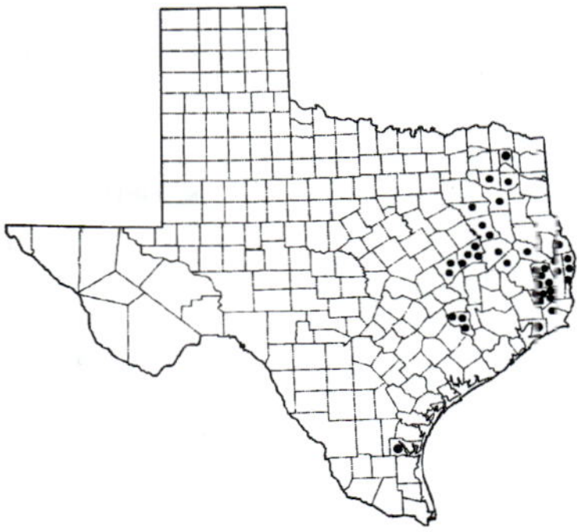

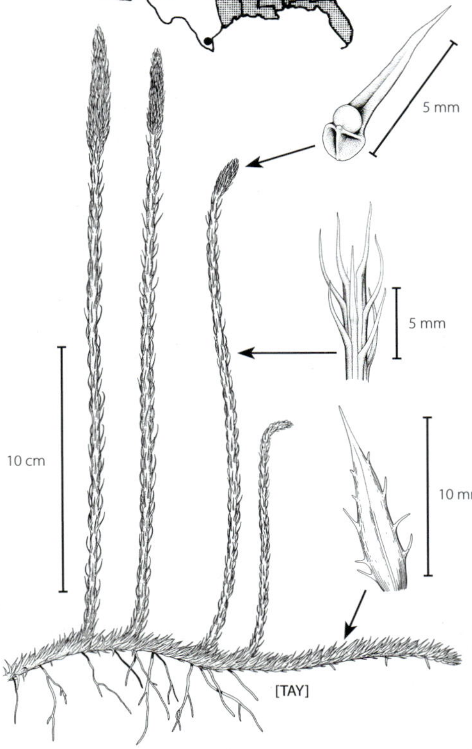

Lycopodiella appressa

APPRESSED BOG CLUB-MOSS

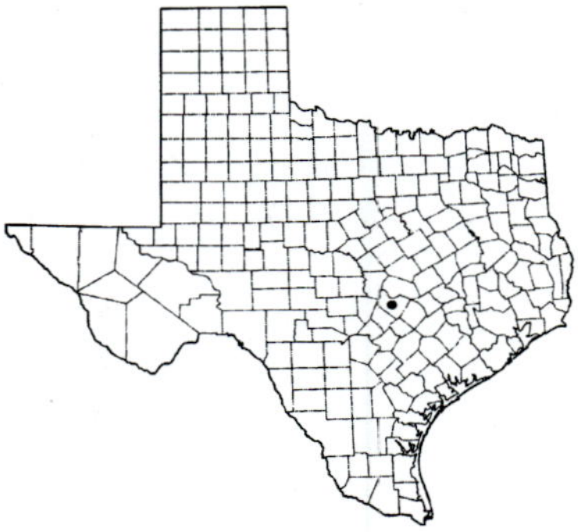

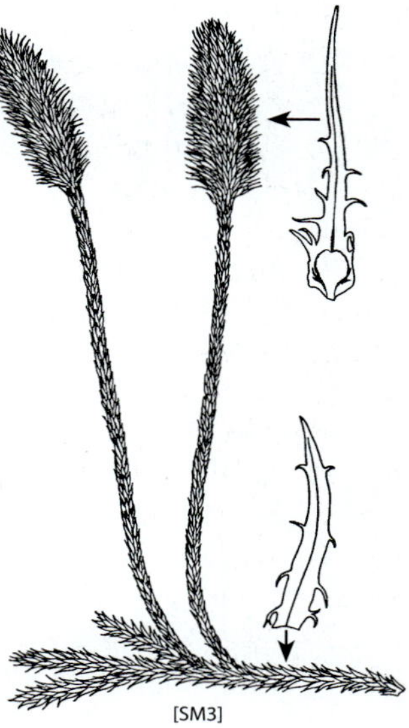

[SM3]

Alan Cressler

Alan Cressler

Lycopodiella prostrata

FEATHER-STEM BOG CLUB-MOSS

cm long, 11–20 mm wide; sporophylls wide-spreading; $2n$ = 156 (Wagner & Beitel 1993). Wet places in savannahs, boggy areas in low open pinelands and seeps, in acidic soils; Hardin (*Singhurst 2240*, BAYLU), Jasper (*Singhurst 3318*, *Singhurst, Bridges, & M. Pagoulat 14620*, BAYLU, *McRoberts & McRoberts 2903*, TAES), Austin, Lamar, Newton, and Orange (Turner et al. 2003) cos., mainly Pineywoods; also Jefferson Co. in n Gulf Prairies and Marshes (Turner et al. 2003); in addition, the range map in Flora of North America (Wagner & Beitel 1993) has a locality mapped from c TX but we have been unable to confirm this location; e U.S. from NY s to FL w to AR and TX; also s Mexico and Cuba. Sporulating Jul–Nov. [*Lycopodium alopecuroides* L.] *Lycopodiella alopecuroides* has an interesting biogeography. It is one of a few species and numerous deciduous forest genera that occur broadly across the eastern United States as far west as TX and then reappear disjunctly in the mountains of Mexico and in some cases even in Guatemala; although its Mexican and Cuban occurrences may be the result of long distance dispersal, see page 37 for a different possible explanation of its occurrence in the highlands of s Mexico. This species can hybridize with *L. appressa*; two sheets at BRIT (duplicates at TEX-LL) labeled as *L.* ×*copelandii* (Eiger) Cranfill [*L. alopecuroides* × *L. appressa*] (*Orzell & Bridges 5679*, Jasper Co. and *Bridges & Orzell 8584*, Newton Co.), do seem somewhat intermediate between the two purported parents; however, they could be just unusual specimens of *A. alopecuroides*. TX populations represent the extreme sw occurrence of this species in the U.S. It is rare and of very limited distribution in the state. We thus consider it to be of conservation concern in TX. (TOES 1993: IV) ⚠

Lycopodiella appressa (Chapm.) Cranfill, (Latin: *appressus*, lying flat or pressed closely against something, appressed, in reference to the appressed leaves), APPRESSED BOG CLUB-MOSS, SOUTHERN BOG CLUB-MOSS, CHAPMAN'S CLUB-MOSS, SOUTHERN CLUB-MOSS, TIGHT-LEAF CLUB-MOSS, SLENDER CLUB-MOSS, APPRESSED CLUB-MOSS, SOUTHERN APPRESSED CLUB-MOSS. Horizontal stems long-creeping, flat on ground, rooting throughout from lower surface, 1.5–2 mm in diam. (excluding leaves); leaves monomorphic, linear-lanceolate, ca. 5–7 mm long, ca. 0.8–1.2 mm wide, appressed, sparsely toothed or entire; upright stems usually unbranched, 8–30(–40) cm tall, densely covered with leaves; strobili slender, ca. 2.5–7.5(–11) cm long, 3–6(–9) mm wide; sporophylls incurved, appressed; $2n$ = 156 (Wagner & Beitel 1993). Depressions, bogs, moist areas; Pineywoods w to Austin (*H. Scarborough 50-24*), Henderson (*Correll 14196*), Robertson (*J.T. Painter & F.A. Barkley 13483*) (all BRIT, TEX-LL), Anderson, Leon (TAMU, TEX-LL), and Wood (TEX-LL) cos. in the Post Oak Savannah; also n Gulf Marshes and Prairies (Chambers Co., *Brown 20559*, SBSC; Jefferson Co.—Turner et al. 2003) and one collection along the s coast (Kleberg Co., *L.E. Brown 5969*, BRIT); se Canada and e U.S. from ME s to FL w to KS and TX; also Cuba. Sporulating Jun–Oct. [*Lycopodium appressum* (Chapm.) F.E. Lloyd & Underw., *Lycopodium inundatum* L. var. *appressum* Chapm.] There is a history of confusion in the correct epithet to use for this species—it has sometimes been incorrectly spelled as "adpressum." Chapman originally named the taxon in 1878 in Botanical Gazette (Vol. 3, pg. 20), at the level of variety as *Lycopodium inundatum* L. var. *appressum*. However, in his *Flora of the Southern United States* (Chapman 1883), he spelled the name as *L. inundatum* L. var. *adpressum*. This misspelling was subsequently picked up by a variety of botanists (e.g., Palmer 1919a; Correll 1956; Correll & Correll 1972). However, according to the International Code of Botanical Nomenclature (McNeill et al. 2006), the original spelling of a name or epithet is to be retained, except for the correction of problems such as typographical or orthographical errors. The spelling as given in the basionym (*Lycopodium inundatum* var. *appressum*) is thus the one to be used (but corrected for gender as *appressa* when treated in the genus *Lycopodiella*).

Lycopodiella prostrata (R.M. Harper) Cranfill, (Latin: *prostratus*, down flat, laid low, prostrate, in reference to the prostrate stems), FEATHER-STEM BOG CLUB-MOSS, CREEPING CLUB-MOSS, PROSTRATE BOG CLUB-MOSS, SOUTHERN CLUB-MOSS, FEATHER-STEM CLUB-MOSS. Horizontal stems long-creeping, flat on ground, essentially rooting throughout, 1–1.5 mm in diam. (excluding leaves); leaves linear-lanceolate, 3–8 mm long, 0.4–1.8 mm wide, with conspicuous marginal teeth, spreading, nearly perpendicular to stem, slightly dimorphic, those of the upper side slightly smaller (3–5 mm long); upright stems usually unbranched, 15–35 cm tall, densely covered with leaves; strobili 4–8 cm long, 15–20 mm wide; sporophylls wide-spreading; $2n$ = 156 (Wagner & Beitel 1993). Travis Co. (Correll 1956—"*B.C. Tharp s.n.*"; Turner et al. 2003) near the w edge of the Blackland Prairie, also the range map in Wagner and Beitel (1993) indicates occurrence in the extreme se part of the Pineywoods and the n part of the Gulf Prairies and Marshes as well as c TX (while we have followed Wagner and Beitel in mapping this species along the TX Gulf coast on our U.S. map, we have been unable to find supporting specimens); se U.S. from NC s to FL w to AR and TX. Sporulating mainly summer–fall. [*Lycopodium alopecuroides* L. var. *pin-*

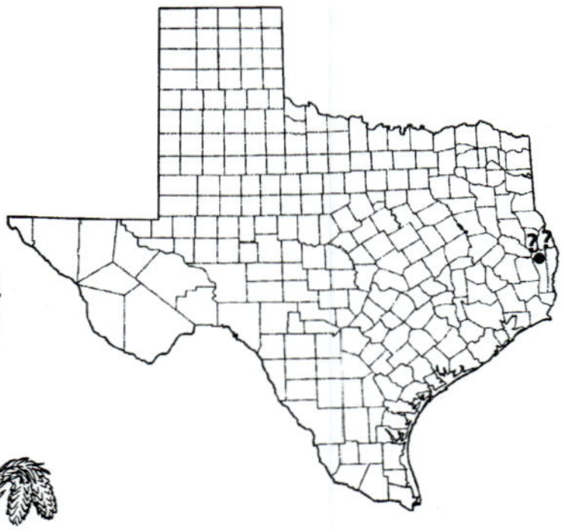

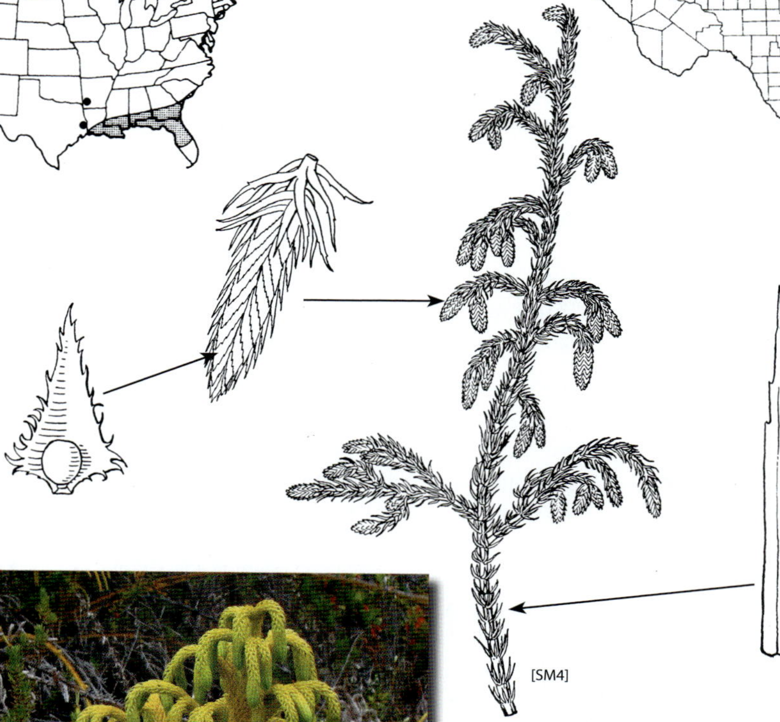

[SM4]

Alan Cressler

Alan Cressler

Palhinhaea cernua

NODDING CLUB-MOSS

natum (Chapm.) J. Lloyd & Underw. ex C.A. Br. & Correll, *Lycopodium inundatum* L. var. *pinnatum* Chapm., *Lycopodium prostratum* R.M. Harper] Correll (1956), who treated this taxon as *L. alopecuroides* var. *pinnatum*, indicated that it was known in TX only from Travis Co., but did not cite the herbarium of the *B.C. Tharp s.n.* collection (dates and herbaria of specimens were not included in Correll's book as part of the listings of specimens). We have not been able to relocate that collection. Carr (2002d) noted, "It is possible that a Tharp specimen at TEX-LL [of *L. appressa*], collected 15 Oct 1939 from Austin County, was mistakenly attributed to the city of Austin and thus to Travis County. *Lycopodiella prostrata* occurs mainly in mesic to wet pine forests of the southeastern United States, and its presence in Travis County seems unlikely." However, Correll (1956) noted that "The apparent isolation of this variety on the western edge of the Blackland Prairies, far from its nearest known stations to the east in Beauregard and Natchitoches parishes, Louisiana, is of uncommon interest." It thus seems unlikely that Correll would not have carefully studied the specimen. (A similar situation involving *Cheilanthes lanosa* was recently resolved with Correll's sole citation for TX being confirmed [Holmes et al. 2011]). Therefore, until further information is available, we are accepting Correll's citation and including *L. prostrata* in the TX flora. Also, although the Travis Co. site is significantly disjunct to the w from most of the range of the species, such a disjunction is not uncommon in homosporous pteridophytes (see pages 2, 30). This species hybridizes with *L. alopecuroides* (Snyder & Bruce 1986) and has sometimes been lumped with it (Radford et al. 1968); however, most recent taxonomic treatments (e.g., Wagner & Beitel 1993; Nauman et al. 2000) recognize both species. It also hybridizes with *L. appressa* (to form *L.* ×*brucei* Cranfill); two specimens at TEX-LL appear to be examples of such a hybrid (Jasper Co., *Correll & Correll 12518*; Jefferson Co., *Mrs. J.L. Hooks s.n.*, "1/51936"). Because of its rareness and limited distribution in the state, we consider this species to be of conservation concern in TX.

PALHINHAEA Vasc. & Franco NODDING CLUB-MOSS

Palhinhaea has traditionally been placed in a more broadly defined *Lycopodium* or by some authorities in *Lycopodiella* (e.g., Mabberley 1997; Wikström & Kenrick 2000). As defined here, following the treatment in *Flora of North America* (Wagner & Beitel 1993), *Palhinhaea* is a subcosmopolitan genus of 10–15 species, widespread mainly in the tropics and subtropics. It has a base chromosome number (*x*) of 55 (Weakley 2010). (Named for Ruy Telles Palhinha, 1871–1950, Azores-born, Portuguese botanist)
REFERENCE: MacRoberts & MacRoberts 1995a; Bray 1996.

Palhinhaea cernua (L.) Vasc. & Franco, (Latin: *cernuus*, nodding, drooping, in reference to the drooping branch tips), NODDING CLUB-MOSS, STAG-HORN CLUB-MOSS. Plant overwintering as buried stem tips, the rest dying; horizontal stems branching, rooting where they touch the ground, with remote leaves; upright stems to 45(–70) cm tall, many-branched (and resembling a miniature tree), with lateral branches drooping at tips; leaves linear-needle-like; strobili nodding, terminating branches, 4–8 mm long; sporophylls triangular-ovate, 1–2 mm long, coarsely toothed, wider than the sterile leaves; sporangia nearly globose, solitary at base of upper side of sporophylls. Hillside pitcher plant bog on wet but not inundated sand in full sun; Jasper Co. (Pineywoods) in the Angelina National Forest—known from two recently discovered plants and thus photographed but not collected (*MacRoberts & MacRoberts 2860,* photos at TEX-LL; MacRoberts & MacRoberts 1995a); Turner et al. (2003) also mapped Angelina and San Augustine cos. (but we have been unable to confirm these records and they are indicated by question marks on the county distribution map); se U.S. from SC s to FL w to AR and TX (Wagner & Beitel 1993); also tropics of Old and New worlds. Sporulating summer–fall. [*Lycopodiella cernua* (L.) Pic. Serm., *Lycopodium cernuum* L.] This species was previously known from the se U.S. w to AR and LA; the TX location is ca. 125 km sw of the nearest known location in LA, and as such is the westernmost station for the species in the U.S. (MacRoberts & MacRoberts 1995a). Weakley (2010) speculated that its occurrence in the se U.S. "may be adventive," but it is considered native by most authorities. This species, widespread in both the Old and New World tropics, is probably the world's most abundant CLUB-MOSS (Wagner & Beitel 1993). However, it is certainly one of the rarest native species in the TX flora. It is a showy species easily distinguished by the many-branched upright stems. The rare (and often transitory) occurrence of certain pteridophytes beyond the margins of their main range (e.g., *Palhinhaea cernua* in TX) has sometimes been attributed to long-distance dispersal of spores by major storms or hurricanes (J. Peck, pers. comm.). Because of its rareness and limited distribution in the state, we consider this species to be of conservation concern in TX.

PSEUDOLYCOPODIELLA Holub BOG CLUB-MOSS

Pseudolycopodiella has traditionally been placed in a more broadly defined *Lycopodium* or by some authorities in *Lycopodiella* (e.g., Øllgaard 1987; Mabberley 1997; Wikström & Kenrick 2000). As treated here, following the treatment in *Flora of North America* (Wagner & Beitel 1993), *Pseudolycopodiella* is a subcosmopolitan genus of 12 species, with only one species in North America. It has a base chromosome number (*x*) of 35 (Wagner & Beitel 1993). (Name derived from Greek: *pseudo*, false, and the genus *Lycopodium*, from Greek: *lykos*, wolf, and *pous* or *podium*, foot; in reference to the resemblance of the bristly, almost furry-appearing branch tips to a wolf's paw, plus the Latin diminutive suffix, *-ella*)
REFERENCES: Bruce 1976; Holub 1983.

Pseudolycopodiella caroliniana (L.) Holub, (of Carolina, because its type specimen came from the "Carolinas," a region which historically included much of the se U.S.—Nelson 2000), SLENDER BOG CLUB-MOSS, CAROLINA CLUB-MOSS, SLENDER CLUB-MOSS. Plant perennial; horizontal stems evergreen at least at apex, short-creeping, flat on ground, rooted throughout from lower surface, densely covered with leaves, 8–12 mm wide including leaves; horizontal stem leaves dimorphic, the lateral leaves spreading and appearing 2-ranked, lanceolate to lanceolate-ovate, 3.5–7 mm long, 1.2–2.1 mm wide, entire; median leaves smaller, ascending; upright stems (bearing strobili) unbranched, scattered along stems, 5–30 cm long, with only scattered, scale-like, subulate (= awl-shaped), typically appressed, entire leaves ca. 3–5 mm long and 0.5–0.7 mm wide; strobili solitary, terminating upright stems, slender, 9–80(–120) mm long, 2.5–5(–8) mm wide including sporophylls; sporophylls diverging, broadly ovate to deltate basally narrowing to acuminate apically, marginally erose, ca. 3–5 mm long, differently shaped (much wider basally) than the leaves of the supporting upright stems; sporangia reniform (= kidney-shaped), solitary at base of upper side of sporophylls; $2n = 70$, 140 (Wagner & Beitel 1993). Depressions in savannahs and open flat pinelands, in acidic soils, often with sphagnum moss; Angelina (*F.W. Gould & C. Leinweber 6536*, BRIT), Jasper (*W.F. Mahler 8350*, BRIT), Tyler (BRIT), Hardin, Henderson (BAYLU), and San Augustine (TEX-LL) cos. in s Pineywoods and Lee (TEX-LL) and Leon (*Orzell & Bridges 10343*, TEX-LL) cos. in the Post Oak Savannah; e U.S. from PA s to FL w to AR and TX and disjunct to MA and KY; also West Indies, Mexico to South America, Asia, Africa. Sporulating Jul-Sep. [*Lycopodium carolinianum* L.] The slightly disjunct population in Lee Co. is presumably a relict of more humid conditions in the past that has been able to survive since Ice Age times due to unique conditions associated with porous sandstone layers. Here water percolates through acidic sand derived from Eocene strata (roughly 50 million years old), reaches an impenetrable layer beneath, and moves laterally, eventually seeping out to the surface as permanent springs, acidic seeps, and wetlands. Numerous other mesic species are disjunct to similar such wet microhabitats in otherwise rather dry inhospitable environments in c TX (see page 36). This species is extremely widespread, being known in both the New and Old Worlds. Its TX populations represent the extreme sw occurrence of the species in the continental U.S.

SELAGINELLACEAE Willk.

SPIKE-MOSS FAMILY

An ancient, cosmopolitan but primarily tropical and subtropical family currently treated as a single genus with > 700 (Valdespino 1993) to perhaps 750 (Weakley 2010) species. Based on molecular studies, some authors speculate the genus will eventually be subdivided into a number of segregates (e.g., Weakley 2010; see discussion under genus). The earliest fossil evidence for the family is from the Carboniferous 350–330 million years ago (Rowe 1988; Korall & Kenrick 2002). However, Wikström (2001) noted that fossils in the isoetalean group from the late Devonian suggest the minimum age for the split between the lineages leading to Isoëtaceae and Selaginellaceae is 377 mya. As noted earlier, along with other Lycopodiophyta, they are part of an ancient evolutionary lineage that has been separate from all other vascular plants for around 400 million years (Pryer et al. 2004a). According to Korall et al. (1999), "Greatest diversity occurs in lowland to midmontane primary tropical rain forest, but this cosmopolitan family is also widely distributed in subtropical, temperate, montane, and rarely subarctic regions." Selaginellaceae are usually terrestrial or epiphytic, superficially moss-like vascular plants bearing spores differentiated into microspores and megaspores

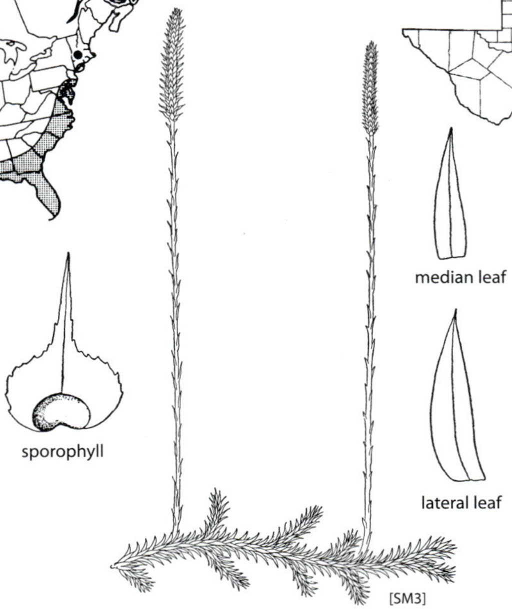

Alan Cressler

Alan Cressler

Pseudolycopodiella caroliniana

SLENDER BOG CLUB-MOSS

(plants heterosporous). The leaves (lycophylls) usually have a single vein and a ligule (= minute, tongue-like basal protuberance on a leaf; the function of ligules is uncertain; however, they only occur on heterosporous species—*Isoëtes* and *Selaginella*). Recent molecular evidence supports the monophyly of the Selaginellaceae (Korall et al. 1999) and the family is apparently only distantly related to the Lycopodiaceae and Isoëtaceae. (Order Selaginellales)

FAMILY RECOGNITION IN THE FIELD: superficially somewhat moss-like, small herbs with numerous, scale-like, 1-veined leaves; stems terminating in ± 4-angled, spore-producing cones.

REFERENCES: Alston 1955; Correll 1956, 1966; Jermy 1990b; Thomas & Quansah 1991; Webster 1992; Valdespino 1993; Korall et al. 1999; Yatskievych & Windham 2009b.

SELAGINELLA P. Beauv. SPIKE-MOSS

Texas species small (usually ca. 12 cm or less tall), terrestrial or lithophytic (= growing on rocks) (plants epiphytic elsewhere) perennial herbs; stems prostrate to erect, leafy; rhizophores (= modified leafless root-like stems) present, producing small roots near their tips; vegetative leaves small, somewhat scale-like, with ligule on upper (= adaxial) side near base, all alike or of 2 main kinds (technically, a third kind of leaf is axillary between branches), in 4 ranks or spirally arranged; sporophylls (= spore-bearing leaves) modified, in strobili (= cones) at branch tips; sporangia solitary in axils of sporophylls, of 2 kinds, the microsporangia with numerous microspores and megasporangia usually with 4 relatively large megaspores.

Selaginella, the only living genus in the family, has an extremely long history in the fossil record—fossils resembling *Selaginella* are known from the Carboniferous Period onwards (Collinson 1991; Thomas 1992); they are thus at least 300 million years old. In fact, the morphology of the group has changed little since that time (Thomas 1992; Korall & Kenrick 2002). The genus is currently most diverse in the tropics. However, Korall and Kenrick (2002) pointed out that it "contains frost-tolerant, arctic-alpine species, delicate terrestrial rainforest species, and physiologically robust, drought-adapted xerophytes of desert scrub and heathland." Most TX species have xerophytic (= drought tolerant) adaptations and some are well known as "resurrection" plants, capable of reviving after long periods of desiccation. Because of its heterogeneity, some authorities (e.g., Skoda 1997) recommend splitting *Selaginella* into several smaller genera. Small (1938), for example, separated *S. apoda* and similar species into the genus *Diplostachyum*, and Thomas (1992) argued that, "The presence of both isophyllous and heterophyllous *Selaginella*-like plants in the Carboniferous [ca. 300 mya] supports the idea that the genus should be divided into at least two genera." However, splitting the genus would require hundreds of name changes (J. Peck, pers. comm.), and we are following most recent authors (e.g., Valdespino 1993; Korall et al. 1999; Korall & Kenrick 2002; Mickel & Smith 2004) who continue treating all species in a single genus. Tropical species are known for their unusually colored leaves—reddish or bronze or iridescent blue-green (Hoshizaki & Moran 2001); the iridescence can be shockingly bright. Although high chromosome numbers are characteristic of pteridophytes in general (due to polyploidy and hybridization), they are rare among the heterosporous pteridophytes including *Selaginella* (Marcon et al. 2005). The key to species is modified in part from those in Valdespino (1993), Yarborough and Powell (2002), and Mickel and Smith (2004). (Name derived from *Selago*, an ancient name for *Lycopodium*, a genus resembling *Selaginella*, and the Latin diminutive suffix, *-ella*, signifying tiny or small)

REFERENCES: Morton 1939; Weatherby 1944; Clausen 1946; Tryon 1955; Somers & Buck 1975; Filippini-de Giorgi et al. 1997; Skoda 1997; Korall & Kenrick 2002; Marcon et al. 2005; Wilbur & Whitson 2005; Schulz 2010a, 2010b; Banks et al. 2011; Reveal & Gandhi 2012.

1. Plants of moist often shady habitats, delicately thin-herbaceous; stem leaves not overlapping or only slightly so, in 4 obvious ranks, 2 lateral and spreading, 2 smaller and appressed-ascending along the upper surface of the stem; lower surface of the stems easily visible; plants annual ____ **S. apoda**
1. Plants of xeric habitats, rather rigid; stem leaves crowded, conspicuously overlapping, appressed to stem, either not in 4 distinct ranks or if in 4 ranks then obviously overlapping; lower surface of the stems not visible (well concealed by leaves); plants perennial.
   2. Plants forming distinct rosettes; main branches (from base of plant) much branched, strikingly flattened in overall aspect when moist (almost as if ironed); leaves in 4 distinct ranks (two medial, two lateral).
      3. Medial leaves with apex acuminate to obtuse (lacking bristle), broadly triangular to ovate in over-

all shape; lateral and medial leaves essentially alike; main branches dramatically curling inward upon drying, and together transforming the rosette into a ball-like appearance ______ **S. lepidophylla**

3. Medial leaves with apex bearing a long bristle (⅓–½ as long as leaf blade), rather narrow (lanceolate and slightly falcate) in overall shape; lateral and medial leaves different in shape and size; main branches remaining almost flat or merely forming a somewhat erect cluster even when dry ______ **S. pilifera**

2. Plants not forming distinct rosettes; main branches often much branched, but not flattened in overall aspect; leaves not in 4 ranks.

4. Plant with a definitely upwardly growing appearance, the aerial stems mostly erect or ascending (= growing upward at an angle); lower part of aerial stems often each with 1 bud-like arrested branch; both underground (rhizomatous) and aerial stems present; rhizophores (= modified root-like leafless stems that produce roots) usually only on lower ½ of stems.

5. Leaf apex without terminal bristle (merely acute to obtuse); leaf margins with only sparse very fine teeth (minutely denticulate) ______ **S. viridissima**

5. Leaf apex bearing a terminal bristle; leaf margins ciliate, variously ciliate but at least with short cilia at base.

6. Leaves and sporophylls with apical bristle usually 1 mm or more long (0.65–1.85), the bristles visually very obvious when viewed with a 10× lens (and ⅓ to ½ as long as the leaf); aerial stem leaves 3–4.7 mm long, the leaf margins with long, spreading cilia ± from base to apex (use magnification) ______ **S. rupincola**

6. Leaves and sporophylls with apical bristle less than 1 mm long (0.25–0.85), the bristles visible, but not obvious when viewed with a 10X lens; aerial stem leaves 1.9–3 mm long, the leaf margins with long or short, spreading cilia at base but the cilia grading to very short, minutely tooth-like and ascending towards apex.

7. Species known in TX only from eastern ½ of state (East TX w to e Edwards Plateau); leaves and sporophylls green, only loosely appressed, the tips often angled away from the stems; leaf margins with the cilia very short even towards the leaf base, those cilia towards the leaf apex minutely tooth-like and ascending; aerial stems lacking rhizophores ______ **S. corallina**

7. Species known in TX only from Trans-Pecos and adjacent counties; leaves and sporophylls often glaucous, tightly appressed; leaf margins with long, spreading cilia at base but the cilia grading to very short, minutely tooth-like and ascending towards apex; aerial stems often with rhizophores on lower ½ of stems ______ **S. ×neomexicana**

4. Plant with the overall appearance of growing on or near the ground, the aerial stems prostrate, creeping, or decumbent (= reclining but with tips upturned); lower part of aerial stems without a bud-like arrested branch; underground (rhizomatous) stems not present; rhizophores usually present throughout length of stem.

8. Leaves with appearance of being strongly upswept or upwardly curved when stem viewed from side, making the views of the leaf-covered stems from above and below distinctly different.

9. Leaves gradually tapering to a yellowish somewhat hardened bristle 0.2–0.5 mm long (the "bristle" essentially just the narrowed apex of the blade tissue; this bristle sometimes breaking off but more persistent on underside leaves); leaves on underside of stem usually abruptly adnate to stem (visible as abrupt color change where attached) ______ **S. wrightii**

9. Leaves often slightly more abruptly tapered apically, with a translucent to whitish bristle of varying lengths (sometimes very short) or merely acute; leaves on underside of stem decurrent to stem (no color change apparent where attached).

10. Plants forming flat cushion-like mats (in TX); leaf apices with bristle 0.5–1.1 mm long; sporophylls usually lanceolate to ovate-lanceolate; species known in TX only from one location in extreme s Brewster Co. in the Trans-Pecos ______ **S. scopulorum**

10. Plants forming loose mats; leaf apices with bristle variable, from lacking to 0.8 mm long; sporophylls ovate-deltate to ovate; including species widespread in the Trans-Pecos and Edwards Plateau.

11. Leaf apices merely acute or with very short flattened bristle 0.3 mm or less long (bristles mostly only on leaves at branch tips or buds); sporophylls merely acute to acuminate, lacking apical bristle; leaves on underside of stems widest at middle (lanceolate in outline), definitely narrowed at base; species known in TX only from the Trans-Pecos ______ **S. arizonica**

11. Leaf apices with an elongate rounded bristle 0.3–0.8 mm long; sporophylls with distinct apical bristle; leaves on underside of stems widest at base (linear to linear-lanceolate in outline) or at least definitely not narrowed at base; species known in TX both in the Trans-Pecos and scattered further e ______ **S. peruviana**

8. Leaves usually without strongly upswept or upwardly curved appearance, the leaf-covered stems therefore appearing radially symmetrical.
   12. Leaf apices with bristles usually absent or < 0.1 mm long, rarely to 0.45 mm; bases of leaves on main stem usually rounded and adnate to stem (visible as abrupt color change) **S. mutica**
   12. Leaf apices with obvious bristles 0.25–1.85 mm long; bases of leaves on main stem EITHER cuneate or oblique and decurrent (no color change apparent) OR (in *S. rupincola* with long apical leaf bristles) rounded and adnate to stem (visible as abrupt color change).
      13. Leaf margins long-ciliate, the cilia spreading ± from leaf base to apex (cilia nearly as long as distance from leaf midrib to leaf margin); bases of leaves on main stem rounded and abruptly adnate to stem (visible as abrupt color change); sporophylls without prominent auricles at base **S. rupincola**
      13. Leaf margins entire to short-ciliate, the cilia sometimes spreading at leaf base but grading to very short, minutely tooth-like and ascending towards apex; bases of leaves on main stem cuneate or oblique and usually decurrent (no color change apparent); sporophylls with OR without prominent auricles at base.
         14. Mats usually rather loose; leaf bases cuneate; leaf apex with usually greenish or greenish-yellow bristle; sporophylls with prominent auricles at base; species scattered in the Trans-Pecos and adjacent Val Verde Co. **S. underwoodii**
         14. Mats flat and cushion-like; leaf bases oblique; leaf apex with whitish bristle; sporophylls without prominent auricles at base; species only known in TX only from one location in extreme s Brewster Co. in the Trans-Pecos **S. scopulorum**

Selaginella apoda (L.) C. Morren var. **apoda,** (Greek, a-, without, and *pous* or *podium*, foot, = footless, in reference to "the stalkless strobili, which were noteworthy in *Lycopodium*, the genus in which Linnaeus originally place this species"—Thieret 1980), MEADOW SPIKE-MOSS, BASKET SELAGINELLA. Plant prostrate-creeping or ascending annual (all other TX species are perennial), often forming mats, delicately thin-herbaceous (our only such species); leaves in 4 ranks, of 2 distinct kinds, without significant hyaline margins, the two types of leaves different in size and shape—lateral leaves ovate to ovate-elliptic, asymmetrical, ca. 1.35–2.25 mm long, 0.75–1.35 mm wide; appressed-ascending (medial) leaves lanceolate, smaller, to ca. 1.2(–1.6) mm long; strobili solitary or paired, obscurely quadrangular (= 4-sided)-flattened, 0.5–2 cm long; 2–4 mm in diam.; sporophylls apically acute to acuminate. Moist areas, ditches, stream banks, seeps, low fields, and low woods; widespread in Pineywoods (e.g., *E. Whitehouse 23088*, BRIT), Post Oak Savannah, n Gulf Prairies and Marshes and disjunct to e Edwards Plateau (e.g., Burnet Co., *Correll & Correll 12746*, BRIT, TEX-LL); e U.S. from ME s to FL w to IL, OK, and TX; also disjunct to the mts. of s Mexico and Guatemala. Sporulating May–Dec. [*Diplostachyum apodum* (L.) Beauv., *Lycopodium apodum* L., not *S. apoda* (L.) Spring (a superfluous name—see discussion below)] In flood-prone bottomlands in East TX this species is sometimes epiphytic on the often moist lower parts of tree trunks (J. Stanford, pers. comm.). A variety of *S. apoda*, var. *ludoviciana* (A. Braun) B.F. Hansen & Wunderlin (based on *S. ludoviciana* (A. Braun) A. Braun), occurs in the Gulf coastal plain (Hansen & Wunderlin 1998). That variety, which differs in minor ways (e.g., hyaline leaf margins) from var. *apoda*, is known from GA, FL, AL, MS, and se LA. *Selaginella apoda* has an interesting biogeography. It is one of numerous TX species that have widespread populations in East TX and isolated disjunct populations on the Edwards Plateau (see discussion on page 37). The range map in Valdespino (1993; and on U.S. map provided here) also shows a disjunct occurrence in s TX (which we have not been able to confirm). Even more interesting, it is one of a few species and numerous deciduous forest genera that occur broadly across the eastern United States as far west as TX, and then reappear disjunctly in the mountains of Mexico and in some cases even in Guatemala; see Fig. 40 and page 37 for a possible explanation of this distribution. *Selaginella apoda* is being considered as a model species for experimental work on lycophytes because of its short life cycle (85 days from spore to spore) and ease of cultivation (Schulz et al. 2010a, 2010b). Reveal and Gandhi (2012) recently published an article discussing technical nomenclatural problems associated with the name *Selaginella apoda* (L.) Spring—there has been disagreement about the correct name for this species for decades (e.g., Morton 1967). They argued, based on rules of the International Code of Botanical Nomenclature, that the correct name is *Selaginella apoda* (L.) C. Morren. Because of their arguments and because the name *S. apoda* (L). C. Morren is now being accepted by most botanical authorities (e.g, IPNI [International Plant Names Index] 2012; TROPICOS 2012), we are following their recommendation.

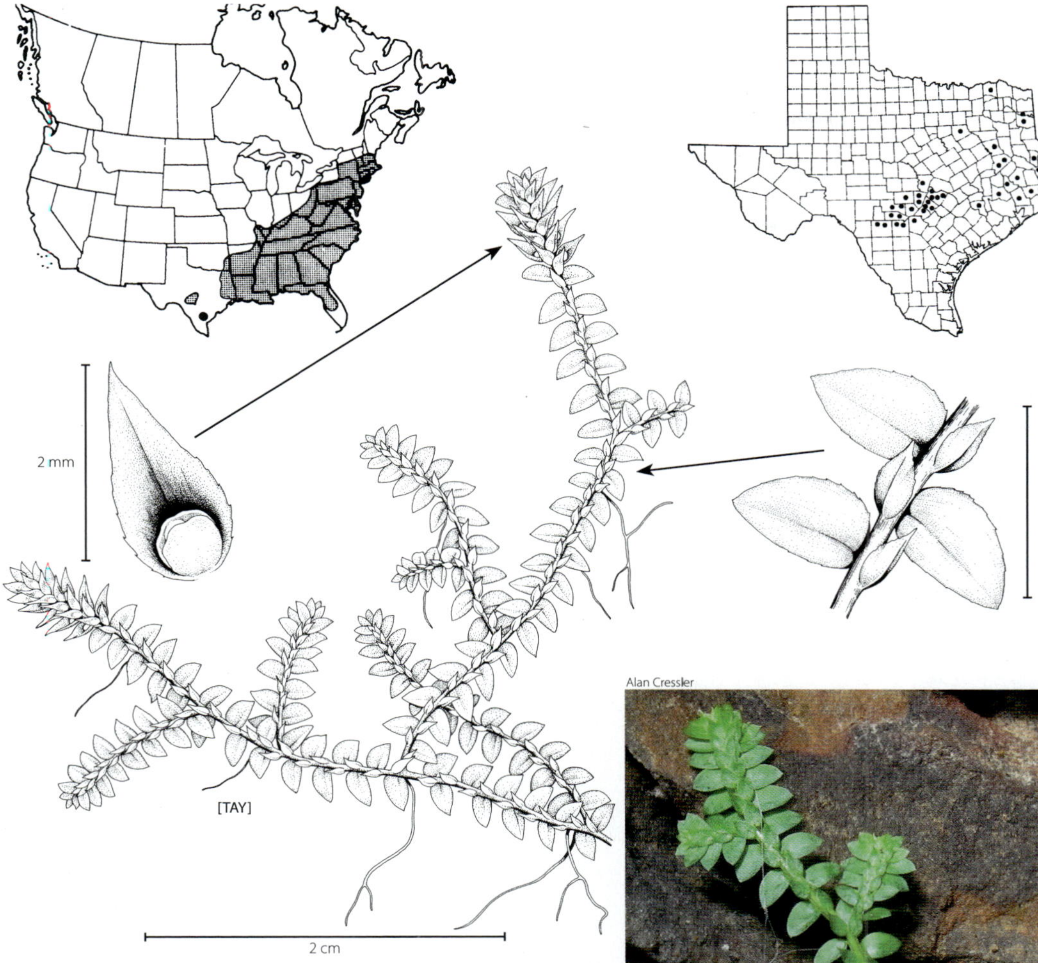

Alan Cressler

Alan Cressler

Selaginella apoda

MEADOW SPIKE-MOSS

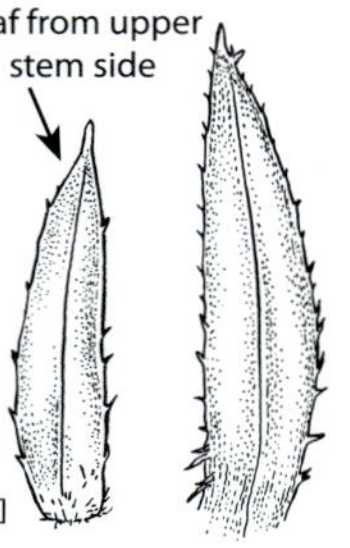

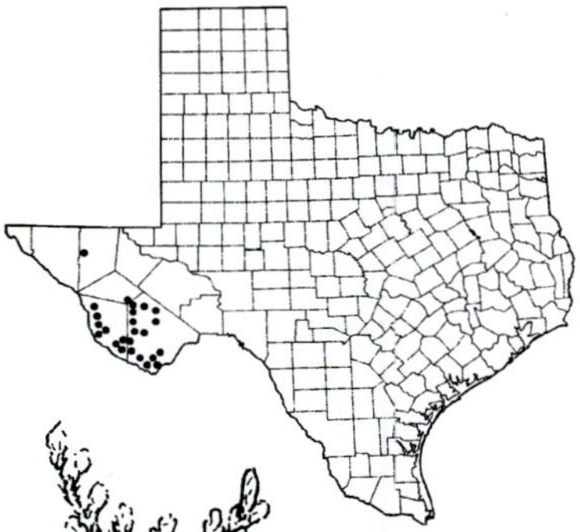

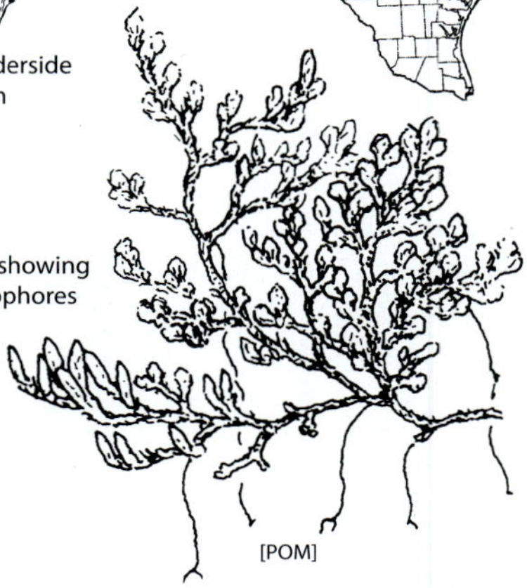

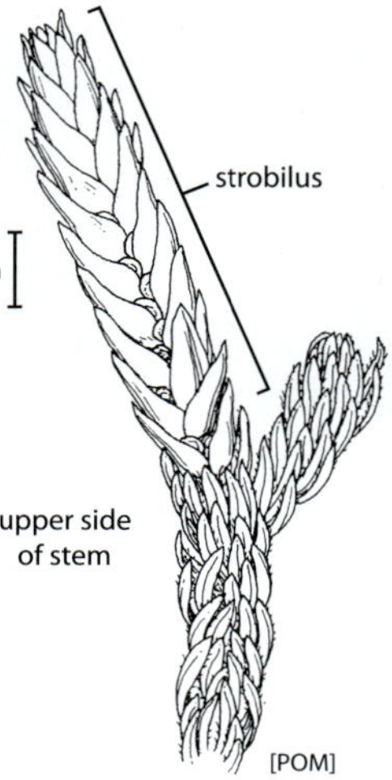

Selaginella arizonica

ARIZONA SPIKE-MOSS

Selaginella arizonica Maxon, (of Arizona, where the type specimen was collected), ARIZONA SPIKE-MOSS. Plant forming loose mats; stems prostrate, with upper and lower surfaces different in appearance (leaves upswept or curving upward making the views of the leaf-covered stems from above and below distinctly different); rhizophores scattered along entire length of stem; leaves tightly appressed to ascending, apically merely acute or with very short flattened bristle 0.3 mm or less long (bristles mostly only on leaves at branch tips or buds), all of one kind though leaves on underside of stems widest at middle (lanceolate in outline), glabrous and decurrent at base, marginally ciliate, those on upper side slightly shorter and narrower, abruptly adnate at base, marginally ciliate; strobili solitary, quadrangular, 5–10 mm long, sporophylls ovate-deltate, apically merely acute to acuminate, lacking bristle. On rock or terrestrial, slopes, gravel, crevices, and ledges, on various substrates (e.g., sandstone, igneous, rarely limestone); mainly Brewster (e.g., *Correll 13594, 29752*, TEX-LL) and Presidio (e.g., *Correll 13741*, TEX-LL) cos. in sc Trans-Pecos, also Sierra Diablo Mts. in Culberson Co. (Yarborough & Powell 2002) and s Jeff Davis Co. (Turner et al. 2003); AZ, TX; also n Mexico. Sporulating Jul–Oct. [*Bryodesma arizonica* (Maxon) Soják] This species is similar to *S. peruviana*, but is "most easily distinguished by the thin, lanceolate leaves on the underside of the stem and the lack of bristles on the sporophylls" (Yarborough & Powell 2002). In comparison, *S. peruviana* has fleshier, narrower (linear to linear-lanceolate) leaves and sporophylls with bristles. ARIZONA SPIKE-MOSS can sometimes be found growing "intertwined with *S. rupincola*" (Yarborough & Powell 2002). Yatskievych and Windham (2009b) noted that, "Where *S. arizonica* co-occurs with the morphologically distinctive *S. rupincola*, occasional putative hybrids also are encountered with intermediate morphology and abortive spores." Given its rareness and limited distribution in the state, we consider this species to be of conservation concern in TX. ⚠

Selaginella corallina (Riddell) Wilbur & Whitson, (Latin: *corallum* or *corallinum* from Greek: *korallion*, red coral, the rationale for the derivation is unclear—the plant has no obvious coral-red color; it is possibly due to the erect habit which resembles in some respects a branched coral), RIDDELL'S SPIKE-MOSS, RIDDELL'S SELAGINELLA, SAND SPIKE-MOSS. Plants forming clumps, the aerial stems erect to ascending, to ca. 12 cm tall, usually smaller, radially symmetrical, without rhizophores; underground (rhizomatous) stems present, with rhizophores; rhizomes and aerial stems often with a bud-like arrested branch near base; leaves of aerial stems essentially of 1 kind, narrowly triangular-lanceolate to linear-lanceolate, ca. 1.2–3 mm long, 0.4–0.5 mm wide, marginally short ciliate towards base but merely minutely denticulate toward apex, apically with whitish bristle; strobili solitary, sometimes with apical vegetative growth, quadrangular, ascending, (0.5–)1–3(–3.5) cm long and ca. 1.2 mm in diam., essentially the same diam. as stems; sporophylls often with a bristle. On rocks (often granite) or terrestrial on sandy or gravelly soils, longleaf pine sand ridges; widespread in e ½ of TX n to Wood (*K. Norton et al. 1031*, BRIT), Palo Pinto (*R.J. O'Kennon 22178*, BRIT), and Cooke (Turner et al. 2003) cos. and w to e Edwards Plateau; AL, AR, GA, LA, OK, and TX. Sporulating throughout the year. [*Lycopodium corallinum* Riddell, *S. riddellii* Van Eselt., *S. arenicola* var. *riddellii* (Van Eselt.) R.M. Tryon, *S. arenicola* subsp. *riddellii* (Van Eselt.) R.M. Tryon] In the past, TX plants of this species were typically treated as *S. riddellii* (Correll 1956) or *S. arenicola* subsp. *riddellii* (e.g., Diggs et al. 2006). However, there has long been disagreement over the classification and correct name of this and related species (the "arenicola complex"). Tryon (1955) recognized three subspecies, although other authors (e.g., Clausen 1946; Snyder & Bruce 1986) treated the three taxa as separate species. Valdespino (1993) considered the complex to have two species, but maintained subsp. *riddellii* as a subspecies of *S. arenicola*. Nauman et al. (2000), did not recognize infraspecific taxa within a broadly defined *S. arenicola*. Recent research by Wilbur & Whitson (2005) has shown that there is an older name (*S. corallina* based on *Lycopodium corallinum*) that has nomenclatural priority and thus must be used according to the International Code of Botanical Nomenclature. The type of *S. corallina* was collected in TX by Riddell in 1839 (near the San Saba River), and a GH specimen has been identified by Wilbur and Whitson (2005). They also noted that "the rank of species, the first ranking provided for each of the three taxa of the *arenicola* complex, serves present needs best and is nomenclaturally simplest." We are therefore following the classification and nomenclature of Wilbur and Whitson (2005) and treating as *Selaginella corallina* the entity that has been going as *S. arenicola* subsp. *riddellii*.

Selaginella lepidophylla (Hook. & Grev.) Spring, (Greek, *lepis* or *lepidos*, scale, and *phyllon*, leaf), RESURRECTION PLANT, SIEMPRE VIVA, FLOWER OF STONE. Plant forming striking open rosettes (to ca. 25 cm in diam.); rhizophores only on basal part of rosette; main branches much branched, flattened in overall aspect when

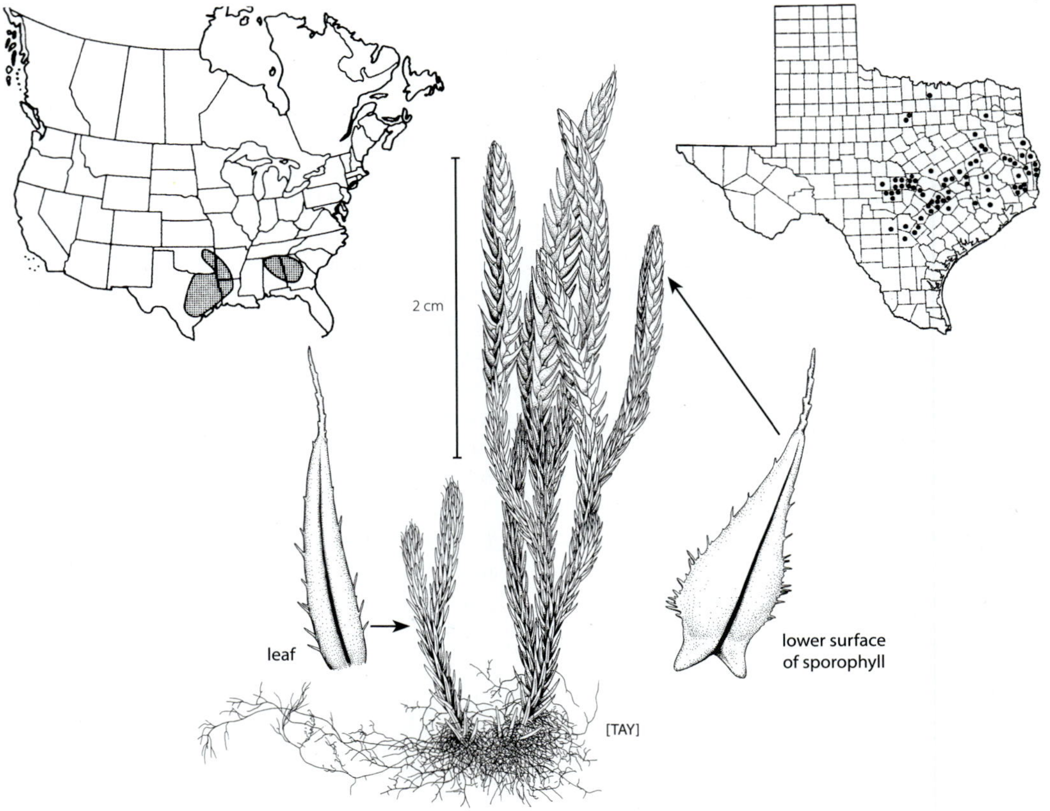

Alan Cressler

Selaginella corallina

RIDDELL'S SPIKE-MOSS

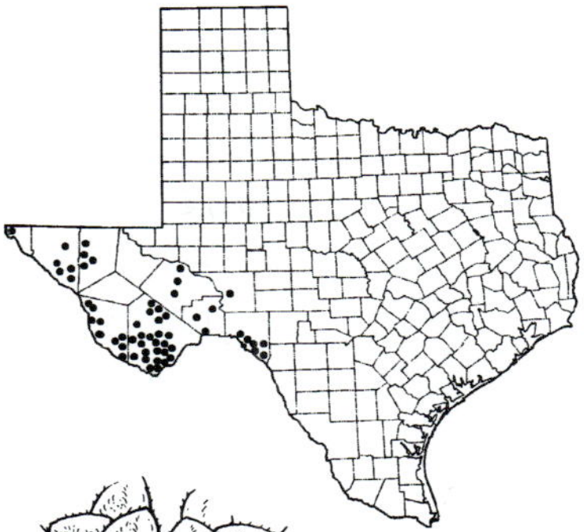

branch upper surface

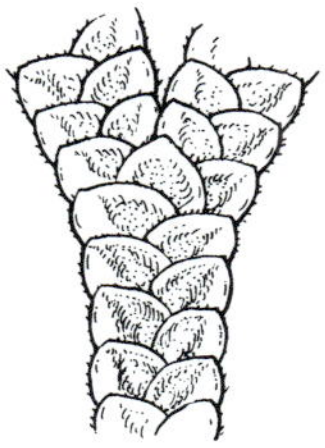

branch lower surface

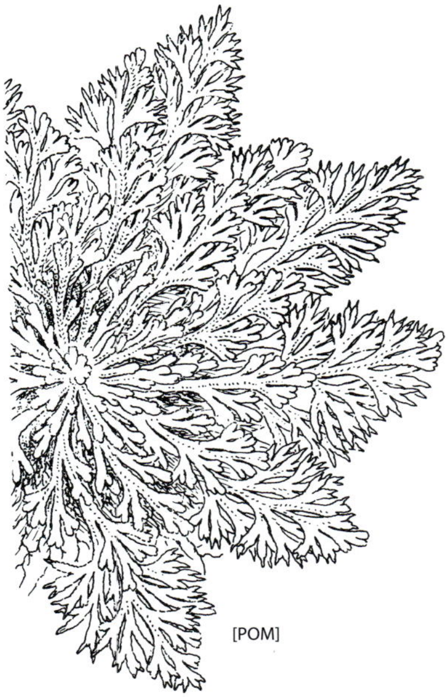

[POM]

Alan Cressler

Peter Guth

Selaginella lepidophylla

RESURRECTION PLANT

moist, curling inward upon drying, giving the rosette a brownish ball-like (fist-sized) appearance; leaves in 4 distinct ranks, 2 lateral, 2 medial, conspicuously overlapping, all without apical bristle and with conspicuous whitish transparent margins, the two types of leaves similar—the lateral leaves deltate to deltate-ovate, apically rounded, the medial broadly ovate, apically abruptly acuminate to obtuse; strobili solitary, quadrangular, 3–12 mm long; sporophylls apically acuminate to acute. On rock or soil, dry slopes, talus, most commonly on limestone, but sometimes on other substrates, typically in situations where excessive direct sun is minimized; widespread in the Trans-Pecos and in adjacent Val Verde (*J.L. Blassingame 2819*, BRIT, 25 mi w of Loma Alta [e of the Pecos River]; *Correll 14919*, TEX-LL, near mouth of Devil's River [e of the Pecos River]) and Crockett (Turner et al. 2003) cos. in the far w Edwards Plateau; in the U.S. known only from NM and TX; also widespread in Mexico. Sporulating Apr–Jan. [*Lycopodium lepidophyllum* Hook. & Grev.] When moist, the rosettes of this species are flat and quite striking (bright green). Yarborough and Powell (2002) noted that because "this species occurs in abundance only in north-facing habitats, it is a good indicator of direction to those traveling in desert areas." It is called RESURRECTION PLANT or incorrectly (because it is not a fern) RESURRECTION FERN since when dry the flat main branches curl inward and the entire plant becomes ball-like and brownish. However, when rehydrated, uncurling occurs and the plant resumes normal growth "even after years of being dry" (Velaspino 1993) (G. Yatskievych [pers. comm.], however, noted that the uncurling is mechanical and there is no evidence supporting the contention that plants can survive for years in a curled state). Because of this ability, it is grown as a curiosity and house plant and is cultivated in greenhouses. Lebkuecher and Eickmeier (1991, 1993) noted that curling reduced light exposure of the still-living internal tissues by more than 99.7% and protected against damage caused by the combination of bright light, high temperature, and desiccation stress (see page 28). Jim Blassingame (pers. comm.), who has a great deal of experience with this species in the field, pointed out that it can develop a "fairy ring" effect with the central area dying out and peripheral portions continuing to grow. Although there are numerous localities in the Trans-Pecos, Velaspino (1993) considered the species to be of conservation concern. Because it is widely known as "RESURRECTION PLANT," its populations are potentially threatened by over-collecting. ⚠

Selaginella mutica D.C. Eaton ex Underw., (Latin: *muticus*, shortened, curtailed, = blunted, without a point), BLUNT SPIKE-MOSS. Plants mat-forming, the mats usually rather loose; stems prostrate-creeping with leaves without strongly upswept or upwardly curving appearance, the leaf-covered stems therefore appearing radially symmetrical; rhizophores scattered along entire length of stem; leaves tightly appressed, all of one kind, linear-lanceolate to lanceolate or lanceolate-elliptic, marginally ciliate to minutely denticulate, basally usually rounded and adnate, apically with bristle usually absent or < 0.1 mm long, rarely to 0.45 mm; strobili solitary, quadrangular, (6–)10–30 mm long; sporophylls ovate-lanceolate to deltate-ovate, apically without bristle or bristle very short. On rock or terrestrial; rocky slopes, cliffs, ledges, canyons, often on igneous rocks, but sometimes on other substrates (e.g., limestone, sandstone); known in TX only from scattered localities in the w Trans-Pecos; AZ, CO, NM, UT, TX, WY; also state of Chihuahua in n Mexico. Sporulating Apr–Oct. [*Bryodesma muticum* (D.C. Eaton ex Underw.) Soják] *Selaginella mutica* is often found growing with *S. underwoodii* and *S.* ×*neomexicana* (and has thus been hypothesized to be a parent of *S.* ×*neomexicana*). It is sometimes confused with *S. viridissima* because they both typically have leaves without bristles at the apex; however, the creeping habit distinguishes *S. mutica* from the erect *S. viridissima*. Velaspino (1993) noted that there are "two rather distinct, morphological extremes" within *S. mutica* which are often recognized as varieties (as done here); the two varieties are not distinguished on the county distribution map. Some specimens have leaves with long spreading cilia and a short often broken apical bristle (treated by Correll as var. *texana*); however, Velaspino (1993) noted that they belong in var. *mutica*. Mickel and Smith (2004) described *S. mutica* as "a very distinct taxon" with "tightly appressed, ciliate, mucronate leaves. The tightly appressed leaves give the stems the appearance of a woven rope."

1. Leaves with apical bristle short but definitely present, the bristles 0.2–0.45 mm long; leaf and sporophyll margins mostly with very short, ascending cilia (cilia rarely longer) or minute teeth _____ var. **limitanea**
1. Leaves with apical bristle lacking or very short (< 0.06 mm long); leaf and sporophyll margins usually with long, spreading cilia (use magnification) _____ var. **mutica**

var. **limitanea** Weath., (Latin: *limitatus*, bounded, limited, or bordered, in reference to its occurrence along the U.S.-Mexican border, the type description saying that it is a "very appropriate epithet for a population

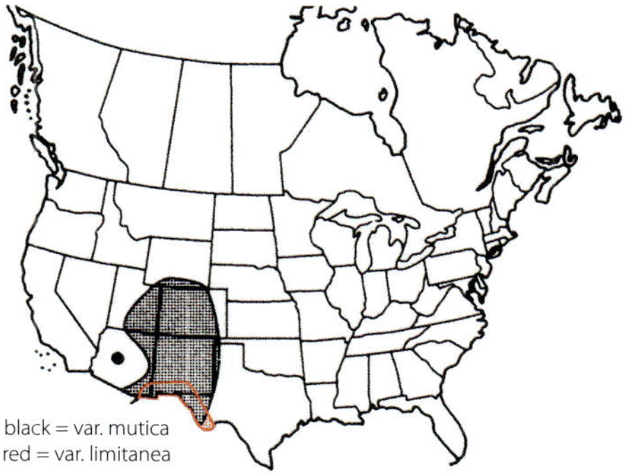

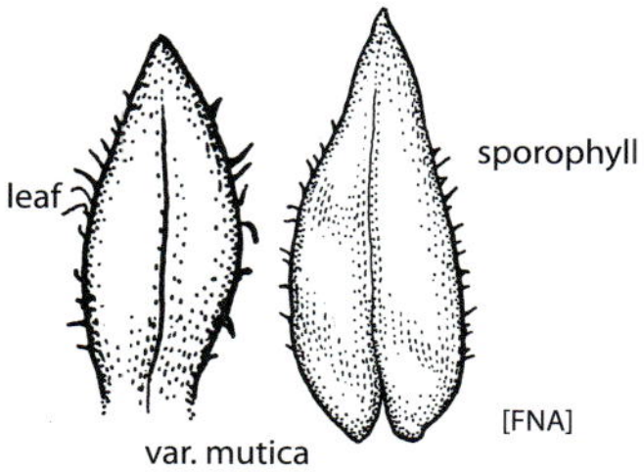

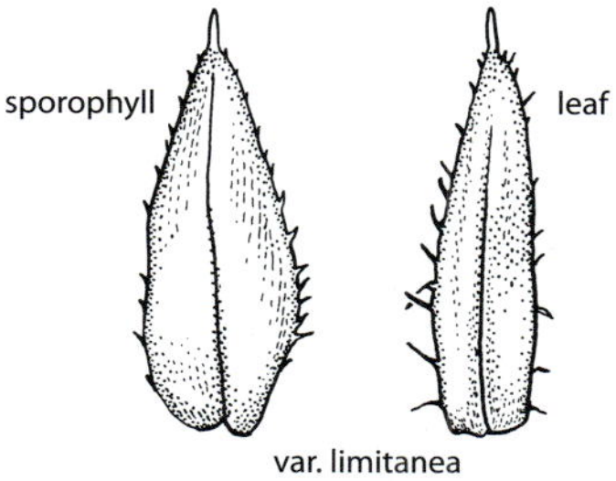

Selaginella mutica

BLUNT SPIKE-MOSS

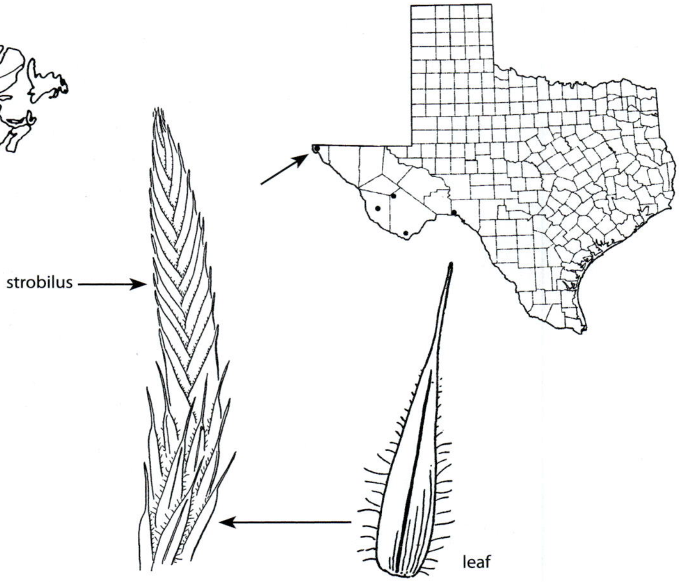

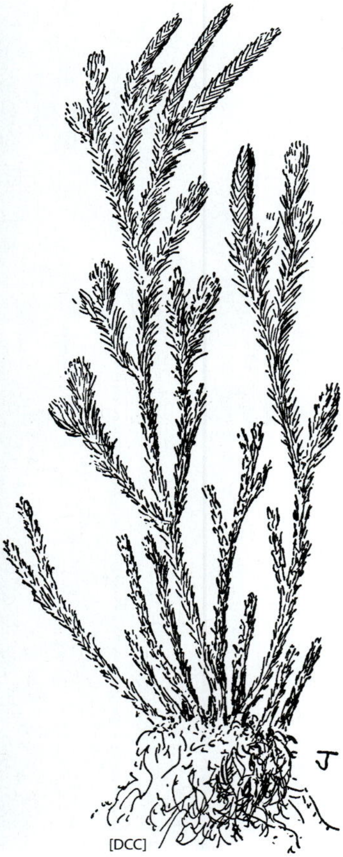

Selaginella ×neomexicana

NEW MEXICO SPIKE-MOSS

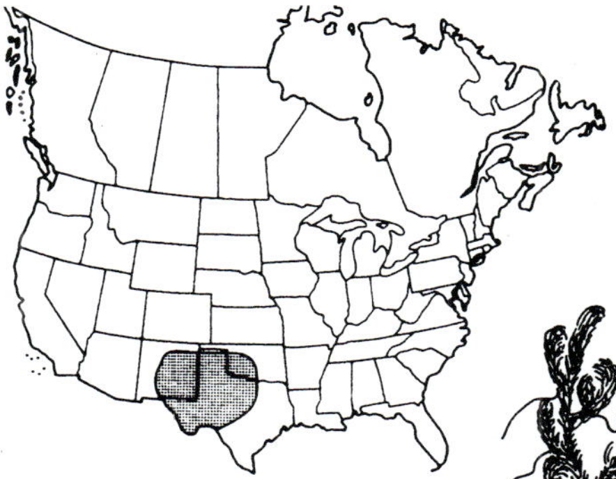

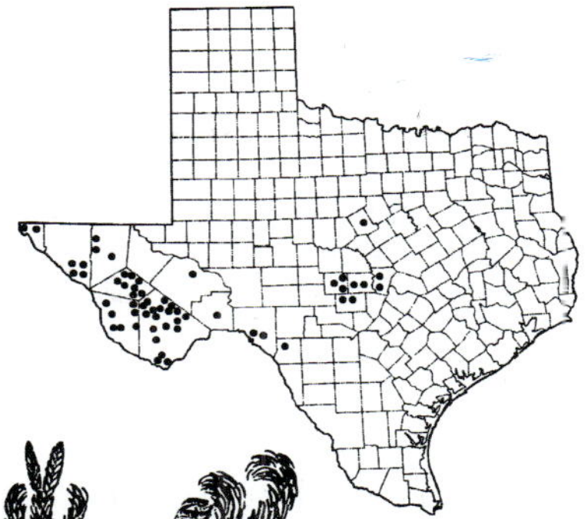

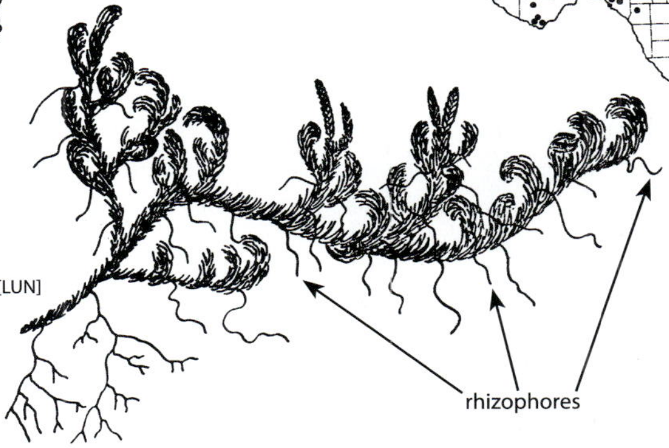

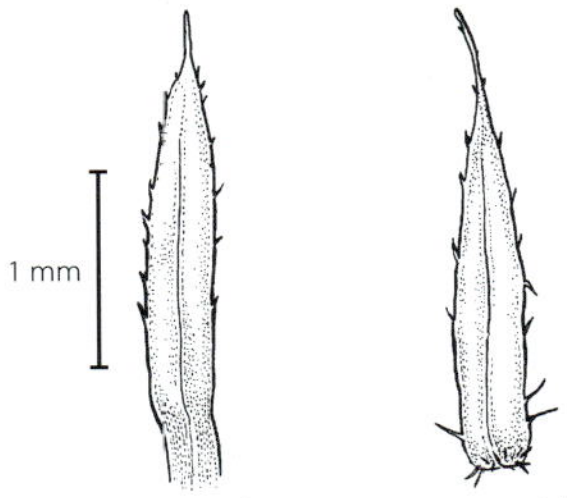

P. B. Pelser

P. B. Pelser

Selaginella peruviana

PERUVIAN SPIKE-MOSS

occurring in a narrow strip of territory along the Mexican border"). Differing from var. *mutica* as given in the key; leaves with apical bristle definitely present, the leaf and sporophyll margins mostly with very short, ascending cilia, but the cilia rarely longer in forms sometimes recognized as var. *texana*. The type description (Weatherby 1944) cites specimens from Mt. Franklin, El Paso Co. (*Slater s.n.*, 1924, US [also *Correll 15027*, TEX-LL; *Warnock 23890*, SRSC]) and the Davis Mts., Jeff Davis Co. (*Moore & Steyermark 2046*, in part, G, *E.J. Palmer 31951*, US [also *Correll 13514*, TEX-LL]), and the type of the synonym *S. mutica* var. *texana* is from the Chicos Mts., Brewster Co. (*Moore & Steyermark 3196*, US, MO). Yarborough and Powell (2002) noted that TX collections of this variety are "mainly from the Mt. Livermore area of the Davis Mts., Jeff Davis Co. where the variety is often found in mixed mats with *S. underwoodii*"; AZ, NM, TX. [*S. mutica* var. *texana* Weath.] Weatherby (1944), who named *S. mutica* var. *texana* (which we are treating as a synonym of var. *limitanea*), noted that "This is a rather indefinite variety, combining the long, spreading cilia of typical *S. mutica* and the terminal seta of var. *limitanea*, and known only from west Texas where the ranges of the two meet. Morphologically, it is little more than a series of intermediates between them...." Carr (2002a) listed var. *limitanea* as one of only two *Selaginella* taxa on his list of rare plants of TX; we thus consider it to be of conservation concern. (Carr 2002a: G4T3S3) ⚠

var. **mutica.** Scattered in the w Trans-Pecos (e.g., Culberson Co., *J.L. Blassingame 3212*, BRIT); AZ, CO, NM, UT, TX, WY; also n Mexico. [*S. watsonii* Underw. var. *mutica* (D.C. Eaton ex Underw.) Clute] Yarborough and Powell (2002) noted that two Sul Ross (SRSC) collections from the Chisos Mts., Brewster Co. "show intermixing with *S. underwoodii*."

Selaginella ×neomexicana Maxon, (of New Mexico, the species occurring in NM as well as AZ and TX), NEW MEXICO SPIKE-MOSS. Plants forming clumps, the aerial stems erect to ascending (can be pendant in large plants when on a nearly vertical surface—see photo), radially symmetrical, with rhizophores mainly on lower ½ of erect aerial stems; underground (rhizomatous) stems present; rhizomes and aerial stems often with a bud-like arrested branch near base; leaves of aerial stems essentially of 1 kind, appressed, linear-lanceolate, 1.9–2.7 mm long, marginally long-ciliate at base but merely minutely denticulate toward apex, apically keeled, with whitish bristle; strobili solitary, quadrangular, usually 1–3 cm long; sporophylls apically keeled, bristled. On rocks of canyons, often on granite: rare in the Trans-Pecos and adjacent Val Verde Co.—in the Franklin Mts., El Paso Co. (e.g., *Correll 15025*, *15035*, TEX-LL), and Chisos Mts. and Sunny Glen, Brewster Co. (*P. Zelazny 93*, SRSC; Yarborough & Powell 2002), and reported for Presidio and Val Verde cos. (Correll 1956; Yarborough & Powell 2002) (we have been unable to relocate the Val Verde Co. collections cited by Correll; therefore the location for that county is mapped in the sw part of Val Verde w of the Pecos River since Correll described the species as being "rare in the Trans-Pecos region of west Texas"); endemic to AZ, NM, and TX. Spores lacking or malformed. This taxon, which is clearly similar in many ways to *S. rupincola* (e.g., long, spreading, marginal leaf cilia, rhizomatous underground stems) has long been treated as a hybrid (e.g., Tryon 1955; Velaspino 1993); megaspores and megasporangia are lacking and the microsporangia are misshapen. It has been proposed that possible parents are *S. rupincola* and *S. mutica* (Tryon 1955) (like *S. ×neomexicana*, the sympatric *S. mutica* has strongly keeled leaf apices). Velaspino (1993) suggested *S. rupincola* × *S. underwoodii* (also sympatric) as another possible parentage or even that *×neomexicana* is an asexual race of *S. rupincola*. Further research is clearly needed to clarify its relationships. Given its very limited range (endemic to extreme s AZ, NM, and TX) and rareness, we consider this taxon to be of conservation concern. ⚠

Selaginella peruviana (Milde) Hieron., (of Peru, the species ranging to South America), PERUVIAN SPIKE-MOSS. Plant forming loose mats; stems prostrate, with upper and lower surfaces different in appearance (leaves upswept or curving upward making the views of the leaf-covered stems from above and below distinctly different); rhizophores scattered along entire length of stem; leaves tightly appressed to ascending, apically with an elongate rounded bristle 0.3–0.8 mm long, all of one kind though leaves on underside of stems widest at base (linear to linear-lanceolate in outline) or at least definitely not narrowed at base, usually pubescent (sometimes glabrous), decurrent at base, marginally ciliate, those on upper side slightly smaller, abruptly adnate at base; strobili solitary, quadrangular, 5–20 mm long; sporophylls ovate-deltate to ovate, with distinct apical bristle. Usually on rock, slopes, crevices, ledges, on sandstone or igneous substrates, less commonly on soils (sandy or clay); primarily Trans-Pecos (e.g., Brewster Co., *Warnock 9171*, BRIT, TEX-LL; El Paso

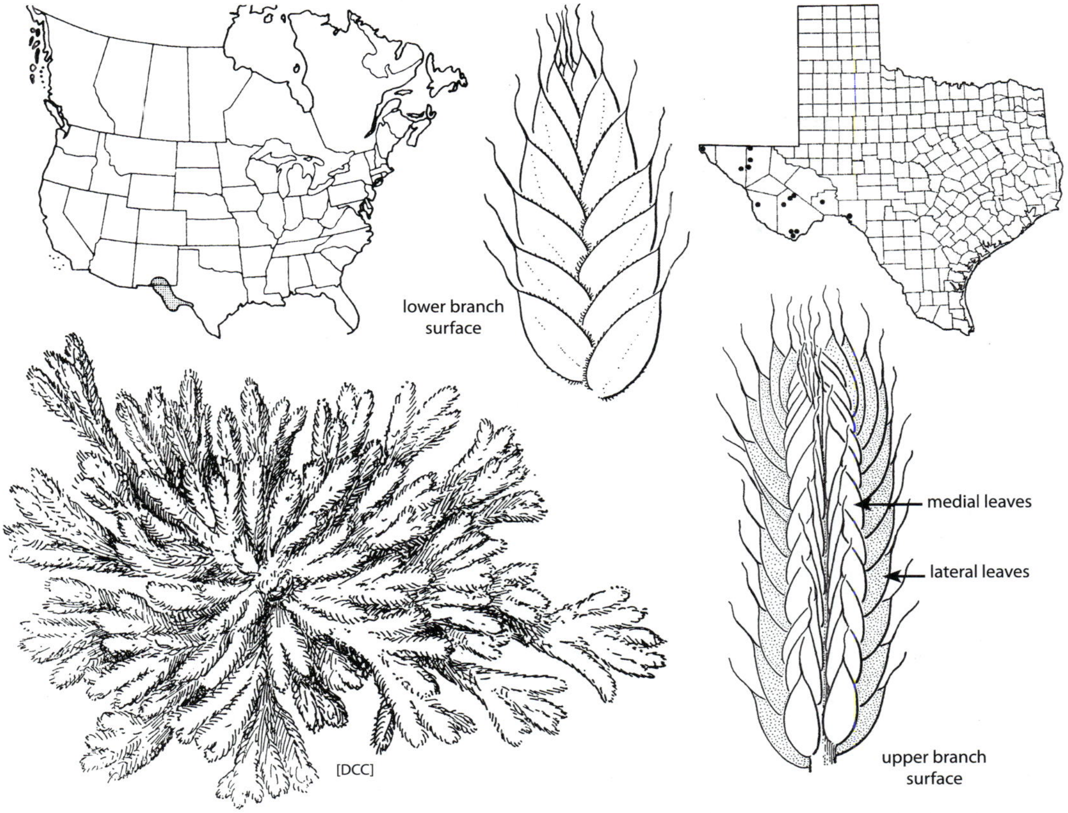

Selaginella pilifera

HAIRY RESURRECTION PLANT

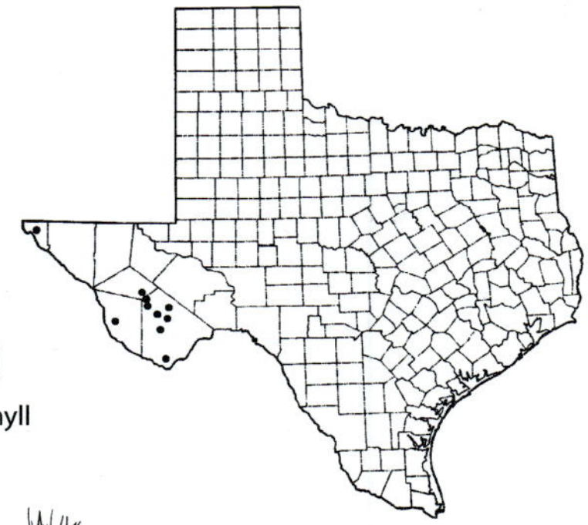

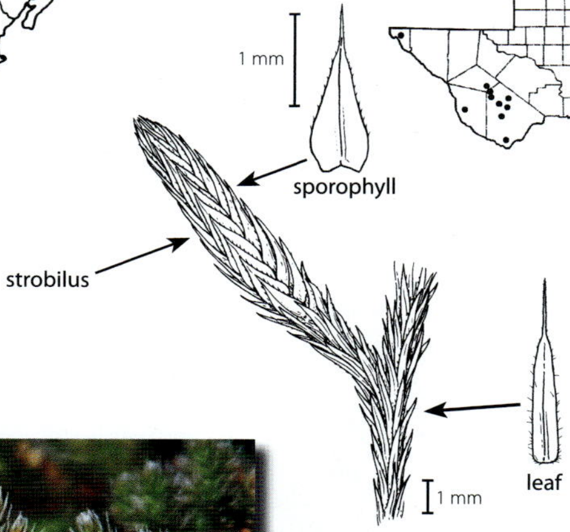

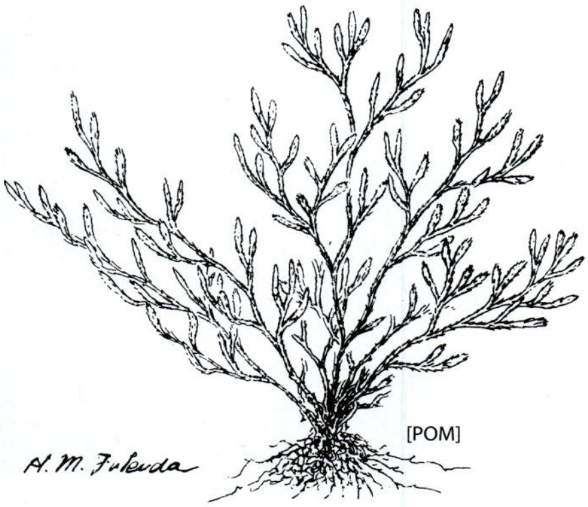

Selaginella rupincola

LEDGE SPIKE-MOSS

Co., *Correll & Johnston 24390-B*, BRIT, TEX-LL) and e Edwards Plateau (e.g., Burnet Co., *G.L. Webster 1982*, TEX-LL) with a slightly disjunct report from Comanche Co. (Stanford 1971) in the Cross Timbers & Prairies; NM, OK, TX; also Mexico and South America. Sporulating Jun–Oct. [*Bryodesma peruviana* (Milde) Soják; *S. rupestris* (L.) Spring forma *peruviana* Milde; *S. sheldonii* Maxon] PERUVIAN SPIKE-MOSS has an interesting biogeography—it occurs in the sw U.S. and Mexico and is distantly disjunct to Argentina, Bolivia, and Peru (Tryon 1969; Mickel & Smith 2004). Yarborough and Powell (2002) noted that this species "occasionally occurs intertwined in the same mixed mats with *S. rupincola* in the Caballos Novaculite area S of Marathon and with *S. underwoodii* or *S. mutica* in the Davis Mountains." It is similar to *S. arizonica*, but can be distinguished by the characters given in the key to species.

Selaginella pilifera A. Braun, (Latin: *pilus*, ball, hair, and *-ifera*, a suffix from Latin *fero*, to bear, = hair-bearing), HAIRY RESURRECTION PLANT, RESURRECTION PLANT, DORADILLA. Plant forming striking open rosettes; rhizophores only on basal part of rosette; main branches much branched, flattened in overall aspect when moist and even when dry still flattened or only somewhat erect; leaves in 4 distinct ranks, 2 lateral, 2 medial, conspicuously overlapping, all with obvious bristles ca. ⅓–½ as long as leaf (0.5–1.5 mm long) and with only inconspicuous whitish transparent margins, the two types of leaves different in size and shape—the lateral leaves usually 3–3.5 mm long, elliptic to elliptic-ovate, the medial smaller, usually 2–3 mm long, lanceolate and slightly falcate; strobili solitary, 5–10 mm long, quadrangular; sporophylls apically with long bristle. Terrestrial or on rock, sheltered areas of rocky slopes, cliffs, canyon floors, typically on limestone, but sometimes on igneous substrates; Brewster (*Correll 13593*, BRIT, TEX-LL), Culberson (*McVaugh 8006*, BRIT, TEX-LL), El Paso (*Wright s.n.*, 1849, isotype: US), Hudspeth (BRIT), Presidio, and Terrell (Turner et al. 2003) cos. in the Trans-Pecos and Val Verde Co. (Seminole Canyon [e of the Pecos River]—Yarborough & Powell 2002) in the extreme w part of the Edwards Plateau; known in the U.S. only from NM and TX; also n Mexico. Sporulating Jun–Oct. [*S. pilifera* var. *pringlei* (Baker) C.V. Morton; *S. pringlei* Baker] Like the similar but more common *S. lepidophylla*, this species is a "resurrection plant" capable of drying out and then resuming growth when rehydrated. The obvious long bristles at the leaf apices, which give the plants a "bristly" or "hairy" appearance, easily distinguish it from *S. lepidophylla*, the only other rosette-forming SPIKE-MOSS in TX. Also unlike that species (whose rosettes become ball-like when dry), in HAIRY RESURRECTION PLANT the lateral stems comprising the rosettes are flattened not only when moist, but remain almost flat to merely forming a somewhat erect cluster even when dry. It is interesting to note that molecular studies (Korall & Kenrick 2002) show that the "resurrection" type morphology has evolved independently a number of times in the genus and that the two resurrection plants in TX, *S. lepidophylla* and *S. pilifera*, are only distantly related. Velaspino (1993) considered this species to be of conservation concern. ⚠

Selaginella rupincola Underw., (Latin: *rupes*, rock, and *cola*, inhabitant, dweller), LEDGE SPIKE-MOSS, ROCK-LOVING SPIKE-MOSS. Plant forming loose clumps, the aerial stems usually suberect to erect, rarely decumbent, to ca. 10 cm tall, with leaves without strongly upswept or upwardly curving appearance, the leaf-covered stems therefore appearing radially symmetrical, usually with rhizophores only on lower ½ (less commonly scattered along entire length); underground (rhizomatous) stems present, with rhizophores; rhizomes and aerial stems often with a bud-like arrested branch near base; leaves of aerial stems tightly appressed, of 1 kind, linear-lanceolate, 3–4.7 mm long, marginally long ciliate ± from base to apex, apically with obvious bristle 0.65–1.85 mm long; strobili solitary, quadrangular, 5–25(–35) mm long; sporophylls apically bristled. On rock or soil, ledges, cliffs, rocky slopes, typically on igneous substrates; Brewster (*Warnock 20893*, BRIT; *W.D. Pohl & J.L. Blassingame 2944*, BRIT; *Warnock 21208*, TEX-LL), Jeff Davis (*O.E. Sperry T738*, SRSC; *Warnock 20273B*, TEX-LL), Presidio (*J. Lott et al. 5378*, SRSC, TEX-LL), and El Paso (Yarborough & Powell 2002) cos.; AZ, NM, and TX; also widespread in Mexico. Sporulating Jun–Oct. [*Bryodesma rupincola* (Underw.) Soják, *S. rupestris* var. *rupincola* (Underw.) Clute] This species is presumed to be one of the parents of the somewhat similar *S.* ×*neomexicana* (Tryon 1955)—see that taxon for further discussion. It can sometimes be found in the same mat as *S. arizonica* or *S. peruviana* (both of which, however, have leaves that are upswept or curving upward—Yarborough & Powell 2002). Because of its rareness and limited distribution in the state, we consider this species to be of conservation concern in TX. ⚠

Selaginella scopulorum Maxon, (Latin: *scopula*, rock, cliff, or crag, the type specimen described as grow-

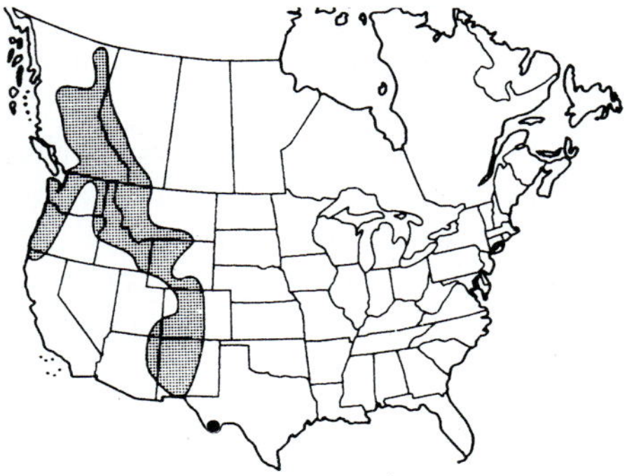

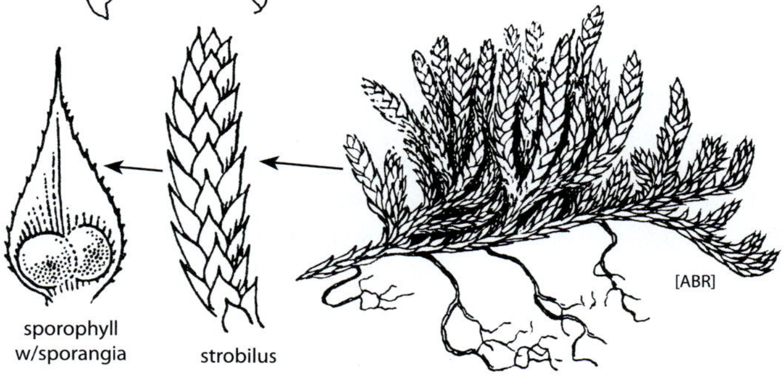

Tanya Harvey

Selaginella scopulorum

ROCKY MOUNTAIN SPIKE-MOSS

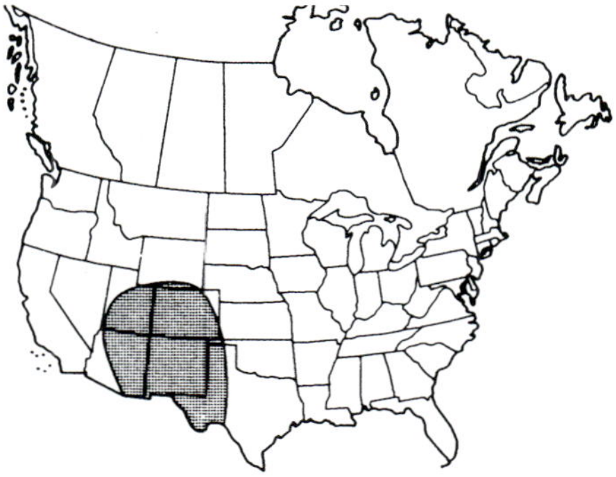

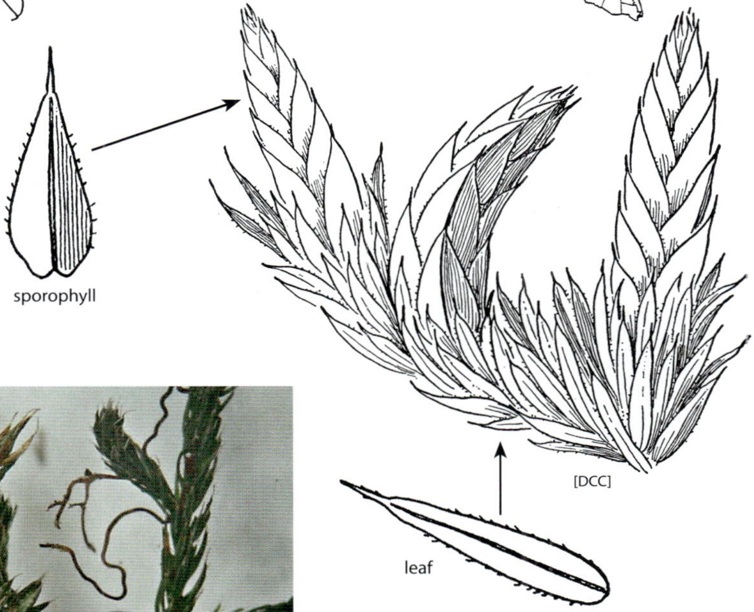

Selaginella underwoodii

UNDERWOOD'S SPIKE-MOSS

ing on "rocks"), ROCKY MOUNTAIN SPIKE-MOSS. Plant forming flat cushion-like mats (in TX, in other parts of the range the mats can be loose); stems prostrate-creeping, with upper and lower surfaces slightly different in appearance, without budlike arrested branches near base; rhizophores scattered along entire length of stem; leaves tightly appressed, linear-lanceolate to linear and sometimes falcate, marginally short ciliate at base but merely minutely denticulate toward apex, the apices bristled, the bristles 0.5–1.1 mm long, whitish, the leaves all of one kind though those on underside of stems glabrous and decurrent at base, marginally short ciliate, those on upper side of stem slightly smaller, glabrous or rarely pubescent and slightly decurrent to adnate at base; strobili solitary, quadrangular, (5–)10–30(–45) mm long; sporophylls lanceolate to ovate-lanceolate, rarely ovate, apically with short bristle. On rock or terrestrial, typically in rather dry habitats (outcrops, rocky slopes, ledges, talus slopes), often on igneous or sandstone substrates; in TX known only from sandstone outcrops 3 mi e of Study Butte in s Brewster Co. (*J.A. Moore & J.A. Steyermark 3245*, US [xerox of specimen at SRSU], identified as *S. densa* var. *scopulorum* by R. Tryon & T. Reeves; Tryon 1955; Correll 1956 [in Addenda];Yarborough & Powell 2002); n British Columbia s to CA (along the Cascades) and s to TX along the Rocky Mts. Sporulating ?–Jun–Jul–? [*S. densa* var. *scopulorum* (Maxon) R.M. Tryon, *S. engelmannii* var. *scopulorum* (Maxon) C.F. Reed] This species has a very broad range along the Rocky Mts. and occurs in a variety of habitats—alpine tundra to mountains in the Chihuahuan Desert of sw TX—its TX occurrence is the most southerly record known. It could be referred to as a "Rocky Mt. disjunct" since its TX location is disjunct by approximately 300 mi/480 km from its nearest other known occurrence in New Mexico. It is part of the *S. densa* Rydb. complex, which is in need of taxonomic work (Valdespino 1993). Yatskievych and Windham (2009b) treated this taxon as *S. densa* var. *scopulorum*. Specimens from the TX population "are yellow-green in color and have very short, tufted branches that form a flat, cushiony mat" (Yarborough & Powell 2002). Given its rareness and limited distribution in the state, we consider this species to be of conservation concern in TX. ⚠

Selaginella underwoodii Hieron., (for L.M. Underwood, 1853–1907, American fern specialist, first fern curator at the New York Botanical Garden, and professor at Columbia Univ.) UNDERWOOD'S SPIKE-MOSS. Plants mat-forming, the mats usually rather loose; stems prostrate-creeping or occasionally pendant, without budlike arrested branches near base, with leaves without strongly upswept or upwardly curved appearance, the leaf-covered stems therefore appearing radially symmetrical; rhizophores scattered along entire length of stem; leaves loosely appressed, apically with bristles 0.25–1 mm long, all of one kind, linear to narrowly triangular-lanceolate, marginally entire to denticulate to minutely short ciliate, basally mostly cuneate and decurrent; strobili solitary or sometimes paired, quadrangular, 5–35 mm long; sporophylls lanceolate to ovate-lanceolate, apically with short or long bristle. On rock, shaded ledges, cliffs, rocky slopes, on a variety of substrates (e.g., granite, limestone, sandstone); known in TX from Brewster (Chisos Mts., *Correll 13668*, BRIT, TEX-LL), El Paso (Franklin Mts., *E. Whitehouse s.n.* 1931, TEX-LL), Jeff Davis (Davis Mts., *Correll 34972-A*, *Warnock 7588*, BRIT, TEX-LL), and Val Verde (*Z. Labus 56*, TEX-LL, Seminole Canyon State Historical Park [e of the Pecos River]; Yarborough & Powell 2002, mapped w of the Pecos River) cos. in the Trans-Pecos and extreme w edge of the Edwards Plateau; AZ, CO, NM, OK, TX, UT, and WY; also n Mexico. Sporulating Jul–Oct. [*Bryodesma underwoodii* (Hieron.) Soják, *S. fendleri* (Underw.) Hieron.] This species can sometimes be found growing in mixed mats with *S. mutica* and *S. peruviana* (Yarborough & Powell 2002). It somewhat resembles *S.* ×*neomexicana*, "but is distinguished by the leaves being more lax and in fewer ranks than in *S.* ×*neomexicana*" (Yarborough & Powell 2002). The prostrate-creeping habit of *S. underwoodii* also differentiates it from the erect to ascending *S.* ×*neomexicana*. Mickel and Smith (2004) described *S. underwoodii* in Mexico as "a rather distinct species; however, it shows some variation in the number and length of the marginal cilia (from none to many), leaf base pubescence, and arista length." Because of its rareness and limited distribution in the state, we consider this species to be of conservation concern in TX. ⚠

Selaginella viridissima Weath., (Latin: *virdis*, green, and *-issimus*, superlative, very), GREEN SPIKE-MOSS, SLENDER SPIKE-MOSS. Plant forming clumps or mounds (to 1 m in diam.), the aerial stems usually erect, radially symmetrical, with rhizophores limited to lower ¼ of stems; underground (rhizomatous) stems present, with rhizophores; rhizomes and aerial stems often with a bud-like arrested branch near base; leaves of aerial stems of 1 kind, loosely appressed, linear-lanceolate to narrowly lanceolate, 1.8–2.1 mm long, marginally with only sparse very fine teeth (minutely denticulate), basally cuneate and decurrent to slightly rounded and adnate, apically acute to obtuse, without terminal bristle; strobili solitary, quadrangular, 5–12(–25) mm long;

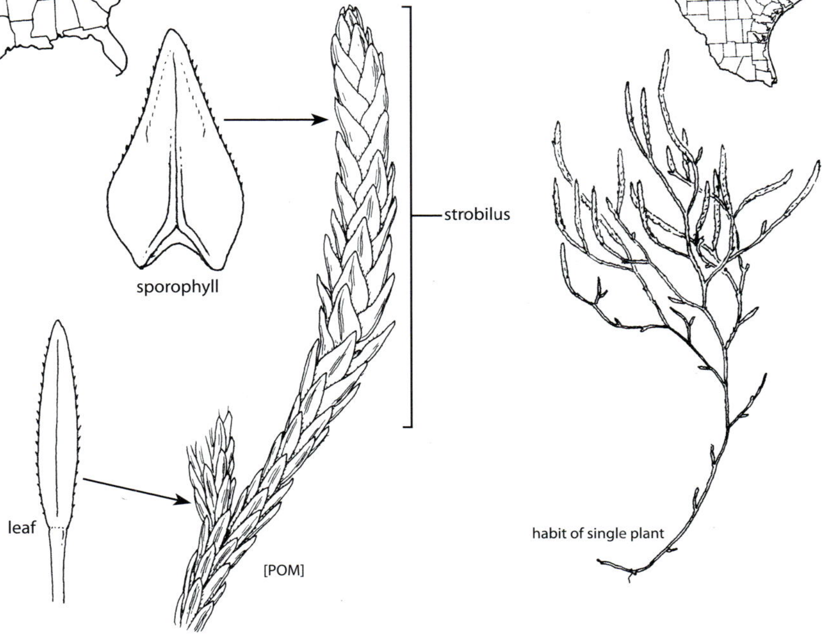

Selaginella viridissima

GREEN SPIKE-MOSS

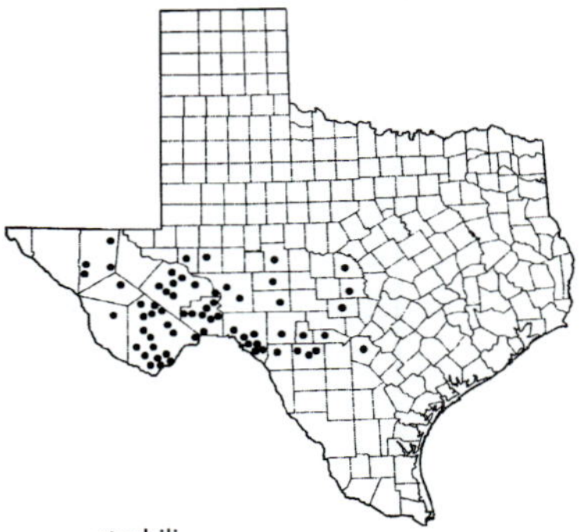

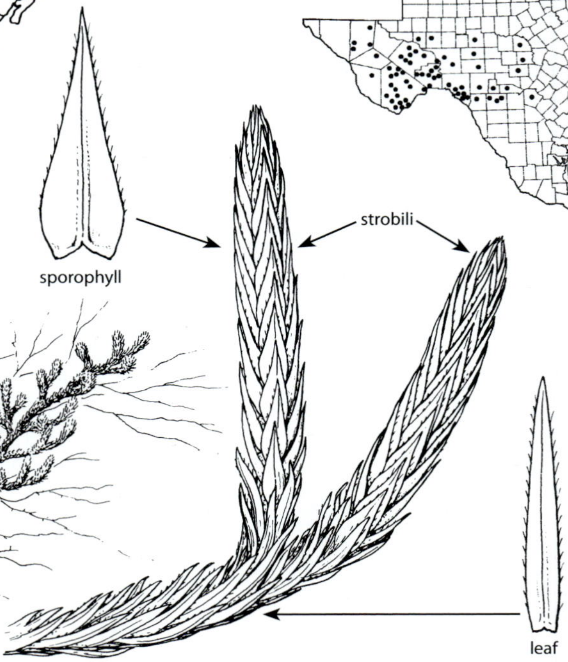

Roy Morey

Selaginella wrightii

WRIGHT'S SPIKE-MOSS

sporophylls apically acute to obtuse. On rock (often igneous), shaded or sheltered ledges, cliffs, crevices, and rocky slopes; in TX known only from the Chisos Mts. of Brewster Co. (*Correll 13627*, TEX-LL; *Warnock 21263*, SRSC; Valdespino 1993) and the "Fern Canyon" (*Warnock 21694*, TEX-LL, US; Carr 2002a) in the Davis Mts. of Jeff Davis Co. (Yarborough & Powell 2002); in the U.S. known only from TX, also Coahuila in n Mexico. Sporulating Jun–Aug. [*Bryodesma viridissimum* (Weath.) Soják, *S. coryi* Weath.] This Chihuahuan Desert endemic (Carr 2002a) is apparently quite rare—it is known from only two counties in the U.S. and a few collections from the n Mexican state of Coahuila (based on Mickel & Smith 2004). The combination of erect habit and leaves without a bristle at the tip is distinctive. Valdespino (1993) noted that the species is of conservation concern and it was one of only two pteridophytes included in *Rare Plants of Texas* (Poole et al. 2007a); it certainly warrants such designation. (Carr 2002a; Poole et al. 2004, 2007a, 2007b: G2S1)

Selaginella wrightii Hieron., (for Charles Wright, 1811–1885, TX collector), WRIGHT'S SPIKE-MOSS. Plant mat-forming, the mats loose to dense; stems prostrate, with upper and lower surfaces different in appearance (leaves upswept or curving upward, making the views of the leaf-covered stems from above and below distinctly different); rhizophores scattered along entire length of stem; leaves tightly appressed, all ± of one kind though leaves on underside of stems narrowly linear-lanceolate and falcate, usually pubescent and usually abruptly adnate at base or slightly decurrent, marginally ciliate, those on upper side slightly shorter, linear-lanceolate, pubescent and abruptly adnate at base, marginally ciliate; leaf apices gradually tapering to a yellowish somewhat hardened bristle 0.2–0.5 mm long (this sometimes breaking off but more persistent on underside leaves), the bristle sometimes poorly differentiated from blade tissue; strobili solitary per branch tip, quadrangular, erect, (7–)15–20(–26) mm long, often abundant; sporophylls lanceolate, with obscure bristle. On rock, typically on limestone, ledges, cliffs, and rocky slopes; widespread in the Trans-Pecos and Edwards Plateau (e.g., Schleicher Co., *J.W. Stanford 3543*, BRIT) e to Bexar (*Carr & P. Turner 12360*, TEX-LL) and San Saba (*Correll & Correll 12671*, TEX-LL) cos.; NM and TX; also widespread in Mexico. Sporulating Mar–Dec. [*Bryodesma wrightii* (Hieron.) Soják, *S. rupestris* forma *wrightii* (Hieron.) Clute] Correll (1956) noted that it superficially resembles *S. peruviana* (treated by him as *S. sheldonii*) in the prostrate habit and upturned stem tips, but *S. wrightii* can be distinguished in various ways including the longer leaves (3.3–5 mm long vs. 2.3–4 mm) with longer marginal cilia and the short (0.2–0.5), yellowish, somewhat hardened bristle at the leaf tips. Velaspino (1993) considered this species to be of conservation concern. However, since it is so widespread and common we are not treating it as such.

DIVISION POLYPODIOPHYTA
FERNS (INCLUDING PSILOPHYTES & HORSETAILS)

A group of ca. 9,000–11,500 species in 231 genera arranged in 33–37 families and 11 orders, with some authorities estimating as many as 12,000 species (Moran 2004; Pryer et al. 2004b; Schneider et al. 2004b; Smith et al. 2006; Mabberley 2008). About 100 new species are described each year, and the group represents about 3.6% of the world's vascular plant species (Kessler 2010). It is divided very unequally among five major lineages—ophioglossoids (grape ferns), whisk ferns, marattioids (a small tropical group), horsetails, and leptosporangiates (the vast majority of living species) (Pryer & Schuettpelz 2009)—four of these five occur in TX. As noted previously, there was a very early evolutionary split in the vascular plants about 400 million years ago with one branch leading to the lycophytes (still in existence) and the other branch diversifying into all other vascular plants including ferns, gymnosperms, and ultimately flowering plants (Pryer et al. 2004b). This second branch is known as the Euphyllophytes, signifying that they had a special complex type of leaf known as a euphyll or megaphyll, in contrast to the simpler lycophyll of the lycophytes. Thus, although morphologically more similar in some ways to lycophytes (e.g., lacking seeds, having an independent gametophyte generation, etc.), ferns are evolutionarily more closely related to seed plants (see Fig. 1 on page 4). Within Euphyllophytes, ferns make up one branch and all seed plants are on a second branch.

The fossil record of ferns dates to ca. 359–345 million years ago, early in the Lower Carboniferous Period (359–299 mya), and related groups with many of the characteristics of ferns are known from as early as the

Middle Devonian 390 million years ago (Devonian Period 416–359 mya) (Rothwell & Stockey 2008). All five of the major lineages of living ferns were present by the end of the Carboniferous (Pryer & Schuettpelz 2009). Although ferns are thus an ancient group and some individual living species have extremely long fossil histories (see further discussion under *Osmunda*), recent molecular evidence suggests that the leptosporangiate ferns (> 95% of living ferns), particularly the polypods (Order Polypodiales) which make up ca. 80% of living fern species, had major diversifications during the Cretaceous Period beginning ca. 100 mya and more recently in the Cenozoic Era. These recent radiations are thought to be a result of ecological opportunities associated with the rise and increasing dominance of the angiosperms—e.g., modern polypods thrive in the low light conditions of angiosperm forests and as epiphytes in the canopy of tropical forests (Eriksson 2004; Schneider et al. 2004b; Torsten 2004; Schuettpelz & Pryer 2008, 2009). This view, that the majority of modern fern diversity results from relatively recent diversifications, is very different from the previous belief that ferns "succumbed to the ecological predominance of angiosperms and are mostly evolutionary holdovers from the late Palaeozoic/early Mesozoic era" (Schneider et al. 2004b). Indeed, it now seems likely that although ferns are an ancient lineage, they more recently underwent extensive adaptive radiation "in the shadow of the angiosperms" (Lovis 1977; Schneider et al. 2004b). Kawai et al. (2003) suggested that a mutation resulting in a unique type of light receptor in polypod ferns may have given them an advantage in the dense shade created by angiosperm forests.

Ferns have also apparently been incredibly successful for brief periods following episodes of severe ecological stress. There is abundant evidence of devastating effects on ecosystems worldwide from the impact of an extraterrestrial object (e.g., asteroid or comet) with the earth ca. 65 million years ago at the Cretaceous-Tertiary (K-T) boundary. Although approximately 65–70% of the world's species became extinct (e.g., dinosaurs, pterosaurs, ammonites), ferns show something surprising—a temporary "fern spike" following the catastrophe. Prior to the impact, microfossils (spores and pollen combined) indicate there was a diverse flora with fern spores making up only ca. 15–30% of the total. However, in sediments from just after the impact the percentage of fern spores spiked, reaching 90% to even 99% of microfossils (Graham 1999a; Vajda et al. 2001; Moran 2004). In other words, a flora composed of numerous species, mostly angiosperms, was suddenly but only temporarily replaced by one dominated by a few ferns. The reasons are not completely clear, but one possibly is that the ease and rapidity of dispersal of large numbers of tiny spores over long distances was a contributing factor in this sudden though transitory dominance of ferns. Another possible explanation is that the spore banks in soil could have been richer than seed banks, favoring ferns.

The leaves of ferns are euphylls or megaphylls (= with branched veins), which apparently are derived from modified branch systems, have an associated "leaf gap" in the stem vascular tissue, and have marginal or apical meristems (all in contrast to the lycophytes). Ferns are "seed-free," reproducing by spores, and the spores are of one type (plants homosporous—most living ferns) or two (plants heterosporous—Azollaceae, Marsileaceae, and Salviniaceae, < 1% of living ferns). Modern species range from tree ferns up to 24 m tall to tiny free-floating aquatics, but most are rhizomatous perennial herbs. They were traditionally divided into a eusporangiate group (only 2 families, Ophioglossaceae and Marattiaceae) and a leptosporangiate group. The eusporangiate ferns have large thick-walled sporangia, each containing from several hundred to thousands of spores, with the sporangia developing from several initial cells. Recently, molecular evidence has shown that two other groups, Psilotaceae and Equisetaceae, which are very different in appearance than most ferns but have eusporangiate type sporangia, are actually ferns. The four eusporangiate families, which together make up < 5% of all living ferns, represent the basal branches of the fern evolutionary tree (see Fig. 5) with living species being the descendants of ancient lineages. The leptosporangiate group, which includes all other ferns and accounts for > 95% of modern fern diversity, is now thought to be a single evolutionary lineage (monophyletic group). They are characterized by small delicate sporangia, with each sporangium usually developing from a single cell and with sporangial walls only one cell thick. Further, each sporangium usually contains 128 spores or fewer (Jones & Luchsinger 1986; Camus 1990; Wagner 1990; Doyle 1998; Moran 2004). Ferns are currently recognized as comprising 11 orders, 7 of these occurring in TX (see page 5). The word fern is "adapted from the Anglo-Saxon *fearn* and the German *farn* with a lineage that goes back to the Sanskrit *parna*, which originally meant a wing or feather" (Durant 1976).

REFERENCES: Eaton 1878; Bush 1903; Reverchon 1903; Small 1938; Correll 1949, 1956, 1966; Jermy et al. 1973; Thieret 1980; Tryon & Tryon 1982; Bell & Woodcock 1983; Taylor 1984; Lellinger 1985; Raven et al. 1986;

Bold et al. 1987; Flora of North America Editorial Committee 1993; Wagner & Smith 1993; Peck & Taylor 1995; Pryer et al. 1995; Raubeson & Stein 1995; Smith 1995; Nauman et al. 2000; Nelson 2000; Hoshizaki & Moran 2001; Pryer et al. 2001, 2004b; Schneider et al. 2004a, 2004b, 2009; Moran 2004; Schuettpelz et al. 2006, 2008; Smith et al. 2006, 2008; Schuettpelz & Pryer 2007, 2008, 2009; Ranker & Haufler 2008; Pryer & Schuettpelz 2009; Schneider et al. 2009; Mehltreter et al. 2010.

ANEMIACEAE Link

ANEMIA OR FLOWERING FERN FAMILY

A family of 1 or 2 genera and ca. 119–124 species (Mickel 1993; Roux 1995) widespread in the tropics and subtropics, with most species in *Anemia*. The group has a long fossil history with *Anemia* being widespread by the early Cretaceous (Skog 2001). It is most closely related to the Schizaeaceae and then to Lygodiaceae, these two families and the Anemiaceae making up Order Schizaeales. This order is characterized by a unique type of sporangium—completely encircled near the tip by the annulus (= the group of cells involved in opening the sporangium) (Moran 2004). As a group, the Schizaeales are sometimes referred to as the "schizaeoid" ferns and are a sister group to the "so-called 'core-leptosporangiates'—a large clade composed of heterosporous, tree, and polypod ferns" (Smith et al. 2008; see Fig. 1 on page 4). The "schizaeoid" lineage is quite old, originating "possibly in the Triassic and definitely in the Jurassic" (Tidwell & Ash 1994). Fossil evidence suggests that the three main living genera of Schizaeales (*Anemia*, *Lygodium*, and *Schizaea*) are ancient, with these genera already established by the late Jurassic/early Cretaceous (Wikström et al. 2002; Hernandez-Castillo et al. 2006). Anemiaceae are sometimes included in the Schizaeaceae (e.g., Kramer 1990f). However, given the group's long history and morphological distinctiveness, we are following Smith et al. (2006) in recognizing it as a distinct family. In addition to *Anemia,* the only other genus in the family, *Mohria*, has only 7 species (Roux 1995) and is restricted to Africa, Madagascar, and Réunion Island; it includes *M. caffrorum* (L.) Desv. (FRANKINCENSE FERN), a cultivated ornamental with scented fronds. Recent molecular evidence (e.g., Wikström et al. 2002) suggests that *Mohria* is a lineage within *Anemia* and should thus be included in that genus. The common name, FLOWERING FERN, is said to be used because the fertile structures of some species are "held erect above the sterile fronds, and are conspicuously covered with yellow to golden brown sporangia which can be quite showy ..." (Nelson 2000). (Order Schizaeales)

FAMILY RECOGNITION IN THE FIELD: the single local species has ± 1-pinnate leaves with 2 conspicuously different types of pinnae: 4–6 pairs of sterile pinnae and below these a pair of very long-stalked, bipinnate, fertile pinnae.

REFERENCES: Mickel 1993; Roux 1995; Wikström et al. 2002.

ANEMIA Sw. FLOWERING FERN, PINELAND FERN

A genus of ca. 120 species of tropical and subtropical regions of the world, especially Brazil (70 species) and Mexico (20 species) but also in Africa and India; only 3 species occur in the U.S.—2 in FL and 1 in TX (Mickel 1993; Mickel & Smith 2004). Mickel (1962) advocated including *Mohria* (the only other genus in the family) in *Anemia* and preliminary molecular data (Skog et al. 2002; Wikström et al. 2002, based on 1 species of *Mohria*) indicate that *Anemia* is paraphyletic if *Mohria* is not included. However, until more information is available we are treating *Anemia* in the traditional sense. Although *Anemia* is sometimes placed in the Schizaeaceae (Kartesz 1994), we are following Mickel (1993—*Flora of North America*) and Smith et al. (2006) in treating it in the Anemiaceae. The genus has a long and rich fossil record, going back to the Lower (early) Cretaceous (Mickel 1972; Skog 1992; Hernandez-Castillo et al. 2006). Some species are known to "hybridize freely and frequently" (Mickel & Smith 2004). The 2 basal pinnae are fertile, long-stalked, erect, and have almost no green tissue, a situation unique among ferns (Moran 2004). This unusual morphology probably caused *Anemia* species to be "popular in cultivation in England during the 19th century" (Mickel 1972). (Greek: *aneimon*, without clothing or naked, referring to the absence of blade protection for the sporangia, the fertile pinnae lacking blade tissue or nearly so)

REFERENCES: Correll 1956, 1966; Mickel 1962, 1981, 1982; Duek 1980; Skog 1992; Carlquist & Schneider 1998; Skog et al. 2002.

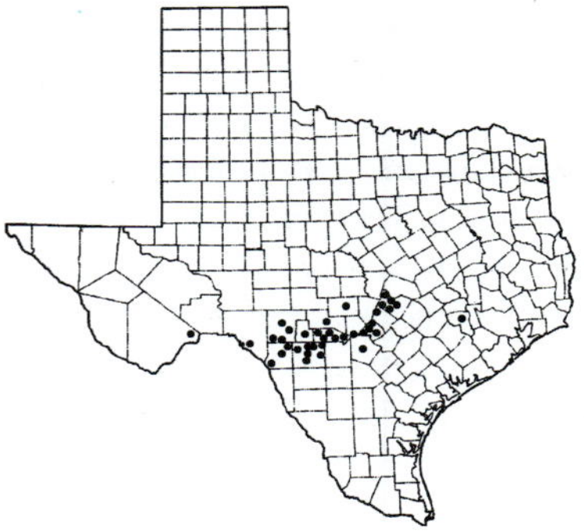

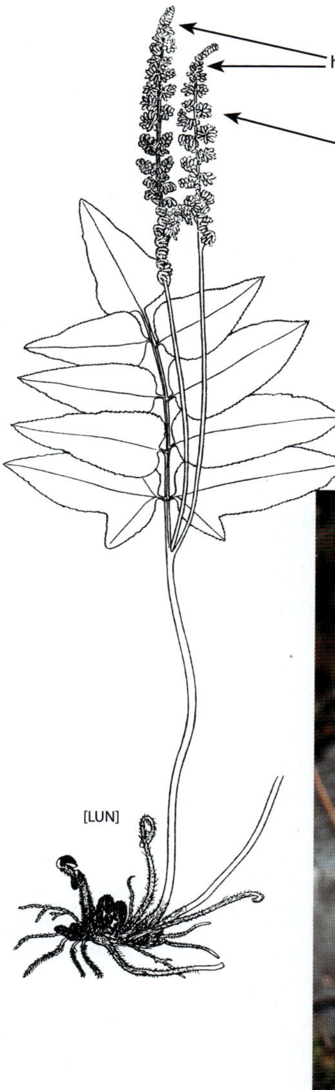

sporangia on ultimate segments

Anemia mexicana

MEXICAN FERN

Anemia mexicana Klotzsch, (of Mexico, where the type specimen was collected), MEXICAN FERN, MEXICAN FLOWERING FERN. Plant terrestrial or on rocks, to ca. 50 cm tall; rhizomes short-creeping, horizontal, covered with coarse dark hairs; leaves 1-pinnate (except for fertile pinnae), partially dimorphic, with 4–6 pairs of sterile pinnae distally and with the lowermost pair of pinnae (fertile) highly modified and extremely different in appearance; fertile pinnae bipinnate, with only a minute amount of blade tissue, very long-stalked, to 30 cm long, usually exceeding the sterile portion of the leaf in length; sterile pinnae triangular-ovate to lanceolate, basally truncate, apically subobtuse to acute or acuminate, marginally serrulate, the lowermost sterile pinnae often lobulate; sporangia in 2 rows on ultimate segments (= smallest subdivisions) of fertile pinnae; indusia absent; $2n = 76$ (Mickel & Smith 2004). Outcrops, under overhanging ledges, rocky partially shaded slopes, banks of ravines; found primarily on limestone outcrops on the Edwards Plateau e and n to Bexar (BAYLU, TEX-LL), Hays (TAES), Comal (*S. Barber 416*, BRIT), and Travis (*M.S. Young s.n.*, "4/6/1918," BRIT, TEX-LL) cos., also Austin Co. in se Blackland Prairie (Correll 1956) and Brewster Co. (*P. Manning 3401*, TEX-LL) in the Trans-Pecos; also Mexico; this is one of 19 pteridophyte species known in the U.S. only from TX (see page 40 and Table 5). Sporulating Feb–Oct. [*Ornithopteris mexicana* (Klotzsch) Underw.] The fertile leaves, with 2 long-stalked, fertile, basal pinnae with almost no green tissue, make this species unmistakable.

ASPLENIACEAE Newman

SPLEENWORT FAMILY

A cosmopolitan family of ca. 700+ species (Wagner et al. 1993). Generic boundaries within the family are controversial—although some authorities treat all as members of a single diverse genus *Asplenium* (e.g., Wagner et al. 1993), others recognize as many as 10 genera. Molecular and anatomical evidence indicate that most segregate genera are lineages within *Asplenium,* and a division of the family into two genera, *Asplenium* (with the vast majority of the species) and *Hymenoasplenium,* seems to be the developing consensus (Schneider 1996; Murakami et al. 1999; Schneider et al. 2004a; Schuettpelz & Pryer 2008; Smith et al. 2008). (Order Polypodiales)

FAMILY RECOGNITION IN THE FIELD: leaves 1-pinnate (or simple or merely forked in 1 rare species), all alike or the fertile slightly smaller; sori elongate along the veins; indusia attached along one side of the sori.

REFERENCES: Kramer & Viane 1990; Wagner et al. 1993; Schneider 1997; Murakami et al. 1999; Schneider et al. 2004a; Schuettpelz & Pryer 2008.

ASPLENIUM L. SPLEENWORT

TX species terrestrial or on rocks; rhizomes short-creeping to erect, scaly; leaves clustered, 1-pinnate, or sometimes simple or merely forked in *A. septentrionale*, monomorphic or slightly dimorphic, mostly evergreen; pinnae auricled basally or not so, the auricle on the acroscopic side (= side toward the leaf apex) of the pinnae; sori elongate along veins; indusia attached along one edge of the sori.

A large and diverse, cosmopolitan genus of ca. 700 species (following Wagner et al. 1993 & Mabberley 2008) of terrestrial, epipetric (= on rocks), and epiphytic species, with centers of diversity in the Appalachians, Central American mts., Andes, and Himalayas. This is one of the largest genera of ferns, and the species range in size from those with leaves only a few centimeters long to the giant BIRD'S-NEST FERNS, with leaves more than 1 m long (Hoshizaki & Moran 2001). The genus is well known for its interspecific hybridization and complex polyploid series with numerous allopolyploids. Ploidy levels range from diploid to hexaploid, and three-fifths of the species are thought to be of hybrid, allopolyploid origin; *A. platyneuron* is part of a polyploid complex centered in the e U.S. (Wagner 1954). A number of species are cultivated as ornamentals (e.g., *A. nidus* L.—BIRD'S-NEST FERN). (Greek: *splen*, spleen; thought by Dioscorides, Greek naturalist of the first century A.D., to be useful for treating spleen diseases)

REFERENCES: Dittmer et al. 1954; Wagner 1954; Correll 1956, 1966; Taylor et al. 1976; Wagner & Johnson 1981, 1983.

1. Plant almost grass-like in appearance (due to linear leaves); leaf blades simple or forked, with only 2–3 (–5) pinnae per leaf, the simple blades or pinnae linear, many times longer than wide; rhizomes much branched; species known in TX only from one collection in Brewster Co. in the Trans-Pecos ________ **A. septentrionale**

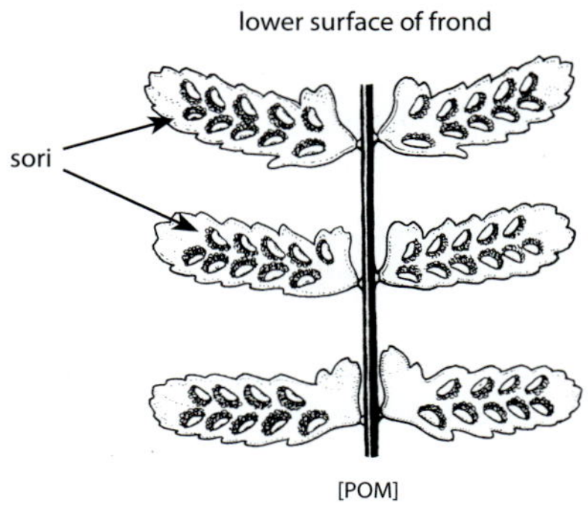

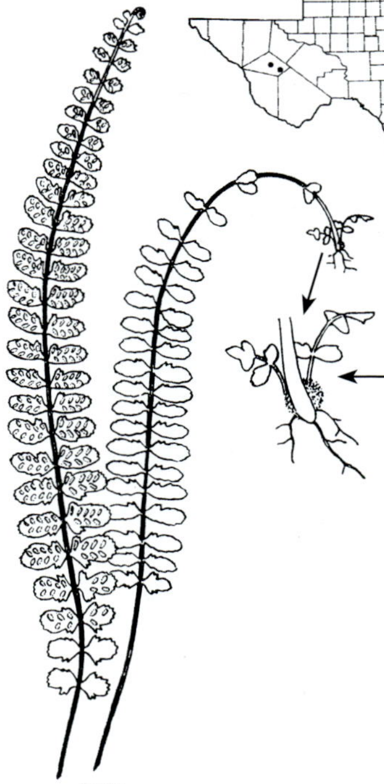

Asplenium palmeri

PALMER'S SPLEENWORT

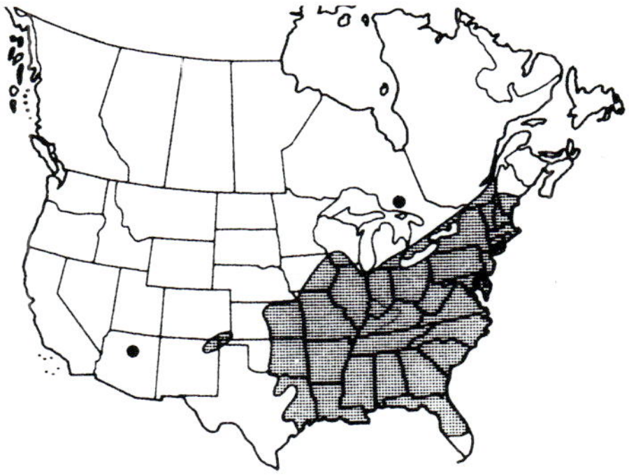

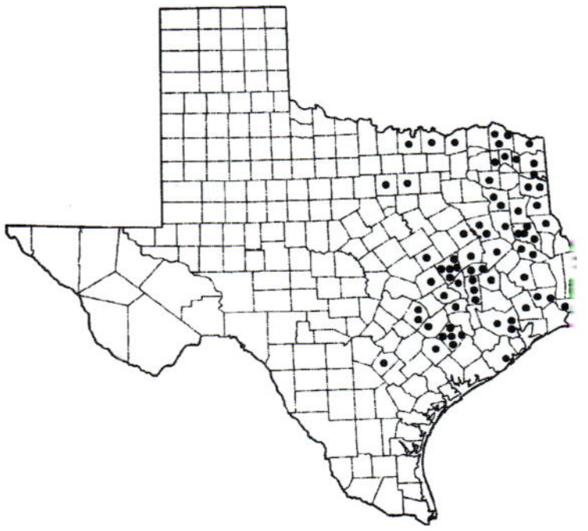

Alan Cressler

lower surface of pinnae showing sori

Alan Cressler

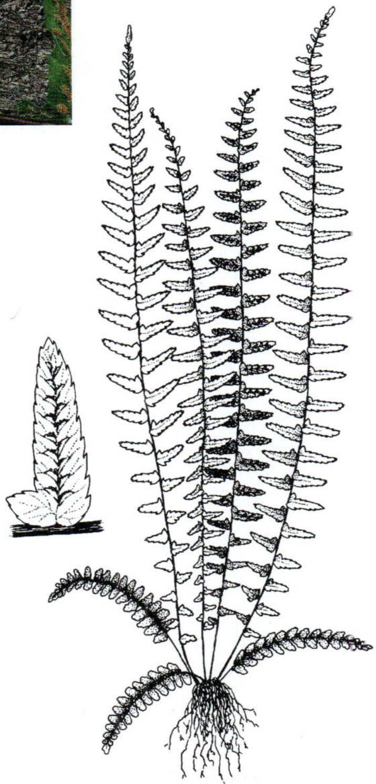

[LUN]

Asplenium platyneuron

EBONY SPLEENWORT

1. Plant definitely fern-like rather than grass-like in appearance; leaf blades with numerous pinnae per leaf, the blades oval to oblong, 1–4 times as long as wide; rhizomes not branched except sometimes branched in *A. trichomanes;* including species widespread in TX.
   2. Leaf blades with rachis elongated into a naked whip-like extension ending in a bud capable of rooting and giving rise to a new plant; species known in TX only from Jeff Davis Co. & Edwards Co. ____ **A. palmeri**
   2. Leaf blades terminating in photosynthetic tissue (leaf apex either pinnatified or with a leaflet); including species widespread in TX.
      3. Pinnae oval to broadly oblong (somewhat rounded in appearance), usually < 2 times as long as wide, the medial ones 8 mm or less long; species documented from TX only in Brewster & Jeff Davis cos. in the Trans-Pecos ____ **A. trichomanes**
      3. Pinnae usually oblong, usually 3–4 times as long as wide, the medial ones usually 10–25 mm long; species widespread in TX.
         4. Pinnae usually alternate, with their basal auricles overlapping the rachis, their margins subentire to deeply serrate or incised; plants terrestrial or growing on rocks; leaves slightly dimorphic, the fertile erect, the sterile smaller and spreading ____ **A. platyneuron**
         4. Pinnae opposite or nearly so, usually not overlapping the rachis, their margins subentire to crenulate; plants usually growing on rocks; leaves monomorphic, all fertile, erect or ascending ____ **A. resiliens**

Asplenium palmeri Maxon, (for its collector, Edward Palmer, 1831–1911, British born botanist who collected in the sw U.S., Mexico, and South America). PALMER'S SPLEENWORT. Rhizomes short-creeping, not branched; leaves monomorphic; leaf blades linear to linear-lanceolate in outline, usually recurved, with rachis elongated into a naked whip-like extension ending in a bud capable of rooting and giving rise to a new plant; petiole and rachis purplish black, shiny; pinnae opposite or nearly so, oblong, somewhat coriaceous, irregularly crenate-serrate to serrate, marginally ± inflexed, the medial ones 6–9 mm long. Dry rock crevices of granite cliffs; in TX known only from Jeff Davis Co. (*Correll 13489*, *Soxman 469*, *Warnock 21781*, TEX-LL; *Warnock 46560*, SRSC) in the Trans-Pecos and disjunct to the Edwards plateau in Edwards Co. (Pulliam Creek, Blue Hole, *Cory 37695*, GH, verified by T. Wendt); AZ and TX; also Mexico, Guatemala, and Belize (a previous report from NM is thought to be erroneous—Alexander 2006). Sporulating summer–fall. The relatively small pinnae, 6–9 mm long, distinguish this species from both *A. platyneuron* and *A. resiliens* (which typically have pinnae 10 mm or more long), while its more oblong pinnae and purplish black rachises distinguish it from *A. trichomanes* (with oval to broadly oblong pinnae and reddish brown rachises). Given its rareness and limited distribution in the state, we consider this species to be of conservation concern in TX. ⚠

Asplenium platyneuron (L.) Britton, Sterns, & Poggenb., (Greek: *platys*, broad, wide, or flat, and *neuron*, nerve, = broad-nerved, the name given by Linnaeus based on an inaccurate early drawing—Fernald 1950a; Thieret 1980; Nelson 2000), EBONY SPLEENWORT, BROWN-STEM SPLEENWORT, INDIAN-HAIR FERN. Rhizomes typically short-creeping, not branched; leaves slightly dimorphic, to 50 cm tall; leaf blades linear-lanceolate to narrowly elliptic-lanceolate or linear-oblanceolate in outline; petiole and rachis usually reddish brown to dark brown (rarely nearly black), shiny; pinnae usually alternate and oblong, usually crenate to serrate, sometimes prominently so to incised, usually with conspicuous acroscopic basal auricle which overlaps the rachis; $2n = 72$ (Wagner et al. 1993). Sandy, moist, wooded banks and slopes, terrestrial or on rocks (it is the only TX species often seen growing terrestrially rather than on rock); widespread in the e ⅓ of TX w to Parker Co. (*J. Quayle & O'Kennon 537*, BRIT) in the West Cross Timbers; se Canada and throughout e U.S. w to NE, KS, OK, and TX and disjunct to isolated areas in AZ, e CO, and e NM; also disjunct to s Africa. Sporulating Apr–Dec. [*A. platyneuron* var. *bacculum-rubrum* (Fernald) Fernald, *A. platyneuron* var. *incisum* (Howe ex Peck) B.L. Rob.] Varieties are sometimes recognized in this species (e.g., Taylor et al. 1976; Kartesz 1999). However, due to the lack of consistent variation, we are following Wagner and Johnson (1983), Wagner et al. (1993), and Yatskievych (1999) in not recognizing infraspecific taxa. This is a generalist species and is according to Correll (1956), "... one of the commonest woodland ferns occurring in eastern Texas. It may be found not only as a solitary plant but also in extensive stands." Wagner et al. (1993) noted that this species "... is remarkable in that it occurs in southern Africa as well as in North America. No other North American fern has this distribution." This disjunct distribution (a distance across the Atlantic Ocean of approximately 12,000 km or roughly 7,500 mi) is almost certainly the result of the long-distance dispersal by spores (Moran & Smith 2001; Moran 2008). Perhaps not surprisingly, Peck et al. (1990) found that this species is "well equipped" for long distance

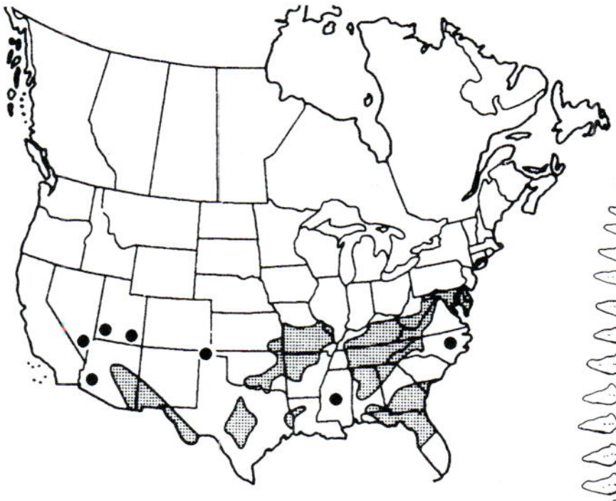

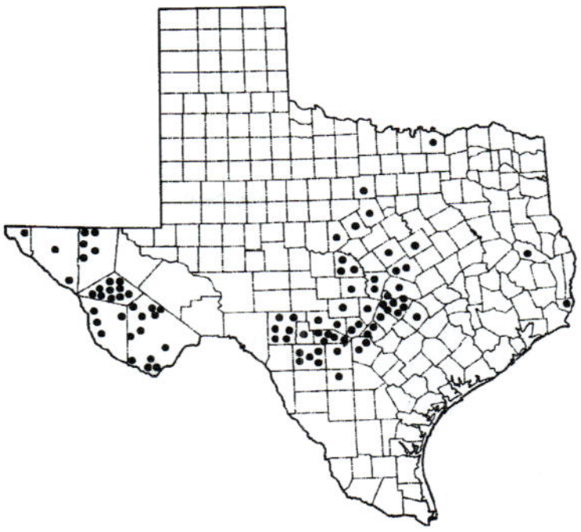

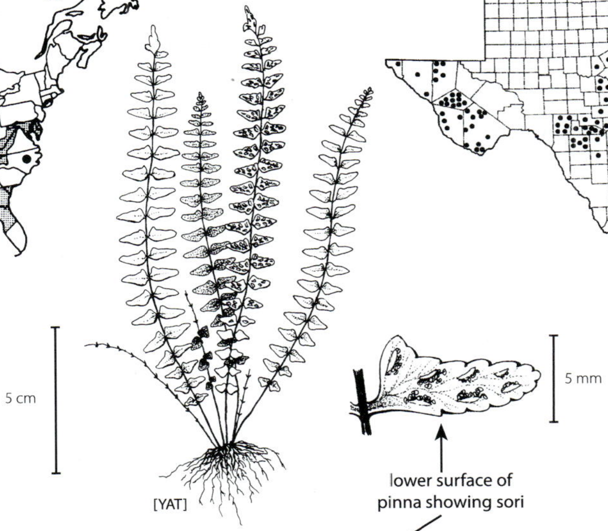

Dale A. Zimmerman Herbarium, Western N. Mex. Univ.

frond with immature sori

Dale A. Zimmerman Herbarium, Western N. Mex. Univ.

Alan Cressler

Asplenium resiliens

BLACK-STEM SPLEENWORT

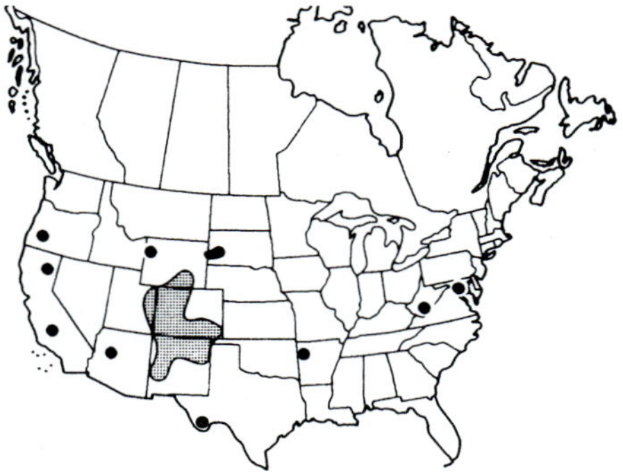

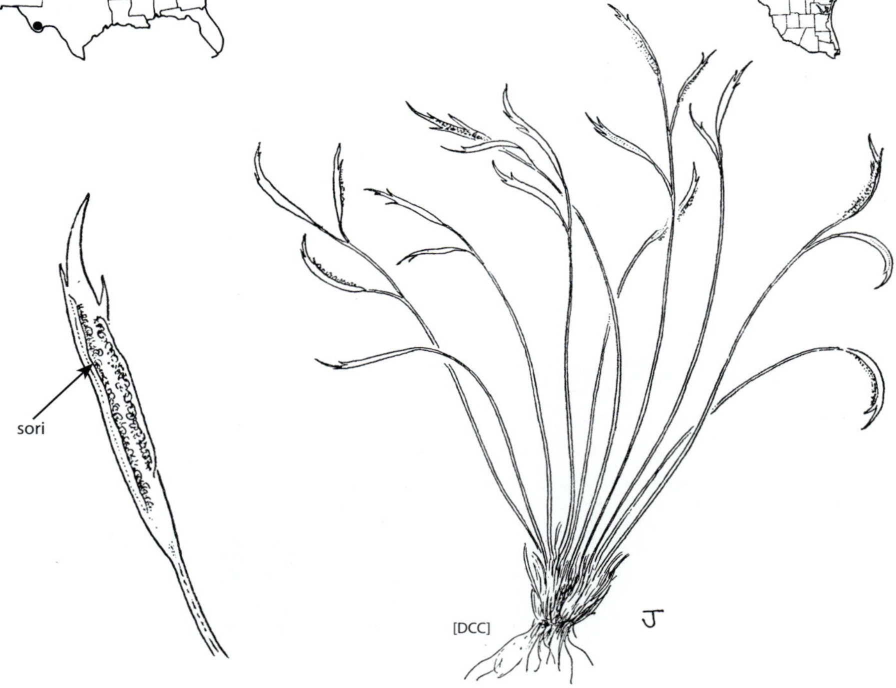

Richie Steffen, Elizabeth C. Miller Botanical Garden

Asplenium septentrionale

FORKED SPLEENWORT

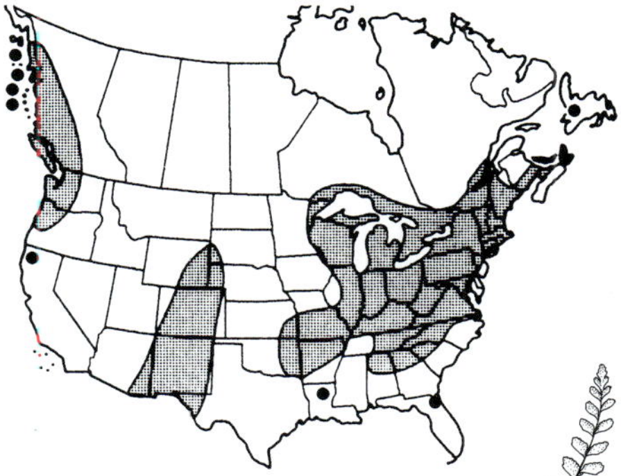

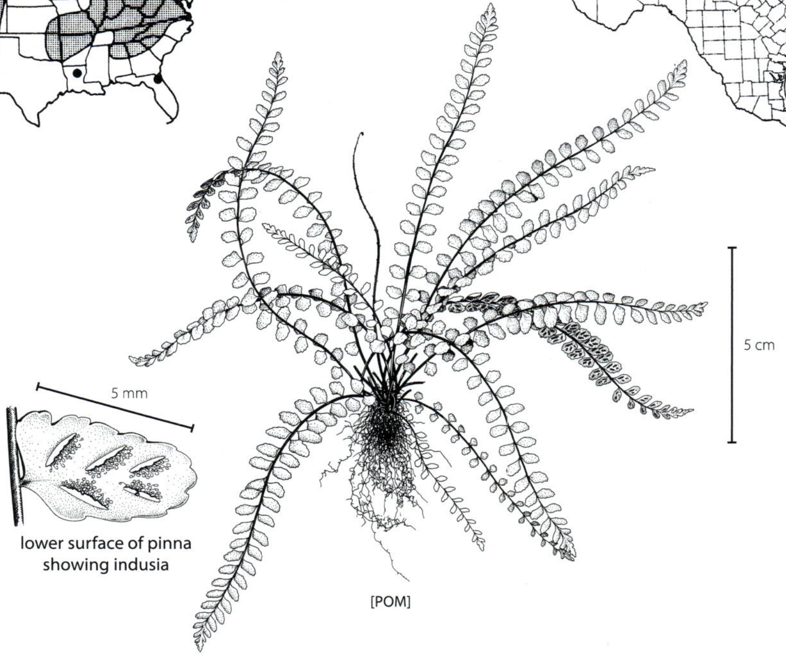

large photo: Alan Cressler; inset photo: Lorena Babcock Moore

Asplenium trichomanes subsp. **trichomanes**

MAIDENHAIR SPLEENWORT

dispersal (e.g., self-fertile gametophytes, low levels of genetic problems). The species is also unusual in possessing trophopods; these modified petiole bases accumulate food reserves and persist after the leaf blade has withered (Wagner & Johnson 1981, 1983; Nauman et al. 2000). Wagner et al. (1993) pointed out that "proliferous buds on the lowest pinnae allow formation of clumps with stems at several layers in the litter." Correll (1956) also noted that if grown in the sun under poor nutrient conditions, the plants are dwarfed, yellowish green, and have nearly entire or only slightly toothed pinnae. In the e U.S. *A. platyneuron* hybridizes with a number of other *Asplenium* species and is involved in a reticulate hybrid complex involving polyploidy (Wagner 1954).

Asplenium resiliens Kunze, (Latin: *resiliens*, jumping, springing back, or bending back, resilient, probably in reference "to the flexible petioles of this species"—Thieret 1980), BLACK-STEM SPLEENWORT, LITTLE EBONY SPLEENWORT. Rhizomes erect, not branched; leaves monomorphic, to ca. 35 cm tall, the blades linear to linear-oblong or linear-lanceolate in outline, usually more coriaceous than in *A. platyneuron*; petiole and rachis black, shiny; pinnae opposite or nearly so, subentire to crenulate, the basal auricle usually not overlapping the rachis; n = $2n$ = 108 (Wagner et al. 1993). Usually growing on rocks, typically limestone or granite; widespread in c TX and the Trans-Pecos, and disjunct n to Grayson Co. (*W.F. Mahler et al. 10102*, BRIT) and e to Orange (*W.L. Bray 62*, TEX-LL) and Angelina (Turner et al. 2003) cos.; across s ½ of U.S. from PA s to FL w to NV and AZ; also Mexico, Guatemala, and South America. Sporulating Apr–Nov. This species has an interesting distribution in TX; it is widespread in the Trans-Pecos and the Edwards Plateau (and adjacent areas) and then disjunct to 3 outlying locations. This distribution is presumably a reflection of the rarity of the necessary rocky substrates in large portions of TX. *Asplenium resiliens* is apogamous (= a type of asexual reproduction that does not involve fertilization; the sporophyte is formed directly from the gametophyte without gamete production—no swimming sperm are formed), which is often considered an adaptation for xeric environments (J. Peck, pers. comm.).

Asplenium septentrionale (L.) Hoffm., (Latin: *septentrionalis*, north or northern, in the type description Linnaeus mentioning its citation in his *Flora Lapponica* and *Flora Suecica*), FORKED SPLEENWORT, NORTHERN SPLEENWORT. Rhizomes erect, much branched; leaves usually numerous, densely tufted, monomorphic, often simple or 1-pinnate (perhaps better described as forked) with only 2–3(–5) strongly ascending pinnae, to only ca. 15 cm tall; leaf blades linear; petioles much longer than blades, dark reddish brown, becoming green distally; rachises green; pinnae linear, 3 cm or less long, 3 mm or less wide, often curved, basally and apically acute, marginally entire or with a few teeth near apex, glabrous; sori 2–3 per pinna, very elongate; $2n$ = 144 (Wagner et al. 1993). Rock crevices and ledges on noncalcareous rocks; in TX known only from one collection at 7,000 ft elevation at Boot Springs in Chisos Mts., Brewster Co. (*J.L. Blassingame 1043*, Jun 1969, HPC, OKLA, xerox SRSC; Correll & Johnston 1970); in North America primarily in the Rocky Mts. from NM n to UT and WY, with scattered localities in AZ, CA, CO, OK, OR, SD, TX, UT, and WY and disjunct e to isolated populations in AR (3000 ft., Mt. Magazine—J.H. Peck, pers. comm.), WV (Emory 1970; Bush 1986), and D.C.; also nw Mexico, Europe and Asia. ?–Jun–Jul–?. [*Acrostichum septentrionale* L.] The isolated Monroe and Hardy county WV populations represent a disjunction of about 700 mi (1,127 km) to the e of the single AR location and that AR population is more than 500 mi (800 km) e of the nearest station in w OK and about 700 mi (1,127 km) nw of the lone TX site (Emory 1970; Wagner 1972; Bush 1986; Wagner et al. 1993; J.H. Peck, pers. comm.). It is one of a number of ferns found primarily in the sw or w U.S. with isolated disjunct populations in the southern Appalachians (see discussion on page 31). This species has a widespread overall distribution in the northern hemisphere (also in Europe and Asia), possibly the result of vicariance (remnants of a past more continuous distribution that has become fragmented) or of long-distance dispersal (Holderegger & Schneller 1994; Weber 2003). FORKED SPLEENWORT resembles a tuft of grass and is thus easily overlooked in the field. Hybrids with *A. trichomanes* (*A.* ×*alternifolium* Wulfen) are known (Wagner et al. 1993). Given its rareness and limited distribution in the state, we consider this species to be of conservation concern in TX. ⚠

Asplenium trichomanes L. subsp. **trichomanes,** (Greek: *thrix*, trichos, hair, and *manes*, a kind of cup, = hairy cup, and also the ancient Greek name for a fern that may have been in some way similar; before Linnaeus botanists placed this species in the unrelated genus *Trichomanes*—Thieret 1980), MAIDENHAIR SPLEENWORT. Rhizomes short-creeping, sometimes branched; leaves monomorphic, usually to only 25 cm tall; leaf blades linear to narrowly linear-oblong in outline; petiole and rachis reddish brown or blackish brown, shiny; pinnae subopposite to alternate, oval to broadly oblong (somewhat rounded in appearance), usually < 2 times as long as wide, the medial ones 8 mm or less long, minutely crenulate to serrulate to subentire; $2n = 72$ (Wagner et al. 1993 for subsp. *trichomanes*). In sheltered crevices of cliffs and ledges on noncalcareous rocks; in TX known only from Sawtooth Mt. (*Correll 15005*, TEX-LL) and Mt. Livermore (Yarborough & Powell 2002) in the Davis Mts. of Jeff Davis Co. and also in Glass Mts. of Brewster Co. (*Warnock 19069*, SRSC; Turner et al. 2003); widespread in the e ½ of the U.S. and scattered in the w ½ of the country; also se and w Canada, n Mexico (Chihuahua), Asia, Europe, Africa, and Australia. Sporulating Mar–Oct. [*A. melanocaulon* Willd.] Correll (1956) noted that the "small roundish pinnae distinguishes this species from the other Aspleniums found in Texas." The plants in TX are subsp. *trichomanes*; three other subspecies are recognized as occurring elsewhere in the world, these sometimes differing in chromosome number and habitat; for example, subsp. *quadrivalens* D.E. Mey., which occurs in the n U.S., is tetraploid and occurs on calcareous rocks (Wagner et al. 1993). Correll's collection from Sawtooth Mt. noted that the plant was "Very rare!" Given its rareness and limited distribution in the state, we consider this species to be of conservation concern in TX. ⚠

ATHYRIACEAE Alston

LADY FERN FAMILY

A subcosmopolitan but especially temperate and montane family of ca. 5 genera and 600 species of mostly medium-sized, understory, terrestrial ferns with most of the species in two large genera, *Athyrium* and *Diplazium* (Rothfels et al. 2012b). In relatively recent treatments *Athyrium* has often been included in a very broadly interpreted Dryopteridaceae (e.g., Smith 1993b; Yatskievych 1999; Diggs et al. 2006) or in a somewhat broadly circumscribed Woodsiaceae containing *Athyrium*, *Cystopteris*, and *Woodsia* as well as a number of small genera (Smith et al. 2006, 2008; Mabberley 2008). However, recent DNA evidence (Schuettpelz & Pryer 2008; Rothfels et al. 2012a) suggests *Athyrium*, *Cystopteris*, and *Woodsia* are all phylogenetically quite remote from Dryopteridaceae, are morphologically similar to each other due to convergent evolution, are not closely related, and should be in completely different families. For further discussion and a key separating plants previously included in a more broadly conceived Woodsiaceae, see the family treatment of Woodsiaceae on page 307. (Order Polypodiales)

FAMILY RECOGNITION IN THE FIELD: sori on veins on lower leaf surfaces, elongate and straight to hooked or curved, somewhat resembling those of *Asplenium*, the indusia with elongate attachment along one side, the attachment as long as the elongate sori; leaf blades all alike, appearing much dissected, 2-pinnate-pinnatifid (rarely sub-3-pinnate), the pinnae usually short-stalked, the lower surfaces without transparent needle-like hairs (distinguishing *Athyrium* from Thelypteridaceae, which have such hairs), the ultimate leaf segments not entire; petiole bases with 2 vascular bundles (distinguishing *Athyrium* from Dryopteridaceae, which have more than 3).

REFERENCES: Wang et al. 2003, 2004; Smith et al. 2006, 2008; Schuettpelz & Pryer 2008; Rothfels et al. 2012a, 2012b.

ATHYRIUM Roth LADY FERN

A cosmopolitan genus of ca. 180–220 species, especially in e and se Asia (Kato 1993a; Mabberley 2008; Rothfels et al. 2012b). Recent molecular data (Sano et al. 2000) suggest the genus is polyphyletic. All members have trophopods, which are persistent swollen petiole bases that function in storing starch (Johnson 1986). (Greek: *athyros*, doorless; the sporangia only tardily push back the outer edge of the indusium)

REFERENCES: Butters 1917a; Benedict 1934; Liew 1972; Schneller 1979; Kato 1993a; Kelloff et al. 2002; Sciarretta et al. 2005.

Athyrium filix-femina (L.) Roth subsp. **asplenioides** (Michx.) Hultén, (sp.: Latin: *filix*, fern and *femina*, female or woman, = lady fern, in reference to the more delicate or graceful appearance of this species in comparison with the superficially similar *Dryopteris filix-mas*, MALE FERN; subsp.: resembling *Asplenium*—spleenwort, and Greek: -*eides*, -oides, like, in reference to the elongate *Asplenium*-like sori), SOUTHERN LADY FERN, LOWLAND LADY FERN, SPRING FERN. Plant terrestrial; rhizomes short-creeping; leaves monomorphic, deciduous, clustered,

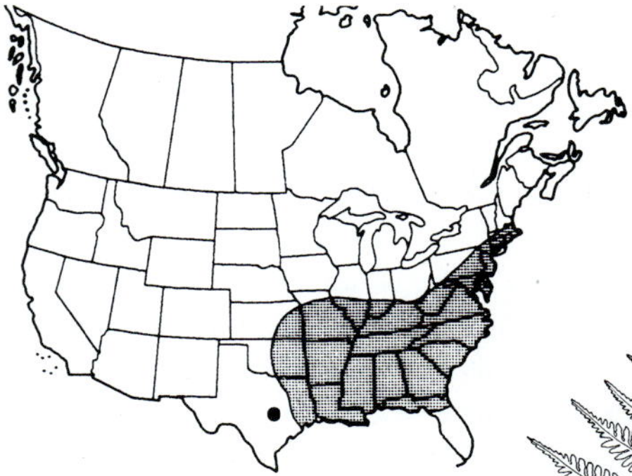

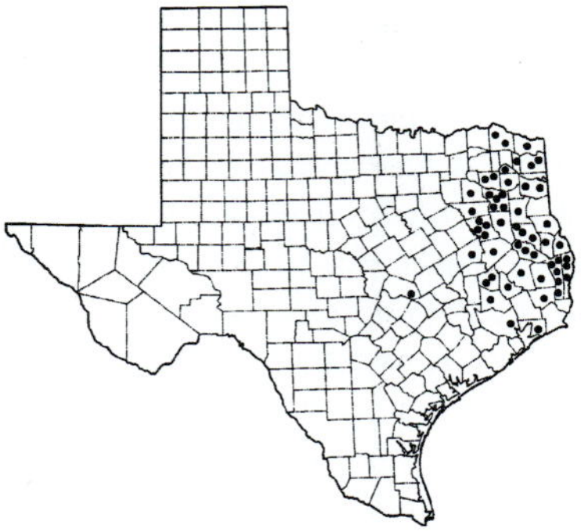

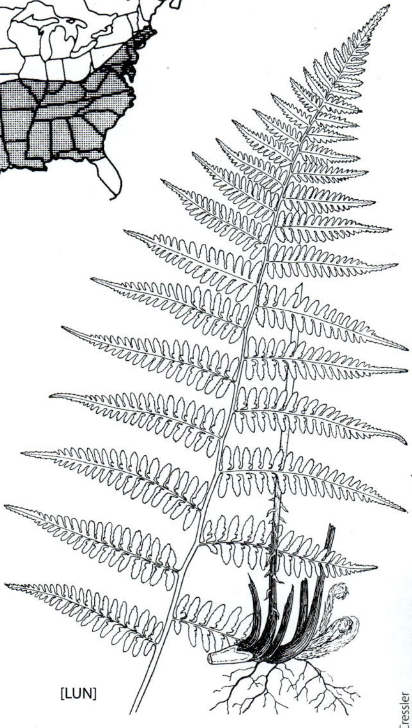

[LUN]

Alan Cressler

Alan Cressler

Athyrium filix-femina subsp. **asplenioides**

SOUTHERN LADY FERN

typically (30–)60–120 cm tall; leaf blades 2-pinnate-pinnatifid (rarely sub-3-pinnate), usually widest near base (only basal 1 or 2 pairs of pinnae shorter), the pinnae usually short-stalked, the ultimate segments (= smallest subdivisions of leaf) with veins free, simple or forked, marginally serrate; petioles greenish or straw-colored, often red- or brown-tinged, with light brown to brown scales (these often early deciduous), the petiole bases swollen (functioning as trophopods), dark reddish brown or blackish, persistent, with 2 vascular bundles; sori elongate along veins, straight to hooked or curved, somewhat resembling those of *Asplenium*, in a single row on each side of the midvein, ca. midway between midvein and margin of ultimate leaf segments; indusia membranous, with elongate attachment along one side, the attachment as long as the elongate sori, opening facing midvein; spores brown to dark brown; 2*n* = 80 (Kato 1993a). Moist woods, thickets, swamps, stream banks; Pineywoods (e.g., Bowie Co., *E. Whitehouse 20364*, BRIT), n Gulf Prairies and Marshes, and w Post Oak Savannah (Leon Co., Turner et al. 2003) and disjunct w to Williamson Co. (Correll 1956 cited *E.N. Plank s.n.*; Turner et al. 2003) in the Blackland Prairie; e U.S. from NY s to FL w to KS and TX. Sporulating May–Nov. [*A. asplenioides* (Michx.) A.A. Eaton, *A. filix-femina* var. *asplenioides* (Michx.) Farw.] Although this taxon has traditionally been treated at the varietal level (e.g., Correll 1956), we are following a number of recent authors (e.g., Yatskievych 1999; Nauman et al. 2000) in recognizing it as distinct at the subspecific level from the more northerly subsp. *angustum* (Willd.) R.T. Clausen. Even though hybrids are known between the two, Kelloff et al. (2002) noted "...substantial differentiation between the *Athyrium* taxa *angustum* and *asplenioides* and the consistent characters uniting a vast number of individuals north and south of their hybrid zone suggest that the taxa should be ranked at least at the level of subspecies. Ranking at the species level would not be inconsistent with the treatment of such taxa in the broader plant literature." It is thus possible that the two taxa will be consistently recognized as distinct species in the future. In fact, both subspecies occur in AR, and J. Peck (pers. comm.) considers them to be distinct species. Weakley (2010) also treated them at the species level. Sciarretta et al. (2005) summed it up nicely when they said subsp. *angustum* and subsp. *asplenioides* are "very recently diverged taxa" and that the *Athyrium filix-femina* complex "seems to represent a lineage 'caught in the act' of speciating." Subspecies *angustum* can be distinguished by its leaf blades usually being widest near the middle, its brown to dark brown petiole scales, and its yellow to light brown spores. This species is sometimes cultivated as an ornamental, and is said to be "one of the most dependable and often-used in fern gardening" (Nelson 2000). It is quite variable with numerous horticultural forms. The cultivars are thought to be derived from the European taxon, subsp. *filix-femina* (Hoshizaki & Moran 2001). The species as a whole occurs broadly across the n hemisphere and has a classic Arcto-Tertiary distribution (see page 33) (Kato 1993b).

AZOLLACEAE Wettst.

MOSQUITO FERN OR WATER FERN FAMILY

A cosmopolitan family of a single genus and only ca. 6 species (Lumpkin 1993; Reid et al. 2006) of tiny floating aquatics (sometimes stranded on mud) with highly reduced vegetative morphology. Such extreme vegetative reduction is often seen in free-floating aquatics (Saunders & Fowler 1992, 1993). Because of their minute size and numerous leaves, the plants do not look like other ferns, and superficially resemble mosses or liverworts. Fossil Azollaceae are known from the Cretaceous Period 89 million years ago (Collinson 1991; Archangelsky 1999), and molecular evidence also suggests a similar age for the family (Metzgar et al. 2007). It is sometimes included in the Salviniaceae (Nagalingum et al. 2008) while recognized as distinct by others (Lumpkin 1993). The Azollaceae and Salviniaceae are both composed of floating aquatics and are each other's closest relatives. However, because the *Azolla* lineage is monophyletic (Nagalignum et al. 2008), has obvious morphological distinctiveness and unique ecology (nitrogen fixation), and is of very ancient origin, we are recognizing it at the family level. This approach was described as "a perfectly acceptable alternative, given the significant differences between the two genera [*Azolla*, *Salvinia*]" (Smith et al. 2008). Three families of heterosporous (= with 2 spore types, megaspores and microspores) water ferns occur in TX (Azollaceae—*Azolla*, Marsileaceae—*Marsilea*, *Pilularia*, and Salviniaceae—*Salvinia*). These groups are all quite distinct morphologically, and traditionally the marsileaceous (*Marsilea* and *Pilularia*) and salviniaceous (*Azolla* and *Salvinia*) lines were considered to have evolved independently from different homosporous fern ancestors.

However, based on recent morphological, fossil, and molecular evidence, all three living heterosporous water fern families appear to comprise a monophyletic group with Marsileaceae most distantly related (Rothwell & Stockey 1994; Hasebe et al. 1995; Pryer 1999; Nagalingum et al. 2008). Heterospory is extremely rare within ferns—it is known from only 5 genera (all in Order Salviniales) and < 1% of living ferns (Pryer et al. 2004b; Nagalingum et al. 2008). (Order Salviniales)

FAMILY RECOGNITION IN THE FIELD: tiny, moss- or liverwort-like, free-floating or mat-forming plants that sometimes form conspicuous velvet-like, green to red mats on the surface of still, stagnant to slow-moving waters.

REFERENCES: Stergianou & Fowler 1989, 1990; Van Cat et al. 1989; Schneller 1990a; Collinson 1991; Lumpkin 1993; Archangelsky 1999; Nagalingum et al. 2006, 2008; Metzgar et al. 2007.

AZOLLA Lam. WATER FERN, MOSQUITO FERN, FAIRY-MOSS

Plants very small (whole plant usually from < 1–3(–5) cm long), aquatic, free-floating or mat-forming, superficially resembling some mosses or liverworts, monoecious; stems branching, usually prostrate; leaves minute, sessile, in 2 ranks on upper side of stem, numerous (roughly 100–200 per plant), often somewhat imbricate, deeply bilobed, the upper/dorsal lobe photosynthetic, deep green to reddish (under stress), the lower lobe typically colorless and translucent; sporocarps (which are specialized sori enclosed by modified indusia), in pairs, the members of a pair of same or opposite sex, arising from lower lobe of first leaf of a lateral branch, of two kinds, the megasporocarps with 1 megasporangium producing 1 megaspore, the larger microsporocarps with numerous microsporangia containing numerous microspores, the microspores aggregated in clusters or masses (= massulae), these covered with barbed hairs (= glochidia, with which the free-floating massulae attach to megaspores); megaspores spherical, covered by a perispore layer (which is sometimes distinctively marked, pitted, etc.) and topped with triangular cap consisting of a collar (involved in attaching the floats), 3 (in TX species) floats, and a blue-green bacterial colony covered by remnants of the indusium (the whole unit collectively referred to as the megaspore apparatus or complex).

Azolla species are considered the world's smallest ferns—whole plants are often only the diameter of a dime and the leaves are ca. 1/32 of an inch long (ca. 1 mm) (Moran 1997, 2004). Buoyancy is achieved by the extreme reduction in size and also by minute pockets of air that become trapped around leaf hairs (which are found in all known *Azolla* species) (Saunders & Fowler 1993). The upper emergent leaf lobes are hollow and inhabited by a mutualistically symbiotic, nitrogen-fixing, filamentous cyanobacterium (= blue-green bacterium), *Anabaena azollae* Strasb., that combines, or "fixes" atmospheric nitrogen with hydrogen to make ammonia, which can be utilized by the ferns (Moore 1969; Shi & Hall 1988; Hoshizaki & Moran 2001; van Hove & Lejune 2002). Because of the resulting high usable nitrogen content, *Azolla* species have been widely used agriculturally as a green manure fertilizer, and are economically valuable. "Cultivating *Azolla* as an organic fertilizer for rice has been a centuries-old practice in China and Vietnam ..." (Moran 1997), dating back perhaps to the 11th century or even before. Wagner (1997) reviewed the extensive literature on the uses of *Azolla* (e.g., biofertilizer, animal feed, human food, medicine, water purifier). Because of these uses, particularly as a biofertilizer, *Azolla* is considered by some to be the world's most economically important fern (Moran 1997). It is also reported to be "the most frequently studied genus of ferns in the world because of its economic importance" (Lumpkin 1993). *Azolla* species, presumably in part due to the presence of the symbionts, have the capacity to grow at remarkable rates, under ideal conditions doubling in biomass in as little as two days (Wagner 1997); it is thus one of the world's fastest growing plants. The symbiosis is permanent—it is automatically transmitted from generation to generation (via both asexual and sexual reproduction) and there are no confirmed reports of cultivation of the symbionts without the host *Azolla* (van Hove & Lejune 2002; Papaefthimiou et al. 2008). In fact, according to Perkins and Peters (1993) the symbiosis is "unique among plant-prokaryote nitrogen-fixing symbioses in that the continuity of the partners is retained through the life-cycle." Jill Reid (pers. comm.) speculates that this is "an example of catching an organelle in the act of evolving" and that the obligately symbiotic *Anabaena* are "incipient organelles." Thieret (1980) pointed out that *Azolla* is excellent material to use in science labs to demonstrate blue-green bacteria: "simply place a few *Azolla* leaves in a drop of water on a glass slide, crush them with a dissecting needle or a forceps, put a cover slip on them, and tap the slip with the eraser end of a pencil to spread apart and flatten the crushed leaf material. Examination at 100X will reveal the ... filaments—each like a section of a bead necklace—with their characteristic blue-green color." Sometimes one or more "beads" will be larger than the others—these are the heterocysts, the specialized nitrogen-fixing cells.

Azolla is extremely well known from the fossil record, with more than 50 extinct species known dating back to 89 mya during the Cretaceous Period (Collinson 1991; Metzgar et al. 2007). As in the case of a number of other fern genera (e.g., *Onoclea, Osmunda, Woodwardia*), during parts of the Tertiary, *Azolla* is known to have had a circum-Arctic distribution at high latitudes. For example, Brinkhuis et al. (2006) described extensive *Azolla* fossils from the Arctic approximately 50 million years ago, indicating a much warmer period in the Earth's history—in fact, at that time the Arctic is thought to have been experiencing relatively warm conditions with ocean surface temperatures ~10° C (50° F) (Brinkhuis et al. 2006). This information corresponds well with the large body of evidence for a widespread primarily temperate (but with some tropical elements) Arcto-Tertiary or Boreotropical flora whose influences can still be seen in modern floras, including that of TX (see page 33). This flora was widely distributed across a northern hemisphere that at the time was both much warmer and had extensive land connections between the continents.

According to Yatskievych and Windham (2008a), the "megaspores [and associated structures] of *Azolla* species are structurally the most complex of any fern." This complexity, including special floats, is thought to be adaptive in bringing together the megaspores and microspores and increasing the chances of fertilization. The floats help disperse the megaspores, and the glochidia of the microspores become entangled with hairs on the megaspores, ensuring proximity when the spores germinate. Nonetheless, spores are rarely produced, and reproduction is typically asexual by fragmentation with dispersal via the movement of water, waterfowl, or other vectors (Yatskievych & Windham 2008a). Their small size, vegetative plasticity, and microscopic characters (e.g., megaspores) can make species of *Azolla* extremely hard to identify (Zimmerman et al. 1989; Stergianou & Fowler 1990; Hoshizaki & Moran 2001). In fact, there has long been disagreement over species concepts within *Azolla*; Reid et al. (2006) noted "most taxonomic systems focus primarily on reproductive characters of *Azolla*, but these structures are rarely present in nature and are often difficult to induce in culture." A recent multiple gene DNA sequence study (Reid et al. 2006 based on 2 chloroplast sequences and 1 nuclear) from all species in the genus seems to have clarified the situation, supporting the recognition of six species, including the 3 treated here for TX. A subsequent molecular study by Metzgar et al. (2007) reached the same conclusion. Two of the 3 TX species (*A. caroliniana* and *A. microphylla*) are extremely similar and frequently confused; further research is needed to clarify their ranges within the state. It should be noted that definitive identification of TX species depends on microscopic examination of reproductive material (which is rare) under magnification, and that the vegetative characters in the key to species, although useful, should be considered as "trends." Jill Reid (pers. comm.), who extensively studied *Azolla*, agrees, noting that the identification of *A. caroliniana* and *A. microphylla* (and even *A. filiculoides*) "seems to be impossible without reproductive structures." A recent molecular study of the *Anabaena azollae* symbiont strains corresponds roughly to the results of the multiple gene DNA sequence study of the host *Azolla* species (Papaefthimiou et al. 2008). The following key to species is adapted from those in Lellinger (1985), Lumpkin (1993), and Mickel and Smith (2004), with nomenclature following the recommendations of Reid et al. (2006). The common name, MOSQUITO FERN, is said to be "due to the belief that its dense covering on the surfaces of ponds retards or prevents the growth of mosquitoes" (Nelson 2000). (Greek: *azo*, to dry, and *ollyo*, to kill, alluding to death from drought, in reference to its dependence on water)

REFERENCES: Svenson 1944, Correll 1956, 1966; Moore 1969; Fowler 1975; Collinson 1980; Lumpkin & Plucknett 1980; Bates & Brown 1981; Calvert et al. 1983; Perkins et al. 1985; Stergianou & Fowler 1989, 1990; Zimmerman et al. 1989, 1991; Johnson 1990; Saunders & Fowler 1992, 1993; Van Coppenolle et al. 1993; Moran 1997; Wagner (G.M.) 1997; Dickinson & Miller 1998; van Hove & Lejune 2002; Evrard & van Hove 2004; Brinkhuis et al. 2006; Reid et al. 2006; Metzgar et al. 2007; Papaefthimiou et al. 2008; Collinson et al. 2009; Darwin Azolla Project 2009.

1. Largest hairs on upper leaf lobes strictly unicellular; plants usually (0.7–)2–5 cm long; megaspores with raised hexagonal markings/angular bumps; species rare if currently present in TX, reported only from the Trans-Pecos ____ **A. filiculoides**
1. Largest hairs on upper leaf lobes with 2 or 3 cells; plants usually 0.5–1.5 cm long; megaspores without raised hexagonal markings/angular bumps, sometimes pitted or not so; species widespread in TX.

2. Plants usually 0.5–1 cm long, the stems pseudodichotomously branched nearly throughout; megaspores not pitted, densely and uniformly covered with tangled filaments (high magnification required; also note: plants rarely fertile); leaves not or scarcely imbricate, the upper lobes usually 0.5–0.6 mm long _____ **A. caroliniana**
2. Plants usually 1–1.5 cm long, the stems ± pinnately branched from a central axis, pseudodichotomously branched mainly at periphery; megaspores pitted, sparsely covered with a few long filaments; leaves slightly to greatly imbricate, the upper lobes usually at least 0.7 mm long _____ **A. microphylla**

Azolla caroliniana Willd., (of Carolina, a region which historically included much of the se U.S.; early botanists gave many plants this epithet; in this case the type description says "Habitat in aquis Carolinae"), MOSQUITO FERN, WATER FERN, CAROLINA MOSQUITO FERN, EASTERN MOSQUITO FERN. Plant ranging from green to with red margins or dark red; largest hairs on upper leaf lobe with 2 or 3 cells, the broad basal cell often ½ or more height of hair, the short apical cell curved, with tip nearly parallel to leaf surface; megaspores without bumps or pits, but with dense covering of tangled filaments, the megaspores, however, usually not present since plants are rarely fertile; $2n$ = usually 44, rarely 66 (Stergianou & Fowler 1990). Quiet water of ponds, lakes, or slow-moving streams or stranded on mud; Austin (*S.G. Jones 7807*), Brazos, Cherokee, Colorado, Dallas, Franklin, Harrison, Hays, Liberty, Madison, Montgomery (*A.K. Neill 1997*), Newton, Nueces, Tarrant, Travis cos. (all BRIT, specimens annotated by J. Peck), and Jim Wells (*M.C. Johnston 542139*, TEX-LL, identified V.M. Bates using SEM) cos.; sporadically but widely scattered in e ½ of TX; B.C., Ont., and e U.S. from NH s to FL w to SD and TX. Sporulating rarely summer–fall (Fertile specimens are rarely collected—Nauman et al. 2000). This species has long been considered to be the only *Azolla* in the e portion of TX, and it and *A. microphylla* are often nearly impossible to distinguish when sterile. Therefore, numerous herbarium and literature records of *A. caroliniana* may be the result of confusion with *A. microphylla*—these records of uncertain identity are mapped as open circles on the county distribution map; black dots on the county map indicate specimens annotated as *A. caroliniana* by J. Peck (pteridologist with expertise in *Azolla*) or by SEM identification (see further discussion under *A. microphylla*). Although there is "ample morphological and molecular evidence" to recognize *A. caroliniana* as a separate species [from *A. microphylla*], there is a nomenclatural problem—the type specimen is sterile and is thought by some to actually be of a different species (Reid et al. 2006). As a result, the species currently going by the name *A. caroliniana* "may require a nomenclatural change pending the satisfactory identification of the current type." Scanning electron micrographs of the megaspore apparatus of this species and the related *A. microphylla* (as *A. mexicana*) can be seen in Perkins et al. (1985) and Lumpkin (1993). Where found, *A. caroliniana* is often abundant, and huge numbers of individuals can at certain times of the summer turn the surface of ponds a striking red color (G. Diggs, pers. obs.). According to Correll (1956), "Its occurrence in remote locations is doubtless due to its dissemination, in part, by water-fowl." The species is commonly cultivated as an aquatic ornamental in aquaria and decorative pools (Lellinger 1985). It is often found growing unintentionally at nurseries and water garden stores where aquatic plants (e.g., water-lilies) are sold in TX; it grows easily in decorative "ponds" or water gardens in TX and overwinters without problems in such situations (e.g., Grayson Co. near the TX-OK border, G. Diggs, pers. obs.) even in temperatures reaching 10° F (-12° C).

Azolla filiculoides Lam., (Latin: *filicula*, diminutive of *filix*, fern, and Greek: *-eides*, -oides, like or having the form of, the type description saying it has the aspect of a very small fern), WESTERN MOSQUITO FERN. Plant ranging from green or yellowish green to with red margins or dark red; stems ± pinnately branched from a central axis, pseudodichotomously branched only at periphery, prostrate and 1–3 cm long when immature, becoming nearly erect when mature and crowded and sometimes to 5 cm or more long; largest hairs on upper leaf lobe with only 1 cell; megaspores with raised hexagonal markings/angular bumps (papillae), with sparse to moderate tangled filaments; $2n$ = usually 44, rarely 66 (Stergianou & Fowler 1990). Quiet water; Jeff Davis Co. (*Correll 15559*, TEX-LL, annotated by V.M. Bates and T. Reeves); Johnston (1990) included *A. filiculoides* for TX pointing out that "Jeff Davis Co. material was referred to this species by Reeves and by Vernon M. Bates Jr. in 1980." It should be noted that Bates studied *Azolla* and also published a report of *A. filiculoides* from Georgia based on SEM examination of megaspores (e.g., Bates & Browne 1981). B.C. in Canada s through w U.S. from WA to OR, CA, and AZ to extreme w TX; also reported for GA and NY (Bates & Browne 1981), the last two locations possibly the result of artificial introductions; native from w North America (includ-

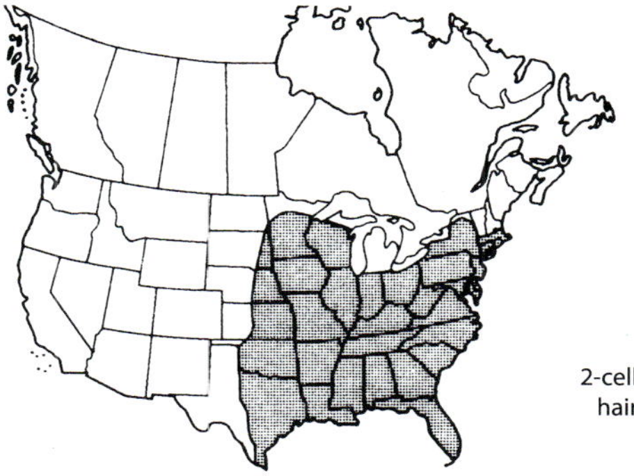

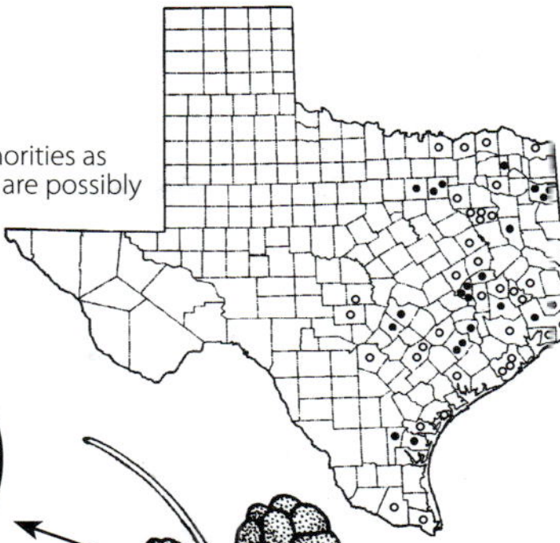

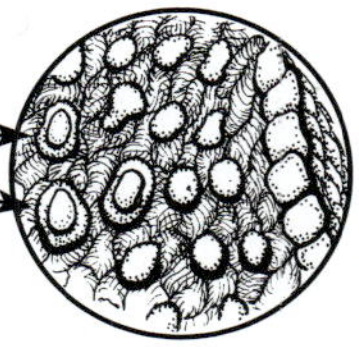

[FNA]

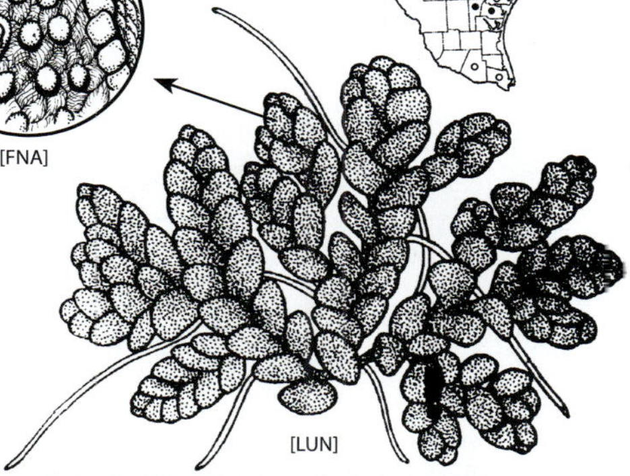
[LUN]

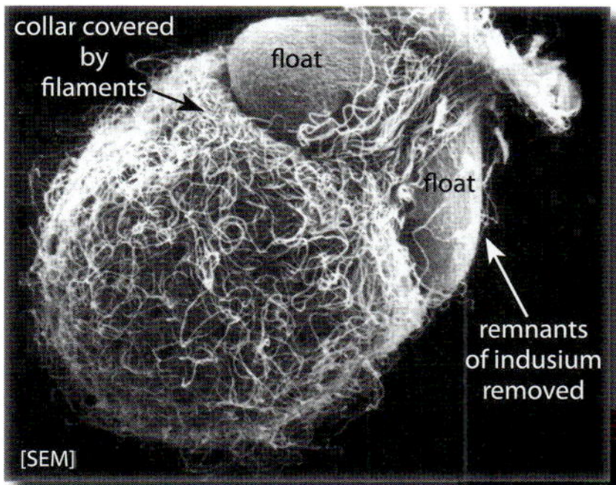

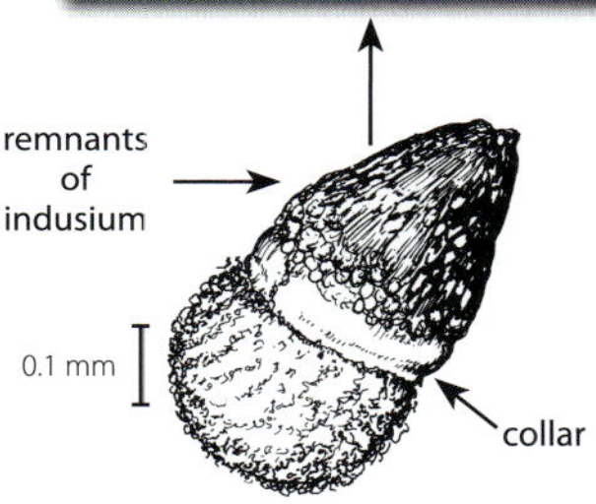

megaspore apparatus
[FNA]

Alan Cressler

pond covered with red
MOSQUITO FERN

Azolla caroliniana

MOSQUITO FERN

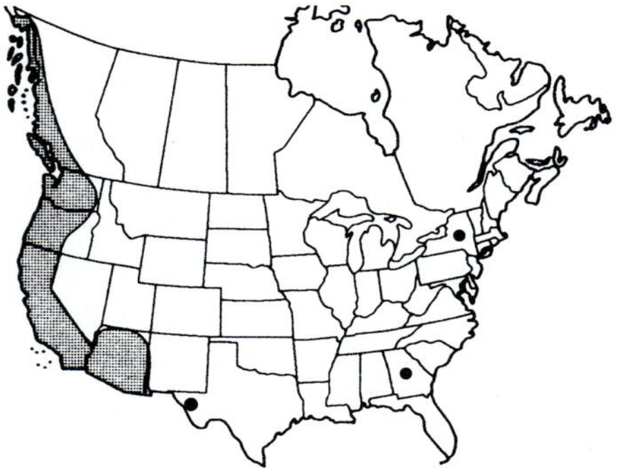

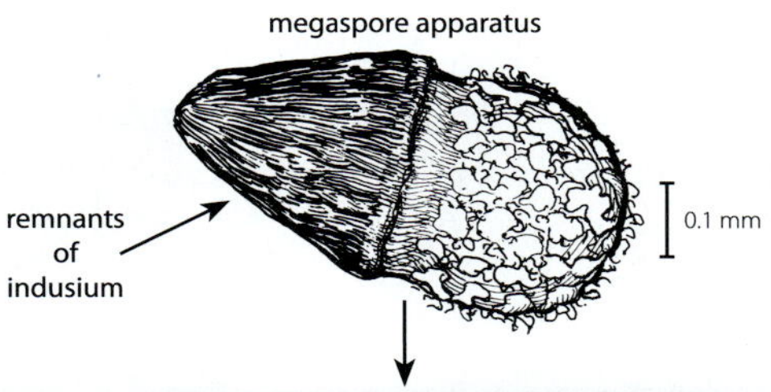

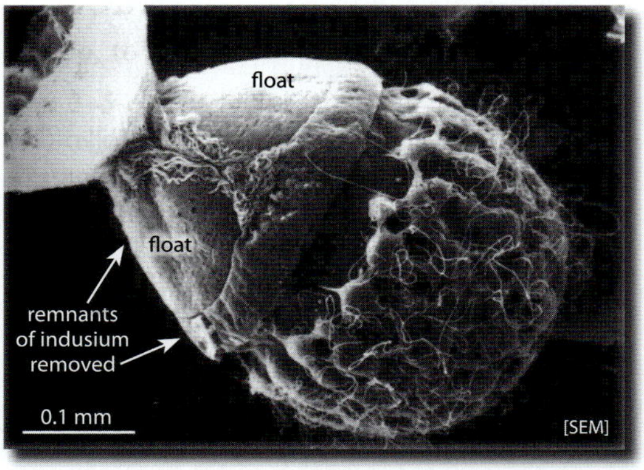

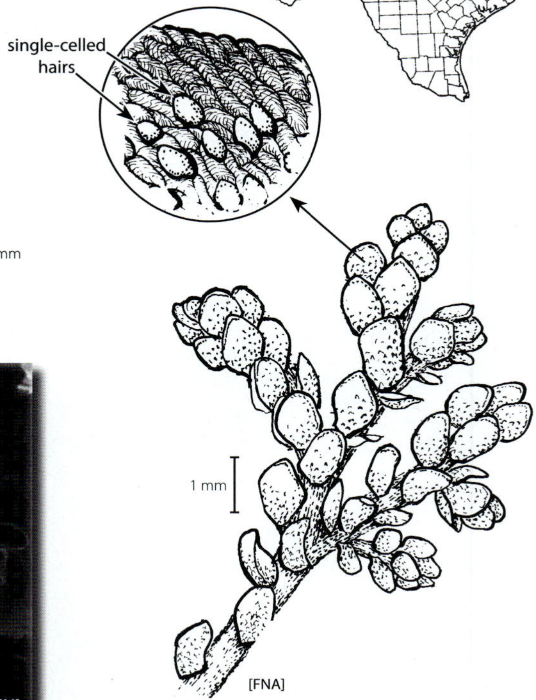

Azolla filiculoides

WESTERN MOSQUITO FERN

ing Mexico) to South America, but widely naturalized in the Old World. Sporulating usually in late spring. Yatskievych and Windham (2008a) noted that "Populations of *A. filiculoides* in Arizona are uncommon and sporadic. Waterfowl move the plants around and eventually it may be discovered at other sites in the state." Given the relative proximity of Jeff Davis Co. to AZ and the fact that *Azolla filiculoides* is transported by waterfowl, it is not surprising that a collection of the predominantly western *A. filiculoides* was made in one of the westernmost counties of TX; collectors should be on the lookout for this species. Velaspino (1993) noted that this species "is cold tolerant, surviving even in fragmented parts under thin ice. It usually reaches a climax population in late spring, becomes fertile, collapses, and is replaced by other more heat-tolerant aquatics such as *Lemna* spp." WESTERN MOSQUITO FERN was introduced into the United Kingdom by at least the late 1800s and became a problematic invasive; the frond-eating weevil, *Stenopelmus rufinasus*, was imported from FL in 1995 and has had "remarkable success" in reducing populations in the British Isles (Weed-Killer.net 2010). Since this species is known from only one collection in TX, we consider it to be of conservation concern in the state. ⚠

Azolla microphylla Kaulf., (Greek: *mikros*, micro, small or little, and *phyllon*, leaf, = small-leaved), MEXICAN MOSQUITO FERN. Similar to *A. caroliniana* and distinguished as in the key above; frequently fertile (Lumpkin 1993); $2n$ = usually 44, rarely 66 (Stergianou & Fowler 1990). Quiet waters; Bowie, Cameron, El Paso, Tyler (BRIT, specimens annotated by J. Peck), Jeff Davis (*W. Hess 3201*, BRIT), Jim Wells (BRIT), Presidio (*Warnock & L.C. Hinckley 46865*, TEX-LL, annotated as *A. mexicana* by T. Reeves), and Brewster (Turner et al. 2003) cos.; B.C. and widespread in c and w U.S. (occasionally introduced in the e U.S.—Lumpkin 1993); also Mexico to South America. Sporulating summer–fall. [*Azolla mexicana* Schlect. & Cham. ex C. Presl] Recent DNA sequence data based on multiple genes support the biological reality of this species as separate from *A. caroliniana* but show that it is conspecific with individuals that had been going under the name *A. mexicana* (a more recent name now placed in synonymy under *A. microphylla*) (Reid et al. 2006; Metzgar et al. 2008). Evrard and Van Hove (2004) raised the possibility that the type specimen of *A. microphylla* is actually of another species—if this turns out to be the case, a nomenclatural change will be necessary. Within the state this species (as *A. mexicana*) has in the past usually been considered to be restricted to far w TX; for example, Turner et al. (2003) mapped it (as *A. mexicana*) in TX only in the Trans-Pecos. However, there has long been confusion between it and the morphologically very similar *A. caroliniana* (J. Peck, pers. comm.) (see more discussion under that species). Peck and Taylor (1995) showed *A. microphylla* (as *A. mexicana*) nearly throughout AR, including adjacent to the extreme ne border of TX, and Lumpkin (1993) mapped it in OK immediately adjacent to the ne border of TX. More recently, Peck (2011a) treated all material occurring in AR as *A. microphylla*. Because of the difficulty in distinguishing it from *A. caroliniana*, the species is probably much more widespread in TX than previously thought. Lumpkin (1993) also stressed that because of inconsistent characters used in the past to identify *Azolla* species, "Literature that attributes a particular species of *Azolla* to a particular state or province must be questioned. ..." Further work thus needs to be done to clarify the distribution of both *A. microphylla* and *A. caroliniana* in TX. *Azolla microphylla* (as *A. mexicana*) was erroneously reported to be "generally less cold tolerant" and to have "a narrower environmental range than *A. caroliniana*" (Lumpkin 1993). However, it occurs in the upper Mississippi River northward to Minneapolis-St. Paul, MN (Peck 1982) and also to the Canadian province of British Columbia. Van Cat et al. (1989) reported hybridizing this species with *A. filiculoides*. Jill Reid (pers. comm.) has suggested that *A. microphylla* may have arisen from a hybridization event between *A. caroliniana* and *A. filiculoides*. She came to that conclusion based mainly on the studies of the reproductive structures—the megasporocarp of *A. caroliniana* is distinctive and non-variable (likewise with *A. filiculoides*), but the megasporocarp of *A. microphylla* is variable, with overlapping morphology of both *A. caroliniana* and *A. filiculoides*. The species is occasionally cultivated (Lellinger 1985).

Azolla pinnata R. Br. (Latin: *pinnatus*, feathered, winged, plumed, in reference to the pinnately arranged branches), FEATHERED MOSQUITO FERN, WATER VELVET, is a widespread Old World native that is listed as a U.S. federal noxious weed (USDA Natural Resources Conservation Service 2010), and as such is prohibited in the U.S. by federal law. There is concern it could be invasive. Yatskievych and Windham (2008a) noted that it "appears sporadically as a contaminant in commercially grown aquatic plants imported from other regions." This

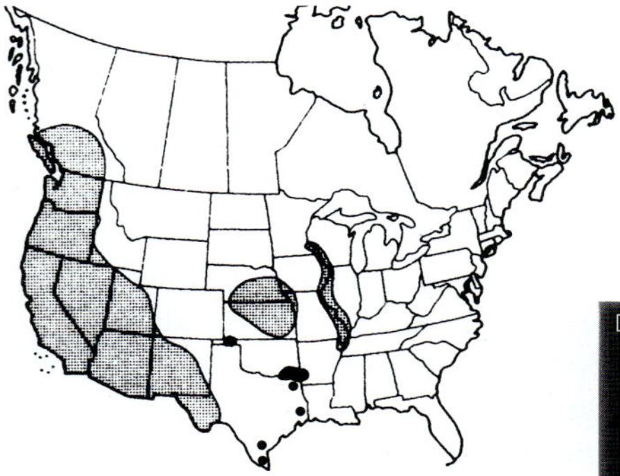

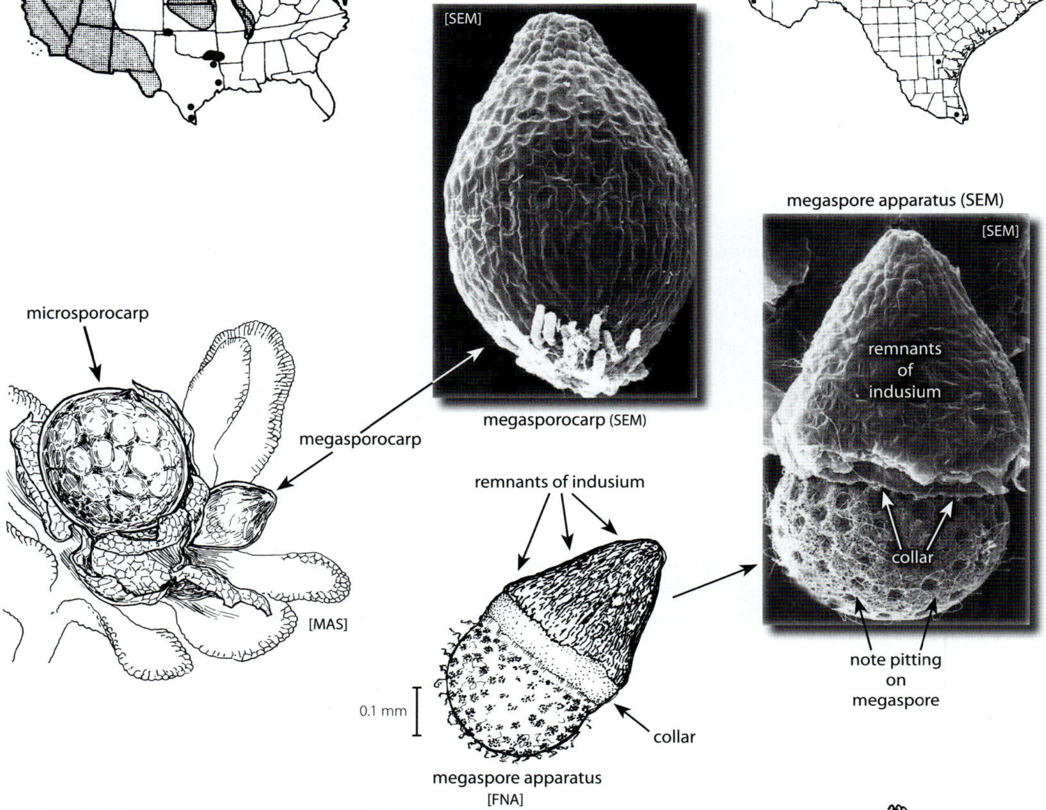

Dale A. Zimmerman Herbarium, Western N. Mex. Univ.

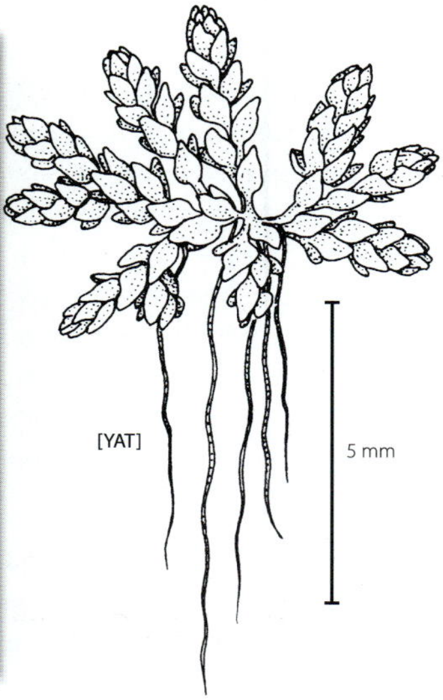

Azolla microphylla

MEXICAN MOSQUITO FERN

species, currently unknown from TX, can be distinguished from all TX species by its 9 megaspore floats (vs. 3 in all TX species). It also has pubescent stems with obviously pinnately arranged branches, the longer basal ones giving the plant a decidedly triangular appearance (vs. glabrous stems at least somewhat dichotomously branched and the plant much less triangular in all TX species). Given the large number of other invasive exotic aquatics that have been introduced into TX, this species is included here as a note to alert collectors and others to be aware of it as a possibility. In southeastern Asia it has been used for centuries as a natural fertilizer in the cultivation of rice (Lumpkin 1993).

BLECHNACEAE Newman

CHAIN FERN, DEER FERN, OR MIDSORUS FERN FAMILY

Terrestrial or in *Blechnum* on rocks; leaves monomorphic or dimorphic; leaf blades 1-pinnate to 1-pinnatifid; sori linear-oblong in chain-like rows or elongate and continuous, parallel to and adjacent to midribs, with linear indusia attached by outer margin and opening towards midrib.

A monophyletic family of ca. 9–10 genera and ca. 200–250 species (Cranfill 1993b; Mickel & Smith 2004; Smith et al. 2008) including some small tree-fern-like species and climbers; it is mostly tropical and s temperate except for the n temperate *Woodwardia*. The family is apparently most closely related to Onocleaceae (Smith et al. 2008). (Order Polypodiales)

FAMILY RECOGNITION IN THE FIELD: sori either discrete, linear-oblong, in a chain-like row along each side of the midvein of a pinna or pinnule OR elongate and continuous along the length of the midveins of the pinna or pinnule; indusia attached by their outer margin, opening toward midvein.

REFERENCES: Kramer et al. 1990a; Cranfill 1993b, 2001; Cranfill & Kato 2003.

1. Sori elongate and continuous along the length of the midveins of the pinnae or pinnules; leaf blades usually 1-pinnate near base, 1-pinnate-pinnatifid distally; veins of sterile leaves free ____ **Blechnum**
1. Sori linear-oblong and distinct from each other, numerous in a single chain-like row along each side of the midveins of the pinnae or pinnules; leaf blades entirely 1-pinnate-pinnatifid; veins of sterile leaves anastomosing into rows of areoles (= small vein-enclosed areas) between midvein and margin ____ **Woodwardia**

BLECHNUM L. DEER FERN

A genus of 180–ca. 220 species, primarily tropical in distribution but with a few n temperate and a few s temperate; it is especially well represented in the s hemisphere; hybrids between species are numerous. The genus includes epiphytic, hemiepiphytic, and terrestrial species with the rhizomes of some forming trunks; many species have the petioles and rachises grooved on upper surface (Nauman 1993e; Mickel & Smith 2004; Mabberley 2009). As presently defined, the genus is paraphyletic, with a number of other genera apparently embedded within it, and it is in need of further study (Cranfill 2001; Schuettpelz & Pryer 2008). Hybrids between species of *Blechnum* are reported to be "numerous" (Mickel & Smith 2004). Within the family, it is related to *Woodwardia* (Mickel & Smith 2004). The young fronds of a number of species are edible or used medicinally; some are cultivated as ornamentals (Mabberley 2009). (Greek: *blechnon*, ancient name for ferns in general)

REFERENCES: Seigler & Lockwood 1975; Nauman 1993e; Mickel & Smith 2004.

Blechnum appendiculatum Willd., (Latin: *appendicula*, diminutive of *appendix*, appendage or addition, and *-atus*, indicating possession, the pinnae sometimes slightly auriculate-appendaged basally), HAMMOCK FERN, NEW WORLD MIDSORUS FERN. Rhizomes elongate, slender, branched; leaves ± monomorphic, usually clumped; petiole grooved on upper (= adaxial) surface, straw-colored to light brown; leaf blades usually 1-pinnate at base, 1-pinnate-pinnatifid distally, sometimes mostly 1-pinnatifid, lanceolate, to ca. 23 cm long, the veins of the sterile leaves free; rachis sparsely or usually densely hairy, the hairs ± gland-tipped; pinnae with margins ± entire, the larger ones curved, to ca. 4(–8) cm long; sori elongate and continuous along most of the length of the midveins of the pinnae or pinnules; $2n = 124$ (tetraploid; Mickel & Smith 2004). In TX in granite crevices (apparently protected from sun for most of year), in other parts of the range terrestrial in forests or among or over rocks; in the state known only from Llano Co. (Enchanted Rock—*Seigler & Lockwood DS-8932*, *J.L. Blassingame 4060*, BRIT, TEX-LL; *C. Ellis et al. 4541*; *J.L. Blassingame & W.D. Pohl 3912*, BRIT; *P. Douglas*

detail of rachis

Harry Guiremand

lower surface of pinnae w/linear sori

Forest & Kim Starr, Creative Commons-BY

Blechnum appendiculatum

HAMMOCK FERN

486, HPC); FL, GA LA, and TX; also Mexico and West Indies to South America. Sporulating Dec–Mar–?. [*B. occidentale* L. var. *minor* Hook., *B. occidentale* forma *pubirachis* (Rosenst.) Lellinger, *B. occidentale* var. *pubirachis* Rosenst.] This is predominantly a tropical species with Seigler and Lockwood (1975) noting, "All of the United States populations probably have arisen from spores blown by the wind from Mexico or the West Indies." It is closely related to and has long been treated as an infraspecific taxon within *B. occidentale* (e.g., Nauman 1993e). That species is not known from the U.S., and differs in having glabrous rachises, larger leaves, and in the tropics, generally growing at lower elevations (Lellinger 1985; Mickel & Smith 2004). Mickel and Smith (2004) noted that morphological differences between *B. appendiculatum* and *B. occidentale* (sensu stricto) "appear correlated with differences in elevation and pubescence, and there are relatively few specimens that are difficult to place in one species or the other." We are therefore following Mickel and Smith (2004) in recognizing this taxon at the specific level. *Blechnum appendiculatum* (as *B. occidentale* var. *minor*) is reported to reproduce extensively by stolons as well as spores (Nauman 1993e). The extremely elongate sori (along most of length of pinnule midvein) is quite distinctive among TX ferns. Given its rareness and limited distribution in the state, we consider this species to be of conservation concern in TX. ⚠

WOODWARDIA Sm. CHAIN FERN

Plants terrestrial; rhizomes in TX species long-creeping, with leaves scattered along the rhizomes; leaves monomorphic or dimorphic, deciduous, the blades 1-pinnate or 1-pinnatifid, the veins anastomosing, forming a regular series of areoles (= small vein-enclosed areas) along costae and costules; sori discrete, linear-oblong, in a single chain-like row along each side of the midvein; indusia attached by their outer margin, opening on side next to midvein, often obscured by dehisced (= opened) sporangia.

A genus of 14 species of North America, Central America, Mediterranean Europe, and e Asia (Cranfill 1993b); the species are distributed in the n hemisphere in a classic Arcto-Tertiary pattern (Cranfill & Kato 2003) (see page 33). Such a modern distribution is probably the result of vicariance (see pages 30, 32) given that fossils of *Woodwardia* are found "throughout the Tertiary of Europe, North America, and Asia" (Pigg & Rothwell 2001). The genus is known in the fossil record beginning in the late Cretaceous and is widespread in the Tertiary (Collinson 2001; Pigg et al. 2006). Although *Woodwardia* (in the broad sense) is considered monophyletic, some authors (e.g., Cranfill & Kato 2003) argue for dividing it into three genera: the monotypic *Anchistea* (for *W. virginica*); *Lorinseria* (also monotypic, for *W. areolata*); and *Woodwardia* (for the remaining species). Indeed, molecular data (Cranfill & Kato 2003) show that *W. areolata* is the basal member of the genus with *W. virginica* the next most basal branch. However, given that there are only 14 species in this monophyletic lineage, and they are very widely known under the genus name *Woodwardia*, we are following most recent authors and maintaining all in that one genus. Some authorities (e.g., Nelson 2000) attribute the name CHAIN FERN "to the conspicuous netted, chain-like areoles that parallel both sides of the mid-vein on the lower sides of the pinnae and most pinnules, and impart a distinctive, chain-like appearance to the sori. These 'chains' are easily seen without magnification, especially when held up to light ..." (Named for Thomas Jenkinson Woodward, 1745–1820, English botanist)

REFERENCES: Correll 1956, 1966; Kitfield 1974; Cranfill 1983; Collinson 2001; Pigg & Rothwell 2001; Cranfill & Kato 2003; Pigg et al. 2006.

1. Leaves conspicuously dimorphic (= of 2 types), the pinnae of fertile leaves contracted and linear in comparison with the much wider lanceolate pinnae of the sterile leaves; sterile blades 1-pinnatifid, with a wing of blade tissue several mm wide along much (at least upper half) of the rachis; pinnae themselves not pinnatifid, sometimes sinuate, the margins serrulate ______ **W. areolata**
1. Leaves monomorphic or nearly so, the pinnae of fertile and sterile leaves essentially the same; blades 1-pinnate, with no leaf tissue along the rachis; pinnae deeply pinnatifid with entire margins ______ **W. virginica**

Woodwardia areolata (L.) T. Moore, (Latin: *areolatus*, with areolae, divided into distinct spaces, networked, checkered, from *areola*, diminutive of *area*, a space, in reference to the veins connecting to form small vein-enclosed spaces), NARROW-LEAF CHAIN FERN, NETTED CHAIN FERN, CHAIN FERN. Sterile leaves few, 40–58 cm long; pinnae of sterile leaves in 7–12 alternate pairs, 1–2.5 cm wide, the veins anastomosing into 2 or more rows of areoles (= small vein-enclosed areas) between midvein and margin; fertile blades with sori nearly completely covering surface of blade; $2n = 70$ (Cranfill 1993b). Low, wet, usually sandy areas; Pineywoods w to Fannin

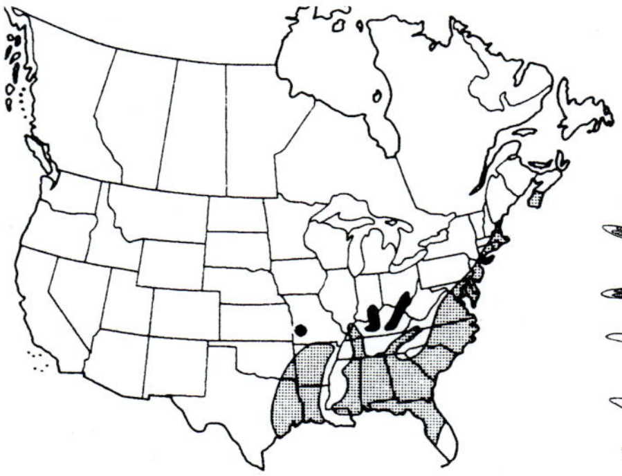

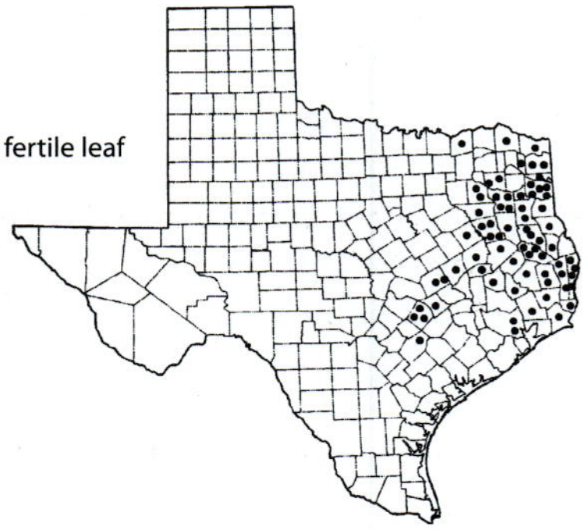

portion of
fertile leaf

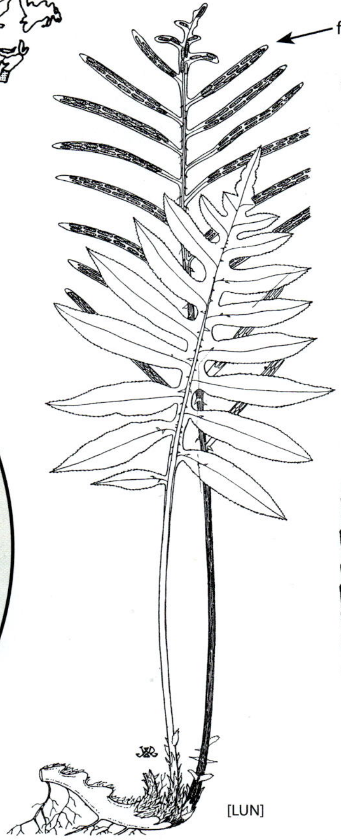

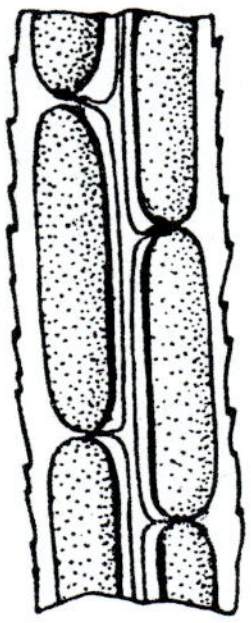

detail of
fertile segment

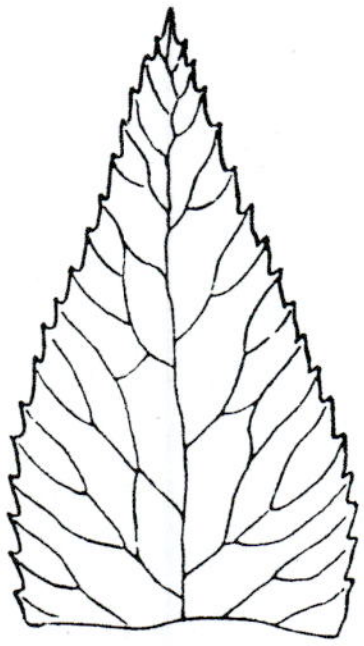

detail of sterile
segment tip

Woodwardia areolata

NARROW-LEAF CHAIN FERN

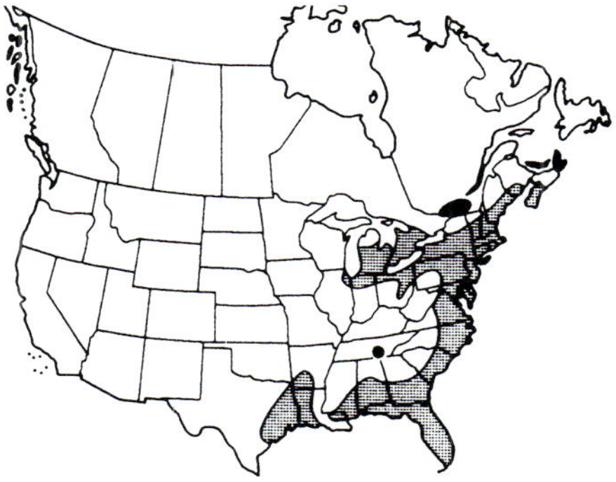

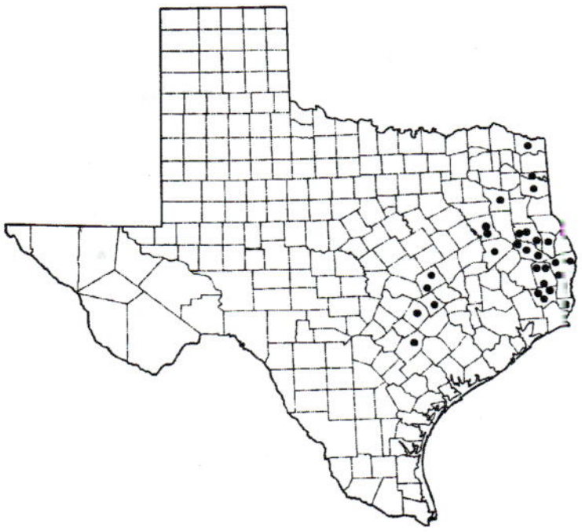

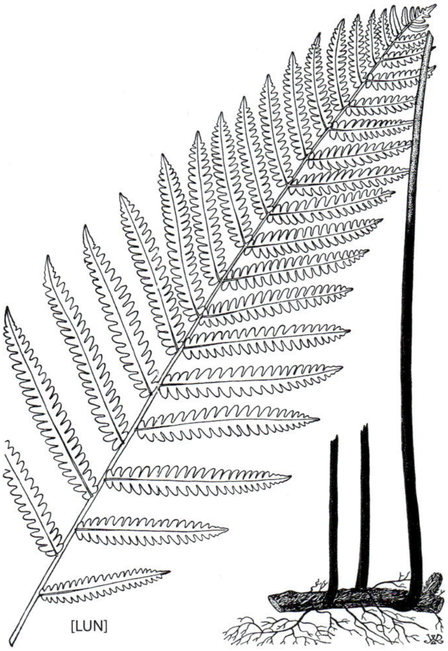

Alan Cressler

Sonnia Hill

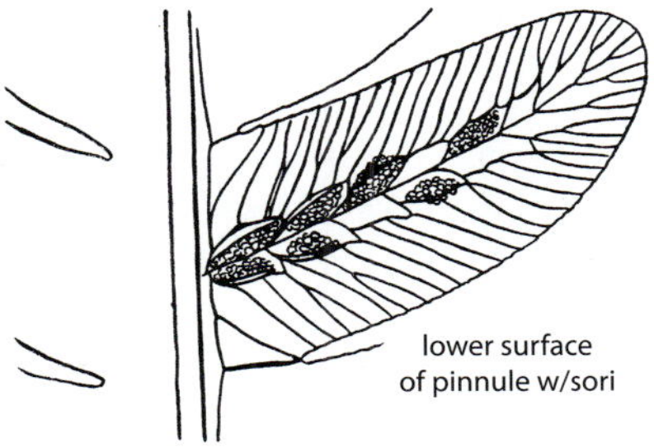

Woodwardia virginica

VIRGINIA CHAIN FERN

Co. (BRIT) in Red River drainage and Bastrop (*W.R. Carr & E.A. Kutac 7741*, BRIT), Freestone (BAYLU), Leon (*Correll 14177*, BRIT), Milam, and Van Zandt (TAES, TEX-LL) cos. near w margin of Post Oak Savannah, and sw to Gonzales Co. (TEX-LL; Fleenor & Tabor 2009); also n edge of Gulf Prairies and Marshes (e.g., Orange Co., *R.L. Vines B15722*, BRIT; Jefferson Co., *Mrs. J.L. Hooks s.n.*, 3/15/1937, TEX-LL); se Canada and through e U.S. from NH (Kitfield 1974) and NY s to FL w to IL, OK, and TX. Sporulating Mar–Nov. This species has sometimes been segregated into the genus *Lorinseria* [as *L. areolata* (L.) C. Presl]. It can rarely produce leaves intermediate between its typical sterile leaves and fertile leaves (Shaver 1970; Thieret 1980). The Gonzales Co. occurrence, which is the southwestern most locality for this widespread species (Cranfill 1983), is in the Ottine wetlands (see note under *W. virginica*). The sterile leaves superficially resemble those of *Onoclea sensibilis*, SENSITIVE FERN (which has subopposite basal pinnae, the pinnae undulate to irregularly deeply lobed, but with the margins entire), except *W. areolata* usually has alternate basal pinnae, the pinnae all lacking lobes (pinnae can be sinuate), and with the margins serrulate.

Woodwardia virginica (L.) Small, (of Virginia, Linnaeus giving many plants from the eastern U.S. this epithet), VIRGINIA CHAIN FERN. Leaves numerous, 50–100 cm long; petioles straw-colored to greenish above, dark reddish brown to purplish or nearly black at base; pinnae in 12–23 pairs, the middle pinnae 1–3.5 cm wide, the veins anastomosing to form a single row of areoles near midvein; rachis straw-colored, sometimes darkly so; sori covering only a small part of the blade surface; $2n = 70$ (Cranfill 1993b). Low areas; primarily Pineywoods and n edge of Gulf Prairies and Marshes (Orange Co.—Turner et al. 2003) disjunct w to Bastrop (*W.R. Carr & E.A. Kutac 7740*, BRIT), Lee (TAMU), Gonzales (*Tharp s.n.* 1935, Ottine, BRIT, TEX-LL), and Milam (TEX-LL) cos. on w margin of Post Oak Savannah; se Canada and e U.S. from ME s to FL w to IL, AR, and TX; also Bermuda. Sporulating Apr–Dec. This species has sometimes been segregated into the genus *Anchistea* [as *A. virginica* (L.) C. Presl]. It has a long fossil history (e.g., Pigg & Rothwell 2001) and fossils from the middle Miocene (ca. 14 mya) are "structurally identical" to modern *W. virginica*. Pigg and Rothwell (2001) note that this species adds "additional data to our growing realization that some homosporous pteridophytes display exceptional species longevity." In TX this species is found primarily in the far e part of the state in the Pineywoods and then disjunct to 4 counties further w. These w populations are presumably relicts of more humid conditions in the past that have been able to survive in isolation since Ice Age times due to unique bogs and wetlands associated with the sandy, porous Carrizo Formation. For example, in Milam Co. (*L.C. Taylor 3574*, TEX-LL) the species was found in a "peat bog." Likewise, the southwestern most location for the species, Gonzales Co., is from the Ottine wetlands—a relict spring-fed ecosystem. Numerous other mesic species are disjunct to similar such wet microhabitats in otherwise rather dry inhospitable environments in c TX, and are thought to be relicts of Ice Age times (see page 36). VIRGINIA CHAIN FERN can be confused when sterile with the superficially similar *Osmundastrum cinnamomeum*, CINNAMON FERN. However, it can be "distinguished at some distance by the dark brown base of the petiole, mostly darker rachis, and by the fronds being well spaced rather than clump forming" (Nelson 2000). When fertile, the two species are dramatically different (in *Osmundastrum* the fertile pinnae are completely different from the sterile—not very leaf-like, with no expanded pinnae, and densely tomentose).

CYSTOPTERIDACEAE (Payer) Shmakov

BLADDER FERN FAMILY

A subcosmopolitan but especially temperate and montane family of ca. 4 genera and 30 terrestrial species. In relatively recent treatments the species have often been included in a very broadly interpreted Dryopteridaceae (e.g., Smith 1993b; Yatskievych 1999; Diggs et al. 2006) or in a somewhat broadly circumscribed Woodsiaceae containing *Athyrium*, *Cystopteris*, and *Woodsia* as well as a number of small genera (Smith et al. 2006, 2008; Mabberley 2008). However, recent DNA evidence (Schuettpelz & Pryer 2008; Rothfels et al. 2012a) suggests *Athyrium*, *Cystopteris*, and *Woodsia* are all phylogenetically quite remote from Dryopteridaceae, are morphologically similar to each other due to convergent evolution, are not closely related, and should be in completely different families. For further discussion and a key separating superficially similar species see the family treatment of Woodsiaceae on page 307. (Order Polypodiales)

FAMILY RECOGNITION IN THE FIELD: plants superficially similar to *Woodsia* (plants relatively small, leaves all alike or nearly so, appearing much dissected, the sori round) but differing in having the indusia attached by one side, hood- or pocket-like, arching over sori towards segment margin, the leaf blades broadest at base or at or just below middle, and the veins of ultimate leaf segments reaching segment margins.
REFERENCES: Smith et al. 2006, 2008; Schuettpelz & Pryer 2008; Rothfels 2012a, 2012b.

CYSTOPTERIS Bernh. BLADDER FERN, BRITTLE FERN

Plants on rock or terrestrial; rhizomes creeping; leaves ± monomorphic (can vary a bit seasonally), dying back in winter, including petioles typically < 50(–75) cm long; leaf blades ovate to ovate-lanceolate, elliptic or deltate, widest point varying from base to middle, 1-pinnate-pinnatifid to 2–3-pinnate or 2–3-pinnate-pinnatifid, pinnatifid apically, with or without glandular hairs, the lower surfaces without transparent needle-like hairs (these present in Thelypteridaceae), the ultimate segments crenate to serrate or dentate, with veins reaching segment margins; petioles glabrous except at very base, the bases with 2 vascular bundles (versus more than 3 in Dryopteridaceae), often persisting over winter as trophopods (= swollen petiole bases that function in storing starch); sori roundish, on veins, in single row between midrib and each margin of ultimate segments; indusia attached by one side, delicate, hood- or pocket-like, arching over sori towards segment margin (sometimes ephemeral or inconspicuous in mature sori).

A subcosmopolitan genus of ca. 12 (Mabberley 2008) to 20 species (Haufler et al. 1993; Rothfels et al. 2012b). Polyploidy and hybridization are common in the genus, and it is taxonomically difficult (Haufler et al. 1993). All members of the genus occurring natively in TX are rare and of limited distribution, and we consider them to all be of conservation concern. In TX they are easily mistaken for members of the superficially similar but much more common genus *Woodsia* (Woodsiaceae). (Greek: *cystos*, bladder, and *pteris*, fern, in reference to the inflated or bladder-like young indusia covering the sori).
REFERENCES: Blasdell 1963; Correll 1965; Reeves 1976; Lellinger 1981; Haufler et al. 1985, 1990, 1993; Haufler & Windham 1991.

1. Leaf blades usually widest at or near base; rachises, costae, indusia, and midribs of ultimate leaf segments with sparse to dense covering of gland-tipped hairs; rachises and costae sometimes with round, pea-like bulblets; species known in TX only from far w part of the state.
   2. Rachises and costae densely glandular-pubescent, often with round, pea-like bulblets; leaf blades long attenuate at tip ____ **C. bulbifera**
   2. Rachises and costae sparsely glandular-pubescent (multicellular gland-tipped hairs particularly in axils of costae), lacking bulblets OR sometimes with misshapen bulblets; leaf blades short attenuate at tip ____ **C. utahensis**
1. Leaf blades usually widest at or just below middle; rachises, costae, indusia, and midribs of ultimate leaf segments usually without glandular hairs; rachises and costae without bulblets; including species of both e and w parts of TX.
   3. Leaf blades 1-pinnate-pinnatifid to 2-pinnate; apices of ultimate leaf segments appearing rounded to slightly pointed; leaves clustered 1–5 cm behind the tip of the rhizome (rhizome tip protruding beyond attachment of leaves); petioles usually green to straw-colored, even near base; rhizomes with tan to golden hairs as well as scales; plants terrestrial, in moist deciduous forests, the species known in TX only from the e ½ of the state ____ **C. protrusa**
   3. Leaf blades 2–3 pinnate to 2–3-pinnate-pinnatifid (appearing relatively more finely dissected); apices of ultimate leaf segments appearing relatively blunt; leaves clustered at tip of rhizome; petioles variable in color, but usually dark purple at least at base; rhizomes without hairs, only scales present; plants often on rocks, the species known in TX only from the w part of the state (Edwards Co. and the Trans-Pecos) ____ **C. reevesiana**

Cystopteris bulbifera (L.) Bernh., (Latin: *bulbus*, a swelling, and *feros*, to bear, bearing bulbs, in reference to the bulblets on the leaves), BULBLET BLADDER FERN. Usually on rocks; rhizomes with scales, but without hairs; leaves clustered at tip of rhizome, monomorphic, to 75 cm long; petioles reddish to greenish or straw-colored; rachis and costae, and midribs of ultimate segments densely glandular-pubescent, the rachis and costae often with round, pea-like bulblets; leaf blades 2-pinnate to 2-pinnate-pinnatifid, usually narrowly deltate, sometimes ± broadly so, widest at base, apically long attenuate, the ultimate segments serrate; indusia usually with glandular hairs; $2n = 84$ (Haufler et al. 1993). Moist calcareous cliffs, in crevices and on ledges; in TX known only from the Guadalupe Mts., Culberson Co. (*Correll & Hanson 29803*, BRIT, TEX-LL, "First Texas collec-

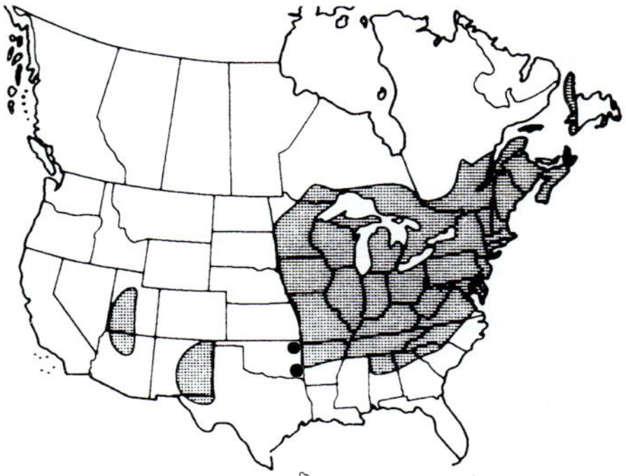

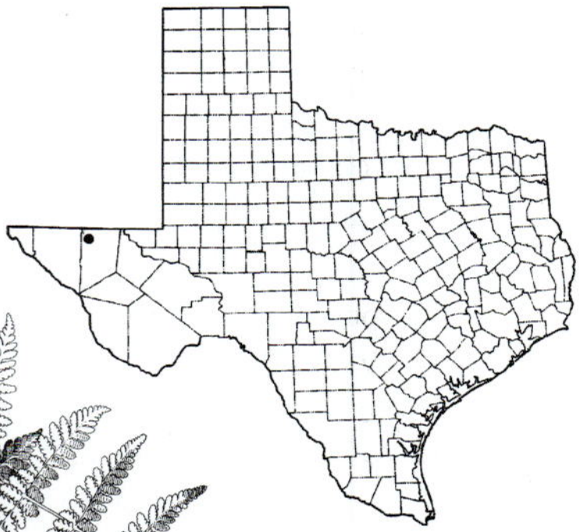

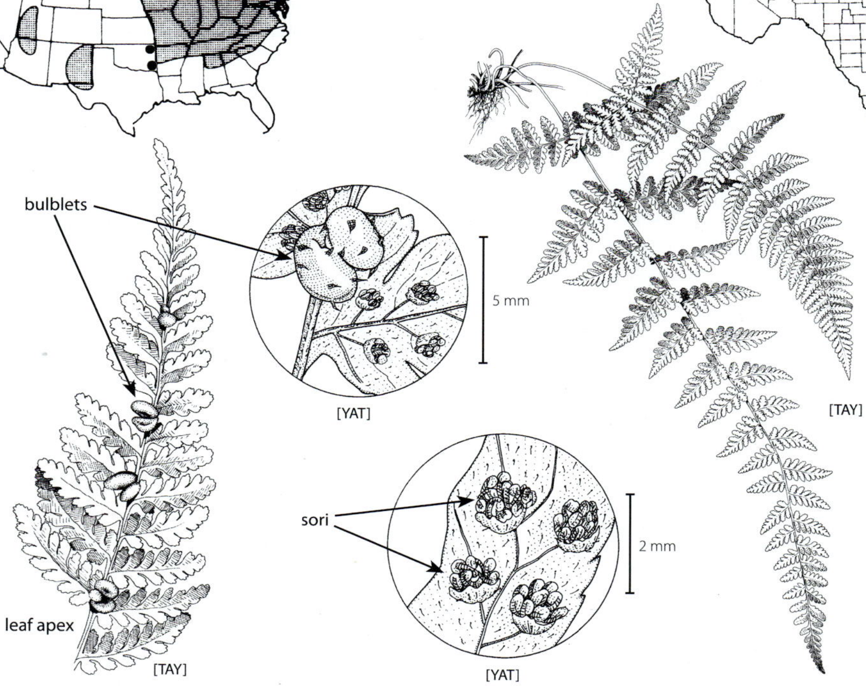

Cystopteris bulbifera

BULBLET BLADDER FERN

tion," BRIT sheet annotated by M.D. Windham; *T.L. Burgess 3718*, TEX-LL); e U.S. (except extreme se) w to AR and WI and se Canada, also disjunct to the w U.S. in AZ, NM, OK, TX, and UT. Sporulating ?–Jun–Aug–?. [*Polypodium bulbiferum* L.] This species was first reported from TX by Correll (1965). He discovered it in McKittrick Canyon in the Guadalupe Mts., whose canyons are "the most mesophytic and florally interesting" in Trans-Pecos, TX (Correll 1965). He also noted that it was "the southernmost locality known for this typically northern species." It is now known that this is a species primarily of the e ½ of North America, with isolated populations in the mountains of AR, NM, OK, TX, and UT. These disjunct populations, which are separated by many hundreds of mi from others of their species, are presumably the result of long-distance dispersal. The combination of leaves usually narrowly deltate, widest at base and long attenuate, with dense glandular pubescence and bulblets, is distinctive. Weakley (2010) noted that, "This species is a diploid involved in the reticulate evolution of *Cystopteris* in e. North America." In the e U.S. *C. bulbifera* is known to hybridize with *C. protrusa* giving rise to the tetraploid *C. tennesseensis* Shaver (Haufler et al. 1990). In the sw U.S. *C. bulbifera* has hybridized with *C. reevesiana* giving rise to the tetraploid *C. utahensis*. The various hybrids usually have less glandular pubescence than *C. bulbifera*, are typically widest beyond the base, and have misshapen bulblets (Haufler et al. 1993). Because of morphological similarities and because both are tetraploids with one parent in common, *C. tennesseensis* (in the e U.S.) has sometimes been incorrectly treated as a synonym of the genetically distinct *C. utahensis* (of the w U.S.) (e.g., The Plant List 2011; Tropicos 2011). Because of its rareness and limited distribution in the state, we consider this species to be of conservation concern in TX. ⚠

Cystopteris protrusa (Weatherby) Blasdell, (Latin: *protrusus*, protruding, pushed out, or exserted, in reference to the rhizome apex extending beyond the point of leaf attachment), SOUTHERN BLADDER FERN, LOWLAND BLADDER FERN, LOWLAND BRITTLE FERN. Terrestrial; rhizomes protruding 1–5 cm beyond attachment of current season's leaves, with tan to light brown or golden hairs as well as scales; leaves seasonally somewhat dimorphic, clustered, erect to erect-spreading, to 45 cm long; petioles usually green to straw-colored, even near base; leaf blades 1-pinnate-pinnatifid to 2-pinnate, ovate-lanceolate to elliptic, widest at or just below middle, to ca. 25 cm long and 12 cm wide, apically broadly acute, the ultimate segments dentate to serrate; earliest leaves small, sterile, coarsely divided, with rounded teeth marginally; later leaves larger, fertile, more finely divided, with sharply pointed teeth marginally; sori on veins of ultimate leaf segments (= smallest subdivisions of leaf); indusia lacking glandular hairs; $2n = 84$ (Haufler et al. 1993). Moist deciduous forests; Gonzales Co. in the s Blackland Prairie (*H.B. Parks s.n.*, Jan 1939, TAES) (cited by Correll [1956] as *C. fragilis* var. *protusa*), Harrison (*Ajilvsgi 4065*, BRIT—annotated by J. Peck) and Houston (Turner et al. 2003) cos. in the Pineywoods; also Victoria Co. (Turner et al. 2003) in the Gulf Prairies and Marshes. *Cystopteris protrusa* is also known from McCurtain Co., OK, just n of the Red River (*Taylor & Taylor 28956*, BRIT). Although the range map in Haufler et al. (1993) does not include TX and the species is apparently quite rare in the state, it is known from several TX localities; e U.S. from NH s to FL w to MN, NE, OK, and TX (the TX populations are disjunct to the sw of most of the range of the species); also Ontario, Canada. Sporulating spring–summer. [*C. fragilis* (L.) Bernh. var. *protrusa* Weatherby] Weakley (2010) noted, "This species is a diploid involved in the reticulate evolution of *Cystopteris* in e. North America." The Gonzales Co. location is from the Ottine wetlands—a relict spring-fed ecosystem. Numerous other mesic species are disjunct to similar wet microhabitats in otherwise rather dry inhospitable environments in c TX, and are thought to be relicts of Ice Age times (see page 36). An Edwards Co. collection of *C. reevesiana* (*Cory 35636*, Devil's Sink Hole, TAES) was incorrectly identified and mapped as *C. protrusa* by Diggs et al. (2006). The terrestrial habit, and rhizomes with tips protruding beyond attachment of leaves and with hairs as well as scales (other TX species usually on rocks and lacking rhizome hairs), distinguish *C. protrusa* from other TX members of *Cystopteris* (Haufler et al. 1993). This species is very similar in general morphology to *Woodsia obtusa*, BLUNT-LOBE WOODSIA, and can be confused with that much more common species in the e ½ of TX. The two can be distinguished by the indusia, by the ultimate leaf segments which are not significantly narrowed at base in *W. obtusa* (versus obviously narrowed towards base in *C. protrusa*), and other characters (e.g., whether veins reach leaf margins) as given in the key separating *Woodsia* and the superficially similar *Cystopteris* in the family treatment of Woodsiaceae on page 307. This species was one of the first ferns for which trophopods (see genus description) were described (Wagner et al. 1970). Because of its rareness and limited distribution in the state, we consider this species to be of conservation concern in TX. ⚠

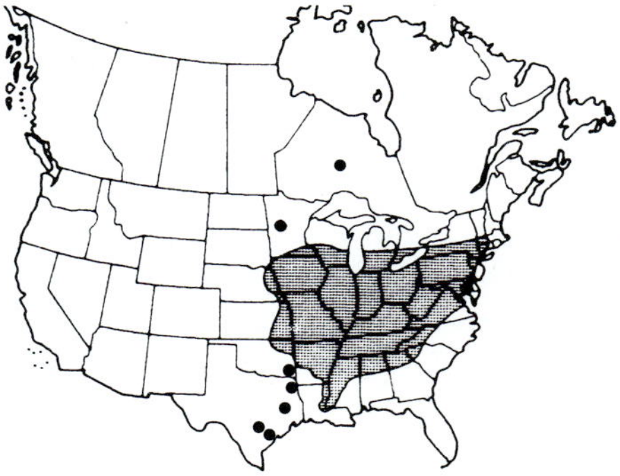

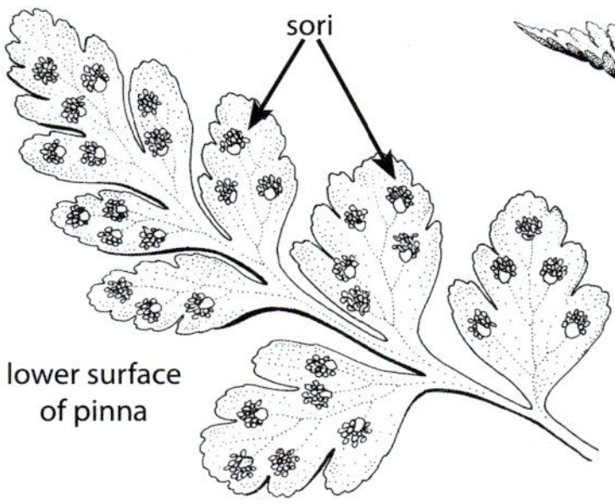

Steve Baskhauf

pinnules w/sori

Alan Cressler

Cystopteris protrusa

SOUTHERN BLADDER FERN

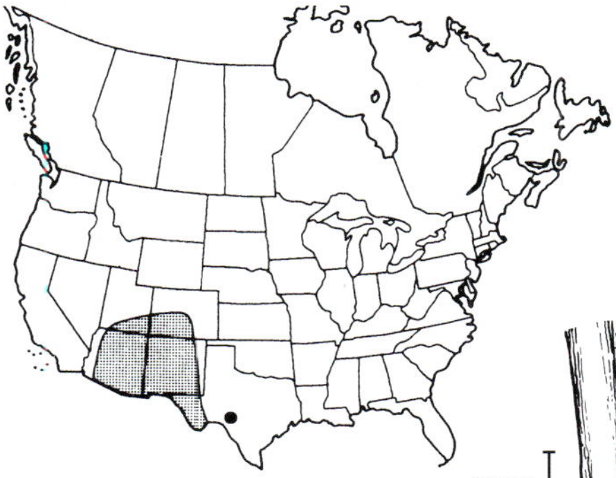

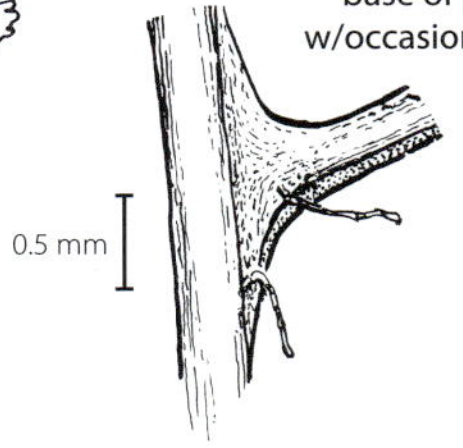

base of pinna
w/occasional hairs

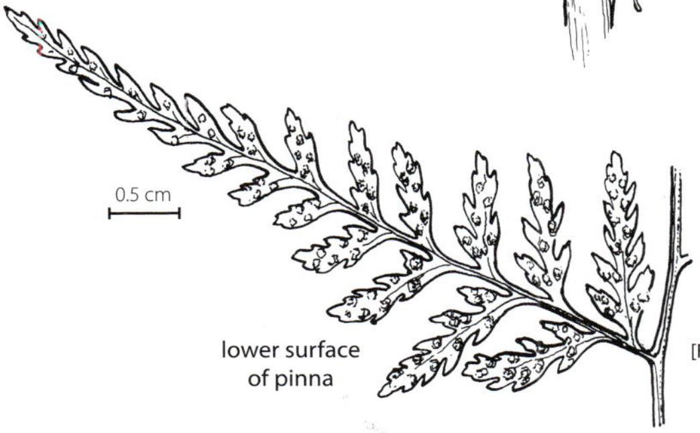

lower surface
of pinna

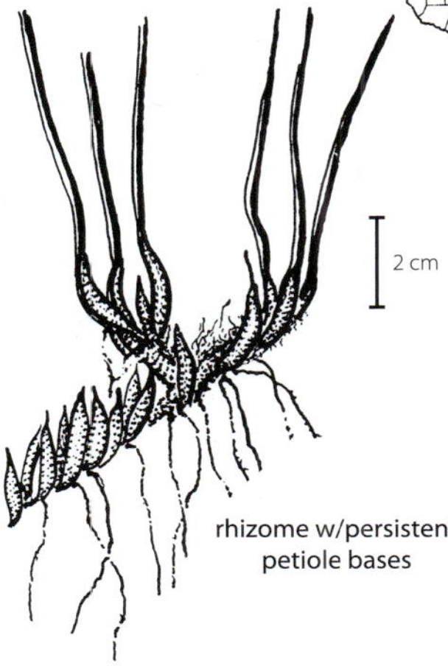

rhizome w/persistent
petiole bases

Dale A. Zimmerman Herbarium, Western N. Mex. Univ.

©Al Schneider, www.swcoloradowildflowers.com

Cystopteris reevesiana

SOUTHWESTERN BRITTLE FERN

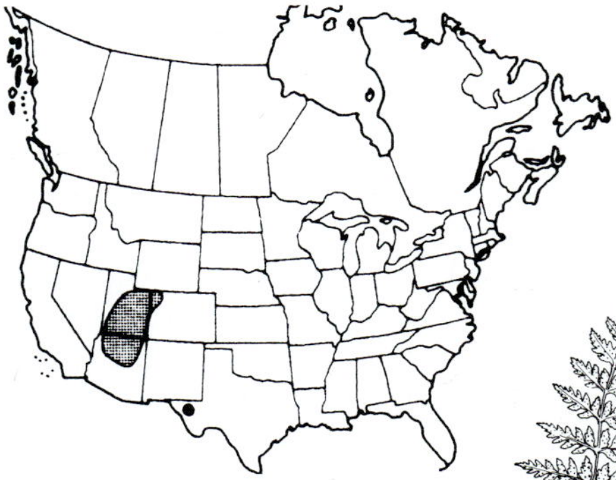

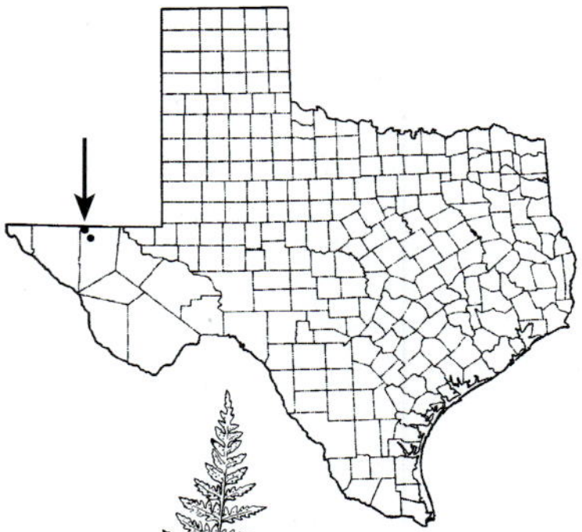

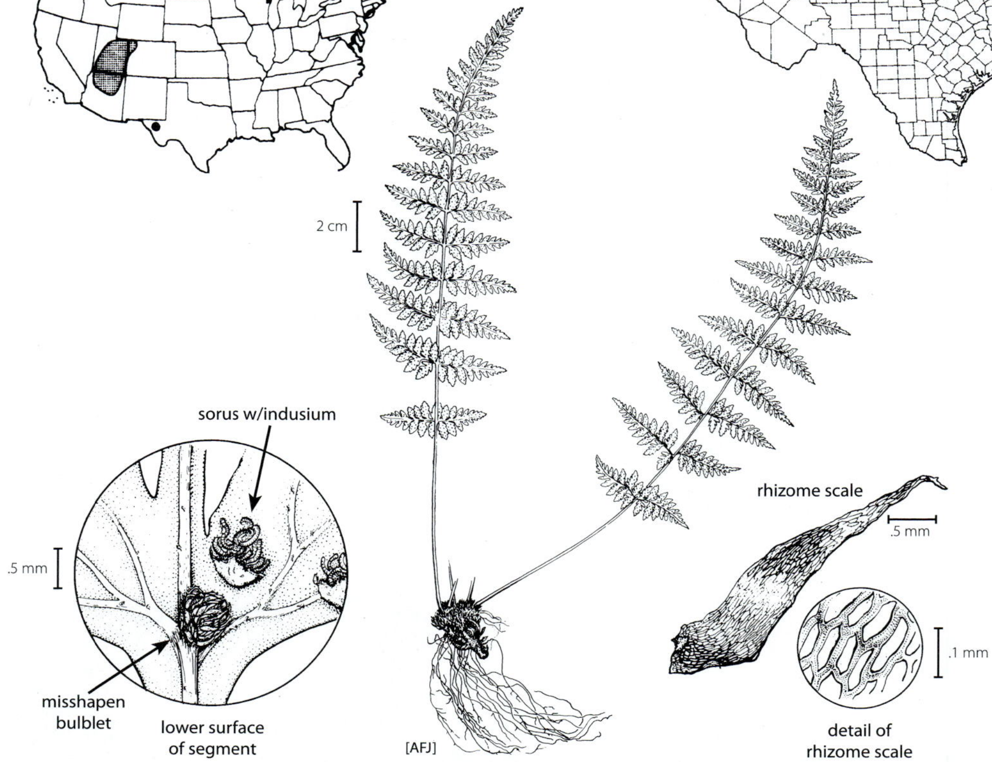

Dan Nickrent

Cystopteris utahensis

UTAH BLADDER FERN

Cystopteris reevesiana Lellinger, (for Dr. Timothy Reeves, 1947–, student and collector of U.S. and Mexican ferns, especially *Cheilanthes*), SOUTHWESTERN BRITTLE FERN, BRITTLE FERN. Terrestrial or on rocks; rhizomes not protruding beyond attachment of leaves, with internodes relatively long, with tan to brown scales but without hairs; leaves clustered at rhizome tips, monomorphic, to 45 cm long; petioles variable in color, but usually dark purple at least at base and sometimes throughout; rachis and costae without glandular hairs (except for occasional gland-tipped hairs in axils of pinnae) or bulblets; leaf blades finely dissected, 2–3-pinnate to 2–3-pinnate-pinnatifid, ovate to elliptic, widest at or just below middle, apically short attenuate, the ultimate segments crenate to dentate; indusia without glandular hairs; $2n = 84$ (Haufler et al. 1993). On rocks (a variety of substrates including limestone or igneous materials) or sometimes terrestrial; Culberson Co. in Guadalupe Mts. (*M.S. Young s.n.* 1916, TEX-LL), Edwards Co. (Punkin Cave, *Schindel s.n.*, 2007, BRIT—identification confirmed D.B. & J.S. Lellinger in 2007; *Cory 35636*, in Devil's Sink Hole, 80 feet below surface, TAES, TEX-LL), and Jeff Davis Co. in Madera Canyon, Mt. Livermore, Davis Mts. (*Hinckley s.n.*, 1935, TEX-LL); AZ, CO, NM, TX, and UT; endemic to the U.S.—not known from Mexico despite earlier erroneous reports of its occurrence there (Mickel & Smith 2004). Sporulating summer–fall. [*C. fragilis* (L.) Berhn. subsp. *tenuifolia* Clute] This species can sometimes cross with *C. utahensis*, producing "sterile triploids of intermediate morphology" (Haufler et al. 1993). Because of its limited distribution in the state, we consider this species to be of conservation concern in TX. ⚠

Cystopteris utahensis Windham & Haufler, (of Utah, where the type specimen was collected), UTAH BLADDER FERN. Similar to *C. reevesiana*; usually on rocks; rhizomes with relatively short internodes, with dark brown scales, but without hairs; leaves clustered at tip of rhizome, monomorphic, to 45 cm long; petioles green to straw-colored or darker near base; rachis and costae with sparse glandular pubescence (multicellular gland-tipped hairs particularly in axils of costae—use high magnification) and sometimes with misshapen bulblets; leaf blades 2-pinnate-pinnatifid, deltate to often narrowly so, widest at or near base, apically short attenuate, the ultimate segments serrate; indusia with scattered glandular hairs; $2n = 168$ (Haufler et al. 1993). Crevices and ledges of limestone cliffs; known in TX only from Culberson Co. (*Warnock 23174*, SRSC, paratype; *Warnock 12000*, BRIT, TEX-LL, McKittrick Canyon; *Johnston et al. 10717*, *Correll & Hanson 29816*, *Riskind 1050*, all TEX-LL); AZ, CO, TX, and UT; also reported from NM (Roth 2001) without location; endemic to the U.S. Sporulating summer–fall. This species is an allotetraploid derived from diploid parents, *C. bulbifera* and *C. reevesiana* (Haufler & Windham 1991); it is somewhat intermediate in morphology between the two. It should thus not be surprising that there is sometimes difficulty in distinguishing *C. utahensis* from *C. reevesiana*. The situation is further complicated by the fact that sterile triploid hybrids are known between *C. utahensis* and *C. reevesiana* (Haufler et al. 1993). The leaves widest at or near base, sparse glandular pubescence, and bulblets lacking or misshapen are helpful in identification of *C. utahensis*. In addition, the tetraploid *C. utahensis* often has somewhat larger sori than do the diploids *C. bulbifera* and *C. reevesiana*. Because of its limited distribution in the state, we consider this species to be of conservation concern in TX. ⚠

DENNSTAEDTIACEAE Lotsy

BRACKEN OR CUPLET FERN FAMILY

Perennials, one TX species terrestrial, the other species rare, only in caves or sinks; rhizomes long-creeping; leaves monomorphic, large, the blades 3–4 pinnate/pinnate-pinnatifid; sori marginal, either linear and continuous along margins of ultimate leaf segments OR globose and discrete at vein tips in the sinuses of the ultimate leaf segments.

The Dennstaedtiaceae are a nearly cosmopolitan but mostly tropical family of ca. 10–20 genera and ca. 170–400 species depending on circumscription (Cranfill 1993a; Mabberley 2008). The group has been variously treated to include as few as 8 genera or in other cases nearly half the genera of higher ferns (Mickel 1973). Smith et al. (2008) in their recent work on fern classification, consider the family to have ca. 11 genera and 170 species. Of these, only *Dennstaedtia* and *Pteridium* are represented in TX. Within the huge polypod order, the family Dennstaedtiaceae is one of the earliest diverging families (Smith et al. 2006; see page 380). (Order Polypodiales)

FAMILY RECOGNITION IN THE FIELD: the two TX species are quite different in distribution and habitat, but both have large, decompound (3–4 times pinnate/pinnate-pinnatifid) leaf blades and marginal sori; one is a widespread colony-forming plant in the e ⅓ of TX with sori linear, along margins of the ultimate leaf segments (= smallest subdivisions of leaf) with the leaf margins recurved over sori to form a false indusium. The other is a rare species known only from caves or sinks in Val Verde Co. in w TX, with discrete globose sori.
REFERENCES: Mickel 1973; Kramer 1990b; Cranfill 1993a; Wolf 1995; Smith et al. 2008.

1. Sori globose, discrete (obviously separate from each other), marginal at vein tips in the sinuses (between teeth/lobes) of the ultimate segments, in cup-like indusia; leaf blades with ultimate segments toothed to lobed with sinuses sometimes reaching to nearly ½ distance to midrib, definitely not appearing marginally entire ______ **Dennstaedtia**
1. Sori linear, continuous along margins of ultimate leaf segments, covered by elongate, linear false indusia formed by the recurved margin of ultimate leaf segments; leaf blades with ultimate segments often pinnately divided, but the segment or its divisions marginally entire ______ **Pteridium**

DENNSTAEDTIA Bernh.
CUP FERN, HAY-SCENTED FERN

A nearly cosmopolitan genus of ca. 60–70 species (Nauman & Evans 1993; Mabberley 2008), but primarily of tropical forests, with 3 in the U.S.—1 in w TX, 1 in FL, and 1, *D. punctilobula*, the HAY-SCENTED FERN, common in the deciduous forests of the e U.S. The genus is most diversified in the Old World and is related to *Pteridium*. Molecular data indicate that *Dennstaedtia* as currently considered is paraphyletic (Schuettpelz & Pryer 2008). BEADED CUPLET FERN has leaves nearly 4 m long, the largest leaves of any TX fern. Some are cultivated as ornamentals. (Named for August Wilhelm Dennstaedt, 1776–1826, German botanist, physician, and director of the Belvedere Garden)
REFERENCES: Correll 1960a; Tryon 1960; Nauman & Evans 1993; Schuettpelz & Pryer 2008.

Dennstaedtia globulifera (Poir.) Hieron., (Latin: *globus*, ball, sphere, or globe, and *-ifera*, a suffix from Latin *fero*, to bear, = globe-bearing, in reference to the globular sori), BEADED CUPLET FERN, SOUTHERN HAY-SCENTED FERN. Plant extremely large for a TX fern (leaves to nearly 4 m long); rhizomes long-creeping; leaves erect or arching; petioles to 1 m long; leaf blades 3(–4) pinnate, 1–3 m long, deltoid to ovate, ca. ⅔ to nearly as wide as long, the tissue soft-herbaceous, glabrous on upper surface, with hairs on the veins on lower surface, the ultimate leaf segments obtusely toothed to lobed with sinuses reaching to nearly ½ distance to midrib; sori distinct, globose, ca. (1–)1.2–1.5 mm in diam., marginal at vein tips in the sinuses (between teeth/lobes) of ultimate segments, with cup-like indusia, these formed by fusion of true indusium and tiny tooth of ultimate leaf segment; $2n$ = ca. 94 (Nauman & Evans 1993). Extremely rare in moist caves or limestone sinks (in other parts of its range, e.g., in Mexico, it is found in wet montane tropical forests); in the U.S. known only from Val Verde Co. (e of the Pecos River in extreme w Edwards Plateau), TX (first TX coll.: *Larry Hoffman s.n.*, 1958, "Fern Cave"; *Correll & Correll 26140*, 1962; *E. Glenn 19280*, 1962, all TEX-LL); also Mexico, West Indies, Central and South America. Sporulating summer. [*Dicksonia globulifera* (Poir.) Kuntze, *Polypodium globuliferum* Poir.] Correll (1960a) reported this species as new to both TX and the U.S. According to label data by Correll (*Correll & Correll 26140*), the plant was growing "On floor of cave about 50 ft. deep, below opening." As in the case of several other TX ferns (Seigler & Lockwood 1975; Seigler 1977), the TX occurrence of this largely tropical species is significantly disjunct from the majority of the species' range. It is able to survive in w TX by occupying an extremely rare microhabitat—humid cave entrances. The extremely large leaves are very surprising to find in a fern in the dry sw U.S. where reduction in leaf size is the norm (see page 27); in fact, Nauman and Evans (1993) noted that this species is "among the largest ferns in the [North American] flora"—the original TX collection cited by Correll (1960a) described "fronds up to 13 ft. [3.96m] long." The combination of the large leaf size and the cup-like indusia and globose sori is quite distinctive among TX ferns. Given its rareness and limited distribution in the state, we consider this species to be of conservation concern in TX. ⚠

PTERIDIUM Gled. ex Scop. BRACKEN FERN

BRACKEN FERNS are among the most widespread and familiar of all ferns—they are known worldwide in diverse habitat with the exception of hot and cold desert areas. *Pteridium* is also a relatively old genus, thought

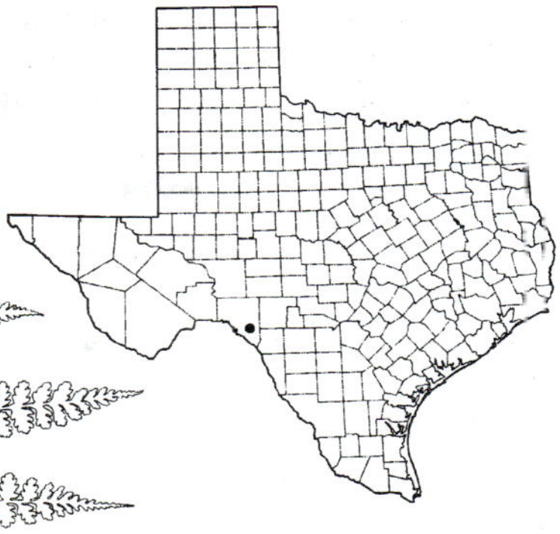

[SCY]

small portion of leaf

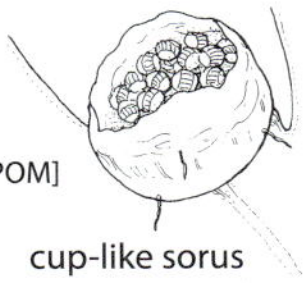

[POM]

cup-like sorus

ultimate segments w/sori

Carlos Velasco

Carlos Velasco

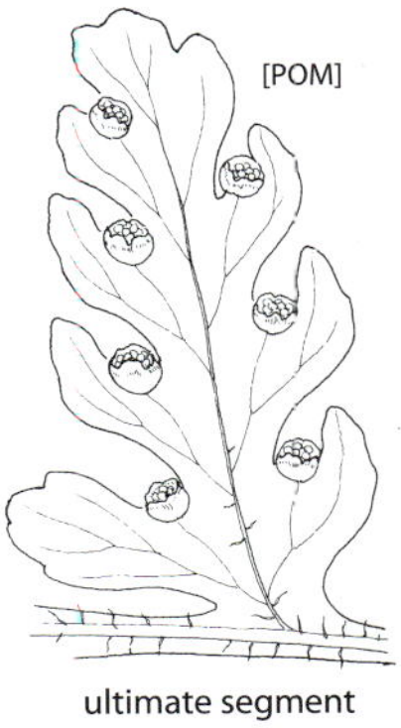

[POM]

ultimate segment w/sori

Dennstaedtia globulifera

BEADED CUPLET FERN

to have separated from *Dennstaedtia* between 114 and 92 mya (Der et al. 2009). Although extremely distinct and easily recognized at the generic level, species concepts within *Pteridium* have been confused and controversial—Thomson (2000b) called the genus "a taxonomic quicksand and an evolutionary conundrum." As Der et al. (2009) emphasized, *Pteridium* has "high levels of phenotypic plasticity, few diagnostic morphological characters, and the presence of intermediate phenotypes where different morphological forms come into contact, demonstrating that reproductive barriers are incomplete." It has traditionally been considered to be a monotypic cosmopolitan genus, with the single species divided into 12 varieties in 2 subspecies: subsp. *aquilinum* (including as varieties *latiusculum, pubescens,* and *pseudocaudatum*) predominantly in the n hemisphere and subsp. *caudatum* (L.) Bonap., largely in the s hemisphere (Tryon 1941; Tryon & Tryon 1982; Jacobs & Peck 1993). However, there have been a variety of taxonomic approaches, and some authorities (e.g., Lellinger 1985; Mickel & Beitel 1988) have recognized a number of the infraspecific taxa as separate species. Recently, Thomson et al. (2008) provided evidence supporting the recognition of 5 species in the genus, with all populations in the U.S. in 3 subspecies of *P. aquilinum* (with the exception of the primarily Central and South American *P. caudatum* (L.) Maxon which reaches s FL). In Mexico, where there are 3 species, the characters separating them remain consistent even when they occur together (Mickel & Beitel 1988; Mickel & Smith 2004); it therefore seems most appropriate to recognize them as separate species. We are following the nomenclature and classification used by Thomson et al. (2008). *Pteridium* has sometimes been placed in the Pteridaceae. (Greek: *pteridon*, a small fern, from *pteron*, feather or wing, due to the shape of the leaves)
REFERENCES: Tryon 1941; Correll 1956, 1966; Cooper-Driver 1976, 1990; Page 1976; Power & Skog 1987; Mickel & Beitel 1988; Jacobs & Peck 1993; Speer & Hilu 1998[1999], Speer et al. 1998[1999]; Smith et al. 2000; Speer 2000, 2008; Taylor & Smith 2000; Thomson et al. 2000, 2008; Thomson 2000a, 2000b, 2004a, 2004b, 2008; Der et al. 2009.

Pteridium aquilinum (L.) Kuhn, (Latin: *aquilinus*, of the eagle, eagle-like), BRACKEN FERN, PASTURE BRAKE. Terrestrial, often in large colonies; rhizomes deeply buried, long-creeping, with hairs but not scales; leaves monomorphic, deciduous, scattered along the rhizomes, to 2 m long; leaf blades broadly triangular to triangular-lanceolate or triangular-ovate or even pentagonal in outline, usually of 3 main divisions, each division 2-pinnate/pinnate-pinnatifid, the pinnae rigidly herbaceous to subcoriaceous, with glands (extrafloral nectaries) at base of some pinnae, glabrous to pubescent; ultimate leaf segments (= smallest subdivisions of leaf) with margins entire; sori marginal, linear, continuous, covered by a false indusium formed by the recurved margin of the ultimate leaf segments and an obscure inner, delicate, true indusium; $2n = 104$ (Thomson et al. 2008). Open woods, wooded slopes, pastures, thickets, often in sandy soils; widespread in e ⅓ of TX with populations of a different subspecies in the Trans-Pecos; predominantly n hemisphere, throughout most of the U.S. and s Canada; also Mexico s to Central America as well as Africa, Asia, and Europe. Sporulating Jun–Nov. Treated broadly (e.g., Jacobs & Peck 1993), this variable species, with a number of infraspecific taxa (depending on the source followed), is virtually worldwide in distribution (all continents except Antarctica) and is the most widely distributed fern. It is considered by some to be the most widespread of all vascular plants (with the exception of a few human dispersed annual weeds) (Page 1976). Recent new evidence, based on morphology, cytology, and molecular studies, suggests that it is perhaps better divided into 5 species, with *P. aquilinum* circumscribed to include only the diploid "predominantly Northern Hemisphere (Laurasian) lineage" (Thomson et al. 2008). Even in this restricted sense it is still extremely widespread, occurring from North and Central America to Africa, Asia, and Europe. It grows aggressively, and its tenacity is shown by regeneration through several meters of volcanic ash on Mt. St. Helens in WA within 1–2 years of the volcanic eruption (Woodland 1997). The species has a deep rhizome, making it well-adapted to fire prone habitats and one of the first species to reappear following a fire (Nelson 2000). In places in the n U.S. "the rhizomatous brackens are the dominant understory, covering thousands of square kilometers. Some clones may be hundreds or even thousands of years old and cover a hectare or more, but because they lack secondary growth, their age is difficult to ascertain" (Wagner & Smith 1993). BRACKEN FERN can sometimes have profound ecological impacts because of its abundance; for example, Pakeman et al. (1996) estimated that in the United Kingdom, *Pteridium* occupies an area of around 1.7 million hectares, which represents 7% of the total land surface.

BRACKEN FERN has a variety of defense mechanisms, some of which have important implications for both animals and humans. For example, in the British Isles it is the cause of "bracken staggers" or "bracken poison-

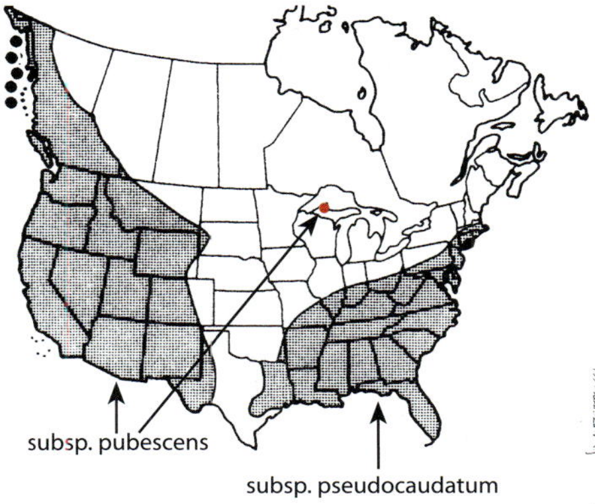

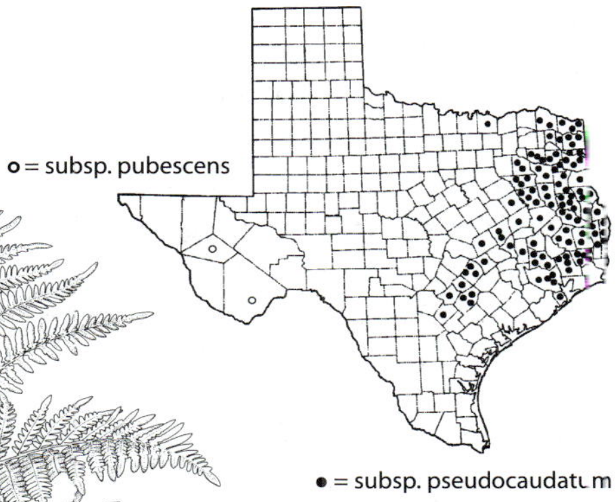

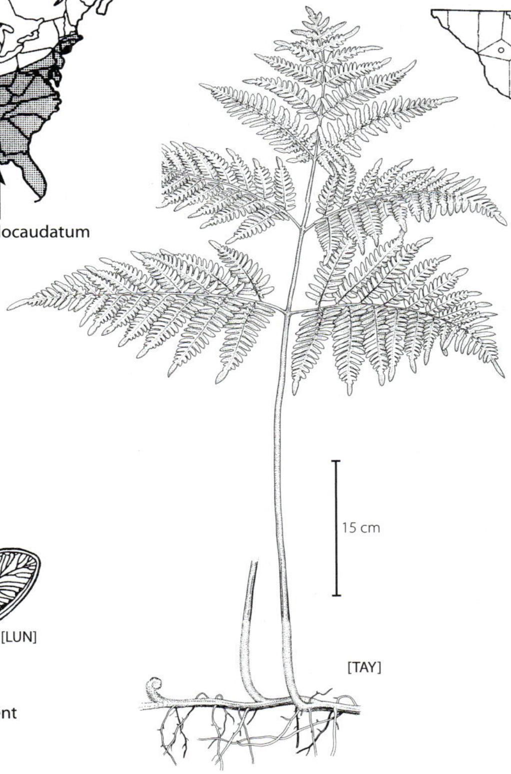

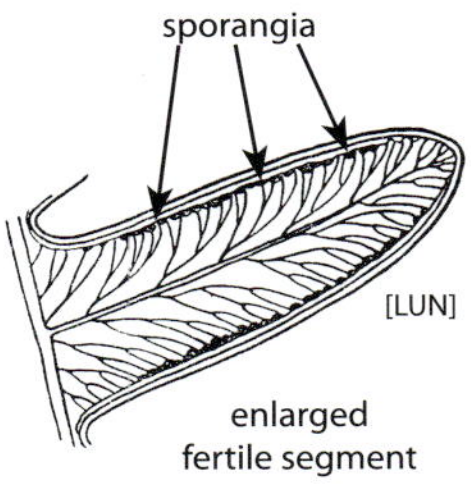

enlarged
fertile segment

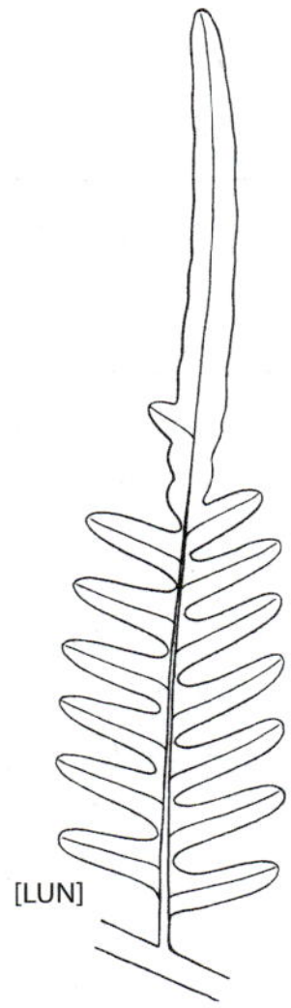

pinnule of
subsp. pseudocaudatum

Robert George

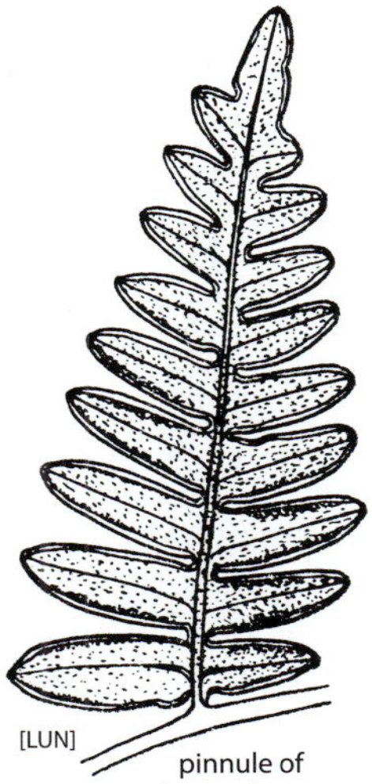

pinnule of
subsp. pubescens

Pteridium aquilinum

BRACKEN FERN

ing," a potentially fatal condition in livestock. Symptoms include a hemorrhagic syndrome, chronic hematuria ("red water"), bone marrow depression, fever, neurological symptoms and staggering, retinal degeneration ("bright blindness"), and cancer. Plant defense chemicals/toxins reported include an enzyme (thiaminase), which can cause fatal thiamine (Vitamin B1) deficiency in livestock, a glycoside (ptaquiloside—mutagenic and carcinogenic), and a cyanide-producing glycoside (prunasin). Ptaquiloside and possibly other carcinogens can be passed to humans via cow's milk. In Asia, particularly Japan and Korea, people consume the fiddleheads as a cooked vegetable and in bars as a salty snack food (Hodge 1973; Moran 2004). However, the fiddleheads unfortunately have high concentrations of ptaquiloside and have been suggested as a cause of stomach and esophagus cancer. Because it is both carcinogenic and mutagenic, BRACKEN FERN should not be eaten. Another chemical defense employed by BRACKEN FERN is the production of phytoecdysones, a class of hormone-like compounds that disrupt ecdysis (= molting) in insects—insects eating the plant thus have their developmental sequence altered. Extrafloral nectaries are yet another possible defense mechanism exhibited by BRACKEN FERN—these leaf-borne nectar-producing glands attract ants which under some circumstances may protect the plant by deterring herbivores (Power & Skog 1987; Cooper-Driver 1990; Rashbrook 1992). BRACKEN FERN is also known to be allelopathic (= the chemical inhibition of one species by another), with toxins leaching from the tissues adversely affecting surrounding plants—e.g., in the western U.S. the toxins can interfere with tree and shrub regeneration after fires or logging and are also known to inhibit the growth and development of potentially competing herbs. The growth of seedlings may be inhibited even after BRACKEN FERN is removed, apparently due to toxins remaining in the soil. In some places BRACKEN FERN is such an effective competitor that it can become dominant and disrupt the procession of ecological succession (Gliessman & Muller 1978; Mabberley 1987; Turner & Szczawinski 1991; Moran 1993a; Foster & Caras 1994; Weathers 1998; Bruneton 1999; Smith et al. 2000; Burrows & Tyrl 2001; Frohne & Pfänder 2005; Yatskievych & Windham 2008b; Robinson et al. 2010).

1. Terminal segments of pinnules ca. 6–15 times longer than wide; lower (= abaxial) surface of leaf blades glabrous or with sparse hairs; central pinnae at 40° to 60° angle to rachis; pinnules at 40° to 70° angle to costa; false indusia glabrous; longest pinnule in 2nd to 4th position from base of pinnae; subspecies widespread in e ⅓ of TX ____ subsp. **pseudocaudatum**
1. Terminal segments of pinnules ca. 4 times longer than wide; lower surface of leaf blades with abundant, spreading hairs; central pinnae at 70° to nearly 90° angle to rachis; pinnules at 75° to 90° angle to costa; false indusia pilose on margin and often on surface; longest pinnule in 1st or 2nd position from base of pinnae; subspecies known in TX only from the Trans-Pecos ____ subsp. **pubescens**

subsp. **pseudocaudatum** (Clute) Hultén, (Greek: *pseudo*, false, and *cauda*, tail or tail-like appendage, this taxon resembling var. or subsp. *caudatum*, now often treated as *P. caudatum*), TAILED BRACKEN FERN, PASTURE BRAKE, BRACKEN FERN. Leaf blades, rachises, and costae with lower surfaces glabrous or with sparse hairs; pinnules with terminal segments much elongated, 6–15 times longer than wide; false indusia glabrous. Open woods, pastures, thickets, often in sandy soils; widespread in Pineywoods and n Gulf Prairies and Marshes w through Post Oak Savannah (e.g., Anderson Co., *C.M. Rowell s.n.*, Sep 1986; Gonzales Co., *W.L. McCart 5870*, both BRIT), also Red River drainage w to Grayson Co. (S. Crosthwaite, pers. comm.); MA s to FL, w to IN, MO, OK, and TX (more abundant in the se U.S.; Thomson et al. 2008). Sporulating Jun–Nov. [*P. aquilinum* var. *pseudocaudatum* (Clute) A. Heller, *Pteris aquilina* var. *pseudocaudata* Clute] It has been suggested that the "more skeletonized" leaf blade of this subspecies may reduce evapotranspiration and thus be "of advantage in the semixeric pinelands of the south-eastern coastal plain" (Speer et al. 1998[1999]).

subsp. **pubescens** (Underw.) J.A. Thomson, Mickel & Mehltreter, (Latin: *pubens*, downy or softly hairy), HAIRY BRACKEN FERN, DOWNY BRACKEN FERN, WESTERN BRACKEN FERN. Leaf blades, rachises and costae typically with lower surfaces with dense covering of contorted, spreading hairs; pinnules with terminal segments not greatly elongated, only ca. 4 times longer than wide; false indusia with hairs at least on margins and sometimes also on surface. Open wooded slopes, near streams, disturbed areas; Jeff Davis (Mt. Livermore, Davis Mts.—*Palmer 32010*, TEX-LL; Palmer 1927) and Brewster (Rosillos Mts.) (Yarborough & Powell 2002) cos.; widespread in w ½ of the U.S. (predominantly w of the Rocky Mts., but extending to their e slopes) and adjacent Canada with disjunct populations e as far as MI and w TX; also Baja California, Mexico (Thomson et al. 2008). [*P. aquilinum* var. *pubescens* Underw., *Pteris aquilina* var. *pubescens* (Underw.) Clute] Though rare in TX, this is the commonly found subspecies in much of the w U.S.

subsp. *latiusculum* (Desv.) Hultén, (Latin: *latis*, wide, breath, broad, and the diminutive *-culum*, = somewhat broad), EASTERN BRACKEN FERN, is known from AR (all 75 counties—Peck 2011a) and LA just to the e of the state border and could possibly be found in TX. It can be distinguished from subspecies *pseudocaudatum* and *pubescens* by its central pinnae borne at 40° to 60° angles to rachis, its pinnules borne at 40° to 70° to costa, the terminal segments of pinnules only 2–4 times longer than wide, and the segment margins and lower surface of rachises and costae shaggy but the false indusia glabrous. [*P. aquilinum* subsp. *latiusculum* (Desv.) W.C. Shieh, *P. aquilinum* var. *latiusculum* (Desv.) Underw. ex A. Heller, *P. latiusculum* (Desv.) Hieron. ex Fries, *Pteris latiuscula* Desv.] This subspecies is widespread in the e ½ of the U.S. and adjacent Canada as well as n Mexico. Speer et al. 1998[1999] suggested that *latiusculum* and *pseudocaudatum* (as varieties) were "geographically overlapping bracken taxa" that "encompass a single uninterrupted gene pool." Based on DNA studies, Speer (2008) argues that *pseudocaudatum* and *latiusculum* "are not two separate taxa, but a single bracken variety" with northern and southern morphotypes. It is thus likely if an individual of the morphology of subsp. *latiusculum* is discovered in TX, it will ultimately not be recognized taxonomically. Jim Peck (pers. comm.) notes that "in AR they are morphologically distinct, with subsp. *latisusculum* more common in the north and subsp. *pseudocaudatum* more common to the south."

DRYOPTERIDACEAE Herter

WOOD FERN FAMILY

TX species usually terrestrial or on rocks; rhizomes short and stout to short-creeping to erect; leaves monomorphic or only slightly dimorphic; leaf blades 1-pinnate to 1–more pinnate-pinnatifid, the lower surfaces without transparent needle-like hairs; petiole bases with vascular bundles more than 3; sori on lower leaf surfaces on veins or vein tips, usually not marginal; indusia round and peltate or round-reniform and attached at a narrow sinus.

The Dryopteridaceae have been treated in a variety of ways in the past. We are following Smith et al. (2006, 2008) in considering it to be a monophyletic family comprising ca. 40–45 genera and 1,700 species (excluding Athyriaceae, Cystopteridaceae, Onocleaceae, Tectariaceae, and Woodsiaceae), with approximately 70% of these in just four genera (*Ctenitis*, *Dryopteris*, *Elaphoglossum*, and *Polystichum*). It is pantropical, and there are also many temperate species. Smith (1993b) treated the family much more broadly (ca. 60 genera and 3,000 species) and included genera here segregated (based on recent molecular data) into six other families (Athyriaceae—*Athyrium*, Cystopteridaceae—*Cystopteris*, Lomariopsidaceae—*Nephrolepis*, Onocleaceae—*Onoclea*, Tectariaceae—*Tectaria*, and Woodsiaceae—*Woodsia*). However, modern knowledge of fern phylogeny shows that such a broad interpretation would require a number of quite well delineated, monophyletic families to also be included (e.g., Aspleniaceae, Blechnaceae, Polypodiaceae, Thelypteridaceae—Smith et al. 2008), a situation that clearly would be neither practical nor desirable. Other authors (e.g., Lellinger 1985) have viewed the family in a much more restricted sense—32 genera and 850 species. Generic concepts within the family, particularly of the polystichoid ferns, currently are in flux (Little & Barrington 2003; Li & Lu 2006; Lu et al. 2007; Liu et al. 2010). The family has sometimes been called the Aspidiaceae, which is an illegitimate name. The key to genera is adapted from Diggs et al. (2006), who treated the family broadly. In order to facilitate identification, a second key, with includes genera previously often treated in the Dryopteridaceae, is provided. (Order Polypodiales)

FAMILY RECOGNITION IN THE FIELD: TX species either 1-pinnate with round sori and peltate indusia OR more than 1-pinnate, the pinnae themselves further divided (either pinnate or pinnatifid) and indusia round-reniform and attached at a narrow sinus; lower surfaces of leaf blades without transparent needle-like hairs (helping to distinguish this family from Thelypteridaceae, which have such hairs), the ultimate leaf segments (= smallest subdivisions of leaf) not entire; petiole bases with more than 3 vascular bundles (distinguishing this family from the Athyriaceae with 2).

REFERENCES: Correll 1956, 1966; Kramer et al. 1990b; Smith 1993b; Sano et al. 2000; Little & Barrington 2003; Li & Lu 2006; Liu et al. 2007b; Liu et al. 2010.

KEY TO GENERA IN DRYOPTERIDACEAE IN THE STRICT SENSE

1. Leaf blades more than 1-pinnate, the pinnae themselves further divided, either pinnate or pinnatifid; indusia round-reniform, attached at a narrow sinus; pinnae without a prominent basal lobe or auricle ________ **Dryopteris**
1. Leaf blades only 1-pinnate, the pinnae themselves not further divided, neither pinnate nor pinnatifid (but large basal lobe or auricle sometimes present on pinnae); indusia round, peltate; pinnae (especially the proximal) often (but not always) with a small to large basal lobe or auricle on the acroscopic side (= side toward the leaf apex).
   2. Terminal pinna essentially like the lateral pinnae except slightly smaller; plants rare in TX and known only from the Trans-Pecos in sheltered areas in canyons and ravines in the mountains ________ **Phaneorphlebia**
   2. Terminal pinna different in shape or size from lateral pinnae; plants known in TX only from the e ½ of the state typically in forest habitats at low elevations.
      3. Pinnae ovate, conspicuously narrowed to an acuminate tip, 4–10(–12) pairs; fertile pinnae essentially like the sterile, with sori scattered broadly over lower surface with obvious space between them; pinnae with veins anastomosing (= veins connecting by cross-veins to form a network) and thus forming areoles (= vein enclosed areas) which then contain free veinlets ________ **Cyrtomium**
      3. Pinnae oblong, parallel-sided (but often with a small basal auricle), acute or blunt at tip, more than 12 pairs (often many more); fertile pinnae reduced in size (limited to distal ⅓ to ½ of leaf), with sori crowded and often covering much of lower surface; veins branching dichotomously (= equally 2-forked) and free (= with their distal ends unconnected to other veins) ________ **Polystichum**

KEY TO PLANTS PREVIOUSLY INCLUDED IN A BROADLY CONCEIVED DRYOPTERIDACEAE

1. Fertile and sterile leaves completely different; fertile leaves without typical blade tissue, the sori in berry-like or bead-like structures; sterile leaf blades 1-pinnatifid (deeply divided but not completely pinnate); rachis with a conspicuous flange of photosynthetic tissue ________ *Onoclea* (see Onocleaceae)
1. Fertile and sterile leaves or portions of leaves similar, the fertile portion never without blade tissue, the sori on blade tissue; leaf blades at least completely 1-pinnate, often more divided; rachis without a flange of photosynthetic tissue.
   2. Leaf blades only 1-pinnate, the pinnae themselves neither pinnate nor pinnatifid (but large basal lobes or auricles sometimes present on pinnae).
      3. Pinnae 1–2(–3) pairs in addition to a single terminal pinna (pinnae sometimes with large basal lobes or auricles); plants primarily of the Edwards Plateau, known as far e as Comal Co. near w margin of the Blackland Prairie ________ *Tectaria* (see Tectariaceae)
      3. Pinnae four or more pairs (usually many more) in addition to a single terminal pinna; plants widespread in East TX and also the Trans-Pecos.
         4. Terminal pinna essentially like the lateral pinnae except slightly smaller; plants rare in TX and known only from the Trans-Pecos in sheltered areas in canyons and ravines in the mountains ________ **Phanerophlebia**
         4. Terminal pinna very different in shape or size from lateral pinnae; plants known in TX only from the e ½ of the state typically in forest habitats at low elevations.
            5. Pinnae ovate, conspicuously narrowed to an acuminate tip, 4–10(–12) pairs; fertile pinnae with sori broadly scattered over lower surface (sometimes appearing as if in numerous rows); pinnae with veins anastomosing (= veins connecting by cross-veins to form a network) and thus forming areoles (= vein enclosed areas) which then contain free veinlets ________ **Cyrtomium**
            5. Pinnae ± oblong, ± parallel-sided (but often with a small basal auricle) and blunt to acute at tip, more than 12 pairs (often many more); fertile pinnae with sori in 2 rows (these sometimes very crowded in *Polystichum*); pinnae with veins forked, free (= with their distal ends unconnected to other veins).
               6. Sori only on the uppermost, reduced, fertile pinnae (limited to distal ⅓ to ½ of leaf); indusia orbicular, not at all kidney-shaped; pinnae margins with bristly teeth; stolons absent ____ **Polystichum**
               6. Sori not restricted to the uppermost pinnae, the fertile pinnae not reduced; indusia orbicular-kidney-shaped; pinnae margins without bristly teeth (small non-bristly teeth may be present); stolons present, wiry and widely creeping ________ *Nephrolepis* (see Lomariopsidaceae)
   2. Leaf blades more than 1-pinnate, the pinnae themselves either pinnate or pinnatifid.
      7. Sori elongate, straight to hooked or curved; indusia with elongate attachment along one side, the attachment as long as the elongate sori ________ *Athyrium* (see Athyriaceae)
      7. Sori round or nearly so; indusia variously attached, but attachment not elongate.
         8. Leaf blades relatively large, 10–30 cm wide, 35–120 cm long; pinnules (= subdivisions of pinnae) of middle pinnae typically 5 mm or more wide at their bases, only shallowly toothed and thus not appearing subdivided; indusia attached at a narrow sinus, the round-kidney-shaped indusia usually obvious with the naked eye or at least with 10X magnification; petiole bases with more than 3 vascular bundles; all species occurring in TX rare ________ **Dryopteris**

8. Leaf blades usually smaller, 3.5–10 cm wide, 8–40(–60) cm long; pinnules of middle pinnae typically less than 5 mm wide at their bases, at least some so deeply toothed or incised as to appear subdivided; indusia not attached at a narrow sinus (rather attached below sori and at first cup-like and completely enclosing sori, but later splitting into several irregular lobes or flaps spreading around sori OR attached along one side of sori, hood- or pocket-like, arching over sori) (NOTE: indusia are often inconspicuous in mature sori); petiole bases with 2 vascular bundles; including species common and widespread in parts of TX.
    9. Petioles with scattered light brown scales and glandular pubescence, glabrate with age; indusia attached below sori, when young cup-like and completely enclosing sori, but later splitting into several irregular lobes or flaps spreading around sori (sometimes inconspicuous in mature sori); leaf blades often broadest above the middle; veins of ultimate leaf segments (= smallest subdivisions of leaf) not reaching segment margins (use magnification; important if sori not available); plants widespread and common in e ½ of TX, and rare in the Trans-Pecos __________ *Woodsia* (see Woodsiaceae)
    9. Petioles glabrous except at very base; indusia attached by one side, hood- or pocket-like, arching over sori (sometimes inconspicuous in mature sori) towards segment margin; leaf blades broadest at base or at or just below middle; veins of ultimate leaf segments reaching segment margins; plants widespread (East TX to Trans-Pecos), but very rare in TX ________ *Cystopteris* (see Cystopteridaceae)

CYRTOMIUM C. Presl

HOLLY FERN, ASIATIC HOLLY FERN, NET-VEIN HOLLY FERN

A taxonomically difficult genus sometimes treated as having only ca. 15 (Yatskievych 1993a) but more recently regarded as having 31 (Lu & Cheng 2003) to ca. 40 species (Lu et al. 2006). It is mainly East Asian (China and Japan) but one species occurs in East Africa and India. A number of the Chinese species are endemic to limestone areas and are mostly apogamous triploids and sexual tetraploids (Lu et al. 2006). Some authorities treat it within *Polystichum* (e.g., Mabberley 2008), and according to Yatskievych (1993a), it "might better be considered a subgenus of *Polystichum*, from which it is poorly differentiated morphologically." However, Little and Barrington (2003) noted that, "Our results lead us to suggest recognizing the clade of earliest diverging *Polystichum* species as a distinct genus (*Cyrtomidictyum*) and to continue to recognize *Cyrtomium* as a separate genus, leaving a monophyletic *Polystichum* sensu stricto (s.s.)." More recent molecular work by Li and Lu (2006), Lu et al. (2007), and Liu et al. (2010) strongly argues that neither *Cyrtomium* nor *Polystichum* as traditionally viewed are monophyletic. Liu et al. (2010), for example, suggest recognizing a number of genera including *Cyrtomium* and *Polystichum* in much restricted senses (see more discussion under *Polystichum*). Thieret (1980) suggested that "with its network of veins—as opposed to free veins in *Polystichum*, the genus seems distinct enough." (Greek: *cyrtoma*, arch, in reference to the arch-like appearance of the anastomosing veins [= veins connecting by cross-veins to form a network])

REFERENCES: Christensen 1930; Degener & Hawkes 1951; MacDougal 1976; Landry et al. 1979; Thieret 1980; Yatskievych 1990, 1993a; Little & Barrington 2003; Lu & Cheng 2003; Peck 2003; Matsumoto 2003; Li & Lu 2006; Lu et al. 2005, 2006, 2007; Oka 2008; Liu et al. 2010.

Cyrtomium falcatum (L.f.) C. Presl, (Latin: *falcatus*, sickle-shaped or hooked, in reference to the curved pinnae), JAPANESE NET-VEIN HOLLY FERN, JAPANESE HOLLY FERN, HOUSE HOLLY FERN. Terrestrial or on rocks or masonry; rhizomes short, stout, conspicuously scaly, the scales orangish brown; leaves monomorphic, evergreen, 28–60(–100) cm long; petioles conspicuously scaly at least near base; leaf blades 1-pinnate, 15–35 cm long, leathery-thickened; pinnae short-stalked, 4–10(–12) pairs, 4–8.5(–11) cm long, 1.5–3 cm wide, obliquely ovate to lanceolate, usually falcate (= sickle-shaped), sometimes with a short basal lobe on the acroscopic side (= side toward the leaf apex) of the pinnae, apically acuminate, marginally variable ranging from nearly entire to undulate or irregularly and coarsely dentate (sometimes even appearing fringed), often nearly entire near the tips, bright green and shiny on upper surface, the terminal pinna typically with more lobing/large teeth than laterals, the veins anastomosing and thus forming areoles which then contain free veinlets; sori round, conspicuous, scattered over whole lower (= abaxial) surface of pinnae (sometimes appearing to be in rows); indusia peltate, at first pale, then becoming brownish or even blackish in center, not shriveled at maturity; $n = 2n = 123$ (Yatskievych 1993a). Widely cultivated and escaped; in the Pineywoods in Houston Co. (Davy Crockett Natl. Forest—*Stotts 38*, HPC) on remnant brick wall of old sawmill site and in San Jacinto Co. (*E.*

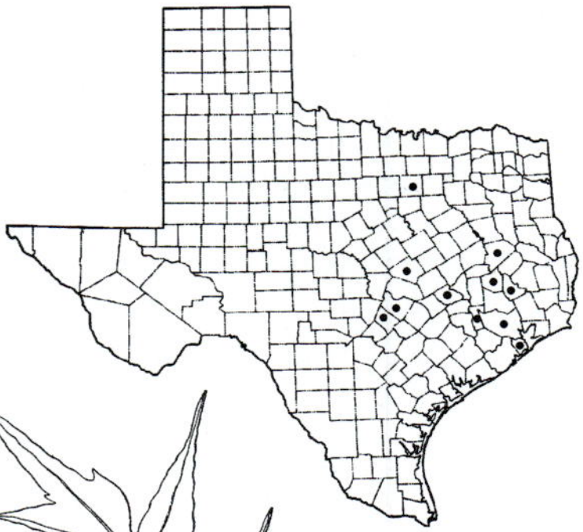

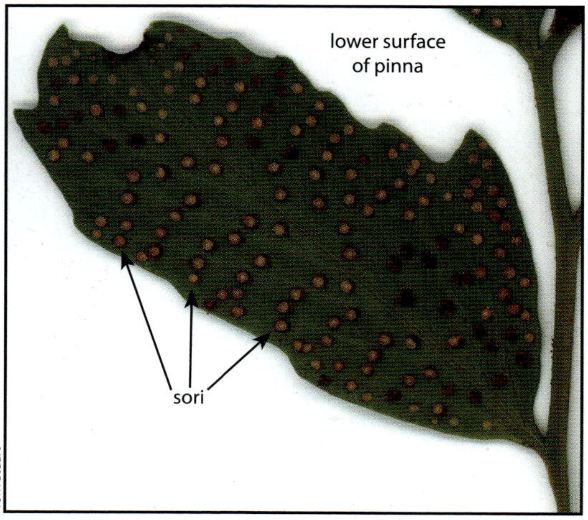

Tom Stuart

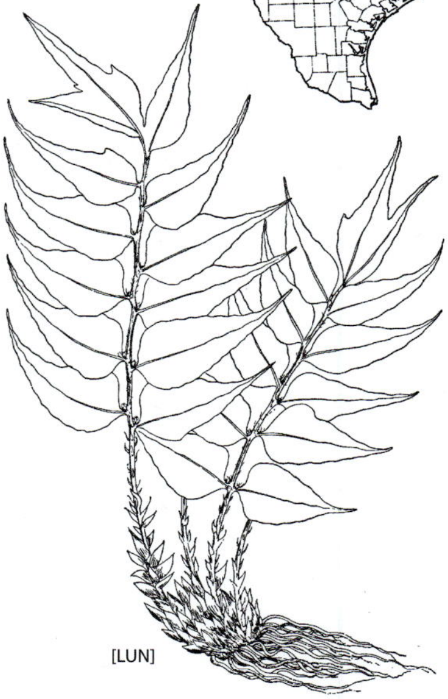

[LUN]

Stan Shebs, cc by-sa

Cyrtomium falcatum

JAPANESE NET-VEIN HOLLY FERN

Keith 3, BRIT) "growing out of the wall of a waterfall," also in the Post Oak Savannah on stream bank in Waller Co. (*E. McWilliams 1995-100*, BRIT) and on bluff in Burleson Co. (*Wilson s.n.*, TAMU), and near w margin of Blackland Prairie in Bell (*Sanchez & Pekins 4651*, TEX-LL on limestone cliff), Hays (*Carr 18255*, TEX-LL on limestone bluff), and Travis (*W.R. Carr, 6019*, BRIT, on moist limestone boulders) cos.; also Jefferson (BAYLU) and Harris (BRIT, TEX-LL) cos. in n Gulf Prairies and Marshes and Tarrant Co. (Turner et al. 2003) in the Cross Timbers and Prairies; naturalized se U.S. from SC s to FL w to AR (Peck 2011a) and TX, also CA and VA (Weakley et al. 2012) and reported from CN, NJ, NY, OH, and OR (USDA PLANTS 2012). Sporulating spring–fall. Native of e Asia. [*Phanerophlebia falcata* (L.f.) Copel., *Polypodium falcatum* L.f., *Polystichum falcatum* (L.f.) Diels] All U.S. plants appear to be apogamous triploids (Yatskievych 1993a). Matsumoto (2003) recognized three subspecies within *C. falcatum*. Hoshizaki and Moran (2001) noted "In wild plants the margins are nearly entire, although a few obscure, minute teeth may be present." This widely cultivated species is naturalized in many parts of the world (e.g., Australia, Azores, Great Britain), including various parts of the U.S. (Yatskievych 1993a; Mabberley 1997). There are a number of cultivars with one of most common being the ROCKFORD FERN (cv. 'Rochfordianum'), which has obviously fringed pinnae margins (Hoshizaki & Moran 2001). While we are not treating infraspecific taxa in this species, J.H. Peck (pers. comm.) indicates that most of the material he has seen is consistent with either subsp. *falcatum* as can be found in the area around Tokyo (see Oka 2008) or with the ROCKFORD FERN cultivar. While in TX this species is typically found growing on rocks (often limestone) or old masonry, it is also known to be halophytic (adapted to live under high salt conditions). In its native range (e Asia) and elsewhere it is often found in maritime environments (Zhao et al. 2002, 2011; Diaz et al. 2005).

Cyrtomium fortunei J. Sm., (for Robert Fortune, 1812–1880, Scottish plant explorer and collector in China, who was responsible for introducing tea from China into India), ASIAN NET-VEIN HOLLY FERN, HARDY JAPANESE HOLLY FERN, FORTUNE'S NET-VEIN HOLLY FERN, FORTUNE'S CYRTOMIUM, native to Japan, Korea, southeast China, Vietnam and eastern Thailand, is known in the U.S. in AR (Peck 2011b), GA, LA, MS, OR, and SC (e.g., MacDougal 1976; Landry et al. 1979). Rhizome scales brown; pinnae sometimes with a short basal lobe on the acroscopic side; indusia tan to pale brown, uniformly colored, shriveling at maturity. [*Phanerophlebia fortunei* (J. Sm.) Copel., *Polystichum falcatum* var. *fortunei* (J. Sm.) Matsum., *Polystichum fortunei* (J. Sm.) Nakai] This is a variable species in which several varieties have been named (Hoshizaki & Moran 2001); plants naturalized in the U.S. are reported to be var. *fortunei* (Yatskievych 1993a). It can be distinguished from *C. falcatum* by its pinnae being in (8–)10–25 pairs, papery, not shiny on upper surface, and with the margins minutely crenulate-denticulate (finely toothed) including close to the tip. It is an apogamous triploid ($n = 2n = 123$) (Yatskievych 1993a). Given that the species is known from LA (e.g., Landry et al. 1979) and AR (Peck 2011b), it is included here as a note to alert collectors and others to be aware of it as a possibility as an escape from cultivation in TX.

DRYOPTERIS Adans. WOOD FERN, SHIELD FERN

Plants of various habitats (terrestrial, on logs, on rocks); rhizomes short-creeping to erect, scaly; petiole bases with more than 3 vascular bundles, these arranged in an arc; leaves 1-pinnate-pinnatifid to 3-pinnate-pinnatifid, monomorphic to slightly dimorphic, in one TX species the fertile leaves sometimes larger than the sterile leaves; pinnae with serrate to crenate to lobed segments; basal pinnae reduced; sori round, in a single row on each side of the midvein of the fertile segments, variously located (in three TX species midway between midvein and margins of ultimate leaf segments, in one species mostly at margins of ultimate leaf segments, and in one species mostly near base of the sinuses between ultimate leaf segments); indusia round-reniform, attached at a narrow sinus.

A widely scattered but largely north temperate genus of ca. 150 (Mabberley 2008) to 250 species (Montgomery & Wagner 1993). A few occur in North America, but most are found in temperate Asia. Hybridization between species is common (Wagner 1970; Montgomery & Wagner 1993; Mickel & Smith 2004) and allopolyploidy is common (Mabberley 2008). The genus is widely cultivated in temperate gardens (Hoshizaki & Moran 2001). Weakley (2010) noted that, "*Dryopteris* and *Athyrium* are often confused when not fertile; they can be easily distinguished by breaking off a leaf and counting vascular bundles (which will appear as thread-like strands)." *Dryopteris* has more than 3 (often 5), while *Athyrium* has 2. All members of the genus occurring natively in TX are rare and of limited distribution, and we consider them to all be of conservation concern. (Greek: *drys,* oak or tree, and *pteris*, fern; several species are associated with oak woodlands) REFERENCES: Walker 1962; Wagner 1970; Thomas et al. 1973; Montgomery & Paulton 1981; Carlson & Wagner 1982; Montgomery 1982; Peck & Peck 1988; Fraser-Jenkins 1986; Werth et al. 1988; Werth 1991; Montgomery & Wagner 1993; Hoshizaki & Wilson 1999; Peck et al. 2000; Peck 2003; Mink et al. 2010, 2011a.

1. Leaf blades 3-pinnate-pinnatifid at least at base; sori mostly EITHER immediately adjacent to base of the sinuses between ultimate leaf segments OR near margins of ultimate segments (if near margins then leaf blades herbaceous); petioles with scales cinnamon-colored (= light reddish brown); species known in TX only from a cave in Val Verde Co. in w TX ______ **D. cinnamomea**
1. Leaf blades 1-pinnate-pinnatifid or 2-pinnate-pinnatifid at base; sori mostly EITHER midway between midvein and margin of ultimate leaf segments OR at margins of ultimate leaf segments (if at margins then leaf blades leathery); petioles with scales various, from cinnamon-colored to light brown, brown or in 1 species sometimes tan with dark central stripe; species known in TX from East TX, the Trans-Pecos, and the Panhandle.
   2. Sori mostly at margins of ultimate leaf segments; petioles with scales tawny or cinnamon-colored; species known in TX only from Hemphill Co. in the Panhandle ______ **D. marginalis**
   2. Sori midway between midvein and margin of ultimate leaf segments; petioles with scales usually brown or light brown or in 1 species sometimes tan with dark central stripe; species known in TX from both extreme East TX and the Trans-Pecos.
      3. Petioles < ¼ total leaf length (petiole and blade combined), with 2 distinct types of scales (1 broad, scale-like and 1 hair-like); species known in TX only from Davis Mts. in the Trans-Pecos ______ **D. filix-mas**
      3. Petioles > ¼ total leaf length, without any hair-like scales; species known in the state only from extreme East TX.
         4. Leaf blades with fertile pinnae occupying at least the distal ½ or sometimes nearly the entire blade; segments of fertile pinnae nearly the same width as those of typical sterile pinnae; teeth along segments of pinnae narrowed to a minute spine-like tip, often incurved; species of extreme ne part of East TX ______ **D. celsa**
         4. Leaf blades with fertile pinnae only towards the tip (distal ⅓ to at most distal ½); segments of fertile pinnae narrower than those of typical sterile pinnae; teeth along segments of pinnae triangular, not incurved; species of se part of East TX ______ **D. ludoviciana**

Dryopteris celsa (W. Palmer) Knowlton, W. Palmer & Pollard, (Latin: *celsus*, raised, held high, elevated, apparently in reference to this species sometimes growing upon logs and stumps), LOG FERN. Rhizomes short-creeping; leaves monomorphic, ± evergreen; petioles ⅓ length of total leaf, scaly at least at base, the scales brown or tan with dark central stripe; leaf blades 1-pinnate-pinnatifid to 2-pinnate-pinnatifid at base, ovate-lanceolate, 35–80 cm long, herbaceous; fertile pinnae in distal ½ of leaf blade to nearly the entire blade; segments of fertile pinnae nearly the same width as those of typical sterile pinnae, not more widely spaced; sori midway between midvein and margin of ultimate leaf segments; $2n = 164$ (Montgomery & Wagner 1993). Rotting logs, stumps, and piles of humus in swamps or seeps, acidic soils; included based on a 1925 Palmer collection (*29404*, MO [annotated by R. Cranfill in 1981 & J. Peck in 1985], GH—Peck & Peck 1988, annotated G. Diggs 2012) from Bowie Co. (cited in Correll 1956 and Correll & Johnston 1970 as *D. cristata*) in the extreme ne corner of e TX and also on the range map in Werth et al. (1988) showing ne TX; e U.S. from PA s to SC w to AR, MO, n LA, and ne TX (the TX location is slightly disjunct from the Ouachita Mts. in AR to the n), also disjunct populations in MI and NY. *Dryopteris cristata* (L.) A. Gray, CRESTED WOOD FERN, a circumboreal species of the ne U.S. and s Canada, was long erroneously reported for TX (Correll 1956; Wherry 1964; Correll & Johnston 1970; Correll & Correll 1972), but does not occur in the state (Thomas et al. 1973; Peck & Peck 1988; Peck 2000; Diggs et al. 2006; Mink et al. 2011a); the erroneous reports were based on incorrect identification of the 1925 Palmer collection. A detailed discussion and photographs of the Palmer specimens involved (some were mixed collections with *Osmundastrum cinnamomeum*) can be found in Mink et al. (2011a). *Dryopteris cristata* is easily recognized in the field because the upper fertile pinnae are held horizontally (like an open venetian blind) in contrast to the situation in *D. celsa*, which has both fertile and sterile pinnae held parallel to the plane of the blade (like a closed venetian blind). Sporulating summer–fall. [*D. goldiana* (Hook. ex Goldie) A. Gray subsp. *celsa* W. Palmer] *Dryopteris celsa* is a fertile allotetraploid resulting from hybridization between *D. goldiana* (Hook. ex Goldie) A. Gray (GOLDIE'S WOOD FERN) and *D. ludoviciana* (Walker 1962; Montgomery & Wagner 1993). It has been proposed that the two parent species, which no longer occur together, were pushed into contact by the climatic change accompanying the last glacial maximum,

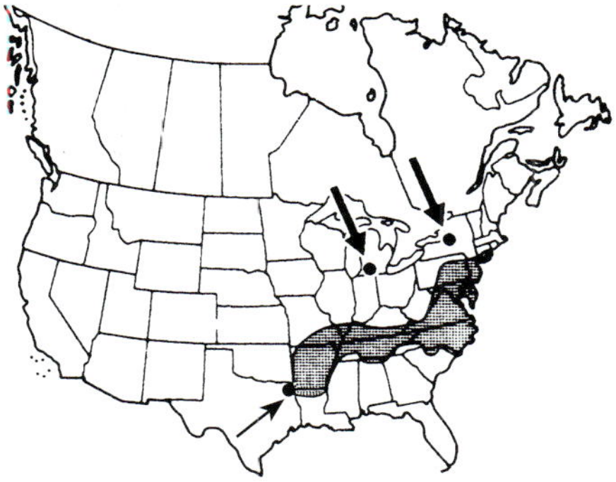

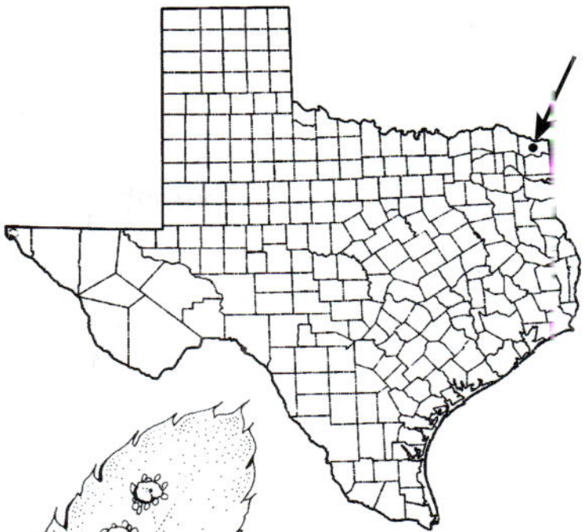

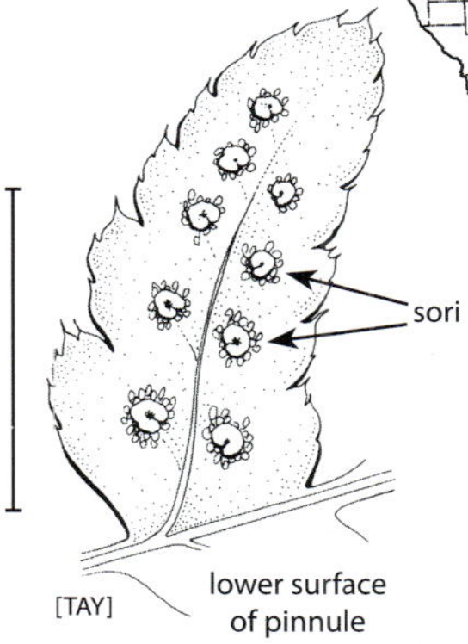

upper surface of pinna

Dryopteris celsa

LOG FERN

only ca. 18,000 years ago. If this is the case the species is quite young in terms of geologic time (Taylor 1984; Moran 2004). In various parts of its range, *D. celsa* is known to hybridize with six other species. The common name, LOG FERN, is derived from the tendency of the species to grow on logs or humus (Snyder & Bruce 1986). Thomas et al. (1973) noted "The plants are best located in midwinter and late winter, when the swamps are relatively clear of foliage. The previous year's fronds lie upon the ground but will usually remain more or less evergreen and excellent specimens may be obtained." Because of its rareness and extremely limited distribution in the state, we consider this species to be of conservation concern in TX. ⚠

Dryopteris cinnamomea (Cav.) C. Chr., (from Latin: *cinnamum*, cinnamon-colored or cinnamon-brown; see *Osmundastrum cinnamomeum* for further discussion of derivation), CINNAMON WOOD FERN. Rhizomes short-creeping to suberect; leaves monomorphic, evergreen; petioles ca. ¼ length of total leaf, with scales at least at base, the scales entire or slightly denticulate marginally, apically acute, cinnamon-colored; leaf blades 3-pinnate-pinnatifid at least at base to 2-pinnate-pinnatifid above, deltate-ovate, to ca. 37 cm long, herbaceous, often with an elongate tip; sori mostly either immediately adjacent to base of the sinuses between ultimate leaf segments or near margins of ultimate segments; $2n = 82$ (Montgomery & Wagner 1993). In TX known only from a cave near Comstock (e of the Pecos River) in Val Verde Co. (*Babcock 100*, 1965, SBSC, TEX-LL) (in Mexico terrestrial or on rocky ledges in light shade of montane forests—Mickel & Smith 2004); in the U.S. known only from TX, but widespread in Mexico. This species was previously cited as occurring in AZ (Montgomery & Wagner 1993), but the AZ specimens have since been determined to be *D. knoblochii* A.R. Sm. (Mickel & Smith 2004; G. Yatskievych pers. comm.). Sporulating ?–Sep–? in TX. [*Tectaria cinnamomea* Cav.] This species is part of the poorly understood *D. patula* complex of Mexico and Central America (Montgomery & Wagner 1993). The TX location is considerably disjunct from other populations. Recently, Mickel and Smith (2004) in discussing *D. cinnamomea*, referred to the TX citation as "uncertain." It is thus possible that the Val Verde Co. Babcock collection is actually another closely related species in the *D. patula* complex that is not currently known from the U.S. Because of its rareness and extremely limited distribution in both TX and the U.S., we consider this species to be of conservation concern. ⚠

Dryopteris filix-mas (L.) Schott, (Latin: *filix*, fern and *mas*, male or masculine, = male fern, in reference to the less delicate or less graceful appearance of this species in comparison with the superficially similar *Athyrium filix-femina*, LADY FERN), MALE FERN. Rhizomes suberect to erect; leaves monomorphic, dying back in winter; petioles < ¼ length of total leaf, with scales at least at base, the scales of 2 distinct types (1 broad, scale-like and 1 hair-like), brown, usually quite obvious; leaf blades 1-pinnate-pinnatifid to 2-pinnate-pinnatifid at base, ovate-lanceolate, ca. 22–90+ cm long, with texture firm but not leathery, basal basiscopic pinnule and basal acroscopic pinnule ca. equal; sori midway between midvein and margin of ultimate leaf segments; $2n = 164$ [?] (Montgomery & Wagner 1993). On ledges, bluffs, cliffs, and moist woods (Yarborough & Powell 2002); in TX known only from the Davis Mts., Jeff Davis Co. (*Correll 15003, 34969; L.C. Hinkley 173; Warnock 7454, 23020, 23094*, all TEX-LL; *L.C. Hinkley s.n.* 1936, *P. Manning et al. 3395*, SRSC; *B. McBride s.n.*, BRIT, HCP, "Dark shaded crevices of large granite boulders, Pile of Rocks"; *D. Benham & C.M. Rowell 172*, BRIT; *E.J. Palmer 32035, 34042, 34262*, MO); ne U.S. and adjacent Canada and widespread in the w U.S. and adjacent Canada; also Mexico (Coahuila), Asia, and Europe. Sporulating May–Nov. [*Polypodium filix-mas* (L.) Schott] Montgomery and Wagner (1993) noted that this species has been considered to be both an auto- and an allopolyploid, that it "may be composed of at least two closely related taxa," and that its taxonomy "is not well understood." They further observed that plants "in the northeast and northwest are tetraploid. These differ morphologically and ecologically from a taxon of unknown chromosome number in the southwestern Rocky Mountains [presumably including the TX collections]. The Rocky Mountain taxon closely resembles the Mexican *D. pseudofilix-mas* (Fée) Rothmaler." Mickel and Smith (2004) observed that plants from the w U.S. "may not be conspecific with those from eastern North America and elsewhere." Thus, this widespread species (North America, Europe, Asia) is in need of detailed study and the exact nature of the TX populations is still unclear. *Dryopteris filix-mas* is one of the oldest known vermifuges (= a medicine that expels intestinal worms)—the drug aspidium, derived from the rhizomes, paralyzes tapeworms (which can then be expelled) but is dangerous because it also paralyzes voluntary muscles of patients, and overdoses are potentially fatal. It has been used since ancient Greek times. Livestock can also be affected upon eating the apical buds of the rhizomes—severe digestive disturbances and neurological problems including blindness can result (Correll

George Yatskievych

Dryopteris cinnamomea

CINNAMON WOOD FERN

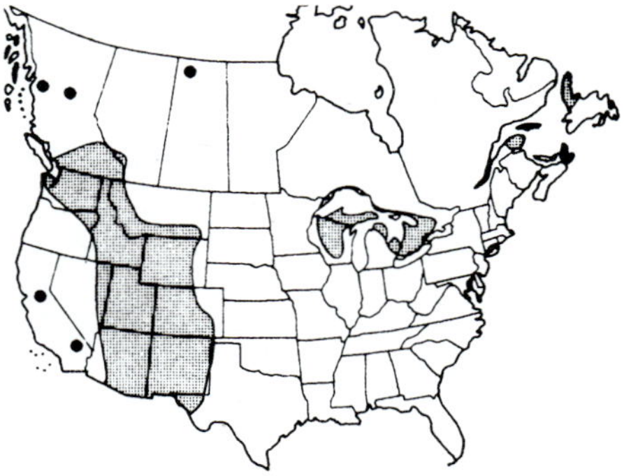

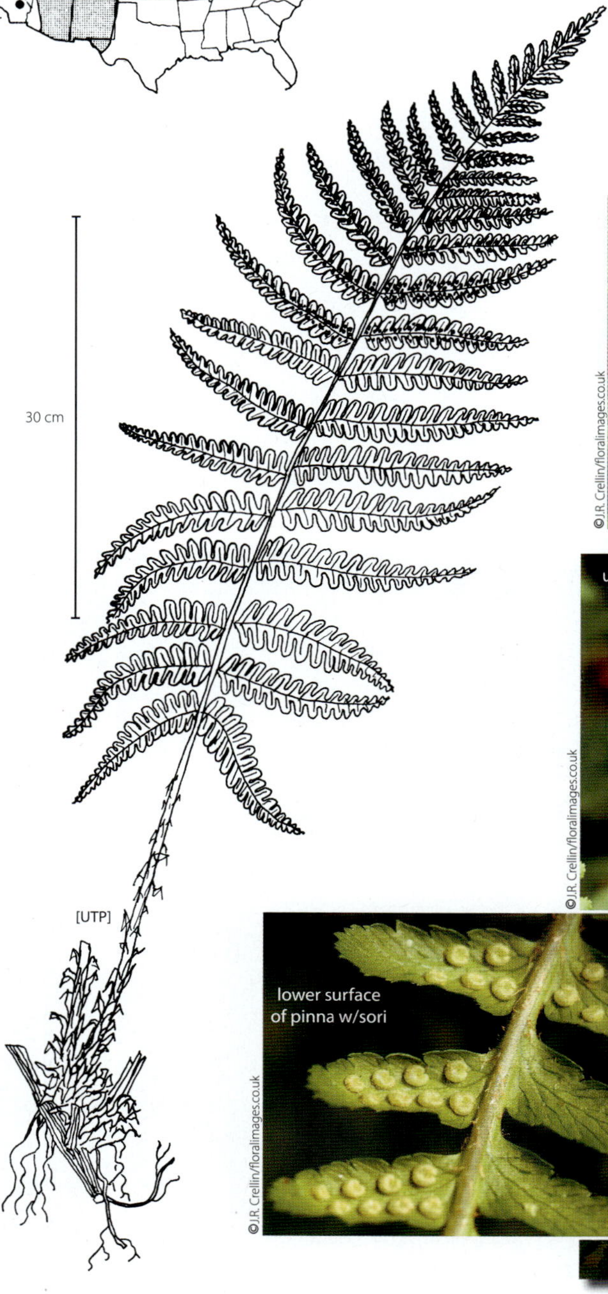

Dryopteris filix-mas

MALE FERN

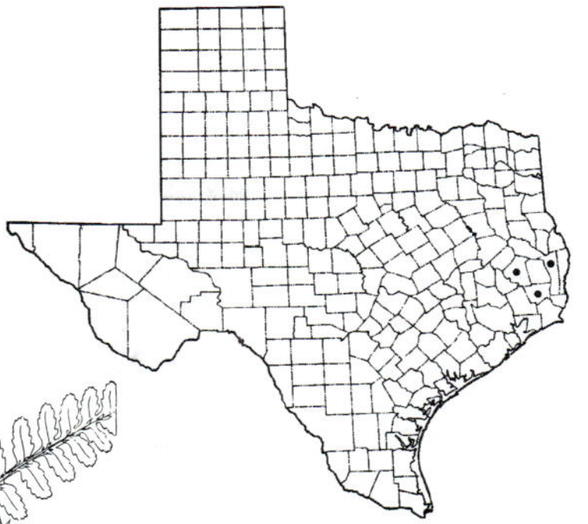

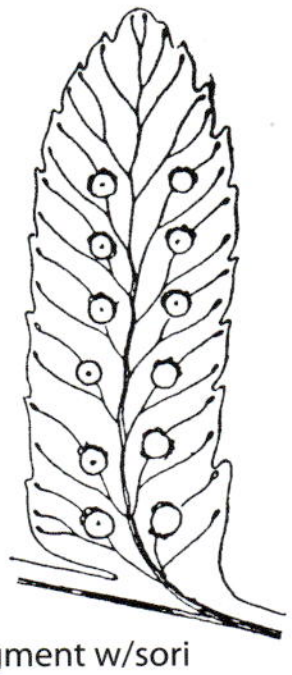

segment w/sori

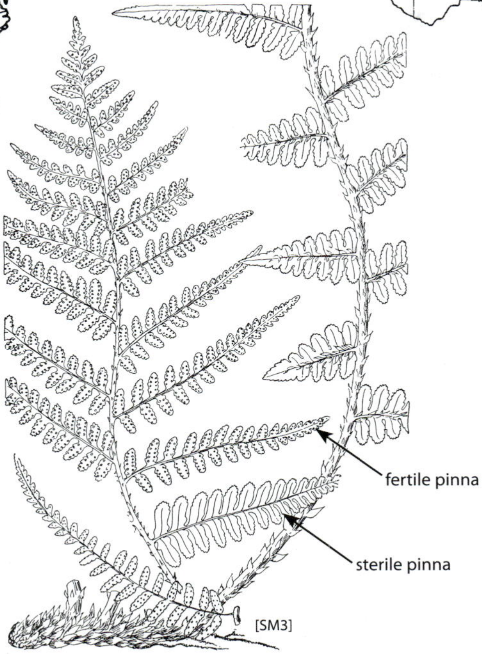

Dryopteris ludoviciana
SOUTHERN WOOD FERN

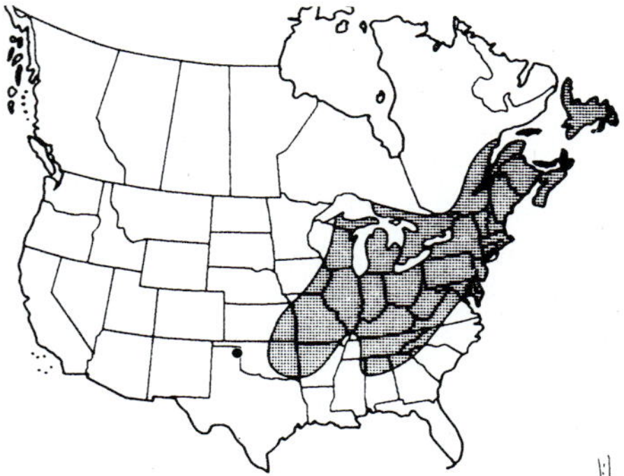

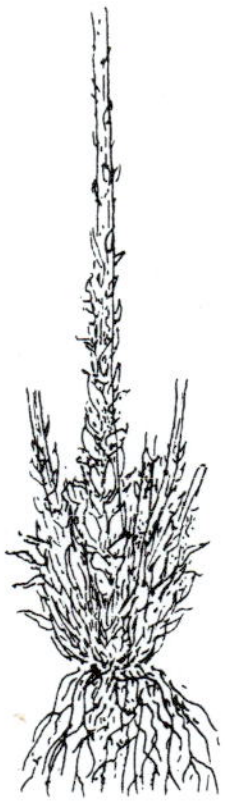

rhizome

[JTM]

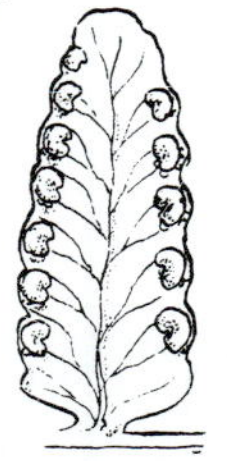

lower surface
of pinnule w/sori

John Hilty, Illinois Wildflowers: www.illinoiswildflowers.info

John Hilty, Illinois Wildflowers: www.illinoiswildflowers.info

Dryopteris marginalis

MARGINAL WOOD FERN

1956; Mabberley 1997; Yarborough & Powell 2002). Because of its rareness and extremely limited distribution in the state, we consider this species to be of conservation concern in TX. ⚠ ☠

Dryopteris ludoviciana (Kunze) Small, (of Louisiana, where the species was first discovered), SOUTHERN WOOD FERN, LOUISIANA WOOD FERN. Rhizomes short-creeping, stout; leaves somewhat dimorphic, evergreen; petioles > ¼ length of total leaf, scaly at base, the scales brown; leaf blades 1-pinnate-pinnatifid, lanceolate, 25–90(–150) cm long, herbaceous, the basal pinnae much reduced, basal basiscopic pinnule slightly longer than basal acroscopic pinnule; fertile pinnae in distal ½ of leaf blade; segments of fertile pinnae distinctly narrower than those of sterile pinnae, often more widely spaced, and sometimes contracted at base; sori midway between midvein and margin of ultimate leaf segments; $2n = 82$ (Montgomery & Wagner 1993). Swamps and wet woods; Hardin (*G.E. Watson s.n.*, 1971, TEX—first confirmed collection for TX; Correll 1972a), Jasper (*S.L. Orzell & E.L. Bridges 5942*, BRIT, MO, TEX-LL), and Polk (*Orzell, Bridges, & G. Watson 5768*, BRIT, MO, TEX-LL) cos. in the Pineywoods; endemic to the se U.S. in AL, AR, FL, GA, LA, NC, SC, and TX. In 1956 Correll said, "This species was reported by Reverchon (1903) as having been collected by J.M. Fetherolf in Newton County in the Timber Belt. It was also reported from TX by Cory and Parks (1937). I have not seen any specimens of this species west of East Baton Rouge Parish, Louisiana." Correll (1972a) subsequently reported it for TX. This diploid species is endemic to the se U.S. (Montgomery & Wagner 1993), most collections being from the extreme southeast with outliers in AR (Peck 2011a), LA, and TX. Sporulating spring–fall. [*Aspidium ludovicianum* Kunze; *Thelypteris ludoviciana* of authors] This species is one of the parents of *D. celsa;* it is also reported to cross with *D. celsa* resulting in sterile hybrids (Werth et al. 1988; Montgomery & Wagner 1993). Weakley (2010) noted that it is "one of the diploid 'parent species' of the e. North American reticulately-evolved *Dryopteris* complex" and is involved in the origin of at least three other species. Because of its rareness and limited distribution in the state, we consider this species to be of conservation concern in TX. (TOES 1993: IV) ⚠

Dryopteris marginalis (L.) A. Gray, (Latin: *marginalis*, bordering or on the periphery or marginal, in reference to the position of the sori typically at margins of ultimate leaf segments), MARGINAL WOOD FERN, MARGINAL SHIELD FERN, LEATHER WOOD FERN. Rhizomes ascending to erect; leaves monomorphic, evergreen; petioles ¼–⅓ length of total leaf, with tawny or cinnamon-colored scales at least at base; leaf blades 1-pinnate-pinnatifid to 2-pinnate-pinnatifid at base, lanceolate, 25–50(–75) cm long, leathery, the basal pinnae slightly reduced, the basal basiscopic pinnule longer than basal acroscopic pinnule; fertile pinnae in distal ⅔ of leaf blade; sori mostly at margins of ultimate leaf segments (but some can be adjacent to base of the sinuses); $2n = 82$ (Montgomery & Wagner 1993). Terraced stream bank adjacent to the Canadian River; in TX known only from the Gene Howe Wildlife Management Area, Hemphill Co. (*J. Singhurst 17614*, 15 Oct 2009, BAYLU) in the Panhandle; ne U.S. and adjacent Canada s to AL and GA and w to WI, OK, and TX. Sporulating in TX ?–Oct–? (summer–fall in other states). [*Nephrodium marginale* (L.) Michx., *Polypodium marginale* L.] This species was reported as new to TX in 2010 (Mink et al. 2010). The TX location is approximately 200 km disjunct to the nw of the nearest location in Comanche County, OK (Mink et al. 2010). Because it is known from only a single station in Hemphill Co., we consider this species to be of conservation concern in TX. ⚠

PHANEROPHLEBIA C. Presl

MEXICAN HOLLY FERN, HOLLY FERN

Plants terrestrial or on rocks; rhizomes short-creeping to erect; leaves monomorphic, evergreen; leaf blades 1-pinnate, with an apical pinna similar to the laterals; pinnae usually falcate, basally lobed or not so, attenuate apically, marginally serrulate or spinulose (rarely deeply incised to lacerate), the proximal usually largest or nearly so, the veins usually free, 1–3-forked; sori round, in 2–4 irregular rows on each side of the midribs of the pinnae, between midrib and margin; indusia peltate, persistent or shriveling.

A neotropical genus of 9 species ranging from AZ and TX to Hispaniola and n South America; 8 of the species occur in Mexico. "Most species grow in mesic forests at middle to higher elevations, often in calcareous rock crevices or calcareous soil" (Mickel & Smith 2004). The relationships of the polystichoid ferns (in TX including *Phanerophlebia*, *Polystichum,* and the introduced *Cyrtomium*) have been unsettled for many decades. For example, *Phanerophlebia* has sometimes been included in *Cyrtomium* (e.g., Morton 1957b),

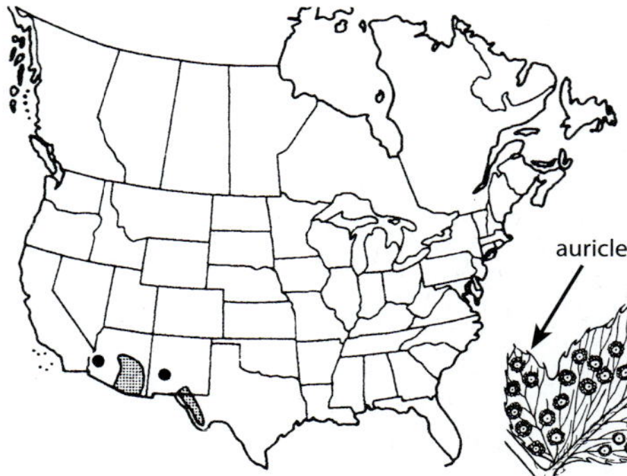

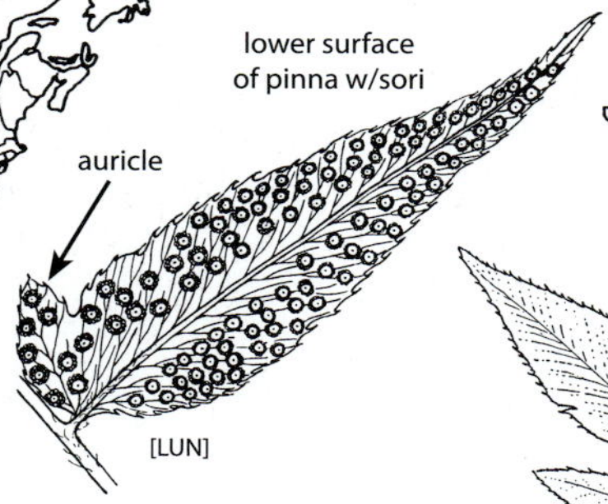

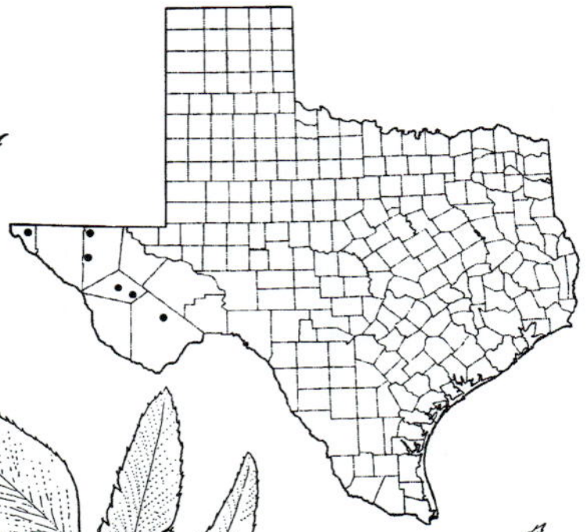

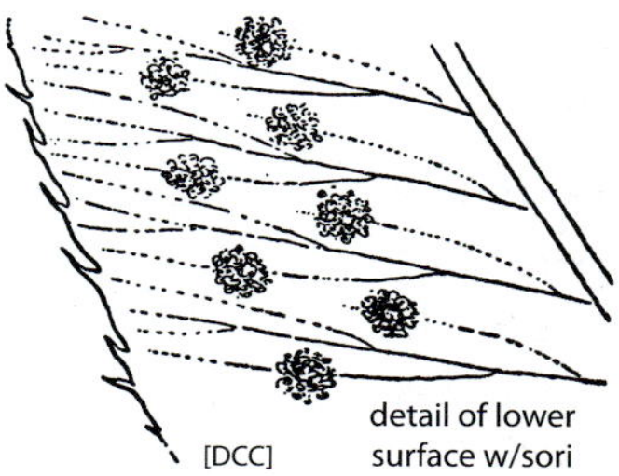

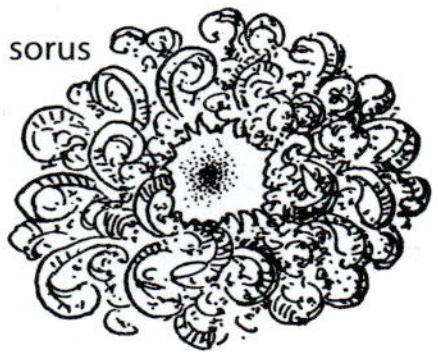

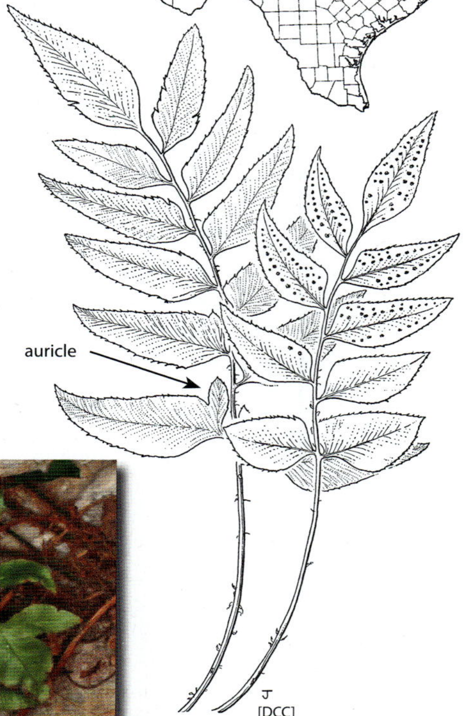

Dale A. Zimmerman Herbarium, Western N. Mex. Univ.

Dale A. Zimmerman Herbarium, Western N. Mex. Univ.

Phanerophlebia auriculata

EARED HOLLY FERN

which it superficially resembles morphologically; however, Yatskievych et al. (1988) and Stein et al. (1989) presented molecular evidence that the similarities are the result of convergent evolution. Yatskievych (1996), in the most recent revision of the taxa, recognized *Phanerophlebia* as a separate genus, and more recently, Little and Barrington (2003) and Liu et al. (2010) obtained molecular evidence showing that *Phanerophlebia* is in a clade that is sister to a clade containing some species of *Polystichum* and some of *Cyrtomium*; molecular evidence thus does not support including *Phanerophlebia* in one of the larger genera. Although the nomenclature is still unsettled, we are following most recent authors including Mickel and Smith (2004) in recognizing *Phanerophlebia* at the generic level. (Greek: *phaneros*, free, and *phlebium*, vein, from the free [nonanastomosing] venation of *P. nobilis*, the type species—Yatskievych 1993b)

REFERENCES: Underwood 1899; Maxon 1912; Morton 1957b; Yatskievych et al. 1988; Stein et al. 1989; Yatskievych 1990; 1992, 1993b, 1993c, 1996; Little & Barrington 2003; Mendoza-Ruiz & Pérez-García 2003; Mickel & Smith 2004; Lu et al. 2007.

1. Leaf blades usually with (2–)5–8(–12) pairs of lateral pinnae (in addition to the morphologically similar terminal pinna); at least some pinnae with a basal, ear-like, acroscopic (= on the side toward the apex) lobe; indusia flat to concave in center, without an umbo (= rounded, knob-like protuberance), shriveling and ± disappearing at maturity _____ **P. auriculata**
1. Leaf blades usually with (8–)10–18 pairs of lateral pinnae; pinnae without basal acroscopic lobes; indusia with a raised, knob-like central umbo, not shriveling at maturity _____ **P. umbonata**

Phanerophlebia auriculata Underw., (Latin: *auricula*, external ear, and *-atus*, indicating possession, = having an ear-like structure), EARED HOLLY FERN. Leaves to 65(–75) cm tall; leaf blades with at least some pinnae on every leaf having an acroscopic basal lobe; pinnae ovate to lanceolate, 2–10 cm long; indusia membranous, flat or concave in center, shriveling at maturity; $2n$ = 164 (Yatskievych 1993b). Moist sheltered areas in canyons and ravines (damp shaded canyon walls, under ledges, soil pockets in crevices); El Paso (e.g., *Correll 13788*, BRIT, TEX-LL, *O'Kennon 10692*, Hueco Tanks, BRIT; *G.M. Soxman 448*, TAES), Culberson (*Johnston et al. 10728*, TEX-LL), Jeff Davis (e.g., *Warnock 21779*, TEX–LL), and Brewster (e.g., *R.A. Studhalter s.n.*, 1921, SRSC) cos. in the Trans-Pecos; AZ, NM, and TX; also n Mexico. Sporulating Jun–Nov. [*Cyrtomium auriculatum* (Underw.) C.V. Morton] Yatskievych (1996) suggested that this species is an allotetraploid with *P. umbonata* as one of the parents. Yarborough and Powell (2002) noted that it is rare. Given its rareness and limited distribution in the state, we consider this species to be of conservation concern in TX.

Phanerophlebia umbonata Underw., (Latin, *umbo*, a rounded knob-like elevation or protuberance, and *-atus*, indicating possession, = having an umbo), CHISOS HOLLY FERN. Leaves to 70(–90) cm tall; leaf blades without basally lobed pinnae; pinnae lanceolate, 5–15 cm long; indusia scarious, with darker, raised, central umbo, persistent; $2n$ = 82 (Yatskievych 1993b). Moist spots in sheltered canyons and forested ravines; 1800–1900 m; in the U.S. known only from the Chisos Mts., Brewster Co. (e.g., *P.D. Whitson 598*, BRIT, TEX-LL; *C. Ellis et al. 299*, BRIT; *Warnock 178*, SRSC, TEX-LL; *Young s.n. 1915*, SRSC, TEX-LL; *Parks & Cory 26487*, TAES; Yatskievych 1996; Yarborough & Powell 2002); also n Mexico. Sporulating Jun–Oct. [*Cyrtomium auriculatum* (Underw.) C.V. Morton] This is the only member of the genus which has persistent indusia with a raised, darkened umbo in the center. Given its rareness and limited distribution in the state, we consider this species to be of conservation concern in TX. (Carr 2002a: G3S1)

POLYSTICHUM Roth

CHRISTMAS FERN, SWORD FERN, HOLLY FERN

A cosmopolitan genus treated by some as having at least 200 to possibly 300 species and one of the 10 largest genera of ferns (Little & Barrington 2003; Mabberley 2008); as currently viewed it is not monophyletic (Liu et al. 2010). The primary center of diversity is in sw China and adjacent regions, with others in tropical America and Malesia (Little & Barrington 2003). The species are typically terrestrial woodland perennials, particularly of montane regions, and some are cultivated as ornamentals (Mabberley 2008). *Polystichum* is difficult taxonomically and "hybridization, allopolyploidy, and apogamy all confound attempts at systematic revision and phylogenetic analysis of the genus" (Little & Barrington 2003). *Polystichum* is one of the polystichoid ferns related to *Cyrtomium* and *Phanerophlebia*, and some authorities lump both of these into *Polystichum* (e.g., Mabberley 1997, 2008). However, recent molecular studies (e.g., Little & Barrington 2003;

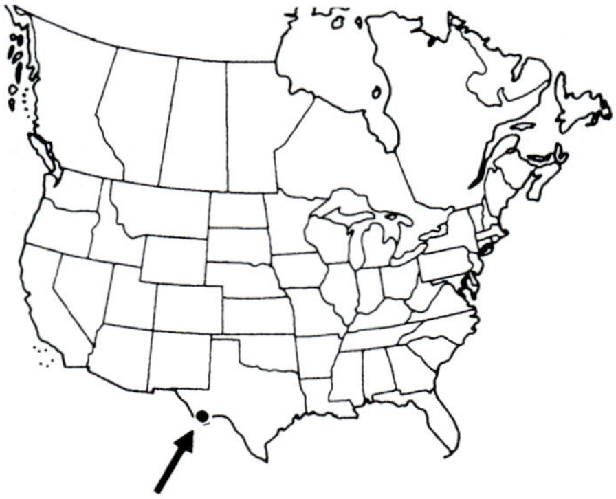

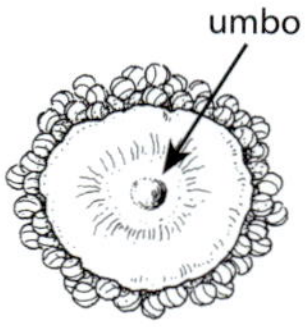

detail of sorus from above

[POM]

Alan Cressler

Alan Cressler

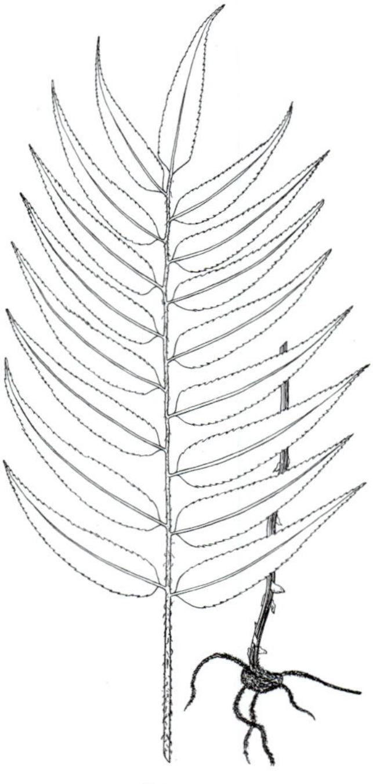

[POM]

Phanerophlebia umbonata

CHISOS HOLLY FERN

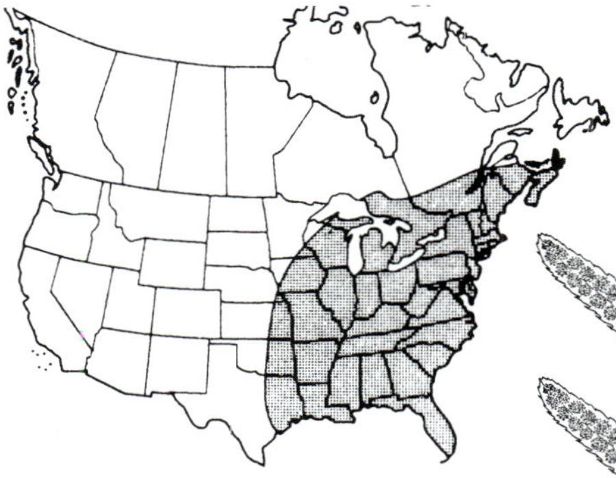

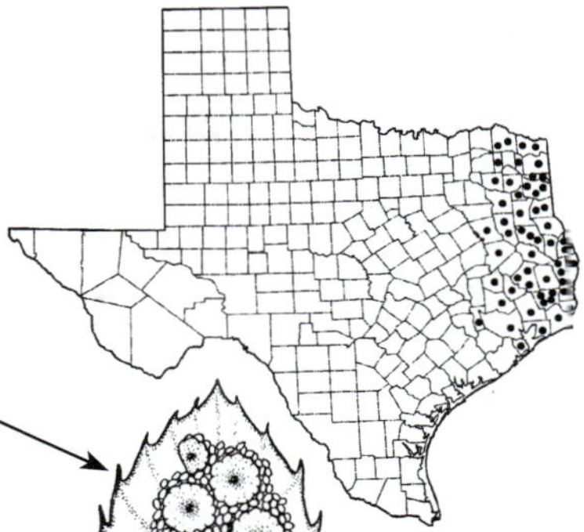

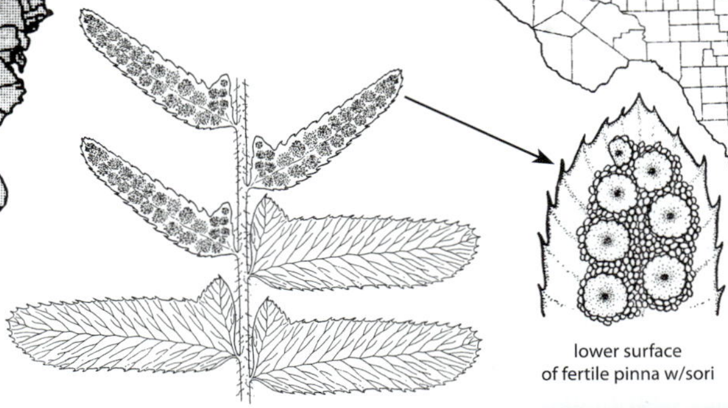

lower surface of leaf showing fertile and sterile pinnae

lower surface of fertile pinna w/sori

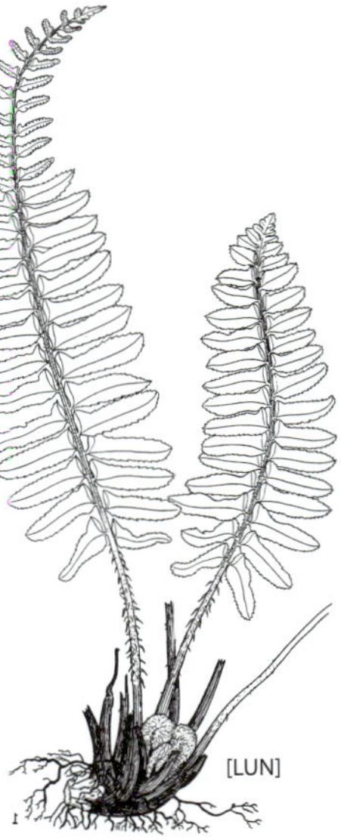

Dan Tenaglia-missouriplants.com

Dan Tenaglia-missouriplants.com

Dan Tenaglia-missouriplants.com

Polystichum acrostichoides

CHRISTMAS FERN

Li & Lu 2006; Lu et al. 2007; Liu et al. 2010) strongly argue that neither *Cyrtomium* nor *Polystichum* as traditionally viewed are monophyletic and that the situation is complicated; Liu et al. (2010), for example, suggest recognizing a number of genera including *Cyrtomium* and *Polystichum* in much restricted senses. Generic boundaries will almost certainly change in the future, but it seems highly likely that *Polystichum* will ultimately be recognized in a more narrow sense. (Greek: *poly*, many, and *stichos*, row, presumably in reference to the several rows of sori on each pinna)

REFERENCES: Wagner 1993; Noodén & Wagner 1997; Schneider & Carlquist 1997; Greer & McCarthy 2000; Little & Barrington 2003; Lu et al. 2007; Liu et al. 2010.

Polystichum acrostichoides (Michx.) Schott, (from *Acrostichum*, leather ferns, another genus of ferns with many crowded sori—Nelson 2000, and *-oides*, like or resembling), CHRISTMAS FERN, DAGGER FERN. Rhizomes erect; leaves evergreen, clustered, to 70 cm long, the blades elliptic-lanceolate to lanceolate in outline, 1-pinnate; pinnae numerous, mostly alternate, auricled basally, the auricle on the acroscopic side (= side toward the leaf apex) of the pinna, the margins bristle-toothed; petioles densely scaly; leaf blades partially dimorphic, the proximal pinnae (those near blade base) sterile, the distal pinnae (those near blade tip) of some blades fertile and conspicuously contracted (but blade tissue still evident); sori round, crowded in usually 2 rows, medial, often confluent at maturity and covering much of lower surface of fertile pinnae; indusia peltate, entire, persistent; $2n = 82$ (Wagner 1993). Rich wooded slopes, moist areas; Pineywoods and Post Oak Savannah w to Anderson (Turner et al. 2003), Waller, Wood (Correll 1956; TAES), and Red River (*Correll & Correll 12380*, BRIT, TEX-LL) cos.; also n Gulf Prairies and Marshes; se Canada and throughout e U.S. w to MN and TX; also disjunct to ne Mexico (Nuevo Leon and Tamaulipas, Mickel & Smith 2004) and naturalized in Europe. Sporulating May–Nov. [*Nephrodium acrostichoides* Michx.] According to Correll (1956), "In some areas the evergreen fronds are gathered for decorative greens at Christmas time, hence the common name." Dunbar (1989) also noted that the prominent "ear" (auricle) at the base of each of the pinnae causes the pinnae, when held vertically, to resemble Christmas stockings. Experimental evidence (Noodén & Wagner 1997) suggests that it is beneficial for the leaves to remain green in winter. Two hypotheses for this benefit are: 1) extension of the period of photosynthesis into the winter, and 2) nutrient storage in the old leaves requiring only a single transfer of nutrients to new leaves in the spring (versus two transfers if storage is in the rhizomes). Additionally, Greer and McCarthy (2000) found that as the sporophyte developed, the "rhizome is superseded by the fronds as the primary storage organs." Wagner et al. (1970) pointed out that plants with deeply incised (vs. toothed) fronds are the result of deviant late season growth developing from cold-damaged leaf primordia.

EQUISETACEAE Michx. ex DC.

HORSETAIL FAMILY

The family is represented in the modern world only by the distinctive and nearly cosmopolitan genus *Equisetum* that contains ca. 15 species (Hauke 1993; Smith et al. 2008). *Equisetum* is immediately recognizable—it is one of the world's morphologically most distinctive plants. As a result of its uniqueness and its extremely long fossil record, the family was often in the past classified as a separate division (the botanical equivalent of an animal phylum), variously called the Equisetophyta, Sphenophyta, or Arthrophyta. Currently the family is usually recognized in its own class, the Equisetopsida, and order, the Equisetales (e.g., Smith et al. 2006). The HORSETAILS are the only surviving group of a very ancient lineage. Fossil forms date to the Late Devonian Period (416–359 mya), and the lineage reached its maximum diversity and abundance in the Paleozoic Era. Relatives of HORSETAILS (the fossil genus *Calamites*) were components of the Carboniferous Period's (359–299 mya) swamp forests that formed present-day coal deposits; some reached the proportions of trees (to 20 or perhaps even 30 m tall—more than 90 ft) and were probably competitors of the tree Lycopodiophyta (Bateman 1991; Moran 2004).

The evolutionary relationships of the group have long been uncertain, and it was not until recently, with the advent of molecular information, that its "intimate" relationship with ferns has been widely accepted. Molecular data indicate that *Equisetum* is embedded within the fern lineage and is apparently most closely related to the Marattiales (an ancient order extending back 340 mya, and often considered to be the second most basal group of living ferns) (Pryer et al. 2001; Moran 2004; Wolfe et al. 2009). The molecular work is supported by morphological evidence (Renzaglia et al. 2000, 2002; Smith et al. 2006). For example, Renzaglia et al. (2002) found that, "*Equisetum* spermatozoids share several critical features with ferns, including *Psilotum*, and support monophyly of a fern-*Equisetum* assemblage." However, Rothwell and Stockey (2008) noted that molecular studies that do not incorporate fossil data yield hypothetical phylogenies that "appear to be increasingly discordant." They and Rothwell and Nixon (2006) thus suggested that the exact phylogenetic position of *Equisetum* is still somewhat uncertain. More recently, a study by Schneider et al. (2009) concluded that both molecular and morphological data are in agreement in placing *Equisetum* within the ferns. Because of the preponderance of evidence and following the consensus of most modern pteridologists, in this book we are considering the Equisetaceae to be ferns. However, future research will hopefully further clarify the exact evolutionary relationships of these unique plants.

The morphologically distinct Equisetaceae are characterized by whorled, small, scale-like leaves with a single vein, hollow, jointed, green stems (which are the primary photosynthetic organs of the plants), and spores borne in cones at the tips of stems (characters shared by the extinct calamites). Some species have numerous small branches and bear a slight resemblance to a horse's tail—this, however, is not the true derivation of the common or scientific names as is often thought (note derivation of scientific name discussed in the generic synopsis of *Equisetum*). The family is unique among living plants in having an apical meristem (= embryonic tissue at the growing tips/apices of stems) that gets smaller each time a new stem segment is produced (a growth pattern known as apoxogenesis). This results in a progressive narrowing upwards of the stems; ultimately the stems cannot keep growing (Moran 2004). (Order Equisetales)

FAMILY RECOGNITION IN THE FIELD: plant body consisting primarily of hollow, jointed, green stems; leaves inconspicuous, scale-like, in whorls at the very distinct nodes; sporangia in small, terminal cones.

REFERENCES: Correll 1949, 1956, 1966; Hauke 1990, 1993; Bell & Woodcock 1983; Raven et al. 1986; Bold et al. 1987; Wagner & Smith 1993; Woodland 1997; Renzaglia et al. 2000, 2001, 2002; Pryer et al. 2001; Bennert et al. 2005; Wikström & Pryer 2005; Brune et al. 2008; Yatskievych & Windham 2008c; Armstrong & Armstrong 2009; Schneider et al. 2009.

EQUISETUM L. HORSETAIL, SCOURING-RUSH

Plants perennial, rhizomatous, the rhizomes subterranean, creeping, branched; aerial stems annual or perennial, monomorphic or dimorphic, erect, hollow in center, with a series of smaller canals, jointed with very distinct nodes, ridged, green, photosynthetic (except fertile stems of the dimorphic *E. arvense*), usually unbranched in TX species (but these can produce lateral branches if the stem apex is injured) or much-branched in *E. arvense*; leaves small, inconspicuous, whorled, scale-like, fused into sheaths surrounding the nodes but with tips free and tooth-like; cones (strobili) discrete, terminal on stems, of tightly arranged, stalked, peltate sporophylls oriented at right angles to the cone axis, the outer surface of the cone appearing as "a series of interlocking hexagonal plates, which separate at maturity through elongation of the central axis" (Yatskievych & Windham 2008c); sporangia eusporangiate, in a ring of 5–10 on the under/inner surface of the sporophylls; spores of 1 kind (plant homosporous), green (except white in hybrids), each spore bearing four strap- or ribbon-like appendages called elaters (this is unique among vascular plants), the elaters hygroscopic—coiling and uncoiling with changes in humidity—and apparently functioning in spore dispersal.

The genus contains ca. 15 species (Hauke 1993) and is nearly cosmopolitan, especially in boreal and n temperate areas (40°–60° n latitude), being absent only from Australia and New Zealand, with most species in habitats that are wet at least seasonally (Hoshizaki & Moran 2001; Des Marais et al. 2003). Although most botanists accept only a single genus (e.g., Hauke 1993), some (e.g., Skoda 1997) would segregate a number of species into the genus *Hippochaete*. That genus, more commonly recognized as *Equisetum* subgenus *Hippochaete* (SCOURING-RUSHES), is distinguished by its sunken stomates, versus stomates flush with the epidermis in subgenus *Equisetum* (HORSETAILS) (Hauke 1963, 1993; Skoda 1997). Recent DNA research (Guillon 2007) suggests two main branches within the genus, the first containing two sub-branches, subgenus *Hippochaete* (7 species) and the unusual *E. bogotense* Kunth (ANDEAN HORSETAIL, COLA DE CABALLO; native of South America and s Central America), and the second main branch being subgenus *Equisetum* (7 species). Two of the TX species are in subgenus *Hippochaete*, and *E. arvense* is in subgenus *Equisetum*. Although the lineage leading to *Equisetum* is very ancient, all living species are apparently the result of a more recent evolutionary radiation. These modern species all have the same chromosome number, with n = 108 (Des Marais et al. 2003), species from the two subgroups have remarkably similar sperm cells (Renzaglia et al. 2002), and molecular data suggest the diversification leading to the modern species occurred in the Tertiary, within the last 50 million years (Des Marais et al. 2003). However, recent fossil evidence (Stanich et al. 2009) shows that *Equisetum* had evolved by at least the early Cretaceous, 136 million years ago. In fact, an older fossil genus, *Equisetites*, which is of Paleozoic (mid-Permian or possibly Carboniferous) origin (Stewart & Rothwell 1993; Des Marais et al. 2003) is sometimes indistinguishable from living horsetails, and *Equisetum* could thus possibly be considered the oldest surviving genus of vascular plants in the world (Guillon 2004; Bennert et al. 2005). Because the genus is ancient (Cretaceous at very latest), its nearly worldwide distribution is not surprising. Further, a number of species have distributions that reflect very different conditions in the past—for example, 8 of the 11 North American species also occur in Japan. Additionally, Tertiary age fossils of *Equisetum* are well known from the high Arctic of extreme northern Canada, reflecting not only the widespread occurrence of this genus but also extremely different climatic conditions in the past at very high latitudes (e.g., McIver & Basinger 1999). (See page 33 for a discussion of such distributions). The coarse, somewhat abrasive stems contain silica, and were used by early settlers to scour pots and pans (Woodland 1997) (hence the common name). They have also been used in polishing fine furniture and pewter (thus an old name, PEWTERWORT), and for honing the reeds of woodwind musical instruments. Herbarium botanist Rebecca Swadek (BRIT) noted that they also make an excellent nail file. The name HORSETAIL is sometimes said to be derived from species with whorled branches, in reference to their bushy appearance (Hoshizaki & Moran 2001), but it is actually derived from a translation of the scientific name, which refers to the roots (see below). The largest living species, *Equisetum giganteum* L., occurs in the Amazon area of South America and can reach 5 m in height (Moran 2004). *Equisetum* is often used in Japan in flower arrangements (Hoshizaki & Moran 2001) and it is sometimes cultivated, including in water gardens; however, it can rapidly spread vegetatively and sometimes become problematic. ☠ Some species are known to contain alkaloids or other toxins such as thiaminase, an enzyme that destroys thiamine and causes Vitamin B1 deficiency; they can be poisonous to livestock (particularly horses) when included in hay, because the enzymes are not destroyed by drying. The resulting condition, referred to as equisetosis, includes neurological symptoms such as loss of muscular control, trembling, and staggering. All species are probably somewhat neurotoxic (Kingsbury 1964; Burlage 1968; Fuller & McClintock 1986; Weathers 1998; Burrows & Tyrl 2001). Species of *Equisetum* are often difficult to distinguish due to several factors—variability within species, state of maturation, the fact that one plant may look quite different when grown under different environmental conditions (Hauke 1963), and "rampant hybridization" between species (Des Marais 2003). Although there is reproduction via spores, according to Yatskievych and Windham (2008c), "much of the reproduction in *Equisetum* species is vegetative from rhizome and aerial stem fragments dispersed by flood-waters or perhaps in mud on the feet and feathers of waterfowl." (Latin: *equis*, horse, and *seta*, bristle, referring to the coarse black roots of *E. fluviatile* L.—Hauke 1993)

REFERENCES: Hauke 1960, 1963, 1978; Rutz & Farrar 1984; Soltis et al. 1988; Korpelainen & Kolkkala 1996; Skoda 1997; Des Marais et al. 2003; Guillon 2004, 2007; Bennert et al. 2005; Brune et al. 2008; Stanich et al. 2009; Mink et al. 2011b.

1. Aerial stems (sterile) much-branched, with regular whorls of branches (these smaller in diameter than main stem and having only 3–4 ridges); aerial stems of two forms, unbranched light brown to tan or pinkish fertile stems and branched green sterile ones; leaves 4–14 per node (number evident as teeth of sheaths); species extremely rare in TX, documented only from Lubbock and Wheeler cos. in the Panhandle ____ **E. arvense**
1. Aerial stems unbranched (except damaged or abnormal stems can have some branches, but these branches have 10 or more ridges), both fertile and sterile ones green, identical; leaves 10–50 per node (number evident as teeth of sheaths); species widespread in TX.
    2. Cones typically not opening (shriveling and drying without releasing spores); spores white, misshapen; hybrid vegetatively intermediate between *E. hyemale* and *E. laevigatum* as contrasted in couplet 3 below (e.g., leaf sheaths with length/width ratio intermediate between 1:1 and 2:1 [see photo], stems only slightly roughened, cone apex somewhat pointed, typically with a small apiculum, teeth of sheaths often at least somewhat persistent, often light in color) ____ **E. ×ferrissii**
    2. Cones opening; spores green, spherical; parental species distinguished as in couplet 3 below.

3. Sheaths (= fused leaves) on mature aerial stems dark-girdled (= with a dark band) at sheath base on most nodes of aerial stems (in addition to thin dark line at sheath apex where dark teeth are attached), whitish to ashy-gray or brownish above dark basal girdle (however, upper few nodes typically not dark-girdled), about as long as wide (length/width ratio 1:1); aerial stems perennial, evergreen through the winter, usually persisting more than one year, 5–10 mm in diam, somewhat roughened (like sandpaper when a fingernail is scraped along the stem) due to minute tubercules on stem ridges; cone apex usually pointed with a noticeable black apiculum; teeth of sheaths often persistent, blackish, or promptly shed _______ **E. hyemale**
3. Most sheaths on mature aerial stems green, with only a thin dark line at sheath apex where teeth are shed, only a few near stem base dark-girdled at sheath base, distinctly longer than wide (length/width ratio ca. 2:1); aerial stems annual, not evergreen, lasting less than a year, only occasionally overwintering, 3–5(–7) mm in diam., relatively smooth due to ± lack of tubercles on stem ridges; cone apex usually rounded or with a very small apiculum; teeth of sheaths promptly shed _______ **E. laevigatum**

Equisetum arvense L., (Latin: *arvum*, cultivated field, applied to plants to mean of cultivated fields or land), COMMON HORSETAIL, FIELD HORSETAIL, BOTTLE-BRUSH. Aerial stems annual, dimorphic, the fertile unbranched, succulent fleshy, light brown to tan or pinkish, non-photosynthetic, ephemeral (produced in spring, withering as the sterile appear), the sterile much-branched, to 60(–100) cm tall, with 4–14 ridges, green, persisting through growing season, the branches numerous in regular whorls, solid, to ca. 25 cm long and ca. 1 mm in diam., with 3–4 ridges; leaves 4–14 per node (number evident as dark teeth of sheaths); cones cylindric to ovoid, to ca. 4 cm long and 10 mm in diam., the apex rounded; $2n$ = ca. 216 (Hauke 1993). Stream banks, lake shores, marshes, other moist areas; in TX known only from the Panhandle in Lubbock (Correll 1956 cited *E.L. Reed 3613*, 1932 at Buffalo Spring), Wheeler (*W.C. Holmes, J.R. Singhurst and J.N. Mink 15094A* [BAYLU]), and possibly Potter (B. O'Kennon, pers. comm.; the Potter locality is not mapped on county distribution map) cos.; nearly throughout the U.S. and Canada except absent from the far se U.S. w through most of TX; also Eurasia. Sporulating early spring. We have been unable to locate or confirm the 1932 Reed collection cited by Correll. Correll (1956) noted that his 1952 attempt to relocate this species at the Buffalo Spring locality was unsuccessful. Jack Stanford (pers. comm.) and Jim Blassingame (pers. comm.) both question whether the identification of the Reed collection is correct—based on their extensive experience in the area, they believe the collection was a specimen of either *E. hyemale* or *E. laevigatum* that was branched due to meristem damage. Marshall Johnston and Bob O'Kennon (pers. comm.) saw a few plants of what they thought was *E. arvense* at the edge of a playa lake in Potter Co. (near Amarillo) in the early 1990s. Mink et al. (2011b) recently reported a 2010 collection from Wheeler Co.; this collection was of a single tiny specimen clinging to the roots of a plant of another species. After examining the specimen, James H. Peck (pers. comm.) and G. Diggs agree that the material is consistent with a very immature specimen of *E. arvense*; additional specimens consisting of more mature material will be important in confirming this species for TX. Although many infraspecific taxa have been named in this species, we are following Hauke (1993) in not recognizing varieties or subspecies. This species occurs broadly across the n hemisphere and has a classic Arcto-Tertiary distribution (see page 33) (Kato & Iwatsuki 1983). Jones (1987) noted that instances of "death are known where hay contaminated with stems of *E. arvense* was fed to horses." In fact, among HORSETAIL species, *E. arvense* is the one that causes most cases of thiaminase poisoning (Vitamin B1 deficiency) in North America. Horses, particularly young ones, eating large amounts of *E. arvense* are most susceptible (Burrows & Tyrl 2001). Given its rareness and limited distribution in the state, we consider this species to be of conservation concern in TX. ⚠ ☠

Equisetum ×ferrissii Clute, (for James H. Ferriss, 1849–1926, Illinois journalist and field biologist with interests in land snails, fossils, ferns, and cacti; at one time president of the Amer. Fern. Soc.; see Clute 1926), FERRISS' HORSETAIL, FERRISS' SCOURING-RUSH, INTERMEDIATE HORSETAIL. Plant vegetatively variable but typically intermediate between the two parental species; aerial stems with basal part over-wintering, monomorphic, unbranched (unless damaged or abnormal), 20–180 cm tall, slightly roughened; leaves 14–32 per node (number evident as teeth of sheaths), the teeth promptly shed or persistent and light in color; cones typically failing to open, the apex somewhat pointed, typically with a small apiculum; spores misshapen, white (in contrast to the green spores of the other three taxa of *Equisetum* in TX). Stream banks, other moist areas; scattered primarily in w ⅔ of TX (solid dots on the county distribution map represent specimens identified by J. Peck and G. Diggs [e.g., Randall Co., *L. Pace 149*; Tarrant Co., *J. Quayle et al. 749*; Uvalde Co., *L. & C. Taylor 33749*,

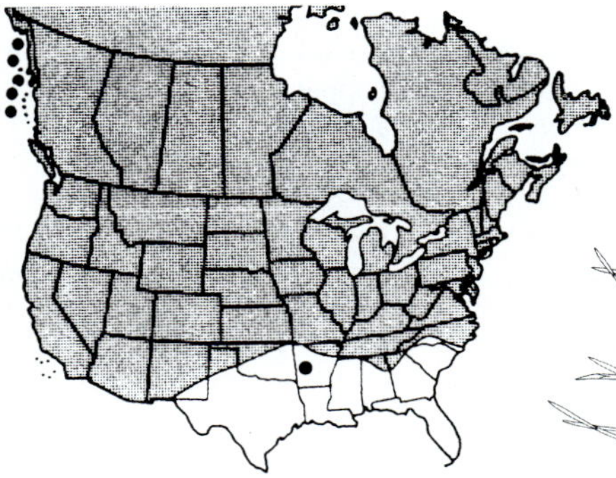

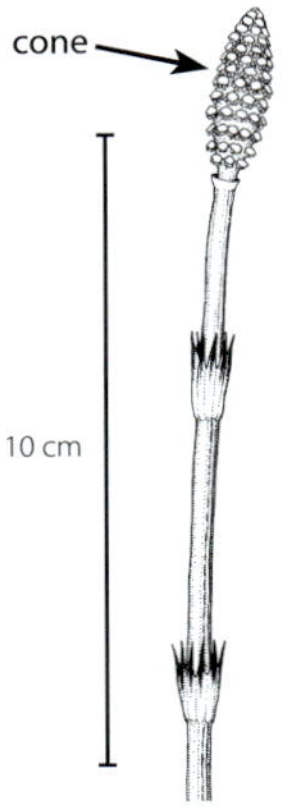

Alan Cressler

Alan Cressler

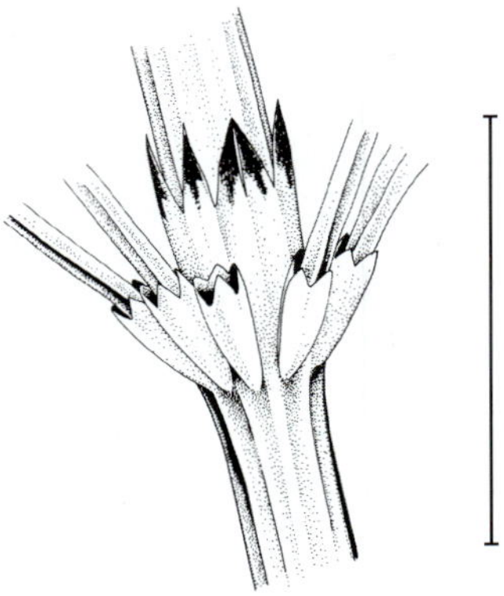

Equisetum arvense

COMMON HORSETAIL

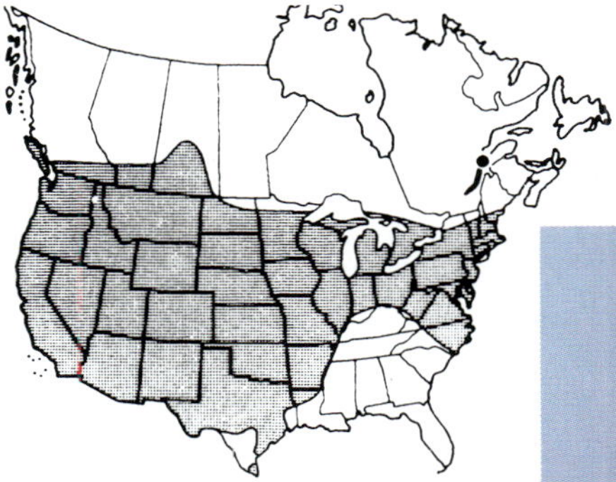

● confirmed collections
○ unconfirmed collections and literature citations

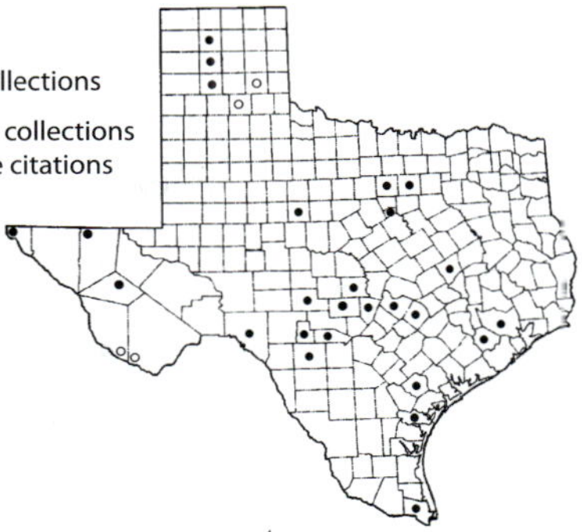

Robin Moran

variation in E. *xferrissii*

l to r: leaf sheaths of *E. hymale*, *E. xferrissii*, and *E. laevigatum*

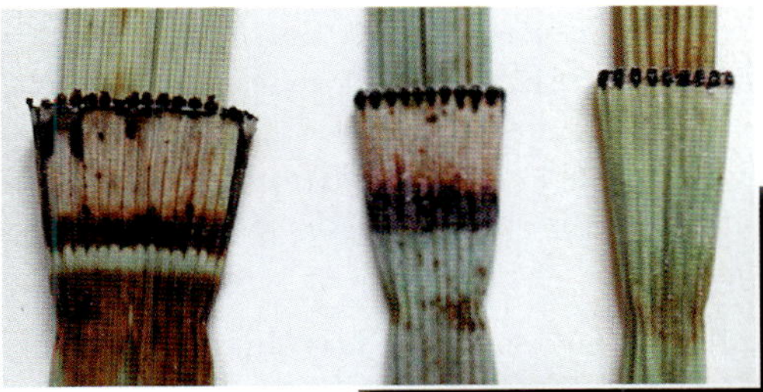

Robin Moran

cones typically failing to open

Jim Peck

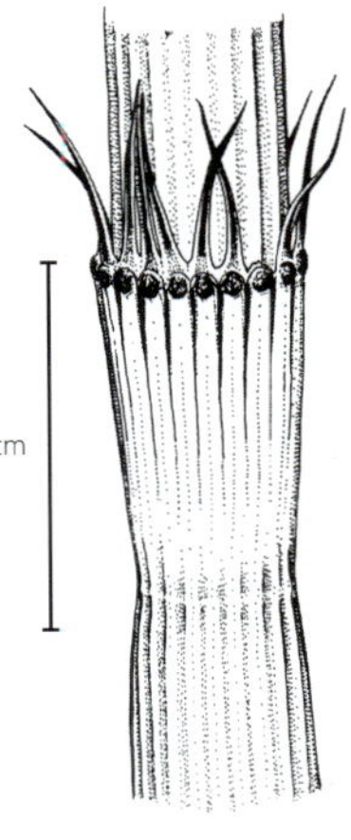

leaf sheath w/persistent teeth

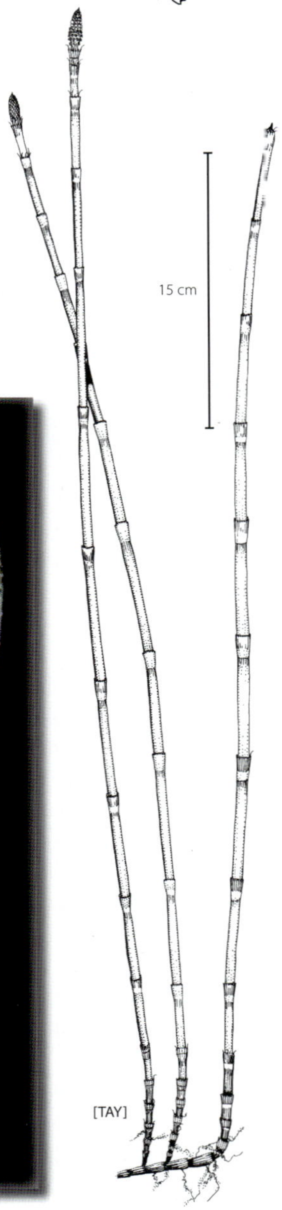

Equisetum ×ferrissii

FERRISS' HORSETAIL

all BRIT] or for which spores are clearly aborted [e.g., Culberson, El Paso, and Jeff Davis cos., TEX-LL], while open circles are herbarium and literature records identified by other authorities—see discussion under *E. hyemale*); widespread in the U.S. and s Canada except absent from much of se U.S.; also n Mexico. Cones present late spring-early summer but not shedding spores. [*E. hyemale* var. *elatum* (Engelm.) C.V. Morton, *E. hyemale* var. *intermedium* A.A. Eaton] Hauke (1960, 1993) considered this taxon to be a hybrid between *E. hyemale* and *E. laevigatum*; molecular evidence confirms its hybrid origin (Des Marais et al. 2003). The cones often fail to open apparently because the central axis does not elongate to expose the sporangia; the sporangia therefore do not dry, split, and release the misshapen, white, abortive spores (PlantSystematics.org 2008). This hybrid has sometimes been treated as a variety of *E. hyemale* or has been mistaken for either *E. laevigatum* or *E. hyemale*; although sterile sexually, it can be found without the parental species and is apparently dispersed vegetatively (Hauke 1963, 1993). Des Marais et al. (2003) described it as "widespread and prolific." In Mexico it is reported to be "nearly as common as its parents" (Mickel & Smith 2004) and in AR it has been found in every county (Peck 2011a). Likewise in TX, *E.* ×*ferrissii* is widespread and common; in fact, we have seen more specimens of what are apparently the hybrid than of the parental species. However, young stems, anomalous regrowth, or incomplete specimens of *E. hyemale* and *E. laevigatum* can often not be definitely separated from *E.* ×*ferrissii* (when lacking spores and cones); populations in the wild are much more easily identified since complete material is available. This hybrid is most easily distinguished from the two parental species (with greenish spherical spores and cones that open) by its white misshapen spores and cones which fail to open; it can also be distinguished by its slightly rough stems (when a fingernail is scraped lengthwise along the stem), sheaths of intermediate length (between the more elongate sheaths of *E. laevigatum* and the shorter "stubby" sheaths of *E. hyemale*), by the somewhat pointed cone apex typically with a small apiculum, and by the often persistent sheath teeth that are rather light in color. Yatskievych and Windham (2008c) noted that "The morphology of the more-or-less evergreen aerial stems is quite variable, but they are generally at least slightly rough to the touch and the stem sheaths sometimes resemble those of *E. hyemale* toward the stem base and those of *E. laevigatum* toward the tip." Both parental species are poisonous, thus this hybrid should be considered so as well. ☠

Equisetum hyemale L. subsp. **affine** (Engelm.) Calder & R.L. Taylor, (sp.: Latin: *hiemalis*, of or pertaining to winter, in reference to the evergreen habit; subsp.: Latin: *affinis*: related to, neighboring, akin to, the type description noting that this plant, which was originally named as a variety of a different species, "approaches very closely, sometimes too much so, to [*E. hyemale*], whence the name"), SCOURING-RUSH HORSETAIL, TALL SCOURING-RUSH, AMERICAN SCOURING-RUSH, COMMON SCOURING-RUSH, GREAT SCOURING-RUSH, ROUGH HORSETAIL, WINTER SCOURING-RUSH, CAÑUELA. Plants often found in large colonies, spreading vigorously by vegetative reproduction via rhizomes (e.g., huge colonies are sometimes seen spreading along creeks forming virtual "thickets"); aerial stems usually perennial, monomorphic, unbranched (unless damaged or abnormal), 18–220(–300) cm tall, typically 5–10 mm in diam. (the plants tending to be more robust than those of *E. laevigatum*), with 14–50 ridges, the stem surface somewhat roughened (like sandpaper), dark green or bluish-green at maturity; leaf sheaths about as long as wide (length/width ratio 1:1), usually with 2 dark girdling bands/lines (sometimes irregular or absent) one at base and a narrow one at apex where teeth are attached and often with a whitish to ashy-gray or brownish band between (however, sheaths at upper few nodes often ± green and lacking basal dark girdle thus somewhat resembling those of *E. laevigatum*), the teeth of sheaths often persistent, blackish, or promptly shed; leaves 14–50 per node (number evident as teeth of sheaths); cones ovoid to cylindric-ellipsoid, (0.8–)1.5–2.5 cm long, 5–10 mm in diam., the apex usually pointed with a noticeable black apiculum; $2n = 216$ (Hauke 1993). Stream banks, wet places, seepage areas, often forming large colonies; widespread from Pineywoods w throughout much of TX (e.g., Cameron Co., *R. Runyon 114*; Grayson Co., *M. Nee & Diggs 43893*; Val Verde Co., *Correll & Correll 12891*, all BRIT) (solid dots on the county distribution map represent specimens identified by J. Peck and G. Diggs, while open circles are herbarium and literature records identified by other authorities—see discussion below); throughout Canada and the entire U.S.; also Mexico, Guatemala, and e Asia. Sporulating Mar–late fall. [*E. hyemale* L. var. *affine* (Engelm.) A.A. Eaton, *E. hyemale* var. *robustum* (A. Braun) A.A. Eaton, *E. prealtum* Raf., *E. robustum* A. Braun var. *affine* Engelm.] Subspecies *affine* is native to both North America and e Asia (Kato & Iwatsuki 1983; Mickel & Smith 2004), while subsp. *hyemale* is European and w Asian in distribution. When the ranges of both subspecies are considered, the species dem-

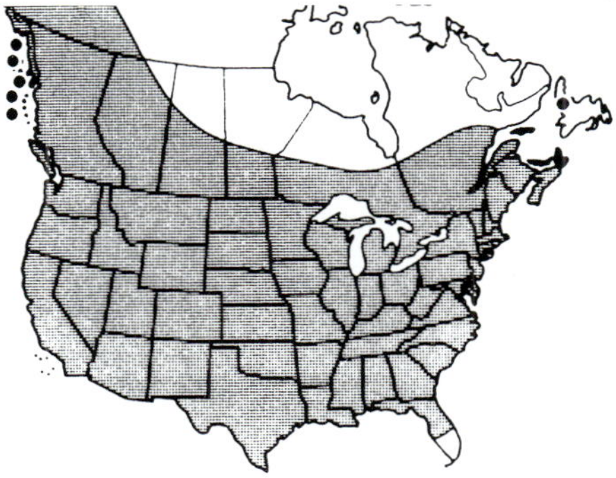

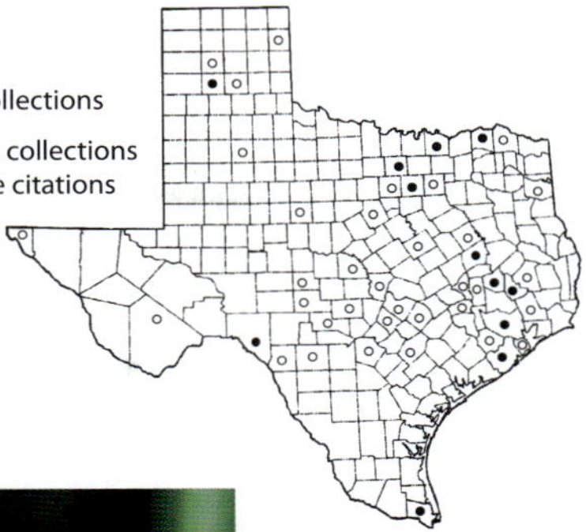

● confirmed collections

○ unconfirmed collections and literature citations

note black apiculum of cone

George Diggs

Alan Cressler

photo above shows old (overwintered) stems on left and new stems on right

Rebecca K. Swadek

photo and drawing at right show typical banding

[LUN]

Equisetum hyemale subsp. **affine**

SCOURING-RUSH HORSETAIL

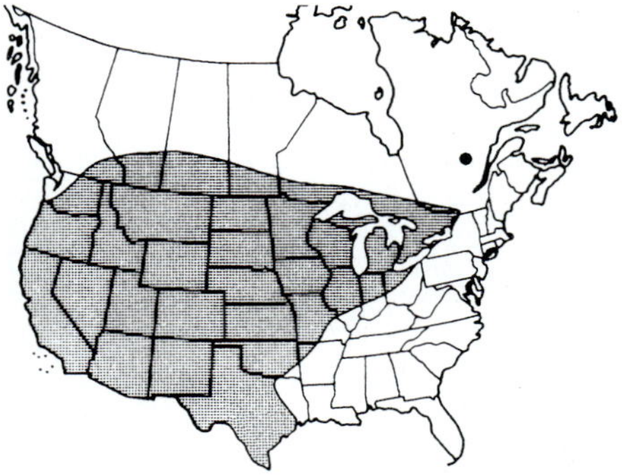

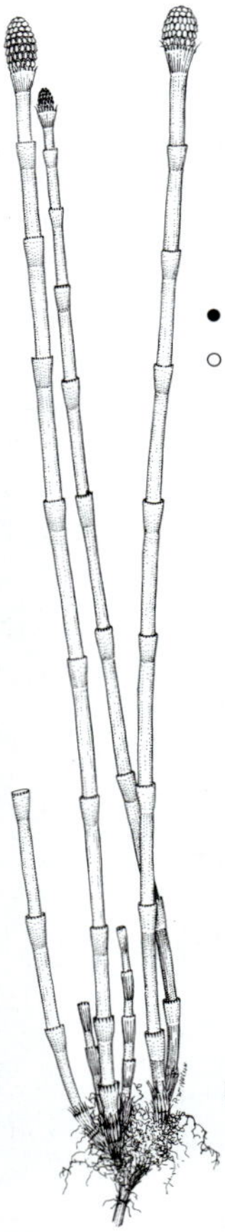

[TAY]

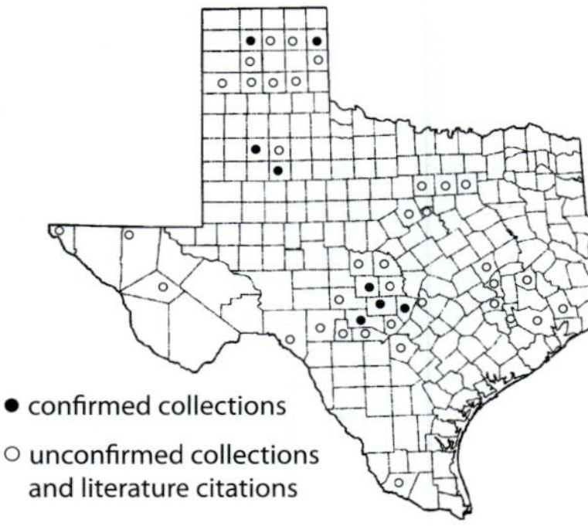

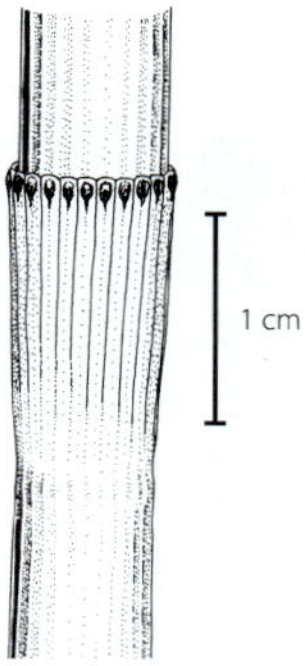

Robert J. O'Kennon

Jim Peck

Equisetum laevigatum

SMOOTH HORSETAIL

onstrates a classic Arcto-Tertiary distribution (Kato & Iwatsuki 1983) (see pages 32 and 33 for a discussion of the North America-e Asia disjunct distribution and the concept of an Arcto-Tertiary flora). Young or damaged stems or fragmentary specimens of this species and *E. laevigatum* can very easily be confused (such as a "top snatch"—specimens consisting of only a stem tip). For example, Mickel and Smith (2004) noted that "Young *E. hyemale* has longer than usual sheaths that often lack the lower dark bands, resembling those of *E. laevigatum*." Indeed, when a plant that has multiple stems (from a rhizome) is available, it is evident that young stems or stems with anomalous regrowth can look remarkably like *E. laevigatum,* even though older stems are obviously typical of *E. hyemale*. Because of this and since the banding of the uppermost or lowermost nodes can often be deceptive, complete, mature stems should be collected whenever possible. Identification is further complicated because the hybrid *E. ×ferrissii* is vegetatively intermediate between its two parents and is morphologically quite variable. Because of the ease with which typical herbarium material can be confused, as noted above, only specimens identified by J. Peck and G. Diggs are represented as solid black dots on the county distribution maps of *E. hyemale*, *E. laevigatum*, and *E. ×ferrissii*. Jack Stanford (pers. comm.) pointed out that he has observed plants of *E. hyemale* with green rhizomes (even when they are underground); this is the only species he has seen exhibiting this phenomenon. This species is sometimes cultivated, including in water gardens in TX, but can spread aggressively through vegetative reproduction and become problematic (G. Diggs, pers. obs.). Poisonous (Burlage 1968; Burrows & Tyrl 2001).

Equisetum laevigatum A. Braun, (Latin: *laevigatus*, smooth, slippery, polished), SMOOTH HORSETAIL, SMOOTH SCOURING-RUSH, BRAUN'S SCOURING-RUSH, KANSAS HORSETAIL, KANSAS SCOURING-RUSH, SUMMER SCOURING-RUSH, COLA DE CABALLO, CAÑUELA. Aerial stems usually annual, monomorphic, unbranched (unless damaged or abnormal), 20–150 cm tall, typically 3–5(–7) mm in diam, (the plants tending to be less robust than those of *E. hyemale*), with 10–32 ridges, the stem surface ± smooth, usually bright green to yellowish green at maturity; leaf sheaths longer than wide (length/width ratio ca. 2:1), usually green except for thin dark band where teeth are attached, the teeth promptly shed (note: the basal few nodes often with sheaths having a basal dark girdle, thus ± resembling those of *E. hyemale*); leaves 10–32 per node (number evident as teeth of sheaths); cones ellipsoid, 1–2 cm long, 7–10 mm in diam., the apex usually rounded or with a very small apiculum; $2n = 216$ (Hauke 1993). Widely scattered in TX, mainly in the w ¾ of the state (e.g., Hemphill Co., *C. Rowell 4043*, Mason Co., *S.E. Wolf 5167*; Potter Co., *G. Nesom & O'Kennon LAMR326*, all BRIT) (solid dots on the county distribution map represent specimens identified by J. Peck and G. Diggs, while open circles are herbarium and literature records identified by other authorities—see discussion under *E. hyemale*); s Canada and through much of U.S. except ne and se; also n Mexico. Sporulating May–Jul. [*E. funstonii* A.A. Eaton, *E. kansanum* J.E. Schaffn., *E. laevigatum* subsp. *funstonii* (A.A. Eaton) Hartman] Hauke (1993) noted that, "Schaffner named this species *Equisetum kansanum* because he applied the name *E. laevigatum* to what we now know is the hybrid *E. ×ferrissii*." This species, *E. hyemale,* and the hybrid *E. ×ferrissii* are often very difficult to distinguish and appear to intergrade when only young, anomalous, or incomplete material is available (see discussions under *E. hyemale* and the hybrid, *E. ×ferrissii*). Smooth horsetail is sometimes sold in nurseries for use in water gardens; in these settings it can be quite aggressive. Poisonous (Burlage 1968; Burrows & Tyrl 2001).

LOMARIOPSIDACEAE Alston

SWORD FERN FAMILY

A family of 4 genera and ca. 70 species (Smith et al. 2008) previously often treated in the Dryopteridaceae. Smith et al. (2008) noted that based on recent molecular data (Schuettpelz & Pryer 2007) "it appears likely that these genera form a monophyletic group." Although the family was not strongly supported in the molecular analyses and was only "tentatively recognized" (Schuettpelz & Pryer 2007), there is also a shared morphological character—the presence of articulate pinnae. The only TX representative is an introduced and naturalized ornamental, *Nephrolepis exaltata*, SWORD FERN. The situation is still not totally clear, the family "requires additional scrutiny," and it is possible that *Nephrolepis* "may eventually require [separate] recognition" (Smith et al. 2008) at the family level. Relationships are thus unsettled at present, but there is strong consensus among leading fern taxonomists that the genera should not be included in the Dryopteridaceae, and we are thus following Smith et al. (2008) in tentatively treating *Nephrolepis* in the Lomariopsidaceae.

rhizome

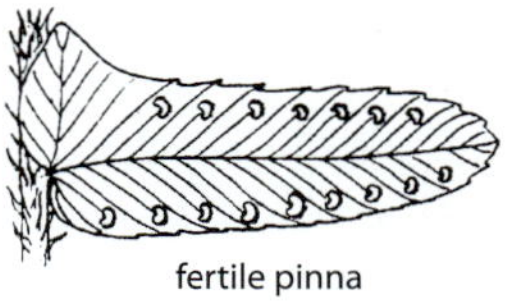

fertile pinna

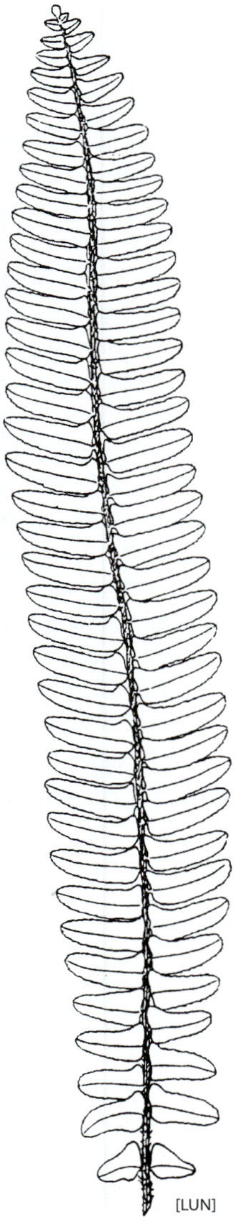

[LUN]

Nephrolepis exaltata
SWORD FERN

Family name from *Lomariopsis*, CLIMBING HOLLY FERN or FRINGED FERN, a pantropical genus of ca. 45 species (15 in the American tropics) of climbers with thick, climbing rhizomes (Moran 2000; Rougan et al. 2007). (From *Lomaria*, a subgenus of the genus *Blechnum*—Greek: *loma*, fringe or border, and *opsis*, resemblance or appearance—Quattrocchi 2000) (Greek: *nephros*, kidney, and *lepis*, scale, in reference to the shape of the indusium) (Order Polypodiales)

FAMILY RECOGNITION IN THE FIELD: the only TX species is rarely naturalized in the eastern ⅓ of the state, has 1-pinnate leaves with numerous pinnae, and has the sori, which are on the lower pinnae surfaces, not restricted to the uppermost pinnae.

REFERENCES: Kaur 1974; Kramer 1990g; Smith et al. 2006, 2008; Schuettpelz & Pryer 2007, 2008.

NEPHROLEPIS Schott

BOSTON FERN, SWORD FERN

A genus of 20–30 species widespread in tropical areas (Nauman 1993c; Mickel & Smith 2004). *Nephrolepis*, whose relationships have been obscure, has often been included in a broadly construed Dryopteridaceae (e.g., Smith et al. 1993b; Diggs et al. 2006) or sometimes placed in the Davalliaceae or Nephrolepidaceae (e.g., Kramer 1990d; Nauman et al. 2000). A number of species are cultivated. (Greek: *nephros*, kidney, and *lepis*, scale, in reference to the shape of the indusium)

REFERENCES: Nauman 1981, 1993c; Schneider & Carlquist 1999b; Peck 2002.

Nephrolepis exaltata (L.) Schott, (Latin: *exaltatus*, raised on high, lofty, in reference to the perched sites where it often grows as an epiphyte in swamps [Peck 2002]), SWORD FERN, BOSTON SWORD FERN, WILD BOSTON FERN, YANKEE FERN. Rhizomes short, ± erect, with wiry, widely creeping stolons; leaves monomorphic, evergreen, clustered, 1-pinnate, usually 0.4–1.5 m long, the blades linear-lanceolate, the pinnae numerous (> 18 pairs), the fertile pinnae not reduced; petiole bases with more than 3 vascular bundles; sori roundish, somewhat closer to margin than to midvein of pinnae, the sori not restricted to the uppermost pinnae, the indusia ± orbicular-reniform to horseshoe-shaped, attached at narrow or broad sinus; $2n = 82$ (Mickel & Smith 2004). Escaping from cultivation; Dallas (persisting and spreading in yard in Highland Park, Dallas, R. O'Kennon, pers. obs.; Peck 2002), Hidalgo (*S. Beadles 1006*, BRIT), Orange (BRIT), Nacogdoches (TAES), Burnet, Henderson, Orange, and Real (Turner et al. 2003) cos.; terrestrial or more often epiphytic in its native habitat; naturalized in scattered locations in TX and in AR and LA; considered by some (e.g., Nauman 1993c, Nauman et al. 2002) to be native to FL, the West Indies, and scattered Pacific Islands; however, Weakley (2010) questioned whether the species was native to FL noting it was "rare, in our area perhaps only introduced." Mickel and Smith (2004) give the following for distribution: "USA (Fla); Mexico; Pan; Gr Ant; Fr Gui, Braz. Native range uncertain, perhaps widely distributed in Old World tropics, but introduced there." [*Polypodium exaltatum* L.] Sporulating throughout most of the year (Correll 1956). It is a commonly cultivated (particularly indoors) and commercially important fern with many cultivars including cv. 'Bostoniensis' (BOSTON FERN) and the locally developed DALLAS JEWEL FERN,™ commonly known as the DALLAS FERN. There is debate over the origin of the name BOSTON FERN, but it apparently originated in the late 1800s or early 1900s when plants of a particular form of *N. exaltata* became popular in nurseries in and around Boston, MA (Benedict 1916). Bailey (1924) noted it was "Introduced 1895." Klingaman (2011) suggested, "Boston fern originated in a shipment of 200 plants sent from a Philadelphia florist to F. C. Becker, a florist in Cambridge, Mass.…Becker recognized one plant in the shipment was faster growing, had wider fronds and an unusual drooping habit, instead of the stiffly upright form of the species. He began to propagate it in 1894. Two years later, botanists in London identified the plant and suggested the Boston name for the variant form."

LYGODIACEAE M. Roem.

CLIMBING FERN FAMILY

A very small family (1 genus, ca. 25–40 species) primarily of tropical regions nearly worldwide, but also temperate areas of North America, Asia, and s Africa, and Pacific Islands (Nauman 1993a; Smith et al. 2008). It is related to the Anemiaceae and Schizaeaceae, both of these families and the Lygodiaceae making up Order Schizaeales—see page 97 under the Anemiaceae for a discussion of this order. Fossil evidence suggests that the Lygodiaceae is old, already being in existence by the late Jurassic/early Cretaceous (Wikström et al. 2002). The family is sometimes included in the Schizaeaceae (e.g., Correll 1956; Radford et al. 1968; Correll & Johnston 1970; Kramer 1990f; Mabberley 1997; Nauman et al. 2000). However, given its long history and morphological distinctiveness, we are following Nauman (1993a) and Smith et al. (2006, 2008) in recognizing it as a distinct family. (Order Schizaeales)

FAMILY RECOGNITION IN THE FIELD: the solitary species in the flora is the only twining/climbing, vine-like fern in TX.

REFERENCES: Nauman 1993a; Wikström et al. 2002.

LYGODIUM Sw. CLIMBING FERN

A primarily tropical and subtropical genus of 25–40 species with the majority in the Old World. They are vine-like ferns with leaves whose midribs twine around and climb other plants; some species can climb to more than 30 m (Nauman 1993a; Wikström et al. 2002; Mickel & Smith 2004; Smith et al. 2006, 2008). The elongating leaf midribs (= rachises) have terminal meristems with indeterminate growth, and it is one of only two fern genera in the world that climb using such twining midribs (Moran 2004). As a result of the continuing growth of the midribs, individual leaves (which arise from subterranean rhizomes) become incredibly long and have the appearance of stems with numerous short "branches" (actually pinnae). The twining stem-like rachises are used for such things as basketry, fish-traps, mats, and yarn. Only one species, *L. palmatum* (Bernh.) Sw., AMERICAN CLIMBING FERN, is native to the U.S., endemic from ME s and w across the e U.S. as far as MI and MS (but not as far as TX). In some parts of the world invasive *Lygodium* species can form "immense, 3-dimensional carpets over 1 m thick that can strangle whole forests" (Robinson et al. 2010). According to Hoshizaki and Moran (2001), "The evergreen fronds [of *L. palmatum*] used to be collected during the winter and used for Christmas decorations. This destroyed so many populations that in 1869 the Connecticut legislature enacted a law to protect the plant—the first plant conservation law passed in the United States." This U.S. native is easily distinguished from *L. japonicum* by its 1-palmately lobed pinnules and petioles borne 10–40 mm apart. The genus has a long fossil history, known from the late Cretaceous/early Tertiary Period and diversifying throughout that geological period (Collinson 2001; Wikström et al. 2002; Rothwell & Stockey 2008). (Greek: *lygodes*, flexible, in reference to the twining rachises—Nauman 1993a).

REFERENCES: Duek 1978; Nauman 1987; Pemberton 1998; Pemberton & Ferriter 1998; Carlquist & Schneider 1998; Lott et al. 2003; Volin et al. 2004; Madeira et al. 2008; Hutchinson & Langeland 2010.

Lygodium japonicum (Thunb.) Sw., (of Japan, in reference to where it is was originally described from by Carl Pehr Thunberg, a student of Linnaeus and author of *Flora Japonica*—Thieret 1980), JAPANESE CLIMBING FERN. Terrestrial; rhizomes subterranean, 2–3 mm in diam., creeping, bearing petioles 2–7 mm apart; leaves with indeterminate growth, climbing, vine-like, to at least 6 m long (reports of leaves to 30 m), the rachis (= midrib) elongate, twining, flexuous, stem-like (true stem/rhizome below ground), bearing numerous pinnae, continuing to elongate and produce additional pinnae; pinnae reduced to short stalks, these short stalks each bearing a pair of opposite pinnules and typically a dormant apical bud; pinnules usually 2–3-pinnate, sparsely to moderately pubescent on the lower (= abaxial) surfaces, the ultimate segments (= smallest subdivisions) of pinnules serrulate; fertile pinnules toward leaf apex, similar to sterile pinnules except having ultimate segments fringed with finger-like fertile lobes; sori (each with 1 sporangium) in 2 rows, 1 on each side of midvein of oblong marginal lobes of ultimate segments, almost completely covered by broad hood-like flaps or flanges of tissue which serve as indusia; $n = 58$, $2n = 58$ (Roy & Manton 2006). Naturalized in low woods, thickets, roadside ditches, circumneutral soils; Hardin (e.g., *V.L. Cory 59153*, BRIT), Harris, Jasper (*Correll & Correll 12529*), Jefferson, Orange, Polk (all BRIT, TEX-LL), Liberty, Montgomery (TAMU, TEX-LL), Tyler (TAES, TEX-LL), Sabine (ASTC), San Jacinto, and Walker (TEX-LL) cos. in s part of the Pineywoods, Madison Co. (TAMU; Neill & Wilson 2001) at the e margin of the Post Oak Savannah, and Brazoria and Chambers (SBSC) cos. in the n Gulf Prairies and Marshes; in some areas (e.g., Jack Gore Baygall Unit of the Big Thicket National Preserve in Hardin Co.) it is so abundant and integrated into the vegetation that it appears native (G. Diggs, pers. obs.). Sporulating Apr–frost. Native of e Asia (China and Japan), India, and ne Australia, now naturalized in the se U.S. from NC s to FL w to AR and TX (Nauman 1993a) and also in Hawaii and Puerto Rico (Robinson et al. 2010). [*Ophioglossum japonicum* Thunb.] This widely cultivated, introduced species has been reported as invasive in a number of countries including some areas of the se U.S.; its dense canopy can prevent

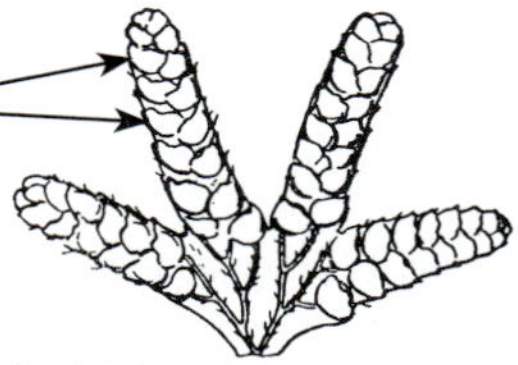

fertile lobes of ultimate segment

Alan Cressler

George Diggs

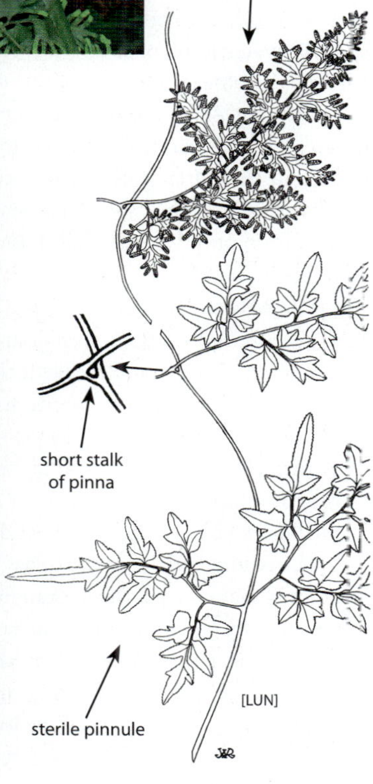

Lygodium japonicum

JAPANESE CLIMBING FERN

the growth of underlying vegetation (Nauman 1993a; Robinson et al. 2010). It escaped from cultivation in the se U.S. prior to 1906 and was first discovered in TX in 1937 (Soxman 1939; Nelson 2000). It is listed as a category I invasive exotic pest species by the Florida Exotic Pest Plant Council (FLEPPC 2009). Volin et al. (2004) listed a number of characteristics that enhance the competitive ability of non-indigenous plants, including "self or wind pollination, rapid growth to reproductive age or size, high and continuous seed or spore production, adaptations for short- and long-distance dispersal, and vegetative as well as sexual reproduction." *Lygodium japonicum* has all of these except vegetative reproduction. (A second Old World species, ⃠ *Lygodium microphyllum* (Cav.) R. Br., OLD WORLD CLIMBING FERN, is also considered a dangerous invasive weed in Florida (Pemberton 1998, 2003; Pemberton & Ferriter 1998; Nauman et al. 2000; Brandt & Black 2001; Volin et al. 2004; Hutchinson & Langeland 2010) and is listed as a category I invasive exotic pest species by the Florida Exotic Pest Plant Council (FLEPPC 2009). It can be distinguished from *L. japonicum* by its 1-pinnate pinnules than are glabrous beneath. Currently it is known in the US only in FL.) ☜ ⃠

MARSILEACEAE Mirb.

WATER-CLOVER OR PEPPERWORT FAMILY

Plants small, rooted, aquatic or of very wet habitats; rhizomes long-creeping, with air channels (an adaptation to an aquatic environment); leaves scattered along the rhizomes, long-petioled and palmately divided into 4 pinnae or else filiform and lacking expanded blades; sori contained in sporocarps (= hard bean- or pea-like structures) on stalks from near base of petiole; sporangia of 2 kinds within the same sorus and sporocarp, the megasporangia with 1 megaspore, the microsporangia with numerous microspores; megasporangium with megagametophyte protruding.

A nearly cosmopolitan family of 3 genera and ca. 75 species of rooted aquatic or wet area plants (Johnson 1993a; Smith et al. 2008). The third genus, *Regnellidium*, is monotypic, occurs only in South America (se Brazil and ne Argentina), and is characterized by having leaves with 2 pinnae/leaflets. All three families of heterosporous water ferns occur in TX (Azollaceae—*Azolla*, Marsileaceae—*Marsilea*, *Pilularia*, and Salviniaceae—*Salvinia*). These groups are quite distinct morphologically; traditionally, the marsileaceous (*Marsilea* and *Pilularia*) and salviniaceous (*Azolla* and *Salvinia*) lines were considered to have evolved independently from different homosporous fern ancestors. However, based on recent morphological, fossil, and molecular evidence, all three living heterosporous water fern families are considered to comprise a monophyletic group (Rothwell & Stockey 1994; Hasebe et al. 1995; Pryer 1999). Heterospory is extremely rare within ferns—it is known from only 5 genera (all in Order Salviniales) and < 1% of living ferns (Pryer et al. 2004b; Nagalingum et al. 2008).

The Marsileaceae have a long fossil history—fossils from 137 mya (early Cretaceous) are part of the Marsileaceae evolutionary line (Yamada & Kato 2002; Schneider et al. 2004c), and it is thought that "the diversification of the Marsileaceae into its extant lineages occurred in the mid-Cretaceous" (Lupia et al. 2000). All living members of the family have reduced leaves with at most four leaflets, and "appear to terminate development at a simplified leaf stage that is similar in complexity to juvenile stages of other ferns" (Pryer & Hearn 2009)—an example of heterochrony (= an evolutionary change in the relative timing of events during development). (Gould 1977). This leaf reduction is least extreme in *Marsilea* (4 leaflets), more so in *Regnellidium* (2 leaflets), and most dramatic in *Pilularia*, which has undifferentiated filiform leaves.

The family is also unique in having sporocarps, which are hard, bean- or pea-like, protective and dispersal structures that contain the sporangia. These sporocarps are apparently highly modified folded pinnae. When the sporocarps are moistened, the spores are released and drift towards the surface of the water. Marsileaceae spores are specially adapted for an aquatic environment, their unique gelatinous outer layer swells upon wetting, serving as a flexible floating organ that allows the spores to float at the water surface. The megaspores "floating at the water/air interface form a concave meniscus" which "creates a vortex-like effect that serves as a trap for free-swimming sperm cells" (Schneider & Pryer 2002). Ecologically, Pryer and Hearn (2009) noted that these ferns are adapted for "seasonally dry areas where water is commonly present only during the rainy season, or where water levels fluctuate considerably from one season to another." They also point out that in contrast to most ferns which take months or even years to produce mature leaves capable of making spores,

Marsileaceae with their combination of highly reduced leaves and spores in specialized sporocarps can reproduce (produce spores) quite quickly ("within a few weeks")—this seems extremely valuable in a seasonal or highly variable environment. Yatskievych and Windham (2009d) also noted the rapid reproduction saying, "less than a week usually is necessary following sporocarp germination for fertilization and the subsequent generation of new sporophytes." (Order Salviniales)

FAMILY RECOGNITION IN THE FIELD: rooted plants of wet areas with leaves resembling a 4-leaf clover or in 1 species apparently rare in TX the leaves are thread-like and ± resemble those of a small grass; sori in hard, bean- or pea-like structures at the base of the petioles.

REFERENCES: Correll 1956, 1966; Kramer 1990c; Johnson 1993a; Lesho 1994; Pryer 1999; Lupia et al. 2000; Schneider & Carlquist 2000b; Schneider & Pryer 2002; Yamada & Kato 2002; Mickel & Smith 2004; Nagalingum et al. 2006, 2008; Pryer & Hearn 2009; Yatskievych & Windham 2009d.

1. Leaf blades resembling a 4-leaf clover, palmately divided into 4 narrowly to broadly cuneate (= wedge-shaped) pinnae __________ **Marsilea**
1. Leaves filiform (= thread-like), very narrow, somewhat grass-like in appearance, without expanded blades __________ **Pilularia**

MARSILEA L. WATER-CLOVER, PEPPERWORT

Small plants, aquatic or of wet habitats, often forming dense colonies, with widely creeping, naked rhizomes; ours rooting only at nodes, or in *M. mollis* occasionally with an internodal root; leaves long-petiolate (the petioles have air channels) with blades palmately divided into 4 pinnae, the leaves produced on land typically smaller, hairier, and with stiffer petioles; pinnae pulvinate at base (the pulvinae are joint-like thickening that "serve to orient the leaflets during the day and cause the leaflets to fold together at night"—Johnson 1993a); sporocarps ovoid, basal, on short peduncle-like stalks above ground, TX species with a portion of the stalk adnate to the sporocarp (this part of the stalk referred to as the raphe), the tip of stalk often protruding as a bump or tooth (proximal tooth), a second distal tooth sometimes present on sporocarps beyond the attachment point of the stalk.

A genus of 45 species of aquatic ferns, nearly cosmopolitan but particularly in temperate and subtropical areas; the area of greatest diversity is Africa (Johnson 1993a; Mickel & Smith 2004; Nagalingum et al. 2007). Within the genus there are two ecological types: "1) true aquatic species with glabrous leaves that inhabit permanent water bodies and 2) semi-aquatic species with hairy leaves that prefer fluctuating wetland habitats and prevail through seasonal extremes in wet and dry periods" (Whitten & Jacono 2009). The leaves, with 4 pinnae (= leaflets), have a superficial resemblance to those of CLOVER; young plants can have unlobed leaves like *Pilularia*. Johnson (1993a) discussed the very "unfernlike" leaves—many of the leaf characteristics of *Marsilea* are well-known among other aquatic plants as adaptations to an aquatic environment. They optimize floatation of the leaf surface, while "minimizing the effects of stresses imposed by water movement." The overall (the 4 leaflets together) circular outline of the leaf, the attachment of the leaflets at a central point, the pliable petiole (allowing adjustment as water level changes), the water-repellent, thickened upper leaf epidermis, buoyancy provided by air-containing cells, and stomates only on the upper leaf epidermis are all examples of aquatic adaptations. The paucity of morphological characters that differ between species has made determining species relationships within *Marsilea* difficult. However, molecular studies (Nagalingum et al. 2007) have recently clarified evolution within the genus—all three TX species fall into the "nodorhizae" subgroup which is restricted to the New World (North and South America, the Caribbean islands, and Hawaii). It is distinguished from other members of the genus by "having roots only at the nodes, sporocarps that are large (2.5–9 mm long) with high numbers of sori (10–23) and lacking transverse veins" (Nagalingum et al. 2007). A number of species have hairs on the leaf surfaces. Wu and Kao (2009) studied an amphibious species, *M. quadrifolia* L., EUROPEAN WATER-CLOVER, that has glabrous submerged and floating leaves, but terrestrial leaves with hairs. They found that when the leaf hairs were removed, the leaves lost significantly more water through transpiration. They hypothesize that leaf hairs (which reduce water loss through boundary layer resistance) are an important adaptation allowing this amphibious species to grow in terrestrial conditions. Some *Marsilea* species can be quite variable morphologically depending on growing conditions—for example J. Stanford (pers. comm.) has observed individuals growing in water to have petioles at least 30 inches long.

☠ An Australian member of the genus, the NARDOO FERN, (*M. drummondii* A. Braun), is particularly rich in the enzyme thiaminase (which breaks down vitamin B1—thiamine) and is reported to have caused the suffering and death of members of the first scientific expedition to cross Australia from south to north (1860–61). Upon running out of food, the explorers ate the fern, which was known to be consumed by Aborigines (but specially prepared by them). The consumption of the inadequately prepared fern resulted in beriberi (vitamin B1 deficiency) affecting almost all members of the expedition. They experienced weakness, emaciation, partial paralysis of the legs, nerve damage, and in some cases death—the lone survivor was saved by Aborigines (but was crippled for life). The fatal poisoning of large numbers of Australian sheep by NARDOO FERN has also been reported (Earl & McCleary 1994; Grant 1995; Moran 1995, 2004; Bruneton 1999). *Marsilea* species are increasingly being found as invasive in Florida and the southeastern U.S., at least in part due to the water garden trade (Whitten & Jacono 2009); therefore, collectors and land managers in TX should be alert to possible invasive exotics. The sporocarps of *Marsilea* are extremely durable and resistant to water loss, and spores may remain viable for up to 130 years (Allsop 1952; Johnson 1985; Moran 1995). Sporocarps can apparently be dispersed on the feet of waterfowl or in their digestive tracks (Malone & Proctor 1965; Johnson 1986; Mickel & Smith 2004). The common name PEPPERWORT is in reference "to the sporocarp, which approximates a peppercorn in size and shape" (Johnson 1993a). Although superficially unlike most ferns, the typical coiled ('fiddlehead') form of young leaves of *Marsilea* species is an excellent clue. (Named for Count Luigi Ferdinando Marsigli, 1658–1730, Italian naturalist and geographer at Bologna and founder of the Institute of Science of Bologna)

REFERENCES: Gupta 1957; Thieret 1977; Landry & Holder 1985; Johnson 1986, 1988; Lesho 1994; Grant 1995; Witsell & Shepherd 1998; Jacono & Johnson 2006; Nagalingum et al. 2007; Whitten & Jacono 2009.

1. Sporocarps with distal tooth (= the one distal to the proximal tooth, which protrudes as a bump or tooth from tip of sporocarp stalk) 0.4–1.2 mm long, acute; pinnae sparsely appressed pubescent or glabrous beneath, the hairs if present not noticeably extending past leaf margin ____ **M. vestita**
1. Sporocarps with distal tooth absent or to only 0.2 mm long, blunt; pinnae densely and conspicuously spreading hairy beneath, the hairs sometimes noticeably extending past leaf margin.
   2. Pinnae 9–35 mm long, 8–39 mm wide; sporocarps 6–9 mm long, 4.5–5.5 mm wide, densely covered with long, matted, often twisted hairs at maturity; sporocarp stalks usually branched, with several sporocarps per stalk; petioles 5–39 cm long; species wide-spread e of the Trans-Pecos ____ **M. macropoda**
   2. Pinnae 2–17 mm long, 1–16 mm wide; sporocarps 5 mm or less long, 3 mm or less wide, glabrate at maturity (but covered with long, needle-like hairs with spreading tips when young); sporocarp stalks unbranched, with 1 sporocarp per stalk; petioles 1–14 cm long; species known in TX only from the Trans-Pecos ____ **M. mollis**

Marsilea macropoda Engelm. ex A. Braun, (Greek: *macro*, long, here meaning large, and *pous* or *podium*, foot, = large-footed), BIG-FOOT WATER-CLOVER, LARGE-FOOT WATER CLOVER, LARGE-FOOT PEPPERWORT, WATER-CLOVER, GOLDEN WATER-CLOVER. Species generally larger (pinnae, petioles, and sporocarps) than *M. mollis* and *M. vestita*, differing as in the key; sporocarps with distal tooth absent or merely a slightly raised area; $2n = 40$ (Lesho 1994). Typically in mud, also shallow water; mainly s and c TX n to Tarrant and Wise (e.g., *O'Kennon & McLemore 18719*, BRIT), e to Harris (SBSC), and w to Brown (*J.W. Stanford 2841, 3482*, HPC) and Val Verde (Correll 1956; Turner et al. 2003) cos.; AL, AR, FL, LA, and TX, also Mexico. Sporocarps produced nearly year round. Johnson (1993a) noted that "*Marsilea macropoda*, with its tall, hairy leaves and branched sporocarp stalks, is a striking and distinctive species, worthy of cultivation." It is sometimes used as an ornamental. Johnson (1993a) also pointed out that "individuals that appear to be hybrids [with *M. vestita*] have been collected in Blanco, Brazos, and Kleberg counties in Texas." These apparent hybrids have large hairy leaves but sporocarps with a relatively large distal tooth.

Marsilea mollis B.L. Robinson & Fernald, (Latin: *mollis*, soft or pliant), MEXICAN WATER-CLOVER. Similar to *M. vestita*, but differing as in the key (size of distal sporocarp tooth and leaf pubescence); petioles 1–14 cm long; pinnae 2–17 mm long, 1–16 mm wide (floating leaves in Mexico reported by Mickel & Smith 2004 to be somewhat larger); sporocarps with distal (sometimes called superior) tooth absent or to 0.2 mm long; $2n = 40$ (Lesho 1994). Mud or in shallow water, desert washes; known from only one TX collection from Limpia Canyon, Davis Mts., Jeff Davis Co. (*Hinkley s.n.*, 1935, TEX-LL, annotated D.M. Johnson; Johnson

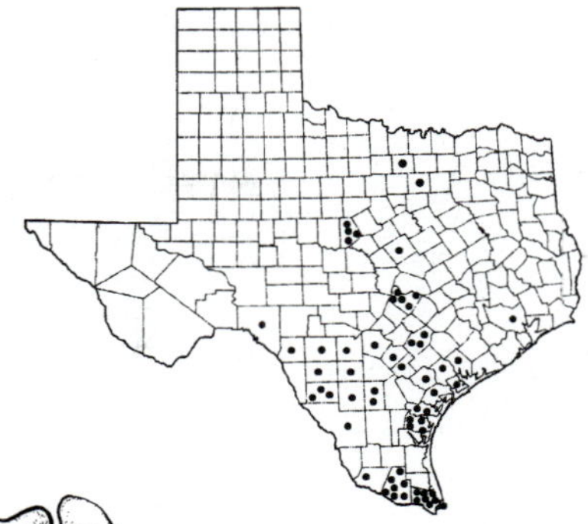

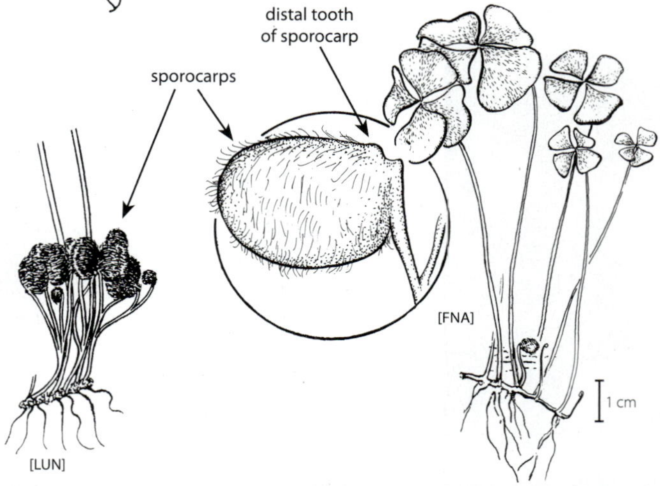

George Diggs

Marsilea macropoda

BIG-FOOT WATER-CLOVER

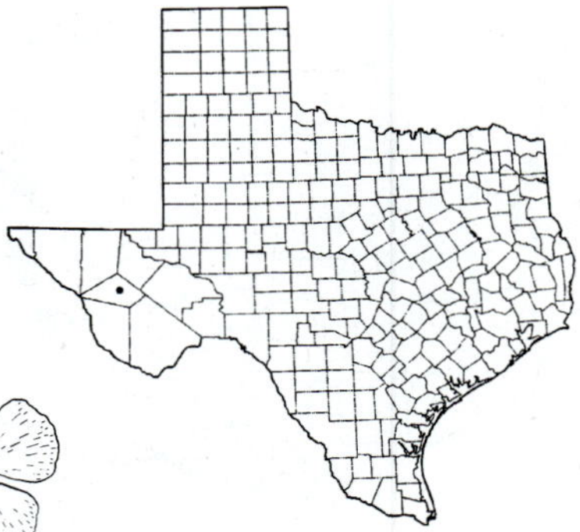

young sporocarps showing covering of hairs

lower surface of leaf showing hairs

Marsilea mollis

MEXICAN WATER-CLOVER

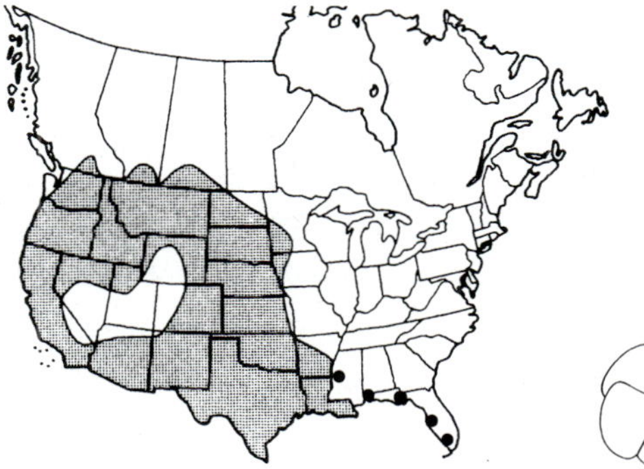

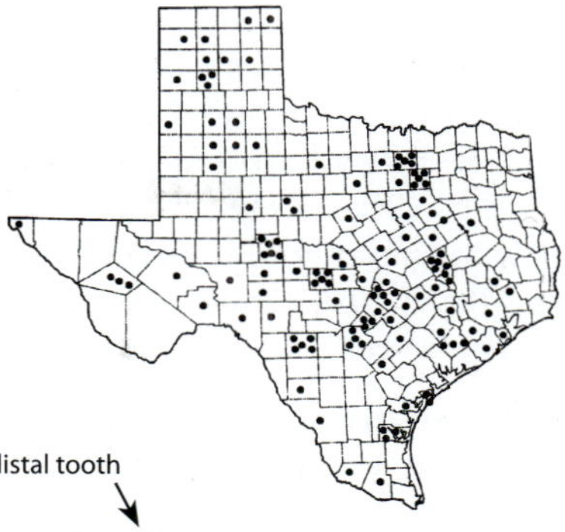

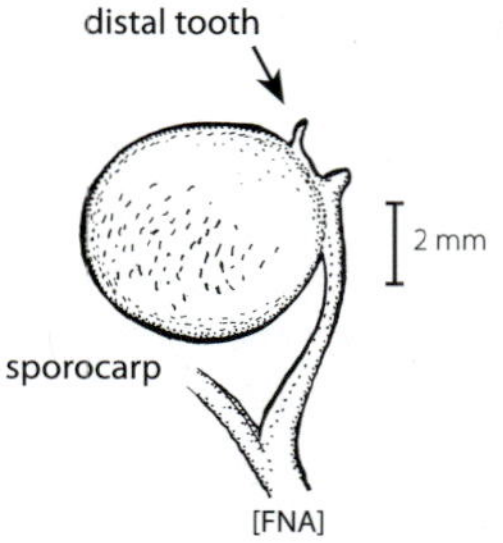

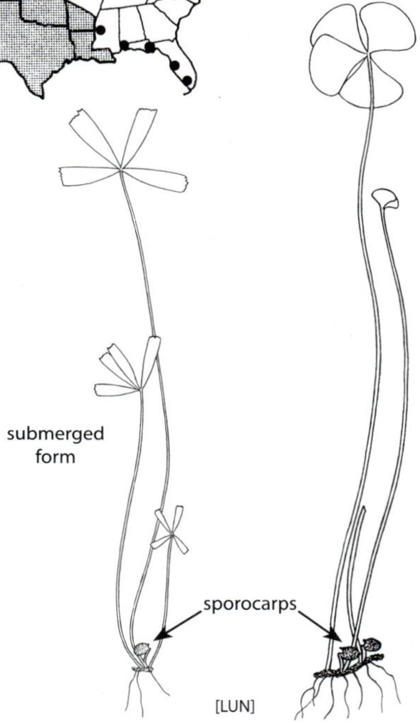

Dale A. Zimmerman Herbarium, Western N. Mex. University

Dale A. Zimmerman Herbarium, Western N. Mex.Universtiy

Marsilea vestita

HAIRY WATER-CLOVER

1986, 1993a); AZ and TX, also Mexico to South America. Sporulating May–Oct. Johnson (1993a) indicated that "the name *M. mexicana* A. Braun has sometimes been misapplied to this species." Johnson (1993a) also noted that the red or brown streaks on the pinnae, supposedly characteristic of this species, actually can be found on floating leaves of many other species. Although known from desert habitats, this species can avoid drought by effectively becoming annual. If drought occurs, the plants die, and after heavy rains their life cycle simply begins again with spore germination—the spores having been protected by the very resistant sporocarps (Heitz 2010). Given its rareness and limited distribution in the state, we consider this species to be of conservation concern in TX. ⚠

Marsilea vestita Hook. & Grev., (Latin: *vestis*, garment, clothing, covering, = covered, apparently in reference to the sparse hairs on the pinnae of some individuals), HAIRY WATER-CLOVER, HOOKED PEPPERWORT, WATER-CLOVER, HAIRY PEPPERWORT, NARROW-LEAF PEPPERWORT. Petioles 2–20 cm long; pinnae 4–19 mm long, 4–16 mm wide, varying in appearance, typically 1–2 times as long as wide, fan-shaped or broadly cuneate, apically entire or undulate-crenulate, but sometimes (when growing submerged) narrow in appearance, 3–7.5 times as long as wide, narrowly and obliquely cuneate, apically irregularly toothed or crenulate; petioles 2–20 mm long; sporocarp stalks unbranched, with 1 sporocarp per stalk; sporocarps 3.6–7.6 mm long, 3–6.5 mm wide, pubescent with appressed hairs or soon typically glabrate (leaving purple or brown speck-like scars), with noticeable distal tooth; $2n = 40$ (Lesho 1994). Ponds, wet depressions, along streams and rivers; widespread in TX except absent from most of the Pineywoods (e.g., Burnet Co., *L.L. Sanchez 4038*; El Paso Co., *R.D. Worthington 4951*, both BRIT); sw Canada throughout w U.S. e to MN, AR, and LA and extending in the se U.S. e to FL; also Mexico and Peru in South America. Sporocarps produced Mar–Oct. [*M. mucronata* A. Braun, *M. uncinata* A. Braun, *M. tenuifolia* Engelm. ex A. Braun, *M. vestita* subsp. *tenuifolia* (Engelm. ex A. Braun) D.M. Johnson] Plants with narrow pinnae are usually easily distinguished in the field and have sometimes been treated as a separate species (*M. tenuifolia*—e.g., Correll & Johnston 1970) or subspecies (*M. vestita* subsp. *tenuifolia*—e.g., Johnson 1986; Kartesz 1994; Diggs et al. 1999). However, this variant was not formally recognized by Johnson (1993a), Jones et al. (1997), or Kartesz (1999), and according to J. Peck (pers. comm.), it is an environmentally induced form that occurs when leaves develop submerged. We are therefore including it in *M. vestita*. Such variation is not surprising since the leaves of *Marsilea* species are known to "show great plasticity of form in response to a wide range of physical and chemical stimuli" (Johnson 1993a). Bob O'Kennon (pers. comm.), however, has noted populations in Gillespie Co. with very narrow leaves in locations where the plants are never submerged. Correll (1961) wrote a paper indicating that he had found *Marsilea mexicana* A. Braun (now recognized as *M. ancylopoda* A. Braun) new for TX, based on a specimen from Aransas Co. (*Correll & Johnston 17623*, TEX-LL); however, that specimen has since been identified as *M. vestita* (by D. Lemke in 2007), and *M. ancylopoda* is not known for TX.

Marsilea mutica Mett., AUSTRALIAN WATER-CLOVER, VARIEGATED WATER-CLOVER, or BANDED NARDOO, native to Australia and New Caledonia, while currently unknown for TX, is known from AR and OK and widely across the se U.S. (Jacono & Johnson 2006). It should be looked for in TX by collectors. This species can be distinguished from all native TX *Marsilea* species by its pinnae being two-toned green, with a lighter color in the basal ⅓–⅔ and in having sporocarps lacking teeth (vs. pinnae evenly green and sporocarps with 1 or 2 ± prominent teeth). The two-tone pattern does not preserve well when specimens are heat-dried, but is still evident to a lesser degree (J. Peck, pers. comm.). It is sold in TX at nurseries and water garden stores for use in decorative "ponds" or water gardens. However, it is potentially invasive and is officially considered a noxious aquatic plant in OK (USDA 2009). ☜ ⃠

PILULARIA L. PILLWORT

A genus of ca. 5–6 very similar species of inconspicuous grass-like or sedge-like plants of North America, South America, Europe, Pacific Islands, Australia, and New Zealand (Johnson 1993a; Mickel & Smith 2004). Mickel and Smith (2004) noted that the "discreteness and interrelationships of species within *Pilularia* need re-investigation using modern methods." It is sometimes placed in its own family. (Latin: *pilula*, a little ball, in reference to the spheric sporocarps)

REFERENCES: LaMotte 1940; Hill 1980; Dennis & Webb 1981; Culwell 1994.

Pilularia americana A. Braun, (of America), AMERICAN PILLWORT, WATER-PEPPER. Small inconspicuous aquatic, submersed or infrequently persisting on bare mud; leaves filiform, 1.6–10.2 cm long, lacking expanded blades; sporocarps produced just below ground surface, globose, 2–6(–10) mm long, 2–3 mm in diam.; 2*n* = 20 (Chopra 1960). Temporary pools, ponds, tanks, and reservoir margins; J. Peck (pers. comm.) indicated that the best time to find this species is after reservoir drawdowns in fall. According to the range map in Johnson (1993a), *P. americana* occurs rather broadly in nc TX, the n part of e TX, and the n part of c TX; the only confirmed records we know of are from Burnet (*R. McVaugh 7656*, Granite Mt.—BRIT, TEX-LL), Colorado (*J.&C. Taylor 32117*, *J. Taylor 32127*, BRIT), Llano (*D. Seigler et al. 2016*, Enchanted Rock, TEX-LL), Mason (*S.R. Hill 8654*, BRIT, TEX-LL, *T. Walters 795*, BRIT), Smith (*O'Kennon 19989*, BRIT), Tarrant (*O'Kennon 22097*, Fort Worth Nature Center & Refuge, BRIT), and Wise (*O'Kennon & McLemore 18341*; BRIT) cos.; AR, CA, GA, KS, MO, NE, OK, OR, TN, and TX; also Baja California and Durango in Mexico and South America (Mickel & Smith 2004). Sporocarps produced late spring–fall. Weakley (2010) noted that, "This peculiar plant has a puzzling distribution, being known from several disjunct regions: OR to s. CA; NE to TX; AR; TN; and GA to SC." It is so inconspicuous that it is rarely recognized or collected. A large population (ca. 150 sq. m) was discovered in AR (Culwell 1994), prompting the comment that the species "may be more common than herbarium specimens indicate" (Culwell 1994). The population was growing in shallow water in a large lake; Culwell (1994) in describing the AR population noted "after the lake had undergone a drawdown, I noticed a large area..., usually under approximately 8–16 cm of water, that appeared to be a bluegrass lawn. This expanse extended about 6 m from the original shoreline at a gradual slope and stretched approximately 25 m along the shore. The population was quite dense and a relatively pure stand, with few invading species." Mickel and Smith (2004) also suggest that because of "its inconspicuous and unfern-like appearance, probably it has been overlooked by collectors." This species was long known in TX only from the Central Mineral Region (e.g., Correll 1956; Turner et al. 2003). Recently, however, Bob O'Kennon (pers. comm.) has found extensive colonies in Lake Tyler (Smith Co. in East TX) and in Tarrant and Wise cos. (in the Cross Timbers and Prairies), and specimens are known from Colorado Co. The Smith, Tarrant, and Wise populations were found on exposed flats during a period of extreme low water following prolonged drought. The plants had presumably been living for many years entirely underwater. Extensive searching in similar nearby habitats did not reveal additional colonies. A reasonable hypothesis for this sporadic appearance is long-distance dispersal of the sporocarps or spores inside of or on the feet of waterfowl. Such dispersal can apparently occur for related *Marsilea* species (Johnson 1986; Mickel & Smith 2004). Recent molecular research (Nagalingum et al. 2008) has found that this species consists of two geographically and genetically distinct groups and needs further study. Weakley (2010) pointed out that the plant "may be recognized as a 'fern' by the typical coiled ('fiddlehead') development of young leaves."

ONOCLEACEAE Pic. Serm.

SENSITIVE FERN FAMILY

A very small, largely n temperate, monophyletic family of 4 genera and 5 species that is most closely related to the Blechnaceae (Schuettpelz & Pryer 2008; Smith et al. 2008). In the past it has been treated as a tribe of the Dryopteridaceae (e.g., Smith 1993b; Gastony & Ungerer 1997; Diggs et al. 2006), but recent molecular analyses support its recognition as a separate family (Smith et al. 2008). The species are terrestrial perennials with strongly dimorphic leaves and sori enclosed by the inrolled leaf margins. One of the best known species in the family is *Matteuccia struthiopteris* (L.) Tod., OSTRICH FERN, native to n North America and Eurasia. It is widely cultivated as an ornamental and is well known for its "edible fiddleheads, the canning of which is a local industry in New England and adjacent Canada" (Johnson 1993c); it could possibly escape from cultivation in East TX. This species resembles *Onoclea sensibilis*, SENSITIVE FERN, in having conspicuously dimorphic leaves, but can be distinguished by its 1-pinnate-pinnatifid sterile leaves with veins free and by its 1-pinnate fertile blades which have the sori enclosed by the hardened revolute margins of the pinnae. (Order Polypodiales)

FAMILY RECOGNITION IN THE FIELD: the only TX species has conspicuously dimorphic leaves with unmistakable fertile fronds; the blade tissue is rolled into globular, berry-like or bead-like structures concealing the sori, the whole fertile leaf superficially resembling a narrow panicle of small round fruits.

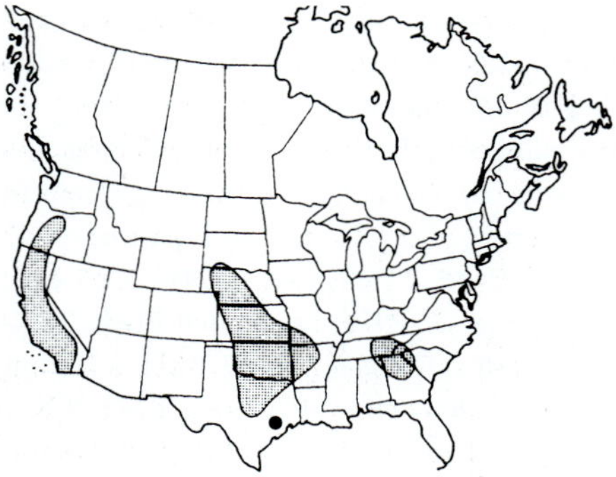

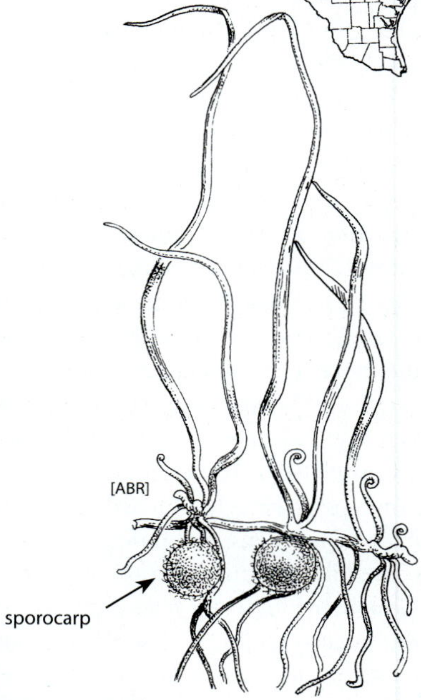

Pilularia americana

AMERICAN PILLWORT

REFERENCES: Gastony & Unger 1997; Smith et al. 2006, 2008; Schuettpelz & Pryer 2007, 2008.

ONOCLEA L. SENSITIVE FERN

A monotypic genus of n temperate areas (Johnson 1993b); sometimes cultivated as an ornamental. (Greek: *onos*, vessel, and *cleisto*, closed, in reference to the "segments of the fertile frond that roll up into hard, bead-like structures enclosing the sori"—Moran 2004)

REFERENCES: Lloyd 1971; Beitel et al. 1981; Rothwell & Stocky 1991; Johnson 1993b; Gastony & Ungerer 1997; Reudink et al. 2005.

Onoclea sensibilis L., (Latin: *sensibilis*, sensitive, perceptible), SENSITIVE FERN, BEAD FERN. Rhizomes creeping; petiole bases with 2 vascular bundles; leaves conspicuously dimorphic, of 2 very different types, scattered along the rhizome, erect, glabrous; sterile leaves to ca. 1(–1.3) m tall, thin herbaceous, deciduous, broadly triangular to ovate in outline, deeply pinnatifid, the pinnae few, subopposite (especially the lowermost), undulate to irregularly deeply lobed, the margins entire, the veins reticulate and forming a series of linear to oblong areoles along the midvein (= costa), the rachis winged; fertile leaves persistent over winter, 2-pinnate, the blades greatly reduced, the ultimate blade segments (= smallest subdivisions of blade) rolled into globular, berry-like or bead-like structures concealing the sori, the whole fertile leaf superficially resembling a narrow panicle of small round fruits; $2n = 74$ (Johnson 1993b). Swamps, low woods, and wet areas; Pineywoods and n edge of Gulf Prairies and Marshes w through Post Oak Savannah to Milam, Upshur (BRIT), Red River, Henderson, Limestone, Falls (Turner et al. 2003), and Wilson (TAES; Correll 1956) cos.; also disjunct w to Gillespie (*O'Kennon 8367*, Bear Mt., wet depression e of granite mine, BRIT), Burnet (Correll 1956), Uvalde, and Wilson (Turner et al. 2003) cos.; se Canada and throughout e U.S. w to ND, WY, CO, and TX; also e Asia (the Asian plants are often recognized as var. *interrupta* Maxim.). Sporophylls produced Apr–Oct, persisting through the winter and releasing the green spores in spring before expansion of the sterile leaves (Johnson 1993b). Both the specific epithet and the common name derive from the sensitivity of this species to cold—the sterile leaves die back with even a light frost (Johnson 1993b). The sterile leaves superficially resemble those of *Woodwardia areolata*, but that species has alternate basal pinnae, the pinnae lacking lobes (pinnae can be sinuate) but with minutely serrulate margins. The disjunct TX populations may represent survival since Ice Age times in tiny pockets of appropriate moist microhabitats; numerous TX plants display such disjunctions (see page 36). SENSITIVE FERN is native to both e North America and e Asia (Hoshizaki & Moran 2001); see Fig. 34 and page 32 for a discussion of this interesting disjunct distribution pattern. Molecular evidence obtained by Gastony and Ungerer (1997) supported the recognition of North American plants as *O. sensibilis* var. *sensibilis* and those of e Asia are as *O. sensibilis* var. *interrupta* Maxim. Fifty-seven million year old Paleocene Epoch fossils virtually identical to modern members of this species provide evidence that fern species can remain essentially unchanged over millions of years (Rothwell & Stockey 1991; Serbet & Rothwell 1999). Unusual leaves, intermediate in form between typical sterile and fertile leaves, are known in this species (Thieret 1980), possibly as a result of some type of injury (e.g., frost, defoliation) (Lloyd 1971; Beitel et al. 1981). Because of the unusual appearance of the fertile leaves (greatly reduced blade tissue and bead-like pinnules), they are sometimes gathered for use in dried arrangements. The species has trophopods—petiole bases that are "specialized to accumulate food and to persist as a storage organ long after the frond to which it belonged has withered and decayed away" (Wagner & Johnson 1983). SENSITIVE FERN is reported to be poisonous; horses are said to become unsteady and collapse upon ingesting large amounts of the plant. Other symptoms include walking in circles, difficulty in chewing (with food sometimes falling from the mouth), seizures, and death. The specific cause is not known with certainty but is thought to be thiaminase, an enzyme that destroys thiamine and causes Vitamin B1 deficiency (A number of other pteridophytes are well known to cause thiamine depletion including species of *Cheilanthes*, *Equisetum*, *Marsilea*, and *Pteridium* [Hodgdon 1951; Burlage 1968; Turner & Szczawinski 1991; Burrows & Tyrl 2001]). SENSITIVE FERN also contains phytoecdysones, a class of hormone-like plant defense compounds that disrupt ecdysis (= molting) in insects—insects eating the plant thus have their developmental sequence altered (Swain & Cooper-Driver 1973).

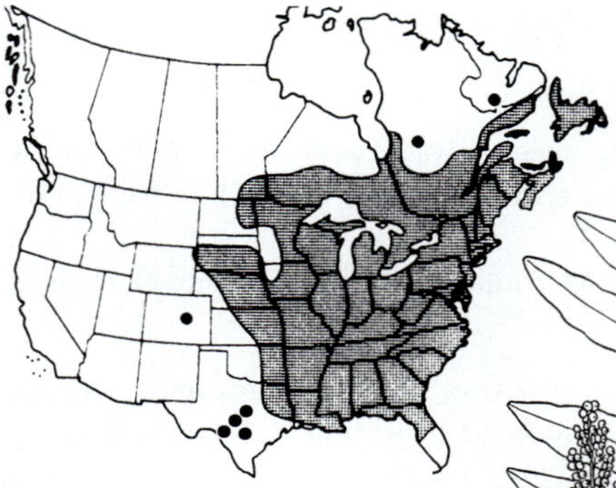

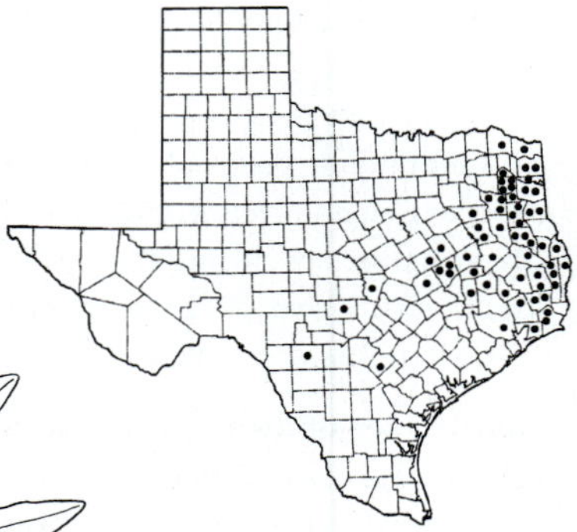

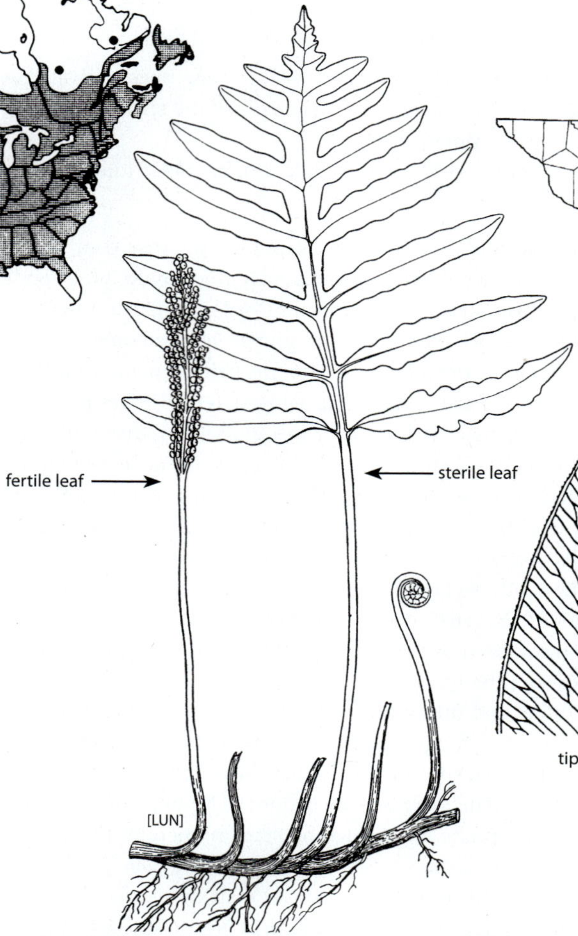

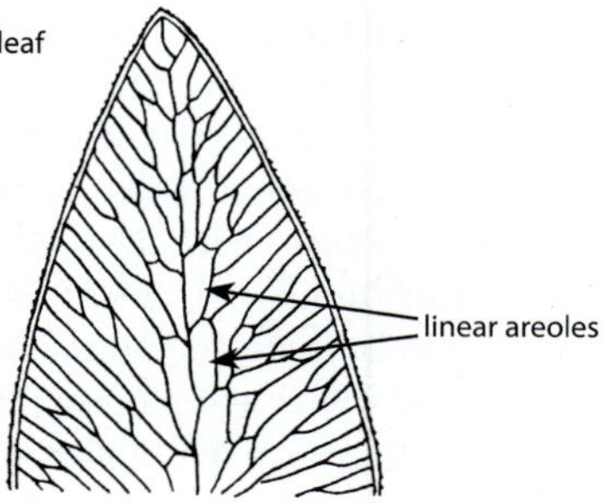

tip of pinna

Sally and Andy Wasowski, Courtesy LBJ Wildflower Center

Onoclea sensibilis

SENSITIVE FERN

OPHIOGLOSSACEAE Martinov

ADDER'S-TONGUE FAMILY

Texas species terrestrial; stems ± subterranean, simple, unbranched, short, upright, fleshy, not rhizomatous; leaves 1 or less commonly 2(–3) per stem, often ephemeral, divided into a blade portion (= trophophore) and a fertile sporangia-bearing portion (= sporophore), the vernation (= the way a new leaf emerges from bud) conduplicate (not circinate as in most ferns); blade portion simple, divided, or compound; fertile portion lacking blade-like tissue, typically consisting of a long stalk with a terminal, branched or unbranched, sporangia-bearing area; sporangia eusporangiate, large (in comparison with those of most other ferns), spherical, thick-walled, lacking an annulus, embedded in 2 rows on each branch or on the unbranched sporangia-bearing area; gametophytes subterranean, with mycorrhizal fungi.

A nearly cosmopolitan but mostly temperate and boreal, morphologically distinctive family of 5–8 genera, depending on taxonomic viewpoint, and ca. 70–80 species (Wagner & Wagner 1993; Hauk et al. 2003; Smith et al. 2008). They are eusporangiate ferns, which means having large, thick-walled sporangia with hundreds to thousands of spores per sporangium—this condition is known in only four ancient fern lineages, Ophioglossaceae, Equisetaceae, Marattiaceae, and Psilotaceae. Molecular evidence shows the family is monophyletic (Hauk et al. 2003). Hauk et al. (2003) argue that several ancient and very distinctive (both at the molecular level and morphologically) lineages traditionally treated in *Botrychium* should be recognized as separate genera.

The Ophioglossaceae have for a considerable period of time been thought to be only distantly related to other ferns (e.g., Kato 1988). Wagner (1990), for example, suggested that "Of modern plants normally classified as 'ferns,' the Ophioglossaceae are the most isolated." Bell and Woodcock (1983) noted they are apparently relicts of an ancient lineage (Bell & Woodcock 1983), and Wagner (1990) further indicated that a number of characteristics "... suggest closer affinities to progymnosperms or cycadophytes than to typical modern ferns." However, modern molecular evidence clearly links them with other ferns. They and the Psilotaceae (WHISK FERNS) are most closely related and together are the sister group to all other ferns. In other words, they are the basal branch of the fern lineage and are thus living representatives of the oldest surviving branch of ferns. It is interesting that while the family thus appears quite ancient, the oldest known fossils date only to the earliest Tertiary (Rothwell & Stockey 1989). The family is made up of two clearly defined subfamilies, Botrychioideae and Ophioglossoideae, sometimes recognized as distinct families. The young leaves unfold (technically conduplicate) rather than having the form of an unrolling (circinate) fiddlehead typical of most ferns (Kato 1988; Wagner 1990; Nelson 2000). Overall leaf structure in the family can also be confusing—unlike in other ferns, the leaves are divided into a photosynthetic blade portion (= trophophore) and a fertile sporangia-bearing portion (= sporophore). The sporophore arises upon the axis of the trophophore and dies after the spores are shed. Wagner (1990) noted that the homologies of this unique sporophore "are not fully understood, but it is commonly believed to be the fusion product of the two basal pinnae (analogous to the fertile basal pinnae of *Anemia*, but with fusion)." In other words, the sporophore is thought to represent two fused basal pinnae.

Mankyua, a new genus of this ancient family, was recently found on an island off the s coast of the Korean Peninsula (Sun et al. 2001). This is an example of the exciting new botanical discoveries continually being made. The sporophytes of members of this family obtain minerals through mycorrhizal fungi associated with their fleshy roots (which lack root hairs); in essence, these ferns trade carbon compounds for minerals (J. Peck, pers. comm.). The Ophioglossaceae are unusual in another way. In contrast to the terrestrial photosynthetic gametophytes typical of most ferns, their gametophytes and (very young sporophytes prior to producing an above ground leaf) are subterranean and non-photosynthetic; instead, they are completely dependent on the fungal symbionts for their energy needs. In other words, at these stages they are mycoheterotrophic: unable to grow without the fungi (Winther & Friedman 2007); the related Psilotaceae have similar subterranean mycoheterotrophic gametophytes. The following treatment draws heavily on Wagner and Wagner (1993). (Order Ophioglossales)

FAMILY RECOGNITION IN THE FIELD: often small plants with only 1 or sometimes 2(–3) leaves; leaves with a blade portion (simple to compound) and an erect, spike-like, fertile portion consisting of an elongate stalk and a terminal, fertile, sporangia-bearing area.

REFERENCES: Clausen 1938; Correll 1956, 1966; Thomas 1972, 1980; Kato 1988; Wagner 1990; Wagner & Wagner 1993; Vangerow et al. 1999; Pryer et al. 2001; Sun et al. 2001; Barker & Hauk 2003; Hauk et al. 2003; Sun et al. 2009.

1. Sterile portion of leaf blades ternately-pinnately compound, divided, or lobed, the margins usually denticulate to serrate or lacerate; veins of leaf blades dichotomous (= equally 2-forked) and free (= with their distal ends unconnected to other veins); sporangia in a pinnately branched, panicle-like arrangement ____ **Botrychium**
1. Sterile portion of leaf blades simple, the margins entire; veins of leaf blades reticulate and anastomosing (= veins connecting by cross-veins to form a network); sporangia in an unbranched, linear, spike-like arrangement ____ **Ophioglossum**

BOTRYCHIUM Sw. GRAPE FERN, MOONWORT

Leaves divided into a blade portion and a fertile sporangia-bearing portion, the blade portion compound, divided or lobed, ovate to triangular or broadly triangular in outline, ± ternate (= divided into three equal parts); fertile portion of leaf consisting of an elongate stalk terminated by a 1–2-pinnate, panicle-like, sporangia-bearing region.

A nearly cosmopolitan genus of as many as 50–60 species (if segregate genera are included—e.g., Wagner & Wagner 1993; Mickel & Smith 2004) or according to some authors only 25–30 species (excluding segregates—e.g., Mabberley 2008; Weakley 2010). The greatest diversity is at high latitudes and high elevations, particularly in North America and e Asia. Most species are quite variable vegetatively, resulting in considerable taxonomic confusion. Plants typically produce only one leaf each year (Nauman et al. 2000). Although molecular studies by Hauk et al. (2003) showed *Botrychium* in the broad sense to be monophyletic, they divided it into several segregate genera. However, because the genus is monophyletic and there is a long tradition of treating it broadly, we are following Mickel and Smith (2004) in maintaining all species within a diverse *Botrychium*. If divided into segregates, the four TX species would be in two different genera: *Botrypus* and *Sceptridium*. (Latin: *botry*, bunch (of grapes), in reference to the sporangial clusters)
REFERENCES: Butters 1917b; Wagner 1960, 1992; Thomas 1979; Thomas et al. 1981; Watano & Sahashi 1992; Hauk 1995; Do et al. 1996; Holmes et al. 1996; Skoda 1997; Hauk & Haufler 1999; Barker & Hauk 2003; Small et al. 2005; Rothwell & Karrfalt 2008; Hauk et al. 2012; Williams & Waller 2012.

1. Plants small, to only ca. 12 cm tall; blade portion of leaf prostrate on ground, small, only 3–8 cm long, short-stalked (petiole-like stalk 1.5–3 cm long); ultimate segments (= smallest subdivisions) of blade portion of leaf fan-shaped, their tips broadly rounded; leaves commonly 2 per plant ____ **B. lunarioides**
1. Plants usually larger, 8–75 cm tall; blade portion of leaf raised above the ground, not prostrate, usually larger, 4–30 cm long, either long-stalked (petiole-like stalk 3–20 cm long) OR sessile (petiole-like stalk absent) in *B. virginianum*; ultimate segments of blade portion of leaf not fan-shaped, their tips usually pointed or obtuse; leaf usually 1 per plant.
   2. Blade portion of leaf sessile (fertile stalk originating at very base of blade portion; NOTE: blade & fertile portions share a lengthy common stalk); blade finely divided, the ultimate segments numerous, small, with coarsely serrate to lacerate (= irregularly cut) margins; leaves absent in winter ____ **B. virginianum**
   2. Blade portion of leaf long-stalked, appearing to have a long petiole (blade portion well-separated from origin of fertile stalk); blade coarsely divided, the ultimate segments relatively few, large, with denticulate margins; leaves present in winter.
      3. Blade portion of leaf 2–3-pinnate (often 2-pinnate), ± papery/herbaceous; ultimate segments of blade portion mostly oblong to obliquely lanceolate to narrowly lanceolate, the margins nearly parallel, basally cuneate, apically relatively blunt (can be somewhat pointed) ____ **B. biternatum**
      3. Blade portion of leaf mostly 3-pinnate, relatively leathery; ultimate segments of blade portion usually ± trowel-shaped (= wider basally), rarely linear, the margins usually not parallel, basally obliquely truncate or obtuse, apically relatively pointed ____ **B. dissectum**

Botrychium biternatum (Savigny) Underw., (Latin: *bi*, two, and *ternarius*, by threes, = twice-ternate, the leaves often divided into three major parts, each of these again often divided into three parts), SPARSE-LOBE GRAPE FERN, SOUTHERN GRAPE FERN, WINTER GRAPE FERN. Plant ca. 10–35 cm tall; roots usually 10 or fewer, blackish; leaves present over winter, mostly green but rarely bronze in winter if exposed; new leaves appearing in late spring to early summer; sterile blade portion green to dark green, long-stalked (stalk 3–20 cm long), ± papery/her-

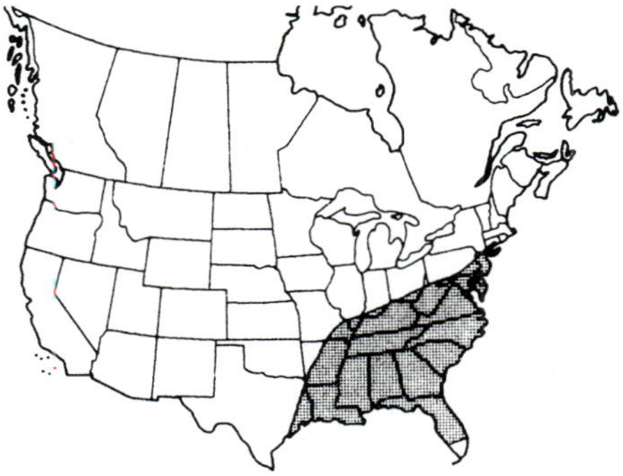

ultimate segment of leaf showing marginal teeth and venation

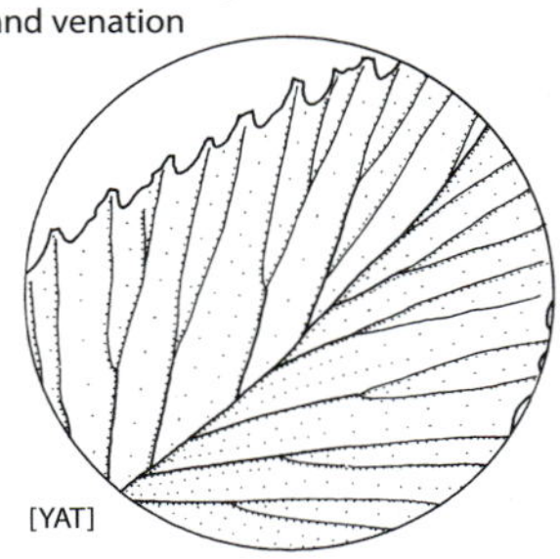

Botrychium biternatum

SPARSE-LOBE GRAPE FERN

baceous, to 18 cm long and 28 cm wide, usually smaller, 2–3-pinnate (often 2-pinnate), the ultimate segments (= smallest subdivisions) elongate, obliquely lanceolate to narrowly lanceolate, the margins nearly parallel, finely denticulate (marginal teeth finer and more uniform than in *B. dissectum*), the apices obtuse to short-acuminate; 2*n* = 90 (Wagner & Wagner 1993). Low woods; Angelina, Bowie, Cass, Hardin (*E. Whitehouse 11191*), Harrison, Montgomery (*V.L. Cory 50680*), Nacogdoches, Newton, Sabine, Titus (all BRIT—identified J. Peck), Tyler (TEX-LL) Harris, and Liberty (Brown 2008) cos. in the Pineywoods and Brazoria Co. (Brown 2008) in the n Gulf Prairies and Marshes; se U.S. from PA s to FL w to IL, OK, and TX. Spores maturing in the fall. [*B. tenuifolium* Underw., *B. dissectum* Spreng. var. *tenuifolium* (Underw.) Farw., *Sceptridium biternatum* (Savigny) Lyon] If *Botrychium* is divided into segregate genera, as argued by Hauk et al. (2003), *B. biternaturm* would be recognized in the genus *Sceptridium*. This species and *B. dissectum* are similar morphologically and are often confused (see discussion under *B. dissectum*).

Botrychium dissectum Spreng., (Latin: *dissectus*, cut up, deeply divided, dissected), CUT-LEAF GRAPE FERN, OBLIQUE GRAPE FERN, DISSECTED GRAPE FERN. Plant extremely similar to *B. biternatum*; leaves present over winter, often bronze or reddish bronze in winter if exposed (but more so in colder areas of the n U.S.); new leaves appearing in late spring; sterile blade portion shiny green, relatively leathery, to 20 cm long and 30 cm wide, mostly 3-pinnate, the ultimate segments usually ± trowel-shaped (rarely linear), the margins usually not parallel, denticulate to coarsely denticulate, the marginal teeth coarser and less uniform than in *B. biternatum*; 2*n* = 90 (Wagner & Wagner 1993). In various habitats from open grassy areas to deep forests; Anderson, Bowie, Fannin (*Diggs & S. Beach s.n.*, 1999), Leon (*E.S. Nixon 18512*), Titus, and Walker (all BRIT—identified by J. Peck) cos., Pineywoods w to Post Oak Savannah and Red River drainage; Hatch et al. (1990) also cited the Blackland Prairie; while the range map in Wagner and Wagner (1993) shows this species in TX only on the extreme e margin of the Pineywoods, it occurs considerably further w; se Canada and throughout e U.S. w to KS, MN, OK, and TX. Spores maturing in the fall. [*B. dissectum* var. *obliquum* (Muhl. ex Willd.) Clute, *B. dissectum* var. *oblongifolium* (Graves) Broun, *B. obliquum* Muhl. ex Willd., *B. obliquum* var. *elongatum* Gilbert & Haberer, *Sceptridium dissectum* (Spreng.) Lyon] If *Botrychium* is divided into segregate genera, as argued by Hauk et al. (2003), *B. dissectum* would be recognized in the genus *Sceptridium*. Barker and Hauke (2003) noted that this is the "most variable North American grapefern species." One form (forma *dissectum*—northern entity occurring as close to TX as c AR—J. Peck, pers. comm.) has highly dissected and quite distinctive leaves. However, molecular studies (Barker & Hauke 2003) do not show the forms to be genetically distinct and thus do not support recognizing the forms as formal varieties or subspecies. The only form occurring in TX (forma *obliquum*, previously treated as a variety) closely resembles *B. biternatum*. Thomas (1980) indicated that the two species intergrade and that some individual specimens puzzle even experts. As a result, *B. dissectum* and *B. biternatum* have often been confused (particularly immature plants). According to Taylor (1984) "... there do not appear to be any stable characters which will always clearly distinguish these two taxa." Nauman et al. (2000) lumped *B. dissectum* into *B. biternatum* saying, "they intergrade in Florida to such an extent that two species cannot be maintained." However, according to W.H. Wagner Jr. (pers. comm.) and Wagner and Wagner (1993), *B. dissectum* has leaves that are more dissected and the ultimate segments usually trowel-shaped (rarely linear), apically more pointed, and with the margins more lacerate. Likewise, J. Peck (pers. comm.) indicates that two species are distinct and that mature (reproductive) plants can usually be distinguished without difficulty. We are therefore following W.H. Wagner, Jr. (pers. comm.), Wagner and Wagner (1993) and J. Peck (pers. comm.) in recognizing *B. dissectum* as a distinct species.

Botrychium lunarioides (Michx.) Sw., (resembling moonwort—*Botrychium lunaria*, a widespread species common in both the Old and New Worlds), WINTER GRAPE FERN, PROSTRATE GRAPE FERN. Roots often 20–30 in number, yellow to brown; leaves appearing in late fall, overwintering and then dying in early spring (plant wholly underground and dormant for 8–9 months per year); sterile blade portion usually pale green, short-stalked, fleshy, 3–8 cm long, 5–12 cm wide, 2–3-pinnate-pinnatifid; ultimate leaf segments fan-shaped, with midvein absent, denticulate, rounded at apex; 2*n* = 90 (Wagner & Wagner 1993). Open closely cut grassy areas, often in cemeteries; n Pineywoods and Post Oak Savannah w to Bastrop (*J. Singhurst 4806*), Falls, Kaufman, Lee (*Singhurst 4773*), Limestone, Navarro (all BAYLU), Hunt (TEX-LL), and Milam (Holmes et al. 1996) cos. on e edge of Blackland Prairie; se U.S. from NC s to FL w to OK and TX. Spores maturing Feb–Apr.

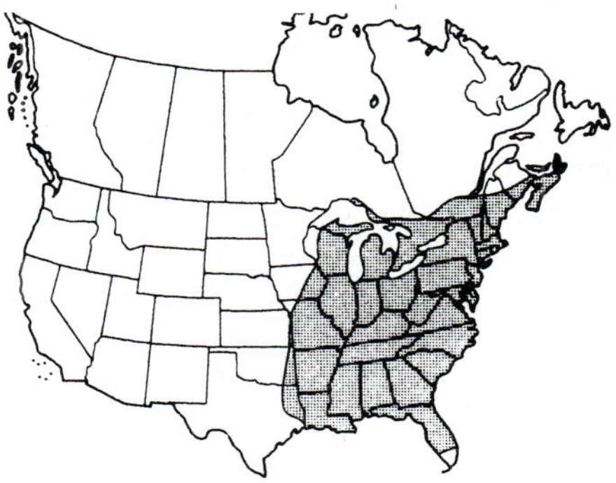

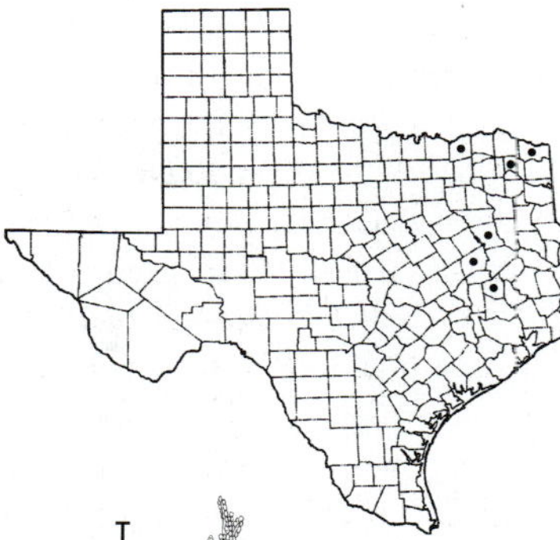

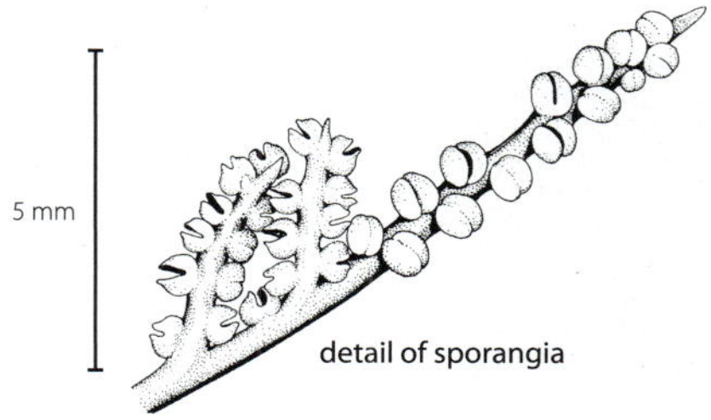

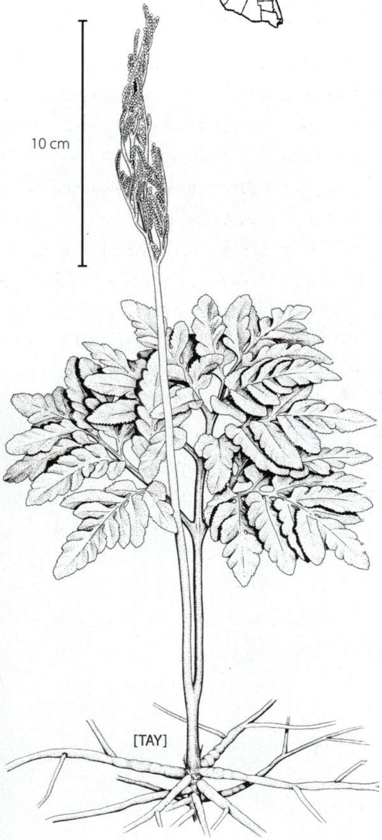

Alan Cressler

Botrychium dissectum

CUT-LEAF GRAPE FERN

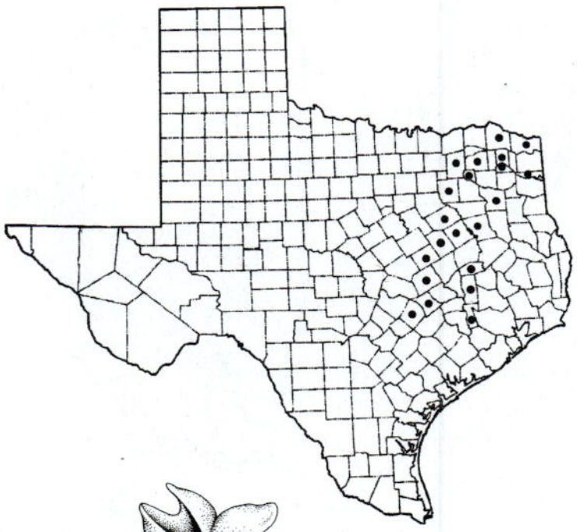

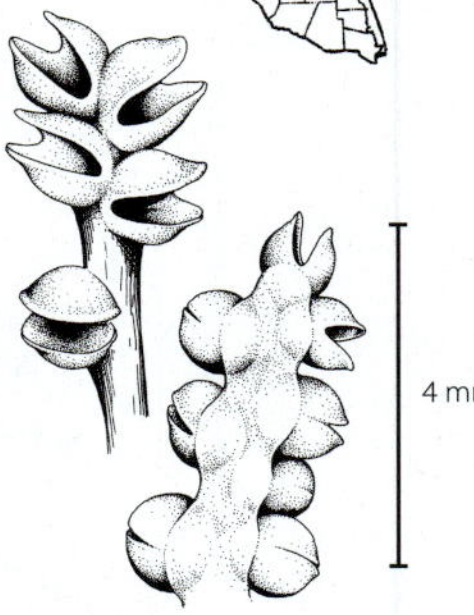

detail of sporangia

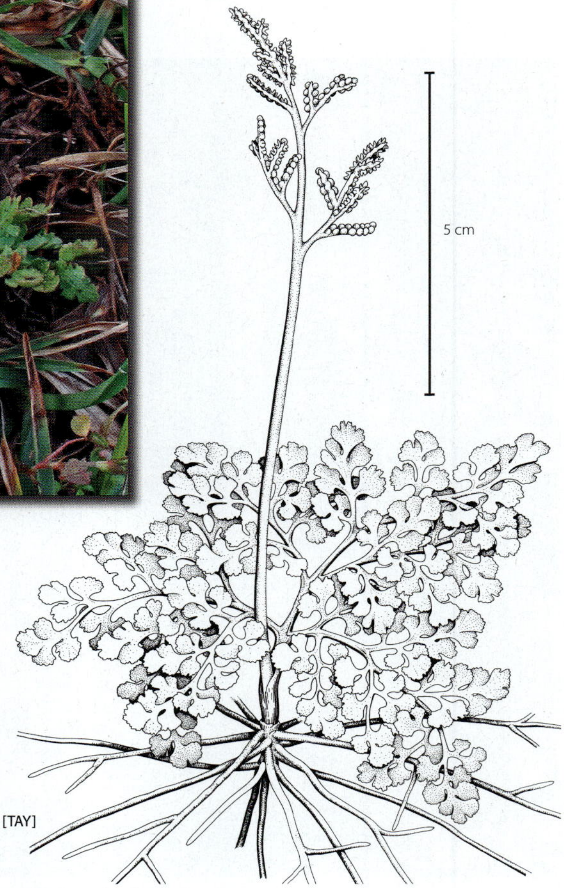

Botrychium lunarioides

WINTER GRAPE FERN

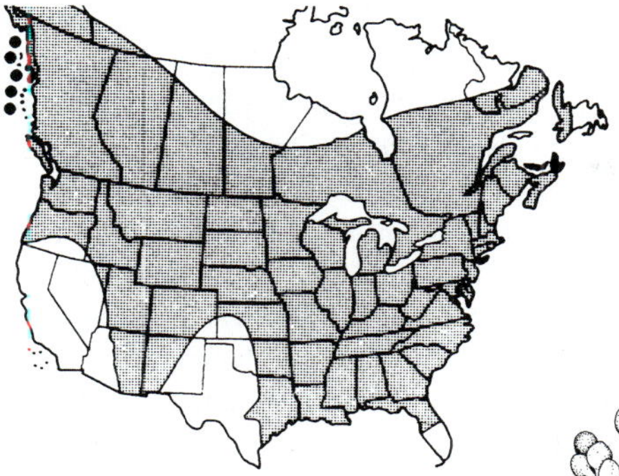

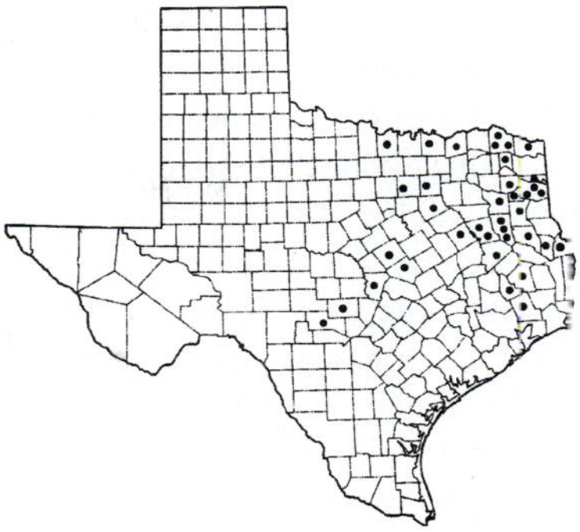

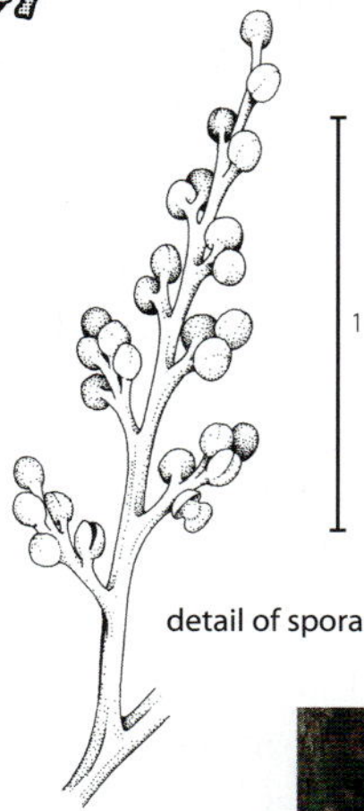

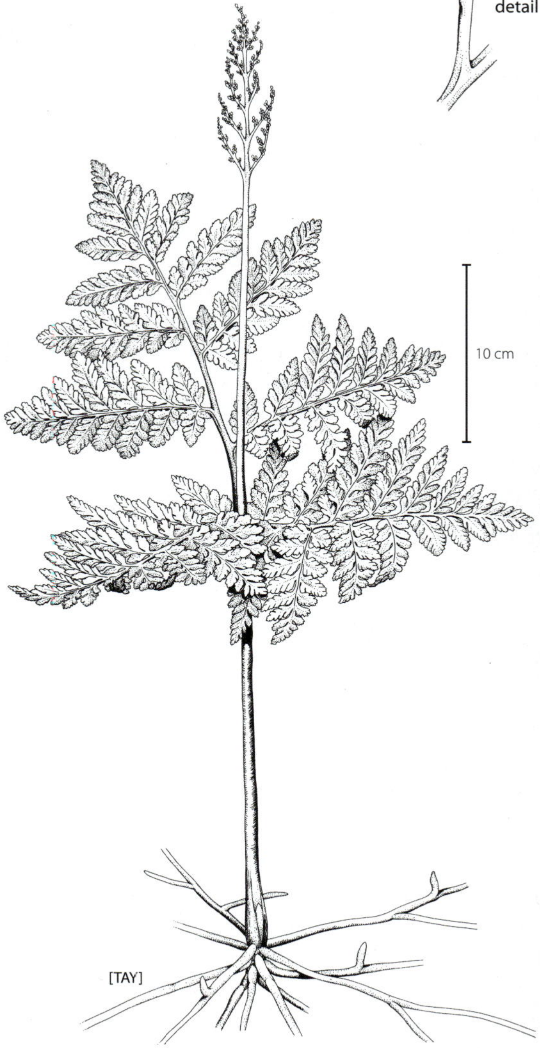

Alan Cressler

Botrychium virginianum

RATTLESNAKE FERN

[*Holubiella lunarioides* (Michx.) Skoda, *Sceptridium lunarioides* (Michx.) Holub] If *Botrychium* is divided into segregate genera, as argued by Hauk et al. (2003), *B. lunarioides* would be recognized in the genus *Sceptridium*. According to Wagner and Wagner (1993), a "... peculiarity of this species is the tendency for the sporophores to remain curled in late fall and early winter and to become erect in February." Based on extensive field experience in AR, LA, and TX, Thomas et al. (1981) noted that this species "will remain dormant during unusually dry springs." It was first collected in TX (San Augustine Co.) in 1972 (Thomas 1979). It was only recently reported from the Blackland Prairie by Holmes et al. (1996), who greatly expanded its known distribution within TX; it is now widely known in the e ⅓ of TX. Wagner (1992) considered *B. lunarioides* to be distinct enough to place it in its own section. This species is easily overlooked, both because the plant is small and because the vegetative portions are prostrate and hidden by grass. Wealkey (2010) also noted that the leaves only appear in late fall and die by early spring (though G. Yatskievych, pers. comm., noted that he has collected the species in wc AR in May). However, when fertile, the vertically oriented sporangia-bearing part sticks well above the prostrate blade portion that is often interwoven in the turf. Thus, as pointed out by Nelson (2000), the "most effective way to find this fern is to search for it in closely cut lawns, especially cemeteries, by crawling on hands and knees or by laying one's face close to the ground, then looking laterally across the top of the grass for the fertile frond segments."

Botrychium virginianum (L.) Sw., (of Virginia, Linnaeus giving many plants from the eastern U.S. this epithet, for this species in the type description he cited Gronovius' *Flora Virginica*, published in two volumes, in 1739 and 1742), RATTLESNAKE FERN, VIRGINIA GRAPE FERN, COMMON GRAPE FERN. Plant erect, 8–75 cm tall; roots 15 or fewer, yellow to brown; leaves seasonal, appearing in early spring and dying in summer; sterile blade portion pale green, sessile, thin, herbaceous, 4–30 cm long and wide, 3–5-pinnate-pinnatifid; ultimate leaf segments linear, with midvein present, serrate to lacerate, pointed at apex; $2n$ = 184 (Wagner & Wagner 1993). Moist, rich woods and thickets; Pineywoods w to Cross Timbers and Prairies (e.g., Montague and Tarrant cos., BRIT; Coryell Co., TEX-LL); also e Edwards Plateau (e.g., Gillespie Co., *O'Kennon 11288*, Bear Mt., BRIT; Kerr Co., Palmer 1919b); throughout most of Canada and the U.S.; also Mexico to South America, Europe, and Asia (Mickel & Smith 2004). Spores maturing Apr–Jun. [*Botrypus virginianus* (L.) Holub] If *Botrychium* is divided into segregate genera, as argued by Hauk et al. (2003), *B. virginianum* would be recognized in the genus *Botrypus*. The common name is probably derived from a resemblance of the clusters of sporangia to the rattles of a rattlesnake. Alternatively, it has "been reported that a salve made by boiling the roots of the plant was used by the Cherokee Indians in the treatment of snake bites" (Nelson 2000).

OPHIOGLOSSUM L. ADDER'S-TONGUE

Plants small, TX species to ca. 25 cm tall; leaves divided into a blade portion and a fertile sporangia-bearing portion, the blade simple; fertile portion of leaf consisting of an elongate stalk terminated by an unbranched, linear, spike-like, sporangia-bearing region.

A nearly cosmopolitan, but mainly tropical and subtropical genus of 25–35 species; India and Africa are areas of high diversity. A number of species are found in open arid areas, appear above ground only after seasonal or sporadic rains, and are ephemeral (Wagner & Wagner 1993; Zech et al. 1998; Mabberley 2008). Of the 8 species known from North America n of Mexico, 6 occur in TX. *Ophioglossum* species have the highest chromosome numbers known for vascular plants, with numbers as high as $2n$ = ca. 1,320 (for *O. vulgatum*, Wagner & Wagner 1993) or even $2n$ = 1,440 (Khandelwal 1990; Barker & Wolf 2010) being reported. The genus is noted for its taxonomic problems due to such factors as the inadequacy of herbarium collections, rarity of populations, subtlety of characters, and variability (Wagner et al. 1984). Many species are small and easily overlooked. They also superficially resemble other plants including *Plantago* species (PLANTAINS), but *Ophioglossum* species can be distinguished by the lack of a midvein on the leaf blades (Hoshizaki & Moran 2001). Wagner (1979) suggested that this lack of midrib and the reticulate venation pattern seen in *Ophioglossum* leaf blades probably arose by "loss of the true blade and amplification of the midrib." Leaf venation is particularly important for identification of some *Ophioglossum* species. Because this character is often difficult to observe in dried specimens, Wagner et al. (1984) suggested wetting the leaf by putting a few drops of 95% ethanol directly on the leaf surface and then observing using transmitted light. *Ophioglossum polyphyllum* A. Br., widespread in the Old World but not previously known for North America, was reported from the Trans-Pecos of TX (Zech et al. 1998) only in the late 1990s; it had previously been confused with *O. engelmannii* (Zech & Manning 1996). Turner et al. (2003) mapped a Winkler Co. location for *O. californicum* Prantl based on an error; we have no evidence that species (endemic to CA, NM, and Baja CA) is a member of the TX flora (see discussion under *O. petiolatum*). (Greek: *ophis*, snake, and *glossa*, tongue, in reference to the tip of the sporangia-bearing structure)

REFERENCES: Thomas 1979; Wagner 1979; Wagner et al. 1984; Shmida & Liston 1987; Thomas et al. 1987; Khandelwal 1990; Zech & Manning 1996; Zech et al. 1998; Schneider & Carlquist 1999a; Münster et al. 2002; Whittier 2003, 2008.

1. Stems (± subterranean and sometimes called rootstocks) globose-bulbose, 3–12 mm diam., nut-like; leaves emerging from cavity in top of stem, the blade portion near ground, spreading or nearly flat on ground, usually roughly triangular to orbicular-ovate or cordate, to only 3.5 cm long; sporangial clusters < 1 cm long; common stalk (to where blade portion and fertile stalk separate) usually < 3 cm long _____ **O. crotalophoroides**
1. Stems cylindric, upright, to ca. 5 mm diam.; leaves developing at top of stem, the blade portion usually well above ground, erect to spreading, usually ovate to lanceolate, to 12 cm long; sporangial clusters 0.5–4 cm long; common stalk varying from < 1 to as much as 10 cm long.
   2. Blade portion of leaf with distinct and sometimes prominent apiculate tip, ± folded when alive OR not so; principal veins of blade forming large primary areoles (= vein-enclosed areas) within which numerous veinlets form secondary areoles, which can then enclose free veinlets.
      3. Plants widespread in TX (Pineywoods and Gulf Prairies and Marshes w to West Cross Timbers, e Edwards Plateau, Panhandle, and Trans-Pecos); blade portion of leaf (when alive) commonly folded, to 10 cm long and 4.5 cm wide, when dried uniformly green without pale central band; roots 0.4–1.5 mm in diam.; fertile stalk 1.3–2.5 times length of blade portion; leaves (blade portion and fertile portion combined) 1–2 per stem.
         4. Plants of desert grasslands of far w TX; roots ca. 25 per plant (varying from 15–30), dark in color, often blackish, relatively thin, 0.4–0.7 mm thick and usually crooked when dried; blade portion of leaf linear-lanceolate to narrowly ovate; persisting somewhat ribbon-like old leaf bases typically about 5 (varying from 2–10) _____ **O. polyphyllum**
         4. Plants of wooded slopes primarily of e ½ of TX; roots ca. 15 per plant (varying from 6–24), varying in color from yellowish to tan, brown or dark brown, relatively thicker, mostly 0.9–1.5 mm thick and usually straight when dried; blade portion of leaf broadly ovate to narrowly ovate, elliptic, or oblong; persisting somewhat ribbon-like old leaf bases relatively fewer, typically 2, varying from 0–4 _____ **O. engelmannii**
      3. Plants likely in TX only from the Pineywoods in far e part of the state (this large form of *O. nudicaule* may not occur within the state); blade portion of leaf (when alive) not folded, to 4.5 cm long and 1.7 cm wide, when dried commonly with a pale central band; roots 0.2–0.8 mm in diam.; fertile stalk 2–6 times length of blade portion; leaves 2–3 per stem _____ **O. nudicaule** (large form)
   2. Blade portion of leaf without apiculate tip, usually rounded to acute at apex OR with small apiculate tip (in small form of *O. nudicaule*), commonly plane (= not folded) when alive; principal veins of blade forming areoles but these including only free veinlets (= with one end not touching another vein).
      5. Blade portion of leaf usually rounded at apex, without apiculate tip, (2.5–)5–10 cm long and 1.5–4(–5.5) cm wide; leaves (blade portion and fertile portion combined) usually 1 per stem, appearing in a single flush once per year _____ **O. vulgatum**
      5. Blade portion of leaf acute at apex, with OR without apiculate tip, less than 4.5(–6) cm long and 1(–3) cm wide; leaves commonly 2–3 per stem, appearing in 1 or more flushes per year, depending on rains.
         6. Blade portion of leaf typically much more than 1 cm long, usually without apiculate tip, the base cuneate to truncate or nearly cordate, usually with relatively coarse venation; roots (at least the major ones) generally straight, 0.8–1.3 mm in diam, usually dark brown _____ **O. petiolatum**
         6. Blade portion of leaf often less than 1 cm long (in the small form of *O. nudicaule*), with short apiculate tip, the base gradually tapered to cuneate, usually with fine intricate venation; roots often wavy, 0.2–0.8 mm in diam., yellowish to pale brown _____ **O. nudicaule** (small form)

Ophioglossum crotalophoroides Walter, (from Greek: *krotalon*, a rattle, *phoros*, bearer, and *-oides*, like or resembling, meaning "like a rattle bearer", due to the resemblance of the sporangial clusters to the rattles of a rattlesnake—Thieret 1980), BULBOUS ADDER'S-TONGUE, DWARF ADDER'S-TONGUE FERN. Plant usually to only 15 cm tall; stems (± subterranean) globose-bulbose, 3–12 mm diam., nut-like or pea-sized; leaves (blade portion and fertile portion combined) 1–3 (often 2) per stem; blade portion to 3.5 cm long and 2.5 cm wide, usually smaller, pale green, spreading or nearly flat on ground; fertile stalk 1–5 times as long as blade portion; sporangia

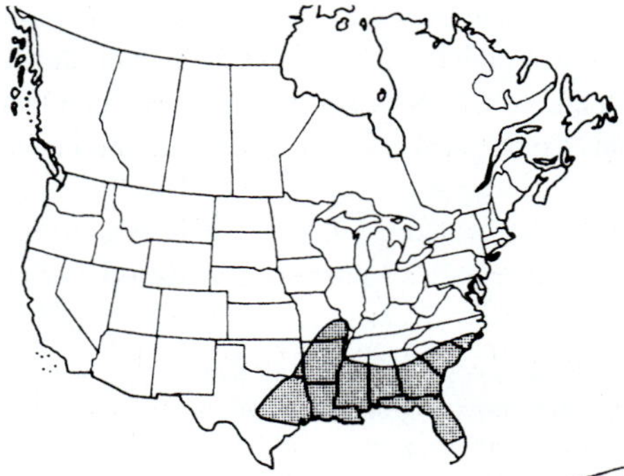

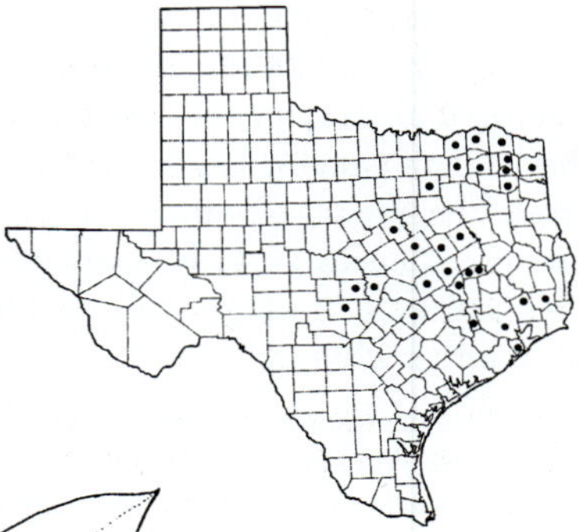

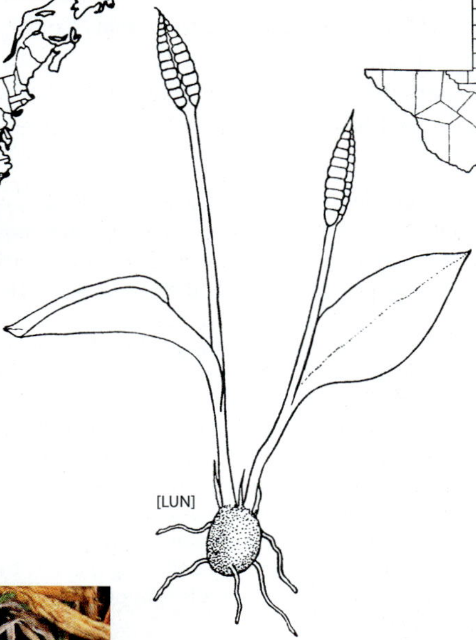

Ophioglossum crotalophoroides

BULBOUS ADDER'S-TONGUE

4–8(–12) on each side of fertile stalk. Usually observed in moist sand, ditches, lawns, cemeteries, areas where grass is short; Pineywoods and Post Oak Savannah w to Fannin (*J.&C. Taylor 4832*, BRIT), Bastrop (BAYLU), Milam (TEX-LL), McLennan (*W.C. Holmes 7613*, BAYLU), and Hunt (Turner et al. 2003) cos. and from lawn in Dallas Co. (TAMU); also Bosque Co. (BAYLU) in Cross Timbers and Prairies, n Gulf Prairies and Marshes, and e Edwards Plateau in Gillespie (*O'Kennon 10456,* Bexar Mt., BRIT), Burnet (TEX-LL), and Llano (Turner et al. 2003) cos.; se U.S. from NC s to FL w to OK and TX; also Mexico to South America and West Indies. Leaves appearing late winter and early spring, sometimes later in season after heavy rains; usually sporulating Mar–May. The globose-bulbose stem is unique among TX members of the genus and Wagner and Wagner (1993) noted the species "is very remarkable morphologically for its highly modified stem and threadlike nonproliferous roots." According to Thomas (1979), "This species is probably the most common fern in East Texas and is found abundantly in early spring in almost every sandy area where grass is short, such as school lawns and cemeteries." Thomas et al. (1981) noted that this species "will remain dormant during unusually dry springs." Nelson (2000) commented that, the "bulbous, rounded, field-pea-size stem distinguishes this species from other members of the genus." The shape and size of the blade portion of the leaf can vary significantly depending on soil texture and depth (Thieret 1980). Seigler (1977) reported that in the Central Mineral Region it occurs "in shallow pools and depressions in granite and sandstones."

Ophioglossum engelmannii Prantl, (for George Engelmann, 1809–1884, German-born American botanist, "who first recognized the plant as distinct"—Nelson 2000), LIMESTONE ADDER'S-TONGUE, ENGELMANN'S ADDER'S-TONGUE. Plant to 25 cm tall; leaves (blade portion and fertile portion combined) 1–2 per stem; blade portion to 10 cm long and 4.5 cm wide, broadly ovate to narrowly ovate, elliptic, or oblong, commonly folded when alive, apically apiculate, when dried uniformly pale green without pale central band, dull, principal veins of blade forming large primary areoles within which are numerous secondary areoles; fertile stalk 1.3–2.5 times as long as blade portion; sporangia 20–40 on each side of fertile stalk. Usually in thin black soils on limestone, wooded rocky slopes; Pineywoods and Gulf Prairies and Marshes w to West Cross Timbers (e.g., Brown Co., *E.J. Palmer 11432,* 1917, photocopy of MO sheet at HPC), also e Edwards Plateau, and disjunct to Bailey Co. in the Panhandle (*T.C. Rosson 1480B*, BRIT); e U.S. from PA s to FL w to NE, AR, and TX and in the sw to NM and AZ; also Mexico and Central America. Leaves appearing from early to late spring, often with a second flush of leaves following summer rains; usually sporulating Dec–Jun. Thieret (1980) noted, "The entire plant of this fern has a very distinctive odor by which it can be identified even after 5 years of storage."

Ophioglossum nudicaule L., (Latin: *nudus*, bare, naked, and *caulis*, stem, in reference to the stalk of the fertile portion which appears "leafless" [i.e., naked] because the blade portion usually arises at or near its base—Thieret 1980), SLENDER ADDER'S-TONGUE, LEAST ADDER'S-TONGUE. Plant to ca. 12 cm tall, variable, ranging from very small-leaved forms to large-leaved forms; leaves (blade portion and fertile portion combined) commonly 2–3 per stem; blade portion to 4.5 cm long and 1.7 cm wide but often very small (< 4 mm long and 3 mm wide), green, dull, the largest with pale central band, flat or nearly so when alive, apically with short apiculum; fertile stalk 2–6 times as long as blade portion, arising at or near ground level; sporangia 5–12 on each side of fertile stalk; $2n = 240$ (Mickel & Smith 2004). Disturbed places (e.g., cemeteries, mowed areas around motels), ditches, grassy slopes, wet meadows, damp depressions in pinelands, moist open woods, bog margins; Shelby Co. (*Thomas 27493*, 16 Feb 1972—BRIT, according to R.D. Thomas' annotation this is actually the first TX collection given that the *Cory 52663* sheet at BRIT, originally thought to be the first TX collection, is actually *O. petiolatum*) near the LA border, also Hardin (BRIT), Orange, San Augustine (Thomas 1979), and San Jacinto (E. Keith, pers. comm., BAYLU) cos. in the Pineywoods, surprisingly disjunct to Gillespie Co. (*O'Kennon 11355*, BRIT, Bear Mt., wet area below granite outcrop, growing with *O. crotalophoroides*) in Central Mineral Region of the Edwards Plateau; se U.S. from VA s to FL w to OK and TX; also Mexico and Central America. Leaves appearing in late winter and early spring, sometimes with a second flush of leaves after heavy rains; usually sporulating Dec–Jun. [*O. dendroneuron* E.P. St. John, *O. ellipticum* Hook. & Grev., *O. mononeuron* E.P. St. John, *O. nudicaule* var. *minus* Clausen, *O. nudicaule* var. *tenerum* (Mett. ex Prantl) Clausen, *O. tenerum* Mett. ex Prantl] According to Wagner et al. (1984), this is the most variable and taxonomically confused species of *Ophioglossum* in the se U.S., with a complete transition series from small- to large-leaved (previously called *O. ellipticum*) forms. All of the specimens we have been able to confirm from TX are of the small-leaved form. Colonies of the small-leaved form typically have the blade portion quite small (< 0.4 cm long and 0.3

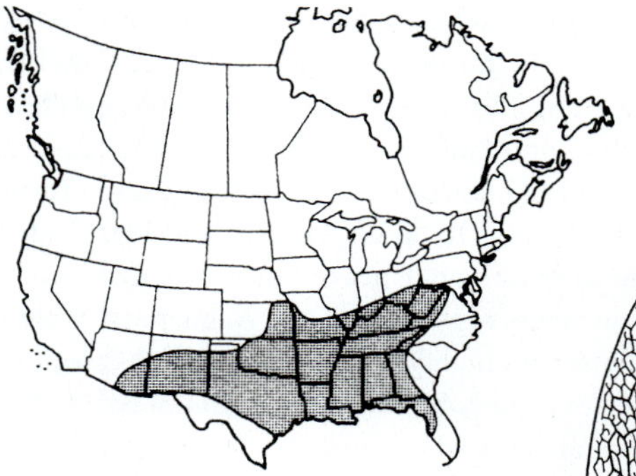

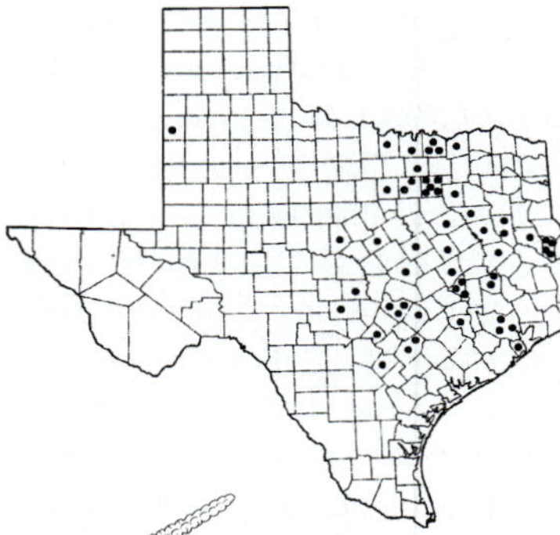

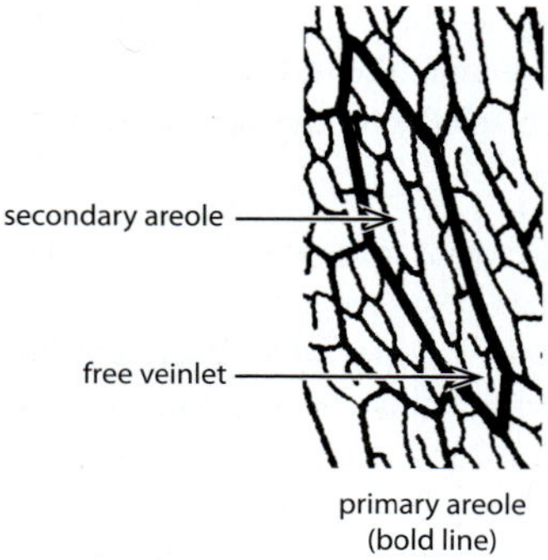

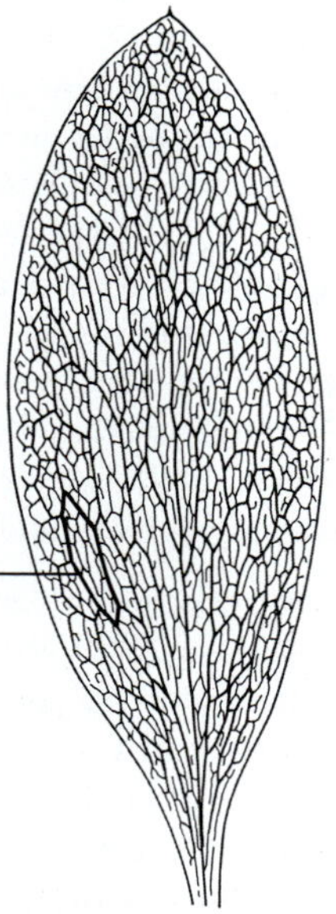

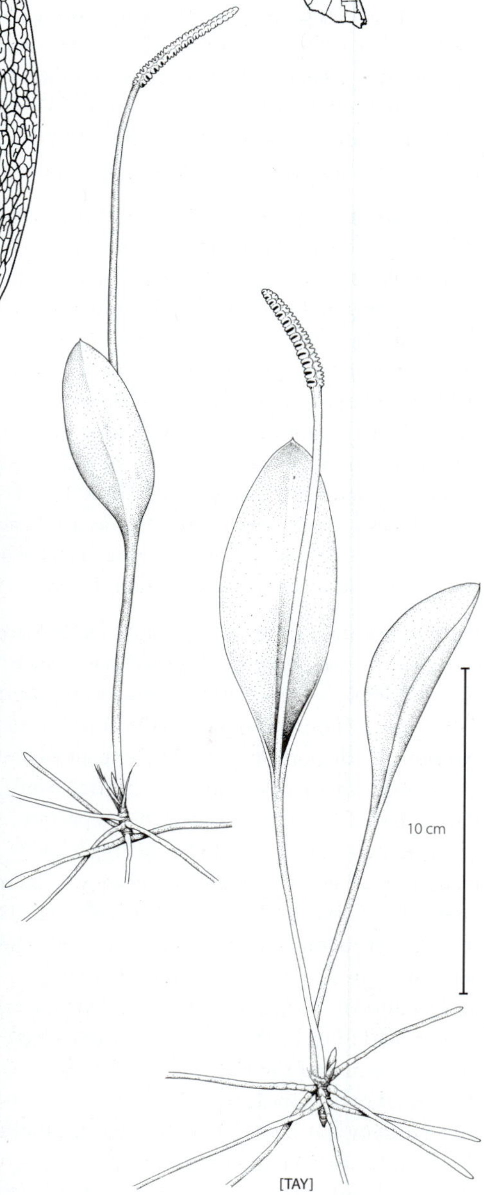

George Diggs

Ophioglossum engelmannii

LIMESTONE ADDER'S-TONGUE

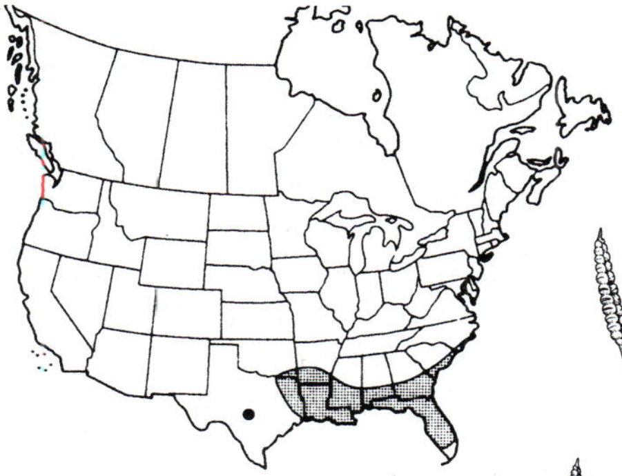

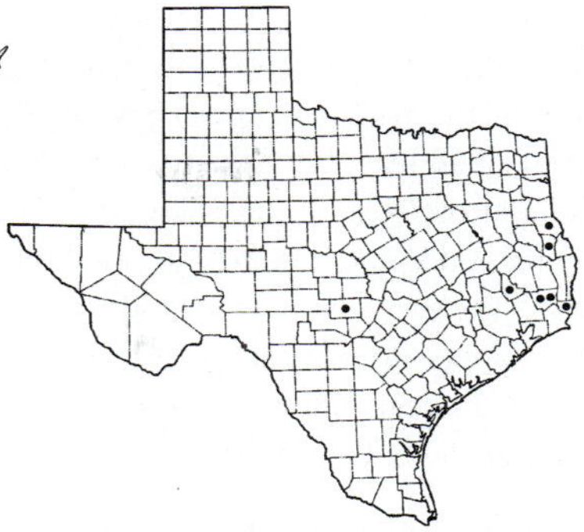

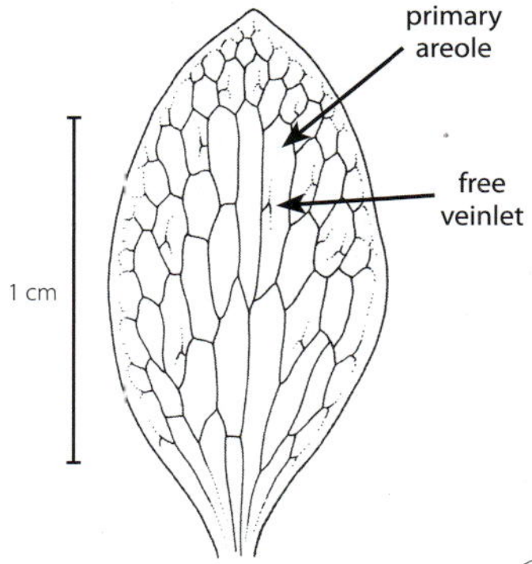

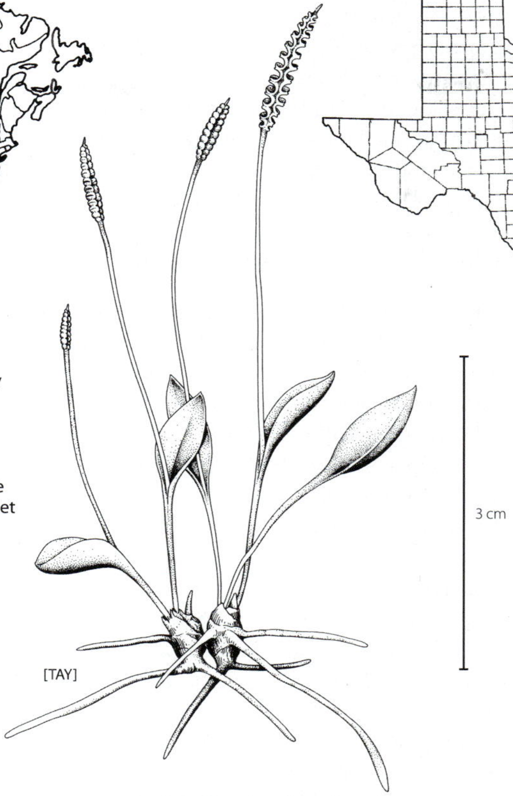

Ophioglossum nudicaule

SLENDER ADDER'S-TONGUE

cm wide), while those of the large-leaved form can have much larger blade portions. The large-leaved form has bireticulate veins (areoles enclosing secondary areoles), but the areoles are "coarser and more open" than those in *O. englemannii* and *O. polyphyllum* (Zech et al. 1998). Nelson (2000) commented that, "The combination of small leaf size (leaves generally less than about 1.5 cm long), flat leaves with netted venation, leaves arising at the base of the plant, and non-bulbous stem distinguishes this species from others in the genus." Wagner and Wagner (1993) pointed out that although this species "is much less common than *O. crotalophoroides*," the two "often occur together and are found in the same or similar habitats." *Ophioglossum nudicaule* is a striking example of a species that is known in East TX and disjunct to the Central Mineral Region of the Edwards Plateau. Other pteridophytes with similar distribution patterns are *Isoëtes butleri*, BUTLER'S QUILLWORT, *O. crotalophoroides*, BULBOUS ADDER'S-TONGUE, and *Onoclea sensibilis*, SENSITIVE FERN.

Ophioglossum petiolatum Hook., (Latin: *petiolus*, little foot, stalk, or stem, and *-atus*, indicating possession, = having a petiole or leaf stalk, in reference to the usually prominent stalk of the blade portion), STALKED ADDER'S-TONGUE, LONGSTEM ADDER'S-TONGUE. Plant to ca. 21 cm tall; leaves (blade portion and fertile portion combined) commonly 2–3 per stem; blade portion to 1.5–6 cm long and 0.6–3 cm wide, gray-green, dull, flat or nearly so when alive, apically acute; fertile stalk 0.8–7 times as long as blade portion; sporangia to 30 on each side of fertile stalk; $2n = 960$ (Yatskievych 1999). Wet woods, disturbed places (e.g., cemeteries, mowed areas), ditches, moist meadows, depressions; Hardin (*Cory 52663*, BRIT), Orange (*Thomas et al. 39021*, BRIT), Jasper, Liberty (Thomas 1979; Thomas et al. 1987), Harris, and Montgomery (Brown 2008) cos. in the Pineywoods; also Jefferson Co. (*J.L. Hooks s.n. 1937*, TEX-LL; Thomas 1979; Turner et al. 2003) in the n Gulf Prairies and Marshes and an isolated occurrence from Winkler Co. in far w TX (*McVaugh 10732*, SRSC, TEX-LL); naturalized se U.S. from NC s to FL w to OK and TX. Leaves appearing during wet periods; usually sporulating Feb–Jun. Native of West Indies, Central America, n South America, Asia (including India), and Pacific Islands including Hawaii (previously thought to occur in Mexico, now considered doubtful there—Mickel & Smith 2004). Wagner and Wagner (1993) indicated that the earliest North American records date from 1900 to 1930, suggesting that the species is probably introduced. The first TX collection was from Jefferson Co. in 1937 (Thomas 1979). The isolated w TX Winkler Co. collection was from an area of high sand dunes "In moist sand, locally abundant in a hollow between dunes; not seen in other similar hollows." Thomas (1979) noted that "This specialized sand dune habitat of Wink[ler] Co. represents the westernmost distribution of this species in the United States except for Hawaii." The collection is the source of an erroneous report of *O. californicum* from TX by Turner et al. (2003). In cultivation (as an ornamental), this plant "is particularly known for developing root buds and a single plant can fill a pot quickly through root proliferation" (Nelson 2000). Thomas (1979) noted that it is "usually abundant where found." This species was apparently spread across the se U.S. in rural cemeteries through contaminated soil in pots that contained decoration plants (J. Peck, pers. comm.); STALKED ADDER'S-TONGUE can also be found in nurseries in the pots of shrubs and trees. In parts of Asia (e.g., Taiwan) the leaves of this species are used in traditional Chinese medicine; they are dried, ground into a powder, and made into a tea (Moran 2004).

Ophioglossum polyphyllum A. Braun ex Schub., (Greek: *poly*, many, and *phyllon*, leaf; an unfortunate epithet since the species usually has only 1 or 2 leaves), LARGE ADDER'S-TONGUE, UPRIGHT ADDER'S-TONGUE. Plants morphologically similar to *O. engelmannii* (distinguished as in the key) but remarkably different in range; leaves (blade portion and fertile portion combined) 1–2(–3) per stem; blade portion linear-lanceolate to narrowly ovate, apically apiculate; principal veins of blade forming large primary areoles within which are numerous secondary areoles; $2n = 240$ (Mickel & Smith 2004). Ditches and low areas in Chihuahuan desert grassland; Brewster (*Manning 896, 897*, SRSC), Jeff Davis (*Manning 915*, SRSC), Presidio (*Manning 925*, SRSC), Pecos, and Reeves (*Manning & Zech 1092*, SRSC) cos. in the Trans-Pecos; AZ and TX; also Mexico (Chihuahua s to Oaxaca), Australia (Barker et al. 2005), Africa, Asia, and Polynesia (e.g., Hawaii). Leaves appearing Jun-Jul, but other times following rains. This native species was first reported for the U.S. (and the New World) in 1998 (Zech et al. 1998); it had previously been interpreted as a narrow-bladed form of *O. engelmannii* (Zech & Manning 1996). It is apparently most closely related to *O. engelmannii*—the two species both have finely bireticulate (= venation pattern where major areoles include minor areoles) blades. This species has an extremely widespread distribution, occurring on four continents and in Hawaii.

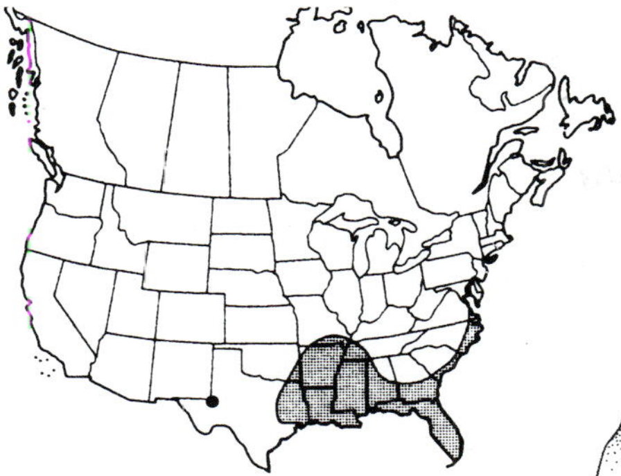

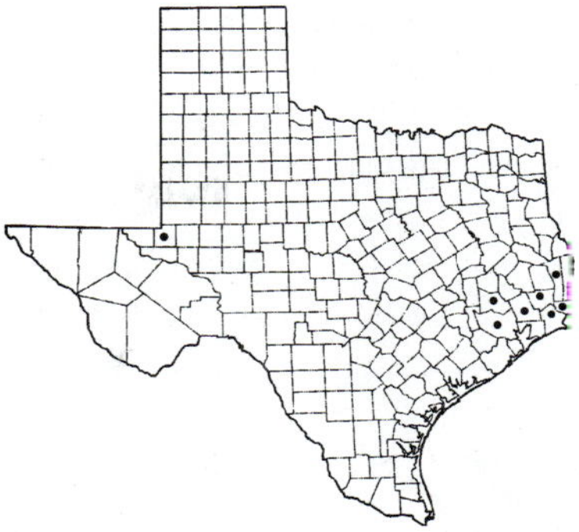

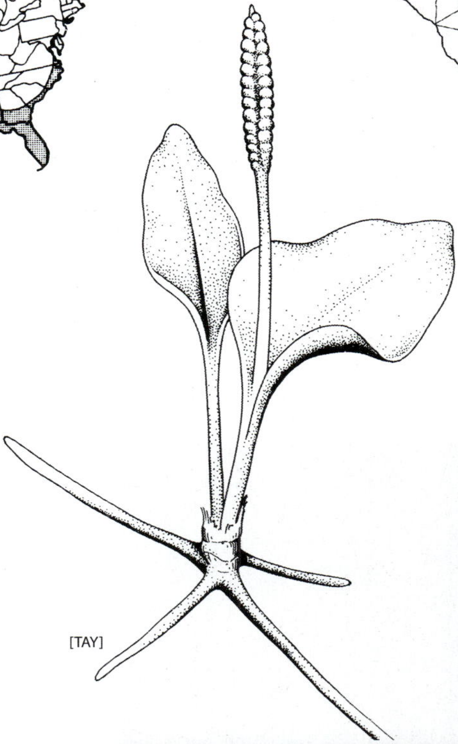

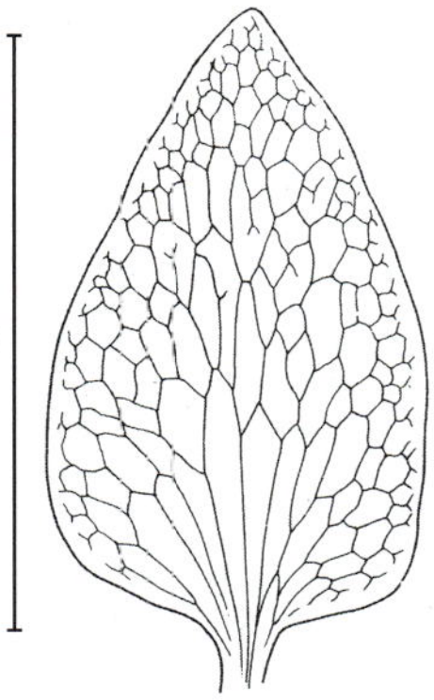

**detail of leaf
w/o secondary areoles**

Bruce Holst

Bruce Holst

Ophioglossum petiolatum

STALKED ADDER'S-TONGUE

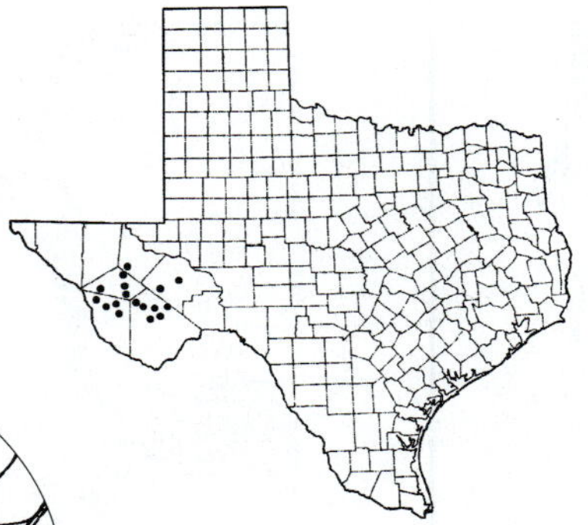

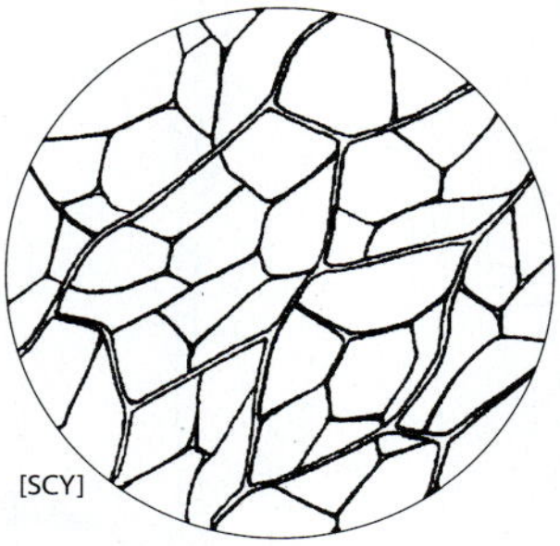

[SCY]

leaf surface with
primary and secondary areoles

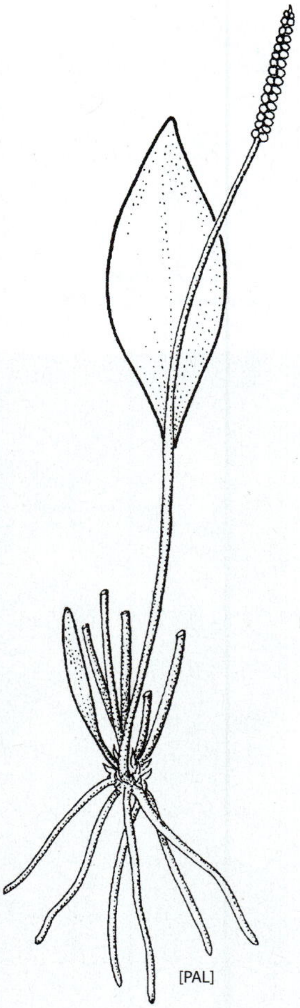

[PAL]

Ophioglossum polyphyllum

LARGE ADDER'S-TONGUE

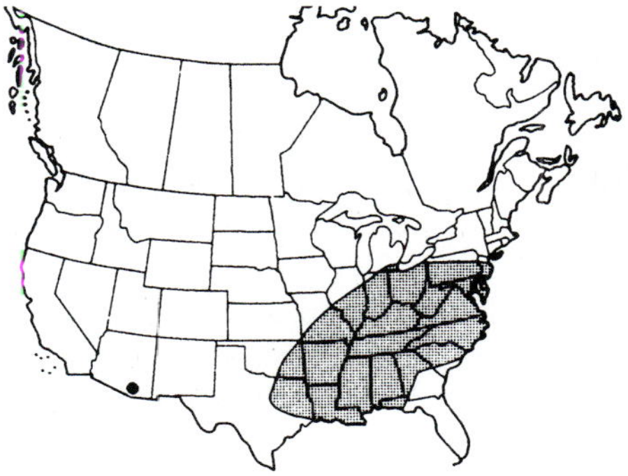

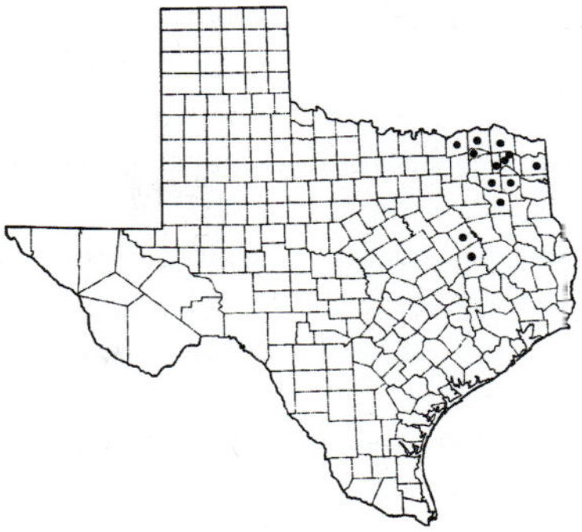

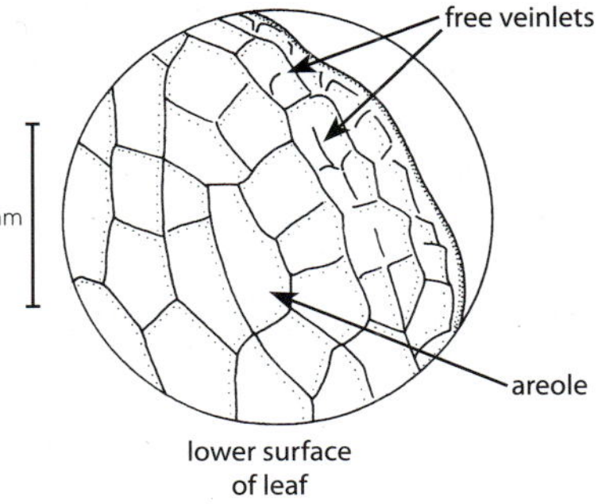

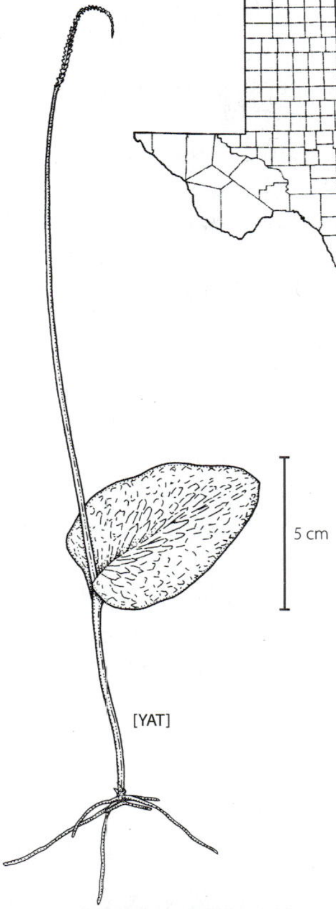

George Yatskievych

Ophioglossum vulgatum

SOUTHERN ADDER'S-TONGUE

Ophioglossum vulgatum L., (Latin: *vulgaris*, common, commonplace, ordinary), SOUTHERN ADDER'S-TONGUE, ADDER'S-TONGUE. Similar to *O. engelmannii*; leaf (blade portion and fertile portion combined) 1 per stem; blade portion (2.5–) 5–10 cm long and 1.5–4(–5.5) cm wide, dark green, somewhat shiny, flat or nearly so when alive, apically rounded; fertile stalk 2–4 times as long as blade portion; sporangia 10–35 on each side of fertile stalk; "2*n* = ca. 1320" (Wagner & Wagner 1993), "2*n* = 960, about 1320" (Yatskievych 1999). Moist woods, meadows, swamps, usually in sandy soils; n Pineywoods, n Post Oak Savannah (e.g., Freestone Co., *Carr & Singhurst 14452*, TEX-LL; Leon Co., *Carr 26629*, TEX-LL), and w in Red River drainage to Fannin Co. (BRIT); also reported for the n Gulf Prairies and Marshes (Jefferson Co.—Correll 1956; but apparently based on a *Mrs. J.L. Hooks s.n.* collection of *O. petiolatum* at TEX-LL) and the Cross Timbers and Prairies (Denton Co.—Turner et al. 2003)—we have been unable to confirm the Denton and Jefferson collections and question their correctness—they are not mapped on the county distribution map; throughout most of the e U.S. w to IL, OK, and TX, also disjunct to s AZ (Wagner & Wagner 1993); also s Mexico and Eurasia. Leaves appearing spring to early summer; usually sporulating Mar–Jun. [*O. pycnostichum* (Fernald) A. Löve & D. Löve, *O. vulgatum* var. *pycnostichum* Fernald] Weakley (2010) and Peck (2011a) treated this species as *O. pycnostichum*, with Weakley (2010) noting, "*O. vulgatum* (defined narrowly) is Eurasian. The best treatment of this complex is uncertain." As there is disagreement about appropriate nomenclature for this species, we are following Mickel and Smith (2004) in continuing to include the New World material in *O. vulgatum*. Wagner and Wagner (1993) pointed out shockingly high chromosome numbers, noting that in the Appalachian Mts. "a distinctive large-spored form has a chromosome number of 2*n* = ca. 1320." This is primarily a species of the e ½ of the U.S. with a disjunct population in s AZ—this isolated occurrence is presumably the result of long-distance dispersal. Also see page 37 for a possible explanation of the disjunct distribution of this species from TX to s Mexico. *Ophioglossum vulgatum* can be confused with *O. petiolatum* but, in addition to the characters in the key, it has unusually persistent leathery basal leaf sheaths (these ephemeral in *O. petiolatum*) (Thomas 1979; Wagner & Wagner 1993).

OSMUNDACEAE Martinov

CINNAMON OR ROYAL FERN FAMILY

Terrestrial; leaves wholly or partly dimorphic, erect to spreading, in a vase-shaped cluster from a stout, woody, creeping to suberect rhizome; old petiole bases persistent; sterile blades 1-pinnate-pinnatifid or 2-pinnate; fertile blades either strikingly different from sterile, lacking expanded pinnae altogether, or similar to sterile except with fertile pinnae borne towards tip; discrete sori absent; sporangia clustered, on contracted non-photosynthetic pinnae; indusia absent; spores green.

An ancient nearly cosmopolitan family of 4 genera (some authorities would say 3 based on differing taxonomic approach) and 20–36 species (Whetstone & Atkinson1993; Smith et al. 2008), with some cultivated as ornamentals. The 2 exclusively Old World genera not occurring in TX are *Leptopteris* and *Todea*. *Todea barbara* (L.) T. Moore, CRAPE FERN or KING FERN, native to e Australia, New Zealand, and South Africa, is a huge tree fern-like plant cultivated as an ornamental; the massive erect rhizomes can sometimes reach 3 m tall, the leaves can be 2.5 m long, and individual plants can weigh up to 1.5 tons (Mabberley 2008). The Osmundaceae are one of the oldest living fern families with an extensive fossil history extending back to the late Permian Period (299–251 mya); the family was widespread by the Triassic (Skog 2001; Judd et al. 2008). In fact, the family has the richest fossil record of any fern family, with over 200 fossil species known (Tidwell & Ash 1994; Pigg et al. 2006; Judd et al. 2008; Rothwell & Stockey 2008).

Osmundaceae are classified as leptosporangiate ferns (the Greek root *lepto-* means thin or slender, in reference to the one-cell-thick walls of the sporangia in Osmundaceae and other leptosporangiates). In contrast, the eusporangiate ferns, lycophytes, and seed plants have sporangial walls several cells thick (Moran 2004). In some respects Osmundaceae are intermediate between eusporangiate and leptosporangiate ferns. The eusporangiate ferns, consisting of only four families (Ophioglossaceae, Marattiaceae, Psilotaceae, and Equisetaceae), normally have thousands of spores per sporangium, the Osmundaceae typically have hundreds (128–512), and the other leptosporangiate ferns usually have 64 or 32 (Lellinger 1985; Jones & Luchsinger 1986; Smith et al. 2006). Morphological and molecular evidence indicates the phylogenetically isolated and monophyletic Osmundaceae is the most basal leptosporangiate fern family (or alternatively stated, it is the sister group of a clade containing all other leptosporangiate ferns) (Li & Haufler 1994; Pryer et al. 1995; Hasebe et al. 1995; Doyle 1998; Yatabe et al. 1999; Metzgar et al. 2008). Generic boundaries in the family have changed in recent years, with the CINNAMON FERN now recognized in its own genus, *Osmundastrum* (see that genus for further discussion).

The wide geographic range of today is probably reflective of an even wider distribution in the past—for example, *Osmunda* (in the broad sense including *Osmundastrum*) is characteristic of Tertiary (Paleocene to Eocene) floras of high, often polar latitudes in the n hemisphere—e.g., Alaska, far n Canada, n Japan, Kamchatka (nw Russia), Spitsbergen (between Greenland and Norway), etc. These Circum-Arctic locations probably result from the time when Tertiary forests extended around the n polar area via land connections during periods of high global temperatures (Collinson 2001; also see page 33). Likewise, the Australia-New Zealand-South Africa distribution of *Todea* suggests it has survived from the time when these land masses were together as parts of the supercontinent Gondwana. Moran (2004) noted that a specialty of this group is "stem armor"—during the time of dinosaurs, now extinct tree-like relatives with sturdy trunk-like stems had a series of extremely hard overlapping leaf bases which protected the stem and presumably provided defense from damage by large herbivores. Modern *Osmunda* and *Osmundastrum* rhizomes, though small, are still protected by tough protective leaf bases. The roots of some species form a dense mass "that may be cut up and used as substrate ('osmundine') for growing orchids or other epiphytes" (Wagner & Smith 1993). (Order Osmundales)
FAMILY RECOGNITION IN THE FIELD: large conspicuous ferns with usually large, wholly or partly dimorphic (fertile leaves or pinnae conspicuously different from sterile) leaves; sporangia not in discrete sori.
REFERENCES: Correll 1956, 1966; Hewitson 1962; Kramer 1990e; Whetstone & Atkinson 1993; Li & Haufler 1994; Phipps et al. 1998; Yatabe et al. 1999, 2005; Pigg et al. 2006; Kato 2007; Schölch 2007; Metzgar et al. 2008; Smith et al. 2008; Judd et al. 2008.

1. Fertile leaves not very leaf-like, without expanded blade tissue, completely different in appearance from sterile leaves; smallest subdivisions of sterile leaves not narrowed at base, the area of attachment as broad as segment; mature sterile pinnae with a persistent tuft of tomentum at base ____ **Osmundastrum**
1. Fertile leaves similar in appearance to sterile leaves except with greatly reduced, sporangia-bearing pinnae at tip; smallest subdivisions of leaves greatly narrowed at very base, attached at one stalk-like point only; mature sterile pinnae glabrous ____ **Osmunda**

OSMUNDA L. ROYAL FERN

A nearly cosmopolitan ancient genus of 6–9 species (Whetstone & Atkinson 1993; Metzgar et al. 2008), excluding *Osmundastrum cinnamomeum*. Metzgar et al. (2008) noted, "A fossil species, *O. claytoniites* Phipps et al. is known from the Triassic [of Antarctica] with gross leaf morphology that is remarkably similar to *O. claytoniana* [a species of the e U.S.], suggesting that *O. claytoniana* has perhaps been in morphological stasis for at least 200 million years and also that the genus *Osmunda* is at least 200 million years old." Note: Yatabe (1999) considered O. *claytoniites* to be most similar to *Osmundastrum cinnamomeum*. (Saxon: *Osmunder*, name for Thor, Norse god of thunder and war, though some authorities (e.g., Moran 2004) note that the origin is unclear and may be from the Latin: *os*, bone, and *munda*, cure, the roots having been used as a cure for rickets or possibly from the Latin: *mundae*, to clean, since it was used medically to clean bones)
REFERENCES: Miller 1967; Vavreck et al. 2006; Metzgar et al. 2008; Schuettpelz & Pryer 2008; Magrini et al. 2010; Tsutsumi 2011.

Osmunda regalis L. var. **spectabilis** (Willd.) A. Gray, (sp.: Latin: *regalis* or *regius*, royal, from *rex*, *regis*, king; var.: Latin: *spectabilis*, notable, showy, spectacular), ROYAL FERN, FLOWERING FERN. Rhizome stout, woody, creeping to suberect; leaves erect to spreading, in a vase-shaped cluster, hemidimorphic (= partly dimorphic, with very different sterile and fertile pinnae on the same leaf blade), 2-pinnate; sterile leaves ca. 0.75–1(+) m long; pinnules lanceolate, mostly ca. 4–6 cm long, the margins subentire to remotely dentate, apically acute to rounded; pinnae without a persistent tuft of tomentum at base, essentially glabrous; fertile leaves similar to sterile except with greatly reduced, sporangia-bearing pinnae at tip; sori absent, the sporangia clustered, initially green, brown at maturity; $2n = 44$ (Whetstone & Atkinson 1993). Wet areas; mainly Pineywoods and Post Oak Savannah w to Lamar Co. (*W.R. Carr & D. Wolfe 13739*, TEX-LL; Carr 1994) in Red River drainage and Travis Co. (*Bray & Long s.n.*, Feb 1901, TEX-LL) on e edge of the Edwards Plateau; also slightly disjunct

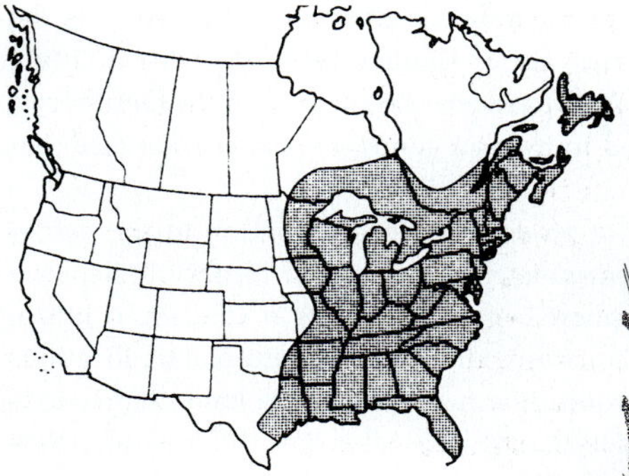

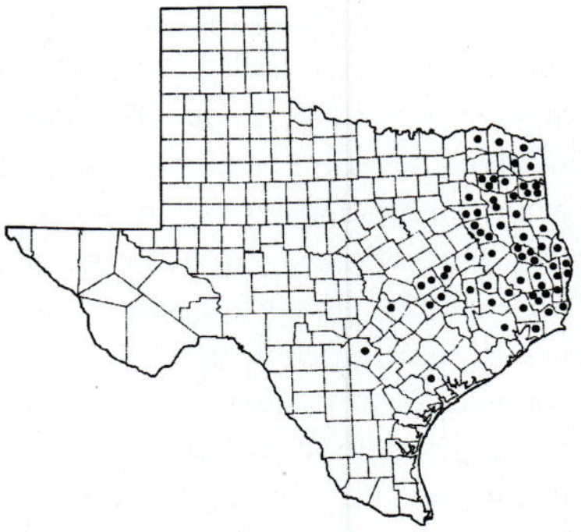

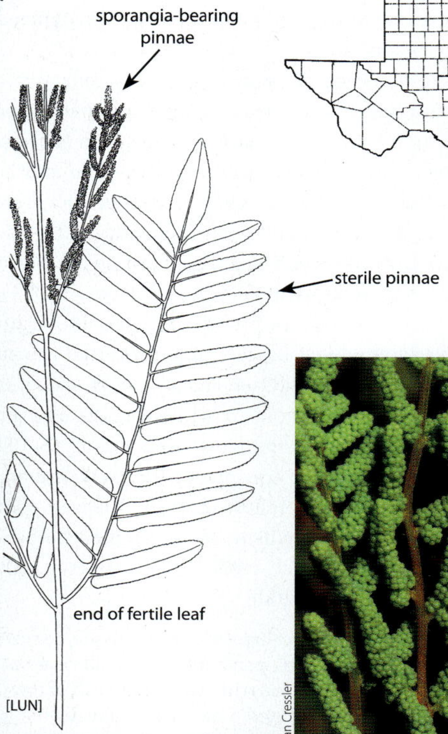

George Diggs
immature fertile leaf tip

Alan Cressler
detail of fertile pinnae

Alan Cressler

Osmunda regalis var. **spectabilis**

ROYAL FERN

to Bexar Co. (Turner et al. 2003) to the sw and to Victoria Co. (Turner et al. 2003) in the Gulf Prairies and Marshes; se Canada and through e U.S. w to MN, OK, and TX; also West Indies, Bermuda, s Mexico to South America. Sporulating Mar–Jul. Variety *regalis* is native to Eurasia and Africa and is distinguished by black hair-like scales along the leaf rachis and a more robust habit (Nauman et al. 2000). There is debate regarding the correct taxonomy for *O. regalis* and its close relatives. For example, Mickel and Smith (2004) noted that "New World material looks distinct enough to be treated as a different species from *O. regalis sensu stricto* (Old World), but more study of both herbarium and living material is needed to properly delimit these taxa." Metzgar et al. (2008) found evidence suggesting that *O. regalis* var. *spectabilis* (North America) might be more closely related to *O. japonica* and *O. lancea* (Asian species) than to the European *O. regalis* var. *regalis*. Further study is clearly needed to clarify these relationships. When this species complex is considered as a unit it demonstrates an Arcto-Tertiary distribution (see pages 12 and 33 for a discussion). Recently, a new tetraploid species of hybrid origin between *O. regalis* and *O. japonica* was described from Asia (Tsutsumi 2011).

The Bexar and Victoria populations are somewhat disjunct from the remainder of the TX populations and may represent remnant populations surviving in isolated favorable microhabitats since Ice Age times (see page 36 for a discussion). Numerous such examples of disjunct species are known in TX (Kral 1966; MacRoberts & MacRoberts 1997a; Diggs & O'Kennon 2003; Taber 2008). Also see page 37 for a possible explanation of the TX-s Mexico disjunct distribution of both this species and *Osmundastrum cinnamomeum*. Whetstone and Atkinson (1993) noted that, "The chloroplasts within the spores give the young sporangia their green color. As the spores mature and are shed, the sporangia change color to a distinctive rusty brown." This species is extremely distinct, particularly when in fertile condition. The fertile leaves are similar to the sterile except for the visually very obvious, greatly reduced, sporangia-bearing pinnae at the blade tip.

OSMUNDASTRUM C. Presl CINNAMON FERN

An ancient monotypic genus of e North America, s Mexico s to South America, and Asia. Although this species has traditionally been included in *Osmunda*, molecular evidence (Yatabe et al. 1999; Metzgar et al. 2008) shows *Osmundastrum cinnamomeum* to be the basal branch of the Osmundaceae (as shown on the phylogenetic tree), with the other branch containing everything else in the family (the Old World genera *Leptopteris* and *Todea* and the remaining species of *Osmunda*). An additional recent analysis combining molecular and morphological characters, including fossil data, agreed with this conclusion (Judd et al. 2008). Thus, in order for *O. cinnamomeum* to be treated in *Osmunda*, everything else in the family would have to be included as well. Yatabe et al. (1999) noted, "When the *rbcL* trees [molecular data], the fossil and morphological evidences are all taken into account, it can be concluded that the extant *Osmunda cinnamomea* has no closely related living species in Osmundaceae...." They also commented, "Thus we can call extant *Osmunda cinnamomea* a 'living fossil'." In fact, the ancient split between *Osmundastrum* and the other genera has been estimated at 294 million years ago (Yatabe et al. 1999; Kato 2007). Given the evidence from multiple sources, we are following Metzgar et al. (2008) and most other recent pteridologists (e.g., Smith et al. 2008) in recognizing the monotypic genus *Osmundastrum*. (name from the related genus *Osmunda*, in which *Osmundastrum* was previously classified, and Latin, *-astrum*, resembling.)

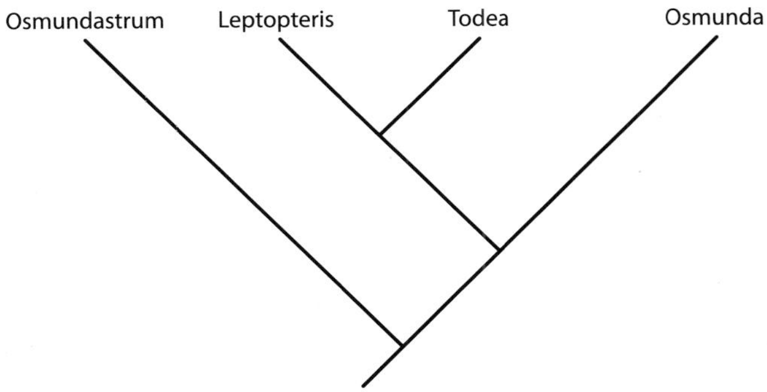

REFERENCES: Miller 1967; Serbet & Rothwell 1999; Yatabe et al. 1999, 2005; Metzgar et al. 2008.

Osmundastrum cinnamomeum (L.) C. Presl, (from Latin: *cinnamum*, which was from Greek: *kinnamon*, *kinnamomon*, and apparently ultimately from Hebrew/Phoenician: *kinamon* or *qinamon*, possibly meaning Chinese amomum or spice, China being from where it was thought to have come—Bashalon 2008; in reference to the bark of *Cinnamomum verum* (Lauraceae), source of the spice cinnamon, meaning cinnamon-colored or cinnamon-brown, a lightish brown mixed with yellow and red; in this species in reference to the color of the hairs on the young petioles, rachis, and pinnae, these usually persisting into maturity only as a tuft at

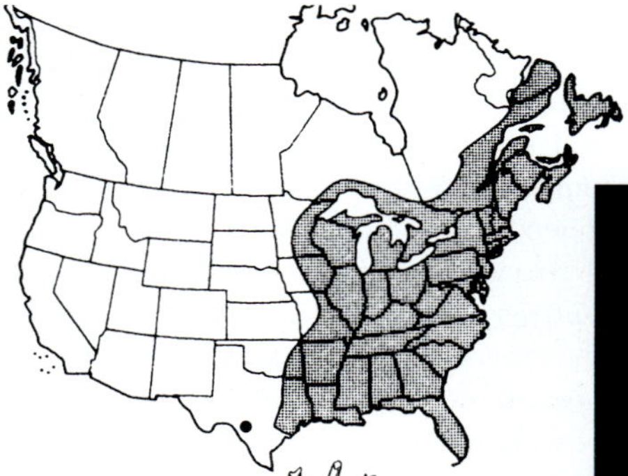

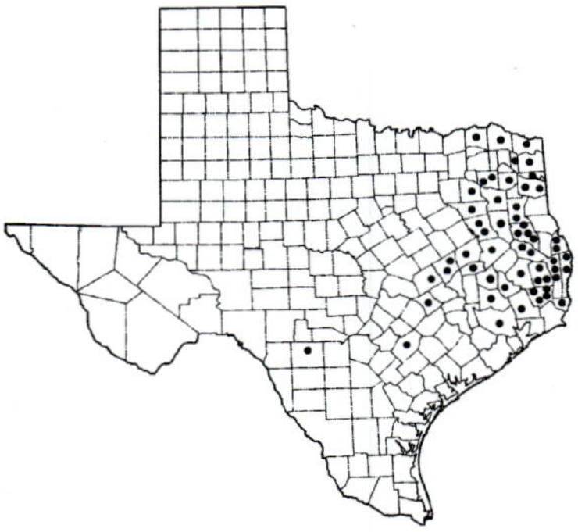

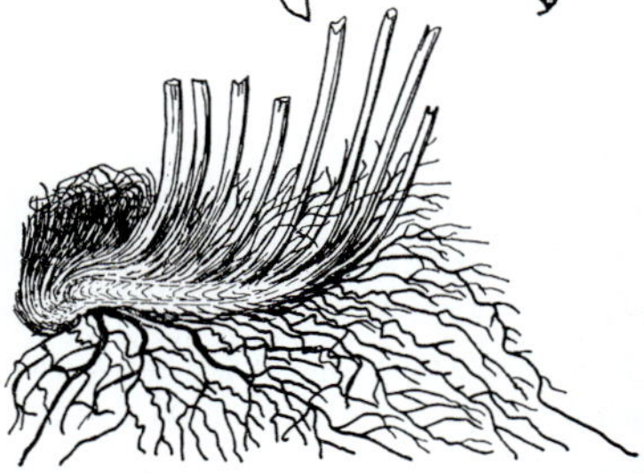

detail of fertile leaf

Robert George

George Diggs

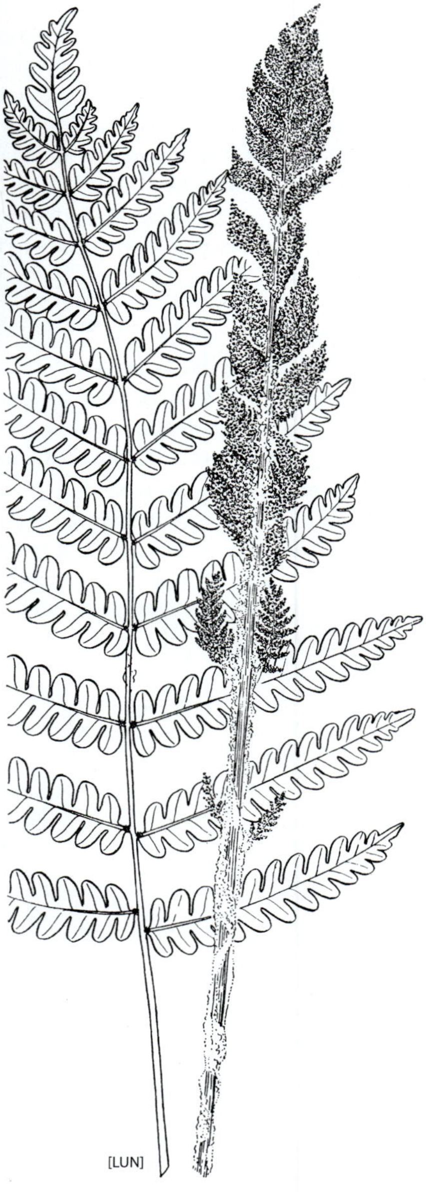

[LUN]

Osmundastrum cinnamomeum

CINNAMON FERN

base of pinnae), CINNAMON FERN, BUCKHORN FERN, BUCKHORN BRAKE, FLOWERING FERN. Leaves dimorphic, erect to spreading, in a vase-shaped cluster from a stout, woody, creeping rhizome; petiole straw-colored to greenish, with hairs when young; sterile leaves 1-pinnate-pinnatifid, ca. 0.3–1.5(–2) m long, the smallest subdivisions with margins entire and usually apically mucronate; pinnae with a persistent tuft of usually cinnamon-colored tomentum on abaxial surface at base; rachis greenish, with hairs when young; fertile leaves strikingly different, not very leaf-like, with no expanded pinnae, densely tomentose, much narrower and shorter than sterile leaves; sori absent, the sporangia clustered, cinnamon-colored; $2n = 44$ (Whetstone & Atkinson 1993). Wet areas; Pineywoods and Gulf Prairies and Marshes w to Leon (BAYLU), Milam (Correll 1956; Turner et al. 2003), Lee (BRIT), Henderson, and Van Zandt (Turner et al. 2003) cos. in the Post Oak Savannah and to Lamar Co. (BRIT) in Red River drainage; also disjunct sw to Gonzales Co. (*Parks & Cory 5630, 5634*, TAES; *Tharp s.n.* 1935, TEX-LL;) and to Uvalde (Turner et al. 2003) Co. on the Edwards Plateau; se Canada and throughout e U.S. w to MN and TX; also s Mexico and s to South America and Asia. Sporulating Mar–Jul or later. [*Osmunda cinnamomea* L.]

This is one of the most interesting TX ferns from the standpoints of biogeography (= scientific study of the geographic distribution of organisms) and evolution. Within TX, CINNAMON FERN is widespread in the Pineywoods of East TX and parts of the adjacent Post Oak Savannah; isolated populations are also disjunct sw to Gonzales Co. in the s Blackland Prairie and even farther to the w to Uvalde Co. in the w Edwards Plateau. The occurrence in Gonzales Co. is in the Ottine wetlands—a unique, isolated, spring-fed, relict wetland ecosystem on the floodplain of the San Marcos River. CINNAMON FERN and a number of other species disjunct to the Ottine wetlands are relicts of more humid conditions in the past. Because of unique hydrological circumstances, they have been able to survive since Ice Age times in a tiny island of appropriate habitat surrounded by a sea of now inhospitable environment. The present occurrence of CINNAMON FERN in the Ottine wetlands is not too surprising, since peat cores from the area dated to the Pleistocene contain CINNAMON FERN spores (see page 36 for discussion of this disjunction). Its broader modern distribution, North America to South America and disjunct to Asia, is also interesting, suggesting an ancient origin and a much more widespread range in the past (Kato & Iwatsuki 1983) (see page 33 for a discussion of such distributions). Recent fossil discoveries from the Upper Cretaceous (Serbet & Rothwell 1999) of w North America provide evidence that *O. cinnamomeum* was more widespread in the past, that it has been in North America "for at least 70 million years." and suggest that fern species can remain virtually unchanged over millions of years (an example of evolutionary stasis). Further, this would mean *O. cinnamomeum* has the greatest known longevity of any individual vascular plant species. Based on the longevity of this and other ferns, Rothwell and Stockey (2008) suggest that there is a "a dramatically different mode and tempo of evolution for some homosporous plants than for seed plants." Old and New World populations of CINNAMON FERN have apparently been separated for a very long time (Yatabe et al. 1999 estimated 25–35 million years) and their relationships are unclear. Metzgar et al. (2008) wrote that, "Because they are clearly distinct based on nucleotide data, it is possible that New World and Asian individuals of *O. cinnamomeum* represent distinct species, as suggested by Yatabe et al. (1999)." Further study will be needed to determine the appropriate taxonomic rank for these widely separated populations. CINNAMON FERN can be confused when sterile with the superficially similar, but only very distantly related *Woodwardia virginica*, VIRGINIA CHAIN FERN, in the Blechnaeae. However, as noted by Weakley (2010), CINNAMON FERN "is coarser (to 2 m tall, vs. to 1 m tall), has cinnamon tufts of tomentum present in the axils of the pinnae (vs. absent), has the rachis greenish and rather fleshy in texture (vs. brown and wiry), and bears fronds clumped or tufted from a massive, woody, ascending rhizome covered with old petiole bases (vs. fronds borne scattered along a thick, horizontal, creeping rhizome)." When fertile, the two species appear dramatically different (in *Woodwardia virginica* the fertile pinnae are essentially like the sterile).

POLYPODIACEAE J. Presl & C. Presl

POLYPODY FERN FAMILY

This family has been treated taxonomically in radically different ways. In the past it was considered to be a huge family encompassing ca. 7,000 species or nearly two-thirds of the living ferns. Correll (1956, 1966), for example, put 20 of the 28 fern genera he included for TX in the Polypodiaceae. More recently, Smith (1993c) treated it much more narrowly, including only ca. 40 genera and 500 species (Smith 1993c). Based on molecular studies it is now considered to include ca. 56 genera and 1,200 species (Smith et al. 2006, 2008). When treated in this way, it includes the grammitid ferns, a group of ca. 20 genera and 600 species often segregated as Grammitidaceae (Ranker et al. 2004; Schneider et al. 2004d), which "clearly nest within Polypodiaceae" (Smith et al. 2008). The Polypodiaceae are largely pantropical with a few represented in temperate areas; most species are epiphytic or on rock, with a few terrestrial. In comparison with many fern groups, it is thought to have evolved relatively recently. Molecular research indicates that polypod ferns (Order Polypodiales, made up of Polypodiaceae and related families) had major diversifications during the Cretaceous Period beginning ca. 100 mya and more recently in the Cenozoic Era. These bursts of speciation are possibly the result of ecological opportunities associated with the rise and increasing dominance of the angiosperms—e.g., modern polypods thrive in the low light conditions of angiosperm forests and as epiphytes in the canopy of tropical forests (Schneider et al. 2004b; Schuettpelz & Pryer 2008, 2009). Generic limits within the family are still being clarified (e.g., within the genus *Polypodium*) (Smith et al. 2008). Family name from *Polypodium*, POLYPODY, a cosmopolitan genus of ca. 100 species, currently more narrowly defined than previously. It is part of the Order Polypodiales, the largest order of ferns, which is characterized by having an annulus only partially encircling the sporangium, the annulus being interrupted by the stalk of the sporangium (the annulus is thus "incomplete") (Moran 2004; see Fig. 13 on page 13). (Greek: *poly*, many, and *pous* or *podium*, foot, referring to the rhizomes which "bear on their upper surface two rows of slightly elevated leaf bases known as phyllopodia. This imparts the appearance of an upside- down creeping caterpillar with its two rows of feet . . ."—Moran 2004) (Order Polypodiales)

FAMILY RECOGNITION IN THE FIELD: the TX species are typically either epiphytic or on rocks; the discrete, large, round or oval sori without indusia are found in single rows on each side of the midvein of the blade or the lobes; leaf blades are deeply pinnatifid or in one species simple; the lower leaf surfaces are conspicuously covered with peltate scales.

REFERENCES: Weatherby 1939; Correll 1956, 1966; de la Sota 1973; Hennipman et al. 1990; Smith 1993c; Schneider et al. 2004d; Jansson et al. 2007; Yatskievych & Windham 2009a.

PLEOPELTIS Humb. & Bonpl. ex Willd.

SCALY-POLYPODY, SHIELD-SORUS FERN

Usually epiphytic or on rocks; rhizomes creeping, scaly; leaves monomorphic, simple to deeply 1-pinnatifid, variously scaly with peltate scales, the scales often bicolored; sori round to oval, discrete, large and obvious, in single rows on each side of the midvein of the blade or of the lobes; indusia absent.

A widespread but primarily neotropical genus of ca. 50 species of mostly epiphytic ferns (Andrews & Windham 1993). Some of the species now treated in *Pleopeltis* were formerly included in *Polypodium* ("Early pteridologists placed any fern with round, non-indusiate sori in *Polypodium*"—Hoshizaki & Moran 2001). Yatskievych & Windham (2009a) noted that, "Until recently, *Pleopeltis* was restricted to a small group of mostly simple-leaved ferns. Phylogenetic research on the Polypodiaceae has indicated that most of the groups of New World polypodies with peltate leaf scales are more closely related to *Pleopeltis* than *Polypodium* sensu stricto and an expanded circumscription of the genus has become accepted by many pteridologists." In fact, *Pleopeltis* species can be readily distinguished by having scales (hence the common name SCALY-POLYPODY) of a distinctive type scattered over the lower surface of the leaf blade (as well as many other characters differentiating them from *Polypodium*) (Windham 1993e). Both Otto et al. (2009) and Sprunt et al. (2011) emphasize that the presence of persistent pelatate laminar scales is a "taxonomically distinctive character." Some authorities (e.g., Mickel & Smith 2004) still treat a number of the species in *Polypodium* (e.g., *P. polypodioides* and *P. riograndensis*) but note that they will ultimately need to be transferred to *Pleopeltis* "pending future publications." However, while we are following Yatskievych & Windham (2009a) in treating the TX species in *Pleopeltis*, recent molecular research (Otto et al. 2009; Sprunt et al. 2011) suggests that *Pleopeltis* and related genera are still not completely understood and that many taxonomic problems (including the exact circumscription of *Polypodium* and *Pleopeltis*) are still unresolved. (Greek: *pleos*, many, and *pelte*, shield, in reference to the shield-like peltate scales covering immature sori)

REFERENCES: Pessin 1925; Wendt 1980; Andrews & Windham 1993; Windham 1993e; Moran 1998; Hooper 1995; Bush et al. 1999; Mickel & Smith 2004; Otto et al. 2009; Sprunt et al. 2011.

1. Leaf blades simple, not divided into lobes/segments _______ **P. polylepis**
1. Leaf blades deeply pinnatifid, the lobes/segments obvious.
   2. Petioles with the scales often overlapping and marginally entire or nearly so; sori so sunken into leaf blade tissue as to form distinct bumps on upper blade surface; leaf blades with mature scales bicolored (dark brown in center, tan to silvery gray transparent margins), circular to deltate-ovate, the margins entire or merely erose; plants usually epiphytic on trees (less commonly on rocks), occurring in e ⅔ of TX _______ **P. polypodioides**
   2. Petioles with scales usually not overlapping and minutely toothed or ciliate marginally; sori not sunken into leaf blade tissue and usually not visible on upper blade surface; leaf blades with mature scales reddish brown to dark reddish brown, not bicolored or sometimes with narrow transparent margins, broadly ovate-lanceolate, the margins fringed with short cilia; plants usually growing on rocks, occurring in TX only in the Trans-Pecos _______ **P. riograndensis**

Pleopeltis polylepis (Roem. ex Kunze) T. Moore var. **erythrolepis** (Weath.) T. Wendt, (sp: Greek: *polys*, many, and *lepis*, scale, in reference to the leaf scales; var.: Greek *erythro*, red, and *lepis*, scale), RED-SCALE SCALY-POLYPODY, RED-SCALE POLYPODY. Growing on rocks; rhizomes slender, widely creeping, scaly, the scales lanceolate-acuminate to broadly lanceolate; leaves monomorphic, usually spaced 1 cm or less apart; petioles ca. ½ blade length; leaf blades simple, lanceolate to narrowly oblanceolate, 5–7 times longer than wide, to ca. 7 cm long, usually 0.8–1 cm wide, coriaceous, weakly hygroscopic, the margins thickened, the midribs abaxially green, the surfaces weakly scaly on upper surface, densely covered with overlapping peltate scales on lower surface, the scales broadly ovate-lanceolate, irregularly fringed-ciliate, with dark reddish brown center, margins pale brown to whitish; sori round to oval, discrete, not or only slightly sunken into blade tissue. Mats on porphyritic rocks (in Mexico on limestone); in the U.S. known only from upper Limpia Canyon in the Davis Mts., Jeff Davis Co. (Palmer 1930 cited *Palmer 34378* which was collected in 1928; Correll 1956; Andrews & Windham 1993; Yarborough & Powell 2002), 2300 m; also n Mexico. Sporulating May–Sep. [*Phlebodium erythrolepis* (Weath.) Conz., *Pleopeltis erythrolepis* (Weath.) Pic. Serm., *Polypodium erythrolepis* Weath.] More recently (1984), Jackie Poole (pers. comm.) saw in upper Limpia Canyon a dense colony of "...hundreds (or perhaps thousands) of individuals"...[that] "only occupied about a 10 square meter area on a N-facing vertical cliff within a coniferous evergreen woodland. ..." The more common, morphologically similar var. *polylepis*, endemic to Mexico, differs in a number of ways including having entire to erose, spheric scales on the lower surface of the blades, the scales not densely overlapping, petioles ca. ⅕ blade length, leaf blades 8–10 times as long as wide, with midribs abaxially green or blackened, rhizome scales ovate to broadly lanceolate, and in nearly always being epiphytic (Wendt 1980; Andrews & Windham 1993; Mickel & Smith 2004); while the n Mexico and TX populations are easily distinguished as var. *erythrolepis* and the c and s Mexico populations are clearly var. *polylepis*, there are intervening areas where "a broad range of intergradation between the varieties occurs" (Wendt 1980). Because of its rareness and limited distribution in the state, we consider this species to be of conservation concern in TX. ⚠

Pleopeltis polypodioides (L.) E.G. Andrews & Windham var. **michauxiana** (Weath.) E.G. Andrews & Windham (sp: resembling *Polypodium*, POLYPODY, a widespread related genus; var.: for André Michaux, 1746–1803, French botanist and explorer of North America), RESURRECTION FERN, GRAY POLYPODY, TREE FERN. Usually epiphytic or sometimes growing on rocks; rhizomes slender, widely creeping, much branched, densely scaly; leaves monomorphic, evergreen, widely spaced; leaf blades deeply 1-pinnatifid, oblong to triangular-oblong in outline, to 18+ cm long and 5 cm wide, thick, opaque, strongly hygroscopic, involute upon drying, glabrous on upper surface except for a few scales along midvein, densely covered with peltate scales on lower surface, the scales obscurely clathrate (= net- or lattice-like in appearance); sori round, so sunken into leaf blade tissue as to form conspicuous bumps on the upper surface of leaves; $2n = 74$ (Andrews & Windham 1993). Usually growing on various species of trees, especially oaks, sometimes on rocks, usually in shady damp situations; widespread in e ½ of TX and slightly disjunct to Uvalde Co. (Correll 1956) (e.g., Anderson Co., *C.M. Rowell & J.L. Blassingame 5038*, BRIT); e U.S. from DE s to FL w to KS and TX, also Mexico and Central America. Sporulating May–Nov. Previously included in the genus *Polypodium* [as *P. polypodioides* (L). Watt var. *michauxianum* Weath.]. The common name, RESURRECTION FERN, refers to the leaves "... which become brown and appear dead during dry periods, but 'resurrect' themselves after rains and rapidly become green and lush" (Nelson 2000). An astonishing 97% of the plant's water can be lost without harm. The "resurrection" occurs

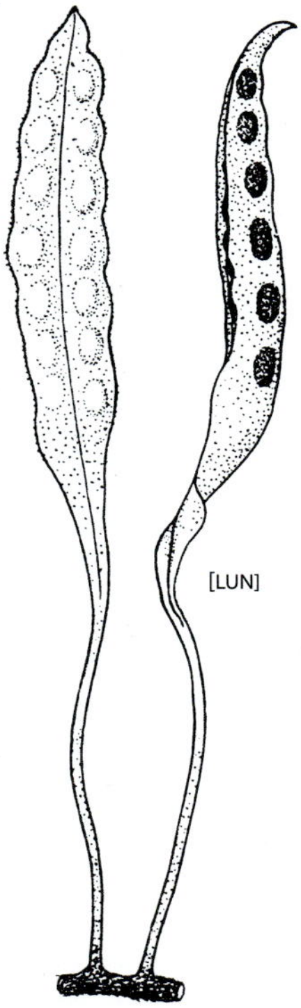

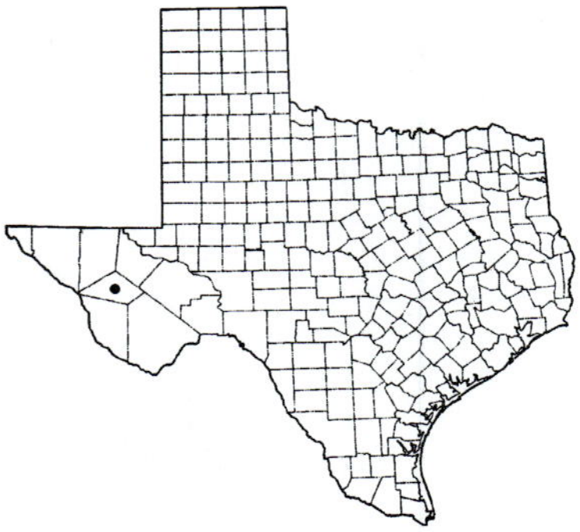

[POM]
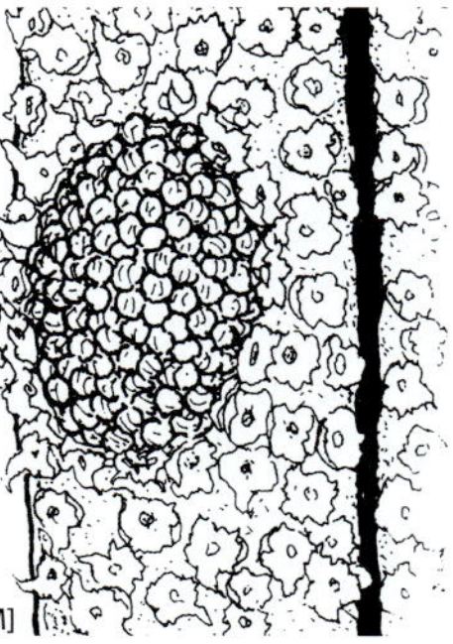

detail of lower
leaf surface w/sorus

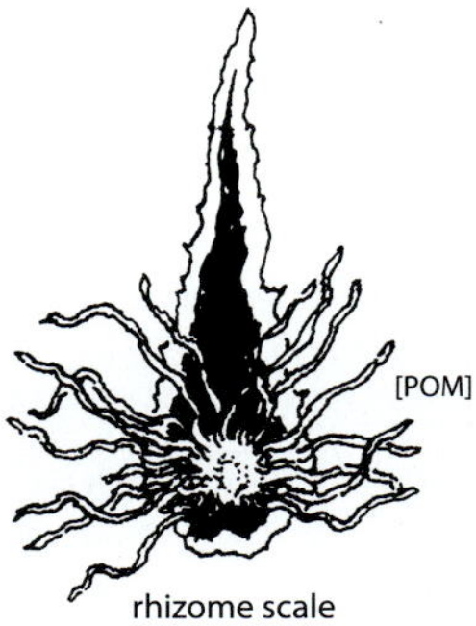

rhizome scale

Carlos Velasco

Pleopeltis polylepis var. **erythrolepis**

RED-SCALE SCALY-POLYPODY

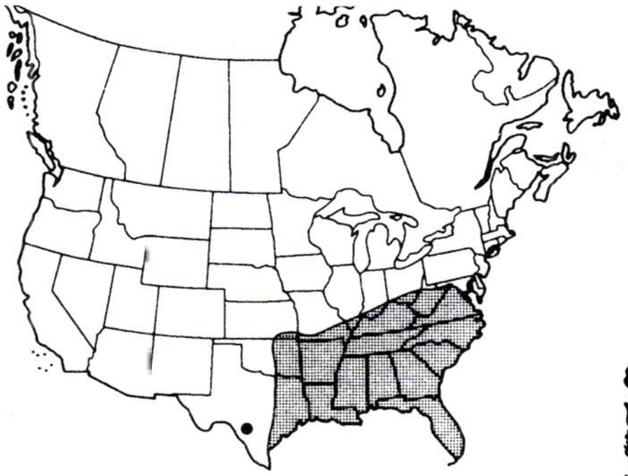

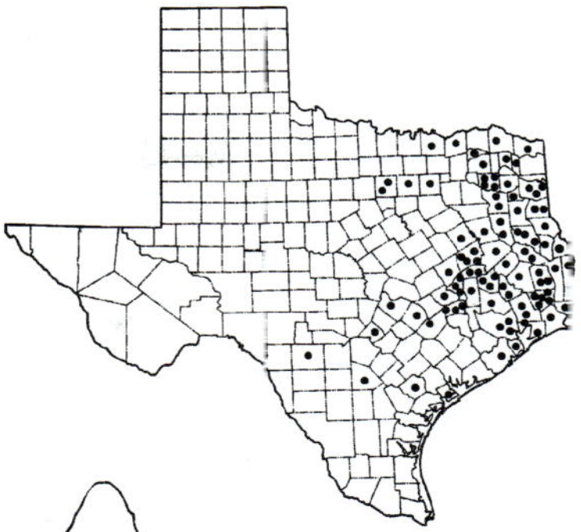

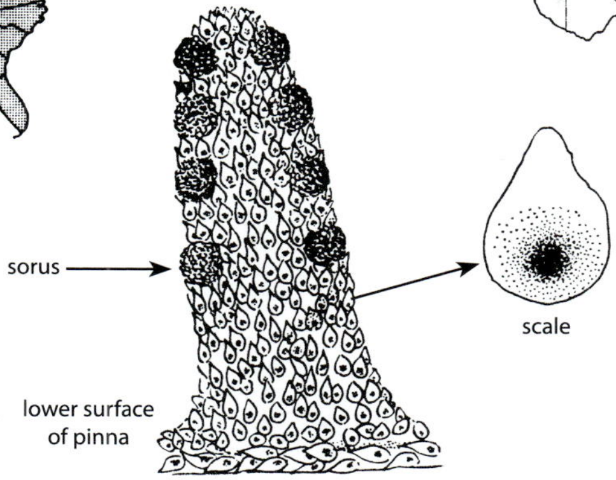

George Diggs

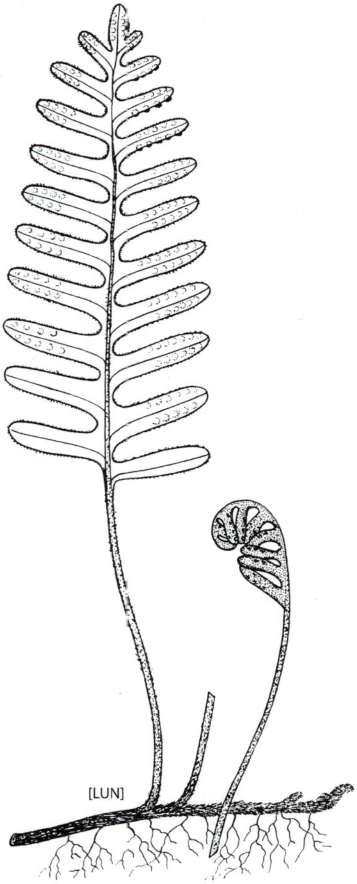

Pleopeltis polypodioides var. **michauxiana**

RESURRECTION FERN

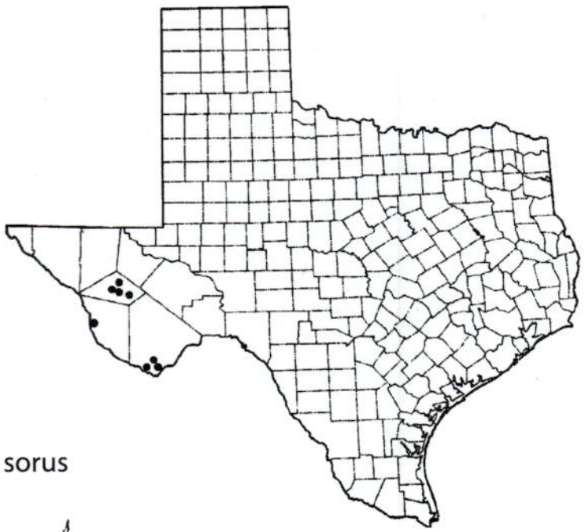

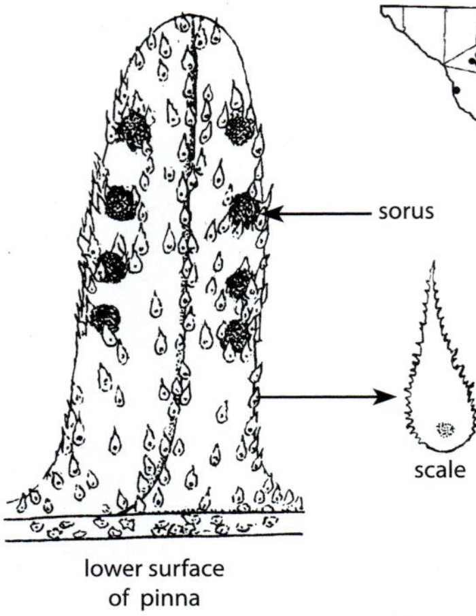

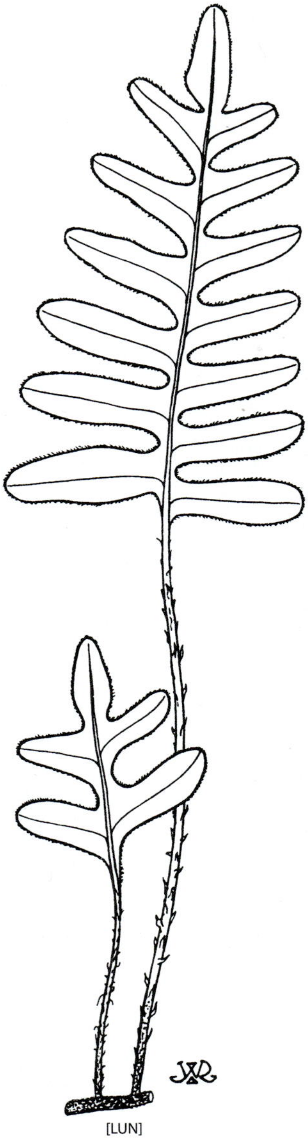

Pleopeltis riograndensis

RIO GRANDE SCALY-POLYPODY

because special peltate scales on the lower surface of the leaf are able to absorb water and quickly channel it into the leaf—and it is the lower leaf surface that is exposed when the leaf curls up upon drying. Each scale is made up of a disk-like portion composed of numerous dead cells and a stalk of living cells. When water is present, it is drawn into the dead cells by capillary action, then taken up by the living stalk cells, and subsequently transported into the leaf tissue (Moran 1998, 2004). Nauman et al. (2000) noted, "The species is represented by six varieties which range from the southeastern United States southward throughout the American tropics to Argentina and in Africa." More recently, an additional variety was named from Mexico (as *Polypodium polypodioides* var. *knoblochianum* Mickel—Mickel & Smith 2004). Sprunt et al. (2011) suggested that recent nucleotide sequence data supports the elevation of var. *michauxiana* to the level of species.

Pleopeltis riograndensis (T. Wendt) E.G. Andrews & Windham, (of the Rio Grande, in reference to its proximity to the Rio Grande River), RIO GRANDE SCALY-POLYPODY, RIO GRANDE POLYPODY. Usually growing on rocky soil or in crevices; rhizomes slender, short-creeping, somewhat branched, scaly, the scales tan marginally with black centers; leaves monomorphic, closely spaced; leaf blades deeply 1-pinnatifid, the basal segments alternate, oblong to narrowly triangular in outline, to 20 cm long and 4 cm wide, subcoriaceous, strongly hygroscopic, glabrous on upper surface, the venation mostly free, sparsely scaly on lower surface with scattered to only slightly overlapping scales (blade surface not concealed), the scales ovate to ovate-lanceolate, distinctly clathrate (= net- or lattice-like in appearance) under magnification; sori round, not or only slightly sunken into blade tissue; $2n = 148$ (Andrews & Windham 1993). Moist areas of canyons (rocky slopes, ledges, crevices), at 1500–2500 m; Boot Springs in Chisos Mts. of Brewster Co. (e.g., *C.H. Mueller s.n., 1932*, BRIT, GH, NY, TEX-LL; *Warnock 1095*, BRIT, SRSC, TEX-LL; *O.E. Sperry 1095*, TAES), Davis Mts. of Jeff Davis Co. (e.g., *Palmer 32196*, TEX-LL, US) and Chinati Mts., Presidio Co. (*M.L. Butterwick & E.J. Lott, 3897*, GH, MEXU, TEX-LL—holotype, Wendt 1980); AZ and TX; also n Mexico (Coahuila, Chihuahua, and Sonora). Sporulating summer–fall. [*Polypodium riograndense* (T. Wendt) Mickel, *Polypodium thyssanolepis* A. Braun ex Klotzsch var. *riograndense* T. Wendt] This species in the past was considered a variety of *P. thyssanolepis* (e.g., Wendt 1980); however, that species, which is widespread in Mexico, has densely scaly petioles and leaves, the blade scales nearly spherical, mostly areolate venation, and the basal leaf segments opposite or nearly so (Andrews & Windham 1993). Because of its rareness and limited distribution in the state, we consider this species to be of conservation concern in TX. ⚠

PSILOTACEAE J.W. Griff. & Henfr.

WHISK FERN FAMILY

The Psilotaceae are a very small group (2 genera, ca. 12 species) found worldwide in tropical regions (Thieret 1993; Smith et al. 2008) and extending to warm areas. In addition to *Psilotum*, the only other genus is *Tmesipteris*, which occurs from se Asia to Australia, New Zealand, and some Polynesian islands. Psilotaceae lack roots (instead they have rhizomes with absorptive rhizoids and mycorrhizal fungi), and in the TX species have essentially naked, dichotomously branching stems with only small, veinless, scale-like outgrowths (= enations; these enations are possibly highly reduced leaves, but they have often not been considered to be true leaves, which by definition have vascular tissue; it has also been suggested that the enations could be highly-reduced lateral stems). Their simple structure makes Psilotaceae morphologically the least complex of all living terrestrial vascular plants. They resemble (at least superficially) fossils of some of the earliest land plants (Mabberley 1997), and the group has traditionally been thought to be direct descendants of the earliest known vascular plants from the Silurian and Devonian periods—e.g., the fossil genus *Rhynia* (Woodland 1997). As a result, the Psilotaceae were long considered to be the most primitive of living vascular plants (Tryon & Tryon 1982) and traditionally have been classified as a separate division (the botanical equivalent of an animal phylum), the Psilophyta. Alternatively, it was suggested (e.g., Bierhorst 1977) that the simple morphology of Psilotaceae may instead be the result of reduction from an ancestral fern. The sporangia are eusporangiate—large, with numerous (> 1000) spores, and walls > one cell thick. This connects the Psilotaceae with the eusporangiate fern family Ophioglossaceae (*Botrychium* and *Ophioglossum*) in the Ophioglossales. A number of early molecular studies (e.g., Manhart 1995; Wolf 1997; Vangerow et al. 1999) also linked Psilotaceae with the Ophioglossaceae, which have often been considered the most isolated and basal among the modern plants traditionally classified as ferns (Wagner 1990). More recent phylogenetic research (Pryer et al. 2001; Smith et al. 2006; Roper et al. 2007), based on molecular data from a number of different genes and morphology, strongly supports the link of Psilotaceae with the Ophioglossaceae and supports their inclusion within the Polypodiophyta (ferns). As a result, most pteridologists (e.g., Moran 2004; Smith et al. 2006, 2008; Schneider et al. 2009) now consider the Psilotaceae to be a family of ferns. However, the family is so distinctive that it is still recognized in its own order, the Psilotales (e.g., Smith et al. 2006). In addition to molecular data, the subterranean, non-photosynthetic, mycoheterotrophic (= obtaining food from decaying organic material via a special relationship with a symbiotic fungus) gametophytes are similar to those of Ophioglossaceae, and the multiflagellate sperm and spore structure and formation also link the Psilotaceae to ferns (Tryon & Tryon 1982; Renzaglia et al. 2000, 2001). However, Renzaglia et al. (2001) noted, "the less streamlined architecture of *Psilotum* gametes and the lack of architectural features shared with any specific taxon examined to date suggest that *Psilotum* is an early divergent fern, with relatively remote affinities to Ophioglossaceae and Equisetaceae." Reduction of the root system may also link Ophioglossaceae and Psilotaceae—"ophioglossoids have simple unbranched roots that lack root hairs, and whisk ferns lack roots altogether" (Schuettpelz & Pryer 2008). *Psilotum nudum* is the only species in the family known to occur in North America. (Order Psilotales)

FAMILY RECOGNITION IN THE FIELD: the only species in the TX flora is a dichotomously branched, essentially leafless, spore-bearing perennial with green photosynthetic stems 0.5 m or less tall.

REFERENCES: Bierhorst 1977; Cooper-Driver 1977; Gensel 1977; Kaplan 1977; Wagner 1977; White 1977; Kramer 1990a; Thieret 1993; Manhart 1995; Pryer et al. 1995; Wolf 1997; Vangerow et al. 1999; Rothwell 1999; Schneider & Carlquist 2000a; Pryer et al. 2001; Smith et al. 2006; Gutierrez 2007.

PSILOTUM Sw. WHISK FERN, FORK-FERN

A genus of 2 or 3 species (Thieret 1993; Smith et al. 2008) with green, photosynthetic stems. The common name, WHISK FERN, comes from the much-branched essentially leafless stems which give the plant the appearance of a whisk broom. (Greek: *psilos*, naked, smooth, or bare, either in reference to the essentially leafless stems or to the "naked" sporangia which are not covered by indusia)

REFERENCES: Correll 1960b; Philips 1970; Lodwick 1975; Renzaglia et al. 2001.

Psilotum nudum (L.) P. Beauv., (Latin: *nudus*, bare, naked), WHISK FERN. Terrestrial, sometimes epiphytic perennial with short-creeping, rhizoid-bearing, branched rhizomes; roots absent (but rhizoids absorptive); aerial stems erect to ascending or arching, to 25(–50) cm tall (greenhouse material often more robust), photosynthetic, dichotomously branched 3–5 times, 3(–several)-ridged; stem appendages (= enations) veinless, scale-like, minute, 0.7–3 mm long (apparently reduced leaves but veins absent); spores all of 1 kind (plants thus homosporous), produced in eusporangiate sporangia, these fused in groups of (2–)3 to form ± globose, (2–)3-lobed synangia; synangia 2–3 mm wide, solitary in axils of minute stem appendages; gametophytes subterranean, with mycorrhizal fungi; chromosome number variable, n = 46–56, 104, 210 (Thieret 1993). Low woods, swamps, wet areas, wet peaty humus, and around bases of trees and stumps; Hardin Co. in s part of Pineywoods (*Correll, Correll, & Rosier 123516*, TEX-LL; Correll 1960b—this collection from the Big Thicket was the first in the state; Lance Rosier led Correll to the location) and Freestone Co. in the Post Oak Savannah (*L.N. Lodwick 216*, TEX-LL; Lodwick 1975); se U.S. from NC (Perry & Musselman 1994) s to FL w to AR, LA, and 2 disjunct locations in TX, also disjunct to AZ (Gutierrez 2007) and nw Mexico (Sonora); extremely widespread in the Americas, Africa, Asia, Australia, New Zealand, and Pacific Islands. Sporulating summer. [*Lycopodium nudum* L.] Grown as an ornamental in some areas, particularly Japan (Hyam & Pankhurst 1995); it can be a minor weed in greenhouses (Thieret 1993). Although extremely rare in TX, in some areas of the U.S. (e.g., FL) it can appear as a weed. The species can be epiphytic in situations such as on tree bases, stumps, old logs, or tree forks with accumulated debris (Nauman et al. 2000; Nelson 2000). The whisk broom or "leafless twig" appearance of the WHISK FERN is nearly unique among pteridophytes (Nelson 2000). The rare (and often transitory) occurrence of certain pteridophytes beyond the margins of their main range (e.g., *Psilotum* in TX) has sometimes been attributed to transport of spores by major storms or hurricanes (J. Peck, pers. comm.). Because of its rareness and limited distribution in the state, we consider this species to be of conservation concern in TX. ⚠

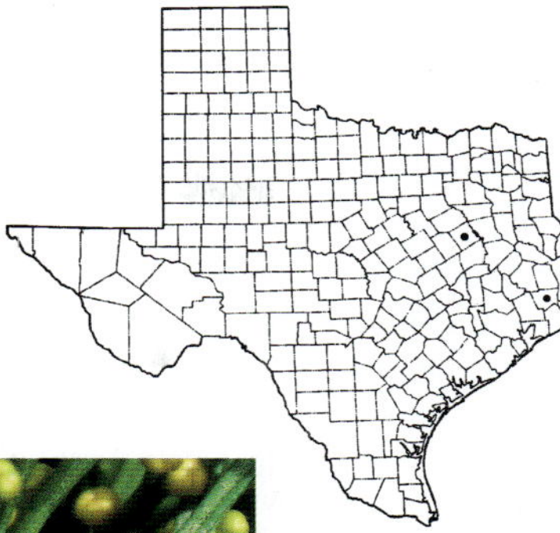

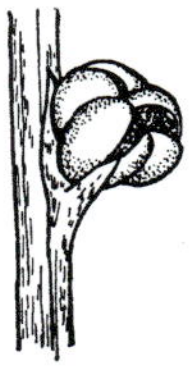

synangium

Alan Cressler

detail of synangia

Forest & Kim Starr, Creative Commons-by

[KNO]

Psilotum nudum

WHISK FERN

PTERIDACEAE E.D.M. Kirchn.

MAIDENHAIR FERN OR BRAKE FAMILY

Texas species mostly xerophytic, on rocks, sometimes terrestrial (or in the introduced *Ceratopteris* aquatic or semiaquatic); rhizomes compact to long-creeping; leaves usually monomorphic (rarely somewhat dimorphic or in *Ceratopteris* distinctly dimorphic and in *Llavea* hemidimorphic); leaf blades 1–4(–5) pinnate or 1–more pinnate-pinnatifid, usually with hairs, glands, and/or scales, sometimes with farina (= a wax-like flavonoid material produced by short, capitate, glandular hairs), or rarely glabrous; sporangia on lower surface of leaf blades, marginal or submarginal; margins of ultimate segments (= smallest subdivisions) of leaf blades recurved to form false indusia (except in *Astrolepis* and *Bommeria*).

The taxa included here in the Pteridaceae have been variously treated at the family level. For instance, some authorities (e.g., Carlquist & Schneider 2000a) segregated *Cheilanthes* and close relatives into the Cheilanthaceae, although others (e.g., Lellinger 1985) recognized some of the genera as Sinopteridaceae. Windham (1993a) in *Flora of North America* treated the Pteridaceae broadly, including ca. 40 genera and 1,000 species. Most recently, based on a variety of molecular studies, the family is considered to be a monophyletic group containing ca. 50 genera and 1,000 species—about 10% of all fern species worldwide (Prado et al. 2007; Schuettpelz et al. 2007; Smith et al. 2008). However, Weakley (2010) noted that the family "may be further subdivided" in the future. Although subcosmopolitan, it is most numerous in the tropics and in dry areas. The newer name Adiantaceae has sometimes been applied to the family. One very unusual genus, the aquatic *Ceratopteris*, has often been recognized in its own family, the Parkeriaceae; however, all recent molecular analyses place it in the Pteridaceae, and its unusual characteristics "are probably a consequence of the shift to aquatic habitats" (Smith et al. 2008). Of the approximately 1,000 species in the family, about 400 are in a Cheilanthoid clade (which corresponds to the subgroup sometimes treated as Sinopteridaceae) (Schuettpelz et al. 2007; Rothfels et al. 2008) including species of *Argyrochosma*, *Astrolepis*, *Bommeria*, *Cheilanthes*, *Notholaena*, and *Pellaea* which occur in TX. These are mostly dry-adapted ferns (see discussion on page 208) that have evolved the ability to colonize, survive, and thrive in seasonally dry habitats—this has resulted in convergent evolution, superficial morphological similarities, and consequently, great taxonomic difficulty. Tryon and Tryon (1982), for example, called the cheilanthoids "the most contentious group of ferns with respect to a practical and natural generic classification." Rothfels (2009) noted that "there seems to be a limited number of morphologies that allow ferns to survive and reproduce in extremely xeric conditions (basically the hairy, waxy, and thick-leaved morphologies ...), and cheilanthoid ferns have explored each of them multiple times independently." Hybridization, polyploidy, and apomixis occur widely within the cheilanthoid ferns, which has further hindered attempts to reach a stable system of classification. Molecular studies have recently been extremely valuable in clarifying evolutionary relationships in the family. Within the huge polypod order, the Pteridaceae are one of the early diverging families (Smith et al. 2006; see page 380). This is the largest family of ferns in TX, with 44 native species, which is more than the total number of pteridophytes in some states. Because of its size and because the genera can usually be recognized when a few differences are understood, a "QUICK TIPS FOR GENUS ID" section is included after the references for each genus. (Order Polypodiales)

FAMILY RECOGNITION IN THE FIELD: usually dry-adapted plants typically growing on rocks (except 1 species aquatic or semi-aquatic); sporangia at or near margins of the ultimate leaf segments with the leaf margins usually recurved over sporangia to form false indusia.

REFERENCES: Correll 1956, 1966; Tryon et al. 1990; Windham 1993a; Gastony & Rollo 1995, 1998; Carlquist & Schneider 2000a; Windham & Yatskievych 2003b; Prado et al. 2007; Schuettpelz et al. 2007; Rothfels et al. 2008.

1. Plants aquatic or semiaquatic; leaves distinctly dimorphic (the fertile quite different than the sterile); leaf blades with adventitious buds or small plantlets in notches along margins __________ **Ceratopteris**
1. Plants mostly on rocks, sometimes terrestrial; leaves usually monomorphic or in *Pellaea* somewhat dimorphic or in *Llavea* strikingly hemidimorphic (= with very different sterile and fertile pinnae on the same leaf blade); leaf blades without adventitious buds or plantlets.
   2. Leaves strikingly hemidimorphic, the fertile pinnules only in the distal ⅓ of the leaf blade, linear,

2.5–8.5 cm long, the sterile pinnules ovate-deltate, 2–3 cm long; scales on rhizomes blackish, those at bases of petioles yellowish; genus known in the U.S. only from a single collection in Presidio Co., TX _______ **Llavea**

2. Leaves not hemidimorphic, the pinnules various; scales various; including species widespread in TX.
   3. Leaf blades broadly pentagonal (= 5-sided) in outline, about as wide or very slightly wider than long, pedately divided into 3 main segments, these deeply pinnate-pinnatifid, without farina (= wax-like, mealy, or powdery material) on lower surfaces; margins of ultimate leaf segments (= smallest subdivisions of leaf) not recurved to form false indusia, the sporangia occupying ⅔–¾ distance from margin of blade to midvein of each ultimate segment; species known in TX only in the Trans-Pecos _______ **Bommeria**
   3. Leaf blades not broadly pentagonal (except in the broadly pentagonal *Notholaena standleyi*, which has the lower blade surfaces with obvious yellowish farina); margins of ultimate leaf segments recurved to form false indusia covering the marginal or submarginal sori (except false indusia not present in *Astrolepis*); including species widespread in TX.
      4. Margins of ultimate leaf segments not recurved to form false indusia; leaf blades 1-pinnate to 1-pinnate-pinnatifid throughout; lower (= abaxial) leaf surfaces densely covered with coarsely ciliate and/or stellate scales; upper (= adaxial) leaf surfaces usually with coarsely ciliate or stellate scales _______ **Astrolepis**
      4. Margins of ultimate leaf segments recurved to form false indusia; leaf blades usually 2–5 pinnate at least at base OR at least partially 2-pinnatifid (except 1-pinnate in 1 species of *Pteris* and 1-pinnate-pinnatifid in 1 species of *Cheilanthes*); lower leaf surfaces scaly, pubescent, or glabrous; upper leaf surfaces without coarsely ciliate or stellate scales.
         5. Petioles greenish to light brown near leaf blade, with upper surface 2–3-grooved; rachis conspicuously winged in at least the distal ½ of leaf blade OR not so; leaf blades 1-pinnate or divided at most so that they are 2-pinnatifid (in these usually at least some pinnae are deeply palmately 3-divided) _______ **Pteris**
         5. Petioles variously colored, but often black to brown throughout, with upper surface rounded, flattened, or 1-grooved; rachis not winged; leaf blades more divided, usually 2–5-pinnate at least at base.
            6. Ultimate leaf segments with only the apical margin recurved (the false indusium thus appearing "interrupted" rather than continuous); sporangia borne directly on underside of recurved apical margins of ultimate leaf segments (= borne on the false indusium); veins of ultimate leaf segments prominent, dichotomously branched (= equally 2-forked), essentially parallel near the margins _______ **Adiantum**
            6. Ultimate leaf segments with apical and lateral margins recurved over sporangia; sporangia borne on lower leaf surface and covered by the recurved margins (= borne on blade tissue beneath the false indusium); veins of ultimate leaf segments obscure or, if prominent, pinnately branched and more divergent near the margins.
               7. Leaf blades with conspicuous whitish to yellowish farina on lower surfaces (this can be hidden by dense scales in *Notholaena aschenborniana*).
                  8. Ultimate leaf segments sessile to subsessile, often fused to costae or costules, the base rounded to truncate or cuneate; rhizome scales black or bicolored with dark central strip and lighter margins _______ **Notholaena**
                  8. Ultimate leaf segments usually distinctly stalked or sometimes subsessile, not fused to costae, the base often ± cordate; rhizome scales tan to brown, uniformly colored or weakly bicolored _______ **Argyrochosma**
               7. Leaf blades lacking conspicuous whitish to yellowish farina on lower surfaces.
                  9. Leaf blades usually 1-pinnate distally (2-pinnate proximally), usually glabrous on lower surface or nearly so; largest ultimate leaf segments variable in size, but sometimes > 4 mm wide; rhizome scales strongly bicolored with dark central stripe OR if uniformly colored then largest ultimate leaf segments > 4 mm wide. _______ **Pellaea**
                  9. Leaf blades usually 2–more pinnate or 2–more pinnate-pinnatifid nearly throughout, often with either pubescence on lower surface or lower surface obscured by large scales of costae (except only sparsely pubescent to nearly glabrous in *Cheilanthes alabamensis*, *C. aemula*, *C. wrightii*, and *Argyrochosma microphylla*); rhizome scales usually uniformly colored to weakly bicolored; ultimate leaf segments usually less than 4 mm wide.
                     10. Lower surface of leaf blades glabrous; most distal ultimate segments of leaf blades ± cordate at base and attached only by distinct dark-colored stalk _______ **Argyrochosma**

10. Lower surface of leaf blades usually with conspicuous pubescence or scales or at least appearing so OR if glabrous then the most distal ultimate segments of leaf blades not cordate at base and ± sessile (attached at least partially by greenish stalks or blade tissue) ____ **Cheilanthes**

ADIANTUM L. MAIDENHAIR FERN

Plants delicate, terrestrial or on rocks; rhizomes short-creeping, ascending, or suberect; leaves monomorphic, clumped or closely spaced; petioles brownish to blackish, glabrous; leaf blades 2–4(–more)-pinnate, the ultimate segments (= smallest subdivisions) variously shaped, cuneate to fan-shaped, irregularly rhombic, transversely oblong or nearly round, only the apices recurved to form false indusia (the false indusium thus appearing "interrupted" rather than continuous), stalked (the stalks 0.5–4 mm long), the veins prominent, dichotomously branched (= equally 2-forked), essentially parallel near the margins, a distinct midvein absent; sporangia submarginal along or rarely between the ends of the ultimate veins, appearing marginal on the back (abaxial surface) of the recurved apices of the ultimate segments (= false indusia), the false indusium appearing "interrupted" rather than continuous since only the apical (but not the lateral) margins of the ultimate leaf segments are recurved.

A genus of 150–200 species, nearly worldwide in distribution except at higher latitudes (> 60°), but particularly common in Andean South America and Mexico (Paris 1993; Mickel & Smith 2004). It is sometimes placed in the Adiantaceae. Some species are used medicinally, and a number are cultivated as ornamentals for their delicate, beautiful foliage. The position of the sporangia is definitive for identification—in other related ferns with false indusia (e.g., *Cheilanthes*, *Pellaea*), the sporangia are borne on the blade tissue beneath the false indusium, rather than on the false indusium itself as in *Adiantum*. The leaves "have the unusual property that, when wetted, water beads-up into silvery droplets that quickly roll off" (Hoshizaki & Moran 2001)—hence the scientific name. Although the origin of the common name, MAIDENHAIR FERN, is unclear, it possibly results from an inexact translation of a very old usage of what is now the specific epithet of *A. capillus-veneris*, (literally, "Venus' hair"), a species widespread in both the Old World and the New World. The name has been used for a very long time—Arber (1912) pointed out that in one of the earliest herbals printed in England (by Rycharde Banckes published in 1525), under the heading "Capillus venerus" the plant was described in part as, "This herbe is called Mayden heere or waterworte." (Greek: *adiantos*, unwetted, for the glabrous leaves, which shed raindrops)

REFERENCES: Fernald 1950b; Paris 1993; Wolf et al. 2003; Mickel & Smith 2004.

QUICK TIPS FOR GENUS ID: Very delicate, non-leathery ferns whose leaves have numerous small separate segments (2–4 pinnate) with veins that are easily seen and equally 2-forked and with only the margin at the segment tips curved over and protecting the sori.

1. Leaf blades glabrous; ultimate segments of leaf blades shallowly or deeply incised into lobes (giving a glove-like effect with short "fingers" present); dark color of segment stalks extending slightly along edge of base of segments (use lens); petioles and rachises usually blackish or purplish black, occasionally somewhat dark brownish; species widespread in TX ____ **A. capillus-veneris**
1. Leaf blades minutely hairy (use lens); ultimate segments of leaf blades entire, to crenulate or serrulate (giving a mitten-like effect without "fingers"); dark color of segment stalks ending ± abruptly at base of segments; petioles and rachis dark reddish brown; species rare in TX, known only from 2 counties on Edwards Plateau ____ **A. tricholepis**

Adiantum capillus-veneris L., (Latin: *capillus*, hair, and *veneris*, pertaining to Venus, Roman goddess of love, beauty, and fertility, = Venus' hair), SOUTHERN MAIDENHAIR, VENUS'-HAIR FERN, CULANTRILLO, MATTRESS FERN. Terrestrial or on rocks; rhizomes short-creeping; leaves weakly deciduous, clumped or closely spaced, numerous, lax-arching or pendulous, 15–75 cm tall; petioles and rachises usually blackish, or purplish black, occasionally somewhat brownish; leaf blades 2–3(–more) pinnate, glabrous, membranous to thin-herbaceous, bright green, the ultimate segments usually wedge or fan-shaped to irregularly rhombic (= 4-sided, diamond-shaped), ca. as long as broad; $2n = 120$ (tetraploid; Paris 1993). Continuously moist calcareous areas, particularly limestone bluffs, rocks and ledges along streams; widespread in TX except far s, common in the Edwards Plateau, with somewhat disjunct populations in East TX in Harris (*J.W. Kessler 3451*, BRIT), Newton (TAES), Orange (*Correll 13336*, TEX-LL), Washington (Turner et al. 2003), and Webb (Rob Fondren, pers. comm.) cos.; s ½

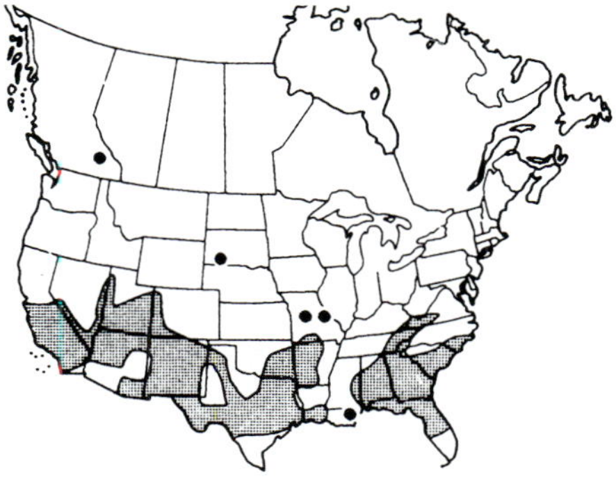

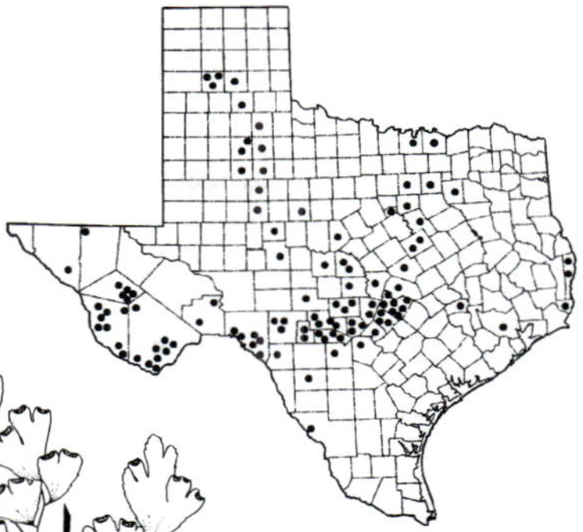

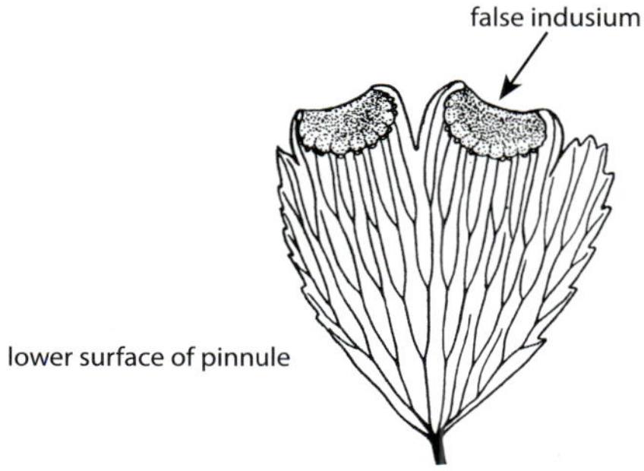

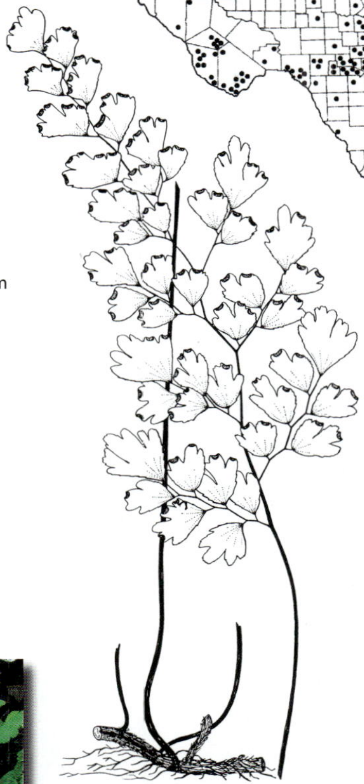

[LUN]

Adiantum capillus-veneris

SOUTHERN MAIDENHAIR

Jim Conrad

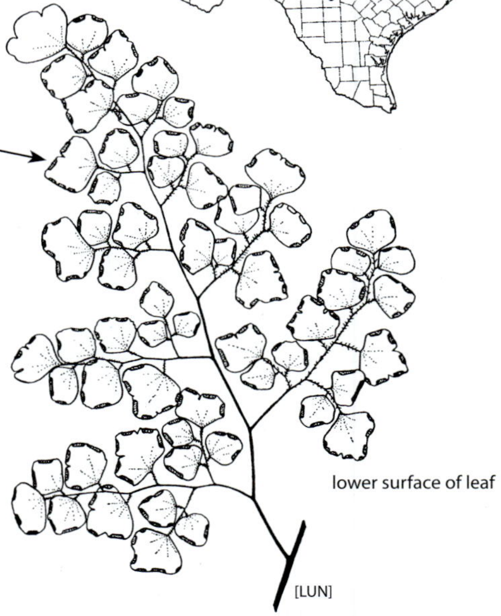

Jim Conrad

Adiantum tricholepis

HAIRY MAIDENHAIR FERN

of U.S. from VA s to FL w to CA, also disjunct to SD and B.C.; also Mexico, West Indies, Central and South America, Eurasia, and Africa. Sporulating May–Dec. [*A. capillus-veneris* var. *modestum* (Underw.) Fernald, *A. capillus-veneris* var. *rimicola* (Sloss.) Fernald] Although the species is variable, recognition of infraspecific taxa does not seem warranted (Mickel & Smith 2004). VENUS'-HAIR FERN has long been used medicinally for conditions of the skin, scalp, and internal organs (Cheatham & Johnston 1995). Hoshizaki and Moran (2001) considered this species, widespread in warm-temperate to subtropical areas, to be one of the most widely distributed ferns in the world. However, the Old World representatives are apparently diploid, with $2n = 60$ (Manton 1950; Paris 1993), and it is possible that North American and Old World plants are actually different species (Paris 1993). The distribution in North America is interesting—the species is widespread across the s ½ of U.S., with an outlying population in the Black Hills of SD and a significantly disjunct population in British Columbia, Canada (Wagner 1972). This isolated Canadian occurrence is presumably the result of long-distance dispersal. Wagner (1972) noted that the British Columbia population of this "south-temperate to subtropical fern is flourishing in the runnels of a system of hot springs, an obviously unusual site." Jim Peck (pers. comm.) pointed out that "likewise, it occurs on tufa of hot springs in Hot Springs, AR."

Adiantum tricholepis Fée, (Greek: *thrix*, trichos, hair, and *lepis*, scale, = hairy scale), HAIRY MAIDENHAIR FERN, FUZZY MAIDENHAIR FERN. Rhizomes short-creeping to ascending or suberect; leaves clumped, lax-arching, to 70 cm tall; leaf blades 3–4-pinnate, minutely hairy on both upper and lower surfaces with numerous stiff whitish hairs to 0.5 mm long, the ultimate segments fan-shaped to transversely oblong to nearly round; petioles and rachis dark reddish brown, somewhat stouter than in *A. capillus-veneris*. Limestone cliffs along streams, on boulders in creeks, and among rocks on steep slopes; Medina (e.g., *J. Stanford 5002*, BRIT, HPC; *Correll 14112*, BRIT, TEX-LL; *G.M. Soxman 89*, TEX-LL) and Bandera (Correll 1956; Paris 1993; no recent collections known in Bandera) cos. on the Edwards Plateau; in the U.S. known only from these 2 counties in TX; also Mexico, Belize, and Guatemala. Sporulating spring through Nov, depending on rains. Correll (1956, 1966) and Paris (1993) noted that specimens from the mouth of the Pecos River are actually *A. capillus-veneris*. Seigler and Lockwood (1975) emphasized that the Edwards Plateau is "well known for its unusual ferns" pointing out that a number of tropical ferns, *Adiantum tricholepis*, *Dennstaedtia globulifera*, *Blechnum appendiculatum* (as *B. occidentale* var. *minor*), and *Tectaria heracleifolia*, have been found in the area, "which is the extreme northern limit of their ranges." This is one of 19 pteridophyte species known in the U.S. only from TX (see page 41). Because of its rareness and limited distribution in the state, we consider it to be of conservation concern in TX (and in the U.S.).

ARGYROCHOSMA (J. Smith) Windham FALSE CLOAK FERN, SILVER FERN

Plants usually on rocks; rhizomes compact, erect to ascending, the scales usually uniformly tan or brown; leaves monomorphic, clustered, to only 30 cm long (usually smaller); petioles, rachises, and stalks of ultimate leaf segments dark brown, lustrous; leaf blades 3–5 pinnate, deltate to ovate, oblong, or lanceolate, usually somewhat leathery; the lower surfaces glabrous or covered by a whitish, wax-like, mealy or powdery material (= farina); ultimate leaf segments (= smallest subdivisions of leaf) small (< 4 mm wide), their margins recurved to revolute, forming rather poorly defined false indusia, these often concealing or partially concealing the sporangia; sporangia on the lower leaf surfaces, along the veins, submarginal.

A New World genus of ca. 30 species (Windham et al. 2009). Although the species have been variously treated in a variety of genera (e.g., *Cheilanthes*, *Notholaena*, and *Pellaea*), Windham (1987b) segregated *A. dealbata*, *A. limitanea*, *A. microphylla*, and related species as the genus *Argyrochosma*. Morphological and chromosome evidence ($x = 27$—unique among cheilanthoid ferns) supports separate recognition of the genus (Windham 1993b), as does more recent molecular evidence (Gastony & Rollo 1998; Windham et al. 2009; Sigel et al. 2011). Molecular studies show the genus is more closely related to *Astrolepis*, *Pellaea*, and other members of the "pellaeid" group than to *Cheilanthes* or *Notholaena* (Windham et al. 2009). The presence of farina has been an important character in the taxonomy of *Argyrochosma* and related groups. For example, Tryon and Tryon (1982) and Tryon et al. (1990) included the species treated here as *Argyrochomsa* in their otherwise exclusively farinose *Notholaena*. Further, the presence of farina in some species of *Argyrochosma* but not others has caused considerable taxonomic diasagreement. Based on recent molecular evidence, Sigel et al.

(2011) noted that "*Argyrochosma* comprises two major monophyletic groups: one exclusively non-farinose and the other primarily farinose. Within the latter group, there has been at least one evolutionary reversal (loss) of farina and the development of major chemical variants that characterize specific clades." (Greek: *argyros*, silver, and *chosma*, powder, referring to whitish farina covering the lower surface of leaf blades in most species)

REFERENCES: Tryon 1956; Knobloch et al. 1973; Windham 1987b, 1993b, 1993e; Windham et al. 2009; Sigel et al. 2011.

QUICK TIPS FOR GENUS ID: Dry area ferns with grayish green to bluish green leaves, with numerous, tiny, separate, stalked leaf segments (usually 3–5 pinnate), without obvious covering of scales or hairs, but lower leaf surfaces (only) often with a striking whitish farina.

1. Lower (= abaxial) surfaces of leaf blades without whitish covering; ultimate leaf segments articulate (the dark brown color of the segment stalks ending abruptly at the segment bases); rachises flattened or shallowly grooved on upper (= adaxial) side ____ **A. microphylla**
1. Lower surfaces of leaf blades obscured by a covering of a whitish powdery or wax-like material (= farina); ultimate leaf segments not articulate (the dark brown color of the segment stalks continuing into the segment bases on the upper side); rachises rounded to slightly flattened on upper side.
   2. Leaf blades deltate (= triangular) in outline, somewhat herbaceous in texture; petioles reddish brown, 0.5–0.75 mm in diam.; ultimate leaf segments with veins often visible on lower surfaces; species of the e Edwards Plateau and Cross Timbers and Prairies e to near w margin of the Blackland Prairie ____ **A. dealbata**
   2. Leaf blades lanceolate to oblong in outline, leathery in texture; petioles reddish brown to black, 0.75–2 mm in diam.; ultimate leaf segments with veins obscure on lower surfaces; species known in TX only from mts. of the Trans-Pecos ____ **A. limitanea**

Argyrochosma dealbata (Pursh) Windham, (Latin: *dealbatus*, white-washed or whitened, in reference to the conspicuous whitish material on the lower leaf surfaces), POWDERY FALSE CLOAK FERN, POWDERY CLOAK FERN, FALSE CLOAK FERN. Leaves to only ca. 15 cm long, evergreen; leaf blades 3–5-pinnate, less divided distally, somewhat herbaceous, the upper surface bluish green, glabrous or with a few glands, the lower surface with very conspicuous whitish powdery or wax-like material; $2n = 54$ (Windham 1993b). Crevices of limestone and other calcareous rocks; mainly Cross Timbers and Prairies and e Edwards Plateau n to Wise Co. (*O'Kennon & S. Lusk 18673,* BRIT) and e to Bell (TAES), Travis (BRIT), Comal (TEX-LL), Hays (Turner et al. 2003), and Ellis (Correll 1956) cos. near w margin of Blackland Prairie; AR, IL, KS, MO, NE, OK, and TX and slightly disjunct to the e in KY. Sporulating summer–fall. [*Cheilanthes dealbata* Pursh, *Notholaena dealbata* (Pursh) Kunze, *Pellaea dealbata* (Pursh) Prantl] This species has a striking color contrast between the upper and lower surfaces of the leaf blades—bluish green above and with obvious whitish farina below.

Argyrochosma limitanea (Maxon) Windham subsp. **mexicana** (Maxon) Windham, (sp.: Latin: *limitatus*, bounded, limited, or bordered, in reference to its occurrence along the U.S.-Mexican border; subsp.: of Mexico, where the type specimen was collected), MEXICAN SILVER FERN, SOUTHWESTERN FALSE CLOAK FERN. Leaves to 30 cm long, evergreen; leaf blades 4–5 pinnate, leathery, the upper surface bluish green, glabrous or with a few glands, the lower surface with very conspicuous whitish powdery or wax-like material; $n = 2n = 81$ (Windham 1993b). Limestone cliffs or rocky slopes; Brewster (*O'Kennon 10952*, BRIT, *E.G. Marsh 245a*, Chisos Mts., TEX-LL, *R.T.Clausen & H. Trapido 5305*, TAMU, and Dead Horse Mts.), El Paso (*R.D. Worthington 3463.5*, Franklin Mts., TEX-LL), and Culberson (*W.D. Pohl & J.L. Blassingame 3518*, BRIT; Sierra Diablo and Guadalupe Mts.—Yarborough & Powell 2002) cos.; AZ, NM, and TX; also n Mexico. Sporulating summer–fall. [*Cheilanthes limitanea* (Maxon) Mickel var. *mexicana* (Maxon) Mickel, *Notholaena limitanea* Maxon subsp. *mexicana* Maxon, *N. limitanea* var. *mexicana* (Maxon) M. Broun] Although this taxon is similar to *A. dealbata*, according to Windham (1993b) the "distinctions between *Argyrochosma dealbata* and *A. limitanea* are subtle but apparently absolute. Although both occur in TX, their ranges do not overlap. Despite the morphologic similarities, isozyme analyses indicate that *A. dealbata*, a diploid species, was not involved in the origin of polyploid *A. limitanea*." Windham (1993b) also noted that the species is an asexual triploid that reproduces via apogamy (= a type of asexual reproduction that does not involve fertilization; the sporophyte is formed directly from the gametophyte without gamete production). The taxonomy of this plant is not yet settled—although it and *A. limitanea* subsp. *limitanea* (AZ, CA, NM, UT, and n Mexico) are similar and are

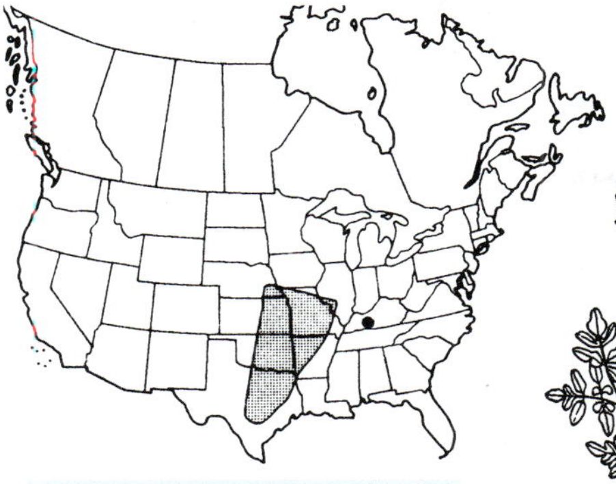

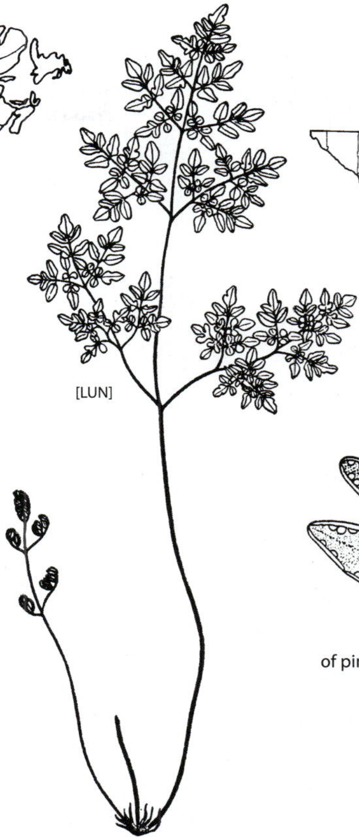

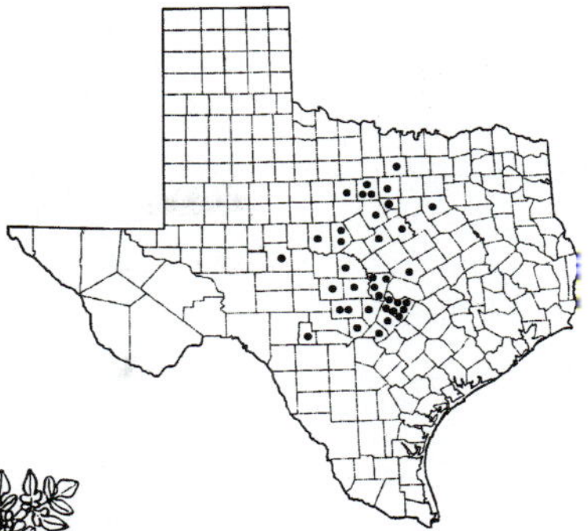

Botanical Research Institute of Texas

lower leaf surfaces
showing whitish farina

lower surface
of pinna showing sporangia

George Yatskievych

Argyrochosma dealbata

POWDERY FALSE CLOAK FERN

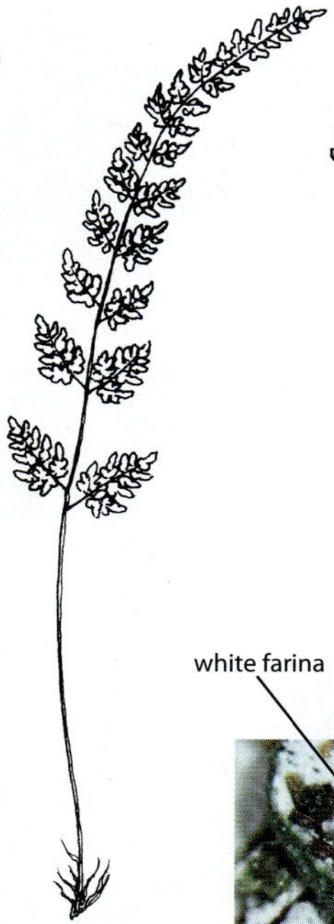

[POM]

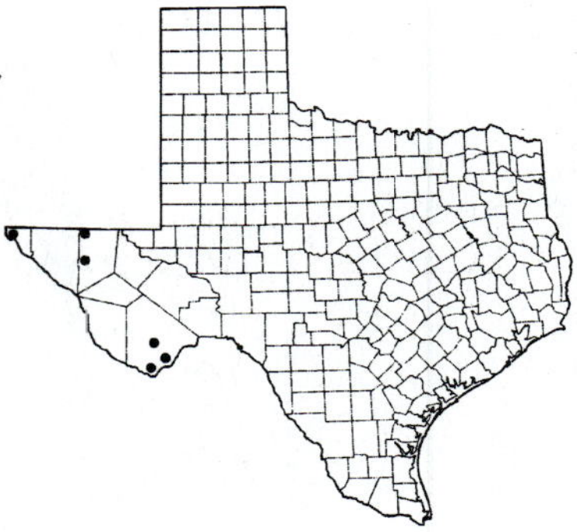

Dale A. Zimmerman Herbarium, Western N. Mex. Univ.

Ann Larson

Dale A. Zimmerman Herbarium, Western N. Mex. Univ.

Argyrochosma limitanea subsp. **mexicana**

MEXICAN SILVER FERN

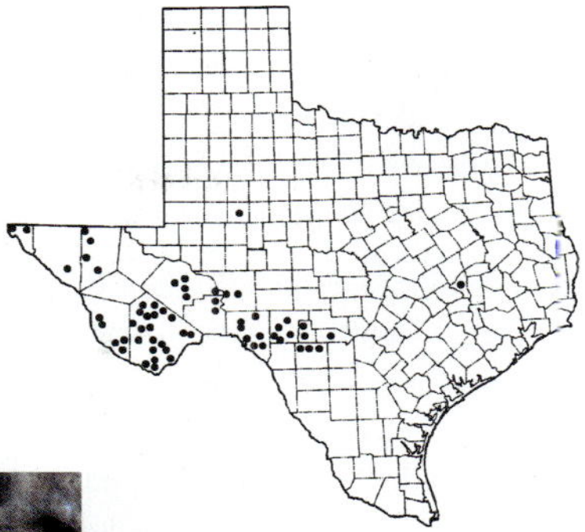

lower surface of pinnule showing sporangia

[LUN]

[LUN]

Argyrochosma microphylla

SMALL-LEAF FALSE CLOAK FERN

both triploids, molecular studies show they have different polyploid origins (Windham 1993b); changes in nomenclature may be necessary in the future if their evolutionary relationships are clarified. As with *A. dealbata,* the leaf blades are bluish green above and with obvious whitish farina below; however, the two can be distinguished by the key to species. Because of its rareness and limited distribution in the state, we consider this species to be of conservation concern in TX. ⚠

Argyrochosma microphylla (Mett. ex Kuhn) Windham, (Greek: *mikros*, micro, small or little, and *phyllon*, leaf, = small-leaved), SMALL-LEAF FALSE CLOAK FERN, SMALL-LEAF CLOAK FERN, SMALL-LEAF SILVER FERN. Leaves 7–25 cm long; leaf blades 3–4-pinnate basally, 2-pinnate distally, leathery, glabrous on both surfaces, often glaucous; $2n$ = 54 (Windham 1993b). Hillsides, cliffs, and canyons, on ledges and talus, and in crevices, both in shade and exposed, apparently limited to limestone substrates; mainly occurring in Trans-Pecos and the Edwards Plateau, this species is disjunct in East TX in Brazos Co. (*H.B. Parks s.n.*, 14 Dec 1938, TAES; Correll 1956), and Howard Co. (*Correll 15177*, TEX-LL) in the s Plains Country; NM and TX; also n Mexico. Sporulating summer–fall. [*Pellaea microphylla* (Mettenius ex Kuhn) Windham, *Cheilanthes parvifolia* (R.M. Tryon) Mickel, *Notholaena parvifolia* R.M. Tryon] This species is easily distinguished from both *A. dealbata* and *A. limitanea* by the lack of whitish powdery or wax-like material on the lower leaf surfaces. Work by Knobloch et al. (1973) suggests that the species is made up of both diploids and tetraploids. It is reported to be poisonous to sheep (Correll 1956). ☠

ASTROLEPIS D.M. Benham & Windham

STAR-SCALED CLOAK FERN, SCALY CLOAK FERN

Plants usually on rocks; rhizomes compact to short-creeping; leaves monomorphic, evergreen, densely clustered; leaf blades 1-pinnate to 1-pinnate-pinnatifid throughout, with 20–60 pairs of pinnae, leathery, the lower leaf surfaces with overlapping, ovate to lanceolate ciliate scales and usually an underlying layer of stellate scales concealing the surface, the upper surfaces sparsely to densely covered with stellate or coarsely ciliate scales to glabrescent with age, the pinnae stalked or subsessile; sporangia marginal along the veins or nearly so, forming a ± continuous band; false indusium absent.

A New World genus of ca. 7–10 species (Benham & Windham 1993; Mickel & Smith 2004; Beck et al. 2010), 4 of which occur in the U.S. (and in TX); 2 diploid species were only recently recognized (Beck et al. 2010). They are often found in full sun and are considered by some to be the most sun-tolerant of the xerophytic ferns (Moore 2008). The taxa treated here as *Astrolepis* have been previously lumped into various genera including *Notholaena* or *Cheilanthes*. However, Benham and Windham (1992) convincingly argued that these and several related species are a monophyletic group worthy of recognition as the genus *Astrolepis*. In addition to morphological characters, the base chromosome number of *Astrolepis*, $x = 29$, distinguishes it from *Notholaena* and the majority of species of *Cheilanthes*. Treatment at the generic level is supported by molecular evidence (Gastony & Rollo 1998; Rothfels et al. 2008; Windham et al. 2009; Beck et al. 2010), which shows that *Astrolepis* is more closely related to a "pellaeoid" clade (containing part of *Pellaea*, *Argyrochosma*, and several other small genera) than to either *Cheilanthes* or *Notholaena*. The key to species is adapted from Benham and Windham (1993). (Greek: *astro*, star, and *lepis*, scale, in reference to the star-like scales on the upper surfaces of the leaf blades)

REFERENCES: Tryon 1956; Benham 1989; Benham 1992; Benham & Windham 1992, 1993; Carlquist & Schneider 1997; Beck et al. 2010, 2011, 2012.

<u>QUICK TIPS FOR GENUS ID</u>: Dry area ferns with narrow leaves, only 1-pinnate or 1-pinnatifid, with obvious distinctive scales—the scales on the upper surface tiny, whitish, star-shaped (with numerous branches) and those on the lower surface fringed, white to brownish, and ribbon-like; segment margins not recurved to protect sori.

1. Leaf blades with pinnae relatively small, the largest usually 4–7 mm long; scales on upper (= adaxial) surfaces of pinnae peltate (= attached near middle), circular to slightly elongate in outline (including marginal projections/cilia), those on the lower surfaces ovate, 1 mm or less long ____ **A. cochisensis**
1. Leaf blades with largest pinnae usually 7–35 mm long; scales on upper surfaces of pinnae if present attached at their base, elongate (= longer than wide) in outline (including marginal projections/cilia), usually those on the lower surfaces lanceolate, 1–1.5 mm long.
   2. Upper surfaces of pinnae densely scaly, particularly near margins, the scales usually persistent, the

body of the scales 5–7 cells wide (base of scale body clearly wider than length of marginal cilia); largest pinnae entire or slightly asymmetrically lobed ____ **A. integerrima**

2. Upper surfaces of pinnae only sparsely scaly to glabrescent, the scales often deciduous with age, the body of the scales 1–4 cells wide (base of scale body narrower than to slightly wider than length of marginal cilia); largest pinnae slightly to conspicuously symmetrically lobed.
   3. Largest pinnae shallowly crenately lobed, the lobes usually broadly rounded; pinnae with upper surfaces sparsely scaly, usually having at least some scales persistent, the scales with bodies 2–4 cells wide; scales on lower pinnae surfaces with relatively coarse cilia (magnification required) ____ **A. windhamii**
   3. Largest pinnae typically deeply lobed, the lobes often somewhat acute; pinnae with upper surfaces sparsely scaly to nearly glabrous, most scales deciduous with age, the scales if present with bodies 1–2 cells wide; scales on lower pinnae surfaces with comparatively delicate cilia ____ **A. sinuata**

Astrolepis cochisensis (Goodd.) D.M. Benham & Windham, (from the type locality in Cochise County, AZ), COCHISE SCALY CLOAK FERN, JIMMY FERN. Leaves 7–40 cm long; longest pinnae usually 4–7 mm long, entire or asymmetrically shallowly lobed with 1–4 broadly rounded lobes, the lower surface completely concealed by scales, the upper surface with scales sparse, deciduous, the scales mostly circular to slightly elongate, stellate to coarsely ciliate. Rocky slopes and cliffs usually on limestone; in TX mostly in Trans-Pecos (e.g., Brewster Co., *Correll 14709*, BRIT, TEX-LL), with a few localities in w Edwards Plateau (e.g., Uvalde Co., *O.E. Sperry 3069*, BRIT; Kimble Co., *L.L. Hansen 4894*, TEX-LL; Turner et al. 2003); AZ, CA, NM, OK, and TX; also n Mexico. Sporulating summer–fall. [*Notholaena cochisensis* Goodd., *Cheilanthes cochisensis* (Goodd.) Mickel, *C. sinuata* (Lag. ex Sw.) Domin var. *cochisensis* (Goodd.) Munz, *Notholaena sinuata* (Lag. ex Sw.) Kaulf. var. *cochisensis* (Goodd.) Weath.] Benham and Windham (1993) treated this species as being composed of three subspecies, two of which occur in TX (a sexual diploid and an apogamous triploid). They distinguished the three subspecies based on chromosome number, number of spores per sporangium, spore size, and the type of substrate the plants grow on. Mickel and Smith (2004) did not recognize the subspecies, merely viewing the different forms as cytotypes. We are continuing to follow Benham and Windham (1993) in recognizing the taxa at the subspecific rank because of their apparent reproductive isolation (based both on genetic and ecological isolating mechanisms). They separated the two TX taxa as in the following key. However, the two are not readily recognizable morphologically and are not distinguished on the county distribution map. The third subspecies (subsp. *arizonica* D.M. Benham, known only from AZ) is usually found on noncalcareous substrates and is an apogamous tetraploid, apparently with a complex origin possibly involving other taxa (Benham & Windham 1993). This species is known to be poisonous to livestock. An unidentified poison causes incoordination, arched back, rapid pulse, gasping, trembling (known as "the jimmies"), and prostration; in severe cases death can result from respiratory paralysis. Sheep are apparently most susceptible, followed by goats and cattle. A dosage of 0.8–3% of body weight for 2–3 days has been shown to be toxic. (Weathers 1998; Burrows & Tyrl 2001). ☠

1. Sporangia containing 64 spores; spores averaging 39–46 μm diam. ____ subsp. **chihuahuensis**
1. Sporangia containing 32 spores; spores averaging 59–70 μm diam. ____ subsp. **cochisensis**

subsp. **chihuahuensis** D.M. Benham, (of Chihuahua, reflecting its distribution primarily in the Chihuahuan desert), CHIHUAHUA SCALY CLOAK FERN. $2n = 58$ (Benham & Windham 1993). Rocky slopes and cliffs usually on limestone; this subspecies is mapped in Benham and Windham (1993) as occurring widely in the Trans-Pecos area; presumably most TX specimens of this species (which is widespread in Trans-Pecos, with a few localities in w Edwards Plateau) are of this subspecies (e.g., Brewster Co., *Correll 13577*; Hudspeth Co., *Correll 15067*; Pecos Co., *McVaugh 7320*, all TEX-LL, all annotated D.M. Benham); NM and TX; also ne Mexico. Sporulating summer–fall. This subspecies is reported to be a sexual diploid (Benham & Windham 1993). ☠

subsp. **cochisensis,** COCHISE SCALY CLOAK FERN. $n = 2n = 87$ (Benham & Windham 1993). Rocky slopes and cliffs usually on limestone; El Paso Co. (*Correll 13832, 15039*, TEX-LL, annotated D.M. Benham); Benham and Windham (1993) mapped this subspecies only in the extreme westernmost tip of the Trans-Pecos in the area of El Paso; AZ, CA, NM, OK (Panhandle), and TX; also n Mexico. Sporulating summer–fall. This subspecies is reported to be an apogamous triploid (Benham & Windham 1993). Because of its rareness and limited distribution in the state, we consider this subspecies to be of conservation concern in TX. ⚠ ☠

Astrolepis integerrima (Hook.) D.M. Benham & Windham, (Latin: *integerrimus*, very entire or whole, an

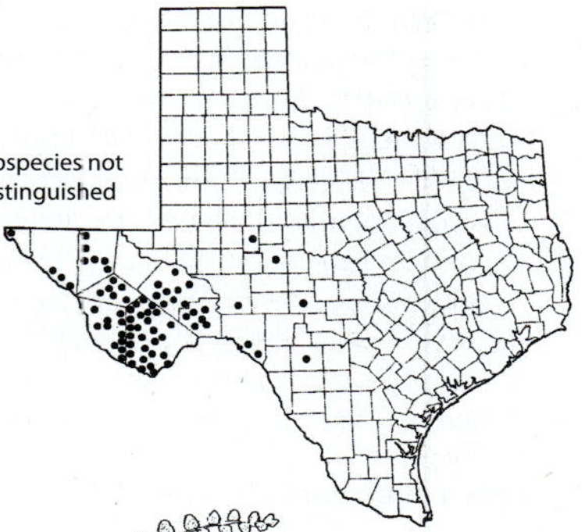

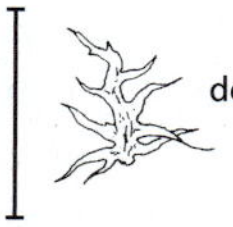

detail of upper scale

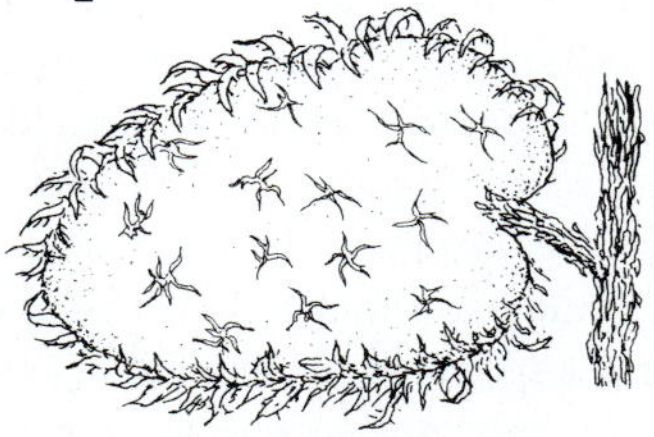

upper surface of pinna

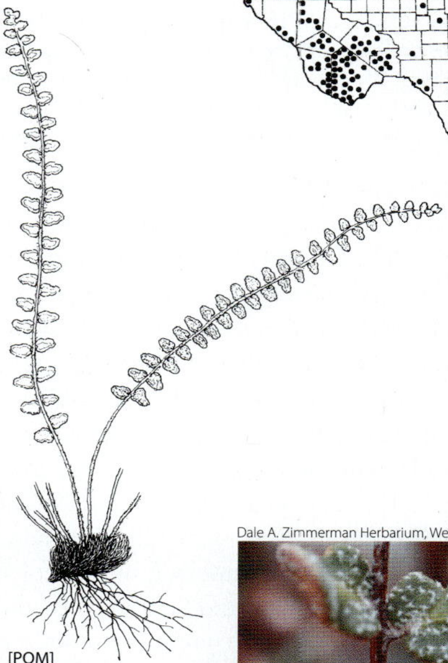

[POM]

Dale A. Zimmerman Herbarium, Western N. Mex. Univ.

Petei Guth

Astrolepis cochisensis

COCHISE SCALY CLOAK FERN

emphatic assertion of the entirety of an organ), HYBRID SCALY CLOAK FERN, HYBRID CLOAK FERN, SOUTHWESTERN CLOAK FERN, WHOLE-LEAF CLOAK FERN. Plants morphologically heterogeneous; leaves 8–45 cm long; longest pinnae usually 7–12(–15) mm long, entire or shallowly asymmetrically lobed with 2–7 broadly rounded lobes, the lower surface completely concealed by scales, the upper surface densely scaly, particularly near margins, the scales usually persistent, elongate, stellate to coarsely ciliate; *n* = 2*n* = 87 (Benham & Windham 1993). Rocky slopes, outcrops, or cliffs, usually limestone or other calcareous substrates. Widely scattered in the w ⅔ of TX e to Palo Pinto, Burnet (*Correll & Correll 12728*), Starr (*J. Ideker BA165*) (all BRIT), Bell, Comal, Travis (TEX-LL), and Bexar (Turner et al. 2003) cos.; mostly sw U.S. (AZ, NM, NV, OK, and TX), and remotely disjunct to AL (*J. Allison 6695*, UNA, VDB; Allison 2001; Allison & Stevens 2001; in Bibb Co. on glades on an ancient [upper Cambrian] rock formation known as the Ketona Dolomite); also Mexico and the Carribean. Sporulating summer–fall. [*Cheilanthes integerrima* (Hook.) Mickel, *Notholaena integerrima* (Hook.) Hevly, *Notholaena sinuata* (Lag. ex Sw.) Kaulf. var. *integerrima* Hook.] Morphologically this variable species is somewhat intermediate between *A. cochisensis* and *A. sinuata*. However, it was hypothesized to be an apogamous triploid hybrid between *A. cochisensis* and an unnamed Mexican taxon related to *A. crassifolia* (of Mexico and Central America) (Benham 1989; Benham & Windham 1993). Subsequently, the "missing" parent was identified as *Astrolepis obscura* J. Beck & Windham (Beck et al. 2010). Recent research (Beck et al. 2011, 2012) has shown that *A. integerrima* has multiple independent origins. In other words, this asexual taxon is an assemblage of many lineages that originated independently at least 10 different times from the same diploid ancestors (with four of the lineages occurring in TX). The morphological variability seen in *A. integerrima* is more understandable given its numerous independent origins. Beck et al. (2012) suggest that such multiple orgins may be the "rule, rather than the exception" among asexual polyploids and that the phenomenon warrants further study. Individuals with shallowly lobed leaves are superficially similar to *A. windhamii*, but can be distinguished by "the abundance and greater width of adaxial scales and the asymmetrical lobing of the pinnae." Individuals with entire pinnae could be confused with *A. cochisensis*, but can be separated by the leaf blades with abundant basally attached scales on the upper surfaces and the larger pinnae (Benham & Windham 1993). This is one of a number of ferns found primarily in the sw or w U.S. with isolated disjunct populations in the southern Appalachians (see discussion on page 31). The population in AL is probably an excellent example of long-distance dispersal of spores.

Astrolepis sinuata (Lag. ex Sw.) D.M. Benham & Windham, (Latin: *sinuatus*, curved or wavy, in reference to the wavy-margined pinnae), WAVY SCALY CLOAK FERN, WAVY CLOAK FERN, LONG CLOAK FERN, JIMMY FERN, BULB LIP FERN. Leaves 11–130 cm long (usually smaller); longest pinnae 7–35 mm long, deeply symmetrically lobed with 6- to 14-lobes, the lower surface concealed by scales, the upper surface with scales sparse, deciduous, elongate, stellate. Rocky slopes, outcrops, or cliffs, calcareous or other substrates. Sporulating Mar–Nov. Benham and Windham (1993) treated this species as being composed of two subspecies, both of which occur in TX (a sexual diploid and an apogamous triploid). They distinguished the subspecies based on chromosome number, number of spores per sporangium, spore size, and range. Mickel and Smith (2004) did not recognize the subspecies, merely viewing the different forms as cytotypes. As in the similar case of *A. cochisensis*, we are continuing to follow Benham and Windham (1993) in recognizing the two reproductively distinct cytotypes as subspecies. They separated the taxa as in the following key. However, the two are not readily recognizable morphologically. The symmetrically deeply lobed pinnae readily distinguish this species. Burlage (1968) reported this species as toxic to livestock. Burrows and Tyrl (2001), however, found that *A. sinuata* does not produce adverse effects, and that the toxic species is the related *A. cochisensis* (sometimes in the past treated as *Notholaena sinuata* var. *cochisensis* (Goodd.) Weatherby).

1. Sporangia containing 64 spores; spores averaging 37–44 µm diam.; plants apparently restricted to w TX and se NM ______ subsp. **mexicana**
1. Sporangia containing 32 spores; spores averaging 50–65 µm diam.; plants widespread in sw U.S. and disjunct in GA ______ subsp. **sinuata**

subsp. **mexicana** D.M. Benham, (of Mexico), MEXICAN SCALY CLOAK FERN. 2*n* = 58 (Benham & Windham 1993). Rocky slopes and cliffs on various substrates; Brewster Co. (*V.L. Cory 40288*, TEX-LL, annotated D.M. Behnam), and Jeff Davis Co. (*Lundell & Lundell 14253-A, 13123*, TEX-LL, annotated D.M. Benham); known in TX only from the Chisos Mts., Brewster Co. and the Davis Mts., Jeff Davis Co. (Benham & Windham 1993), though sev-

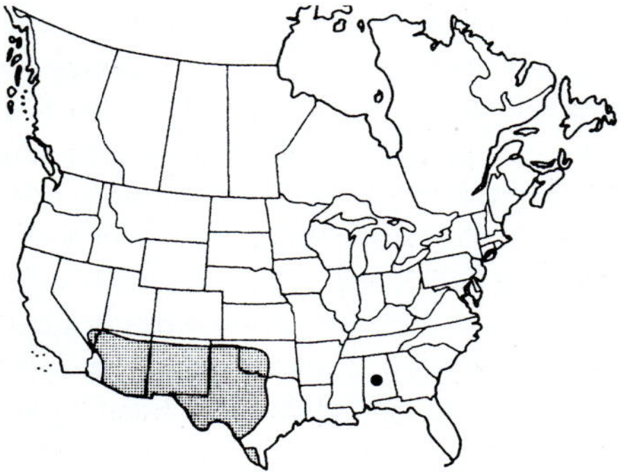

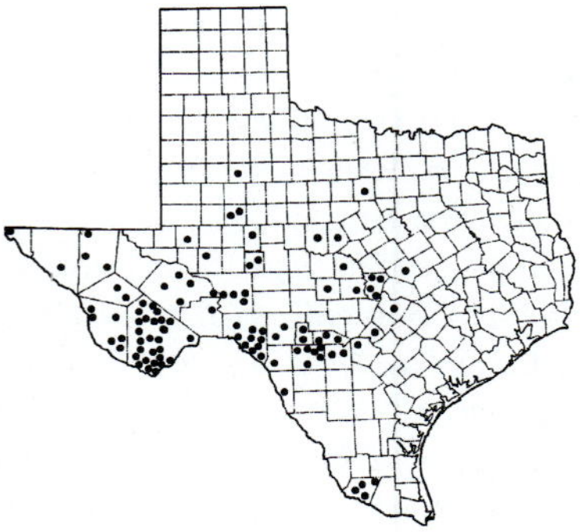

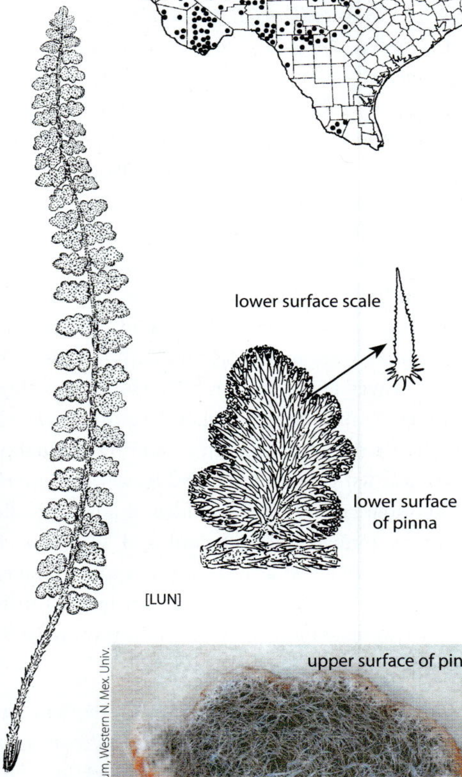

Astrolepis integerrima

HYBRID SCALY CLOAK FERN

subsp. sinuata

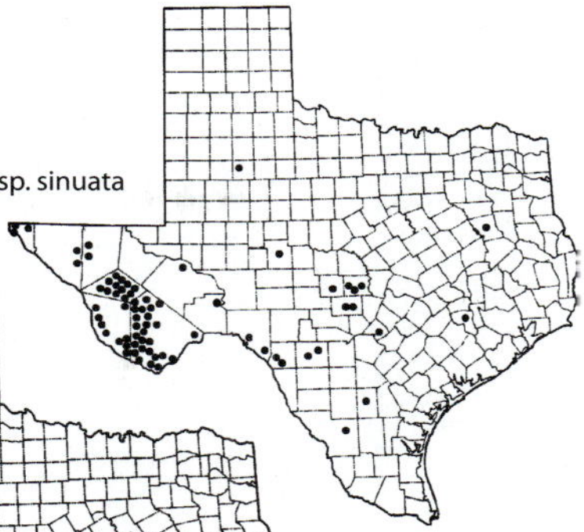

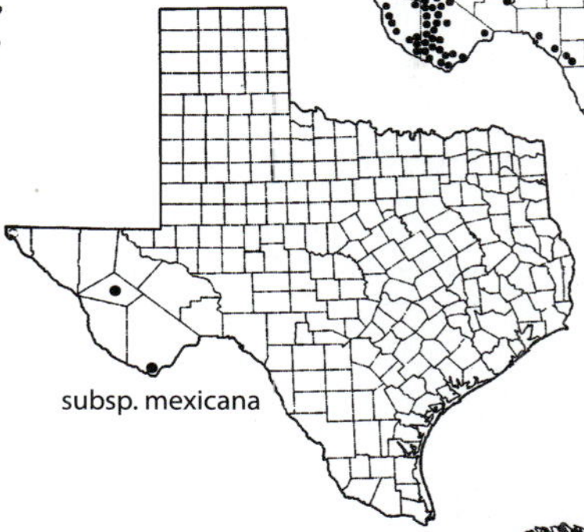

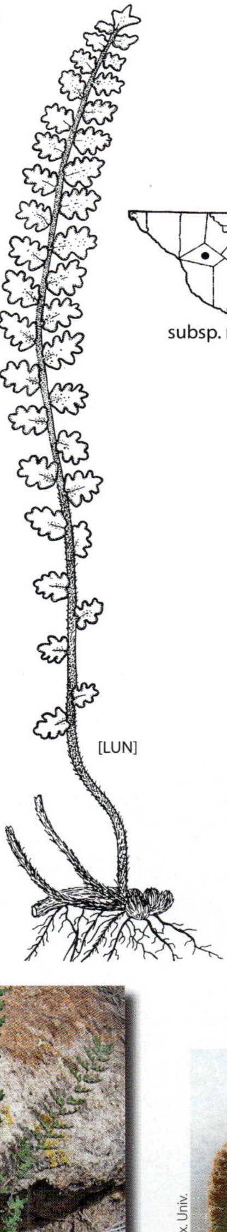

[LUN]

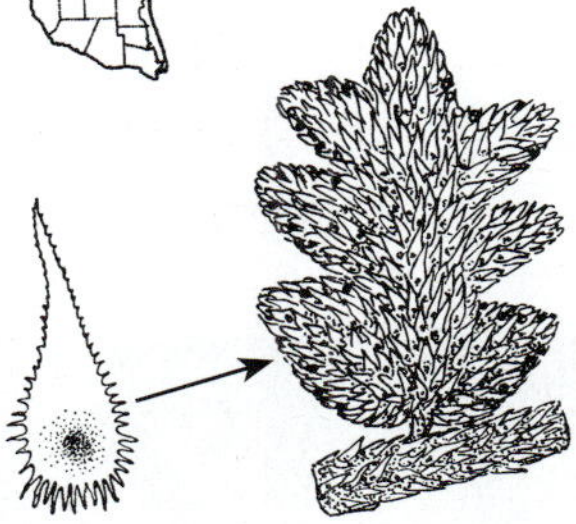

scale

lower surface of pinna

Stan Shebs, cc by-sa

Sharon Yarborough

Dale A. Zimmerman Herbarium, Western N. Mex. Univ.

Dale A. Zimmerman Herbarium, Western N. Mex. Univ.

Astrolepis sinuata

WAVY SCALY CLOAK FERN

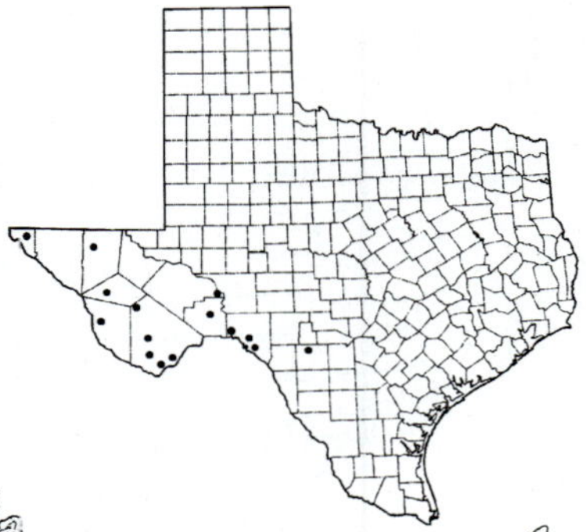

lower surface of pinna
showing scales

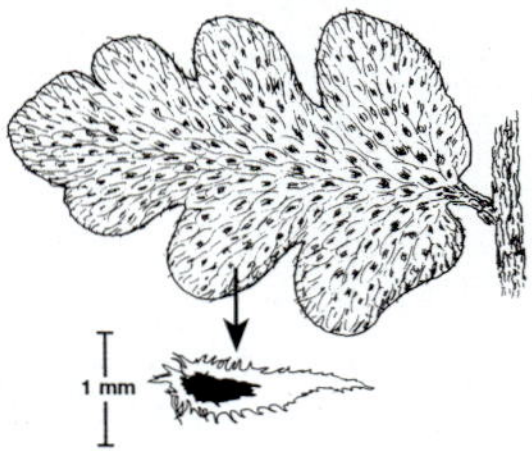

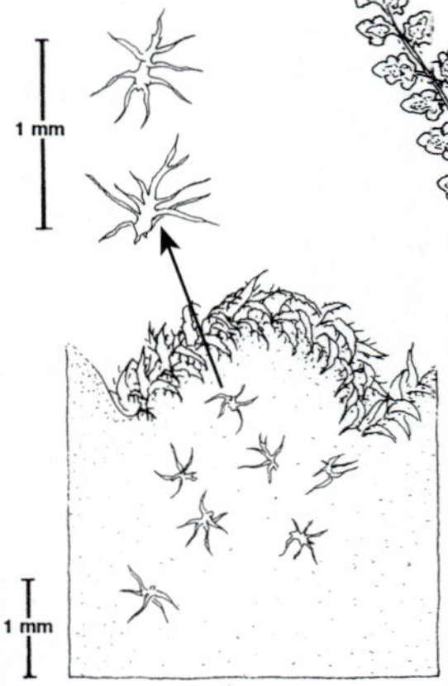

[AFJ, POM]

Dale A. Zimmerman Herbarium, Western N. Mex. Univ.

upper surface of pinna
showing scales

Lorena Babcock Moore

Dale A. Zimmerman Herbarium, Western N. Mex. Univ.

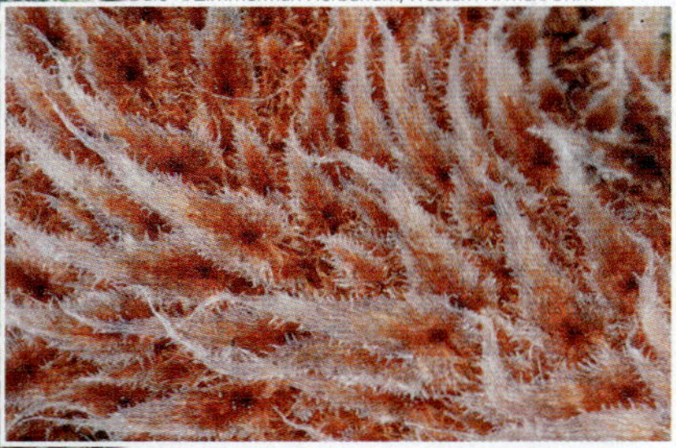

lower surface of pinna
showing scales

Astrolepis windhamii

WINDHAM'S SCALY CLOAK FERN

eral ambiguous specimens from the Sierra Diablo Mts. of Culberson Co. (e.g., *Correll 13757*, TEX-LL) and from Presidio Co. (e.g., *C.L. York 48042*, TEX-LL) are possibly this subspecies; NM and TX; also Mexico and Central America. Sporulating summer–fall. This sexual diploid is apparently quite rare in the U.S. Because of its rareness and limited distribution in the state, we consider this subspecies to be of conservation concern in TX. ⚠

subsp. **sinuata,** WAVY SCALY CLOAK FERN. $n = 2n = 87$ (Benham & Windham 1993). Rocky slopes and cliffs on various substrates; mainly Trans-Pecos and Edwards Plateau but scattered to the s and e in Anderson (*E.N. Plank s.n.*, fide Correll 1956), Austin (*E.H. Boyd s.n.*, fide Correll 1956), Comal, McMullen, and Webb (TEX-LL; Correll 1956) cos.; mostly sw U.S. (AZ, NM, and TX) but disjunct to GA (e.g., Meriwether Co.; Allison & Stevens 2001); also Mexico, West Indies, Central and South America. Sporulating summer–fall. [*Acrostichum sinuatum* Lag. ex Sw., *Cheilanthes sinuata* (Lag. ex Sw.) Domin, *Notholaena sinuata* (Lag. ex Sw.) Kaulf.] This subspecies is reported to be an apogamous triploid derived via autopolyploidy from the diploid subsp. *mexicana* (Benham 1989; Benham & Windham 1993). WAVY SCALY CLOAK FERN is one of a number of ferns found primarily in the sw or w U.S. with isolated disjunct populations in the southern Appalachians (see discussion on page 31). Reports of the species from OK (e.g., USDA PLANTS 2011) are apparently based on nomenclatural confusion involving collections of *Astrolepis integerrima* (previously known as *Notholaena sinuata* var. *integerrima*) that is well known from three localities in OK (Smith 2010).

Astrolepis windhamii D.M. Benham, (for Michael Windham, 1954–, cytogeneticist and pteridologist, formerly Utah Museum of Natural History, currently Curator of Vascular Plants, Duke Univ.), WINDHAM'S SCALY CLOAK FERN, WINDHAM'S CLOAK FERN. Leaves 10–50 cm long; longest pinnae 7–15 mm long, usually symmetrically shallowly lobed with 6–11 broadly rounded lobes, the lower surface concealed by scales, the upper surface with scales sparse, mostly persistent, elongate, stellate; $n = 2n = 87$ (Benham & Windham 1993). Rocky hillsides and cliffs on both limestone and igneous materials; Trans-Pecos (e.g., Brewster Co., *Correll 13482*, BRIT; El Paso Co., *Correll 13832*, BRIT; Val Verde Co., *Correll & Correll 12903*, TEX-LL [w of the Pecos River], annotated D.M. Benham) and extreme w Edwards Plateau (Uvalde Co., *Correll & Correll 12876*, TEX-LL; Val Verde Co., *Z. Labus 37*, TEX-LL, Seminole Canyon State Historical Park [e of the Pecos River], also Turner et al. 2003); AZ, NM, and TX; also n Mexico. Sporulating summer–fall. Mickel and Smith (2004) included *A. windhamii* within *A. integerrima*. However, we are following Benham (1992) and Benham and Windham (1993) in recognizing it as a distinct species. Based on isozyme studies (Benham 1989), they noted that it is an apogamous triploid containing three genomes (from *A. sinuata*, *A. cochisensis*, and an unnamed Mexican taxon related to *A. crassifolia*). In contrast, *A. integerrima*, while also an apogamous triploid, contains only two genomes (from *A. cochisensis* and the unnamed Mexican taxon). Because of their different origins, because they are reproductively distinct lineages, and because they are usually morphologically distinguishable (e.g., lobing of pinnae and scale characters) we believe recognition of *A. windhamii* at the species level is most appropriate. This species can usually be distinguished by its symmetrically shallowly lobed pinnae (vs. pinnae deeply lobed in *A. sinuata* and asymmetrically shallowly lobed or entire in *A. cochisensis* and *A. integerrima*). Benham and Windham (1993) noted that, "Although the features that separate these taxa are subtle, the pinna lobing and scale characteristics of *A. windhamii* ... adequately distinguish them in most cases."

BOMMERIA E. Fourn.

COPPER FERN, BOMMERIA

A New World genus of 5 species ranging from the w U.S. to Costa Rica, typically in dry rocky areas; some are cultivated as ornamentals (Mickel & Smith 2004; Mabberley 2008). Recent molecular research (Windham et al. 2009) has shown that *Bommeria* is one of the most basal lineages among the cheilanthoid ferns. (for Jean Edouard Bommer, 1829–1895, Belgian pteridologist)

REFERENCES: Maxon 1913; Gastony & Haufler 1976; Haufler 1979, 1985, 1993; Haufler & Giannasi 1982; Haufler & Soltis 1984; Ranker 1990; Haufler & Welling 1994; Gastony & Rollo 1995, 1998; Windham et al. 2009.

QUICK TIPS FOR GENUS ID: Dry area ferns with leaves star-shaped (pentagonal/5-sided) in outline, hairy on both upper and lower surfaces and also scaly on the very differently appearing lower surfaces; segment margins not recurved to protect sori.

Bommeria hispida (Mett. ex Kuhn) Underw., (Latin: *hispidus*, hairy, bristly, or rough, in reference to the hairs on the leaves), COPPER FERN, DANCING BOMMERIA, HAIRY BOMMER. Plant terrestrial; rhizomes prostrate, long-creeping, often branched; leaves monomorphic, usually 7–10(–rarely more) mm apart on the rhizome; petioles light grayish brown; leaf blades broadly 5-sided in outline (= pentagonal), small, 7 cm or less long, ca. as wide or very slightly wider, pedately divided into 3 main segments, these deeply pinnate-pinnatifid, dark green but often becoming copper-colored or reddish with age, herbaceous; pinnae with straight needle-like hairs on upper surface, with scales, coiled hairs, and straight needle-like hairs on lower surface (the indumentum very light in color when young, yellowish tan to brownish at maturity), the veins free; rachis chestnut brown; sori along the distal half of the veins; indusia and false indusia absent; $2n = 60$ (Haufler 1993). Rocky soil, rock crevices in mountains and canyons; Brewster (e.g., *Warnock 21559, Correll 13621*, BRIT), El Paso (e.g., *R.D. Worthington 3858*, BRIT), Jeff Davis (e.g., *G.M. Soxman 459*, TEX-LL), Presidio (*Correll 13742*, BRIT, TEX-LL), Culberson, and Hudspeth (Correll 1956) cos. in the Trans-Pecos; AZ, NM, and TX; also Mexico. Sporulating summer–fall. [*B. schaffneri* E. Fourn., *Gymnogramma hispida* Mett. ex Kuhn, *Gymnogramma ehrenbergiana* Klotzsch var. *muralis* Pringle ex Davenp., *Gymnogramma schaffneri* (E. Forn.) Baker, *Gymnopteris hispida* (Mett. ex Kuhn) Underw., *Neurogramma hispida* (Mett. ex Kuhn) Diels] This was one of the first homosporous ferns shown to be obligately outcrossing—in other words, the eggs and sperm from a single gametophyte do not produce a sporophyte. Rather, sporophytes of *B. hispida* in nature are typically the result of outcrossing between genetically different gametophytes (Haufler & Soltis 1984)—this type of sexual reproduction is important because it ensures genetic variation among the offspring. *Bommeria hispida* is reportedly the only member of the genus known to occur n of Mexico (Haufler 1993) and the only one known to be in areas with subfreezing temperatures (Haufler 1979). Haufler (1979) noted that it can withstand more environmental stress than other members of the genus and that its "small leaf size, dissected segments, double palisade layer, and extensive and diverse indument" are adaptations to a high light-low humidity habitat. There are only two members of the family with pentagonal leaf blades in the Trans-Pecos, this species and *Notholaena standleyi*. *Bommeria hispida* has pinnae with scales, coiled hairs, and straight needle-like hairs beneath, while those of *N. standleyi* "have copious yellow or greenish farina but no hairs or scales" (Yarborough & Powell 2002). Also, when dry the leaves of *Bommeria hispida* curl so that the upper leaf surface is exposed; in contrast, leaf curling in *N. standleyi* causes the farina-covered lower surface to be exposed.

CERATOPTERIS Brongn.

ANTLER FERN, WATER FERN, FLOATING FERN

A taxonomically difficult genus of morphologically variable aquatics or semiaquatics widespread in the tropics to warm temperate areas. It is considered to have 3 or 4 species (Lloyd 1993) or perhaps 6 or 7 if cryptic species are included (Masuyama et al. 2002). Some are grown as aquarium plants (Hoshizaki & Moran 2001). The genus is said to have the fastest life cycle of any fern—as little as one month from spore to spore-bearing plant. Because of this, it is used in laboratory genetics studies (Hoshizaki & Moran 2001). *Ceratopteris* has often been treated in the Parkeriaceae, the ANTLER FERN or WATER FERN family (e.g., Correll & Johnston 1970; Lloyd 1993). According to Lellinger (1985), "The Parkeriaceae clearly belongs to the Schizaeales [including Anemiaceae and Lygodiaceae], but its exact position is uncertain. As in some cases of flowering plants, the aquatic habitat has caused great modifications of structure in *Ceratopteris* that obscure the relationships of these plants to the other, non-aquatic families in the order." However, recent molecular studies (Gastony & Rollo 1998), which did not rely on highly modified morphological characters, indicated that the sister group of *Ceratopteris* is *Pityrogramma*, a genus traditionally placed in the Pteridaceae. More recently Smith et al. (2006) noted that it "nests within Pteridaceae in all molecular analyses." We are thus following Tryon (1987), Tryon et al. (1990), Gastony and Rollo (1998), Smith et al. (2006, 2008), and Robbin Moran (pers. comm.) in treating *Ceratopteris* in the Pteridaceae; it is possibly best placed in a separate subfamily, as originally suggested by Tryon (1987). *Ceratopteris* is one of only two genera of homosporous (= with 1 spore type) aquatic ferns; the related genus, *Acrostichum*, is the other; heterosporous aquatic ferns include the Azollaceae, Marsileaceae, and Salviniaceae. The following treatment draws heavily on Lloyd (1993). (Greek: *cerato*, horned, and *pteris*, fern, derived from *pteron*, wing or feather, in reference to the antler-like fertile leaf)
REFERENCES: Benedict 1909; Morton 1967a; Hannan 1969; Lloyd 1974, 1993; Petrik-Ott & Ott 1976; Tryon

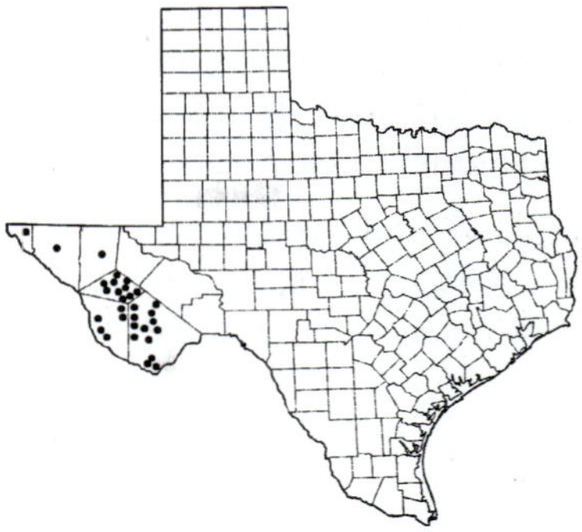

Dale A. Zimmerman Herbarium, Western N. Mex. Univ.

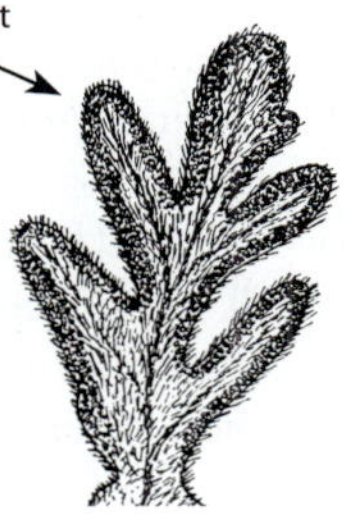

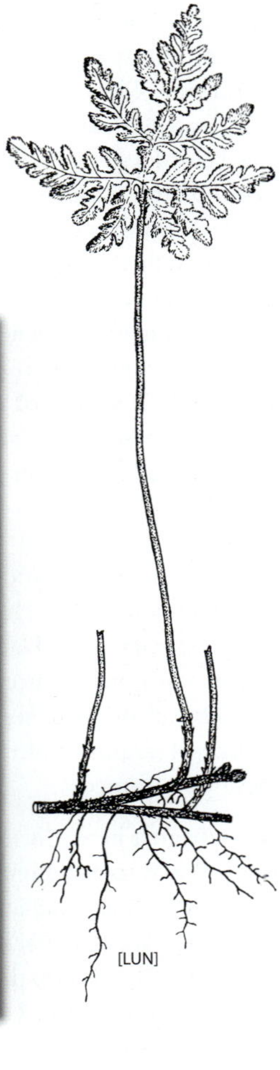

[LUN]

Alan Cressler

Bommeria hispida

COPPER FERN

1987; Watano & Masuyama 1991; Lemke 1994; Gastony & Rollo 1998; Carlquist & Schneider 2000b; Masuyama et al. 2002; Masuyama & Watano 2005; Nakazato et al. 2006; Adjie et al. 2007; Masuyama 2008. QUICK TIPS FOR GENUS ID: Introduced aquatic or semiaquatic fern with buds or small plantlets along margins of 1–3 pinnate leaves.

Ceratopteris thalictroides (L.) Brongn., (resembling *Thalictrum*, meadow-rue, in the Ranunculaceae, and Greek: *-eides*, -oides, like), WATER SPRITE, WATER FERN. Plant short-lived, aquatic or semiaquatic, usually rooting, but sometimes floating; leaves dimorphic, with adventitious buds or small plantlets in notches along margins; sterile leaves 1–3 pinnate, the blades lanceolate to ovate or deltate in outline, 2–41 cm long, 2–20 cm wide, with relatively broad ultimate segments (broader than the linear ultimate segments of the fertile leaf blades), firm, the petioles 1–31 cm long, not inflated; fertile leaves 3–4 pinnate proximally, 2-pinnate distally, larger than the sterile leaves, 2–117 cm long, to 48 cm wide, the blades lanceolate to ovate, deltate, or cordate in outline, finely dissected with the ultimate segments linear, the petioles 1–46 cm long; margins of fertile blades revolute, covering the 1–3 rows of sporangia and forming false indusia; $2n = 154$, 156 (tetraploid) (Masuyama & Watano 2005). Cultivated in aquaria and warm-weather lily or fish ponds and presumably escaping; usually rooted in mud in quiet or moving water of springs and adjacent rivers and lakes; Comal (*P.A. Amerson 2973, 2080*, BRIT, *Petrik-Ott & Ott 1003*, TAES), Hays (*P.A. Amerson 2087*, BRIT; *Correll & Correll 38315*, TEX-LL; *S.A. Conry 41*, BAYLU), and Bell (*Petrik-Ott & Ott 1004*, TAES, US—Petrik-Ott & Ott 1976) cos. along the Balcones Fault at the edge of the Blackland Prairie and the Edwards Plateau. According to D. Lemke (pers. comm. 1998), at that time it was still present in the San Marcos River in Hays Co. Petrik-Ott and Ott (1976) suggested that the constant temperature of the springs along the Balcones Fault may allow this mainly tropical species to persist indefinitely in TX. The species was first reported from TX by Morton (1967a) and was apparently introduced to the state in 1963 (Hannen 1969). Sporulating Mar–Oct. Native to tropical areas worldwide, except Africa, it has been found escaped in the U.S. in CA, FL, LA, TX (Lloyd 1993), and MS (Kartesz 1999); also s Mexico. [*Acrostichum thalictroides* L.] This is an extremely widespread and variable species (Lloyd 1974). Based on molecular work and breeding studies, Masuyama et al. (2002), Masuyama and Watano (2005), Adjie et al. (2007), and Masuyama (2008) found there are at least three morphologically very similar cryptic species hidden within *C. thalictroides* ("north type," "south type," "third type"). Apparently there is reticulate evolution with the three cryptic species having independent allopolyploid origins, with contributions variously from the Old World diploid *C. cornuta* and two hypothetical diploid ancestors. New World populations seem to be of the "south type" (Masuyama et al. 2002). *Ceratopteris thalictroides* is cultivated in some areas (e.g., Japan) as a spring vegetable in flooded rice fields (Mabberley 1997). Tryon and Tryon (1982) suggested that although *Ceratopteris* is usually considered to be an annual, it "... may be more accurately described as short-lived." They noted that many tropical habitats "... lack the seasonality necessary to provide a basis for an annual lifecycle."

CHEILANTHES Sw. LIP FERN

Plants mostly xerophytic, usually growing on rocks or sometimes in soil; rhizomes compact to long-creeping; leaves monomorphic, clustered or scattered along the rhizomes; petioles dark brown to black (straw-colored in one species); leaf blades 2–more-pinnate-pinnatifid (1-pinnate-pinnatified in one species), usually with conspicuous pubescence or apparently so on lower (= abaxial) surface but sometimes glabrous or nearly so, the veins of ultimate segments free or rarely anastomosing, obscure; sporangia marginal on the lower leaf surfaces; margins of ultimate leaf segments (= smallest subdivisions of leaf) recurved, forming false indusia, these well-formed.

As treated here, *Cheilanthes* is a genus of ca. 150 species (Windham & Rabe 1993) of primarily dry area ferns. When thus circumscribed it is subcosmopolitan with many New World species, but it has well-developed evolutionary radiations on six continents (G. Yatskievych, pers. comm.). According to Windham and Rabe (1993), *Cheilanthes* (viewed broadly) is the largest and most diverse genus of xerophytic ferns. With 17 species in TX, it is also the most diverse genus of pteridophytes in the state. However, as Mickel (1979) noted, ever since it was named, "botanists have had great difficulty in delimiting the genus." Even after the removal of segregates including *Argyrochosma* and *Astrolepis*, it is still a heterogeneous and apparently polyphyletic genus. In fact, recent molecular evidence (Gastony & Rollo 1998; Kirkpatrick 2007; Prado et al. 2007; Rothfels

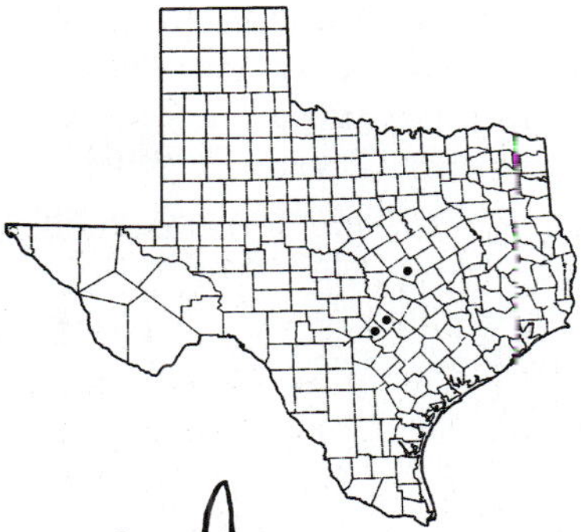

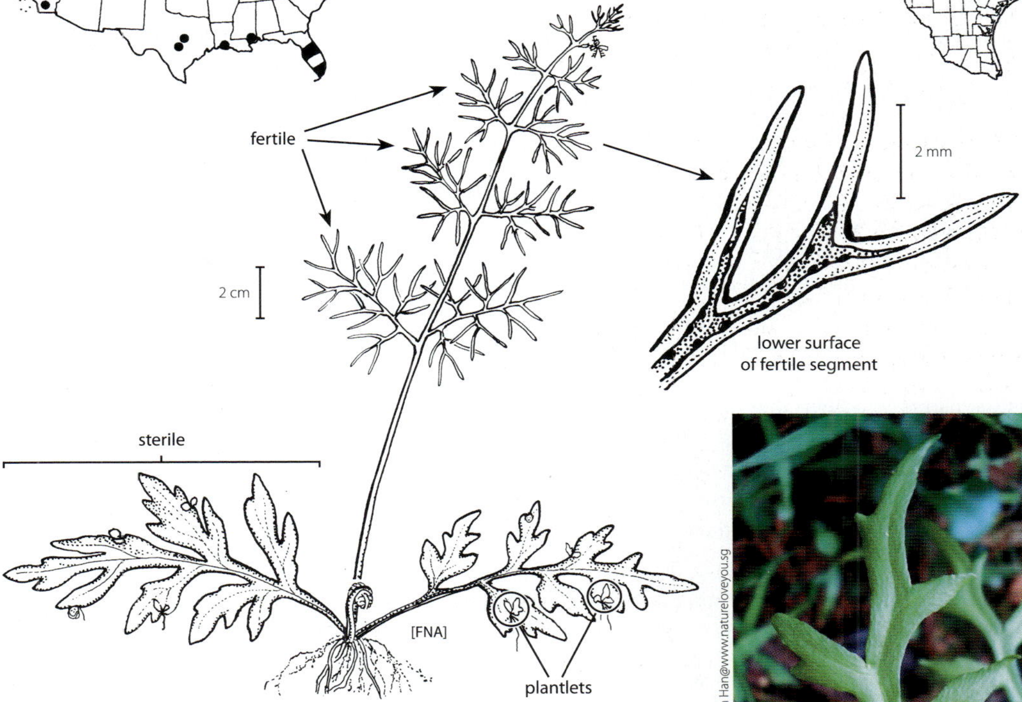

Kwan Han@www.natureloveyou.sg

Kwan Han@www.natureloveyou.sg

Ceratopteris thalictroides

WATER SPRITE

et al. 2008; Windham et al. 2009) supports the hypothesis that it is polyphyletic. For example, a group of "American species" including TX species such as *C. alabamensis* and *C. tomentosa* have been shown to be more closely related to *Argyrochosma*, *Astrolepis*, and *Pellaea* than to other species of *Cheilanthes* (Kirkpatrick 2007), and might warrant recognition as a separate genus. Generic limits within the cheilanthoid ferns (a group of ca. 400 dry-adapted species including such genera as *Argyrochosma*, *Astrolepis*, *Bommeria*, *Cheilanthes*, *Notholaena*, and *Pellaea*) is in flux and the generic delimitation of *Cheilanthes* will almost certainly be changed in the future. Because the situation is not yet settled, we are continuing to recognize *Cheilanthes* in the manner of Windham and Rabe (1993). Grusz et al. (2009) have recently shown that there are very complicated reticulate relationships between the species of one subgroup, the "*C. yavapensis* complex"—see that species for discussion. The scientific and common names for the genus both refer to the position of the sori on the margin or "lip" of the ultimate leaf segments. Because hair-like scales as narrow as 0.1 mm wide can occur, and because the scales are sometimes ciliate, magnification is often necessary for accurate identification. ☠ An Australasian species, *C. seiberi*, is well known as a cause of poisoning in livestock, causing thiamine deficiency, hemorrhaging, hematuria, neoplasia, and death. The compounds thought to be responsible, the enzyme thiaminase and a mutagenic and carcinogenic glycoside known as ptaquiloside, are the same as found in *Pteridium aquilinum*, BRACKEN FERN (Burrows & Tyrl 2001); North American species of *Cheilanthes* are not known to be poisonous. The convex, bead-like shape of the tiny segments of some *Cheilanthes* leaves is probably an adaptation to minimize water loss. This shape hinders the movement of desiccating air across the lower leaf surface and has also been suggested to shade the underside of the leaf (Moore 2008). The key to species is adapted from Windham and Rabe 1993. (Greek: *cheilos*, margin or lip, and *anthus*, flower, referring to the sporangia located near the lip or margin of the leaf and protected by the recurved leaf margin; early botanists apparently thought that the sori were flowers or their equivalent—Thieret 1980)

REFERENCES: Mickel 1979; Gastony & Windham 1989; Windham 1993e; Windham & Rabe 1993; Prado et al. 2007; Schuettpelz et al. 2008; Grusz et al. 2009; Windham et al. 2009; Crow et al. 2011; Holmes et al. 2011.

QUICK TIPS FOR GENUS ID: Dry area ferns with usually obviously hairy and also often scaly, "fern-like" leaves (2 or more pinnate-pinnatifid), the ultimate leaf segments not completely separate from each other and sometimes bead-like.

1. Costae (= midveins of pinnae) and/or rachis (= midvein of leaf) with scales on lower surface (= abaxially) (NOTE: hairs can also be present); young unexpanded leaves hooked but not coiled at tips (vernation not circinate).
    2. Ultimate leaf segments (= smallest subdivisions of leaf) scabrous (= rough to the touch) on upper surface, covered with stiff hairs ____ **C. horridula**
    2. Ultimate leaf segments smooth to the touch on upper surface, lacking stiff hairs (various other hairs can be present).
        3. Scales of costae linear, inconspicuous, only slightly wider than hairs, the largest 0.1–0.4 mm wide ____ **C. tomentosa**
        3. Scales of costae linear to lanceolate to ovate, conspicuous, obviously much wider than hairs, the largest 0.4–1.0 mm wide.
            4. Margins of scales of costae entire to finely toothed, not ciliate (except *C. eatonii* rarely with 1–2 cilia at base) (this character best seen with dissecting scope); rhizomes short and stout with leaves in a dense clump OR long-creeping with leaves closely scattered along rhizomes.
                5. Ultimate leaf segments with upper surfaces completely glabrous; rhizomes long-creeping, slender, 3 mm or less in diam., with leaves closely scattered along the rhizomes; scales of rhizomes mostly uniformly brown in color; species known in TX only from the Trans-Pecos ____ **C. fendleri**
                5. Ultimate leaf segments with upper surfaces having sparse to dense pubescence; rhizomes short, stout, usually 5–10 mm in diam., with leaves in a dense clump; scales of rhizomes with a dark central stripe and light brown margins; including species occurring widely across much of TX.
                    6. Scales of costae linear-lanceolate to lanceolate, loosely overlapping and not concealing the lower surface of ultimate segments; ultimate leaf segments with a dense coating of fine hairs on lower surface; species occurring widely across much of TX ____ **C. eatonii**
                    6. Scales of costae ovate-lanceolate to deltate, strongly overlapping, spreading and nearly completely obscuring the lower surface of ultimate segments; ultimate leaf segments with relatively few coarse hairs on lower surface; species known in TX only from the Trans-Pecos ____ **C. villosa**

4. Margins of scales of costae ciliate with very elongate cilia, particularly near base; rhizomes long-creeping with leaves closely scattered along rhizomes.
    7. Upper surface of pinnae appearing densely tomentose due to the scales of costae having elongate marginal cilia fine, curly, becoming tangled and wrapping around to cover upper surface; largest ultimate leaf segments tiny, 1 mm or less wide; species widespread in w ⅔ of TX ____ **C. lindheimeri**
    7. Upper surface of pinnae appearing glabrous or only slightly pubescent, the scales of costae with marginal cilia coarser, not much tangled nor wrapping to cover upper surface (but still elongate and obvious); largest ultimate leaf segments 1–3 mm wide; species known in TX only from the Trans-Pecos.
        8. Upper surface of pinnae glabrous; scales of costae often with cilia only in basal ½ (but sometimes along entire margin); rhizome scales usually uniformly brown in color, loosely appressed and deciduous on older parts of rhizomes; spores averaging < 64 µm diam. (plants triploid) ____ **C. wootonii**
        8. Upper surface of pinnae *appearing* slightly pubescent (due to cilia of scales of costae slightly wrapping around to upper surface, but actually glabrous); scales of costae with cilia along nearly entire margin; rhizome scales often with dark central stripe and slightly lighter margins, usually tightly appressed and persistent; spores averaging > 64 µm diam. (plants tetraploid) ____ **C. yavapensis**

1. Costae and rachis lacking scales on lower surface or with extremely narrow, inconspicuous, hair-like scales (but can be strikingly pubescent to glabrous); young unexpanded leaves tightly coiled with appearance of a fiddlehead (vernation circinate) OR not so.
    9. Rachis and pinnae glabrous or essentially so; petioles grooved above for majority of length ____ **C. wrightii**
    9. Rachis and often pinnae with glandular or non-glandular hairs; petioles variously shaped but not grooved above in basal half.
        10. Leaf blades essentially glabrous to sparsely and inconspicuously pubescent on lower surface AND not at all scabrous on upper surface; ultimate leaf segments narrowly elliptic to elongate-deltate, not at all suborbicular to bead-like; rachises with different types of pubescence on upper (= adaxial) and lower surfaces, the hairs on upper surface dense, short, twisted, and appressed, the hairs on lower surface sparse, long, and spreading.
            11. Leaf blades lanceolate to oblong, 1–7 cm wide; basal pair of pinnae slightly smaller than adjacent pair; basal pair of pinnules of basal pinnae ± equal in size ____ **C. alabamensis**
            11. Leaf blades broadly triangular to ovate, 5–15 cm wide; basal pair of pinnae slightly larger than adjacent pair; basal pair of pinnules of basal pinnae conspicuously unequal in size ____ **C. aemula**
        10. Leaf blades EITHER with conspicuous pubescence on lower surface OR scabrous (= rough to the touch) on upper surface; ultimate leaf segments suborbicular to bead-like OR not so; rachises with hairs essentially the same on upper and lower surfaces.
            12. Ultimate leaf segments nearly glabrous above, somewhat rounded, bead-like, with coarse non-glandular hairs beneath; false indusia well-formed, 2–3 mm wide, not limited to the segment margin, making a pouch with a somewhat constricted opening, much of the lower surface of the ultimate segments covered by the false indusia; species known in TX only from the Trans-Pecos ____ **C. lendigera**
            12. Ultimate leaf segments with non-glandular or glandular hairs above, bead-like OR not so, with pubescence beneath various; false indusia marginal, not well-formed or absent, neither making a pouch nor covering much of the lower surface of the ultimate segments.
                13. Ultimate leaf segments scabrous on upper surface, covered with stiff hairs; including species widespread in TX ____ **C. horridula**
                13. Ultimate leaf segments smooth to the touch on upper surface, lacking stiff hairs (but can be variously pubescent).
                    14. Leaf blades entirely 1-pinnate-pinnatifid (very different in appearance from all other TX *Cheilanthes*); pinnae articulate (dark color of stalk stopping at base of pinna) ____ **C. bonariensis**
                    14. Leaf blades 2–4 pinnate or 2-pinnate-pinnatifid at least basally; pinnae not articulate (dark color of stalk continuing into the pinna on its lower surface).
                        15. Leaf blades usually 5–10 cm wide, elongate-pentagonal (5-sided) in shape due to the basal pinnae being much larger than the adjacent pair; proximal basiscopic (= on side facing base of leaf) pinnules much larger than acroscopic pinnules.
                            16. Ultimate leaf segments resinous/oily due to exudate from short gland-tipped hairs on both upper and lower surfaces; petioles dark brown; species of eastern Edwards Plateau and Trans-Pecos ____ **C. kaulfussii**
                            16. Ultimate leaf segments not resinous/oily, the hairs long and non-glandular on both upper and lower surfaces; petioles straw-colored; species of far w Edwards Plateau ____ **C. leucopoda**

15. Leaf blades usually 5(–8) cm or less wide, linear to ovate-lanceolate in shape, the basal pinnae ± the same size as adjacent pair; proximal basiscopic pinnules ± the same size as the acroscopic.
    17. Ultimate fertile leaf segments elongate, not bead-like, the upper surfaces sparsely hirsute with long, jointed (4–5 segments) hairs; leaf blades 2-pinnate-pinnatifid near base; species known in TX only from historical collection from McLennan Co. on the w margin of the Blackland Prairie ____ **C. lanosa**
    17. Ultimate fertile leaf segments rounded, bead-like, the upper surfaces with dense pubescence, the hairs various; leaf blades 3- or 4-pinnate near base; species widespread in TX.
        18. Petiole and rachis densely tomentose, particularly when young, the hairs loosely appressed and not noticeably jointed (single-celled); leaf blades usually 4-pinnate at base, 1.5–8 cm wide; leaves to 45 cm long ____ **C. tomentosa**
        18. Petiole and rachis not densely tomentose, instead very sparsely to densely hispidulose, the hairs spreading and noticeably jointed (multicellular, can be seen under strong hand lens or dissecting scope); leaf blades usually 3-pinnate at base, 1–3 cm wide; leaves 20 cm or less long ____ **C. feei**

Cheilanthes aemula Maxon, (Latin: *aemula*, rival, emulating or imitating, probably because at the type locality it was growing with, and was initially confused with *C. microphylla*, a species widespread in Latin America), TEXAS LIP FERN, RIVAL LIP FERN. Rhizomes short-creeping; leaves clustered, 10–50 cm long; petioles dark brown to black, rounded on upper surface; leaf blades broadly triangular to ovate, 5–15 cm wide, 3-pinnate to 3-pinnate-pinnatifid basally, the pinnae not articulate, the largest ultimate segments 3–6 mm long, not bead-like, with sparse hairs or glabrous on lower (unusual in that most other species except *C. alabamensis* and *C. fendleri* are hairy) and upper surfaces; costae without scales on lower surfaces; rachises similar to those in *C. alabamensis*; $2n = 58$ (Windham & Rabe 1993). Rocky slopes and ledges, on limestone; Austin Co. in se Blackland Prairie (Correll 1956; Turner et al. 2003), Val Verde (e of the Pecos—Turner et al. 2003), BRIT), Comal, and Uvalde (TEX-LL) cos. on the Edwards Plateau, Val Verde (*Correll & Correll 12910*, BRIT, TEX-LL, *D. Flyr 739*, TEX-LL, both w of Pecos River), Big Bend State Park near border of Brewster and Presidio (Yarborough & Powell 2002) cos. in the Trans-Pecos, and Starr Co. (*R. Runyon 2914*, TEX-LL) in far s TX; in the U.S. known only from TX; also n Mexico. Sporulating May–Nov. This species, which is similar to *C. alabamensis*, is known in the U.S. from only a few localities in e, c, and w TX (Windham & Rabe 1993). Because of its rareness and limited distribution in the state, we consider this species to be of conservation concern in TX. ⚠

Cheilanthes alabamensis (Buckley) Kunze, (of Alabama, where the plant was originally collected in 1843—Nelson 2000), ALABAMA LIP FERN, SMOOTH LIP FERN. Rhizomes short-creeping to compact; leaves clustered, 6–50 cm long; petioles black, rounded on upper surface; leaf blades lanceolate to oblong, 1–7 cm wide, 2-pinnate to 2-pinnate-pinnatifid at base, the pinnae not articulate, the largest ultimate segments (= the smallest subdivisions of the leaf) 3–7 mm long, not bead-like, with sparse hairs or glabrous on lower (unusual in that most other species except *C. aemula* and *C. fendleri* are hairy) and upper surfaces; costae without scales on lower surfaces; rachises rounded on upper surfaces, with different types of pubescence on upper and lower surfaces, the hairs on upper dense, short, twisted, and appressed, the hairs on lower sparse, long, and spreading; $n = 2n = 87$ (triploids; Windham & Rabe 1993). Limestone hillsides, crevices of limestone ledges and cliffs, but sometimes on igneous substrates; widespread across TX, but most common in Edwards Plateau (e.g., Kerr Co., *Correll & Correll 12812*, BRIT, TEX-LL); mostly s ½ of U.S. from VA s to FL w to TX, NM, and AZ; also Mexico. Sporulating nearly throughout the year, especially Mar–Nov. [*Pteris alabamensis* Buckley, *Pellaea alabamensis* (Buckley) Hook.] Populations in the U.S. are apogamous triploids apparently derived via autopolyploidy; sexual diploids have, however, been found in Mexico (Windham & Rabe 1993; Windham & Yatskievych 2003b).

Cheilanthes bonariensis (Willd.) Proctor, (according to the type description "Habitat in Bonaria," apparently in reference to the Caribbean island of Bonaire, located just north of Venezuela), BONAIRE LIP FERN, GOLDEN LIP FERN. Rhizomes short-creeping to compact; leaves clustered, 10–60 cm long; petioles dark brown, rounded on upper surface; leaf blades linear, 1–4 cm wide, entirely 1-pinnate-pinnatifid (unique among TX *Cheilanthes*),

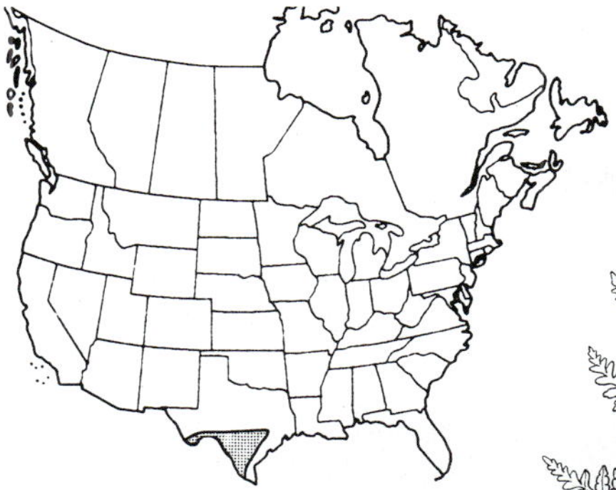

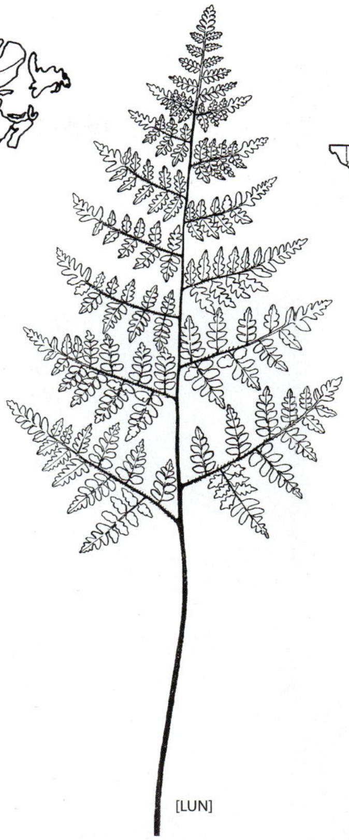

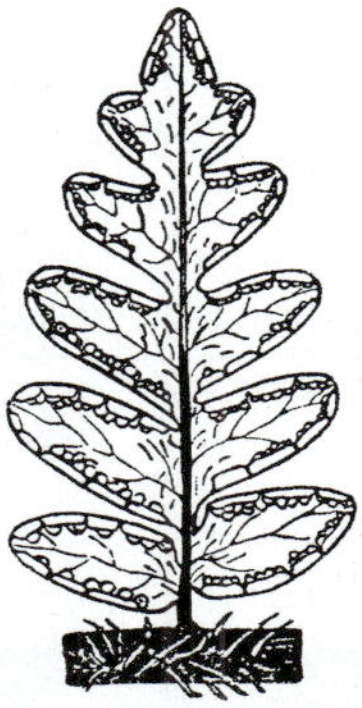

lower surface
of pinnule

Cheilanthes aemula

TEXAS LIP FERN

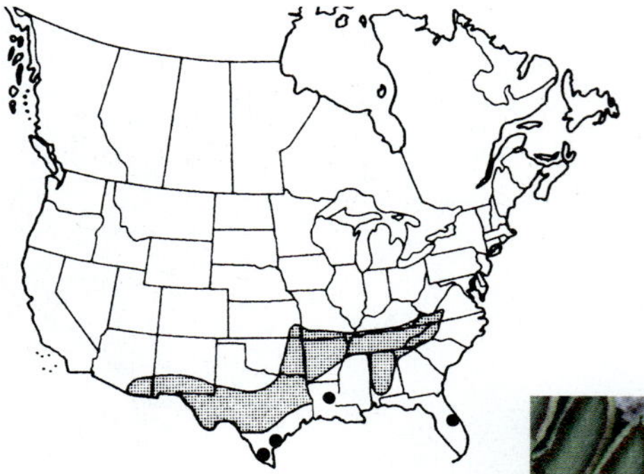

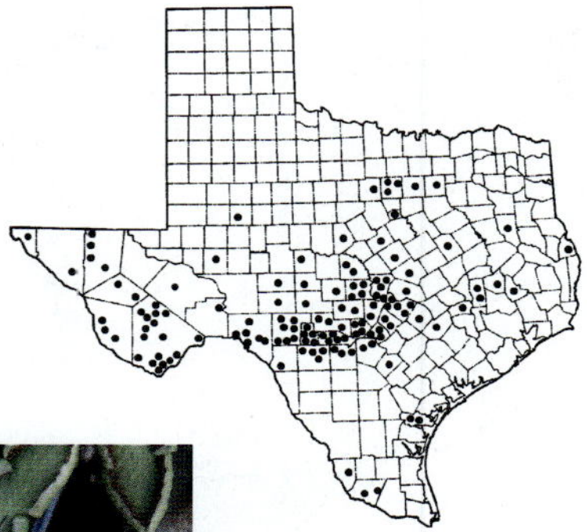

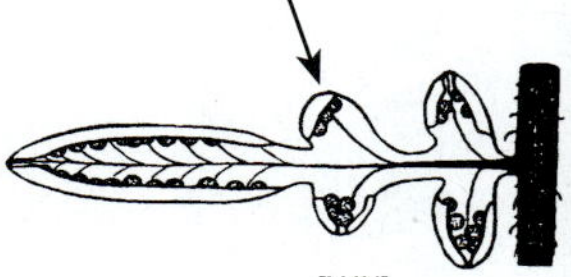

[LUN]

Roy Morey

Alan Cressler

[LUN]

Cheilanthes alabamensis

ALABAMA LIP FERN

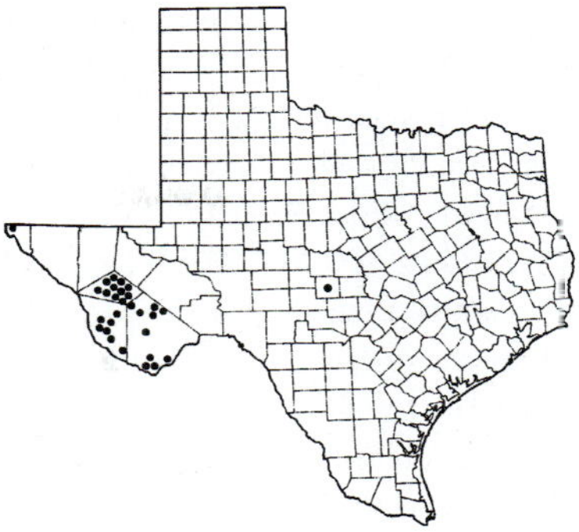

upper surface of pinna

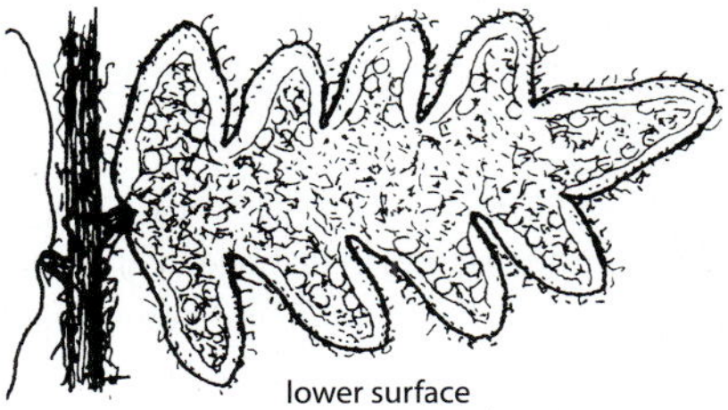

lower surface of pinna

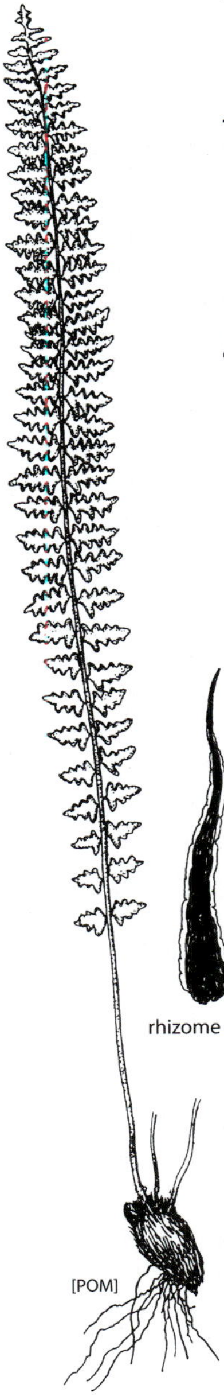

rhizome scale

[POM]

Lorena Babcock Moore

Cheilanthes bonariensis

BONAIRE LIP FERN

the pinnae articulate (dark color of stalk stopping at base of pinnae), the ultimate segments (pinnae lobes) 3–8 pairs per pinna, 1–7 mm long, not bead-like, usually densely white to rusty tomentose on lower surfaces (obscuring surface from view), hirsute above; costae absent; rachises with dense pubescence but without scales; *n* = 2*n* = 90 (Windham & Rabe 1993). Rocky slopes, bluffs, and ledges; primarily in the Trans-Pecos in Brewster (*C.M. Rowell & J.L. Blassingame 5075*, BRIT), Jeff Davis (*J.W. Stanford & J.L. Blassingame 1200*, BRIT), Presidio (BRIT, TEX-LL), and El Paso (e.g., *E.J. Palmer 30608*, TEX-LL) cos., but one collection from the Edwards Plateau (Mason Co., *T. Starbuck 336*, BRIT); AZ, NM, and TX; also Mexico, West Indies, Central and South America. Sporulating summer–fall. [*Acrostichum bonariense* Willd., *Notholaena aurea* (Poir.) Desv., *Notholaena ferruginea* (Willd. ex Link) Hook.] This species, with its 1-pinnate-pinnatifid leaf blades, superficially resembles *Astrolepis sinuata*, but differs in having ± long, matted, white to rusty hairs on the lower leaf surfaces instead of scales (Mickel & Smith 2004). While it was often included in *Notholaena* in the past (e.g., as *Notholaena aurea* by Correll 1956), recent molecular research (Rothfels et al. 2008) clearly shows that it is more closely related to species of *Cheilanthes*. This species is reported to be the "most widespread and abundant pteridophyte species in Mexico" (Mickel & Smith 2004). It is used medicinally in Mexico for stomach aches and coughs (Yarborough & Powell 2002). The species is apparently an apogamous triploid that arose through autopolyploidy (Windham & Rabe 1993).

Cheilanthes eatonii Baker, (for its discoverer, A.A. Eaton, 1865–1908), EATON'S LIP FERN. Rhizomes compact; leaves clustered, 6–35 cm long; petioles dark brown, rounded on upper surface; leaf blades oblong lanceolate, 1.5–5 cm wide, 3–4-pinnate basally, the pinnae not articulate, the ultimate segments oval to round, bead-like, the largest 1–3 mm long, densely hairy to nearly glabrous above (the hairs usually whitish or grayish, but sometimes reddish), densely tomentose on lower surfaces; costae and/or rachis with conspicuous scales on lower surface, but these not concealing the lower surface of ultimate segments; *n* = 2*n* = 90, 120, apogamous (Windham & Rabe 1993). Rocky slopes and ledges; mainly Edwards Plateau and Trans-Pecos with outliers in Morris (*Correll & Correll 12480*, BRIT, TEX-LL) and Cherokee (*Correll 35220*, TEX-LL) cos. in the ne part of East TX and Wilson Co. (Turner et al. 2003) in the sw corner of East TX; also Brown Co. (HPC; Correll 1956) in the sw part of the Cross Timbers and Prairies and Mitchell Co. (Correll 1949, 1956) in the Rolling Plains; mostly sw U.S. (AR, AZ, CO, NM, OK, TX, UT), disjunct from AR (Baxter and Benton cos.—Taylor 1984; where now locally extinct, J. Peck, pers. comm.; Peck 2011b) to isolated stations in VA and WV (Knobloch & Lellinger 1969; a disjunction of ca. 650 mi/1046 km); also Mexico and Costa Rica. Sporulating Mar–Nov. [*C. castanea* Maxon, *C. eatonii* forma *castanea* (Maxon) Correll, *C. tomentosa* var. *eatonii* (Baker) Davenp.] The disjunct occurrences in ne TX are the result of isolated appropriate rocky microhabitats in an area where such environments are quite rare. For example, a Cherokee Co. collection (*D.S. Correll, 35220*, TEX-LL) was from "crevices of iron-ore rock," while the Morris Co. collections (*D.S. Correll & H.B. Correll 12438*, BRIT, TEX-LL; *Correll & Correll 22480*, BRIT) were made on "Slope of mountain, on iron ore rocks" and "On rocks on mountain." This is one of a number of ferns found primarily in the sw or w U.S. with isolated disjunct populations in the southern Appalachians (see discussion on page 31). It is a highly variable species composed of both apogamous triploids and tetraploids "of unknown parentage" (Windham & Rabe 1993). We are following Correll (1956) and Windham and Rabe (1993) in treating *C. castanea* as a synonym of *C. eatonii*. Correll (1956) noted that when extreme forms of typical *C. eatonii* and what he considered to be forma *castanea* are found "they are so distinctive that one would immediately consider them to be specifically different. Unfortunately, very little material of extreme f. *castanea* exists in herbaria. An overwhelming amount of material does exist, however, which grades into one or the other of these two forms. It is simply an arbitrary matter as to which category they should be relegated to." Correll (1956) "placed all plants which have all or part of their fronds tending to be glabrescent [when mature] on the upper surface of their segments into f. *castanea*" and treated "those plants with segments hoary and densely tomentose above as the typical form." Windham and Rabe (1993) noted, "Because there is no clear morphologic break, *C. castanea* is placed here in synonymy under *C. eatonii* pending further study." The species is reported to hybridize with *C. villosa* (Lellinger 1985; Mickel & Smith 2004); individuals that appear to be morphological intermediates between the two are known from w TX and NM (Windham & Rabe 1993); see discussion under *C. villosa*. It is also closely related to and similar to *C. tomentosa*, but with scales of costae linear-lanceolate to lanceolate (versus linear, hair-like and inconspicuous). It is "one of the most common ferns in mountainous areas of the Trans-Pecos" (Yarborough & Powell 2002).

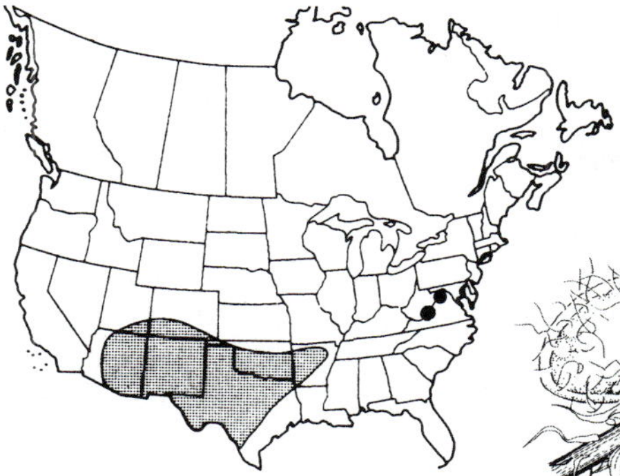

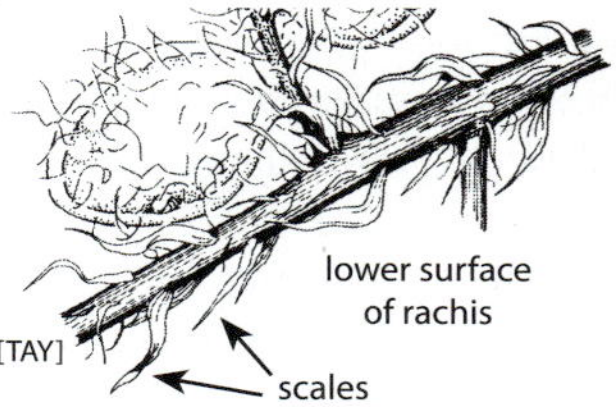

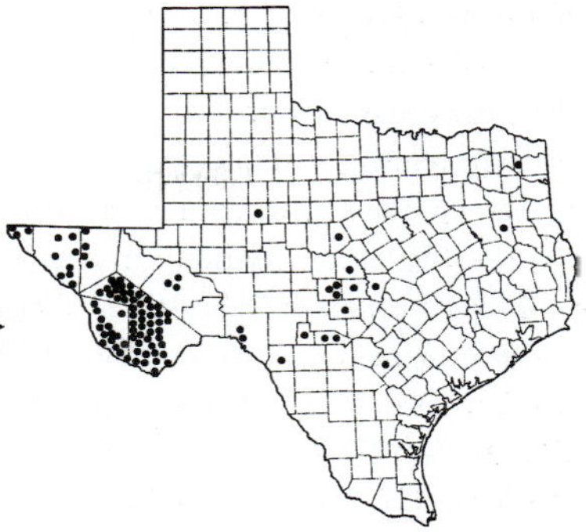

Dale A. Zimmerman Herbarium, Western N. Mex. Univ.

Dale A. Zimmerman Herbarium, Western N. Mex. Univ.

Lorena Babcock Moore

Cheilanthes eatonii

EATON'S LIP FERN

Cheilanthes feei T. Moore, (for A.L.A. Fée, 1789–1874, French botanist), SLENDER LIP FERN, WOOLLY LIP FERN, FEE'S LIP FERN. Plant similar to *C. tomentosa* but with jointed hairs and without tomentum on the petiole and rachis; rhizomes compact to short-creeping; petioles dark brown to black, rounded on upper surface; leaves clustered, relatively small, 4–20 cm long; leaf blades linear-oblong to lanceolate, 1–3 cm wide, 3-pinnate at base, the pinnae not articulate, the ultimate segments 1–3 mm long, rounded, bead-like, densely long hairy on lower surfaces, sparsely hairy to nearly glabrous upper surfaces; costae without scales on lower surfaces; $n = 2n = 90$ (Windham & Rabe 1993). Dry, limestone or calcareous, rocky slopes, ledges, and crevices, usually on limestone; widespread in the w ½ of TX e to Cross Timbers and Prairies—Hamilton and Palo Pinto (Correll 1956; Turner et al. 2003) cos. (e.g., Bandera Co., *C.M. Rowell & J.L. Blassingame 4563*; Culberson Co, *Correll 13755*, both BRIT); sw Canada and much of w U.S. e to WI, IL, MO, AR, and TX, also disjunct to isolated localities in KY and VA—the VA locality is disjunct by 450 km/289 mi from the nearest site in KY (Wieboldt & Bentley 1982); also n Mexico. Sporulating Mar–Nov. This species is a triploid (of unknown parentage) that reproduces via apogamy (= a type of asexual reproduction that does not involve fertilization; the sporophyte is formed directly from the gametophyte without gamete production) (Windham & Rabe 1993, Moran 2004). *Cheilanthes feei* can be confused with *C. lendigera*, but the latter species can be distinguished by its well-formed false indusia forming a pouch and also by its usually larger leaves (Mickel & Smith 2004). It can also be confused with small plants of *C. tomentosa*, but the petiole pubescence in "*C. feei* consists of crisp, multicellular, spreading hairs, with no linear scales, whereas *C. tomentosa* has loosely appressed, single-celled hairs mixed with linear scales" (Yarborough & Powell 2002). This is one of a number of ferns found primarily in the sw or w U.S. with isolated disjunct populations in the southern Appalachians (see discussion on page 31).

Cheilanthes fendleri Hook., (for August Fendler, 1813–1883, one of the first botanists to collect in New Mexico and Venezuela), FENDLER'S LIP FERN. Rhizomes long-creeping; leaves scattered, 7–30 cm long; petioles dark brown, rounded on upper surface; leaf blades lanceolate to ovate-deltate, 1.5–5 cm wide, 3–4-pinnate at base, the pinnae not articulate, the largest ultimate segments 1.5–3 mm long, rounded, bead-like, glabrous on lower (unusual in that most species except *C. aemula* and *C. alabamensis* are hairy) and upper surfaces; scales on underside of costae lanceolate-ovate, conspicuous and often concealing the ultimate segments when viewed from below; $2n = 60$ (Windham & Rabe 1993). Dry wooded banks, ledges, crevices, around boulders, rocky slopes in pine-oak forests; Jeff Davis (e.g., *Correll 13510, 13522*, BRIT, TEX-LL), Presidio (*Warnock & A. Powell 17787*, SRSC), El Paso, Hudspeth, and Brewster (Yarborough & Powell 2002) cos. in the Trans-Pecos; a colletion labelled as coming from Culberson Co. (*Warnock 23704*, SRSC, TEX-LL) is apparently an error because the location given (Eagle Mts.) is in Hudspeth Co; AZ, CO, NM, and TX; also nw Mexico. Sporulating summer–fall. [*Myriopteris fendleri* (Hook.) E. Fourn.] This species is a sexual diploid, and based on molecular evidence it is the maternal parent of the triploid *C. wootonii* and has also contributed genetic material to *C. yavapensis* (Grusz et al. 2009). The combination of ovate non-ciliate costal scales and the complete lack of other leaf pubescence is distinctive for this species (Yarborough & Powell 2002). Because of its rareness and limited distribution in the state, we consider this species to be of conservation concern in TX. ⚠

Cheilanthes horridula Maxon, (Latin: *horridus*, prickly, bristly, or rough, in reference to the stiff hairs on the upper leaf surfaces), ROUGH LIP FERN, PRICKLY LIP FERN. Rhizomes short-creeping; leaves clustered, 5–30 cm long; petioles black to dark brown, rounded on upper surface; leaf blades oblong-lanceolate, 1–4 cm wide, 1-pinnate-pinnatifid to 2-pinnate-pinnatifid at base, pinnae not articulate, the ultimate segments narrowly elliptic to elongate-deltate, not bead-like, the largest 3–5 mm long, scabrous on both upper and lower surfaces due to distinctive stiff hairs, these often inflated basally; scales on underside of costae lanceolate, not conspicuous compared to those of some *Cheilanthes* species; $2n = 58, 116$ (Windham & Rabe 1993). Rock crevices, rocky slopes, and ledges, typically on limestone; mainly Edwards Plateau (e.g., Real Co., *J.L. Blassingame 2906*, BRIT) and Trans-Pecos, but scattered in adjacent areas; in the U.S. primarily in TX but also in extreme s OK (Murray Co.); also n Mexico. Sporulating mainly May–Nov. According to Windham and Rabe (1993), the scabrous pustulose hairs make this "one of the most distinctive species of *Cheilanthes* in North America." It includes both sexually reproducing diploids and tetraploids and needs further study. This is the only member of *Cheilanthes* that has its leaf surfaces roughened by distinctive stiff, prickly, basally inflated hairs.

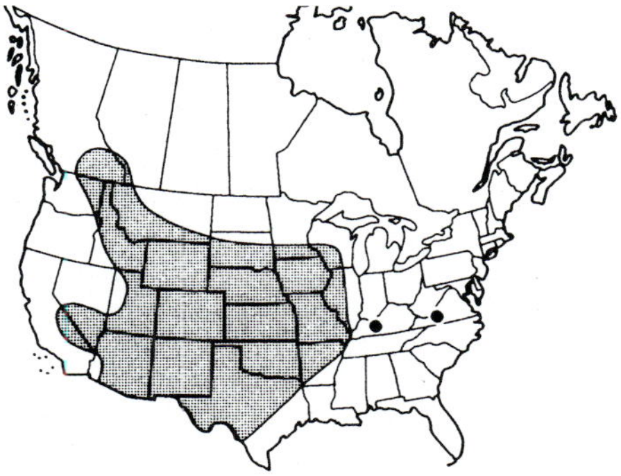

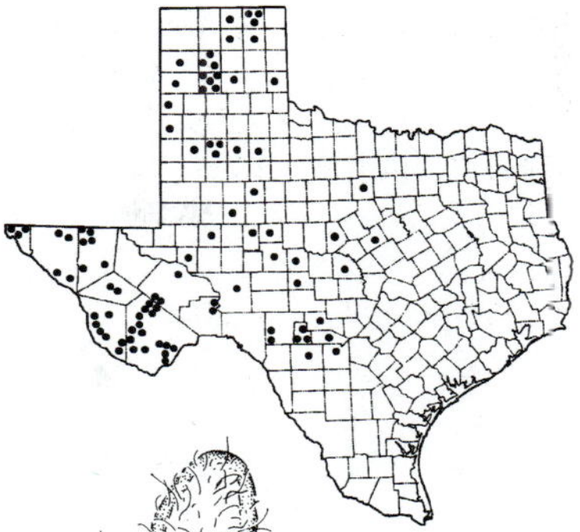

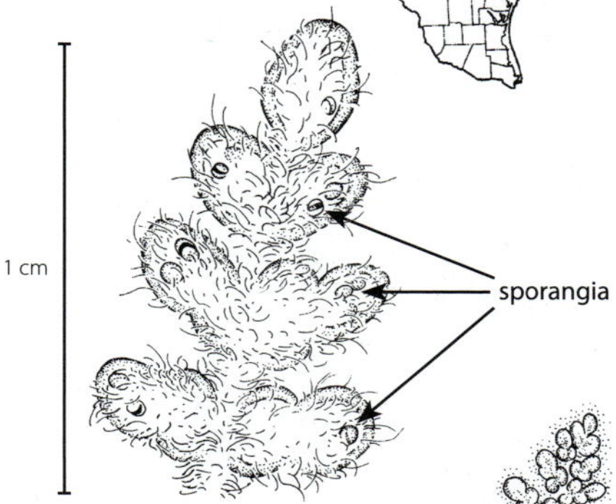

lower surface
of pinna

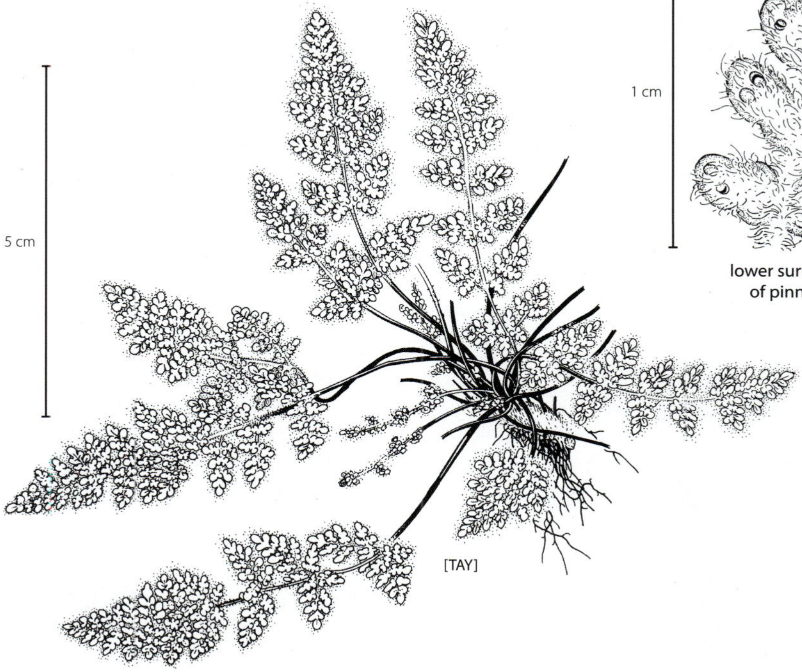

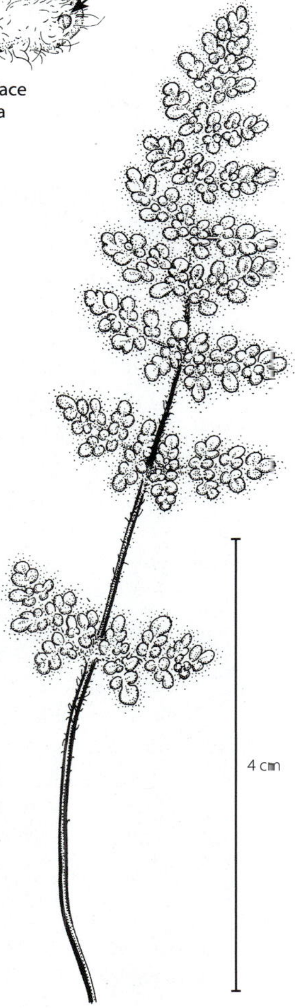

Cheilanthes feei

SLENDER LIP FERN

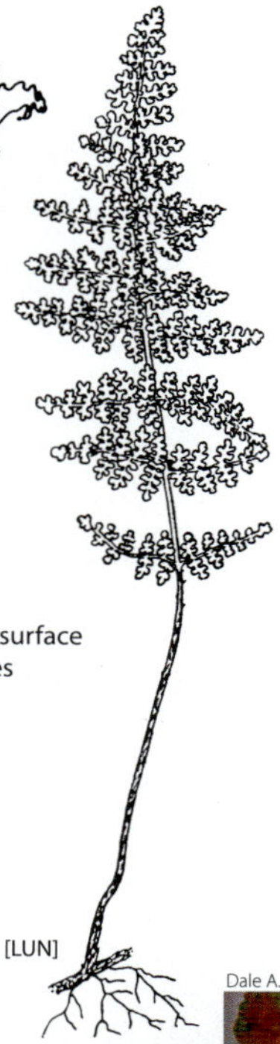

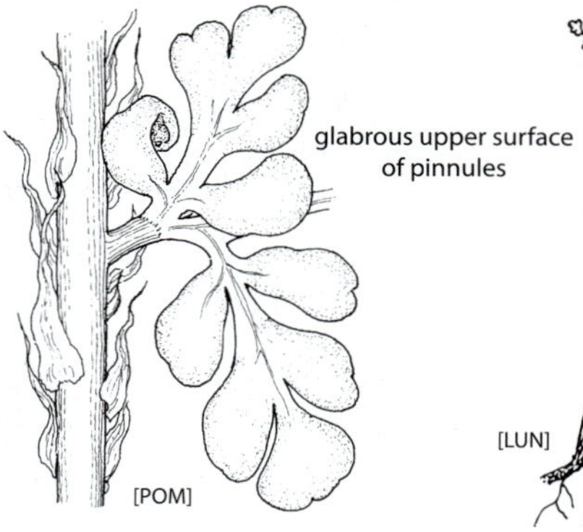

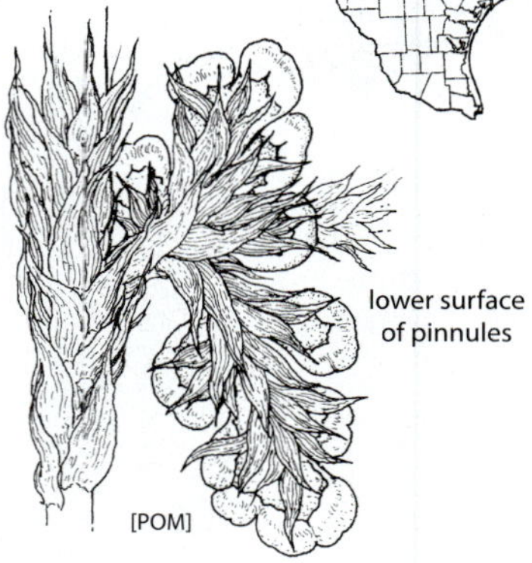

Dale A. Zimmerman Herbarium, Western N. Mex. Univ.

Lorena Babcock Moore

Lorena Babcock Moore

Cheilanthes fendleri

FENDLER'S LIP FERN

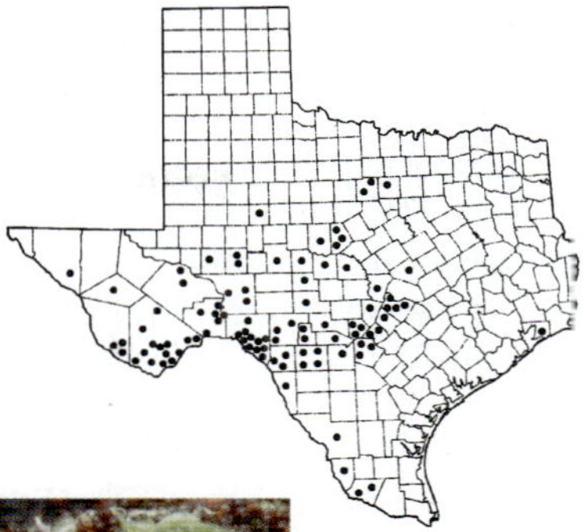

Roy Morey

Roy Morey

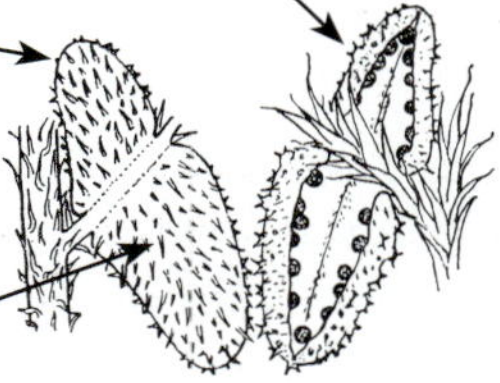

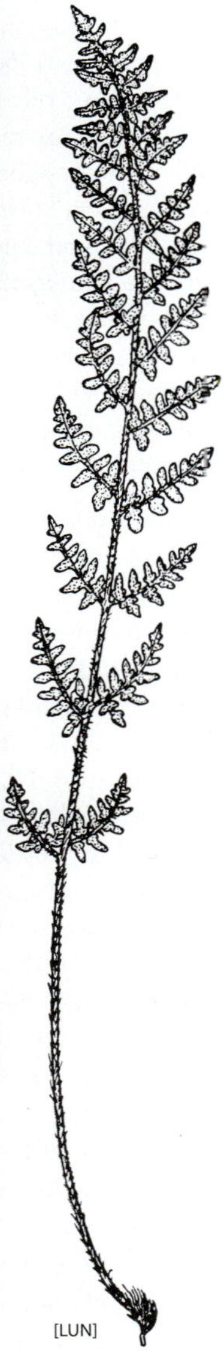

[LUN]

Roy Morey

Cheilanthes horridula

ROUGH LIP FERN

Cheilanthes kaulfussii Kunze, (for G. Kaulfuss, 1786–1830, noted German fern botanist who in 1824 published *Enumeratio Filicum*), KAULFUSS' LIP FERN, GLANDULAR LIP FERN. Rhizomes short-creeping; leaves clustered, 8–35 cm long; petioles dark brown, flattened or slightly grooved distally on upper surface; leaf blades elongate-pentagonal (due to larger basal pinnae with enlarged proximal basiscopic pinnules), (3–)5–10 cm wide, 3-pinnate-pinnatifid to 4-pinnate at base, the pinnae not articulate, the ultimate segments not distinctly bead-like, glandular-pubescent on both lower and upper surfaces, resinous/oily due to exudate from the glandular hairs; costae without scales below. Rocky slopes and ledges, typically on igneous materials; Brewster (Chisos Mts.) (BRIT, TEX-LL) and Jeff Davis (TEX-LL) cos. in the Trans-Pecos and strikingly disjunct to Gillespie (*O'Kennon 8379*, BRIT), Llano (*Correll & Correll 12766*; *Taylor & Taylor 13181*, BRIT; *D. Seigler et al. 2747*, TEX-LL; Seigler 1977), Mason (*C.M. Rowell et al. 2941*, *J.W. Stanford & T.J. Starbuck 476,* BRIT), and Burnet (*J. M. Poole & W. A. Watson 1297*, TEX-LL) cos. much further e; in the U.S. known only from TX; also Mexico, Central and n South America. Sporulating summer–fall. The disjunct occurrence in c TX of this species, which is usually found associated with igneous rocks, is a reflection of soil-related conditions of the Central Mineral Region, an island of igneous materials (Precambrian granites, gneisses, and schists) surrounded by escarpments of Paleozoic and Mesozoic limestone and sandstone (Walters & Wyatt 1982). For example, a Llano Co. collection (*D. Seigler et al. DS-2747*, TEX-LL) was made on Enchanted Rock. Populations in the U.S. appear to be polyploids (derived from diploid Mexican populations) that reproduce apogamously (Windham & Rabe 1993). Mickel and Smith (2004) noted that this is "one of the most widely distributed pteridophytes in Mexico." This species "is covered with conspicuous glandular, stipitate hairs that exude a sticky light brown oil when fresh" (Yarborough & Powell 2002). It is similar to *C. leucopoda* in terms of leaf shape—elongate-pentagonal (= elongate 5-sided) which results from the basal or lowermost pinnae being largest. The basal pinnae have the pinnules closest to the rachis asymmetrically enlarged; the pinnules facing the base of the leaf (= basiscopic) are much larger. *Cheilanthes kaulfussii* can be distinguished from *C. leucopoda* both by its resinous/oily leaves, with short gland-tipped hairs, and by its dark petioles (versus leaves not resinous/oily with long non-glandular hairs and petioles straw-colored in *C. leucopoda*). Because of its rareness and limited distribution in the state, we consider this species to be of conservation concern in TX. ⚠

Cheilanthes lanosa (Michx.) D.C. Eaton, (Latin: *lana*, wool, = woolly, in reference to the leaf hairs), HAIRY LIP FERN, WOOLLY LIP FERN. Rhizomes compact to short-creeping or sometimes long-creeping; leaves clustered or along the rhizome, 7–50 cm long; petioles dark brown, rounded on upper surface, with jointed hairs; leaf blades linear-oblong to lanceolate, 1.5–5 cm wide, usually 2-pinnate-pinnatifid at base, the pinnae not articulate, the ultimate segments oblong to lanceolate, not bead-like, the largest 3–5 mm long, with sparse, long, jointed hairs on both upper and lower surfaces; costae without scales on lower surfaces; $2n = 60$ (Windham & Rabe 1993). Dry rocky slopes and ledges, sandstone, limestone, and granite; in TX only known from McLennan Co. (*E.T. Wherry s.n.*, 1925, US; Correll 1956; Holmes et al. 2011;); the species is also mapped for ne TX by Windham and Rabe (1993); e U.S. from NY s to FL w to MN, KS, OK (e.g., *L. Kelly 36*, BAYLU, Wichita Mts. Wildlife Refuge), and disjunct to McLennan Co. in TX. Sporulating Apr–Oct. While the species has consistently been included in the TX flora (e.g., Cory & Parks 1937; Correll 1956; Correll & Johnston 1970; Hatch et al. 1990; Turner et al. 2003; Diggs et al. 2006), its status in TX has not always been clear (e.g., Diggs et al. 1999, 2006). Correll (1949) noted that only "one locality was known for Texas" and in (1956) cited only the *Wherry s.n.* collection from McLennan Co. (but did not indicate in what herbarium it was deposited). The location and actual identity of the specimen were uncertain until Holmes et al. (2011) relocated it at the U.S. National Herbarium (accession number 1466623; annotated by Correll in 1947 as *C. lanosa*) and confirmed its identity. Holmes et al. (2011) provide a detailed discussion and a photo of the collection. Attempts to relocate the species in McLennan Co. have been unsuccessful (Holmes et al. 2011). Weakley (2010) noted that this is "the most 'eastern' [species] of a predominantly western genus." It is similar in some respects to *C. tomentosa* but with hispidulous jointed hairs instead of tomentum on the petiole and rachis, and with ultimate leaf segments not bead-like. Because of its rareness and limited distribution in the state, we consider this species to be of conservation concern in TX. ⚠

Cheilanthes lendigera (Cav.) Sw., (Latin: *lens* or *lendis*, nit—egg of a louse, or *lens*, lentil—a small lens-shaped fruit, and -ger, from *gero*, to bear or carry, = bearing nits or lentils, presumably in reference to the small, bead-like appearance of the ultimate leaf segments), BEADED LIP FERN, NIT-BEARING LIP FERN. Rhizomes

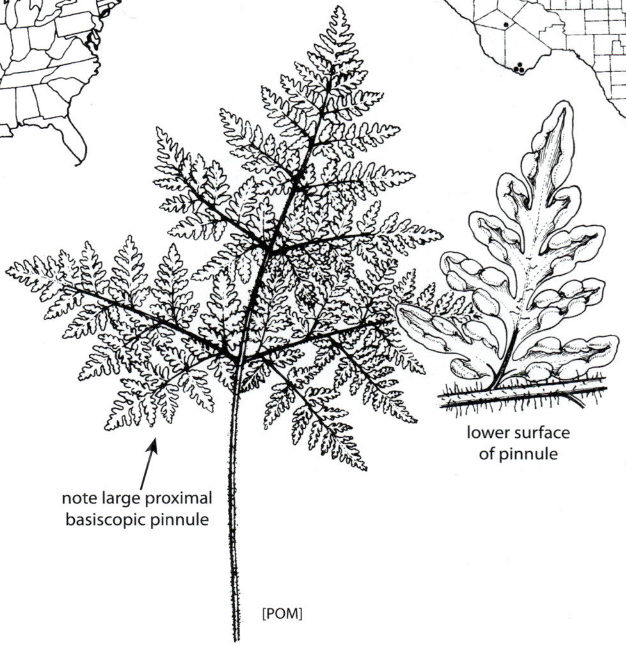

Cheilanthes kaulfussii
KAULFUSS' LIP FERN

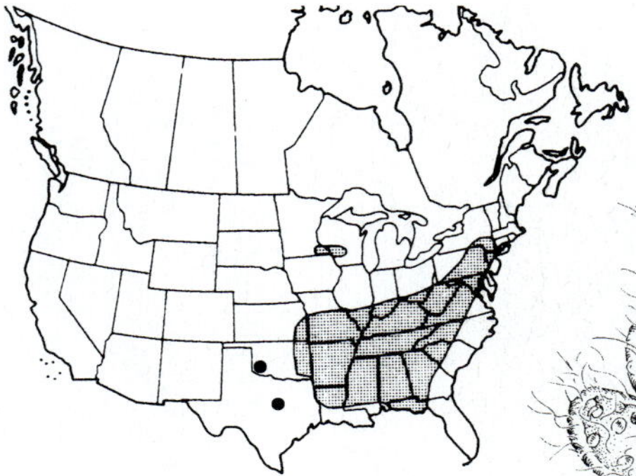

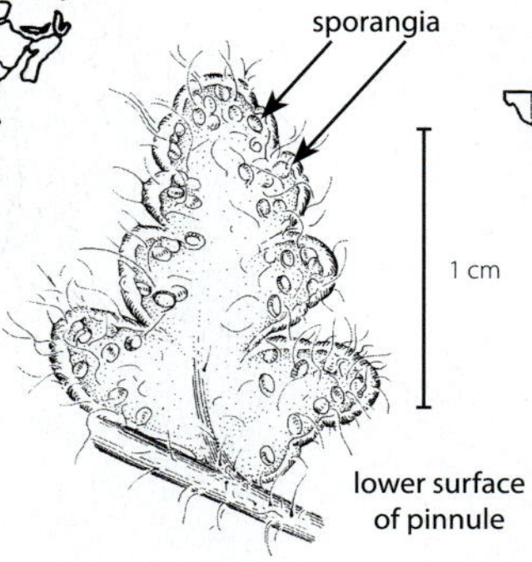

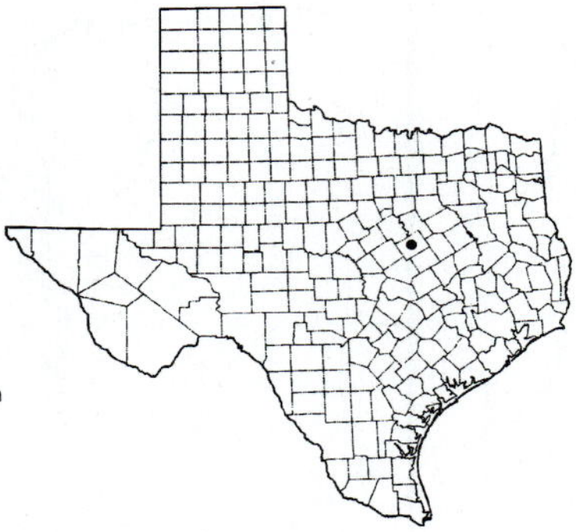

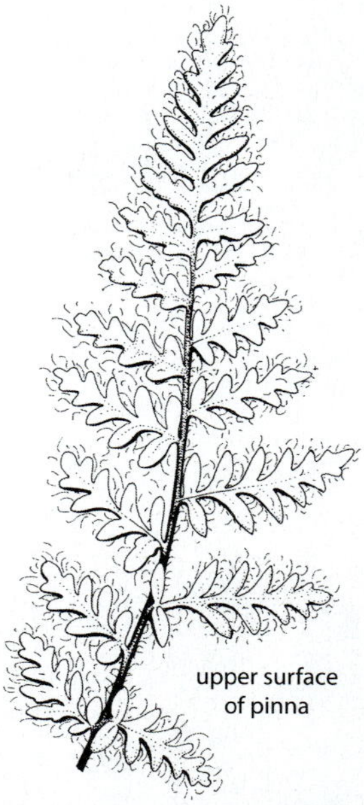

Cheilanthes lanosa

HAIRY LIP FERN

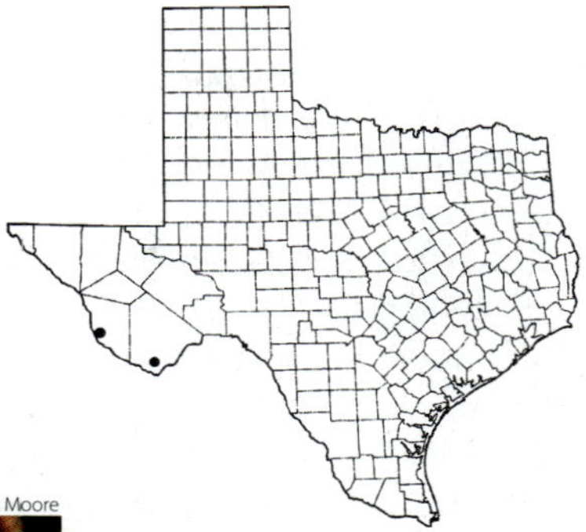

Lorena Babcock Moore

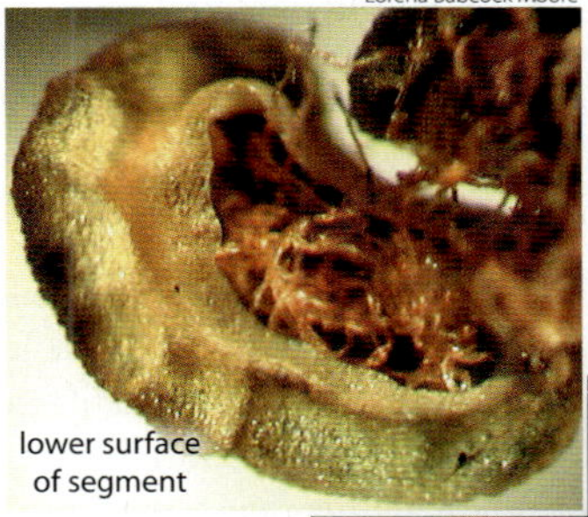

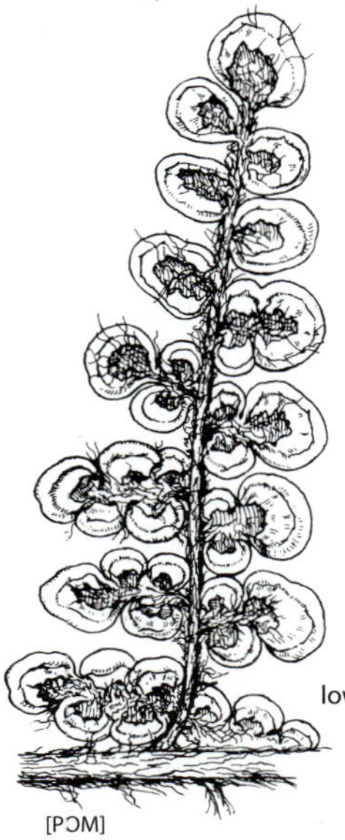

lower surface
of pinna

[PϽM]

Stan Shebs, c c by-sa

Stan Shebs, c c by-sa

Cheilanthes lendigera

BEADED LIP FERN

long-creeping; leaves scattered to clustered, 5–30 cm long; petioles usually dark brown, rounded on upper surface; leaf blades ovate-deltate to oblong-lanceolate, (1.5–)5–8 cm wide, usually 4-pinnate at base, the pinnae not articulate, the ultimate segments rounded, bead-like, the largest 1–3 mm long, with coarse hairs on lower surfaces, glabrous on upper surfaces; scales on underside of costae inconspicuous, linear, hair-like; false indusia well-formed, 2–3 mm wide, not limited to the segment margin, making a pouch with a somewhat constricted opening, much of the lower surface of the ultimate segments covered by the false indusia; $2n = 120$ (Windham & Rabe 1993). Rocky slopes and ledges, typically on igneous substrates; in TX known only from the Chisos Mts. in s Brewster Co. (*B. Warnock 1100*, BRIT, TEX-LL) and from the Chinati Mts. in Presidio Co. (*Warnock 456, 460, 495*, all SRSC); AZ and TX; also Mexico, Central and n South America; all U.S. populations are within about 13 mi (21 km) of the international border. Sporulating summer–fall. [*Pteris lendigera* Cav., *Myriopteris lendigera* (Cav.) J. Smith] The well-formed, broad false indusia making a "pouch" are distinctive—they give the ultimate leaf segments a bead-like appearance (hence the common name). It has been suggested that this species is a fertile allotetraploid resulting from the hybridization of two Mexican species (Mickel & Smith 2004). Because of its rareness and limited distribution in the state, we consider this species to be of conservation concern in TX. ⚠

Cheilanthes leucopoda Link, (Greek: *leukos*, white, and *pous* or *podium*, foot), WHITE-FOOT LIP FERN. Rhizomes compact; leaves clustered, 7–30 cm long; petioles straw-colored, shallowly grooved distally on upper surface; leaf blades elongate-pentagonal (due to larger basal pinnae with enlarged proximal basiscopic pinnules), 3–10 cm wide, 4-pinnate at base, the pinnae not articulate, the ultimate segments oblong to lanceolate, not bead-like, the largest 3–5 mm long, with long non-glandular hairs on both upper and lower surfaces; costae without scales on lower surfaces; false indusia weakly formed; $2n = 60$ (Windham & Rabe 1993). Rocky slopes and ledges, on limestone substrates, sometimes locally abundant; Edwards (e.g., *Correll 13451A*, BRIT, TEX-LL), Uvalde (e.g., *Correll 13459A, H.B. Parks s.n., 6 Jun 1938*, BRIT), Real (e.g., *J.L. Blassingame 2685*, BRIT; *Correll 13422*, TEX-LL), and Kinney (Turner et al. 2003) cos. on the Edwards Plateau; in the U.S. known only from four counties on the Edwards Plateau of TX; also n Mexico. Sporulating summer–fall. Windham and Rabe (1993) noted that among the species covered in the *Flora of North America* treatment, this species is "unique in being a sexual diploid that consistently produces 32 spores per sporangium" [sexual species usually have 64 spores]. Recent molecular studies by Rothfels et al. (2008) show that it is actually more closely related to species of *Notholaena* than to *Cheilanthes* (the confusion caused by convergent evolution, which has resulted in distantly related species having superficially similar appearances)—nomenclatural changes will almost certainly be needed in the future (see further discussion under the genus *Notholaena*). Because of its limited distribution in the state, we consider this species to be of conservation concern in TX. ⚠

Cheilanthes lindheimeri Hook., (for F.J. Lindheimer, 1801–1879, German-born TX collector), LINDHEIMER'S LIP FERN, FAIRY-SWORDS. Rhizomes long-creeping, the scales uniformly brown; leaves closely scattered, 7–30 cm long; petioles usually dark brown, rounded on upper surface; leaf blades oblong-lanceolate to ovate-deltate, 2–5 cm wide, 4-pinnate at base, the pinnae not articulate (= dark color of stalk stopping at base of pinna), the ultimate segments rounded, bead-like, tiny, the largest 0.7–1 mm long, nearly glabrous on upper and lower surfaces (but appearing densely tomentose on upper due to the scales of costae having curly, tangled marginal cilia); scales on underside of costae conspicuous, often concealing ultimate segments, long-ciliate with the curly cilia forming an entangled mass; $n = 2n = 90$ (Windham & Rabe 1993). Rocky slopes and ledges; mainly Edwards Plateau and Trans-Pecos but scattered n to Palo Pinto (*G.M. Soxman 114*, TEX-LL) and Parker (*W.R. Carr 10495*, TEX-LL) cos., e to Burnet Co. (*Whitehouse 18437*, BRIT), and s to McMullen Co. (Turner et al. 2003); AZ, NM, OK (Comanche Co.—OK Native Plant Database 2009), and TX; also Mexico. Sporulating Mar–Nov. [*Myriopteris lindheimeri* (Hook.) J. Smith] Jack Stanford (pers. comm.) noted that this species is found on both granite and limestone. Jim Blassingame (pers. comm.) commented that this species spreads prolifically by rhizomes. It is primarily an apogamous triploid of autopolyploid origin, though 3 diploid populations have been discovered (Schuettpelz et al. 2008). LINDHEIMER'S LIP FERN is similar to *C. wootonii* and *C. yavapensis*, but the broad, uniformly colored rhizome scales, the very small ultimate leaf segments (1 mm or less wide), and the upper leaf surfaces appearing to be densely hairy (actually due to curly, tangled, marginal cilia on the costal scales) are helpful in identification. Although a number of TX *Cheilanthes* species have

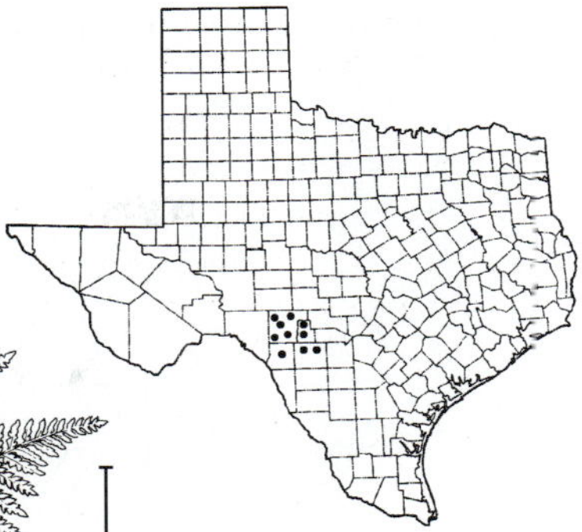

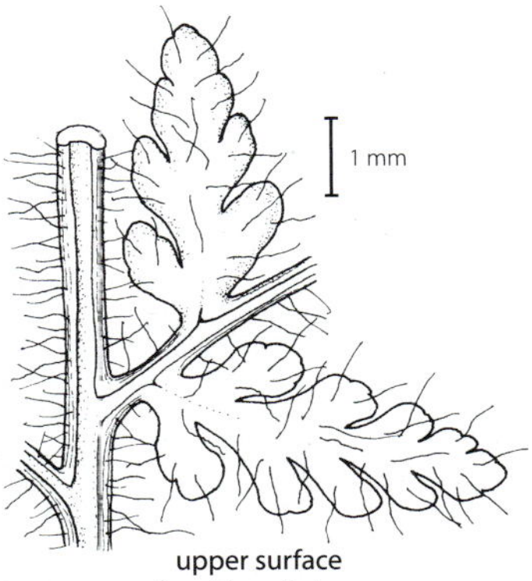

upper surface
of portion of pinnae

Carlos Velasco

Carlos Velasco

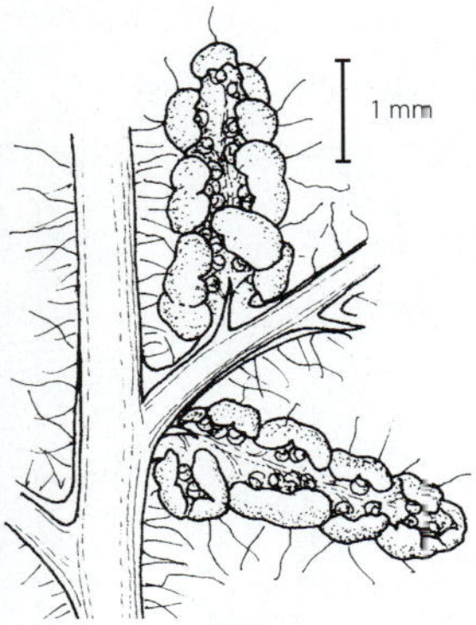

lower surface
of portion of pinnae

Cheilanthes leucopoda

WHITE-FOOT LIP FERN

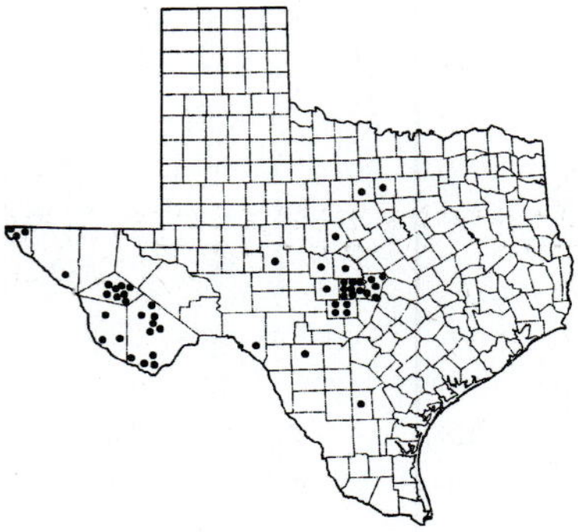

Dale A. Zimmerman Herbarium, Western N. Mex. Univ.

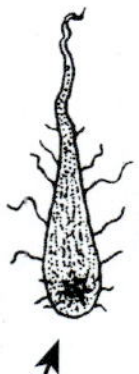

costal scale

Lorena Babcock Moore

[LUN]

Cheilanthes lindheimeri

LINDHEIMER'S LIP FERN

long-creeping rhizomes with the leaves closely scattered along the rhizomes, this is the only such species that occurs e of the Trans-Pecos (with the exception of the extremely rare *C. lanosa* which can sometimes have the rhizomes long-creeping). These long-creeping rhizomes distinguish *C. lindheimeri* from the somewhat similar *C. eatonii* and *C. tomentosa*. LINDHEIMER'S LIP FERN is sometimes called FAIRY-SWORDS because of the upright position of the elongate leaves and their pale appearance due to cilia on the costal scales.

Cheilanthes tomentosa Link, (Latin: *tomentum*, woolly hairs, = densely woolly, covered with short, soft, curly, densely matted or entangled hairs), WOOLLY LIP FERN. Rhizomes compact; leaves clustered, 8–45 cm long; petioles usually dark brown, rounded on upper surface, densely tomentose; leaf blades oblong-lanceolate, 1.5–8 cm wide, usually 4-pinnate basally, the pinnae not articulate, the ultimate segments oval (rarely oblong), bead-like, the largest 1–2 mm long, densely tomentose on lower surfaces, pubescent with fine, unbranched hairs on upper surfaces; scales on underside of costae inconspicuous, linear, only slightly wider than hairs, usually entire; sporangia black at maturity; $n = 2n = 90$ (Windham & Rabe 1993). Rocky slopes and ledges; widely distributed in TX (e.g., Burnet Co., *S.L. Hatch 6238*, BRIT, TEX-LL); across much of s ½ of U.S. from PA and WV s to GA w to TX, NM, and AZ; also Alberta, Canada and Mexico. Sporulating mainly May–Oct. [*Myriopteris tomentosa* (Link) J. Smith] This species is a triploid (parentage unknown) that reproduces via apogamy (= a type of asexual reproduction that does not involve fertilization; the sporophyte is formed directly from the gametophyte without gamete production) (Windham & Rabe 1993; Weakley 2010). It is "closely related to *C. eatonii*, but it is distinguished by having narrower [filiform], less prominent costal scales" [these more evident in *C. eatonii*] (Windham & Rabe 1993). It could also be confused with *C. feei*—see that species for discussion.

Cheilanthes villosa Davenp. ex Maxon, (Latin: *villosus*, softly hairy), VILLOUS LIP FERN. Rhizomes compact; leaves clustered, 7–30 cm long; petioles usually dark brown, rounded on upper surface; leaf blades oblong-lanceolate to ovate, 1.5–5 cm wide, 3–4-pinnate at base, the pinnae not articulate, the ultimate segments round to oval, bead-like, the largest 1–2 mm long, glabrous on lower surfaces except for a few coarse hairs, villous with coarse, unbranched hairs on upper surfaces; scales on underside of costae conspicuous (see below); $n = 2n = 90$, apogamous (Windham & Rabe 1993). Rocky slopes, cliffs, crevices, usually but not exclusively on limestone; widespread in the Trans-Pecos (e.g., *Correll & Correll 30644*, TEX-LL, Brewter Co.) including w of the Pecos River in Val Verde Co.; AZ, NM, and TX; also n Mexico. Sporulating summer–fall. The combination of large ovate-lanceolate to deltate scales on the lower side of the costae and scattered curly hairs on the upper side is distinctive for this species (Mickel & Smith 2004). The scales on the underside of the costae are particularly striking—they are often whitish, very large (2 mm wide), spreading, and nearly cover the underside of the ultimate segments; they are even visible when viewing the leaf from above. This species is an apogamous triploid, but is reported to hybridize with *C. eatonii* (Lellinger 1985; Mickel & Smith 2004), also considered to be apogamous in North America (Windham & Rabe 1993). Several TX specimens (*Correll & Correll 12915*, Val Verde Co., BRIT, TEX-LL; *Correll & Correll 12906*, TEX-LL) seem intermediate between *C. villosa* and *C. eatonii* and such morphological intermediates between the two have been reported from both w TX and NM (Windham & Rabe 1993); they have been referred to as *C. pinkavii* Reeves (but that name is not validly published). Although apogamous triploids are not self fertile (rather, sporophytes are produced through vegetative growth of gametophyte tissue without the fusion of eggs and sperm), the gametophytes can have functional antheridia and sperm can be produced. The sperm from these triploids can fuse with eggs from a different sexually reproducing species to form fertile hybrids (Gastony & Windham 1989). However, as noted by Windham and Rabe (1993), "recent gene exchange [between *C. villosa* and *C. eatonii*] is unlikely because both taxa are [currently] apogamous in North America."

Cheilanthes wootonii Maxon, (for Elmer Ottis Wooton, 1865–1945, pioneer New Mexico botanist, first resident plant taxonomist in the state, and co-author of the 1915 *Flora of New Mexico*; see Allred 1990, 2008), WOOTON'S LIP FERN. Very similar morphologically to *C. yavapensis*; rhizomes somewhat long-creeping, the scales usually uniformly brown or weakly bicolored, loosely appressed and deciduous on older parts of rhizomes; leaves somewhat scattered; petioles usually dark brown, rounded on upper surface; leaf blades oblong-lanceolate to nearly ovate, 2–6 cm wide, 3–4 pinnate at base, the pinnae not articulate, the ultimate segments round to oblong, bead-like, the largest 1–3 mm long, glabrous on lower surfaces or with a few scales

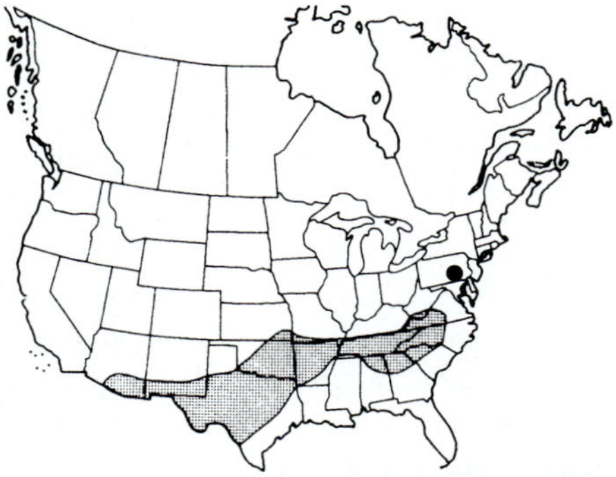

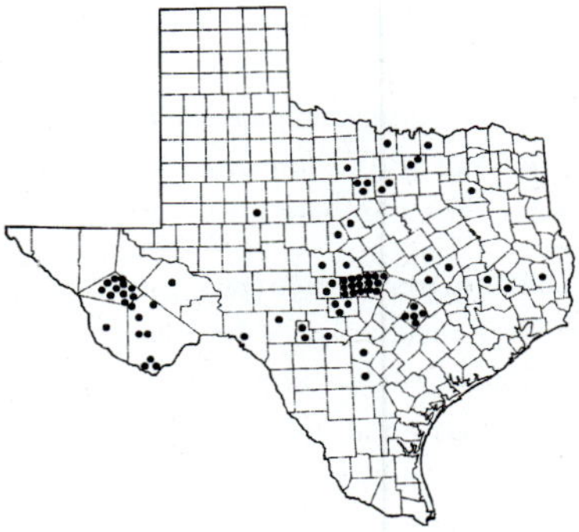

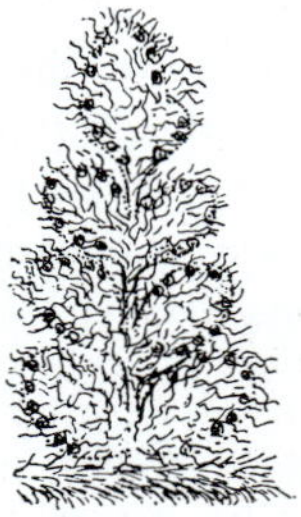

lower surface of pinnule
note black sporangia

Alan Cressler

Alan Cressler

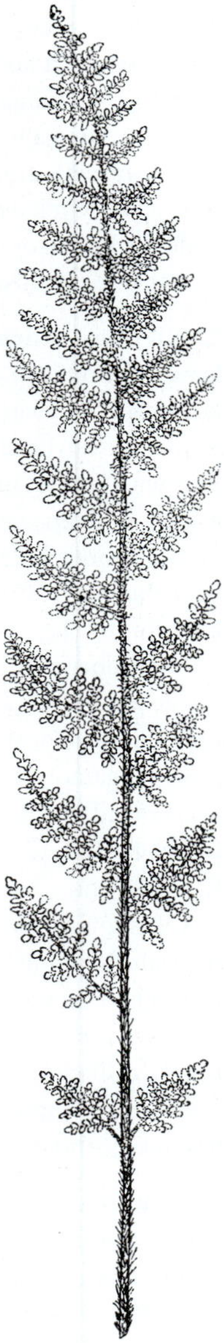

[LUN]

Cheilanthes tomentosa

WOOLLY LIP FERN

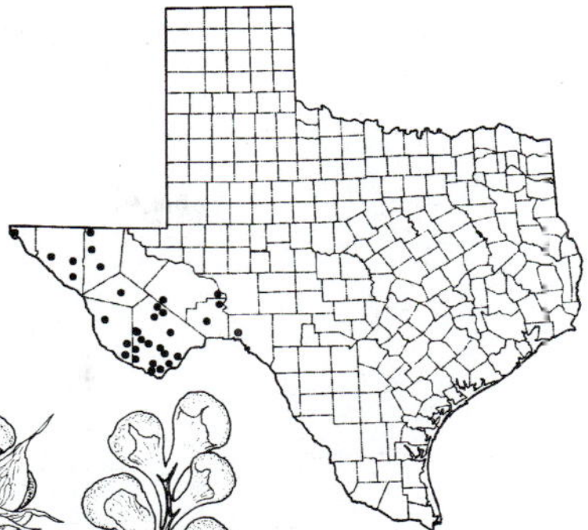

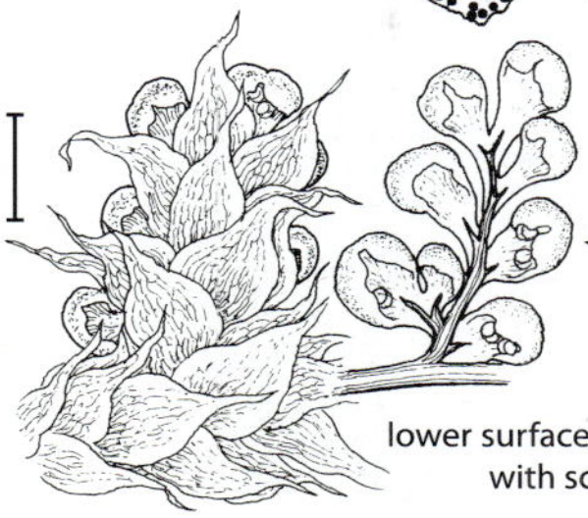

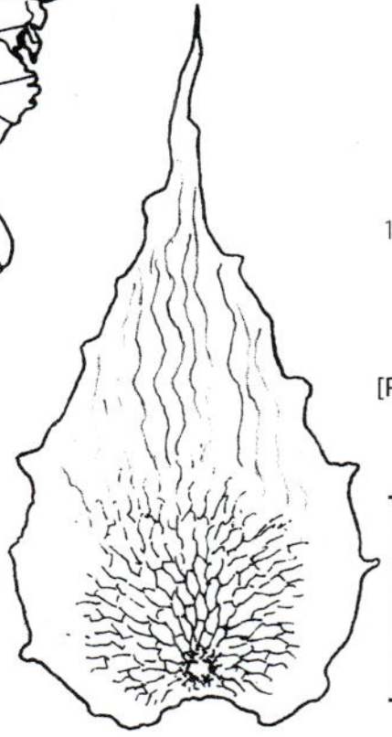

detail of costal scale from lower surface

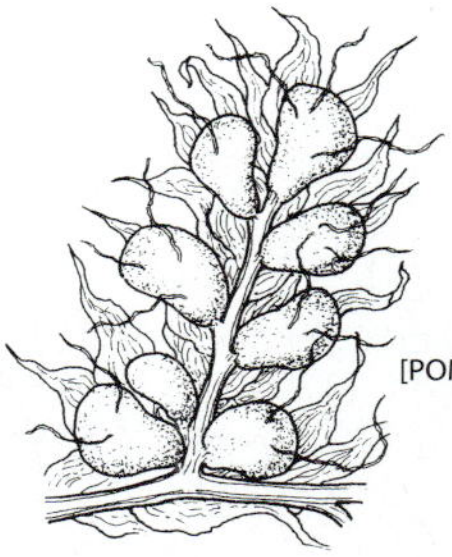

upper surface of pinnule

Cheilanthes villosa

VILLOUS LIP FERN

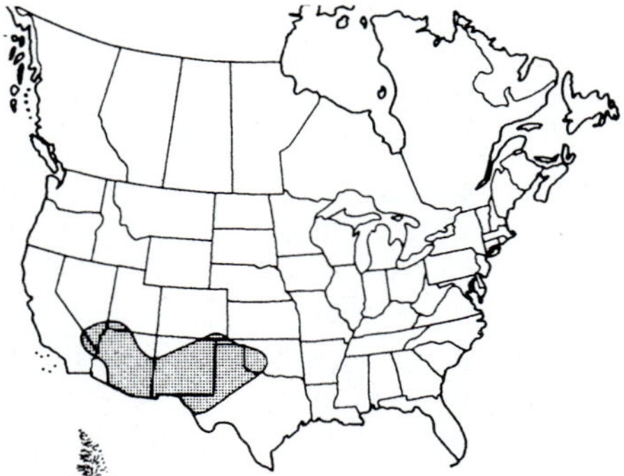

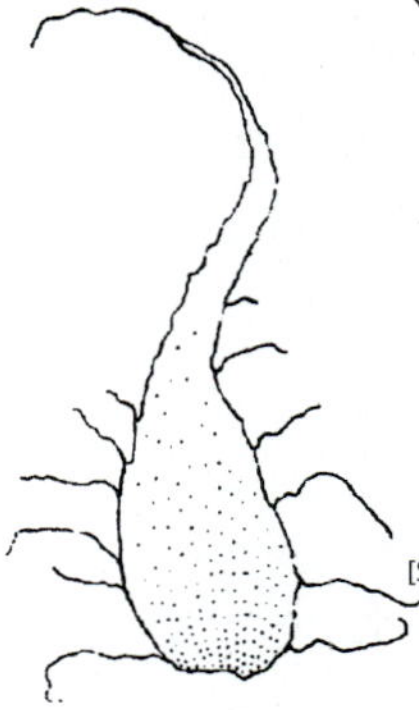

[SCY]

costal scale from
lower surface

Dale A. Zimmerman Herbarium, Western N. Mex. Univ.

Lorena Babcock Moore

Cheilanthes wootonii

WOOTON'S LIP FERN

near base, glabrous on upper surfaces; scales on underside of costae lanceolate-ovate, conspicuous, often concealing ultimate segments when viewed from below; $n = 2n = 90$ (Windham & Rabe 1993); spores averaging < 62 µm diam. (versus > 62 µm in *C. yavapensis*) (Windham 1993e). Rocky slopes, ledges, typically on igneous materials; Hudspeth (*Correll & Johnston 24328*, BRIT, ident. by M.D. Windham, TEX-LL, ident. by T. Reeves), El Paso (*R.D. Worthington 3982*, TEX-LL, annotated by M.D. Windham), Brewster (*P. Zelazny s.n.* 1997, SRSC), Jeff Davis (*J.C. Zech et al. 154*, SRSC), and Presidio (Turner et al. 2003) cos. in the Trans-Pecos; AZ, CA, CO, NV, NM, OK, TX, and UT; also nw Mexico. Sporulating summer–fall. A collection from Llano Co. (Edwards Plateau) previously identified as *C. wootonii* (*Seigler & Crutchfield 1066A,* TEX-LL; Seigler 1977) has since been determined to actually be *C. lindheimeri*. Recent detailed molecular analysis (able to identify genome source) shows that *C. wootonii* is an allotriploid resulting from interspecific hybridization between *C. fendleri* and *C. lindheimeri*, with *C. fendleri* being the maternal parent (Grusz et al. 2009). Further, Grusz et al. (2009) have shown that of the three genomes present, two are from *C. fendleri* and one from *C. lindheimeri*. It is morphologically similar to *C. lindheimeri* and particularly to *C. yavapensis* (Yarborough & Powell 2002); see further discussion under *C. yavapensis*. Yarborough and Powell (2002) indicated that the species is "rare" in TX. Because of its rareness and limited distribution in the state, we consider it to be of conservation concern in TX. ⚠

Cheilanthes wrightii Hook., (for Charles Wright, 1811–1885, TX collector), WRIGHT'S LIP FERN. Rhizomes long-creeping; leaves clustered to somewhat scattered, 4–25 cm long; petioles brown, grooved on upper surface; leaf blades lanceolate to ovate-deltate, 1–4 cm wide, 2-pinnate-pinnatifid at base, the pinnae not articulate, the ultimate segments oblong to linear, not bead-like, the largest 3–7 mm long, glabrous on both upper and lower surfaces; costae without scales on under surfaces; rachises grooved on upper surfaces; $2n = 60$ (Windham & Rabe 1993). Rocky slopes, ledges, typically on igneous materials, sometimes quite common, occasionally mixed with mat-forming *Selaginella* spp.; Brewster (*Correll 13598, 13618*, BRIT, TEX-LL, *R.T. Clausen & H. Trapido 9293*, TAMU), Jeff Davis (*Correll 13503*, BRIT, TEX-LL, *Whitehouse 11209*, BRIT), Presidio (*Correll 13744*, TEX-LL), El Paso (Turner et al. 2003), and Hudspeth (Correll 1956; Yarborough & Powell 2002) cos. in the Trans-Pecos; AZ, NM, and TX; also n Mexico. Sporulating summer–fall. The combination of glabrous petioles grooved on upper side and essentially glabrous pinnae distinguishes this from somewhat similar species (e.g., *C. aemula* and *C. alabamensis* have petioles rounded on upper side). Windham and Rabe (1993) noted that it could also be confused with certain *Pellaea* species, but can be distinguished from all local *Pellaea* "by having both a grooved rachis and a thin, long-creeping stem." This species was the focus of a recent study in which DNA barcoding exposed a case of mistaken identity in the fern horticultural trade. Plants being sold by a horticultural company in Dallas, TX as *C. wrightii* were discovered to actually be *C. distans* (R.Br.) Mett. (BRISTLY CLOAK FERN), native to Australia and adjacent islands. The authors of the paper noted that the molecular technique of DNA barcoding is "a valuable new technology available to the horticulture industry—not only to correct persistent errors in fern identification in the international trade, but also to protect against their overexploitation" (Pryer et al. 2010).

Cheilanthes yavapensis T. Reeves ex Windham, (for the type locality in Yavapai Co., AZ), YAVAPAI LIP FERN, GRACEFUL LIP FERN. Very similar morphologically to *C. wootonii*; rhizomes long-creeping; leaves scattered; petioles dark brown; leaf blades oblong-lanceolate to nearly ovate, 2–6 cm wide, 4-pinnate at base, the pinnae not articulate, the ultimate segments round to oblong, bead-like, the largest 0.4–1 mm long, glabrous on lower surfaces or with a few scales near base, glabrous on upper surfaces (though appearing sparsely pubescent); scales on underside of costae lanceolate, conspicuous, often concealing the ultimate segments when viewed from below; $n = 2n = 120$ (Windham & Rabe 1993); spores averaging > 64 µm diam. (Windham 1993e). Rocky slopes, ledges, typically on igneous materials; known in TX only from El Paso Co. (Franklin and Hueco Mts.) (e.g., *Correll 13790*, BRIT, TEX-LL, *Correll 15096*, TEX-LL, *G.M. Soxman 361*, TEX-LL, *Warnock & Johnston 16179, 16181, 16275*, SRSC, all ident. by M.D. Windham who named the species; Yarborough & Powell 2002); AZ, NM, and TX; also nw Mexico. Sporulating summer–fall. Recent molecular analysis (able to identify genome source) shows that this species is an apogamous tetraploid containing genetic material from three species, with the maternal parent being *C. lindheimeri* and the other two genomes being derived from *C. fendleri*, and *C. covillei*, a species that occurs in Mexico and the sw U.S. to the w of TX. (Grusz et al. 2009). Grusz et al. (2009) also provided "compelling" evidence that *C. yavapensis* has "multiple origins" from these

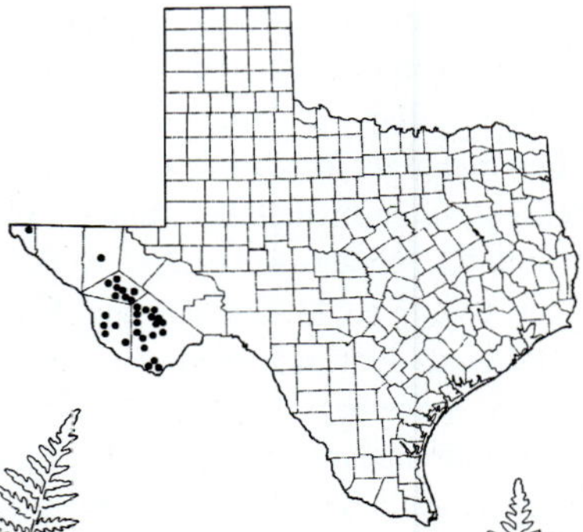

lower surface of pinnule

cross section of rachis

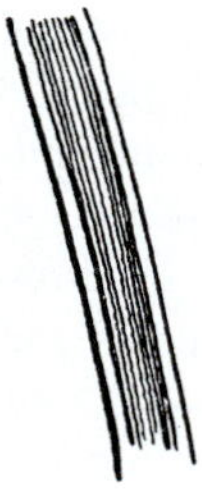

upper surface of rachis

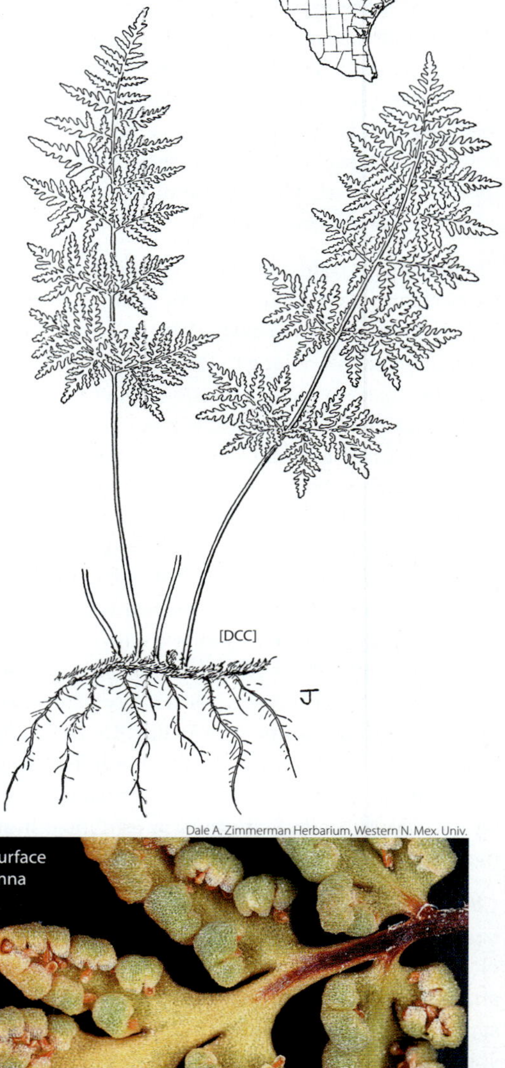

Peter Guth

Dale A. Zimmerman Herbarium, Western N. Mex. Univ.

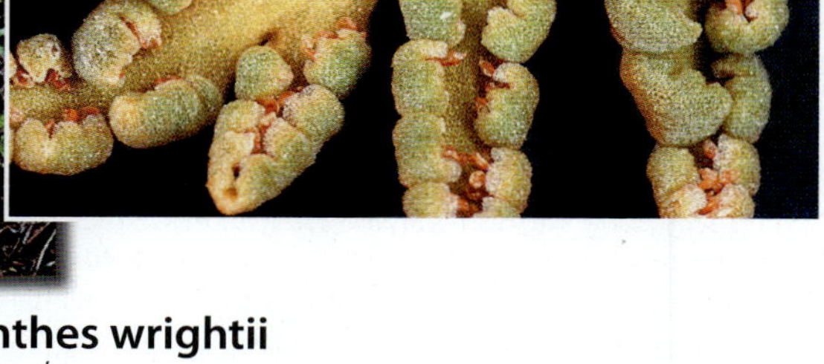

Cheilanthes wrightii

WRIGHT'S LIP FERN

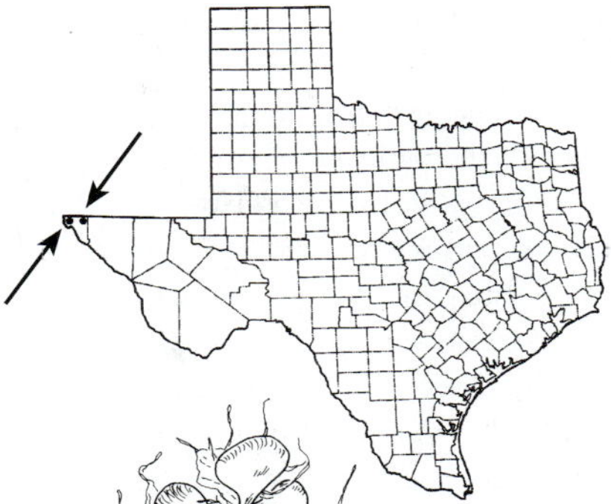

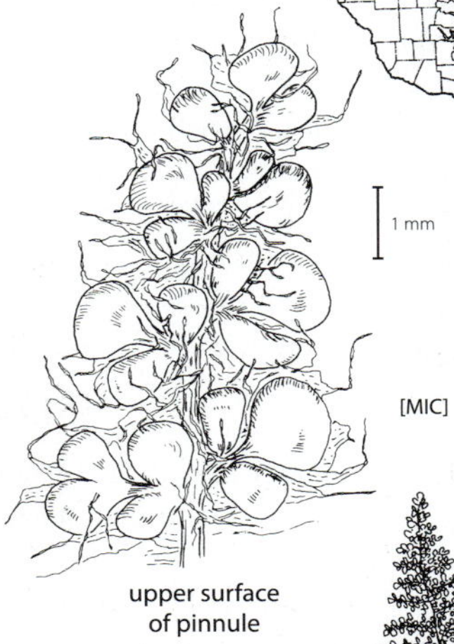

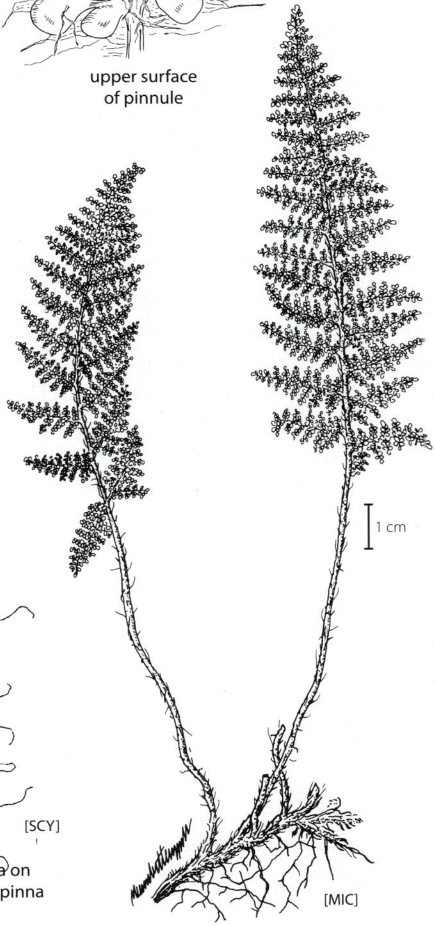

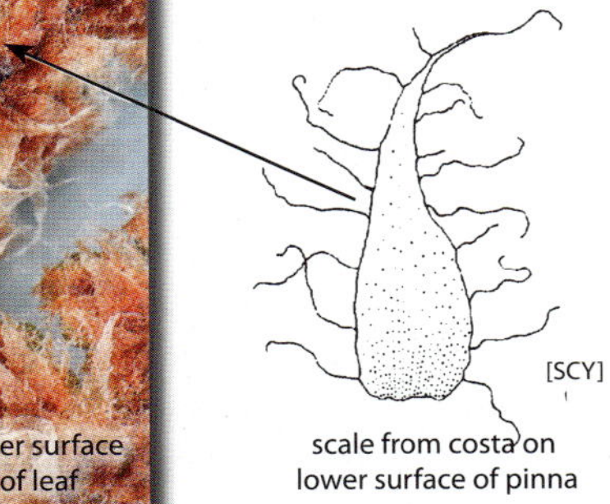

Cheilanthes yavapensis

YAVAPAI LIP FERN

three parental species—in other words, there are two cryptic taxa. Some *C. yavapensis* have two copies of the *C. covillei* genome and one each of the *C. fendleri* and *C. lindheimeri* genomes, but some individuals have two of *C. lindheimeri* and one each of *C. covillei* and *C. fendleri*. Although such a multiple origin might be an argument for recognition of two different species, Grusz et al. (2009) noted that preliminary observations show the two types of *C. yavapensis* to "differ only slightly in morphology. For practical purposes of identification, they may prove impossible to differentiate." We are thus only recognizing a single species for this genetically complicated entity. YAVAPAI LIP FERN is morphologically similar to *C. wootonii* and has sometimes been included as a part of that species. However, Windham and Rabe (1993) have noted that the similarities are due to "convergence rather than common ancestry," and Grusz et al. (2009) have confirmed that *C. wootonii* was not involved in the origin of *C. yavapensis*. Unfortunately, although they have different origins and very different chromosome numbers (120 vs. 90 in *C. wootonii*), Grusz et al. (2009) pointed out that, "Despite evidence that *C. wootonii* and *C. yavapensis* are distinct entities, they remain difficult to distinguish by superficial observation of morphological traits" and that identification can be "particularly difficult" in TX. Likewise Windham and Rabe (1993) noted, "These characteristics [separating *C. wootonii* and *C. yavapensis*] can be subtle, and some specimens will be difficult to place in either *C. wootonii* or *C. yavapensis*." Yarborough and Powell (2002) indicated that the species is "rare" in TX. Because of its rareness and limited distribution in the state, we consider this species to be of conservation concern in TX. ⚠

LLAVEA Lag. LLAVE´S FERN

Llavea is a morphologically distinctive, monotypic genus found primarily in Mexico, but distributed from extreme sw TX to Guatemala and Costa Rica. Its evolutionary relationships have been unclear—Copeland (1947) considered it to be derived from *Pellaea*, Pichi Sermolli (1963) placed it with *Cryptogramma* and *Onychium*, while Tryon and Tryon (1982) suggested there was a distant relationship to *Lygodium*. Family placement has ranged from Adiantaceae to Cryptogrammaceae (e.g., Smith et al. 2006), Llaveaceae, and Pteridaceae. Recent molecular research (e.g. Schuettpelz et al. 2007; Schuettpelz & Pryer 2007, 2008) has shown it to be in the Pteridaceae in a cryptogrammoid clade (= a monophyletic lineage containing a common ancestor and all its descendants) of three genera including *Cryptogramma* (6–11 species of North America, South America, Europe, and Asia) and *Coniogramme* (an Old World genus). This cryptogrammoid clade is the most basal of the five clades comprising the Pteridaceae, making *Llavea* only distantly related to other members of the family. Schuettpelz et al. (2007) noted regarding this clade, "The morphology is highly variable in this group, and a clear morphological synapomorphy is lacking. Although all three genera display sterile-fertile leaf dimorphism, this character is widespread throughout the Pteridaceae." *Llavea* has long been cultivated as an ornamental—Hooker (1860) pointed out that it was "one of the most beautiful, and, in a state of cultivation, rarest of ferns, native of Mexico. ... No fern-collection suitable to a warm greenhouse should be without this charming plant." (Named for Pablo de la Llave, 1773–1833, a traveler in Mexico who discovered the genus—Hoshizaki & Moran 2001; Geldhill 2008)

REFERENCES: Hooker 1860; Stolze 1981; Tryon & Tryon 1982; Mickel & Beitel 1988; Pacheco 1995; Jaramillo et al. 2000; Mickel & Smith 2004; Schuettpelz et al. 2007; O'Kennon & Diggs 2011.

QUICK TIPS FOR GENUS ID: Rare fern of the Trans-Pecos with dramatically different sterile and fertile (much longer and narrower) pinnae on the same leaf and with blackish scales of rhizomes contrasting with yellowish scales at bases of petioles.

Llavea cordifolia Lag., (Latin: *cordis*, heart, and *folius*, leaved, = heart leaf), LLAVE´S FERN, CORDATE-LEAF LLAVEA, HELECHO DE LA LLAVE. Plants terrestrial or on rocks; rhizomes usually horizontal or nearly so, compact, 2–2.5 cm diam., with scales linear-lanceolate, 0.6–1.4 cm long, entire, blackish; leaves clumped, 0.45–1.2(–1.47) m long, hemidimorphic (= partially dimorphic, the fertile leaves similar to sterile except with much narrower, longer fertile pinnae limited to the terminal ⅓ of the blades); petioles ca. equal in length to blade, grooved, glabrous except for large (1–3 cm long), yellowish, linear-lanceolate scales near base; leaf blades broadly ovate, mostly 3-pinnate basally, 2-pinnate apically, subcoriaceous, glabrous, with ultimate segments distinctly stalked, with veins 1–2 forked, free, extending to segment margins, the ends prominent; sterile segments ovate-deltate, 2–3 cm long, 0.8–2.2 cm wide, grayish green, marginally serrulate; fertile segments much narrower, linear, 2.5–8.5 cm long, 1–4(–5) mm wide, with inrolled margins partially covering sori; sporangia

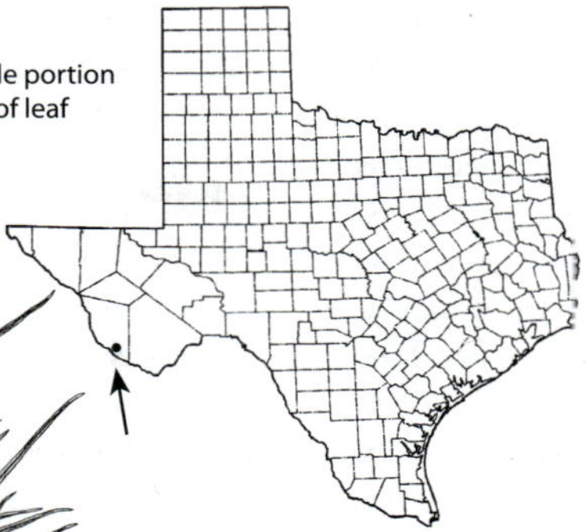

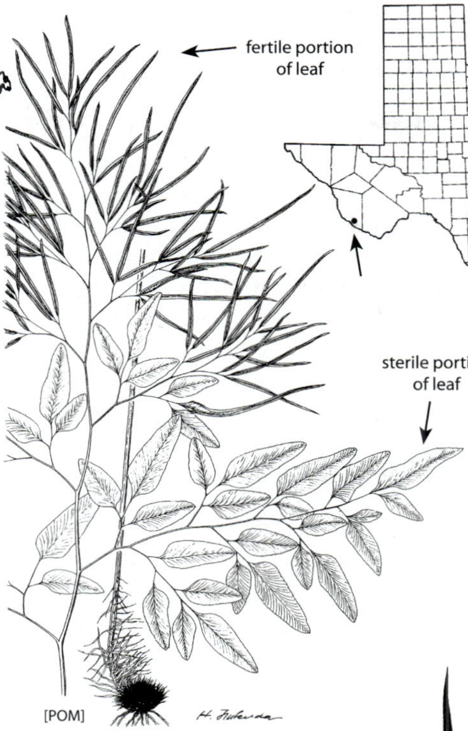

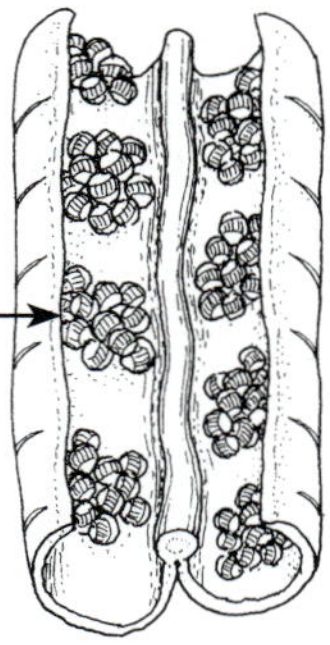

detail of fertile segment

rhizome scale

petiole scale

Llavea cordifolia

LLAVE'S FERN

borne along the veins, so abundant as to often essentially cover the entire abaxial surface of the segments; indusium differentiated, with short glandular hairs on inside; *n* = 29 (Knobloch 1967; Tryon & Tryon 1982; Mickel & Smith 2004). Moist limestone bluff, 1108 m (3635 ft); in TX known only from Big Bend State Park in Presidio Co. (*R.J. O'Kennon & B. Warnock*, 1992, BRIT, SRSC) approx. 6.9 mi from the U.S.-Mexico border; in the U.S. known only from TX; also Mexico, Guatemala, and Costa Rica. The sole U.S. collection was sporulating on 10 May. [*Allosorus karwinskii* Kunze, *Botryogramme karwinskii* (Kunze) Fée, *Ceratodactylis osmundoides* J. Sm.] While the TX collection of this species was made in 1992, at the time it was thought to be an aberrant individual of an *Osmunda* species. The specimen remained in storage until late 2010 when its identity was realized. This represents the first known TX collection and the first report for the U.S. of this primarily Mexican species (O'Kennon & Diggs 2011). In Mexico, it occurs on rocky slopes of moist woods and cliffs (Mickel & Smith 2004). Tryon and Tryon (1982) noted, "*Llavea* grows in mesic canyons, or other rocky places in pine and oak woods, or in tropical forests. Sometimes it occurs on roadsides, on rock walls or in damp soil. It is principally, perhaps always, a calciophile." It superficially somewhat resembles *Osmunda regalis*, ROYAL FERN, (in having very different sterile and fertile pinnae on the same leaf), but is quite different in appearance from all other TX species. Because it is known in this country from only a single station in Presidio Co., we consider this species to be of conservation concern in both TX and the U.S. ⚠

NOTHOLAENA R. Br. CLOAK FERN

Plants xerophytic, usually growing on rocks; rhizomes short-creeping to compact, scaly, the scales black or bicolored with dark central stripe and lighter margins; leaves monomorphic, clustered, small (35 cm or less long); petioles brown or black; leaf blades 1-pinnate-pinnatifid to 4-pinnate, varying from linear-lanceolate to ovate, deltate or pentagonal, leathery, the lower surfaces covered with a whitish or yellowish farina (= a wax-like flavonoid material produced by short, capitate, glandular hairs) as well as scales and/or hairs, the ultimate segments sessile to subsessile, often adnate to costae or costules, the veins of ultimate segments free, obscure; sori on submarginal vein tips on the lower leaf surfaces; margins of ultimate leaf segments recurved to form narrow poorly developed false indusia.

A genus of ca. 35 species, with ca. 28 in Mexico (10 of these in the sw U.S. and 8 in TX); the species are typically found on or among rocks in seasonally dry areas; they range from sw U.S. through Mexico and the West Indies to Central and South America (Windham 1993f; Mickel & Smith 2004; Rothfels et al. 2008; Yatskievych & Arbeláez 2008). The genus has a complex and confusing history (Yatskievych & Smith 2003); for example, it has been broadly treated by some to include species now recognized in *Astrolepis* and *Argyrochosma*, and even some currently in *Cheilanthes*; as such it is exceedingly difficult to distinguish *Cheilanthes* and *Pellaea*. On the other hand, some workers have included the species recognized here in *Notholaena* within a very broadly viewed *Cheilanthes*. However, we are following Windham (1993f) in treating *Notholaena* in a restricted sense, limited to those species with farinose leaves (= covered with a wax-like, mealy, or powdery material); as such it is a nearly monophyletic group closely related to *Cheilanthes* and found only in the w hemisphere (Windham 1993f; Gastony & Rollo 1995). Recent molecular research (Rothfels et al. 2008) suggests a slightly broader circumscription of the genus that would include a few additional New World species, including the taxon that has traditionally (and in this book) been treated as *Cheilanthes leucopoda*—this has long been considered a "morphologically isolated species" within *Cheilanthes* (Tryon & Tryon 1982). Once additional evidence is available, a few species may thus need to be transferred to *Notholaena*. There has also been argument over technical details of nomenclature including what is the correct name to use for the group here treated as *Notholaena*; for example, the generic name *Chrysochosma* is advocated by some. However, Yatskievych and Smith (2003) gave sound arguments for using the name *Notholaena*, and recent molecular results (Rothfels et al. 2008) make their arguments even stronger. The characteristic, usually whitish to yellowish, wax-like, often dense farina is presumably an adaptation in these dry-adapted plants to reduce evapotranspiration from the leaf surfaces, possibly through increasing the reflectance of sunlight (Hevly 1963), or by waterproofing. Surprisingly, as Rothfels et al. (2008) noted, "When suffering water stress, *Notholaena* plants curl their leaves to expose their abaxial (farina-covered) surfaces to irradiation, despite the fact that this surface is where the stomata are situated" (see discussion on page 27). Seigler and Wollenweber (1983) pointed out that *Notholaena* is probably part of an "ancient xeric flora that has evolved in response to the development of aridity in Mexico and the adjacent U.S." This xeric flora is sometimes referred to as the Madro-Tertiary flora (named after the Mexican Sierra Madre Mountains) (see discussion on page 13). Key adapted from Windham (1993f). (Greek: *notho*, false, and *chlaena*, coat, for the margins of the ultimate leaf segments that are reflexed to form false indusia)

REFERENCES: Hall 1950; Tryon 1956; Windham 1986, 1993f; Gastony & Windham 1989; Yatskievych & Smith 2003; Mickel & Smith 2004; Rothfels et al. 2008; Yatskievych & Arbeláez 2008.

QUICK TIPS FOR GENUS ID: Dry area ferns having leaves usually with obvious white or yellowish farina on lower surface and sometimes with scattered farina on upper surface as well, often also with scales, the smallest leaf segments not stalked.

1. Lower (= abaxial) surfaces of leaf blades with scales or multicellular hair in addition to whitish or yellowish farina (= wax-like, mealy, or powdery material).
   2. Lower surfaces of leaf blades in addition to farina mainly having deeply dissected, stellate scales (with obvious scale body, but the scale branches so thin as to be hair-like), the farina completely hidden by the dense overlapping scales ______ **N. aschenborniana**
   2. Lower surfaces of leaf blades in addition to farina mainly having linear to narrowly lanceolate scales and/or multicellular hairs, the farina easily seen.
      3. Lower surfaces of leaf blades in addition to farina mainly having shiny, dark brown, needle-like hairs (these along rachises and costae); sporangia each with 64 spores ______ **N. nealleyi**
      3. Lower surfaces of leaf blades in addition to farina mainly having dull, light brown, linear to narrowly lanceolate scales (these along rachises and costae); sporangia each with 16 or 32 spores.
         4. Upper (= adaxial) surface of leaf blades with sparse covering of long whitish hairs; scales on lower surface of leaf blades ciliate (magnification needed) ______ **N. aliena**
         4. Upper surface of leaf blades without long whitish hairs, glabrous except for scattered, tiny, sessile glands; scales on lower surface of leaf blades entire ______ **N. grayi**
1. Lower surfaces of leaf blades lacking scales or multicellular hairs, with only whitish or yellowish farina.
   5. Leaf blades 2–4 times as long as wide, narrowly deltate; petioles with upper (= adaxial) side flattened or grooved; upper surface of leaf blades with dense whitish glandular pubescence; leaf blades with basal pinnae without greatly enlarged proximal basiscopic pinnules, ± the same shape as more distal pinnae ______ **N. greggii**
   5. Leaf blades only 1–2 times as long as wide, narrowly deltate-pentagonal to pentagonal (= 5-sided in outline) or ovate; petioles with upper side rounded; upper surface of leaf blades glabrous or with sparse glandular pubescence; leaf blades with basal pinnae having proximal basiscopic pinnules being much enlarged (almost like an extra pinna arising at the same point of attachment on the rachis), the basal pinnae thus much differently shaped than the more distal pinnae.
      6. Leaf blades 3–4-pinnate proximally, with at least some ultimate segments constricted near point of attachment and slightly stalked, the farina pale yellowish; ultimate segments of leaf blades with margins strongly recurved, often hiding sporangia ______ **N. neglecta**
      6. Leaf blades 1-pinnate-pinnatifid to 2-pinnate-pinnatifid, the ultimate segments not at all constricted, without any stalk whatsoever, the farina white OR yellowish; ultimate segments of leaf blades with margins only slightly recurved, rarely hiding sporangia.
         7. Lower surface of leaf blades with white farina; leaf blades ovate in outline, 2-pinnate-pinnatifid, the pinnae in middle portion of leaf blade completely distinct from one another, not connected by a wing of blade tissue ______ **N. copelandii**
         7. Lower surface of leaf blades with obvious yellowish farina; leaf blades broadly pentagonal in outline, 1-pinnate-pinnatifid, deeply pinnatifid but not completely pinnate above base (casually appearing more pinnate but segments connected to one another by narrow wings of green blade tissue) ______ **N. standleyi**

Notholaena aliena Maxon, (Latin: *alienus*, foreign, strange, or belonging to another, Maxon [1916] did not give any reason for his choice of epithet; it is possibly in reference to the fact that when named the species was known only from Mexico), MEXICAN CLOAK FERN, FOREIGN CLOAK FERN. Leaves 15 cm or less long; leaf blades linear-lanceolate, 2-pinnate-pinnatifid, 3–6 times longer than wide, the lower surface with cream or pale yellow farina and narrowly lanceolate ciliate scales (magnification needed to see cilia), the upper surface with sparse covering of long lax whitish hairs (villous), the basal pinnae ± the same size as adjacent pair, without greatly enlarged proximal basiscopic pinnules, the ultimate leaf segments sessile, the margins of the ultimate segments only slightly recurved. Rocky areas and cliffs, on igneous rocks (e.g., basalt ledges, among boulders); in the U.S. known only from Brewster Co. (*J.W. Stanford, W.D. Pohl, & J.L. Blassingame 2521*, 1973, BRIT, TEX-LL, *S. Barber 183*, HPC; [Chisos Mts., Black Gap Wildlife Management Area, and lower canyons of the

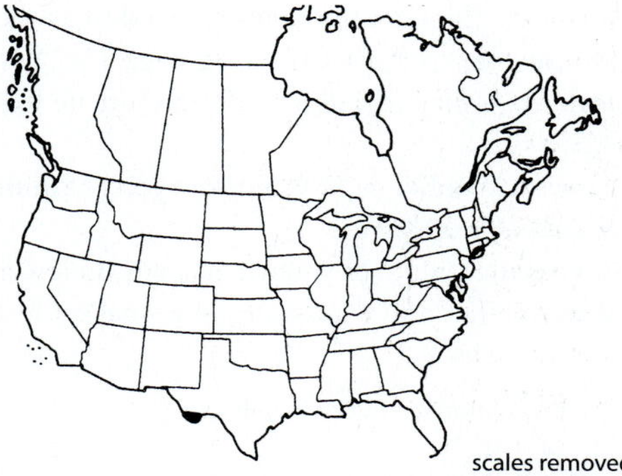

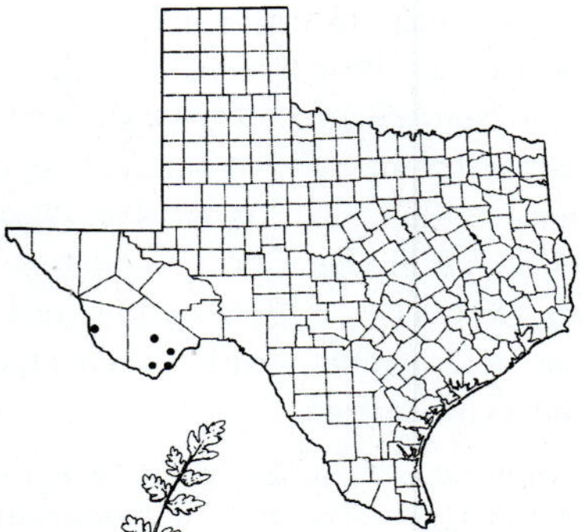

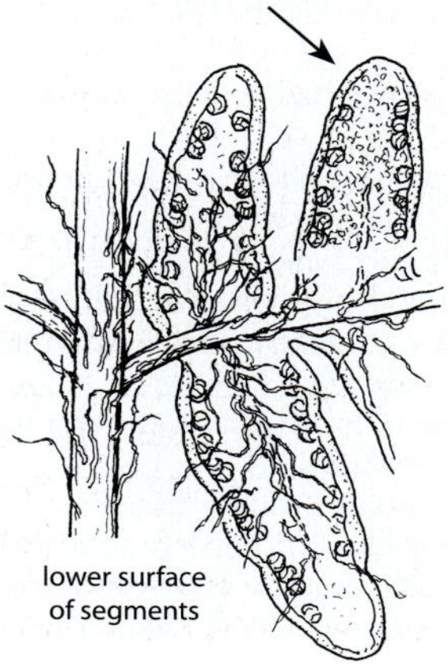

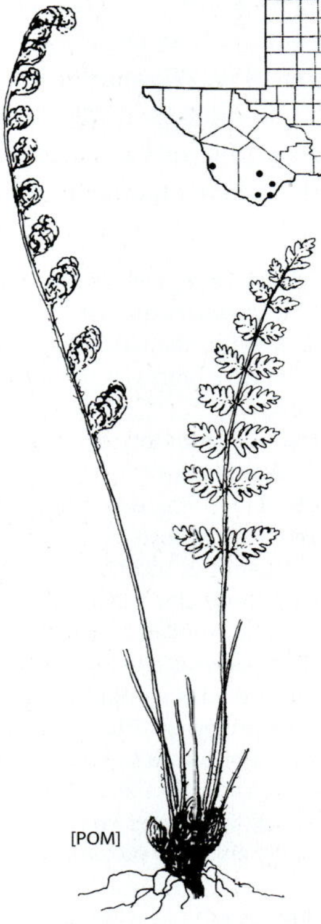

Roy Morey

Notholaena aliena

MEXICAN CLOAK FERN

Rio Grande River—Yarborough & Powell 2002] [Caballos Novaculite s of Marathon, Boquillos Canyon—J. Blassingame, pers. comm.]) and from Presidio Co. (*Warnock 58*, SRSC; *E.J. Lott et al. 5528*, Chinati Mts. State Natural Area, 2005, SRSC, TEX-LL); also n Mexico. Sporulating Feb–Nov. [*Chrysochosma aliena* (Maxon) Pic. Serm.] Windham (1993f) noted that *N. aliena* is closely related to *N. grayi* (distinguished as in the key), that it appears to reproduce by apogamy, and that it is probably triploid. Rothfels et al. (2008) reported their data confirm it to be very closely related to *N. grayi*, and "suggest that *N. aliena* may be an allopolyploid, the first reported for *Notholaena* s.str." Jim Blassingame (pers. comm.), who is very familiar with this species in the field, notes that it and *N. greggii* are two of the most dry-adapted ferns of the Trans-Pecos—they can grow in "desert" areas with only around 8–10 inches of annual rain. *Notholaena aliena* is considered rare or very rare (Windham 1993f; Yarborough & Powell 2002). Because of its rareness and limited distribution in the state, we consider this species to be of conservation concern in TX. ⚠

Notholaena aschenborniana Klotzsch, (for Alwin Aschenborn, 1816–1865, 19th century plant collector who found the type specimen in Mexico), ASCHENBORN'S CLOAK FERN, SCALED CLOAK FERN. Leaves 8–35 cm long; leaf blades lanceolate, 2-pinnate, 2–6 times longer than wide, the lower surface with yellowish or whitish farina and tan to brownish, deeply dissected, stellate scales (with obvious scale body, but the scale branches long and so thin as to be hair-like), the farina completely hidden by the dense overlapping scales, the upper surface with stellate hairs giving the blade a grayish green appearance, the basal pinnae ± the same size as adjacent pair, without greatly enlarged proximal basiscopic pinnules, the ultimate segments sessile or subsessile, the margins of the ultimate segments only slightly recurved; $n = 2n = 90$ (Windham 1993f). Rocky areas on limestone substrates in the mts.; Brewster (Glass and Del Norte Mts.) (*Correll 13588, Correll 14076*, both BRIT, TEX-LL), Pecos (*G. Mckenzie 447*, SRSC), Terrell (*P. Zelazny 181, 183*, SRSC), Val Verde (*Correll & Correll 12904*, TEX-LL), and Culberson (Turner et al. 2003) cos. in the Trans-Pecos and Real (*J.L. Blassigame 2912, 2686*, BRIT) and Val Verde (*J.L. Blassingame 2809* [e of Pecos River], BRIT) cos. in the Edwards Plateau; AZ and TX; also Mexico. Sporulating May–Nov. [*Cheilanthes aschenborniana* (Klotzsch) Mett., *Chrysochosma aschenborniana* (Klotzsch) Pic. Serm., *Notholaena bipinnata* Liebm.] Windham (1993f) noted that all of the individuals tested were apogamous triploids. This species is distinguished by the overlapping scales completely hiding the farina on the lower leaf surfaces. Because of its rareness and limited distribution in the state, we consider this species to be of conservation concern in TX. ⚠

Notholaena copelandii C.C. Hall, (for Edwin Bingham Copeland, 1873–1964, American botanist and fern specialist), COPELAND'S CLOAK FERN. Leaves 25 cm or less long; leaf blades ovate, 2-pinnate-pinnatifid, 1–2 times as long as wide, the lower surface with obvious stark white farina, without scales, the upper surface with glandular pubescence, the basal pinnae having proximal basiscopic pinnules being much enlarged (almost like an extra pinna arising at the same point of attachment on the rachis), the basal pinnae thus much differently shaped than the more distal pinnae, the ultimate leaf segments sessile, the margins of the ultimate segments only slightly recurved; $2n = 60$ (Windham 1993f). Limestone substrates, canyons, ledges, crevices, and rocky slopes; Brewster (e.g., *Correll & Correll 35364*, TEX-LL), Terrell (*Warnock & Johnston 17265*, TEX-LL), and Pecos (Correll 1956) cos. in the Trans-Pecos and scattered in the w Edwards Plateau from Val Verde (*E. Whitehouse s.n.* 1932, SRSC, Del Rio [on e side of Pecos River]; *C.M. Rowell & J.L. Blassingame 3053*, BRIT; *Correll & Correll 12911*, BRIT), Bandera, Edwards, Kerr, Kimble, Real (*J. Blassingame 3111*), Uvalde (all BRIT), and Kinney (TEX-LL) cos. e to Gillespie Co. (*B. O'Kennon 10457*, Enchanted Rock SNA, BRIT; *G. Jermy 349*, US, fide Hall 1950); in the U.S. known only from TX; also n Mexico. Sporulating Mar–Dec. [*Cheilanthes candida* M. Martens & Galeotti var. *copelandii* (C.C. Hall) Mickel, *Chrysochosma candida* (M. Martens & Galeotti) Kümmerle var. *copelandii* (C.C. Hall) Pic. Serm., *N. candida* (M. Martens & Galeotti) Hook. var. *copelandii* (C.C. Hall) R.M. Tryon] The ovate leaf blades with stark white farina are distinctive. While sometimes in the past treated as a variety of the Mexican and Central American *N. candida*, Windham (1993f) discussed a variety of evidence pointing to the two as separate species.

Notholaena grayi Davenp. subsp. **grayi,** (for Asa Gray, 1810–1888, pre-eminent Harvard botanist who was a pioneer in the study of plant geography and author of *Manual of the Botany of the Northern United States*, widely known as "Gray's Manual"). GRAY'S CLOAK FERN. Leaves usually 20 cm or less long; leaf blades linear-lanceolate, 2-pinnate-pinnatifid, 3–6 times longer than wide, the lower surface with whitish farina and lin-

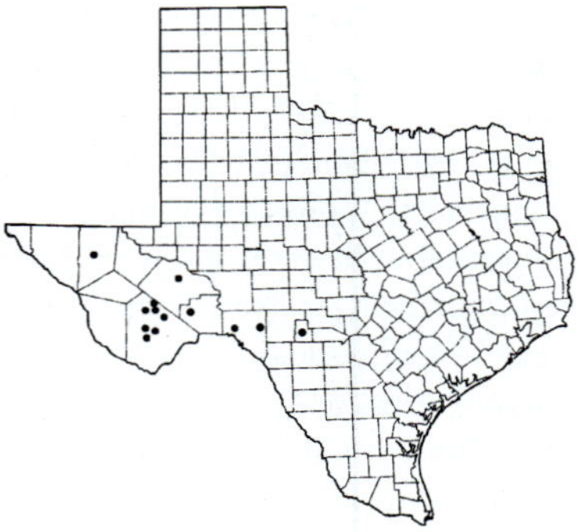

Notholaena aschenborniana

ASCHENBORN'S CLOAK FERN

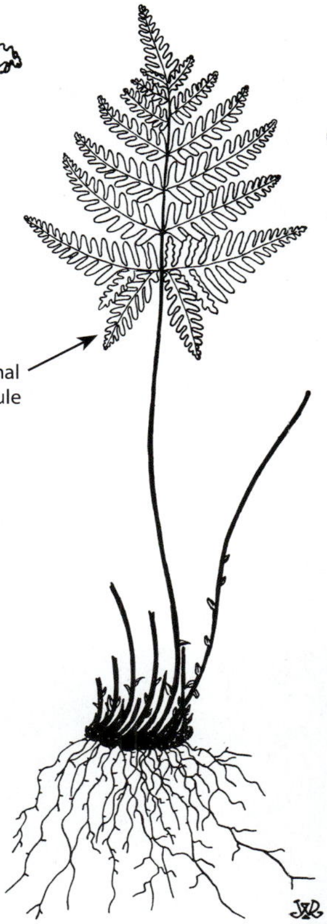

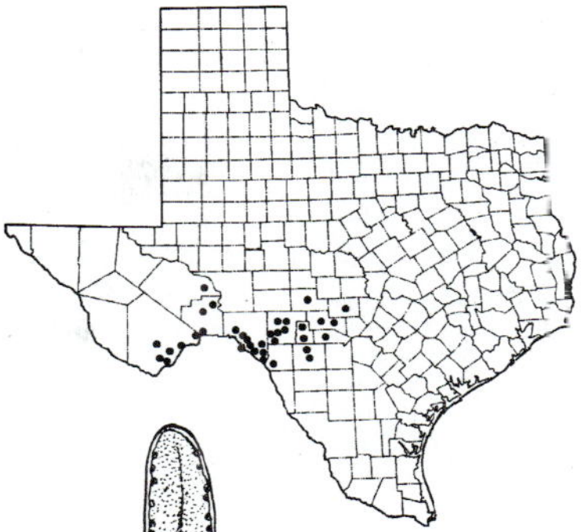

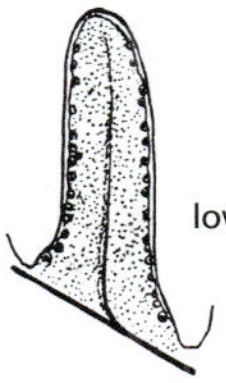

Roy Morey

Roy Morey

Roy Morey

Notholaena copelandii

COPELAND'S CLOAK FERN

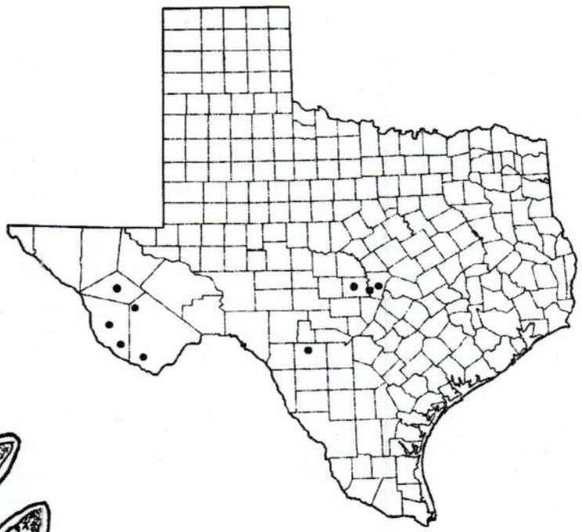

upper surface of pinna [left]
lower surface of pinna [right]

note white farina on
lower surface

Lorena Babcock Moore

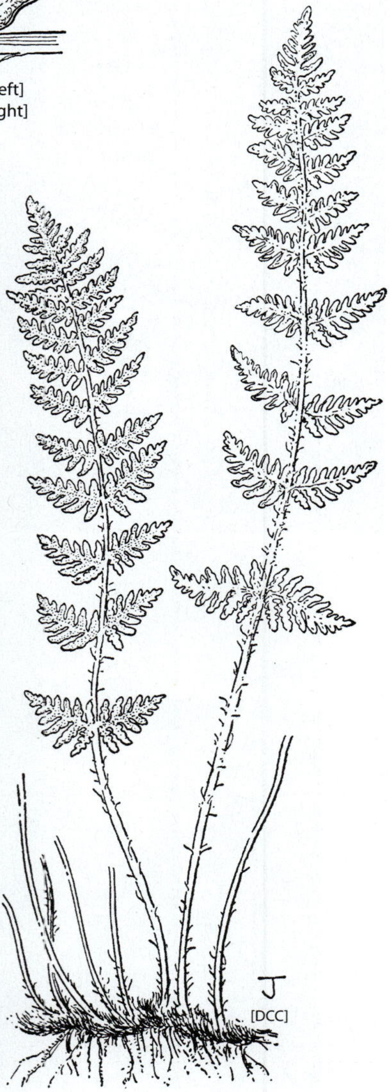

Notholaena grayi subsp. **grayi**

GRAY'S CLOAK FERN

ear to narrowly lanceolate entire scales, the upper surface glabrous except for scattered, tiny, sessile glands, the basal pinnae ± the same size as adjacent pair, without greatly enlarged proximal basiscopic pinnules, the ultimate leaf segments sessile, the margins of the ultimate segments only slightly recurved; $n = 2n = 90$ (Windham 1993f). Granite or limestone substrates, on talus, in crevices, other rocky areas; Brewster (*Warnock 20245, 20247*, TEX-LL), Jeff Davis (Correll 1956), and Presidio (*Warnock 382*, SRSC) cos. in the Trans-Pecos and Burnet (*Correll & Correll 12732, 12734*, BRIT, TEX-LL, Granite Mt.), Llano (Seigler 1977; *Seigler et al. 1514*, cited as being at TEX-LL, but specimen cannot be located), and Uvalde (Correll 1956) cos. on the Edwards Plateau; AZ, NM, and TX; also n Mexico. Sporulating Mar–Oct. [*Cheilanthes grayi* (Davenp.) Domin, *Chrysochosma grayii* (Davenp.) Davenp.] Windham (1993f) indicated that subsp. *grayi*, characterized by 16 spores per sporangium, is made up of apogamous triploids, whereas subsp. *sonorensis* Windham, of AZ and w Mexico and characterized by 32 spores per sporangium, represents sexually reproducing diploids. This species is most similar to *N. aliena* and distinguished as in the key. Because of its rareness and limited distribution in the state, we consider this species to be of conservation concern in TX. ⚠

Notholaena greggii (Mett. ex Kuhn) Maxon, (for Josiah Gregg, 1806–1850, who collected in Mexico and died in the wilderness in n CA), GREGG'S CLOAK FERN. Leaves 20 cm or less long; leaf blades narrowly deltate, 2–3-pinnate, 2–4 times longer than wide, the lower surface with obvious but sometimes sparse whitish farina, but scales absent, the upper surface with dense whitish glandular pubescence, the basal pinnae ± the same size as adjacent pair, without greatly enlarged proximal basiscopic pinnules, the ultimate leaf segments sessile or subsessile, the margins of the ultimate segments strongly recurved and often hiding sporangia; $2n = 60$ (Windham 1993f). Rocky mountain slopes, bluffs, ledges, crevices, and canyons, on limestone or gypsum; in the Trans-Pecos in the Big Bend area of s Brewster Co. (e.g., *J.L. Blassingame 1325*, BRIT, TEX-LL, "lava ledges," Black Gap; *J.W. Stanford et al. 2538*, BRIT, Black Gap; *Correll & Warnock 14978*, TEX-LL, Boquillas Canyon; *S. Adair 91*, HPC, Black Gap) and in s Presidio Co. (e.g., *Correll & Johnston 24461*, SRSC, TEX-LL, Bofecillos Mts.); in the U.S. known only from TX; also n Mexico. Sporulating summer–fall. [*Cheilanthes greggii* (Mett. ex Kuhn) Mickel, *Chrysochosma greggii* (Mett. ex Kuhn) Pic. Serm., *Pellaea greggii* Mett. ex Kuhn] Windham (1993f) noted that all known U.S. populations lie within 25 km of the international border. The combination of densely glandular upper leaf surfaces and petioles with their upper side flattened or grooved is distinctive. This and *N. neglecta* are the only TX species with the margins of the ultimate segments strongly recurved. Jim Blassingame (pers. comm.) notes that it one of the most dry-adapted ferns of the Trans-Pecos—it can grow in "desert" areas (creosote bush flats with only around 8–10 inches of annual rain) versus the seasonally dry habitats of many other Trans-Pecos ferns. It is considered rare (Windham 1993f; Yarborough & Powell 2002). Because of its rareness and limited distribution in the state, we consider this species to be of conservation concern in TX. ⚠

Notholaena nealleyi Seaton ex J.M. Coult., (for Greenleaf Cilley Nealley, 1846–1896, TX botanist and possibly the first botanist at Texas A&M Univ.), NEALLEY'S CLOAK FERN. Leaves 15 cm or less long; leaf blades lanceolate, 2-pinnate-pinnatifid, 3–6 times longer than wide, the lower surface with obvious whitish farina and shiny, dark brown, needle-like hairs (the hairs along the rachises and costae), the upper surface with glandular pubescence, the basal pinnae ± the same size as adjacent pair, without greatly enlarged proximal basiscopic pinnules, the ultimate leaf segments sessile to subsessile, the margins of the ultimate segments only slightly recurved. Cliffs, ledges, crevices, and talus slopes, usually on limestone; Val Verde [including e of the Pecos River in the Edwards Plateau] (*Correll & Correll 12899*, BRIT, TEX-LL), Brewster, (*W.D. Pohl & J.L. Blassingame 2962*, BRIT; *Correll 31568, 31584, M. Butterwick & E. Lott 3707, 3708*, all TEX-LL), Terrell (*P. Zelazny 182*, SRSC and Presidio (Coulter 1890—cited Chenate [Chinati] Mts. as location of original collection; Correll 1956) cos. in or adjacent to the Trans-Pecos and disjunct to Starr Co. (*S.E. Wolf 800-s*, BRIT, *Correll 14890*, TEX-LL; Correll 1956) in far s TX; in the U.S. known only from TX (all collections within ca. 30 km of the international border); also Mexico. Sporulating May–Nov. [*Cheilanthes nealleyi* (Seaton ex J.M. Coult.) Domin, *Chrysochosma schaffneri* var. *nealleyi* (Seaton ex J.M. Coult.) Pic. Serm., *Notholaena schaffneri* (E. Fourn.) Underw. ex Davenp. var. *nealleyi* (Seaton ex J.M. Coult.) Weath.] The shiny dark brown needle-like hairs on the lower surface of the leaves are definitive in distinguishing this species from other members of the genus in TX. Mickel and Smith (2004) included this species in the more widespread *N. schaffneri* (E.

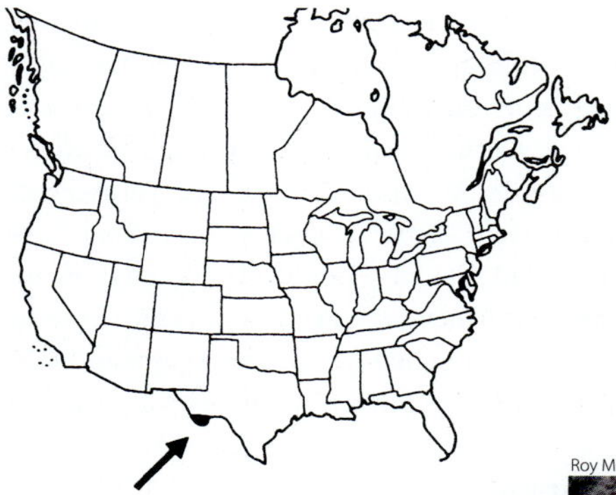

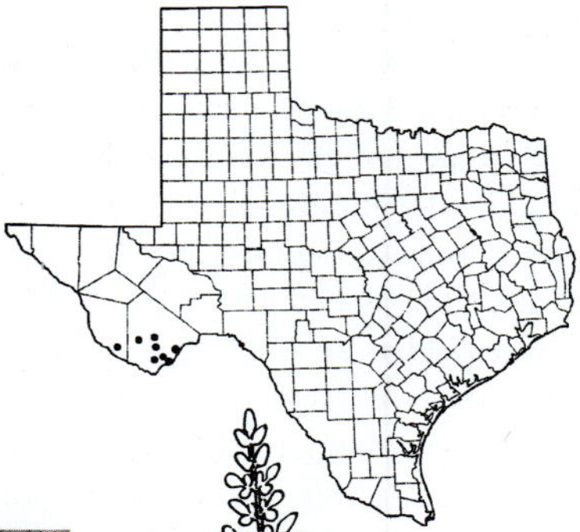

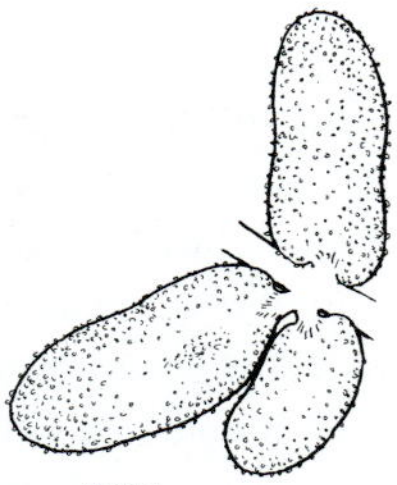

upper surface of leaf
w/whitish
glandular pubescence

[POM]

Roy Morey

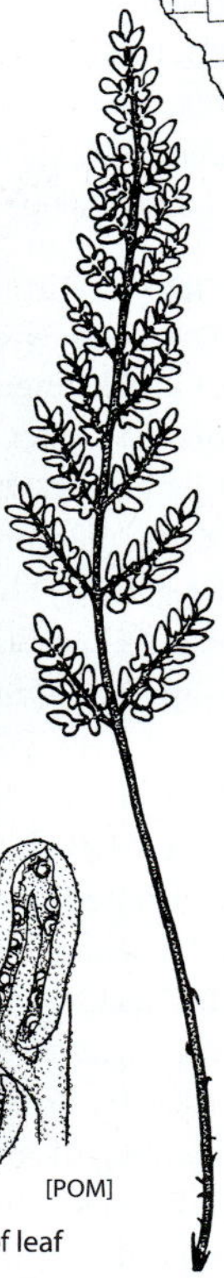

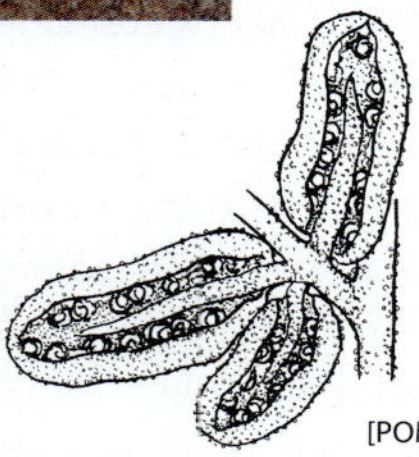

[POM]

lower surface of leaf
w/farina

[LUN]

Roy Morey

Roy Morey

Notholaena greggii

GREGG'S CLOAK FERN

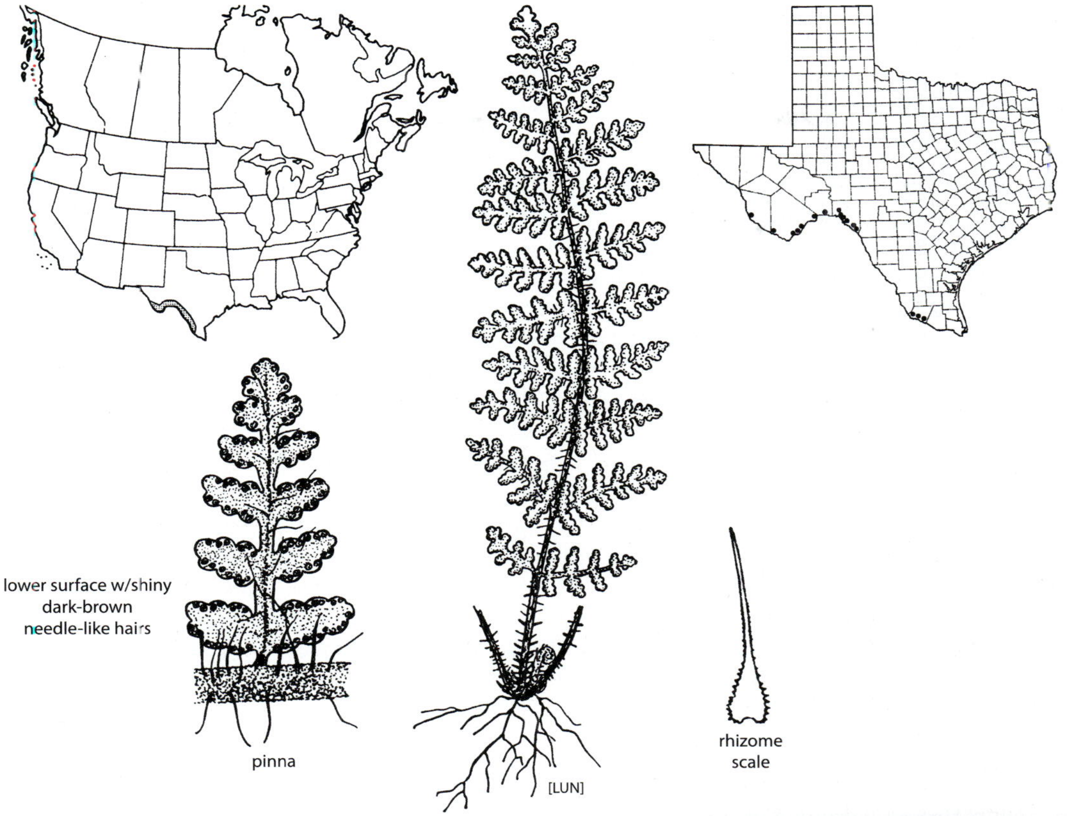

Carl Rothfels

Notholaena nealleyi

NEALLEY'S CLOAK FERN

Fourn.) Underw. ex Davenp., noting that it "appears to be only a diminutive form of *N. schaffneri*." However, we are following Windham (1993f), who recognized it as a distinct species and suggested that it may be an allotetraploid hybrid between *N. schaffneri* and an as yet unidentified species. This species is considered rare (Yarborough & Powell 2002). Because of its rareness and limited distribution in the state, we consider this species to be of conservation concern in TX. ⚠

Notholaena neglecta Maxon, (Latin: *neglectus*, pay no attention, neglected, overlooked), MAXON'S CLOAK FERN. Leaves 15 cm or less long; leaf blades narrowly deltate-pentagonal, 3–4-pinnate, 1–2 times as long as wide, the lower surface with pale yellow farina, without scales, the upper surface glabrous or with sparse glandular pubescence, the basal pinnae much larger than adjacent pair due to the proximal basiscopic pinnules being much enlarged, with at least some ultimate segments constricted near point of attachment and slightly stalked, the margins of the ultimate segments strongly recurved and often hiding sporangia; $n = 2n = 90$ (Windham 1993f). Rocky limestone slopes and cliffs, crevices, canyons; Brewster (*Correll 13583, 14075*, BRIT, TEX-LL), Val Verde (*Correll & Correll 12900*, TEX-LL, Rio Grande Canyon, Langtry [w of the Pecos River]), and Culberson (*J.L. Blassingame 3156*, BRIT; Sierra Diablo Mts., Yarborough & Powell 2002) cos. in the Trans-Pecos, and Uvalde Co. (Correll 1956) in the far w Edwards Plateau; AZ and TX; also n Mexico. Sporulating Apr–Nov. [*Cheilanthes neglecta* (Maxon) Mickel, *Chrysochosma neglecta* (Maxon) Pic. Serm.] *Notholaena neglecta* is similar to *N. copelandii* (though typically smaller) but differs in having leaf blades that are more finely dissected, more pentagonal in outline, and with pale yellow farina. This and *N. greggii* are the only TX species with the margins of the ultimate segments strongly recurved. Windham (1993f) pointed out that all known U.S. populations are apogamous triploids; however, sexually reproducing forms have been found in Mexico (Windham & Yatskievych 2003b). According to Moore (2008) the "dry leaves curl into crescent shapes." Correll noted on some collections that this species is "Rare!" and it is also considered rare by Yarborough and Powell (2002). Because of its rareness and limited distribution in the state, we consider this species to be of conservation concern in TX. ⚠

Notholaena standleyi Maxon, (for Paul Carpenter Standley, 1884–1963, American botanist at Field Museum of Natural History and author of *Trees and Shrubs of Mexico*), STAR CLOAK FERN, STAR FERN. Leaves to 30 cm long; leaf blades broadly pentagonal, 1-pinnate-pinnatifid, deeply pinnatifid but not completely pinnate above base (casually appearing more pinnate but segments connected to one another by narrow wings of green blade tissue), 1–2 times as long as wide, the lower surface with obvious yellowish farina (sometimes pale), without scales, the upper surface glabrous, the basal pinnae much larger than adjacent pair due to the proximal basiscopic pinnules being much enlarged, the ultimate segments sessile, the margins of the ultimate segments only slightly recurved; $2n = 60$ (Windham 1993f). Rocky areas on igneous and limestone substrates; widespread in the Trans-Pecos (e.g., *Correll 13619*, Brewster Co., BRIT; *P. Zelazny 147, 157, 195*, Terrell Co., SRSC; *Correll & Correll 12896*, TEX-LL, "Rio Grande Canyon, Langtry," w Val Verde Co.) and scattered on the Edwards Plateau in Llano (*J.W. Stanford 4514*, BRIT, TEX-LL; *Reverchon 1218*, House Mt., BRIT, TEX-LL; Seigler 1977), Gillespie (*O'Kennon 10643*, BRIT), and Edwards (Turner et al. 2003) cos.; AZ, CO, NM, OK, and TX; also Mexico. Sporulating Mar–Nov. [*Cheilanthes standleyi* (Maxon) Mickel, *Chrysochosma hookeri* Kümmerle, *Notholaena hookeri* D.C. Eaton] A record from Gaines Co. in the High Plains (*Nealley 277*, MSC; Seigler & Wollenweber 1983) is apparently an error based on erroneous location data (J. Blassingame, pers. comm.; J. Stanford, pers. comm.). Recent molecular research (Rothfels et al. 2008) shows that of all the species typically placed in *Notholaena* (e.g., by Windham 1993f), *N. standleyi* is most distantly related to the rest. Windham (1993f) reported that all known plants were sexual diploids, but G. Yatskievych (pers. comm.) noted that this is no longer true. Seigler and Wollenweber (1983), following up on a discussion in Correll (1956, 1966), found that there were three groups within *N. standleyi* that differed in soil-related requirements, were "almost distinct" geographically, and that could be visually recognized based on slight differences in farina color—they referred to them as the "yellow, yellow-green, and gold races." The yellow and yellow-green races occur in TX. No definitive morphological characters could be found between them, and no formal taxonomic recognition has been made; further investigation is probably warranted. In dry weather, the plant appears as a cluster of dark, brittle, often broken petioles with curled leaf blades. Correll (1956, 1966) noted that when dry, the segments of the leaf blade "are rolled inward to form a tight yellow whitish ball" (see discussion on

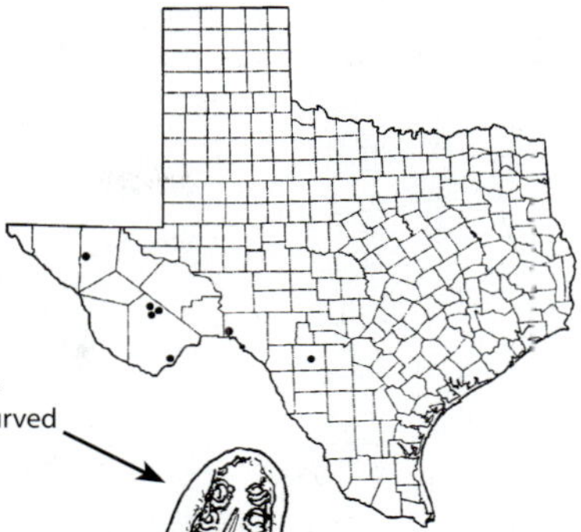

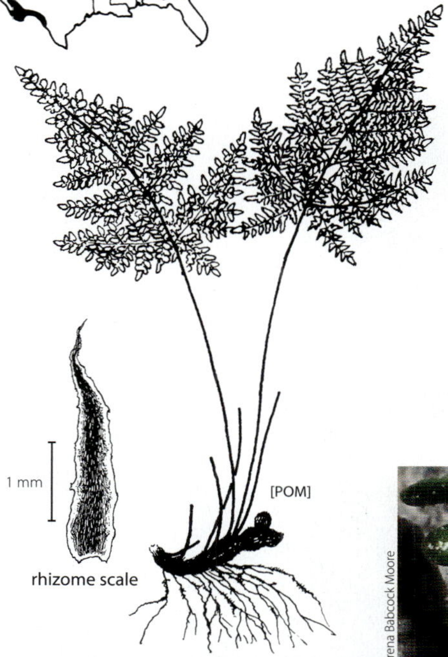

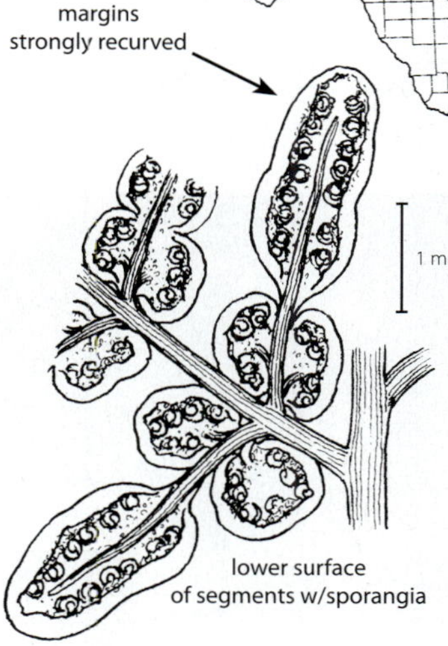

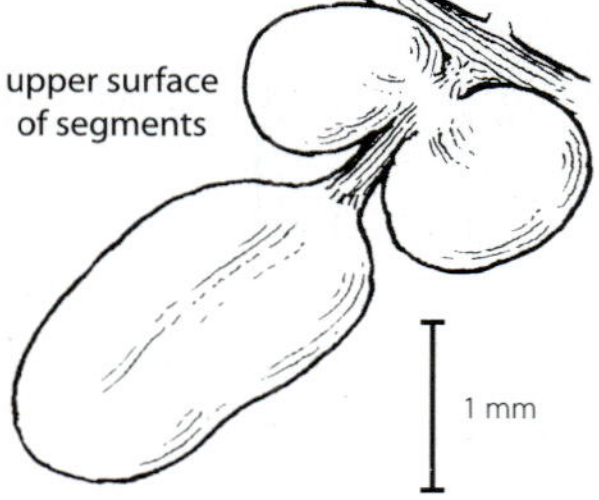

Notholaena neglecta
MAXON'S CLOAK FERN

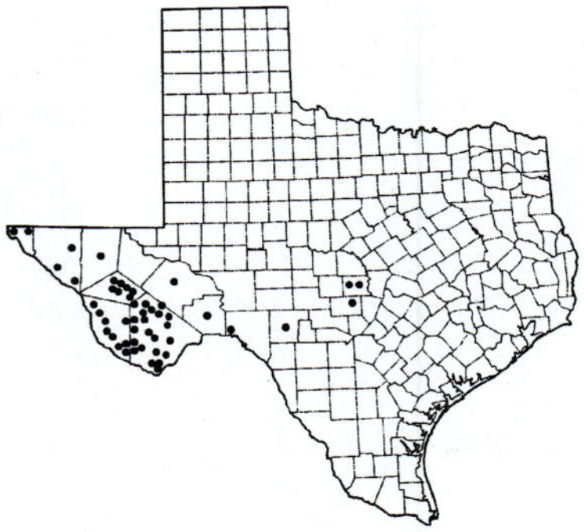

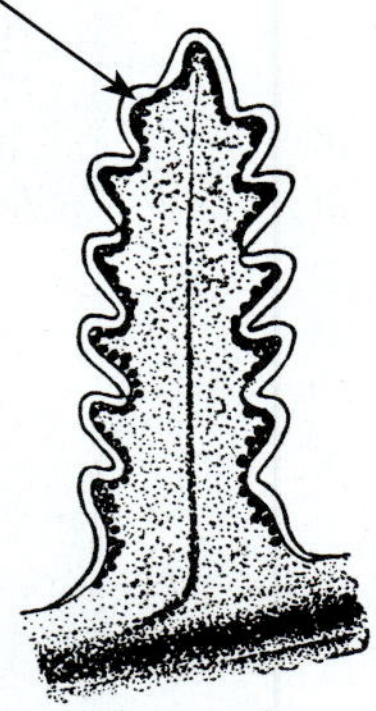

note yellow farina
on lower surface

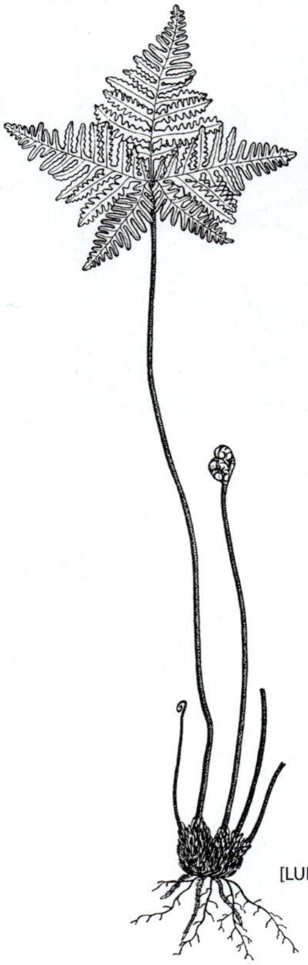

Notholaena standleyi

STAR CLOAK FERN

page 27). The "star-like" pentagonal leaf blades with abundant yellowish farina are distinctive. Although the leaf shape is similar to that of *Bommeria hispida*, that species lacks farina, the lower surface of its leaves merely being hairy (Yarborough & Powell 2002).

PELLAEA Link CLIFF-BRAKE

Xerophytic, usually on or among rocks; rhizomes compact to creeping; leaves monomorphic or somewhat dimorphic, evergreen, clustered to scattered, usually 1–4 pinnate proximally, glabrous or variously pubescent, leathery to thick-herbaceous; ultimate leaf segments (= smallest subdivisions of leaf) usually slightly stalked; sporangia along veins near margins of ultimate leaf segments on the lower (= abaxial) leaf surfaces; margins of ultimate leaf segments recurved forming often rather poorly defined false indusia.

A genus often recognized as having ca. 35–40 species primarily of dry rocky areas distributed mainly in the New World with a few in Asia, Africa, the Pacific Islands, and Australia (Windham 1993c; Mickel & Smith 2004). It has sometimes been circumscribed more broadly, but as such it is probably polyphyletic. In fact, molecular evidence (Gastony & Rollo 1998; Kirkpatrick 2007; Prado et al. 2007) supports the hypothesis that even as presently viewed *Pellaea* is paraphyletic or polyphyletic. Windham (1993c) noted that the species in the U.S. "belong to a closely knit alliance that is usually recognized as a distinct section (sect. *Pellaea*)." As with many Pteridaceae, the generic limits of *Pellaea* and related taxa are still uncertain due to convergent evolution, and changes in generic circumscription should be expected. Based on molecular work, Kirkpatrick (2007) suggested that one possibility would be to divide the "pellaeoid" clade into two genera: *Astrolepis* and *Pellaea*. He noted that another reasonable approach would be to divide this "pellaeoid clade" into four genera: *Pellaea* (limited to the traditional sect. *Pellaea*), *Platyloma* (section *Platyloma*), *Astrolepis*, and a newly circumscribed *Paragymnopteris*. Kirkpatrick (2007) also pointed out that "further taxon sampling and phylogenetic analysis…is needed before formal taxonomic revision." Relative to other TX genera, *Pellaea* is most closely related to *Astrolepis* and then to *Argyrochosma* and *Cheilanthes*. (Greek: *pellos*, dark or dusky, possibly referring to bluish gray leaves or to the dark petiole and rachis of some species)
REFERENCES: Tryon (A.F.) 1957, 1968, 1972; Knobloch & Britton 1963; Gastony 1988; Gastony & Windham 1989; Gastony et al. 1992; Windham 1993c, 1993e; Platt & Townsend 1996; Heafner 2001; Mendoza et al. 2001; Windham & Yatskievych 2003a; Kirkpatrick 2007; Prado et al. 2007.
QUICK TIPS FOR GENUS ID: Dry area ferns with grayish green to bluish green or dark green, often leathery leaves typically without obvious covering of scales, hair, or farina, with separate segments (usually 2-pinnate at base, 1-pinnate towards tip).

1. Petiole and rachis straw-colored, tan or gray, not shiny; ultimate leaf segments usually ± rounded at tip; rhizomes scales narrowly lanceolate to ovate, 0.3–0.8 mm wide.
 2. Rachis uniformly zigzag throughout; pinnae attached to rachis so that their tips point somewhat towards base of leaf (= attachment of pinna costa to rachis is distal to pinna tip) ______ **P. ovata**
 2. Rachis not uniformly zigzag, at most slightly flexuous; pinnae attached to rachis so that their tips point ± straight out or somewhat towards tip of leaf.
  3. Ultimate leaf segments round to deltate, their bases distinctly cordate; rachis glabrous; rhizomes with uniformly orange-brown scales ______ **P. cordifolia**
  3. Ultimate leaf segments elliptic to ovate, not distinctly cordate at base (instead truncate or nearly so); rachis ± pubescent; rhizomes with scales blackish in center and brown along margins ______ **P. intermedia**
1. Petiole and rachis reddish purple to dark brown or black, usually shiny; ultimate leaf segments usually ± pointed at tip; rhizome scales narrowly linear, 0.3 mm or less wide.
 4. Rhizome scales with 2 distinct colors, black in center with brown along margins (look near attachment of petioles); rachis usually glabrous on upper (= on adaxial) surface OR villous (= ± shaggy with long hairs).
  5. Basal pinnae of 3 incompletely divided segments (pinnae pinnatifid); petioles dark purple or black; ultimate leaf segments 10–40 mm long.
   6. Rachis villous (= shaggy, with long, soft, spreading or ascending, unmatted hairs), especially in axils of pinnae; pinnae with hairs scattered along main veins on lower surface; pinnae in distal ⅓ to ½ of leaf usually entire, those in basal part of leaf ternately divided (= into three portions) ______ **P. villosa**
   6. Rachis glabrous or with only a few scattered hairs; pinnae completely glabrous on lower surface; all or almost all pinnae deeply ternately divided (except 1–5+ at leaf tip) ______ **P. ternifolia**

5. Basal pinnae with 3–25 completely separate segments (pinnae truly pinnate; if basal pinnae only have 3 segments, the terminal segment is on a distinct short stalk); petioles dark brown; ultimate leaf segments 4–20 mm long.
   7. Individual pinnae mostly with 9–25 ultimate segments each; length of entire costa (= midvein of individual pinna) of a pinna much longer than longest individual pinnule on that pinna; leaf blades ovate-deltate in outline, 4–18 cm wide ____ **P. truncata**
   7. Individual pinnae mostly with 3–9 ultimate segments each; length of entire costa of a pinna less than or equal to length of longest individual pinnule on that pinna; leaf blades linear-oblong in outline, 1.5–4.5(–5) cm wide ____ **P. wrightiana**

4. Rhizome scales uniformly reddish brown or tan; rachis densely pubescent above with short curly hairs (except glabrous in the very rare *P. glabella*).
   8. Rachis densely pubescent above with short curly hairs; basal pinnae mostly relatively longer-stalked (stalks 5–15 mm); petioles dark reddish purple to nearly black; widespread species known nearly throughout TX ____ **P. atropurpurea**
   8. Rachis glabrous above; basal pinnae sessile to short-stalked (stalks < 5 mm long); petioles dark brown; species mainly of ne U.S., disjunct to TX and known in the state only from Potter Co. in the High Plains ____ **P. glabella**

Pellaea atropurpurea (L.) Link, (Latin: *atropurpureus*, black-purple, = dark purple, in reference to the color of the petiole, which can be so dark as to appear almost black), PURPLE CLIFF-BRAKE, PURPLE-STEM CLIFF-BRAKE, CLIFF-BRAKE, BLUE FERN. Plant to 45(–50) cm tall; rhizome scales uniformly colored, reddish brown or tan; petioles dark reddish purple to nearly black, shiny; leaf blades usually 1-pinnate or 2-pinnate below, 10–30 cm long, 5–20 cm wide, somewhat dimorphic, the sterile ones shorter, less divided, and with relatively wider ultimate segments; rachis and costae densely pubescent on upper surfaces with short curly hairs; pinnae attached to rachis so that their tips point ± straight out or somewhat towards tip of leaf, the basal pinnae mostly relatively longer-stalked (stalks 5–15 mm), the upper pinnae sessile or nearly so; ultimate leaf segments linear-oblong, apically obtuse to very slightly mucronate, leathery, often with a few hairs near midrib on lower surfaces; $n = 2n = 87$ (Windham 1993c). Rocky slopes and woods, cliffs, usually on limestone or calcareous rocks but also on igneous substrates; nearly throughout TX (e.g., *Correll & Correll 12715*, BRIT, TEX-LL); se Canada and most of the U.S. from VT s to FL w to MN, NV, and AZ; also Mexico and Guatemala. [*Pteris atropurpurea* (L.) Link, *Pellaea atropurpurea* var. *cristata* Trel.] Sporulating Mar–Nov. This species is a triploid that reproduces via apogamy (= a type of asexual reproduction that does not involve gamete formation or fertilization; this type of reproduction allows triploids, which would otherwise produce sterile spores and be sterile, to make fertile spores and thus reproduce) (Tryon 1972). Its origin is unclear; Windham (1993c) suggested that "it appears that *P. atropurpurea* is an autopolyploid derivative of a single diploid taxon that has not yet been located." *Pellaea atropurpurea* "is possibly the most widely distributed fern in Texas, growing in many areas where rock outcrops appear" (Yarborough & Powell 2002). Tryon (1972) pointed out that the purple cliff-brake "is the only American species of the large complex of cheilanthoid ferns that has a range extending from the tropics into northern regions, including New England."

Pellaea cordifolia (Sessé & Moc.) A.R. Sm., (Latin: *cordis*, heart, and *folius*, leaved, = heart leaf), HEART-LEAF CLIFF-BRAKE. Plant to 50 cm tall; rhizomes compact, ascending, distinctive in being rather stout (> 5 mm in diam.) and soft or brittle, the scales uniformly colored, orangish brown, relatively long; petioles straw-colored, tan, or gray, not shiny; leaf blades 2 pinnate below, 5–20 cm wide, somewhat dimorphic, the sterile ones shorter; rachis and costae glabrous; pinnae attached to rachis so that their tips point ± straight out or slightly towards tip of leaf; ultimate leaf segments round to deltate, cordate basally, rounded or even retuse apically, herbaceous to leathery, glabrous or puberulent; $2n = 58$ (Windham 1993c). Canyons, cliffs, mtn. slopes; Chinati Mts., Presidio Co. (*McVaugh 7451, Muller 8334*, BRIT, TEX-LL), Chisos Mts., Brewster Co. (*E.G. Marsh, 251a*, TEX-LL), and Davis Mts., Jeff Davis Co. (*J.W. Stanford & J.L. Blassingame 1168*, BRIT; *Correll 34978, A.M. Powell & S.A. Powell 4915*, TEX-LL) in the Trans-Pecos; in the U.S. known only from TX; also Mexico. Sporulating Jul–Nov. [*Adiantum cordifolium* Sessé & Moc., *P. cardiomorpha* Weath., *P. sagittata* (Cav.) Link var. *cordata* (Cav.) A.F. Tryon] This species is a diploid. In the past it was treated as a variety of the Mexican, Central and South American *P. sagittata*; however, the two are separated by a number of morphological differences and "it seems unlikely that they represent cytotypes of a single species" (Windham 1993c). Yarborough and Powell (2002) considered *P. cordifolia* to be rare in TX. Because of its rareness and limited distribution in the state, we consider this species to be of conservation concern in TX. ⚠

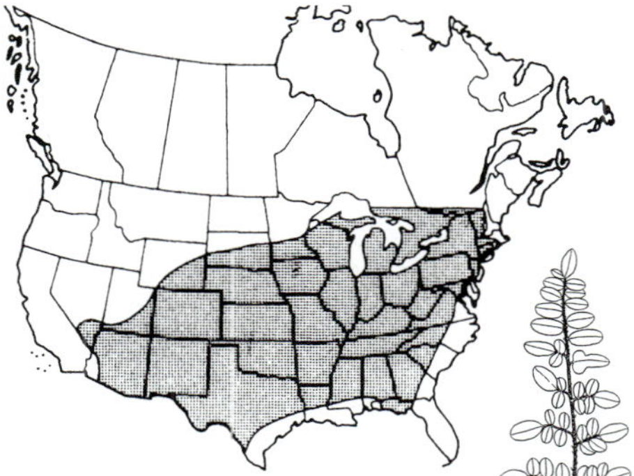

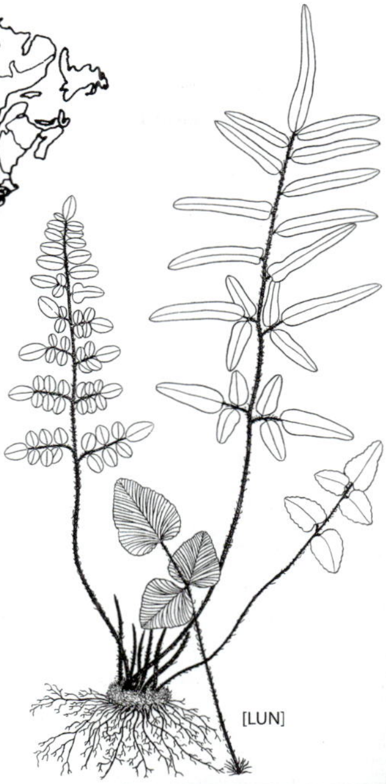

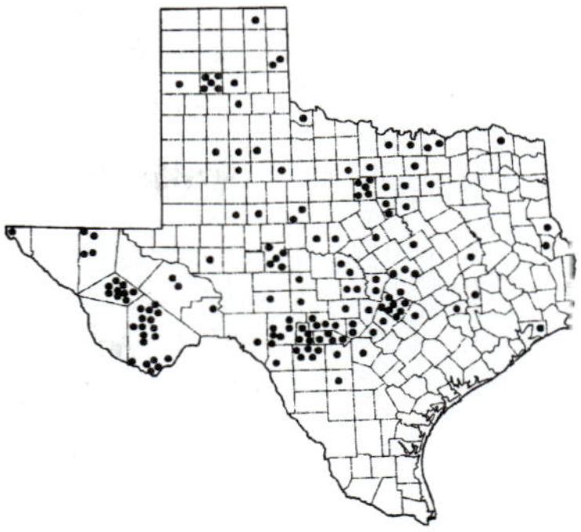

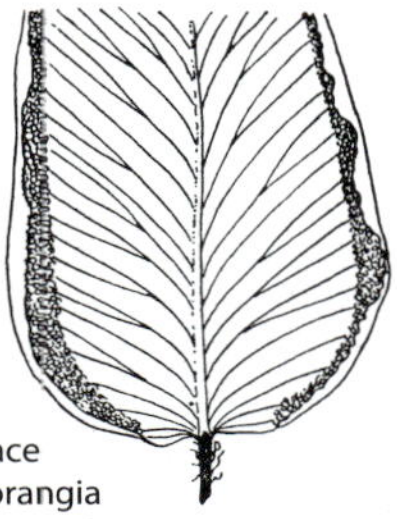

Dale A. Zimmerman Herbarium, Western N. Mex. Univ.

Alan Cressler

Alan Cressler

Pellaea atropurpurea

PURPLE CLIFF-BRAKE

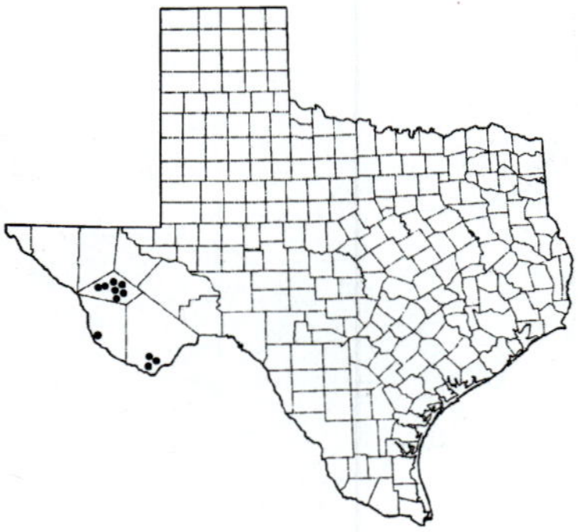

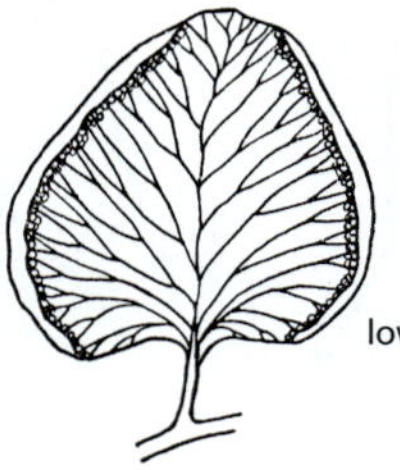

Missouri Botanical Garden

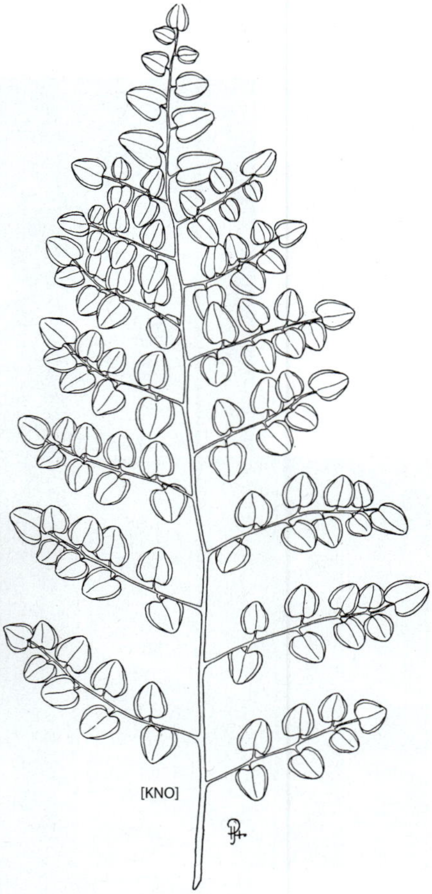

Pellaea cordifolia

HEART-LEAF CLIFF-BRAKE

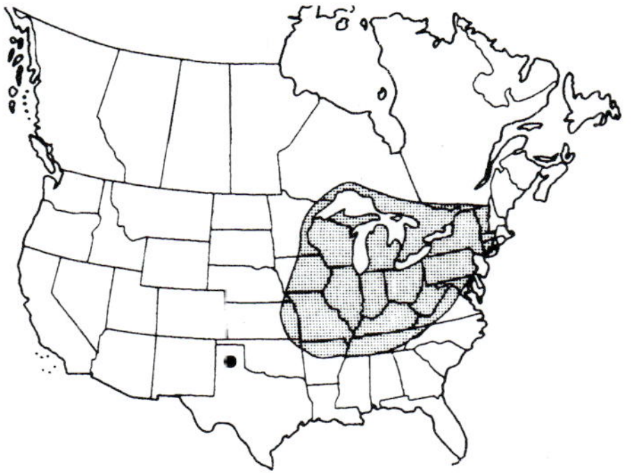

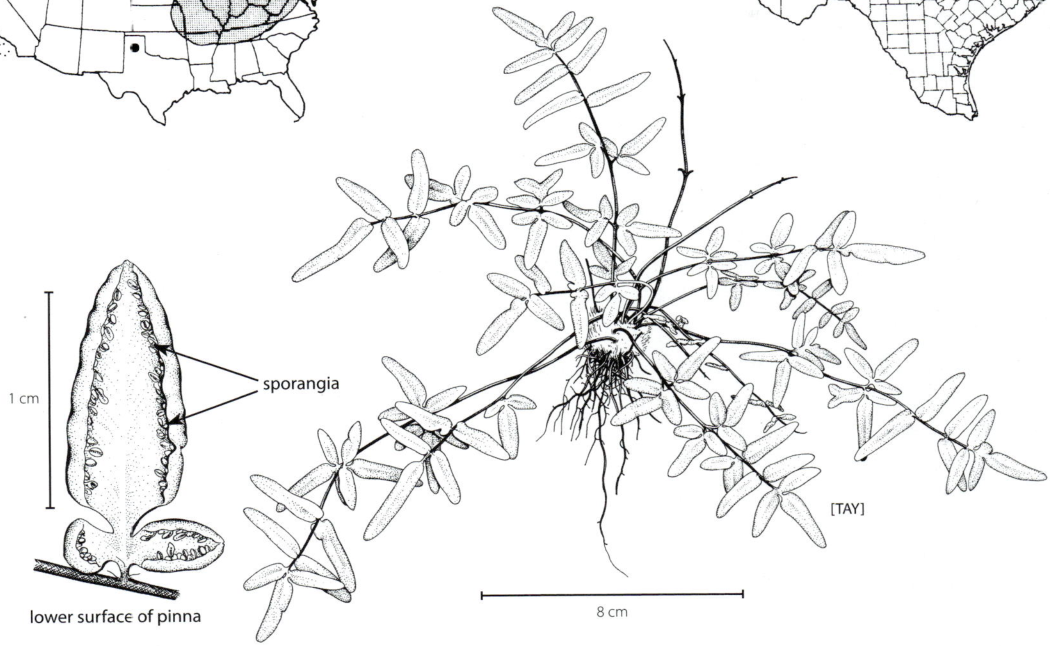

Alan Cressler

Pellaea glabella subsp. **glabella**

SMOOTH CLIFF-BRAKE

Pellaea glabella Mett. ex Kuhn. subsp. **glabella,** (Latin: *glaber*, without hair, plus the diminutive suffix, *-ella*, signifying tiny or small), SMOOTH CLIFF-BRAKE. Plant to 40 cm tall; rhizome scales uniformly colored, reddish brown; petioles dark brown, shiny; leaf blades 1-pinnate or usually 2-pinnate below, 1–8 cm wide, monomorphic; rachis and costae glabrous; pinnae attached to rachis so that their tips point ± straight out or slightly towards tip of leaf, the basal pinnae sessile to short-stalked (stalks < 5 mm long), the upper pinnae sessile; ultimate leaf segments oblong-lanceolate, apically obtuse, leathery to herbaceous, glabrous except some ultimate segments with hair-like scales on lower surface near midrib; $n = 2n = 116$ (Windham 1993c). Cliffs and ledges, in crevices, on limestone or calcareous sandstone; in TX known only from Potter Co. (*Correll & Correll 13085*, along John Rey Creek, in crevices of dry limestone, 1946, BRIT, TEX-LL; Turner et al. 2003, in the High Plains); widespread in the ne U.S. (and adjacent s Canada) s to NC, TN, and AR, and w to MI, KS, and OK, disjunct to the TX Panhandle. Sporulating Jun–Oct. [*P. atropurpurea* var. *bushii* Mack. & Bush] *Pellaea glabella* is currently recognized as being made up of 4 geographically isolated subspecies (Windham 1993c); subspecies *glabella* is an apogamous tetraploid (Windham 1993c). Prior to the use of molecular techniques, it was thought that this tetraploid might have arisen as a hybrid involving *P. atropurpurea*; however, research by Gastony (1988) and Gastony et al. (1992) conclusively showed this was not the case. It appears to be an autotetraploid derived from the diploid *P. glabella* subsp. *missouriensis* (Gastony) Windham, known only from Missouri. Because of its rareness and limited distribution in the state, we consider this species to be of conservation concern in TX. ⚠

Pellaea intermedia Mett. ex Kuhn, (Latin: *intermedius*, that which is in between, = intermediate), INTERMEDIATE CLIFF-BRAKE, CREEPING CLIFF-BRAKE. Plant to 50 cm tall; rhizomes slender (< 5 mm diam.), cord-like, long-creeping, the scales with 2 distinct colors, black in center with brown along margins; petioles straw-colored, tan, or gray, not shiny; leaf blades usually 2-pinnate below, 4–20 cm wide, monomorphic; rachis and costae pubescent with short hairs, or glabrescent; pinnae attached to rachis so that their tips point ± straight out or slightly towards tip of leaf; ultimate leaf segments ovate to elliptic, apically obtuse to very slightly mucronate, leathery, essentially truncate at base, glabrous or puberulent on lower surfaces; $n = 2n = 87$, 116 (Windham 1993c). Canyons, slopes, rocky areas; widespread in the Trans-Pecos and Upton (*Correll 13567*, BRIT, TEX-LL) and Val Verde (*Tharp 4286*, TEX-LL, "Devil's River, Del Rio", which is well to the e of Pecos River) cos. in w Edwards Plateau; AZ, NM, and TX; also n Mexico. Sporulating Mar–Nov. [*P. intermedia* var. *pubescens* Mett. ex Kuhn] Windham (1993c) noted that members of this species in the U.S. are apogamous triploids and tertraploids. However, a sexual diploid population is known from Mexico (Tryon 1968) and Mickel and Smith (2004) suggested that "given the morphological similarity among the three cytotypes" the polyploids probably arose through autopolyploidy.

Pellaea ovata (Desv.) Weath., (Latin: *ovatus*, egg-shaped or ovate, presumably in reference to the shape of the ultimate leaf segments), ZIGZAG CLIFF-BRAKE, OVATE-LEAF CLIFF-BRAKE. Plant usually relatively large, 15 cm to 1 m or more tall; rhizome scales with 2 distinct colors, black in center with brown along margins; petioles straw-colored, tan, or gray, not shiny; leaf blades 2–3-pinnate, usually 3-pinnate basally, 15–70 cm long, 5–25 cm wide, monomorphic, pale green; rachis and costae strikingly flexuous (rachis appearing "zigzag"), usually glabrous; pinnae attached to rachis so that their tips point somewhat towards base of leaf; ultimate leaf segments lanceolate-deltate to sometimes almost ovate, apically obtuse to truncate, leathery, glabrous or with sparse pubescence; $2n = 58$ (Windham 1993c). Rocky slopes and ledges, including limestone; scattered mainly Trans-Pecos and Edwards Plateau e to Bexar (*E.J. Lott & S.P. Rankin 4644*, BRIT) and Travis (*E. McKinney 043*, TEX-LL) cos., also Brown and Palo Pinto (Turner et al. 2003) cos. in the Cross Timbers and Prairies; in the U.S. known only from TX; also Mexico and Central and South America. Sporulating Mar–Nov. [*P. flexuosa* (Kaulf. ex Schltdl. & Cham.) Link, *Pteris ovata* Desv.] Populations in the U.S. are sexual diploids; apogamous triploids are reported to predominate other portions of the species' range south of the U.S. (Tryon 1968; Windham 1993c). The zigzag rachis distinguishes this species from other members of the genus in TX.

Pellaea ternifolia (Cav.) Link, (Latin: *terni*, in three's, and *folius*, leaved, = with three leaves, in reference to the pinnae which are deeply divided into three portions), TRANS-PECOS CLIFF-BRAKE, TERNATE CLIFF-BRAKE. Plant to 50 cm tall; rhizome scales with 2 distinct colors, black in center with brown along margins; petioles dark purple to black, shiny; leaf blades with all or almost all pinnae deeply ternately divided (= into three por-

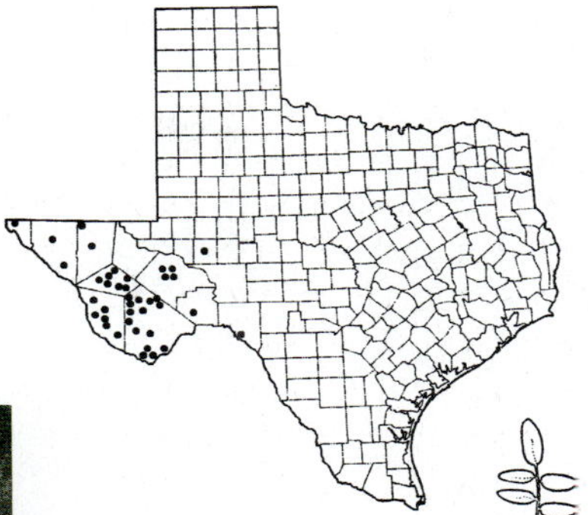

Dale A. Zimmerman Herbarium, Western N. Mex. Univ.

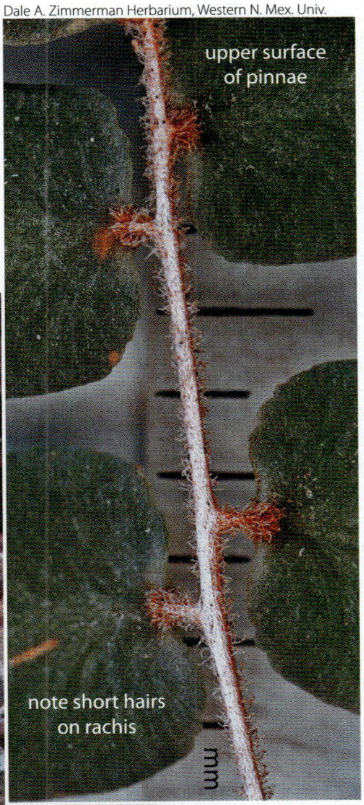

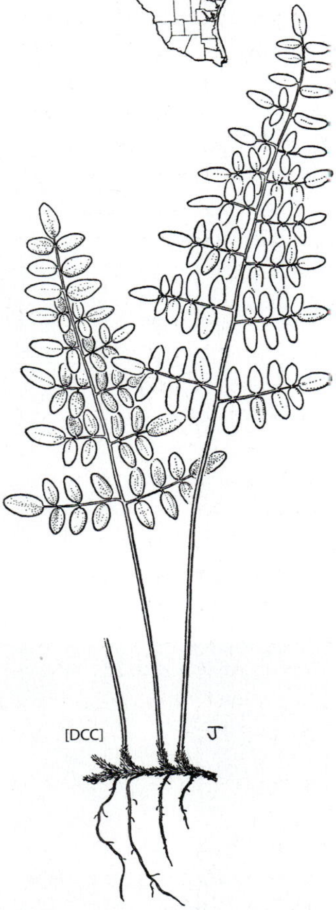

Lorena Babcock Moore

Dale A. Zimmerman Herbarium, Western N. Mex. Univ.

Pellaea intermedia

INTERMEDIATE CLIFF-BRAKE

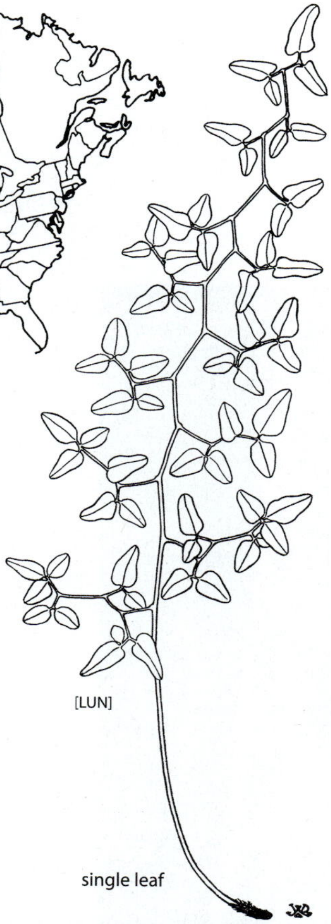

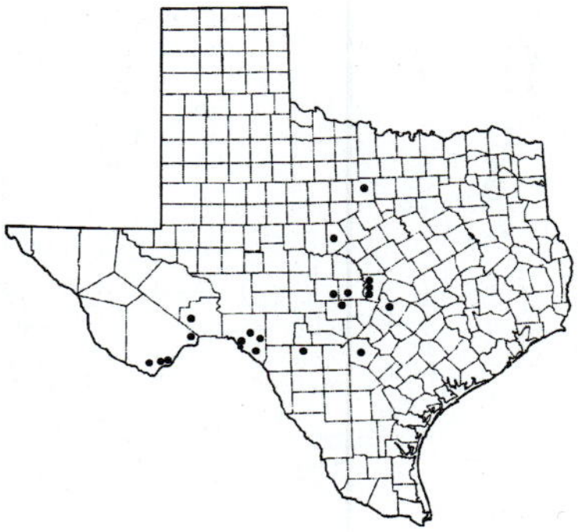

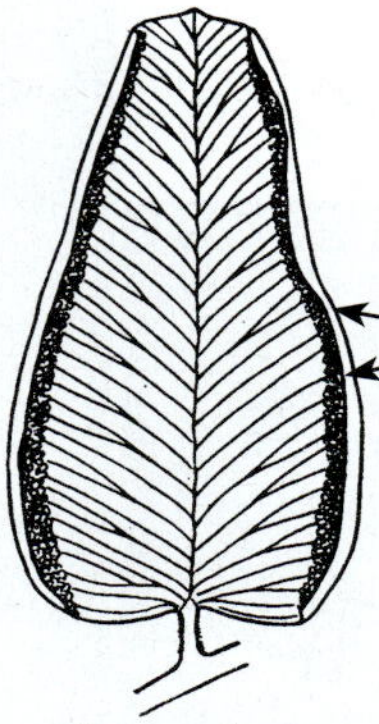

Joseph Marcus, Courtesy LBJ Wildflower Center

Pellaea ovata

ZIGZAG CLIFF-BRAKE

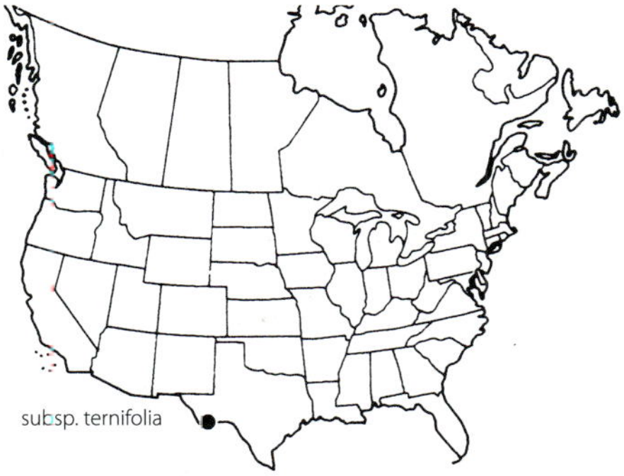

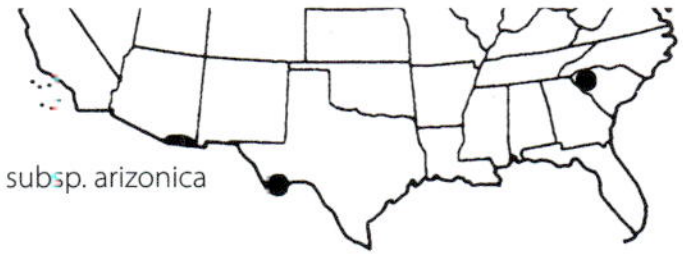

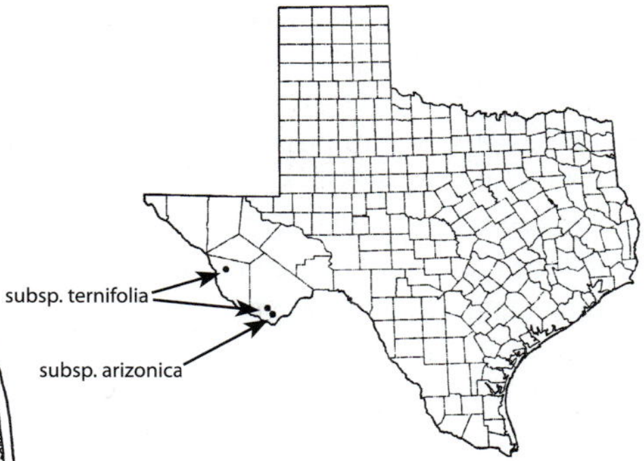

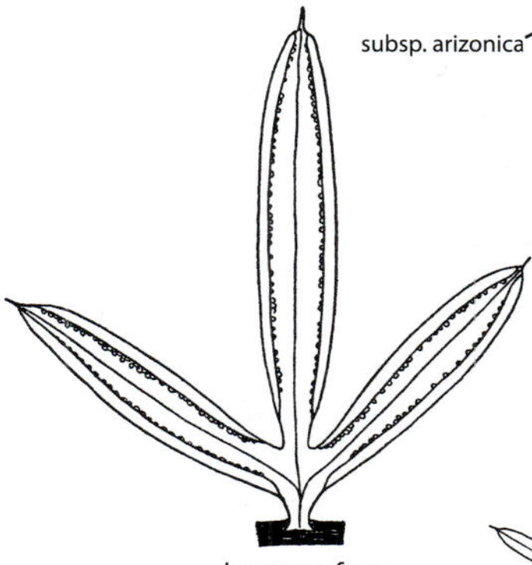

lower surface
of pinna

Roy Morey

Pellaea ternifolia

TRANS-PECOS CLIFF-BRAKE

tions) (1–5+ pinnae at leaf tip can be entire), the blade thus 2-pinnate-pinnatifid throughout, 2–8.5 cm wide, monomorphic; rachis glabrous or with only a few scattered hairs; pinnae attached to rachis so that their tips point ± straight out or slightly towards tip of leaf; ultimate leaf segments linear-oblong, apically mucronate, leathery, completely glabrous on lower surface; 2*n* = 58 or 116. Cliffs, rocky slopes, ledges, on igneous substrates; Trans-Pecos. Sporulating May–Oct. [*Pteris ternifolia* Cav.] This species has the "most extensive distribution of any species in the genus" (Windham & Yatskievych 2003a); it ranges from the sw U.S. and Mexico to Chile, with disjunct populations in Hawaii and Hispaniola (Tryon 1957; Windham & Yatskievych 2003a). Yarborough and Powell (2002) consider this species to be rare in TX. Currently, three subspecies are recognized within *P. ternifolia* (Mickel & Smith 2004); one (subsp. *arizonica*) occurs in the sw U.S. and Mexico and disjunctly in SC, another (subsp. *ternifolia*) in TX, Mexico to South America, and Hawaii, and the third, subsp. *brandegeei* (C.C. Hall) Mickel, in Baja California, Mexico and in Hawaii. Subspecies *brandegeei* differs from the other two in having the pinnae short-stalked (the stalks [technically petiolules] 1–2 mm long) versus sessile. Windham (1993c) noted that the two subspecies occurring in TX are separated by chromosome number and minor morphological differences, and that they "show a tendency toward geographic isolation." At one time Windham (1993c) treated *P. villosa* as a subsp. of *P. ternifolia* (see discussion under *P. villosa*). Windham (1993c) separated subsp. *arizonica* and subsp. *ternifolia* as follows:

1. Largest ultimate segments (excluding terminal pinnae) usually more than 18 mm; distal portion of petioles rounded or slightly flattened adaxially; spores usually 46–53 μm diam. ______ subsp. **arizonica**
1. Largest ultimate segments (excluding terminal pinnae) usually less than 18 mm; distal portion of petioles grooved or flattened adaxially; spores usually 39–45 μm diam. ______ subsp. **ternifolia**

subsp. **arizonica** Windham, (of Arizona, where the type specimen was collected), ARIZONA CLIFF-BRAKE. Plants similar to subsp. *ternifolia*, glabrous or essentially so, the rachis glabrous or with a few widely scattered hairs (in contrast to *P. villosa*); 2*n* = 116 (Windham 1993c). In TX known only from Brewster Co. (*O.E. Sperry s.n.* 1936, Chisos Mts., SRSC, annotated by M.D. Windham and mapped by Windham 1993c); in the U.S. known only from AZ and TX in the sw and also remarkably disjunct to the southern Appalachians in SC (Heafner 2001); also Mexico. The SC population was originally identified as *P. wrightiana* (Platt & Townsend 1996); however, Heafner (2001) has shown that it is actually *P. ternifolia* subsp. *arizonica*. It is one of a number of ferns found primarily in the sw or w U.S. with isolated disjunct populations in the southern Appalachians (see discussion on page 31). This subspecies is represented by tetraploids (Windham 1993c). A hybrid between *P. ternifolia* subsp. *arizonica* and *P. wrightiana* is known from the Chisos Mts., Brewster Co. (*Soxman 246-X*, TEX-LL, identified by M.D. Windham). Because of its rareness and limited distribution in the state, we consider this subspecies to be of conservation concern in TX. ⚠

subsp. **ternifolia,** TRANS-PECOS CLIFF-BRAKE. Plants glabrous or essentially so, the rachis glabrous or with a few widely scattered hairs; largest ultimate segments (excluding terminal pinnae) usually less than 18 mm, but occasionally slightly more (e.g., 21 mm on a BRIT sheet [*Correll 13684A*] annotated by M.D. Windham); 2*n* = 58 (Windham 1993c). Known in the U.S. only from the Chisos Mts., Brewster Co. (*M.S. Young s.n.*, 1915, BRIT—annotated M. Windham; *Correll 13684A*, BRIT, TEX-LL—both annotated M. Windham; Windham 1993c) and the Capote Falls area, Presidio Co. (*Johnston et al. 10668*, TEX-LL); also Mexico to South America and Hispanola and Hawaii. This subspecies is represented by diploids (Windham 1993c). Either of the two subspecies of *P. ternifolia* occurring in TX could be confused with *P. villosa*. However, that species has many pinnae entire (not ternately divided) and rachises that are villous (= with long, soft, spreading hairs). Because of its rareness and limited distribution in the state, we consider this subspecies and the species as a whole to be of conservation concern in TX. ⚠

Pellaea truncata Goodd., (Latin: *truncus*, cut off, = ending abruptly as if cut off straight across, truncate), SPINY CLIFF-BRAKE. Plants to 40 cm tall; rhizome scales with 2 distinct colors, black in center with brown along margins; petioles dark brown, shiny; leaf blades usually 2-pinnate basally, 4–18 cm wide, somewhat dimorphic; rachis and costae glabrous; pinnae attached to rachis so that their tips point ± straight out or slightly towards tip of leaf, mostly with 9–25 ultimate segments each, at least the basal pinnae with segments completely separate (pinnae truly pinnate); ultimate leaf segments narrowly oblong to oblong (variable in shape), apically mucronate (= with a short, small, abrupt pointed tip), leathery, glabrous; 2*n* = 58 (Windham

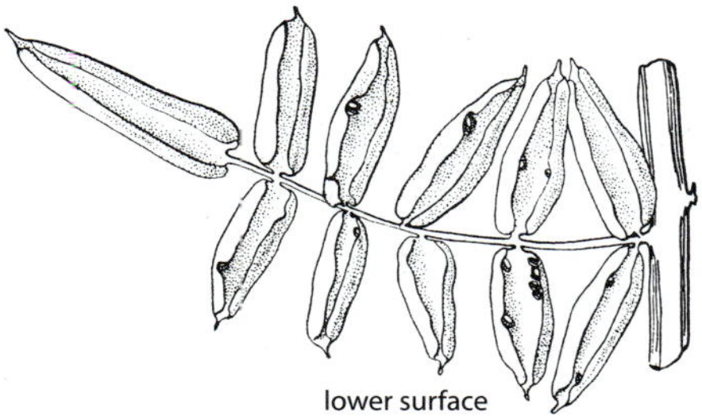

lower surface
of pinna

bicolored
rhizome scale

Pellaea truncata

SPINY CLIFF-BRAKE

1993c). Rocky slopes on a variety of substrates, but usually not limestone; in TX known from the Trans-Pecos in Franklin Mts. (*R.D. Worthington 9475*, BRIT) and Hueco Mts. (*Correll 13799*, BRIT, TEX-LL) in El Paso Co., Cornudas Mts. (*Correll & Johnston 24327*, TEX-LL) and Quitman Mts. (*McVaugh 8033*, BRIT, TEX-LL) in Hudspeth Co., Davis Mts. in Jeff Davis Co. (*T. Simpson 202*, BRIT; *Correll 33665*, TEX-LL), and from the extreme w part of the Edwards Plateau in Val Verde Co. (*Plank s.n.*, Dec 1891, Del Rio [well e of the Pecos River]—Correll 1956; Yarborough & Powell 2002); sw U.S. (AZ, CA, CO, NV, NM, TX, and UT); also n Mexico. Sporulating late spring–fall. [*P. longimucronata* Hook., *P. wrightiana* var. *longimucronata* (Hook.) Davenp.] This species has in the past often gone by the nomenclaturally invalid name *P. longimucronata* (Windham 1993c). Morphologically intermediate hybrids with *P. wrightiana* are reported to be "common in regions where the ranges of the two species overlap, but these [hybrids] are easily identified by their malformed spores" (Windham 1993c). Because of its rareness and limited distribution in the state, we consider this species to be of conservation concern in TX. ⚠

Pellaea villosa (Windham) Windham & Yatsk., (Latin: *villosus*, softly hairy), VILLOSE CLIFF-BRAKE. Plant to 60 cm tall; rhizome scales with 2 colors, black in center with brown along margins; petioles dark purple to black, shiny; leaf blades 2-pinnate-pinnatifid and distally 1-pinnate, monomorphic; rachis villous, especially in axils of pinnae; pinnae attached to rachis so that their tips point ± straight out or slightly towards tip of leaf, at least some and often many distal pinnae marginally completely entire, those in proximal ½–⅔ of leaf deeply ternately divided (= into three portions), the ultimate leaf segments (either portions of the ternately divided proximal pinnae or the undivided, entire, distal pinnae) linear-oblong, apically mucronate, leathery, with hairs scattered along main veins on lower surface; 2*n* = 116 (Windham 1993c). Various rocky substrates; in TX known only from the Davis Mts., Jeff Davis Co. (*R. Magill 101*, TEX-LL [annotated M.D. Windham]; Windham 1993c; *P. Zelazny 217*, SRSC) in the Trans-Pecos; known in the U.S. only from AZ (*Fishbein et al. 2861*, ARIZ) and TX; also rather widespread in n Mexico. Sporulating summer–fall. [*Pellaea ternifolia* (Cav.) Link subsp. *villosa* Windham] When Windham (1993e) named this taxon as a variety of *P. ternifolia* (var. *villosa*) he suggested that *P. ternifolia* was represented by "three genetically distinct taxa characterized by differences in leaf morphology and chromosome number." He further noted that subsp. *villosa* was probably an allotetraploid resulting from hybridization between typical *P. ternifolia* and some other as yet unidentified species. Since 1993, a new species, *P. ribae* Mendoza & Windham, has been discovered in the Mexican state of San Luis Potosí (Mendoza et al. 2001) and appears to be the previously unidentified missing parent (Windham & Yatskievych 2003a). Windham and Yatskievych (2003a) have argued that since the entity going by the name *P. ternifolia* subsp. *villosa* is a morphologically distinct, "reproductively competent" allotetraploid of hybrid origin "it should be recognized as a distinct species." We are thus following Windham and Yatskievych (2003a) in treating it at the species level. They marshalled evidence demonstrating that it is "morphologically intermediate" between the two proposed parents "in nearly every character examined." For example, *P. ternifolia* has all pinnae deeply ternately divided and *P. ribae* has all pinnae entire; *P. villosa*, however, has proximal pinnae ternately divided and distal pinnae entire. Such reproductively isolated allotetraploid species are well known among ferns (Wagner 1954). This taxon was previously known in the U.S. from only a single individual in the Davis Mts. of TX; however, "recent botanical surveys of the Madrean 'sky islands' of southern Arizona have identified a second U.S. population." (These Madrean "sky islands," are isolated high elevation areas [e.g., Chisos Mts., Davis Mts., Guadalupe Peak] similar to habitats found in the Sierra Madre Mountains of Mexico). *Pellaea villosa* is one of a small group of species including *Cheilanthes lendigera, Notholaena aschenborniana, N. neglecta*, and *Pellaea ternifolia* subsp. *arizonica*, "that are known from Arizona and Trans-Pecos Texas but not from the intervening regions of New Mexico" (which has fewer of the Madrean "sky island" habitats) (Windham & Yatskievych 2003a). Because of its rareness and limited distribution in the state, we consider this species to be of conservation concern in TX. ⚠

Pellaea wrightiana Hook., (for Charles Wright, 1811–1885, TX collector), WRIGHT'S CLIFF-BRAKE. Plant 15–30(–50) cm tall; rhizome scales with 2 distinct colors, black in center with brown along margins; petioles dark brown, shiny; leaf blades 2-pinnate basally to 1-pinnate-pinnatifid distally, usually 8–25 cm long, 1–4.5(–5) cm wide, monomorphic; rachis and costae usually glabrous; pinnae attached to rachis so that their tips point ± straight out or slightly towards tip of leaf, mostly with 3–9 ultimate segments each, at least the basal pin-

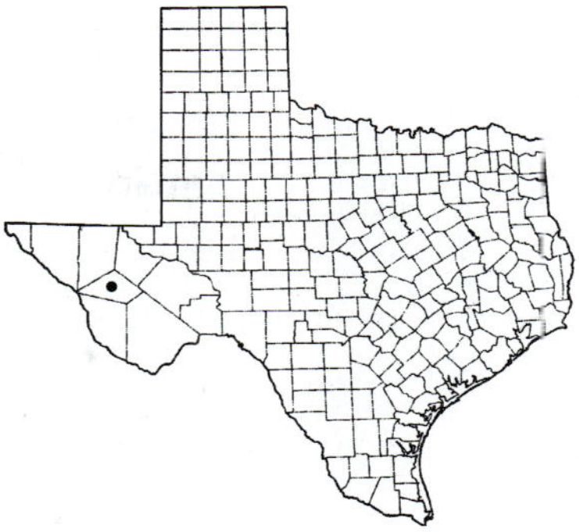

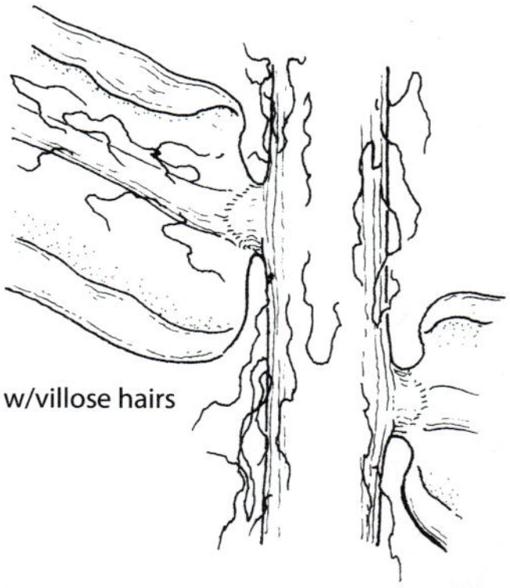

[POM]

Missouri Botanical Garden

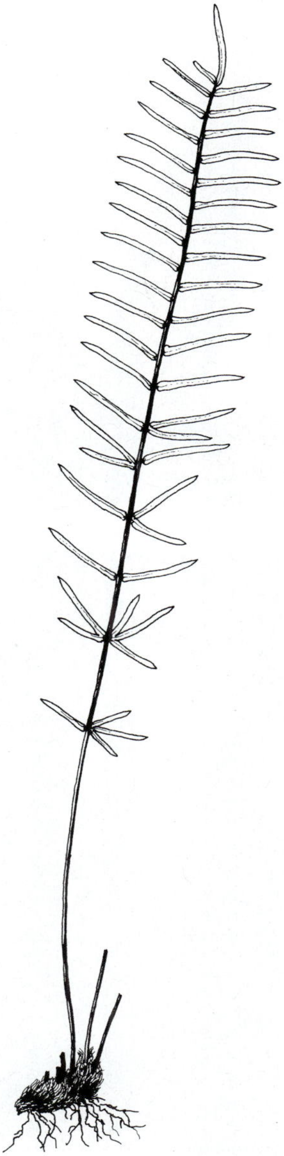

Pellaea villosa

VILLOSE CLIFF-BRAKE

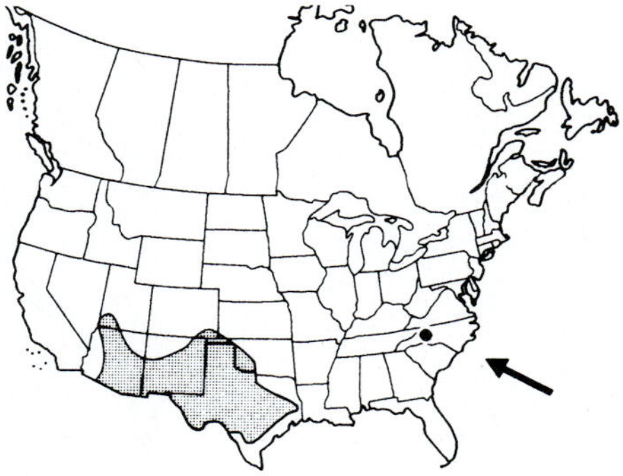

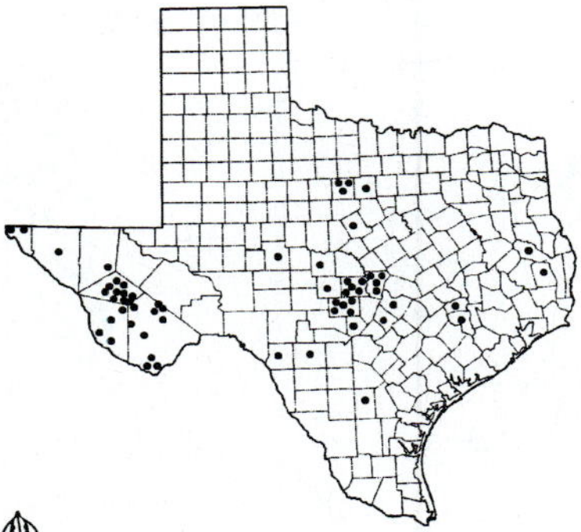

Dale A. Zimmerman Herbarium, Western N. Mex. Univ.

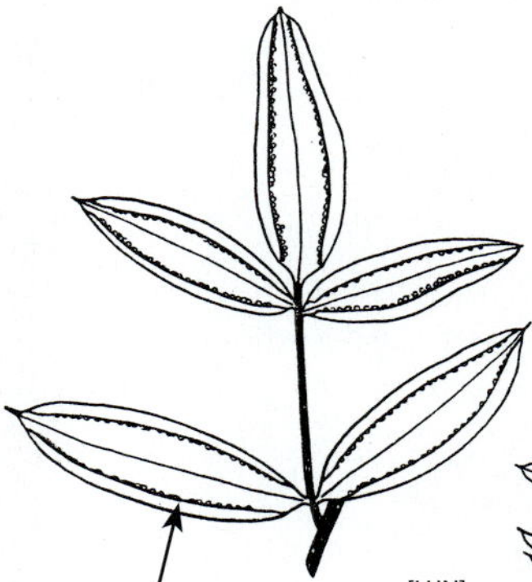

Petei Guth

Pellaea wrightiana

WRIGHT'S CLIFF-BRAKE

nae with completely separate segments (pinnae truly pinnate with the terminal segment on a distinct short stalk); ultimate leaf segments narrowly oblong (occasionally oblong), apically mucronate, leathery, glabrous; $2n$ = 116 (Windham 1993c). Rocky areas and cliffs on various rocky substrates (acidic to mildly basic); primarily Trans-Pecos, Edwards Plateau (e.g., Burnet Co., *W.R. Carr 8495*, BRIT), and w Cross Timbers and Prairies, but also Tyler (*Correll 13295*, "on sandstone boulders," BRIT, TEX-LL), Angelina (Upland Island Wilderness Area—J. Singhurst, pers. comm.), Austin, Hays, Travis, and Washington (Turner et al. 2003) cos. in the s part of East TX; mainly sw U.S. (AZ, CO, NM, OK, TX, and UT) and distantly disjunct in NC; also n Mexico. Sporulating Mar–Nov. [*P. ternifolia* (Cav.) Link var. *wrightiana* (Hook.) A.F. Tryon] This species has been shown to be a fertile allotetraploid derived from *P. truncata* and *P. ternifolia* (Windham 1993c). Hybrids have been reported with *P. atropurpurea* (they are known only from w OK), *P. ternifolia* subsp. *arizonica* (see discussion under that species), and *P. truncata* (see discussion under that species) (Windham 1993c). In the Edwards Plateau, the obviously mucronate ultimate leaf segments of this species quickly distinguish it from other sympatric *Pellaea* species. This is one of a number of ferns found primarily in the sw or w U.S. with isolated disjunct populations in the southern Appalachians (see discussion on page 31). The population in North Carolina (Wagner 1965, 1972) is probably an excellent example of long-distance dispersal of spores.

PTERIS L. BRAKE

Often on rock or terrestrial; rhizomes short-creeping, usually densely scaly, the scales often extending along the rachises; leaves monomorphic, evergreen or not so, clustered; petioles green to light brown, 2–3-grooved on the upper side; leaf blades strictly 1-pinnate to at least partially 2-pinnatifid; pinnae mostly opposite; sori near margins of fertile pinnae or pinnules, usually ± continuous, the pinna or pinnule margins reflexed over sori to form false indusia.

A genus of ca. 250–300 species found worldwide in warm and tropical areas (Nauman 1993b; Mabberley 2008). The young fronds of some species are reported to be edible or useful medicinally, and some are cultivated as ornamentals. Some species are known to accumulate arsenic (up to 1% dry weight), possibly as an adaptation to tolerate arsenic rich environments or as an antiherbivory mechanism (Meharg 2003; Pickering et al. 2006). (Greek: *pteris*, fern, derived from *pteron*, wing or feather, in reference to the closely spaced pinnae which give the leaves a somewhat feather-like appearance—Nauman 1993b).

REFERENCES: Chandler 1941; Soxman 1945; Jones 1955; Morton 1957a; Stevens 1977; Nauman 1993b; Huang et al. 1994; Stanford & Diggs 1998; Ma et al. 2001; Bondada & Ma 2003.

QUICK TIPS FOR GENUS ID: Introduced ferns with 1-pinnate to partly 2-pinnatifid leaves with petioles that are greenish to light brown and 2–3-grooved on the upper side.

1. Leaf blades at least partially 2-pinnatifid (at least the basal pinnae with lobes or divisions), 10–35 cm long; pinnae 3–7 pairs per leaf, the distal pinnae conspicuously decurrent along rachis; scales of rhizomes dark reddish brown to chestnut brown ____ **P. multifida**
1. Leaf blades strictly 1-pinnate (the pinnae without lobes or divisions), (15–)25–50(–80) cm long; pinnae numerous (12 or more pairs per leaf), not decurrent along rachis; scales of rhizomes pale brown ____ **P. vittata**

Pteris multifida Poir., (Latin: *multi-*, many, and *-fidus*, cleft, = many cleft or much divided, in reference to the highly divided leaves), SPIDER BRAKE, HUGUENOT FERN, FERN, CHINESE BRAKE, SAW-LEAF BRACKEN. Leaves evergreen (10–)25–60 cm long; leaf blades 10–35 cm long,10–25 cm wide, essentially glabrous, at least partially 2-pinnatifid; pinnae 3–7 pairs, lanceolate to linear, those in at least the distal (= toward the apex) ½ of mature leaves decurrent on the conspicuously winged rachis (rachis wing constricted distal to each pair of pinnae); basal and sometimes the medial pinnae with 1–2 lobes or pinnules (at least some pinnae are deeply palmately 3-divided, except on the leaves of young plants, which can have palmately compound leaves with three unlobed pinnae), the distal pinnae simple; sterile pinnae wider than fertile pinnae, with margins serrulate to serrate; fertile pinnae entire to serrate near apex; $2n$ = 116 (Nauman 1993b). Masonry of old brick buildings, calcareous sandstone talus blocks, and sandy soils of woods; Fayette (*D.H. Riskind s.n.*, 1977, BRIT, TEX-LL), Jefferson (*Cory 49988*, BRIT), Newton (*C. Allen et al. 19630*, BRIT), Hardin (*J.L. Blassingame 2329*, BRIT; *B.C. Tharp 44408*, SRSC, TAES; *Correll 29389*, TEX-LL; Soxman 1945), and Montgomery (Turner et al. 2003) cos. in East TX and Cameron Co. (*R. Runyon 4265*, "on east side of a brick wall," TEX-LL) at the very s tip of TX; naturalized se U.S. from MD s to FL w to AR (where it occurs on tufa) and TX, also s CA, IL, IN, KY, and NY. Sporulating

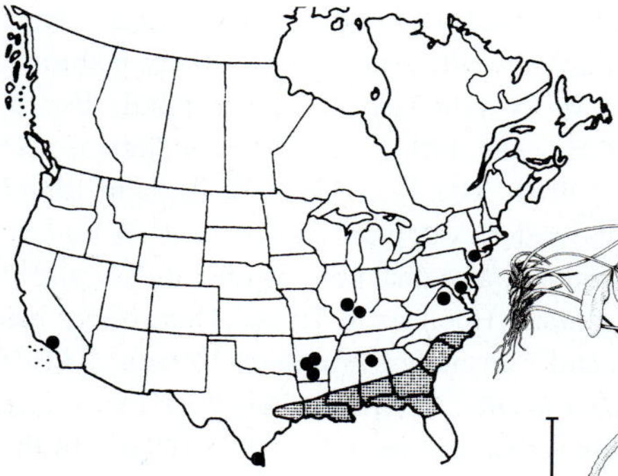

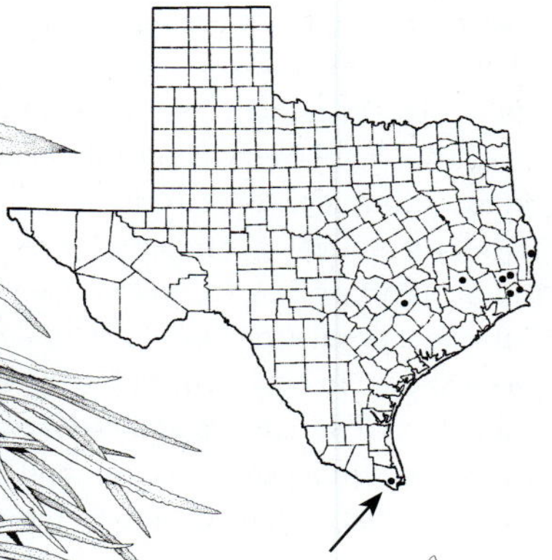

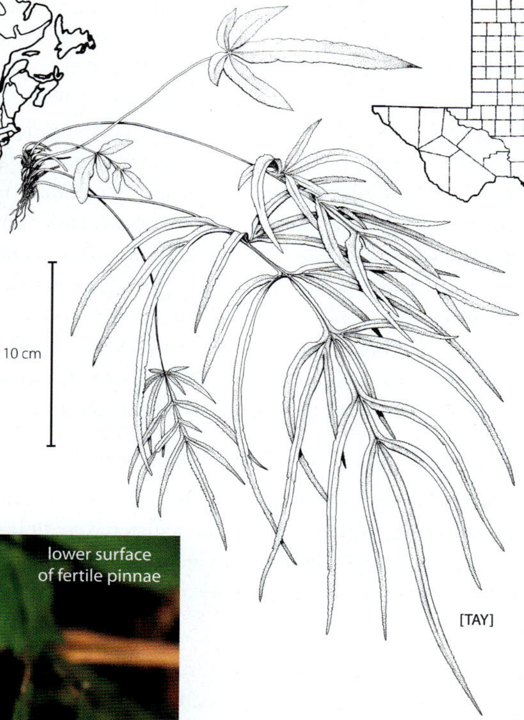

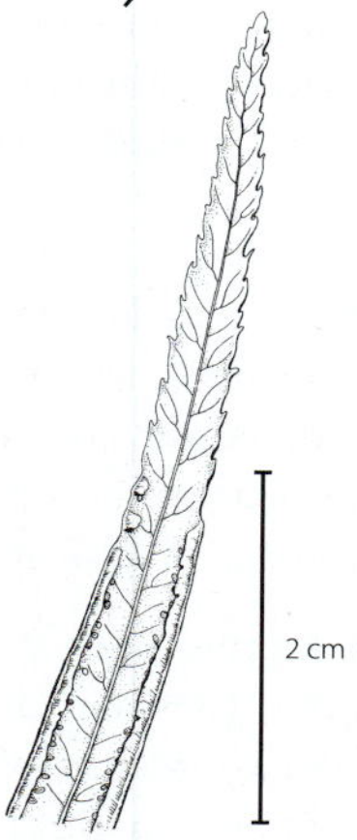

lower surface
of fertile pinna

Mississippi Museum of Natural Science (MS-DWFP)

Mississippi Museum of Natural Science (MS-DWFP)

Pteris multifida

SPIDER BRAKE

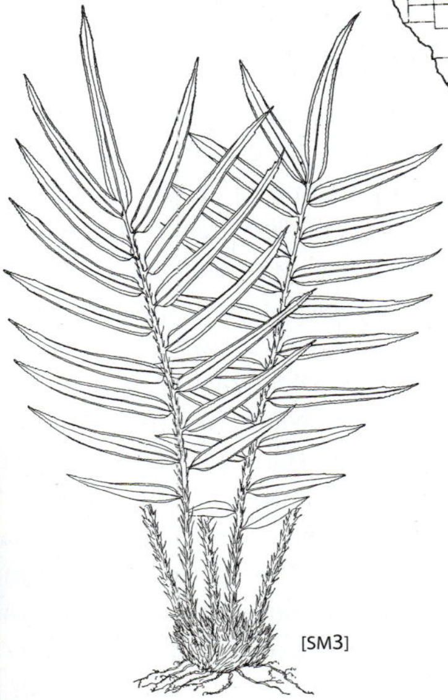

Mississippi Museum of Natural Science (MS-DWFP)

Mississippi Museum of Natural Science (MS-DWFP)

Pteris vittata

LADDER BRAKE

Jun–Dec. Native of e Asia. [*Pycnodoria multifida* (Poir.) Small] According to Correll (1956), this species "... was cultivated on many of the older plantations in the Deep South where it still persists. It has escaped in many areas into the nearby woods and can now be found some miles from its original point of cultivation. It is so completely at home in some places in east Texas that it must be considered as naturalized." According to Nelson (2000), *P. multifida* "was first discovered in the U.S. in 1868 in a Huguenot cemetery in Charleston, South Carolina; hence, one of its common names, Huguenot fern. The common name 'spider brake' is likely in reference to the spiderlike appearance of the deeply-divided leaves." As noted by Peck (2011a), this species is apogamous and "readily capable of producing new sporophytes from single spores."

Pteris vittata L., (Latin: *vitta*, ribbon, band, or stripe, = longitudinally striped—in reference to the linear false indusium), LADDER BRAKE, CHINESE BRAKE, CHINESE LADDER BRAKE. Leaves not remaining green through winter, 0.3–10 dm long; leaf blades oblanceolate or narrowly so, 15–50(–80) cm long, to 16(–25) cm wide, strictly 1-pinnate, glabrous; pinnae 12–20(–30) pairs per leaf, not decurrent on rachis, linear lanceolate, to 18 cm long, serrulate, apically acuminate, basally asymmetrically cordate to truncate; distal pinnae conspicuously longer than proximal pinnae with the terminal pinna typically longest; $2n = 116$ (Nauman 1993b). The only known TX collection is from a stream-side boulder in San Saba Co. (*J.W. Stanford 5308*, 1987, BRIT, HPC, SPLT) on the Edwards Plateau, the site subsequently destroyed by flooding; naturalized (apparently escaped from cultivation); se U.S. from SC s to FL w to TX, also CA and D.C. and invasive in Hawaii; also naturalized in ne Mexico (Nuevo Leon and Tamaulipas) and in various places in the New World subtropics and tropics. Sporulating in TX ?–Nov–?. Native of Asia. [*Pycnodoria vittata* (L.) Small] This species was first reported for the TX flora by Stanford and Diggs (1998). Small (1938) indicated that during the 1930s it appeared in abundance at several Florida localities (e.g., Everglades). It is now widely naturalized in the se U.S. on exposed limestone (e.g., pinelands) and on a variety of man-made calcareous substrates (e.g., sidewalks, buildings, old masonry) (Nauman 1993b). This species has been found to accumulate arsenic. It is a "hyperaccumulator" that is "extremely efficient in extracting arsenic from soils and translocating it into its above-ground biomass" (Ma et al. 2001). It is thus of potential use in remediating arsenic-contaminated soils (Bondada & Ma 2003). Since the species can tolerate high arsenic levels, this is presumably an adaptation for survival on soils naturally high in arsenic or possibly a defense against herbivores. Although not an invasive in TX, it is a prolific spore producer with a known ability to become invasive in other areas (e.g., in Hawaii). Therefore there are concerns about its potential widespread application in bioremediation (Robinson et al. 2010).

SALVINIACEAE Martinov

FLOATING FERN OR WATER-SPANGLE FAMILY

A very small family (1 genus, ca. 10–12 species) of heterosporous (= with 2 spore types), small, floating aquatics found mostly in the tropics, but ranging from s U.S., Mexico, West Indies, Central America, and South America to Eurasia and Africa, including Madagascar (Nauman 1993d). The Azollaceae has often been included in the Salviniaceae (e.g., Nagalingum et al. 2008; Smith et al. 2008) or treated separately (Lumpkin 1993). Molecular evidence indicates that *Salvinia* and *Azolla* are each other's closest living relative (Hasebe et al. 1995), and these two families of floating aquatics are appropriately treated in the same order (Schneller 1990b). However, because of the monophyletic nature of the two lineages (Nagalingum et al. 2008), their obvious morphological distinctiveness, and ancient separation (at least 89 mya) (Collinson 1991; Metzgar et al. 2007), we are recognizing both Azollaceae and Salviniaceae at the family level. This approach was described as "a perfectly acceptable alternative, given the significant differences between the two genera" (Smith et al. 2008). Three families of heterosporous water ferns occur in TX (Azollaceae—*Azolla*, Marsileaceae—*Marsilea*, *Pilularia*, and Salviniaceae—*Salvinia*). These groups are quite distinct morphologically, and traditionally the marsileaceous (*Marsilea* and *Pilularia*) and salviniaceous (*Azolla* and *Salvinia*) lines were considered to have evolved independently from different homosporous fern ancestors. However, all three living heterosporous water fern families appear to comprise a monophyletic group as suggested by recent morphological, fossil, and molecular evidence (Rothwell & Stockey 1994; Hasebe et al. 1995; Pryer 1999). Heterospory is extremely rare within ferns—it is known from only 5 genera (all in Order Salviniales) and < 1% of living ferns (Pryer et al. 2004b; Nagalingum et al. 2008). (Order Salviniales)

FAMILY RECOGNITION IN THE FIELD: small free-floating aquatics with leaves in groups of three (two leaves floating, nearly rounded to elliptic, entire, with conspicuous branched hairs; one leaf submerged, finely dissected, root-like). REFERENCES: Schneller 1990b; Nauman 1993d; Nagalingum et al. 2006, 2008; Yatskievych & Windham 2008d.

SALVINIA Séguier

FLOATING FERN, WATER-SPANGLES, WATER-MOSS

Free-floating aquatics (can be trapped on mud when water levels fall); roots absent; stems branching, not bearing true roots; leaves dimorphic, in groups of three (two floating and one submerged); floating leaves green, simple, unlobed, nearly round to oblong or elliptic, entire, obtuse or notched apically, the upper surface with numerous whitish, multicellular, papillate hairs, these having a single base and four branches, the lower surface with unbranched hairs; submerged leaves finely dissected, root-like in form and function; sporangia (either microsporangia or megasporangia) in sporocarps (formed by membranous indusia surrounding the sori) on submerged leaves, indehiscent, dispersed when the sporocarps decay (but plants usually infertile and reproducing vegetatively).

A genus of ca. 10–12 species of floating ferns (Nauman 1993d; Nagalingum et al. 2008). The hairs on the upper leaf surfaces trap air, causing the plants to right themselves if turned over in the water (Lellinger 1985). This and the air that is trapped in the mesophyll of the leaves make them virtually unsinkable (Moran 2004; Robinson et al. 2010). The leaves are also unusual in that the upper sides of the floating leaves (the sides visible to an observer), which appear to face the stem axis, are actually morphologically the lower (= abaxial) leaf surfaces (Croxdale 1978, 1979, 1981; Nauman 1993d). Some species have become serious aquatic weeds, particularly on man-made reservoirs and in irrigation systems (Schneller 1990b). Because of their potential threat to aquatic ecosystems, all species of the genus are prohibited in TX (Harvey 1998). Although the floating leaves of the two species discussed below are often different in size, crowding or low-light conditions tend to reduce leaf size, making the character less useful in determining species (Lellinger 1985). *Salvinia* species can easily be spread by water flow, human activities (e.g., clinging to boat trailers or boats), or by waterfowl that can "transport pieces of the plant in mud attached to their feet and feathers" (Mora-Olivo & Yatskievych 2009). (Named for Professor Antonio Maria Salvini, 1633–1729, Italian botanist and professor of Greek at Florence) REFERENCES: Herzog 1935; Weatherby 1937; Morton 1967c; Mitchell & Tur 1975; Croxdale 1978, 1979, 1981; Forno 1983; Jacobsen 1983; Montz 1989; Thomas & Room 1986; Moran 1992; Hatch 1995; Jacono 1999a, 1999b, 1999c, 2001a, 2001b; Moran & Smith 1999; O'Kennon et al. 1999; Garbari et al. 2000; Helton & Chilton 2001; Jacono et al. 2001; Wood et al. 2001; Julien et al. 2002; Madeira et al. 2003; McFarland et al. 2004; Tipping & Center 2005; Jacono & Richerson 2005; Tipping et al. 2008, 2009; Al-Hamdani & Ghazal 2009; Mora-Olivo & Yatskievych 2009; Riefner & Smith 2009; Howard 2010; Morgan 2010.

1. Hairs of the upper surface of floating leaves 4-branched, the branches free at their tips (use magnification; sometimes difficult to observe on herbarium specimens); floating leaves usually 6–15 mm long, usually not folded; lower surface of floating leaves ± covered with hairs ____ **S. minima**
1. Hairs of the upper surface of floating leaves 4-branched, the branches joined at their tips (resembling a "cage" or an "egg beater"); floating leaves usually 13–30(–38) mm long or sometimes longer, often folded; lower surface of floating leaves sparsely hairy or ± glabrous except for hairs along the midvein ____ **S. molesta**

Salvinia minima Baker, (Latin: *minimus*, least or smallest, presumably in comparison with the size of other species), COMMON SALVINIA, WATER-SPANGLES, LEAST WATER-SPANGLES, FLOATING FERN. Stems 1–6 cm long; floating leaves usually 6–15 mm long, rounded to cordate basally, obtuse or notched apically, usually not folded; $2n = 4x = 36$, $2n = 6x = 54$ (de la Sota & Cassa de Pazos 2001). Lakes and other aquatic habitats; Jasper (*Holmes et al. 10787*, BAYLU), Jefferson (*Hatch 6403*, TAES), Harris, (*P. Rolling s.n.*, SBSC), Robertson (Turner et al. 2003), Orange (*L.E. Brown 18756*, TAES), Tyler (R. Helton, pers. comm.), Angelina, Sabine, Travis (Morgan 2010), and probably Newton (it is known from the LA side of Toledo Bend—R. Helton, pers. comm.) cos. in East TX and farther w in Wichita Co. in the e Rolling Plains (Morgan 2010); naturalized in scattered localities across se U.S. (AL, AR [Peck 2011a], FL, GA, LA, MS, SC, and TX [Jacono et al. 2001]), also CA (reported 2009 by Riefner & Smith); also Mexico to South America. Sporulating spring and fall (according to Nauman 1993d; however, the spores are apparently functionally sterile and the plants only reproduce vegetatively—

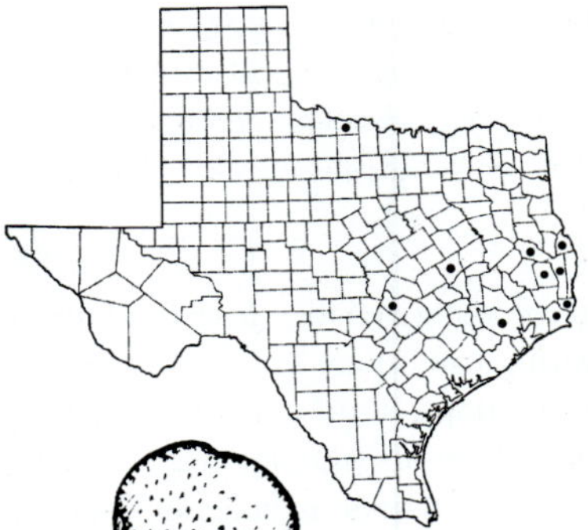

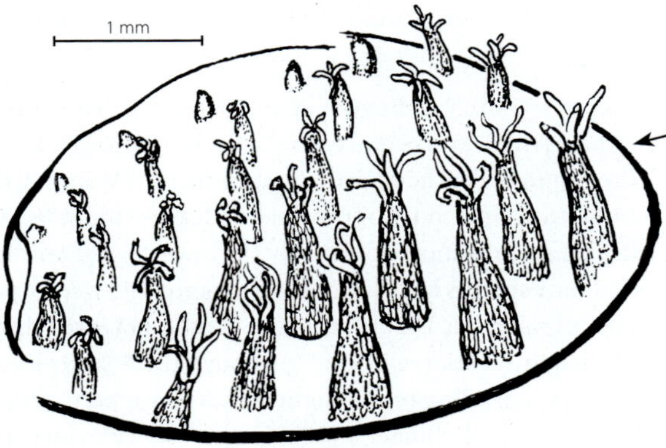

four-branched hairs
of "upper" leaf surface

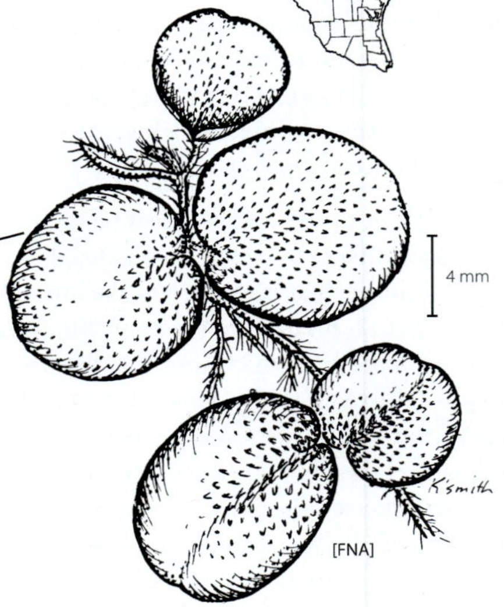

Salvinia minima

COMMON SALVINIA

Riefner & Smith 2009; Morgan 2010). [*S. rotundifolia* of authors, not Willd.] In TX it is considered a "harmful or potentially harmful exotic plant," and it is illegal to release, import, sell, purchase, propagate, or possess this species in the state (Harvey 1998). In certain parts of East TX, populations of this species are so large that recreational access (e.g., boating and hunting) is being impeded (H. Elder, pers. comm.). As is the case with *S. molesta*, this species is capable of forming thick mats (as thick as 20–25 cm) (Montz 1989; Morgan 2010). In the past *S. minima* was considered native to the se U.S. (e.g., Nauman 1993d), but is now thought to have been introduced from tropical America where it is native from Mexico s to Argentina (Jacono et al. 2001; Tipping et al. 2009; Morgan 2010). It was first reported in the U.S. from the St. Johns River, Florida in 1928 (Small 1931 as *S. auriculata*; Tipping et al. 2009). It was probably introduced into FL in the 1920s, and "the plant may have entered natural areas through flooding of water-gardens and ornamental ponds" (Jacono et al. 2001). It was introduced into TX in 1992, apparently from w LA "on a 'marsh buggy,' during geologic exploration" (Jacono et al. 2001; see also Hatch 1995). The spread of this species is apparently limited by a "low tolerance to freezing temperatures and saline waters" (Riefner & Smith 2009). ☞ ⃠

Salvinia molesta D.S. Mitch., (Latin: *molestus*, annoying, troublesome, or burdensome, in reference to problems caused by the rapid vegetative growth), GIANT SALVINIA, KARIBA WEED, AQUARIUM WATER-MOSS, AFRICAN PYLE. Floating leaves often ca. 13–30(–38) mm long but under some conditions up to 60 mm, usually cordate basally, usually notched apically, often folded and compressed into dense chains (flat when young); $2n = 5x = 45$ (pentaploid; Jacono et al. 2001). Lakes and other aquatic habitats, reproducing vegetatively very effectively; primarily in East TX in Robertson (*S.L. Hatch 7301*, BAYLU, BRIT, TAMU, TEX-LL), Harris, Montgomery (SBSC), Liberty (*M. Whitbeck 61*, TAES), Orange (*T. Davern s.n.*, TAES), Houston, Newton, Sabine, Shelby, and San Jacinto (H. Elder, pers. comm.; R. Helton, pers. comm.) cos.; it has also been reported from the Gulf Prairies and Marshes (Brazoria, Chambers, and Fort Bend, cos.—H. Elder, pers. comm.) and as far nw as Flower Mound (Denton Co.) in North Central TX (Jacono & Richerson 2005) and Wichita Co. (H. Elder, pers. comm.) in the Rolling Plains. It was first observed in TX in 1997 (Jacono 1999c), and by summer 2000 it was known in the state from 4 reservoirs and ca. 40 private water bodies (R. Helton, pers. comm.). By 2004 it had been discovered in at least 11 public water sites (H. Elder, pers. comm.) and by 2010 it was known from 17 TX lakes, including some of the state's most popular recreational water bodies: Toledo Bend, Sam Rayburn, Caddo Lake, Sheldon Lake, Lake Texana, and Lake Conroe (TPWD 2010). Particularly large-scale infestations have been noted from Todedo Bend and the Swinney Marsh Complex on the Lower Trinity River, in Liberty County, the latter being one of the most serious in the U.S. (McFarland et al. 2004). It is continuing to spread in the state (H. Elder, pers. comm.), and in 2010 TX initiated a public awareness campaign to alert the public about this potentially devastating invasive (TPWD 2010). Now known to be present across the s U.S. in AL, AR (Peck 2011a), AZ, CA, FL, GA, LA, MS, NC, SC, VA, and TX (Weakley 2010); also introduced to Mexico (Baja Calif., Sonora, Tabasco—Moro-Olivo & Yatskievych 2009). Considered "one of the world's worst weeds" (Jacono 1999c), it is thought to be native to South America (se Brazil—Forno & Harley 1979; Forno 1983; U.S. Geological Survey 2000) and is possibly of hybrid origin. The sporangia abort and the somatic chromosome number is 45 (5 times the base number of $n = 9$), the plant thus being pentaploid (Jacobsen 1983; Jacono et al. 2001b). Jacono (2001b) noted that "*Salvinia molesta* is not known to reproduce by spores. It reproduces vegetatively, that is, new plants develop as fragments break off from mature individuals." It has been introduced by humans to fresh waters of tropical and subtropical areas around the world including Africa, Asia, Australia, s Europe, New Zealand, and the South Pacific, which has resulted in severe economic and environmental problems (Jacono 1999a, 1999b; Garbari et al. 2000). The plants can grow rapidly and under good conditions doubling can occur in ca. one week, with a range of four to ten days (Mitchell & Tur 1975); some authorites indicate that populations can double in a little more than two days (Moran 2004). GIANT SALVINIA covers the surface of lakes and streams, and the floating mats shade and crowd out native plants. Additionally, the mats (sometimes to a meter thick!) reduce oxygen content, degrade water quality, and can cause physical problems including hindering boats, clogging irrigation and drainage canals, and blocking water intakes. The mats can become so dense that they are able to support the weight of a cinder block. Further, in tropical areas the mats provide habitat for mosquitoes, which transmit human diseases including encephalitis, dengue fever, and malaria (Creagh 1991). Plants used in aquaria or water gardens are among the likely original sources for the escaped populations (Thomas & Room 1986; Jacono 1999a, 1999b,

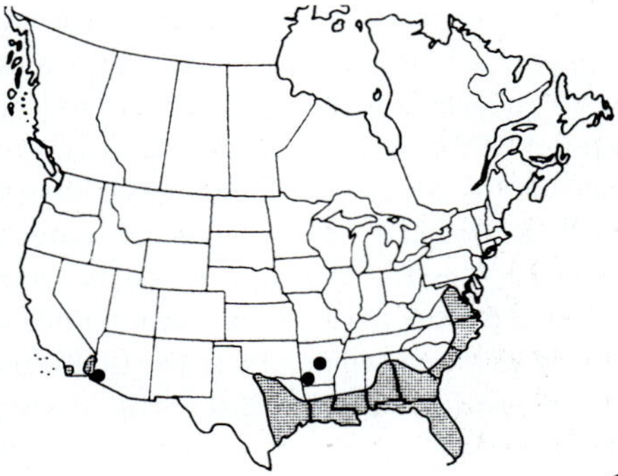

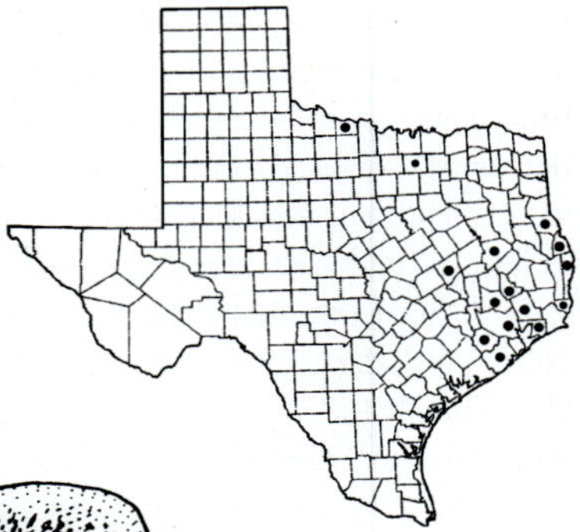

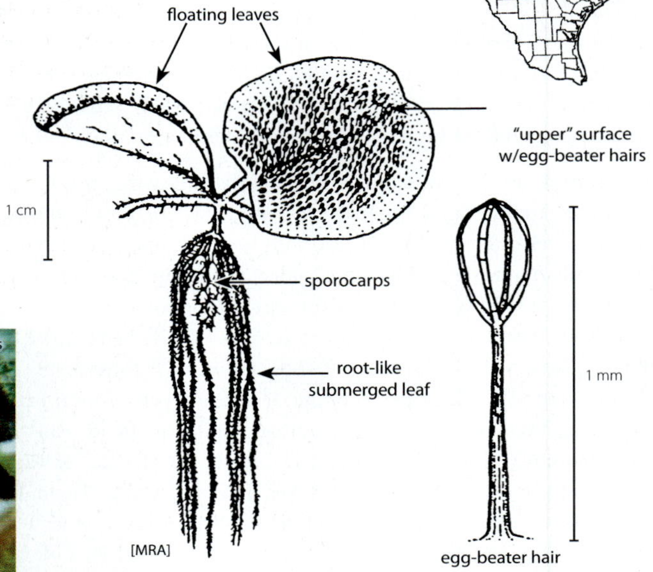

Mississippi Museum of Natural Science (MS-DWFP)

Mississippi Museum of Natural Science (MS-DWFP)

Salvinia molesta

GIANT SALVINIA

1999c; Wood et al. 2001; Moran 2004), but spread by boats or boat trailers also seems likely (as in the case of other invasive aquatics such as *Hydrilla verticillata*, WATER-THYME). It was first reported outside of cultivation in the U.S. in 1995 from a private pond in SC; in 1998 outbreaks were discovered in LA and TX, and since that time it has spread widely across the s U.S. (McFarland et al. 2004). According to R. Helton (pers. comm.), "all reservoirs in East TX are imminently threatened" by this species. In TX, it is considered a "harmful or potentially harmful exotic plant" and it is illegal to release, import, sell, purchase, propagate, or possess this species in the state (Harvey 1998). GIANT SALVINIA has been listed as a federal noxious weed since 1984 (McFarland et al. 2004; USDA Natural Resources Conservation Service 2010), and as such is prohibited in the U.S. by federal law. Jacono (1999a) indicated that, if seen, the species should be eradicated immediately, and that the Texas Parks and Wildlife, Inland Fisheries Division should be contacted at giantsalvinia@tpwd.state.tx.us or (409) 384-9965. Biological control by the salvinia weevil (*Cyrtobagous* spp.) is being tested in East TX; this weevil, native to South America, has been successfully used in control programs in various places in the Old World (Thomas & Room 1986; Moran 1992, 2004; Wood et al. 2001; R. Helton, pers. comm.; H. Elder, pers. comm.). Initial indications are that the weevil is effective in helping to control GIANT SALVINIA in TX (Flores & Carlson 2006; Tipping et al. 2008). Although some authorities have suggested that the ambiguous name *S. adnata* Desv. be used for this species, Moran and Smith (1999) argue convincingly for the continued use of *S. molesta*. The common name KARIBA WEED originated in the 1950s as a result of problems it caused associated with the Kariba Dam project on the Zambezi River between Zambia and Zimbabwe (Robinson 2010).

TECTARIACEAE Panigrahi

HALBERD FERN OR BUCKLER FERN FAMILY

A largely pantropical family of 8–15 genera and ca. 230 species with most in *Tectaria* (in the broad sense) (Smith et al. 2008). As delimited here it is apparently monophyletic (Schuettpelz & Pryer 2007), but Liu et al. (2007a) question the inclusion of some genera. Generic limits within the family, especially within *Tectaria*, have yet to be completely clarified. *Tectaria* used to be treated in Dryopteridaceae (e.g., Smith 1993b; Diggs et al. 2006), but based on recent molecular data does not appear closely related to that group; its inclusion in Dryopteridaceae makes that family polyphyletic (Smith et al. 2008). (Order Polypodiales)

FAMILY RECOGNITION IN THE FIELD: the only TX species has leaf blades with only 1–2(–3) pairs of pinnae in addition to a single terminal pinna (pinnae can be lobed or auricled) and large round sori with peltate indusia. It is found in protected areas on limestone on the Edwards Plateau.

REFERENCES: Liu et al. 2007a; Smith et al. 2008.

TECTARIA Cav. HALBERD FERN

A mostly tropical genus of ca. 200 species, particularly in se Asia and adjacent Pacific Islands, with 7 in Mexico (Moran 1993b; Mickel & Smith 2004); some are cultivated as ornamentals. (Latin: *tectum*, roof, and *aria*, a suffix, in reference to the roof-like indusium of some species—Moran 1993b)

REFERENCES: Morton 1966; Holttum 1986. Moran 1993b; Mickel & Smith 2004.

Tectaria heracleifolia (Willd.) Underw., (*Heracleum*—cow parsnip, in the carrot family, and Latin: *folius*, leaved, = with leaves superficially like those of *Heracleum*), BROAD HALBERD FERN. Rhizomes erect; leaves evergreen, clustered, to 90 cm long; petioles glabrous, sparsely scaly basally, with more than 3 vascular bundles; leaf blades monomorphic, ovate to pentagonal in outline, (12–)20–45(–50+) cm long, 1-pinnate, with 3–5(–7) pinnae (1–3 pairs plus a single terminal pinna), thick-herbaceous to subcoriaceous, glabrous; pinnae, particularly the proximal ones, with large basal lobe(s), the margins with shallow lobes, the apices acuminate to long-attenuate; lateral pinnae ± falcate; sori round, in single rows on both sides of the side veins of the pinnae, to ca. 3 mm in diam.; indusia peltate, entire, persistent; $2n = 160$ (Moran 1993b). Limestone sink-holes and cave entrances, occasionally in shaded canyons or shaded rock outcrops or railroad tunnels; Edwards (Walls of Devil's Sinkhole—*Cory 35624, 35628*, TAES; *Correll 13420, Cory 35630*, BRIT), Uvalde (*Parks & Cory 33414*, BRIT; Limestone sink hole—*Cory & Parks 294*, 1939, TEX-LL; Palmer 1919c, 1934), Irion (*T. Maxwell & R. Ruiz s.n., 1974*, BRIT; *T. Watson 19010*, TEX-LL), Comal (*Lindheimer*, "in the entrance of a limestone cave near New Braunsfels, in 1878"—Palmer 1934), and Mason (Turner et al. 2003) cos. on the

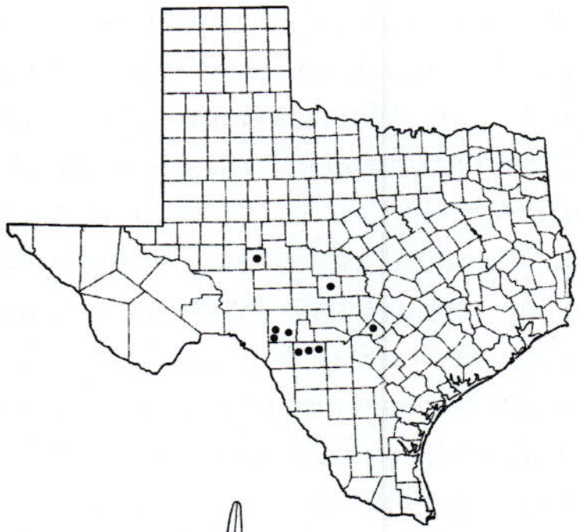

Carlos Velasco

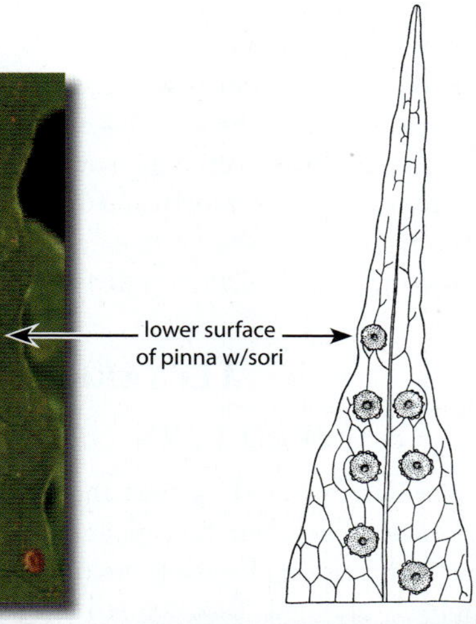

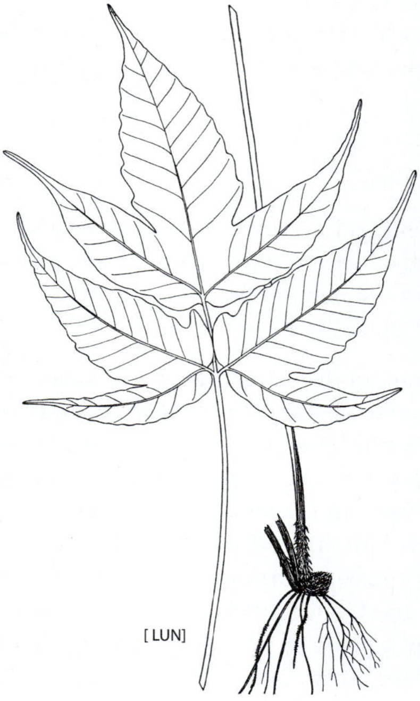

Alan Cressler

Tectaria heracleifolia

BROAD HALBERD FERN

Edwards Plateau; in the U.S. otherwise known only from FL; also West Indies, Mexico, Central and South America. Sporulating Apr–Oct. This is a tropical species whose TX locations are at the extreme n limit of the distribution of the species—it is "one of the commonest ferns at low and middle elevations in southern and eastern Mexico" (Mickel & Smith 2004). It was first collected in TX in 1847 by Lindheimer (Palmer 1934) and is rare in the state, occurring only in particular microhabitats (e.g., sink-holes, caves). An Edwards Co. site is particularly interesting—there *T. heracleifolia* is found on the vertical shaft wall of the 140 ft deep Devil's Sinkhole (a collapsed limestone cavern). The leaf blades can sometimes be huge when growing under very low light conditions (e.g., "ledge 200 feet down in the Devil's Sink Hole"—*Parks & Owens s.n.* 24 Sep 1941, TAES). Other than TX, the species is known in the U.S. only from extreme s Florida. Given this species' rareness and limited distribution in TX and its occurrence in only one other state in the U.S. (FL), we consider it to be of conservation concern in TX. ⚠

THELYPTERIDACEAE Pic. Serm.

MARSH FERN OR MAIDEN FERN FAMILY

Texas species terrestrial; rhizomes short- or long-creeping; leaves usually monomorphic or nearly so (somewhat dimorphic in *Thelypteris dentata*); leaf blades 1-pinnate-pinnatifid, 2-pinnate-pinnatifid, or 2–3-pinnatifid near base, less divided distally, usually with an indument (sometimes sparse) of transparent needle-like hairs on lower surfaces (= abaxially), the ultimate segments (= smallest subdivisions) often but not always entire; sori usually round, on the lower surfaces of the leaf blades (usually on veins), medial to submarginal; indusia round to kidney-shaped or obscure or absent.

A monophyletic family of 900–1000 species of terrestrial ferns (Mabberley 1997; Nauman et al. 2000; Smith et al. 2008). There has been continuing controversy on the number of genera to recognize (Smith 1993a), ranging from 1 to ca. 30 depending on taxonomic approach (Smith 1993a). Yatskievych and Windham (2009c) pointed to the following situation: "with most American authors lumping all of the species into a single genus *Thelypteris*, and most authors working in the Old World splitting the family into about 30 genera. More recently, compromise treatments have been advocated in which 3–6 main groups are recognized as genera." Mabberley (2008), for example, recognized 6 genera, including the 3 known for TX. Although the group is subcosmopolitan overall, most members are tropical. As treated here, only 3 genera occur in North America (2 native), all 3 of these occurring in TX. Members of the family have traditionally been associated with *Dryopteris* (Dryopteridaceae) (e.g., Correll 1956), but there is no close relationship (Smith 1993a); the similarities are apparently due to convergent evolution. Smith and Cranfill (2002) in their recent molecular study of the family have shown that it "comprises two primary lineages, one phegopteroid, the other thelypteroid (including cyclosoroid)." The much smaller phegopteroid lineage (which includes *Macrothelypteris* and *Phegopteris*) is basal in the Thelypteridaceae and "includes those elements that are the most dissected, lack adaxial grooves on the frond axes, and are generally morphologically the most distinct elements within the family." Based on their research, they suggest that, "recognition of an intermediate number of genera may be a reasonable taxonomic course." Smith and Cranfill (2002) also noted that the family is most closely related to a group of terrestrial ferns including the "athyrioid, woodsioid, blechnoid, and onocleoid ferns." (Order Polypodiales)
FAMILY RECOGNITION IN THE FIELD: sori usually on veins (but not marginal) on lower leaf surfaces, round; leaves usually all alike, appearing much dissected, 1-pinnate-pinnatifid to 3-pinnatifid, the lower surfaces with transparent needle-like hairs (distinguishing this family from Dryopteridaceae and Athyriaceae, which have scales and lack such needle-like hairs), the ultimate leaf segments often (but not always) entire or nearly so.
REFERENCES: Smith 1990, 1993a; Smith & Cranfill 2002.

1. Rachis conspicuously and irregularly winged between pinnae (wing tissue extended as obvious lobes in some places); leaf blades ± triangular in outline, ca. as wide as long, usually 7–25(–30) cm long; indusia absent ____ **Phegopteris**
1. Rachis not winged; leaf blades much longer than wide, often > 30 cm long; indusia present, often conspicuous with a hand lens (but in *Macrothelypteris* they are small or obscure in mature sori).
   2. Midveins of pinnae (= costae) grooved on upper (= adaxial) surface; leaf blades 1-pinnate-pinnatifid; veins of ultimate leaf segments reaching segment margins; indusia usually obvious with a hand lens ____ **Thelypteris**

2. Midveins of pinnae not grooved on upper surface; leaf blades mostly 2-pinnate-pinnatifid; veins of ultimate leaf segments not reaching segment margins; indusia small (< 0.3 mm in diam.) or obscure ________ **Macrothelypteris**

MACROTHELYPTERIS (H. Ito) Ching FALSE MAIDEN FERN

A genus of ca. 10 species of tropical and subtropical regions in Asia, Africa, the Pacific Islands, and Australia (Smith 1993a; Mabberley 2008). It has frequently been included in a more broadly circumscribed *Thelypteris* (e.g., Correll & Johnston 1970; Lellinger 1985; Hatch et al. 1990), but most modern authors treat it at the generic level (e.g., Smith 1993a; Peck & Taylor 1995; Kartesz 1999; Nauman et al. 2000). Recent molecular research (Smith & Cranfill 2002) has shown that *Macrothelypteris* (along with *Phegopteris*) is in the small "phegopteroid" lineage that is basal in the family. (Greek: *macro*, large, *thelys*, female, and *pteris*, fern, in reference to the genus *Thelypteris*, FEMALE FERN or MAIDEN FERN)

REFERENCES: Correll 1951; Holttum 1969; Leonard 1972; Gorman et al. 2011.

Macrothelypteris torresiana (Gaudich.) Ching, (for Luis de Torrès of the Marianas Islands, governor of Guam in 1820), TORRES' FERN, MARIANA MAIDEN FERN, FALSE MAIDEN FERN, SWORDFERN. Rhizomes short-creeping; leaves monomorphic, evergreen; leaf blades mostly 2-pinnate-pinnatifid, broadest at base, much longer than wide, to ca. 0.85(–1.5) m long, the lower surface with costal hairs unbranched, septate, mostly over 1 mm; veins of ultimate leaf segments not reaching segment margins; rachis not winged; midveins of pinnae (= costae) not grooved on upper (= adaxial) surface; sori round, medial to submedial; indusia small (< 0.3 mm in diam.) or obscure; 2n = 124 (Smith 1993a). Along streams, damp woods, and moist areas; Newton (*E. Whitehouse 23089*, 1950, BRIT), Liberty (*M. Schedlbauer s.n.*, TAMU), and Hardin (e.g., *L. Rosier s.n.* 1963, TEX-LL) cos. in the Pineywoods, first collected in TX in Newton Co. by Eula Whitehouse in 1950 (Correll 1951; 1956); naturalized in se U.S. from KY (Gorman et al. 2011), TN (Beck 2005), VA (Wieboldt et al. 2011), and SC s to FL w to AR and TX; also naturalized in Mexico and widely in the Neotropics. Sporulating summer and fall. Native to tropical and subtropical Asia and Africa, originally described from the Marianas Islands (Thieret 1980). [*Dryopteris setigera* of authors, not (Blume) Kuntze, *Polystichum torresianum* Gaudich., *Thelypteris torresiana* (Gaudich.) Alston] This species is sometimes cultivated as an ornamental, but in some areas can become invasive (e.g., in AR according to J. Peck [pers. comm.]). The name *Dryopteris setigera* has been misapplied to this species (e.g., Correll 1951). The first known collection of this species in the U.S. was in 1904 from Florida (Leonard 1972) and it is continuing to spread in the se part of the country (e.g., Gorman et al. 2011). At the time of its discovery in TX, Correll (1951) noted that it brought the total number of pteridophyte species known for TX to 103.

PHEGOPTERIS (C. Presl) Fée BEECH FERN

A genus of three (or possibly four) species with one n temperate and boreal, one in e North America, and one in e Asia (Smith 1993a; Mabberley 2008). It has frequently been included in a more broadly circumscribed *Thelypteris* (e.g., Correll & Johnston 1970; Lellinger 1985; Hatch et al. 1990), but most modern authors treat it at the generic level (e.g., Smith 1993a; Peck & Taylor 1995; Kartesz 1999; Nauman et al. 2000). Recent molecular research (Smith & Cranfill 2002) has clarified the situation and shown that *Phegopteris* is in the small "phegopteroid" lineage that is basal in the family. (Greek: *phegos*, beech, and *pteris*, fern, presumably from some species growing under *Fagus* (BEECH) trees)

REFERENCES: Holttum 1969; McNeill & Pryer 1985; Driscoll et al. 2004.

Phegopteris hexagonoptera (Michx.) Fée, (Greek: *hex*, six, *gonio*, angle or corner, and *pteron*, wing or feather, = 6-angled wing, apparently in reference to the shape of the wings along the rachis), BROAD BEECH FERN, SOUTHERN BEECH FERN. Rhizomes long-creeping; leaves monomorphic, dying back in winter; leaf blades 2–3 pinnatifid near base, less divided distally, broadest at base, ± triangular in outline, ca. as wide as long, usually 7–25(–30) cm long, the lower surface with moderate to dense, non-septate short hairs (usually 0.1–0.25 mm long) on costae and veins; proximal pinnae longest, narrowed basally; basal veins of adjacent ultimate leaf segments reaching segment margins or nearly so above the sinus between segments; rachis conspicuously and irregularly winged between pinnae (wing tissue extended as obvious lobes in some places); midveins of pinnae not grooved on upper side (vs. grooved in *Thelypteris*); sori round, submarginal; indusia absent; 2*n* =

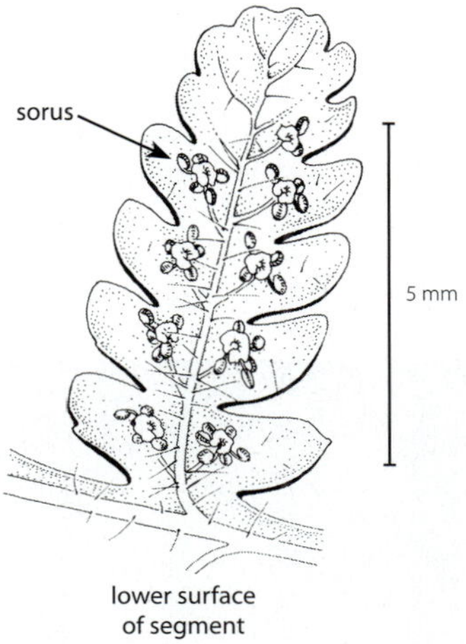

Macrothelypteris torresiana

TORRES' FERN

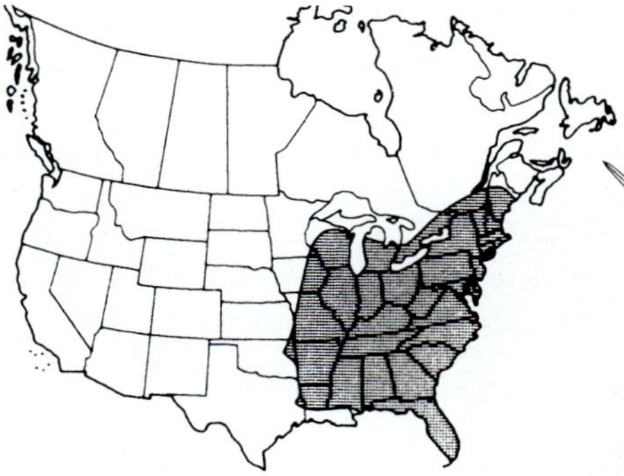

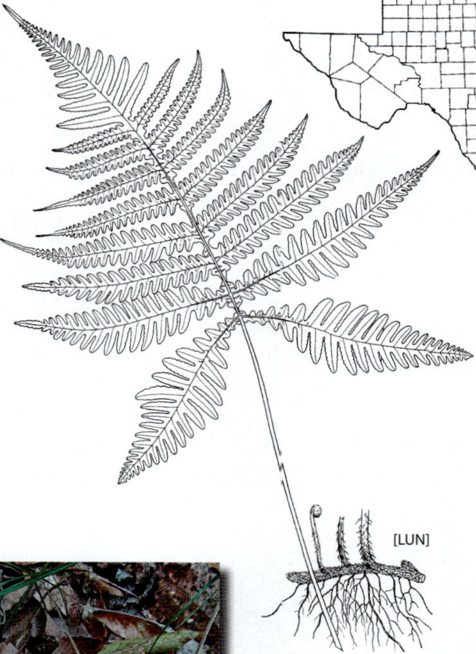

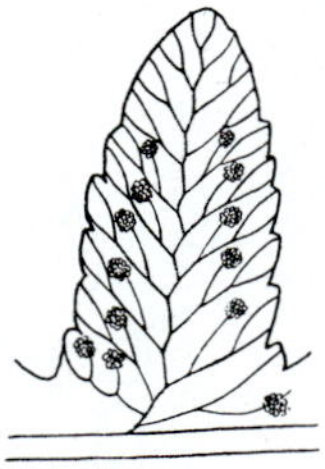

lower surface
of pinnule w/sori

Alan Cressler

Alan Cressler

Phegopteris hexagonoptera

BROAD BEECH FERN

60 (Smith 1993a). Moist wooded areas, margins of bogs, ravines along streams, often in acidic soils; Marion (*Orzell & Bridges 4941*, BRIT, TEX-LL), Newton (*Allen et al. 21363*, BRIT), Sabine (*E. Fritz 326879, McVaugh 8435*, BRIT), Shelby (BAYLU, *Bridges & Orzell 5121*, TEX-LL), Rusk (*Bridges & Ozell 6258*, TEX-LL), and San Augustine (*Correll & Correll 24887*, TEX-LL; Palmer 1919c) cos. in the e Pineywoods; se Canada and throughout the e U.S. w to MN, OK, and TX. Sporulating summer and fall. [*Dryopteris hexagonoptera* (Michx.) C. Chr., *Polypodium hexagonopterum* Michx., *Thelypteris hexagonoptera* (Michx.) Nieuwl.] The winged rachis is very helpful in identification. Correll (1956) considered this species to be "rare" in TX. Given this species' rareness and limited distribution in the state, we consider it to be of conservation concern in TX. ⚠

THELYPTERIS Schmidel FEMALE FERN, MAIDEN FERN, MARSH FERN

Terrestrial; rhizomes horizontal, short- or long-creeping; leaves usually ± monomorphic (somewhat dimorphic in *T. dentata*); leaf blades 1-pinnate-pinnatifid (the pinnae sometimes deeply pinnatifid, within 1 mm of the costae), much longer than wide; ultimate leaf segments (= smallest subdivisions of leaf) entire, with veins reaching segment margins; rachis not winged; midveins of pinnae (= costae) grooved on upper (= adaxial) surface; petioles straw-colored or purplish; sori round, in medial to submarginal position on the leaf segments on the lower (= abaxial) surfaces; indusia round to kidney-shaped, usually obvious with a hand lens, typically with pubescence.

A heterogeneous nearly cosmopolitan genus of ca. 875 species (in the broad sense; Smith 1993a) including the tropical *Cyclosorus*, or ca. 300 species (if considered more narrowly excluding *Cyclosorus*; Mabberley 2008) According to Schuettpelz and Pryer (2008), the genus is "in need of considerable phylogenetic study," and when treated as excluding *Cyclosorus* is paraphyletic. Some of the species are similar and difficult to distinguish without close observation of the venation pattern of the ultimate leaf segments, particularly the basal veins. In some, the basal veins of adjacent ultimate segments join together and run united to the base of the sinus (= space or recess) between the adjacent segments, while in others the basal veins of adjacent ultimate leaf segments are not united, and thus reach margin of segments at or slightly above the sinus. (Greek: *thelys*, female, and *pteris*, fern, derived from *pteron*, wing or feather, perhaps referring to the delicate appearance in contrast to the "male fern," *Dryopteris*—Thieret 1980)

REFERENCES: Smith 1971a, 1971b; Willis & Nester-Hudson 2006.

1. Lateral veins of ultimate leaf segments (= smallest subdivisions of leaf) often forked; leaf blades with **all** lateral veins of ultimate segments extending to margin of blade tissue **above** (= beyond) sinuses between adjacent ultimate segments; margins of fertile ultimate segments usually turned downward ____ **T. palustris**
1. Lateral veins of ultimate leaf segments usually not forked; leaf blades with at least some basal lateral veins of ultimate segments extending to, meeting at, or united **below** the sinuses between the adjacent ultimate segments; margins of fertile ultimate segments usually not turned downward (but can be in old plants and dried specimens of *T. kunthii* and *T. ovata*).
   2. Basal veins of adjacent ultimate leaf segments united below the sinus between the segments, with the resulting single vein extending toward the sinus.
      3. United portion of veins below sinuses usually 2–4 mm long; midveins of pinnae with hairs mostly very short (0.3 mm or less long) on the lower surface (= abaxially); petioles often purplish; leaves usually with more than 2 pairs of greatly reduced basal pinnae ____ **T. dentata**
      3. United portion of veins below sinuses usually < 2 mm long; midveins of pinnae with hairs variable in length (0.3–0.8 mm long) on the lower surface; petioles straw-colored; leaves usually with 0–2 pairs of slightly reduced basal pinnae ____ **T. hispidula**
   2. Basal veins of adjacent ultimate leaf segments mostly not united below the sinus, but with at least some of them extending to or meeting at the sinus (others can separately reach margin of segments slightly above the sinus).
      4. Upper surface of midveins of pinnae glabrous or with a few minute hairs, these never much longer than width of the midveins; a few scales often persistent on lower surface of rachises and midveins of pinnae of mature leaves ____ **T. ovata**
      4. Upper surface of midveins of pinnae with hairs usually longer than width of the midveins (use hand lens); scales absent on lower surface of rachises and midveins of pinnae of mature leaves.
         5. Basal veins of adjacent ultimate leaf segments not united, extending to and meeting at sinus between segments or nearly so; leaf blades usually broadest at or near base, the basal pinnae

as long as or only slightly shorter than more distal pinnae; leaf blades glabrous or sparsely hairy on upper surfaces, except along veins ____ **T. kunthii**

5. Basal veins of adjacent ultimate leaf segments variable, some not united, some meeting at the sinus, some united below the sinus into a single vein; leaf blades broadest above base, the basal and near basal pinnae noticeably shorter than more distal pinnae; leaf blades often somewhat hairy on upper surfaces, on regular blade tissue as well as along the veins ____ **T. hispidula**

Thelypteris dentata (Forssk.) E.P. St. John, (Latin: *dentatus*, toothed, for the leaves which were described as "fronde pinnata, dentata" in the type description, though the "teeth" are actually the very small lobes of the pinnae = the ultimate leaf segments), DOWNY MAIDEN FERN, DOWNY SHIELD FERN. Rhizomes short-creeping; leaves somewhat dimorphic, evergreen; petioles often purplish; leaf blades (25–)40–100 cm long, usually with more than 2 pairs of greatly reduced basal pinnae, the basal acroscopic segment of basal pinnae often auriculate; basal veins of adjacent ultimate leaf segments (= smallest subdivisions of leaf) united below the sinus between the segments, with the resulting single vein extending toward the sinus, the united portion usually 2–4 mm long; midveins of pinnae underneath (= abaxially) with mostly very short (0.3 mm or less long) hairs; sori round, medial to supramedial; indusia pubescent; $2n$ = 144 (Smith 1993a). Wooded slopes, damp woods, swamp hummocks, and along streams; Angelina (*E. Boon s.n. 1934*, TEX-LL), and Sabine (Correll 1956) cos. in the Pineywoods; also Harris Co. (*Tharp 4279*, TEX-LL; Turner et al. 2003) at the n margin of the Gulf Prairies and Marshes; naturalized in the se U.S. from SC (CLEMS, USCH) s to FL w to TX, also KY. Sporulating summer and fall. Native to the tropics and subtropics of Asia and Africa, but widely naturalized in the New World (Mexico, West Indies, Central and South America). This species has apparently spread rapidly in the New World; Smith (1993a) noted that prior to 1900 it was "virtually unknown in the Americas." [*Christella dentata* (Forssk.) Bronsey & Jermy, *Dryopteris dentata* (Forssk.) C. Chr., *Polypodium dentatum* Forssk.] This introduced species "is becoming one of the most common ferns in many areas of the Neotropics" (Mickel & Smith 2004).

Thelypteris hispidula (Decne.) C.F. Reed var. **versicolor** (R.P. St. John) Lellinger, (sp.: Latin: *hispidus*, hairy, bristly, or rough, and *-ula*, diminutive suffix, in reference to the diminutive hairs; var.: Latin: *versicolor*, of various colors or changing color, the botanist who named it [St. John] saying, "Its light-green color often turns to golden or gray as it matures and frequently the leaf-surface acquires rusty-red blotches"), VARIABLE MAIDEN FERN, ST. JOHN'S SHIELD FERN, ROUGH-HAIRY MAIDEN FERN, HAIRY MAIDEN FERN. Rhizomes short-creeping to ascending at tip; leaves monomorphic, evergreen; petioles straw-colored; leaf blades 14–55 cm long, broadest above base, the basal and near basal pinnae usually noticeably shorter than more distal pinnae, the basal acroscopic segment of basal pinnae sometimes auriculate and crenate; basal veins of adjacent ultimate leaf segments variable, some united below the sinus into a single vein (usually for < 2 mm) which then runs to the sinus, some meeting at the sinus, some free; leaf blade tissue (in addition to veins) often somewhat hairy above; midveins of pinnae underneath (= abaxially) with hairs variable in length, 0.3–0.8 mm long; sori medial to supramedial; indusia with pubescence; $2n$ = 72 (Smith 1993a). Wooded slopes and ravines, low woods, along streams; Jasper (*Correll & Correll 12538*, BRIT, TEX-LL), Orange (*Correll 13331*, BRIT, TEX-LL), San Jacinto (*Correll 13376*, BRIT, TEX-LL), Sabine (BRIT), Hardin (*Singhurst 2180*, TEX-LL), and Houston (BAYLU) cos. in the Pineywoods; se U.S. from SC s to FL w to TX; also West Indies in Cuba. Sporulating summer and fall. [*Christella hispidula* (Decne.) Holttum, *Dryopteris versicolor* (R.P. St. John) M. Broun, *T. quadrangularis* (Fée) Schelpe var. *versicolor* (R.P. St. John) A.R. Sm., *T. versicolor* R.P. St. John] This species was at one time considered to be a hybrid between *T. dentata* and *T. kunthii* (e.g., Correll & Johnston 1970). However, Smith (1971a, 1971b) showed this was not possible because the supposed parents were tetraploids and this species is diploid. The species as a whole occurs widely—var. *hispidula* is in the Old World (Africa and Asia). *Thelypteris hispidula* also occurs from Mexico through South America, and Mickel and Smith (2004) noted that, "Mexican and most continental tropical American specimens seem not or only weakly distinguishable from var. *hispidula* in the Old World."

Thelypteris kunthii (Desv.) C.V. Morton, (for Karl Sigismund Kunth, 1788–1850, German botanist), WIDESPREAD MAIDEN FERN, SOUTHERN SHIELD FERN, KUNTH'S MAIDEN FERN. Rhizomes short- to long-creeping; leaves monomorphic, evergreen; petioles straw-colored; leaf blades (9–)30–80 cm long, usually broadest at or near base, the basal pinnae as long as or only slightly shorter than more distal pinnae, the basalmost pinnae varying

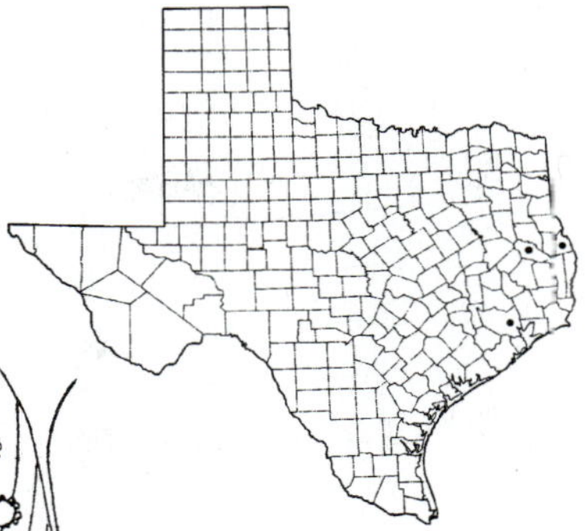

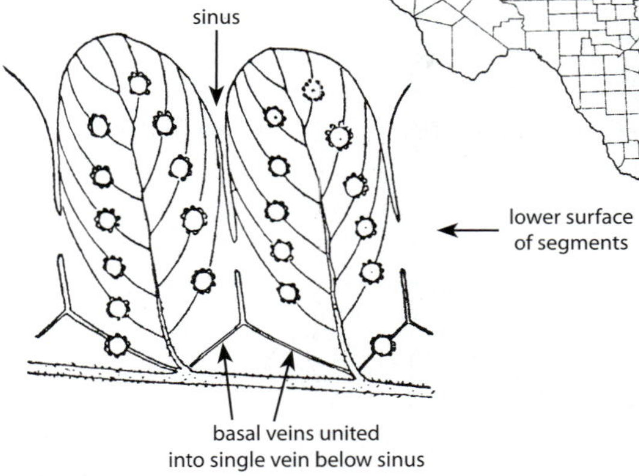

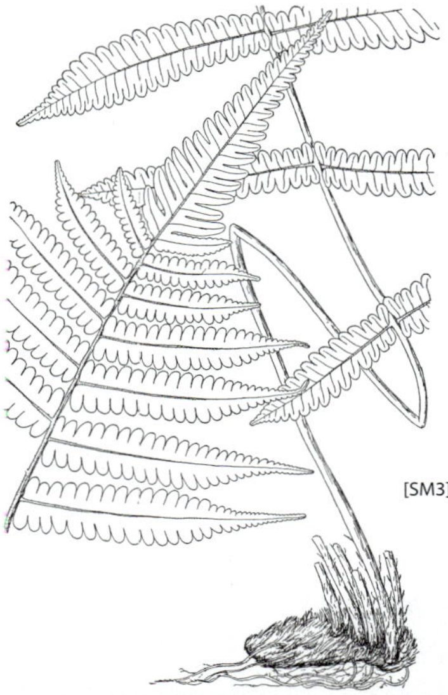

Thelypteris dentata

DOWNY MAIDEN FERN

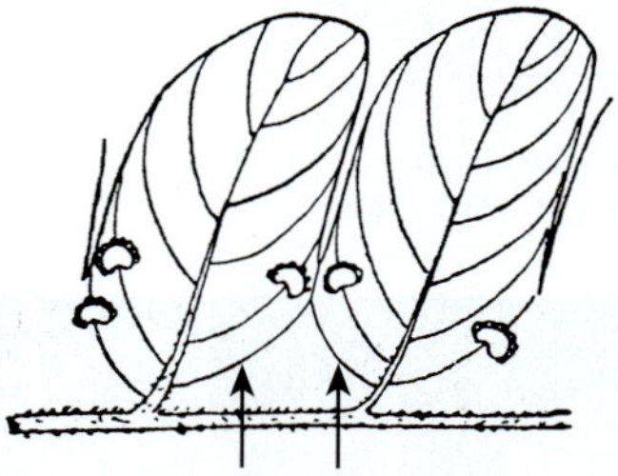

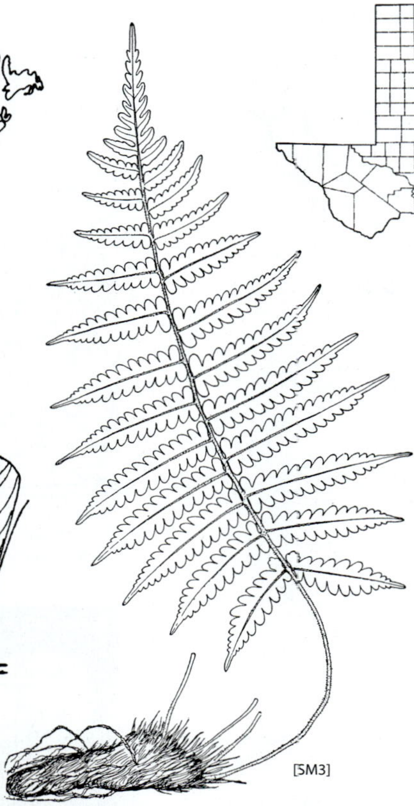

Mississippi Museum of Natural Science (MS-DWFP)

Thelypteris hispidula var. **versicolor**

VARIABLE MAIDEN FERN

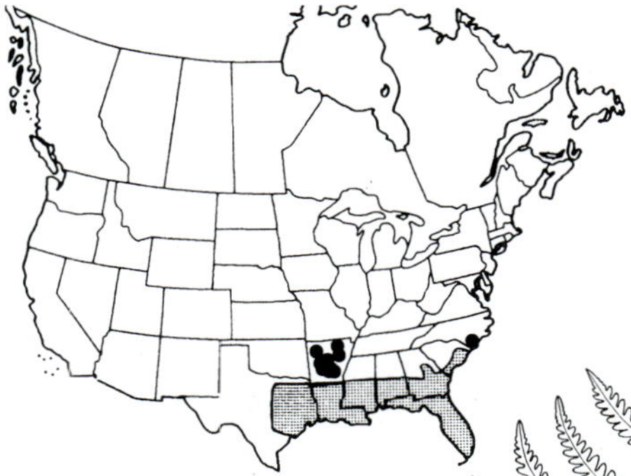

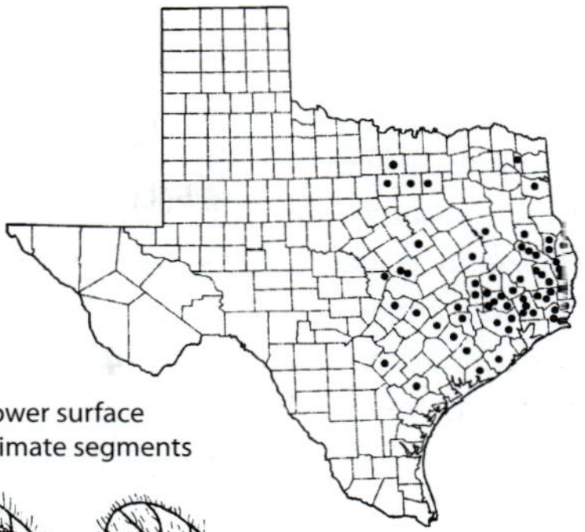

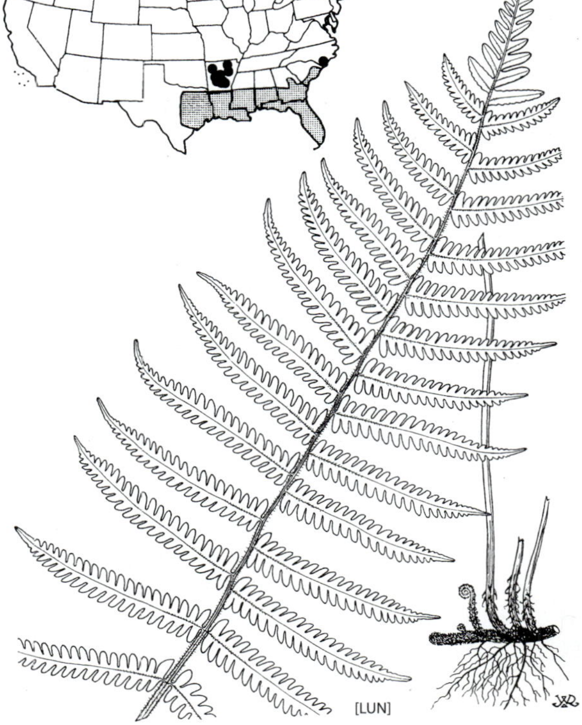

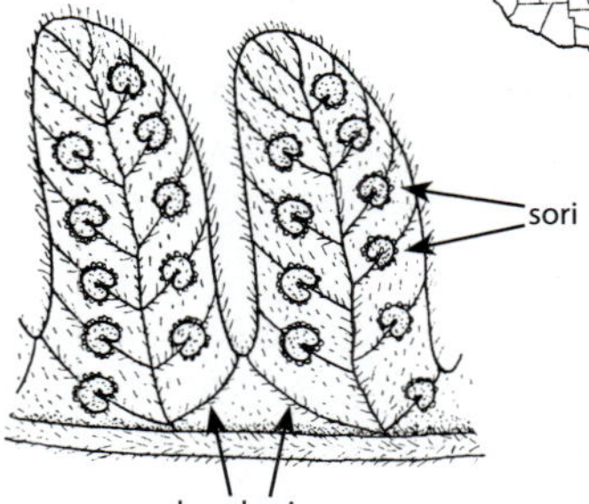

Mississippi Museum of Natural Science (MS-DWFP)

Mississippi Museum of Natural Science (MS-DWFP)H

Thelypteris kunthii

WIDESPREAD MAIDEN FERN

from ± perpendicular to rachis to angled downward, the blades glabrous or sparsely hairy above, except along veins, the lower surface with indument of short hairs on veins and blade tissue; basal veins of adjacent ultimate leaf segments not united, extending to sinus between segments or nearly so and sometimes meeting at the sinus; sori medial to supramedial; indusia with pubescence; 2*n* = 144 (Smith 1993a). Woodlands, streambanks, swamps, ditches; widespread in East TX and adjacent Gulf Prairies and Marshes w to the Blackland Prairie or slightly further w (Bastrop, Dallas, Travis—TEX-LL, Bell, Burnet, Wilson—Turner et al. 2003, and McLennan—J.Mink pers. comm.) cos.; several collections are also known from the Cross Timbers and Prairies, including Tarrant (*O'Kennon 17144*, BRIT) and Wise (*O'Kennon & McLemore 18841*, BRIT) cos. and a Parker Co. collection (*Jeff Quayle s.n.*, 1997, BRIT) at the w margin of the species' range; se U.S. from NC s to FL w to AR and TX; also West Indies and Mexico to South America. Sporulating spring–fall. [*Christella normalis* (C. Chr.) Holttum, *Dryopteris normalis* C. Chr., *T. normalis* (C. Chr.) Moxley] This species has often been confused and lumped (e.g., Correll 1956) with *Thelypteris ovata* var. *lindheimeri*, sometimes under the name *Dryopteris normalis*. *Thelypteris kunthii* has been reported to hybridize with *T. hispidula* and *T. ovata* (Smith 1993a). Willis and Nester-Hudson (2006) reported a *T. kunthii* × *T. ovata* var. *lindheimeri* hybrid from Walker Co., TX.

Thelypteris ovata R.P. St. John var. **lindheimeri** (C. Chr.) A.R. Sm., (sp.: Latin: *ovatus*, egg-shaped or ovate; var.: for F.J. Lindheimer, 1801–1879, German-born TX collector), LINDHEIMER'S MAIDEN FERN, OVATE MARSH FERN, LINDHEIMER'S SHIELD FERN, SAN SABA FERN. Rhizomes usually long-creeping; petioles straw-colored; leaves monomorphic, sometimes evergreen; leaf blades usually broadest between middle and base, tapered, the lower pinnae reduced (leaf ± egg-shaped in outline), the basalmost pinnae angled downward, the basal pair of ultimate segments of the middle pinnae often elongate parallel to rachis, the blades glabrous or with minute pubescence above, with scattered hairs along veins, the lower surface with indument of short hairs on veins and blade tissue; basal veins of adjacent ultimate leaf segments not united, reaching margin at or just above sinus between segments, sometimes meeting at the sinus; sori supramedial to submarginal (sori typically closer to leaf margins than in *T. kunthii*); indusia with pubescence; 2*n* = 72 (Smith 1993a). Low, moist areas, wet bluffs and ledges, including limestone; mainly Edwards Plateau and Trans-Pecos but east to Bell (*S. E. Wolff 3397*, BRIT), Bexar, Comal, Hays, Travis (TEX-LL), and Williamson (Turner et al. 2003) cos. near w margin of Blackland Prairie, also Tarrant Co. (*J. Quayle s.n.* 1996, "possibly escaped from cultivation," BRIT) in the Cross Timbers and Prairies; in the U.S. known only from TX; also Mexico, Central America, and West Indies in Jamaica. Sporulating May–Nov. [*Dryopteris normalis* C. Chr. var. *lindheimeri* C. Chr.] This species has often been confused and lumped (e.g., Correll 1956; Correll & Johnston 1970; Hatch et al. 1990) with *Thelypteris kunthii* (either as *T. kunthii* or under the name *Dryopteris normalis*). Although strikingly similar in overall aspect, the two can be readily distinguished by the characters in the key to species. This is one of 19 pteridophyte species known in the U.S. only from TX (see page 41).

Thelypteris palustris Schott var. **pubescens** (G. Lawson) Fernald, (sp.: Latin: *paluster*, marshy or swampy, presumably in reference to the habitat; var.: Latin: *pubens*, downy or softly hairy, possibly for the hairs on the costae and veins on the underside of the blades or the often pubescent indusia), SOUTHERN MARSH FERN, MARSH FERN, EASTERN MARSH FERN. Rhizomes long-creeping, leaves monomorphic or nearly so, dying back in winter; petioles straw-colored above base; leaf blades 10–40(–55) cm long, the lower surface with sparse to dense hairs on costae (= midveins of pinnae) and sometimes other veins; pinnae deeply pinnatifid, to within 1 mm of the costae (rarely pinnate), the lower pinnae commonly slightly shorter than middle pinnae; lateral veins of ultimate leaf segments often forked, usually extending to margin of blade tissue above sinuses between ultimate segments; margins of fertile ultimate segments entire, usually turned downward; sori medial; indusia often with pubescence; 2*n* = 70 (Smith 1993a). Terrestrial in low woods, along streams, and in swamps and bogs; scattered in the Pineywoods (e.g., Polk Co., *T. Starbuck 403b*, BRIT; Titus Co., *P.A. Amerson 1111*, BRIT) and Post Oak Savannah (e.g., Leon Co., *Orzell & Bridges 6522*, BRIT, TEX-LL—"Acid seep forest…Sparta Sand"); also n Gulf Prairies and Marshes (e.g., Colorado Co., *Correll & J.R. Peace 34950*, TEX-LL); se Canada and throughout e U.S. w to ND, OK, and TX; also e Asia and Europe (circumboreal). Sporulating late summer and fall. [*Dryopteris thelypteris* (L.) A. Gray var. *pubescens* (G. Lawson) Weath., *Thelypteris palustris* var. *haleana* Fernald] The distribution of this fern is a classic example of a disjunction caused by vicariance (= a

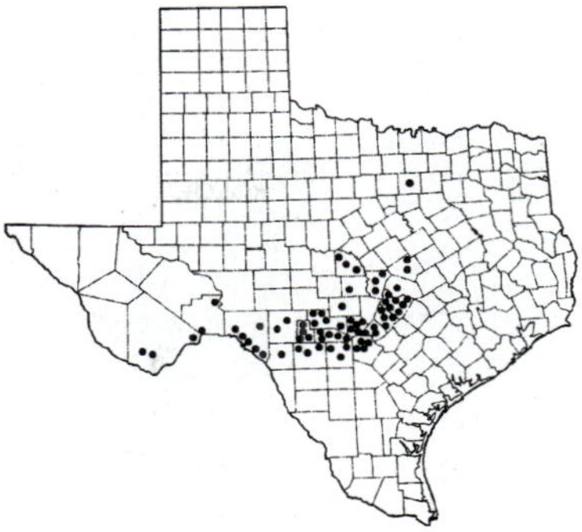

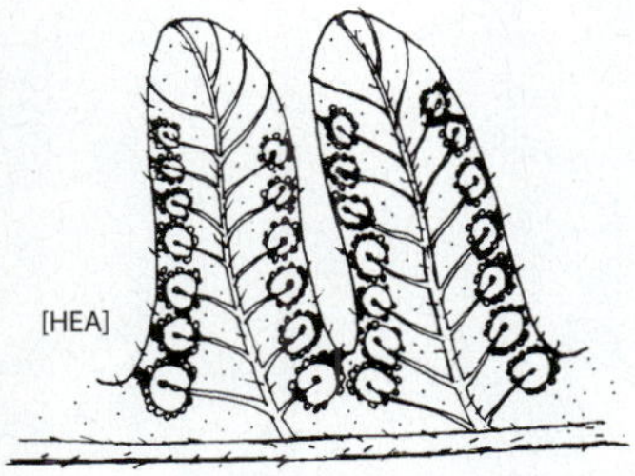

lower surface
of pinnules w/sori

Thelypteris ovata var. **lindheimeri**

LINDHEIMER'S MAIDEN FERN

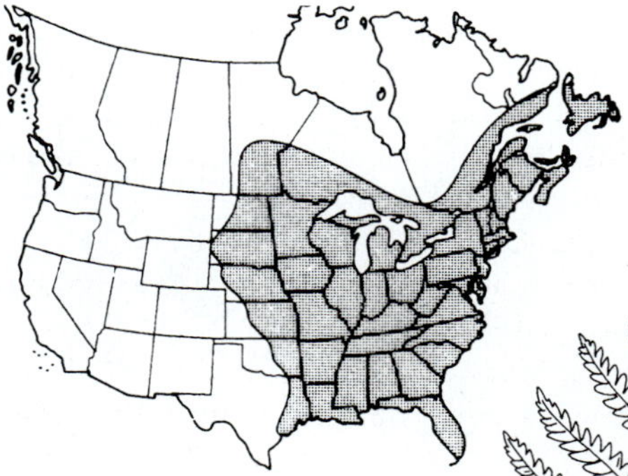

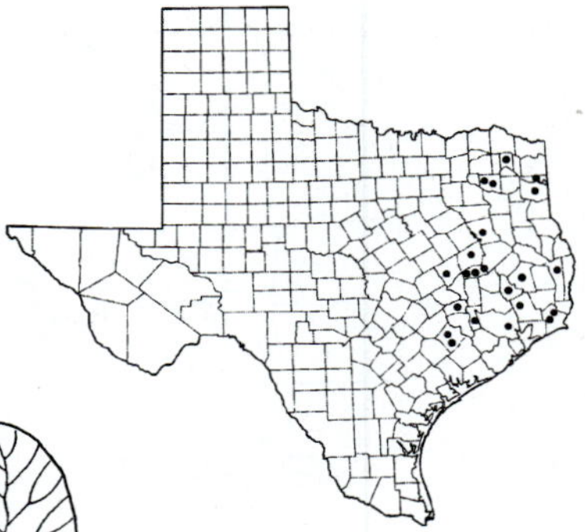

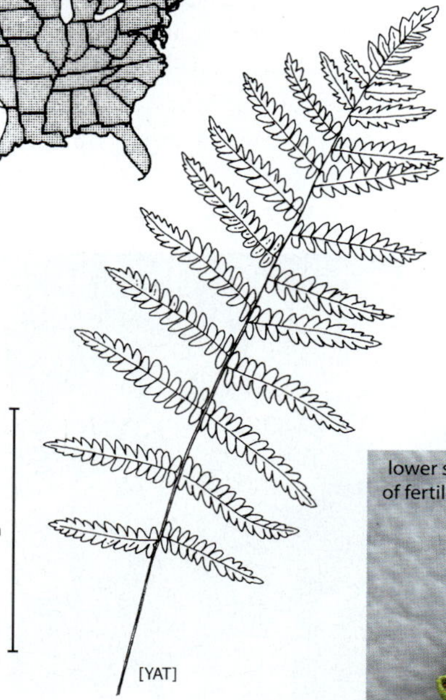

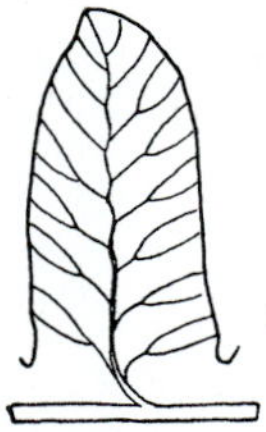

Mississippi Museum of Natural Science (MS-DWFP)

Mississippi Museum of Natural Science (MS-DWFP)

Thelypteris palustris var. **pubescens**

SOUTHERN MARSH FERN

once more widespread species becoming fragmented with survivors in only parts of the former area). It occurs broadly in Europe, e North America, and e Asia (including Japan). In fact, 40% of the fern species of the northernmost island of Japan (Hokkaido) and northeastern North America are common to both regions—a remarkable holdover from the Tertiary Period when the Earth's climate was much warmer, North America, Europe, and Asia were connected by land bridges, and a ± continuous circumboreal forest occurred broadly in the Northern Hemisphere. Much has changed since that time, and *Thelypteris palustris* and a variety of other plants have survived in only limited portions of their ancient range (Kato & Iwatsuki 1983; Moran 2008) (see page 32 and Fig. 35 for further information). The plants in North America and e Asia are var. *pubescens*, while those in Europe and w Asia are var. *palustris*. Sterile specimens of this species can be confused with sterile specimens of CINNAMON FERN (*Osmundastrum cinnamomeum*) because of the similar way in which the lateral veins of the ultimate leaf segments are forked. However, CINNAMON FERN has a persistent tuft of tan to orangish tomentum ("arm pit hairs") where each pinna attaches to the rachis, a feature lacking in *Thelypteris palustris*.

WOODSIACEAE Herter

CLIFF FERN FAMILY

As treated here the Woodsiaceae is a family of a single genus and ca. 35 species. Based on recent molecular evidence (Rothfels et al. 2012a) it is "an isolated lineage, not closely related to the other taxa frequently included in broad concepts of Woodsiaceae" (Rothfels et al. 2012b). In the past *Woodsia* has been treated in a variety of families. Correll (1956, 1966), for example, placed it in his extremely broad Polypodiaceae. In relatively recent treatments the species have often been included in a very broadly interpreted Dryopteridaceae (e.g., Smith 1993b; Yatskievych 1999; Diggs et al. 2006) or in a somewhat broadly circumscribed Woodsiaceae containing *Athyrium*, *Cystopteris*, and *Woodsia* as well as a number of small genera—the family thus conceived has ca. 10–15 genera and 565–700 species (Smith et al. 2006, 2008; Mabberley 2008). In order to facilitate identification, a key including the TX genera in this broad view of Woodsiaceae, is given below. However, Smith et al. (2008) pointed out that the Woodsiaceae (as treated by them) was possibly not monophyletic and might need to be broken up into "numerous small families to house its constituent genera." Indeed, molecular studies have shown that both the classifications involving a very broad Dryopteridaceae and a broad Woodsiaceae are incorrect. Recent DNA evidence (Schuettpelz & Pryer 2008; Rothfels et al. 2012a) suggests *Athyrium*, *Cystopteris*, and *Woodsia* are all phylogenetically extremely remote from Dryopteridaceae, are morphologically similar to each other due to convergent evolution, are not closely related, and should be in completely different families. It is clear that the taxonomy of these and related ferns has been fluid and uncertain. However, molecular research has progressed very rapidly with relationships between these and many other fern families now being much more clearly understood than only a few years ago. As a result, we changed the taxonomy in our manuscript just prior to publication to reflect the very recent and compelling research (Rothfels et al. 2012a, 2012b.) (Order Polypodiales)

FAMILY RECOGNITION IN THE FIELD: plants superficially similar to Cystopteris (plants relatively small, leaves all alike or nearly so, appearing much dissected, the sori round) but differing in having the indusia attached below sori and at first cup-like and completely enclosing sori, but later splitting into several irregular lobes or flaps spreading around sori, the leaf blades often broadest above the middle, and the veins of ultimate leaf segments not reaching segment margins.

REFERENCES: Smith et al. 2006, 2008; Schuettpelz & Pryer 2008; Rothfels et al. 2012a, 2012b.

KEY TO PLANTS PREVIOUSLY INCLUDED IN A MORE BROADLY CONCEIVED WOODSIACEAE

1. Sori elongate, straight to hooked or curved; indusia with elongate attachment along one side, the attachment as long as the elongate sori; leaves (including petioles) typically 60–120 cm long; species of East TX primarily in Pineywoods and Post Oak Savannah _____ *Athyrium* (see Athyriaceae)
1. Sori round or nearly so; indusia variously attached, but attachment not elongate (since the sori are round); leaves (including petioles) typically < 50 cm long (rarely to 75 cm); including species known in various parts of TX from East TX to the Trans-Pecos.
   2. Petioles with scattered light brown scales and glandular pubescence, glabrate with age; indusia attached below sori and at first cup-like and completely enclosing sori, but later splitting into sev-

eral irregular lobes or flaps spreading around sori (sometimes inconspicuous in mature sori); leaf blades often broadest above the middle; veins of ultimate leaf segments (= smallest subdivisions of leaf) not reaching segment margins (use magnification; important if sori not available); plants widespread and common in e ½ of TX, and occasional in the Trans-Pecos ______ **Woodsia**

2. Petioles glabrous except at very base; indusia attached by one side, hood- or pocket-like, arching over sori (sometimes inconspicuous in mature sori) towards segment margin; leaf blades broadest at base or at or just below middle; veins of ultimate leaf segments reaching segment margins; plants widely scattered (East TX to Trans-Pecos), but very rare in TX ______ *Cystopteris* (see Cystopteridaceae)

WOODSIA R. Br. CLIFF FERN

Usually on or among rocks or on well drained sandy or rocky soils; plants usually rather small (leaves typically < 40(–60) cm long); rhizomes compact to short-creeping, erect to horizontal; leaves monomorphic, dying back over winter or in some cases persisting; leaf blades 1–2 pinnate-pinnatifid; pinnae usually with at least some glandular hairs, these sometimes dense, the lower surfaces without transparent needle-like hairs, the ultimate segments dentate, with veins not reaching segment margins, the veins ending in hydathodes (= microscopic epidermal structures that excrete water); petiole bases with 2 vascular bundles; sori round, in 1 row between midvein and each lateral margin of ultimate segments (= smallest subdivisions) of leaf; indusia rather large, attached basally around the sori, at first cup-like and completely enclosing sori, but later splitting into several scale-like or irregular lobes, flaps, or thread-like filaments spreading around sori (often inconspicuous in mature sori).

A genus of ca. 35 species (including the segregate genera *Cheilanthopsis* Hieron., *Hymenocystis* C.A. Mey, and *Protowoodsia* Ching) found mainly in rocky, montane areas of the n hemisphere, particularly the mountains of Eurasia and arid portions of Mexico and the sw U.S. One species extends south through the Andes and also occurs in South Africa (Rothfels et al. 2012b). Rothfels et al. (2012b) also noted that the species "have remarkable ecological and morphological resemblance to members of *Cystopteris* (in Cystopteridaceae), to which they are only distantly related (see key above to distinguish *Cystopteris* and *Woodsia*). Recent molecular research has shown that *Cheilanthopsis*, *Hymenocystis*, and *Protowoodsia* are all nested within *Woodsia* and that the "molecular phylogeny of *Woodsia* is marked by a remarkably deep split between a clade of Old World or holarctic species, and a clade of predominantly New World species" (Rothfels et al. 2012a). The common name, CLIFF FERN, comes from the tendency of a number of species to grow on or among rocks (Hoshizaki & Moran 2001). Key to species adapted from Windham (1993d). (Named for Joseph Woods, 1776–1864, English botanist and architect)

REFERENCES: Brown 1964; Windham 1987a, 1993d, 1993e; Carlquist et al. 1997; Yarborough & Powell 2002.

1. Segments of the indusia relatively broad in at least basal ½, toothed or lacerate into thread-like filaments only towards tip; leaf blades lanceolate to ovate in outline, moderately to densely glandular; species of eastern ½ of TX, the Trans-Pecos, and the Panhandle.
   2. Petioles straw-colored to light brown at maturity; vein tips enlarged to form whitish hydathodes (= water-excreting microscopic epidermal structures) visible on the lower surface of the pinnules; species widespread and common in eastern ½ of TX ______ **W. obtusa**
   2. Petioles reddish brown to dark purple at maturity; vein tips not or only slightly enlarged, barely visible on the lower surface of the pinnules; rare species known in TX only from the Trans-Pecos and Panhandle ______ **W. plummerae**
1. Segments of the indusia divided into narrow, thread-like filaments nearly to the base; leaf blades linear to lanceolate in outline, glabrescent to sparsely glandular; species known in TX only from the Trans-Pecos.
   3. Pinnae relatively pointed, apically often attenuate to narrowly acute; largest pinnae with 7–18 pairs of distinct, well-developed, more widely spaced pinnules; translucent projections on pinnule margins mostly of many cells, usually prolonged to form twisted filaments, the pinnules appearing fringed (under high magnification) ______ **W. phillipsii**
   3. Pinnae relatively blunt, apically abruptly tapered to rounded; largest pinnae usually divided into only 3–7(+) pairs of well-developed, closely spaced pinnules; translucent projections on pinnule margins mostly of only 1–2 cells, only occasionally filamentous, the pinnules not appearing fringed ______ **W. neomexicana**

Woodsia neomexicana Windham, (of New Mexico, where the type was collected), NEW MEXICO CLIFF FERN. An allotetraploid species morphologically similar to the diploid *W. phillipsii* (thought to be one of its par-

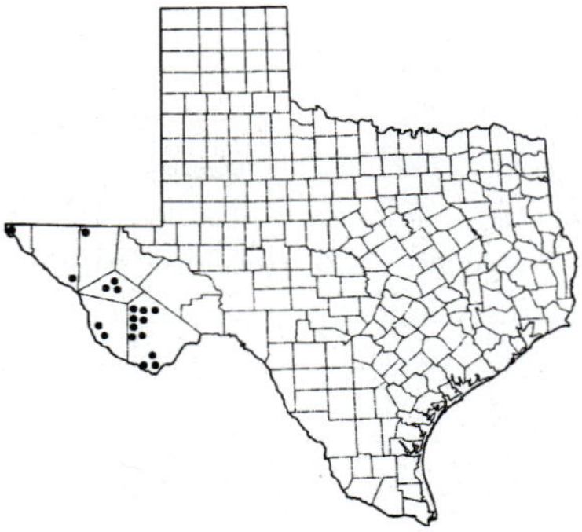

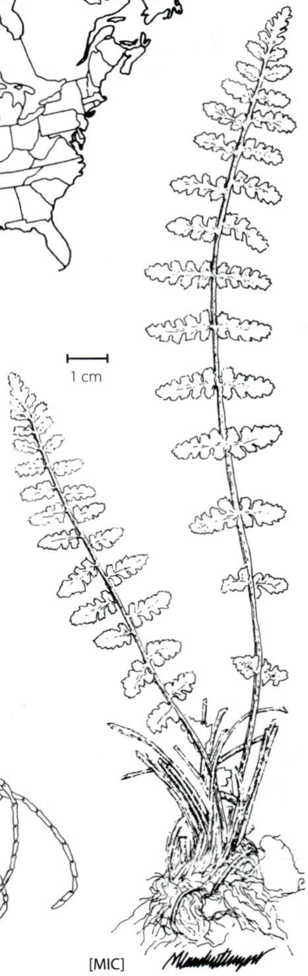

[MIC]

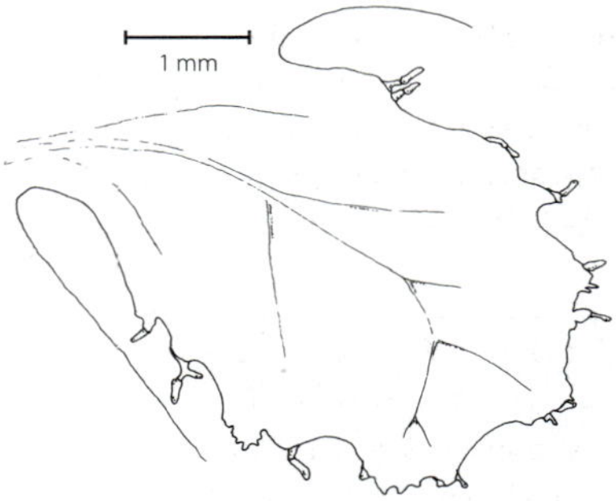

detail of pinnule
w/cellular filaments

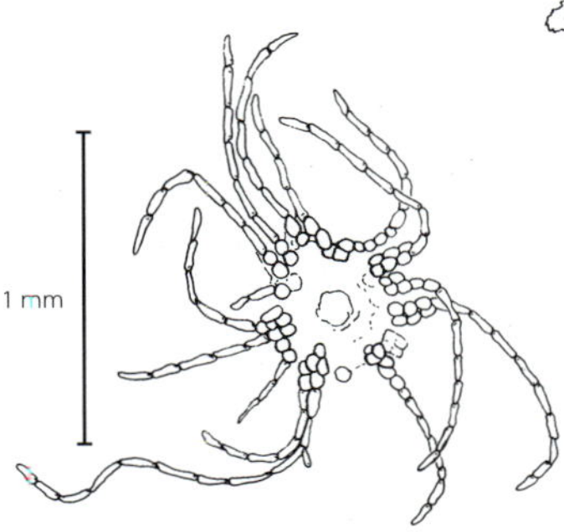

detail of indusium

Dale A. Zimmerman Herbarium, Western N. Mex. Univ.

lower surface of pinnule showing
sporangia and filaments of indusia

Roy Morey

Woodsia neomexicana

NEW MEXICO CLIFF FERN

ents), distinguished as in the key; spores 44–52 µm; 2*n* = 152 (tetraploid) (Windham 1993d). Cliffs and rocky slopes usually in crevices, typically on igneous or sandstone substrates; Brewster (e.g., *Correll 13643A*, BRIT, TEX-LL, both annotated M. Windham), Culberson (*Correll 13911*, BRIT, Guadalupe Mts.), El Paso (*Correll 13791*, BRIT, Hueco Tanks), Jeff Davis (e.g., *Correll 33688*, TEX-LL, *Warnock 15722*, SRSC, both annotated M. Windham), Presidio (*Correll 13746*, TEX-LL, *Warnock 19239*, SRSC, PARATYPE, both annotated M. Windham), and Hudspeth (Yarborough & Powell 2002) cos. in the Trans-Pecos; AZ, CO, NM, w OK, and TX, with disjunct occurrences in c OK and SD; also n Mexico. Sporulating summer–fall. [*W. mexicana* of authors, not Fée] This species was only relatively recently recognized as distinct (Windham 1993e). In the past it and *W. phillipsii* were both often identified as *W. mexicana* (e.g., Correll 1956), which is a related species endemic to Mexico. *Woodsia neomexicana* is an allotetraploid thought to be derived from *W. phillipsii* and another diploid parent. It is morphologically similar to *W. phillipsii*, is sometimes difficult to distinguish without magnification, and can be distinguished most dependably by spore size. Triploid hybrids with *W. phillipsii* are known (Windham 1993d) with several TX collections (*L.C. Hinckley 1122*, Jeff Davis Co.; *E.G. Marsh 259a*, *Correll 13639*, Brewster Co., all TEX-LL; *Warnock 9737*, Brewster Co., *E.J. Palmer 32026a*, Jeff Davis Co., both SRSC) identified as such hybrids by M.D. Windham.

Woodsia obtusa (Spreng.) Torr., (Latin: *obtuse*, blunt or dull, in reference to the tips of the pinnules), BLUNT-LOBE CLIFF FERN, COMMON WOODSIA, BLUNT-LOBE WOODSIA, LARGE WOODSIA. Leaves semi-evergreen, clustered, erect-ascending, to 40(–60) cm tall, often smaller, the blades elliptic-lanceolate to broadly lanceolate, 2-pinnate or 2-pinnate-pinnatifid; indusia rather large, at first enclosing the sporangia and later splitting into several spreading, relatively broad, irregular, non-thread-like segments. Rocky areas, outcrops, well-drained, often sandy areas; Pineywoods w to Montague and Palo Pinto (*W.F. Mahler 8318*, BRIT) cos. in the Cross Timbers and Prairies, to Gillespie (*O'Kennon 8364*, BRIT), Mason (BRIT, TEX-LL), and McCullough (Turner et al. 2003) cos. in the e Edwards Plateau, and s to Gonzales Co. (Turner et al. 2003; Fleenor & Tabor 2009) in the Ottine wetlands (see discussion on page 36); se Canada and throughout e U.S. w to NE, OK, and TX. Sporulating Mar–Nov. Two subspecies of *W. obtusa*, differing in chromosome number, are recognized by Windham (1993d, 1993e) and are separated by him as in the key to subspecies below. Both are widespread in the e ½ of TX with var. *occidentalis* extending further w. Although some mature mid-season material can be clearly and consistently determined to subspecies, it is often difficult to distinguish many specimens. Windham (1993d) further indicated that the 2 subspecies hybridize in the area of sympatry and form sterile triploids with malformed spores. The county distribution map provided for TX does not distinguish subspecies. This species is very similar in general morphology to the rare *Cystopteris protrusa*, SOUTHERN BLADDER FERN, and can be confused with that species in the e ½ of TX. The two can be distinguished by the indusia and other characters (e.g., whether veins reach leaf margins) as given in the key to plants previously included in a more broadly conceived Woodsiaceae (page 307).

1. Spores averaging 42–47 µm [tetraploid]; proximal pinnules of lower pinnae usually shallowly lobed or merely dentate; blades coarsely cut and evidently 2-pinnate; rhizomes compact to short-creeping, individual branches usually 5–10 mm diam. ____ subsp. **obtusa**
1. Spores averaging 35–42 µm [diploid]; proximal pinnules of lower pinnae usually deeply lobed or pinnatifid; blades finely cut, 2-pinnate-pinnatifid; rhizomes short- to long-creeping, individual branches 3–5 mm diam. ____ subsp. **occidentalis**

subsp. **obtusa.** 2*n* = 152 (tetraploid) (Windham 1993d). Widely scattered in East TX (e.g., Grayson, Hopkins, and Milam [*S.L. Orzell & E.L. Bridges 6634*] cos. [all BRIT]—identified J. Peck) w to e Cross Timbers and Prairies (Tarrant Co. [BRIT]); se Canada and e U.S. w to NE and e ⅓ of TX.

subsp. **occidentalis** Windham, (Latin: *occidens*, of the west). 2*n* = 76 (diploid) (Windham 1993d). Widespread in e ½ of TX (based on herbarium collections at BRIT, apparently more common in East TX than subsp. *obtusa*); also Cross Timbers and Prairies and e Edwards Plateau (e.g., Gillespie Co., *Correll & Correll 12753*, BRIT, annotated by M.D. Windham); AR, KS, MO, OK, and e ½ of TX.

Woodsia phillipsii Windham, (for Walter S. Phillips, 1905–1975, collector of the type specimen, "a pioneer in the study of Arizona ferns" [Windham 1993e], and author of *A Checklist of Arizona Ferns*, 1946, 1947; see Mason [1975] for epitaph), PHILLIPS' CLIFF FERN. Leaves to 35 cm long, usually smaller, the blades lanceolate,

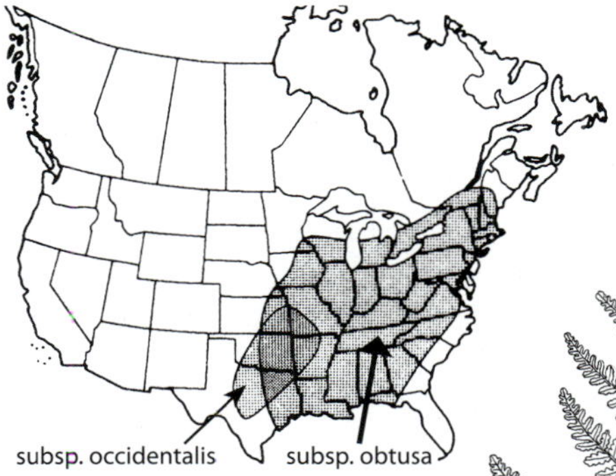

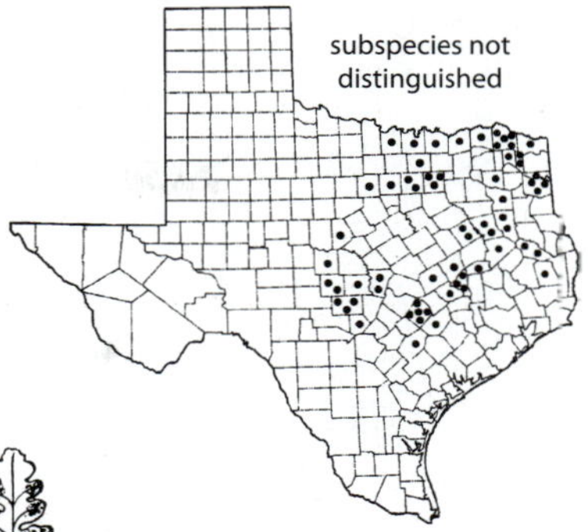

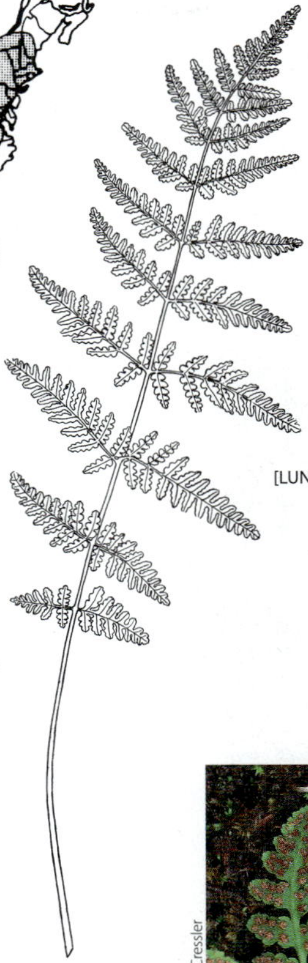

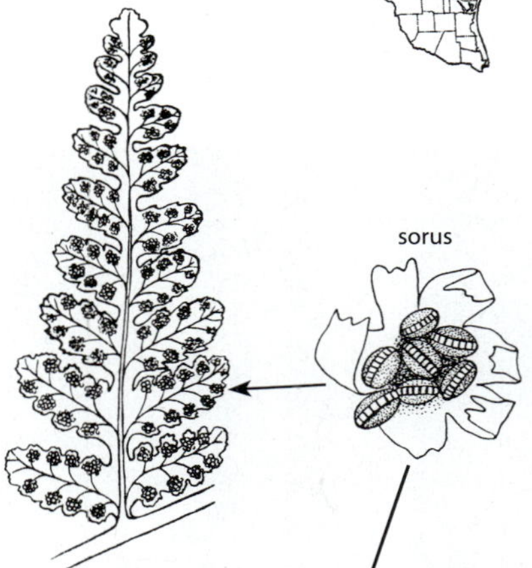

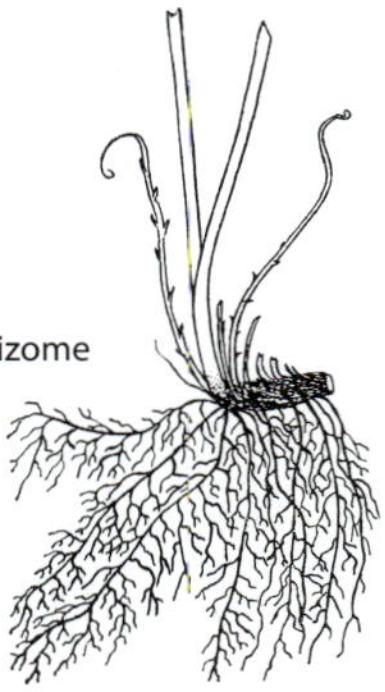

Woodsia obtusa subsp. **obtusa**

BLUNT-LOBE CLIFF FERN

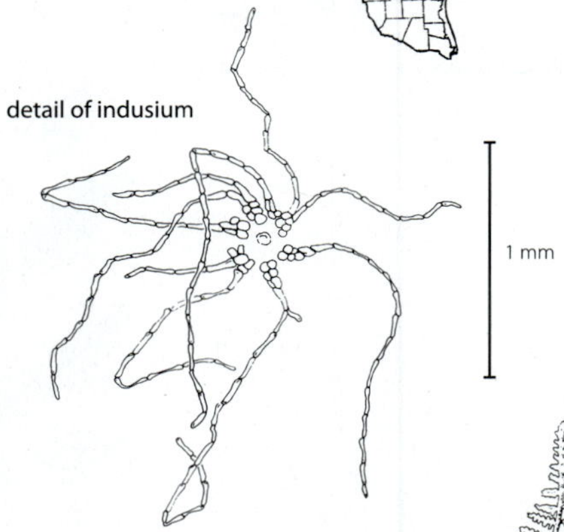

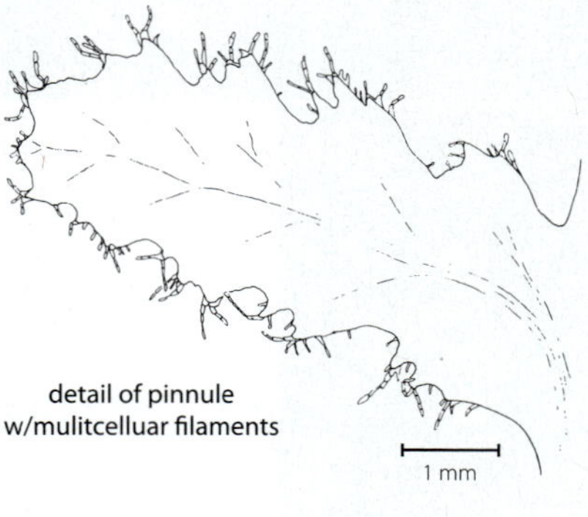

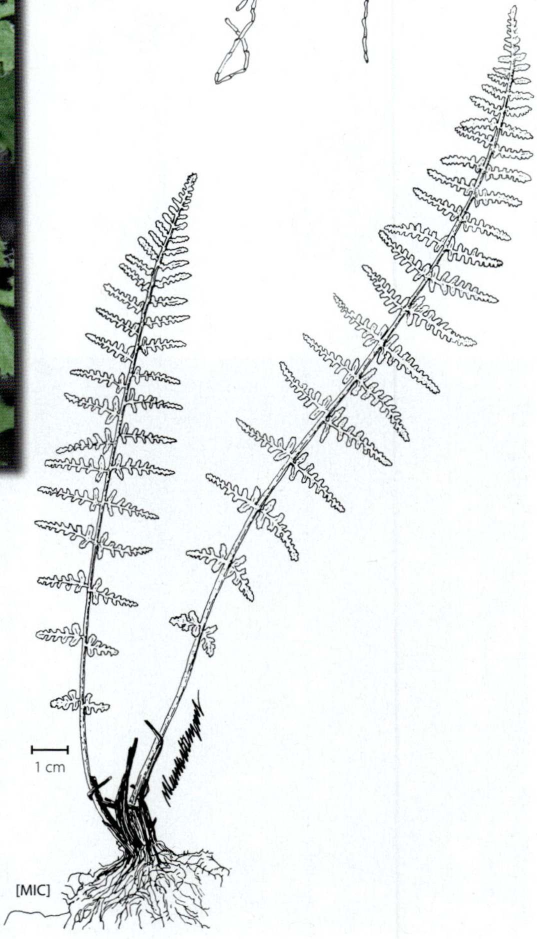

Woodsia phillipsii

PHILLIPS' CLIFF FERN

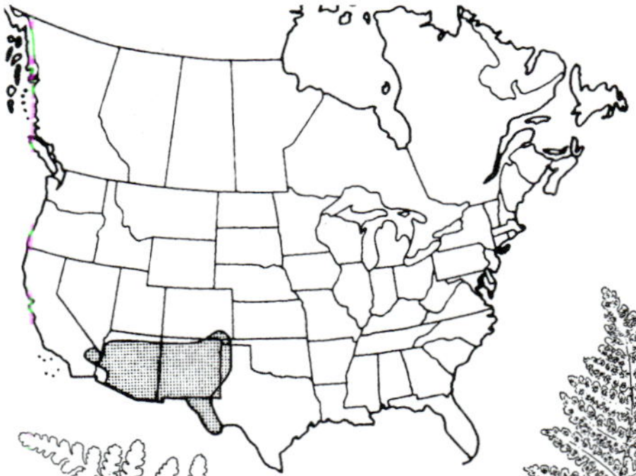

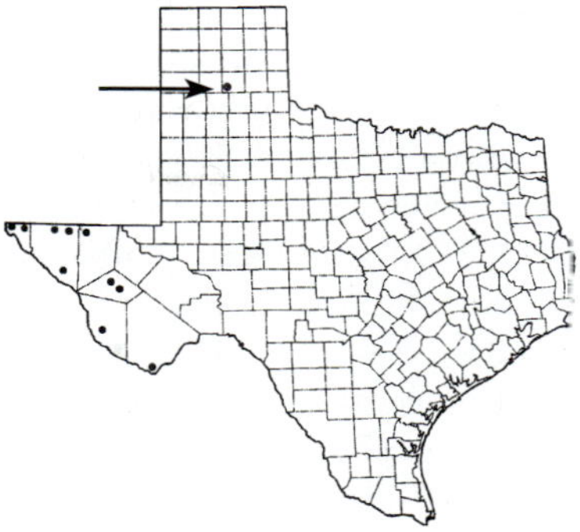

Dale A. Zimmerman Herbarium, Western N. Mex. Univ.

note glandular pubescence

Dale A. Zimmerman Herbarium, Western N. Mex. Univ.

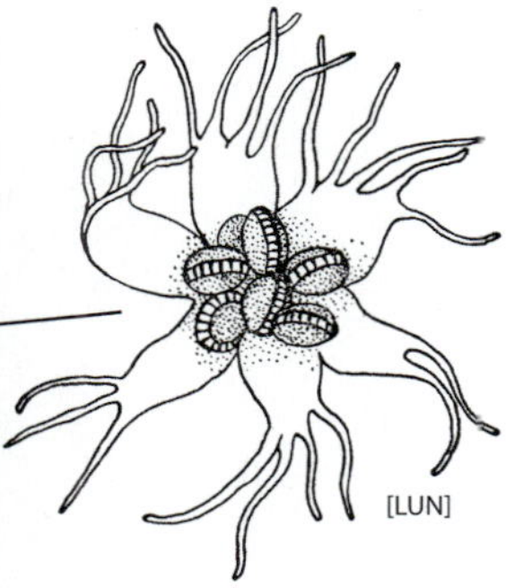

detail of sorus
w/lacerate indusium

Woodsia plummerae

PLUMMER'S CLIFF FERN

2-pinnate-pinnatifid, usually 2-pinnate proximally, occasionally forked at tip, sparsely glandular, the vein tips usually enlarged to form whitish hydathodes visible on the lower surface of the pinnules; indusia with segments divided into narrow, thread-like filaments nearly to the base; spores 37–44 µm, $2n = 76$ (diploid) (Windham 1993d). Cliffs and rocky slopes usually in crevices, typically on igneous substrates; in the Trans-Pecos in Brewster (e.g., *Correll 13681*, BRIT, TEX-LL, both annotated by M.D. Windham), Jeff Davis (*Warnock 23027*, TEX-LL), Presidio (Yarborough & Powell 2002), and Culberson (Turner et al. 2003) cos.; AZ, NM, TX, also n Mexico. Sporulating summer–fall. This species was only relatively recently recognized as distinct (Windham 1993e). In the past it and *W. neomexicana* were both often identified as *W. mexicana* (e.g., Correll 1956), a related species not known in the U.S. *Woodsia phillipsii* sometimes hybridizes with the related and morphologically similar tetraploid *W. neomexicana* to form sterile triploids; it can also hybridize with *W. plummerae* (Windham 1993d). Because of its limited distribution in the state, we consider this species to be of conservation concern in TX. ⚠

Woodsia plummerae Lemmon, (for S.A. Plummer Lemmon, 1836–1923, co-collector of the type specimen and wife of J.G. Lemmon, who described the species), PLUMMER'S CLIFF FERN, FLOWERCUP FERN. Leaves to 18(–25) cm tall, usually smaller, the blades lanceolate to broadly so, 2-pinnate-pinnatifid, usually 2-pinnate proximally, occasionally forked at tip, typically densely glandular and sticky; indusia splitting into several spreading, relatively broad, irregular segments these toothed or lacerate into thread-like filaments only in the distal ½; spores 44–50 µm; $2n = 152$ (tetraploid) (Windham 1993d). Cliffs and rocky slopes typically on igneous materials, but also reported on limestone (Yarborough & Powell 2002); Brewster (*Correll 13639*, BRIT), Jeff Davis (*E. Whitehouse 11224*, BRIT; *J.W. Stanford & J.L. Blassingame 1166*, HPC; *O. Sperry T416*, TAES), Presidio (*R. McVaugh 7448*, BRIT, TEX-LL), Culberson, El Paso, and Hudspeth (TEX-LL) cos. in the Trans-Pecos; and Armstrong Co. (*B.C. Tharp s.n.* 1929, TEX-LL, Palo Duro Canyon) in the Panhandle (a Randall Co. location mapped by Turner et al. [2003] is apparently a mapping error since the Panhandle record at TEX-LL is from Armstrong Co.); AZ, CA, CO, NM, OK, TX, also n Mexico. Sporulating late spring–fall. [*Woodsia obtusa* (Spreng.) Torr. var. *glandulosa* D.C. Eaton & Faxon, W. *obtusa* var. *plummerae* (Lemmon) Maxon, *W. pusilla* E. Fourn. var. *glandulosa* (D.C. Eaton & Faxon) T.M.C. Taylor] This species is a tetraploid whose parentage is unclear; it is known to hybridize with *W. phillipsii* to produce sterile triploids (Windham 1993d).

APPENDIX ONE

Illustration Sources

If there is more than one illustration on a page, only one three-letter code is given for all illustrations unless one or more illustrations is from a different source. In such cases, all illustrations will have a three-letter code so it is clear where each illustration originated.

Ref. Code

ABR Abrams, L. 1923–1960. Illustrated flora of the Pacific states. Stanford University Press, Stanford, CA. Reprinted with permission.

AFJ Benham, D.M. 1992. Additional taxa in *Astrolepis*. Amer. Fern J. 82:59–62. American Fern Society, Washington, DC. Reprinted with permission.

DCC Dittmer, H.J., E.F. Castetter, and O.M. Clark. Drawings by J.R. Janish. 1954. The ferns and fern allies of New Mexico. Univ. New Mexico Publ. Biol. No. 6. The University of New Mexico Press, Albuquerque. Reprinted with permission.

FBC Wiggins, I.L. 1980. Flora of Baja California. Stanford University Press, Stanford, CA. Reprinted with permission of Stanford University Press.

FGP Tryon, A.F. 1957. A revision of the fern genus *Pellaea* Section *Pellaea*. Ann. Missouri Bot. Gard. 44:125–193. Missouri Botanical Garden. St. Louis. Reprinted with permission of the Missouri Botanical Garden.

FNA Flora of North America Editorial Committee, eds. 1993+. Flora of North America. Oxford University Press, New York. Artists: Laurie Lange and Bobbi Angell (*Azolla caroliniana, Azolla filiculoides, Azolla microphylla, Ceratopteris thalictroides, Cystopteris reevesiana, Salvinia minima, Selaginella mutica* var. *mutica*, and *Selaginella peruviana*). From Volume 2. Reprinted with permission of the Flora of North America Association.

FNG McVaugh, R. 1992. Flora Novo-Galiciana: A descriptive account of the vascular plants of Western Mexico. University of Michigan Press, Ann Arbor, Vol. 17, Fig 60A. *Marsilea mollis* © University of Michigan Herbarium. Reprinted with permission.

HEA Diggs, G., B. Lipscomb, and R.J. O'Kennon. 1999. Shinners and Mahler's illustrated flora of North Central Texas. Original illustrations by Linda 'Linny' Heagy, Copyright © 1999. Botanical Research Institute of Texas, Fort Worth.

JTM Mickel, J.T. Illustrated by E.M. Paulton. 1979. How to know the ferns and fern allies. Wm. C. Brown Company, Dubuque, IA. Reprinted with permission.

KNO Knobloch, I.W. and D.S. Correll. With illustrations by Phoebejane Horning and Jane W. Roller. 1962. Ferns and fern allies of Chihuahua, Mexico. Contr. Texas Res. Found. Bot. Stud. Vol. 3. Texas Research Foundation, Renner.

LUN Lundell, C.L. 1961–1969. Flora of Texas. Texas Research Foundation, Renner.

MAS Mason, H.L. 1957. A flora of the marshes of California. University of California Press, Berkeley. Reprinted with permission.

MIC Windham, M.D. 1993. New taxa and nomenclatural changes in the North American fern flora. Contr. Univ. Michigan Herb. 19:31–61. University of Michigan, Ann Arbor. Figs. 1, 5, 6. Reprinted with permission.

MRA Moran, R.C. 1992. The story of the molesting *Salvinia*. Fiddlehead Forum, Bull. Amer. Fern Soc. 19:26–28. The American Fern Society, New York. Reprinted with permission of Robbin C. Moran.

PAL Palmer, D.W. 2003. Hawaii's ferns and fern allies. Contr. Hawaii Biol. Surv. No. 2002-010. University of Hawaii Press, Honolulu. Reprinted with permission.

PHY Singhurst, J.R., A.E. Rushing, C.K. Hanks, and W.C. Holmes. 2011. *Isoetes texana* (Isoetaceae): A new species from the Texas Coastal Bend. Phytoneuron 2011-22:1–6. http://www.phytoneuron.net/PhytoN-Isoetestexana.pdf. Accessed Jun 2011. Reprinted with permission.

POM Mickel, J.T. and A.R. Smith. 2004. The Pteridophytes of Mexico. Mem. New York Bot. Gard. Vol. 88. The New York Botanical Garden, Bronx. Reprinted with permission.

POO Mickel, J.T. 1988. Pteridophyte flora of Oaxaca, Mexico. Mem. New York Bot. Gard. Vol. 46. The New York Botanical Garden, Bronx. Reprinted with permission.

PRC Holmes, W.C., A.E. Rushing, and J.R. Singhurst. 2005. Taxonomy and identification of *Isoetes* (Isoetaceae) in Texas based on megaspore features. Lundellia 8:1–6. University of Texas Plant Resource Center, Austin. Reprinted with permission.

ROL Tryon, R.M., Jr. 1956. A revision of the American species of *Notholaena*. Contr. Gray Herb. 179:1–106. The Gray Herbarium of Harvard University, Cambridge, MA. Reprinted with permission.

RYD Rydberg, P.A. 1932. Flora of the prairies and plains of central North America. Dover Publications Inc., New York.

SCY Yarborough, S.C. and A.M. Powell. Illustrated by Ellen Bergen-Ruggia and Jane W. Roller. 2002. Ferns and fern allies of the Trans-Pecos and adjacent areas. Texas Tech University Press, Lubbock. Reprinted with permission.

SEM Calvert, H.E., S.K. Perkins, and G.A. Peters. 1983. Sporocarp structure in the heterosporous water fern *Azolla mexicana* Presl. Scanning Electron Microscop. 3:1499–1510. Reprinted with permission from Sandy Hancock (S.K. Perkins). Perkins, S.K., G.A. Peters, T.A. Lumpkin, and H.E. Calvert. 1985. Scanning electron microscopy of perine architecture as a taxonomic tool in the genus *Azolla* Lamarck. Scanning Electron Microscop. 4:1719–1734. Reprinted with permission from Sandy Hancock (S.K. Perkins).

SM3 Small, J.K. 1938. Ferns of the southeastern states. The Science Press, Lancaster, PA.
SM4 Small, J.K. 1931. Ferns of Florida: Being descriptions of and notes on the fern-plants growing naturally in Florida. The Science Press, New York, NY.
TAY Taylor, W.C. Illustrated by Paul W. Nelson. 1984. Arkansas ferns and fern allies. Milwaukee Public Museum, Milwaukee, WI. Reprinted with permission of the Milwaukee Public Museum and W.C. Taylor. Reprinted with permission.
UTP Taylor, T.M.C. 1970. Pacific Northwest ferns and their allies. University of Toronto Press, Toronto, Canada. Reprinted with permission.
YAT Yatskievych, G. 1999. Steyermark's flora of Missouri. Missouri Department of Conservation, Jefferson City. Reprinted with permission of the Missouri Department of Conservation.

APPENDIX TWO

A Few Publicly Accessible Locations to See Ferns and Lycophytes in Texas

TRANS-PECOS

Big Bend National Park: http://www.nps.gov/bibe/index.htm [huge park, 801,163 acres in Brewster County and larger than the state of Rhode Island; ferns and lycophytes can be seen in many locations, but especially in high elevation areas like the Chisos Mountains]

Guadalupe Mountains National Park: http://www.nps.gov/gumo/index.htm [86,367 acres in Culberson and Hudsmith counties including Guadalupe Peak, at 2,667 m/ 8,749 ft, the highest point in Texas; ferns and lycophytes are most abundant in such high elevation areas including McKittrick Canyon]

Davis Mountains State Park: https://www.tpwd.state.tx.us/state-parks/davis-mountains [2,708.9 acres with extensive trails in Jeff Davis County in the Davis Mountains; numerous ferns and lycophytes are easily seen]

Chihuahuan Desert Nature Center and Botanical Gardens: http://cdri.org/ [507 acres with botanical gardens and trails located on the grounds of the nonprofit Chihuahuan Desert Research Institute; it is easy to see many species of pteridophytes in a very accessible manner]

EAST TEXAS

Big Thicket National Preserve: http://www.nps.gov/bith/index.htm [National Preserve of just over 100,000 acres in Hardin, Jasper, Jefferson, Liberty, Orange, Polk, and Tyler counties with numerous trails; the Neches River bottomlands and many other areas have huge populations of ferns]

National Forests: http://www.fs.usda.gov/texas/

- **Angelina National Forest:** [153,160 acres in Angelina, Jasper Nacogdoches, and San Augustine counties including the Turkey Hill and Upland Island wilderness areas]
- **Davy Crockett National Forest:** [160,663 acres in Houston and Trinity counties including the Big Slough Wilderness Area]
- **Sabine National Forest:** [160,798 acres in Jasper, Newton, Sabine, San Augustine, and Shelby counties including the Indian Mounds Wilderness Area]
- **Sam Houston National Forest:** [163,030 acres in Montgomery, San Jacinto, and Walker counties including the Little Lake Creek Wilderness Area]

National Wildlife Refuges:

- **Caddo Lake National Wildlife Refuge:** http://www.fws.gov/refuges/profiles/index.cfm?id=21594 [located on Caddo Lake on portions of what was previously the approximately 8,500-acre Longhorn Army Ammunition Plant in Harrison County; Caddo Lake, an extensive swamp ecosystem of approximately 25,000 acres, is the largest naturally occurring body of freshwater in Texas]
- **Trinity River National Wildlife Refuge:** http://www.fws.gov/southwest/refuges/texas/trinityriver/ [approximately 25,000 acres along the Trinity River in Liberty County]

State Parks (there are a number of others[1]):

- **Caddo Lake State Park:** http://www.tpwd.state.tx.us/state-parks/caddo-lake [484 acres in Harrison County on Big Cypress Bayou and the adjacent 8,005-acre Caddo Lake State Wildlife Management Area in Harrison and Marion counties]
- **Dangerfield State Park:** http://www.tpwd.state.tx.us/state-parks/daingerfield/ [507 acres in Harrison County; includes rock outcrops unusual for East Texas]
- **Martin Dies, Jr. State Park:** http://www.tpwd.state.tx.us/state-parks/martin-dies-jr [705 acres in Jasper and Tyler counties on the B.A. Steinhagen Reservoir (15,000 acres)]
- **Palmetto State Park:** http://en.wikipedia.org/wiki/Palmetto_State_Park [270 acres in Gonzales County; includes wet relict habitats more reflective of areas further east; a number of ferns reach their southwestern geographic limits here]

Tyler State Park: http://www.tpwd.state.tx.us/spdest/findadest/parks/tyler/ [985.5 acres in Smith County]
Village Creek State Park: http://www.tpwd.state.tx.us/spdest/findadest/parks/village_creek/ [1,090 acres in Hardin County along Village Creek, a tributary of the Neches River]

Roy E. Larsen Sandyland Sanctuary: http://www.nature.org/wherewework/northamerica/states/texas/preserves/art25155.html [5,654-acre Nature Conservancy preserve in Hardin County with a diversity of habitats and a variety of different pteridophytes]

Jesse H. Jones Park & Nature Center: http://www.hcp4.net/jones/ [With over 300 acres in Harris County, Jones Park is a nature sanctuary to a wide variety of plants, fungi, and wildlife]

Houston Arboretum & Nature Center: http://www.houstonarboretum.org/ [In Harris County, this 155-acre non-profit urban nature sanctuary provides an opportunity to see natural environments. Visitors can explore nature by walking on trails totaling 5 miles]

Mercer Arboretum & Botanic Gardens: http://www.hcp4.net/mercer/index.htm [Located in Harris County, these 300 acres of East Texas Piney Woods showcase the region's largest collection of native and cultivated plants]

EDWARDS PLATEAU

Balcones Canyonlands National Wildlife Refuge: http://www.fws.gov/refuge/Balcones_Canyonlands/ [23,815 acres in Burnet, Travis, and Williamson counties in the Texas hill country]

State Parks & Natural Areas:
Enchanted Rock Natural Area: http://www.tpwd.state.tx.us/spdest/findadest/parks/enchanted_rock/ [1643.5 acres on the border between Gillespie and Llano Counties; a massive granite dome in the Central Mineral Region of central Texas; 29 species of pteridophytes are known from here (some extremely rare)]
Guadalupe River State Park: http://www.tpwd.state.tx.us/state-parks/guadalupe-river [1,939 acres along the Guadalupe River in Comal and Kendall counties]
Lost Maples State Natural Area: http://www.tpwd.state.tx.us/state-parks/lost-maples [2,174 acres in Bandera and Real counties; includes relict populations of various plants including isolated "lost" populations of *Acer grandidentatum*, BIGTOOTH MAPLE]
Perdernales Falls State Park: http://www.tpwd.state.tx.us/state-parks/pedernales-falls [5,211.7 acres in Blanco County in the Texas hill country]

Ladybird Johnson Wildflower Center: http://www.wildflower.org/ [279 acre public botanical garden in Austin focused on native plants]

NORTH CENTRAL TEXAS

Fort Worth Botanic Garden: http://fwbg.org/ [109-acre living museum displaying over 2500 species of native and exotic plants; the oldest botanic garden in Texas, Tarrant County]

Caddo/LBJ National Grasslands: http://www.fs.usda.gov/detail/texas/about-forestdistricts/?cid=fswdev3_008440 [In Wise County, the U.S. Forest Service manages the Caddo-Lyndon B. Johnson National Grasslands in north Texas. The Caddo Grasslands is comprised of 17,785 acres and contains three lakes]

PANHANDLE TEXAS

Wildcat Bluff Nature Center: http://www.wildcatbluff.org/ [More than 600 acres in Potter County of rolling grasslands are threaded with nature trails offering a sense of isolation and tranquility]

SOUTH TEXAS

South Texas Botanical Gardens & Nature Center: http://www.stxbot.org/ [180-acre showcase on Oso Creek in Nueces County]

[1]Texas has 93 state parks, historical sites and natural areas that contain a total of 586,501 acres in 98 counties (Combs 2012).

TEXAS NATURE CENTERS

Nature Centers of Texas are often excellent places to spot ferns. The following Wikipedia link is to over 50 Nature Centers in Texas (Wikipedia). http://en.wikipedia.org/wiki/List_of_nature_centers_in_Texas

APPENDIX THREE

Sources of Additional Information on Ferns and Lycophytes

MAJOR SOURCES OF FERN & LYCOPHYTE INFORMATION ON-LINE

American Fern Society: http://www.amerfernsoc.org/
British Pteridological Society: http://www.nhm.ac.uk/hosted_sites/bps/
Checklist of Ferns and Lycophytes of the World: http://www.rz.uni-karlsruhe.de/~db111/flora/ferns/
Checklist of World Ferns: http://homepages.caverock.net.nz/~bj/fern/
Ferns and Lycophytes of the World: http://www.fernsoftheworld.com/author/msundue/
Ferns of Texas: www.fernsoftexas.org [On-line access to information about Texas ferns and lycophytes through the Botanical Research Institute of Texas]
Hardy Fern Foundation: http://www.hardyferns.org/
Hardy Fern Library: http://hardyfernlibrary.com/ferns/
Illustrated Glossary/Fern Terminology: http://hardyfernlibrary.com/ferns/illustratedGlossary.cfm

A FEW EXCELLENT BOOKS ON FERNS & LYCOPHYTES

Flora of North America Editorial Committee, eds. 1993. Flora of North America north of Mexico, Vol. 2, Pteridophytes and Gymnosperms. Oxford Univ. Press, New York and Oxford. [Technical taxonomic treatment of all North American pteridophytes]

Hoshizaki, B.J. and **R.C. Moran.** 2001. Fern grower's manual: Revised and expanded edition. Timber Press, Portland, OR. [Detailed treatment of all ferns cultivated in the U.S.]

Mehltreter, K., L.R. Walker, and **J.M. Sharpe, eds.** 2010. Fern ecology. Cambridge Univ. Press, Cambridge, England, U.K. [A series of technical chapters on various topics of fern ecology]

Mickel, J.T. 2003. Ferns for American gardens. Timber Press. Portland, OR. [Accessible coverage of fern gardening]

Mickel, J.T. and **A.R. Smith.** 2004. The pteridophytes of Mexico. New York Bot. Gard., Bronx. [Encyclopedic taxonomic treatment of all the pteridophytes in Mexico, including 80 species that occur in TX (63% of the TX species)]

Moran, R.C. 2004. A natural history of ferns. Timber Press, Portland, OR. [Easy to read and enjoyable look at the world of ferns]

Olsen, S. 2007. Encyclopedia of garden ferns. Timber Press, Portland, OR. [Information on a large number of cultivated ferns.]

Ranker, T.A. and **C.H. Haufler, eds.** 2008. Biology and evolution of ferns and lycophytes. Cambridge Univ. Press, Cambridge, England, U.K. [Technical coverage of pteridophyte biology and evolution]

A FEW MAJOR SCIENTIFIC JOURNALS/BULLETINS FOCUSING ON FERNS

American Fern Journal: a scientific journal published by the American Fern Society
Fiddlehead Forum: a member newsletter/bulletin published by the American Fern Society
Pteridologist: a scientific journal published by the British Pteridological Society
The Indian Fern Journal: a scientific journal published by the Indian Fern Society

A FEW OTHER INTERESTING LINKS ON FERNS & LYCOPHYTES

Bibliography of Pteridological Works (1753–2010): http://www.jaknouse.athens.oh.us/ferns/bookfern.html [An extensive list of references on ferns and lycophytes]

Pteridomania–the Victorian Passion for Ferns: http://www.peterboyd.com/pteridomania.htm [A fascinating look at the British obsession for ferns from the 1850s to the 1890s]

Tree of Life: Ferns: http://tolweb.org/Polypodiopsida/20615 [Information on the biodiversity and evolutionary history of ferns from the extensive Tree of Life Project]

Tree of Life: Lycophytes: http://tolweb.org/Lycopodiopsida/20607 [Information on the biodiversity and evolutionary history of lycophytes from the extensive Tree of Life Project]

Desert Fern Guide: http://www.mineralarts.com/ferns/DesertFernsGuide.html [Includes photos and information on many ferns of Trans-Pecos Texas]

Kathryn Pryer Publication Page: http://pryerlab.biology.duke.edu/publications [On-line access to many scientific publications by one of the leading figures in modern fern biology]

Southwestern Fern Society: http://www.voiceplaces.com/southwestern-fern-society-dallas-fort-worth-6543-e/

Texas Gulf Coast Fern Society: http://www.tgcfernsoc.org

A FEW LINKS OF MAJOR GENERAL BOTANICAL IMPORTANCE

Flora of Texas Database (Plant Resources Center): http://www.biosci.utexas.edu/prc/Tex.html [On-line search engine covering the approximately one million herbarium specimens in the Herbarium of the Univ. of Texas at Austin]

Index to American Botanical Literature: http://sweetgum.nybg.org/iabl/ [On-line search engine for a large database on American botanical literature from the New York Botanical Garden]

International Plant Names Database (IPNI): http://www.ipni.org/index.html [On-line database of information on the names and basic bibliographical details of the world's seed plants, ferns and lycophytes]

Kew Databases of Botanical Literature: http://kbd.kew.org/kbd/searchpage.do [On-line search engine to the three major bibliographic databases at the Royal Botanic Gardens Kew; probably the most extensive compilation of botanical literature in the world]

Tropicos: http://www.tropicos.org/ [Huge on-line database on plant names, bibliographic information, and herbarium specimen data hosted by the Missouri Botanical Garden]

Glossary

This glossary is modified from those of Correll (1956), Lellinger (1986), and Diggs et al. (2006).

A

A- A prefix meaning without or not.

Abaxial Located on the side away from axis; e.g., lower leaf surface; contrasting with adaxial.

Abortive Imperfectly or incompletely developed; e.g., spores of hybrids such as *Equisetum ×ferrissii*.

-Aceae Suffix denoting the rank of family in the taxonomic hierarchy.

Acroscopic Located on the side facing or toward the leaf apex.

Acuminate Having a long, tapering point; longer tapering than acute.

Acute Forming a sharp angle of less than 90 degrees; less tapering than acuminate.

Ad- A prefix meaning to or toward.

Adaxial Located on side towards axis; e.g., upper leaf surface; contrasting with abaxial.

Adnate United or fused, when the fusion involves dissimilar structures.

Adpressed (= Appressed) Lying flat against a surface.

Adventitious Referring to structures or organs that develop in an unusual position; e.g., buds or roots that develop out of their usual place; in pteridphytes all roots are adventitious since they arise from rhizomes, rather than from a primary root.

Aerial Above ground level.

-Ales Suffix denoting the rank of order in the taxonomic hierarchy.

Alkaloid Any of a broad class (> 5,000 known alkaloids) of bitter, usually basic (alkaline), organic compounds that contain nitrogen and typically have a ring in their structure. They are often physiologically active in animals; many are poisonous; many affect the nervous system; there are a number of general types based on chemical structure; well-known examples of alkaloids include caffeine, cocaine, morphine, nicotine, theobromine, and strychnine.

Allelopathy, Allelopathic Harmful or detrimental chemical effect by one species upon another; e.g., a plant producing phytotoxic compounds that inhibit the germination or growth of other plants.

Allergen Substance capable of inducing an allergic response.

Allergenic Causing an allergic response or an allergy to become manifest.

Allergy Hypersensitivity of the body cells to specific substances as antigens and allergens, resulting in various types of reactions (e.g., anaphylaxis, contact dermatitis, hay fever).

Allopolyploid An organism with multiple sets of chromosomes derived from different parental species; e.g., allotetraploid.

Alternate Bearing one leaf or other structure at a node; having only one attached at a given point; contrasting with opposite or whorled.

Amphibious Capable of growing on land and in the water.

Anastomosing Net-like; with veins connecting by cross-veins to form a network.

Annual Plant or root system living only one growing season (year); completing the growth cycle within one year.

Annulus A group or ring of thick-walled cells, on the sporangia of some ferns, that are involved in opening the sporangia and allowing spore dehiscence.

Angiosperm (= Flowering plant) Literally, "vessel seed"; a plant having its seeds enclosed in an ovary (= the proximal part of the carpel or "vessel"); a member of Division Magnoliophyta.

Antheridium Male sexual organ borne on the gametophyte and producing male gametes.

Apeture An opening.

Apex (pl. APICES) The tip or summit.

Apical At the tip or apex; relating to the apex.

Apiculate Having a small sharp point formed by blade or other tissue (e.g., of a leaf or cone) rather than by projection of a rib or vein; with an abrupt tip or projection; having an apiculum.

Apiculum A small sharp point; e.g., apiculum on cones of some *Equisetum* species.

Apogamy A type of asexual reproduction that does not involve fertilization; the sporophyte is formed directly from the gametophyte without gamete production.

Apogamous Forming a sporophyte without fertilization (the union of gametes).

Apomixis A collective term for reproduction, including vegetative propagation, that does not involve sexual processes; any form of asexual reproduction.

Appendage Any attached structure that is supplementary or secondary.

Appendiculate With an appendage.

Appressed (= Adpressed) Lying flat against a surface.

Aquatic Living in water.

Archegonium Female sexual organ borne on the gametophyte and producing female gametes.

Areolate Divided into small angular spaces; marked with areolae.

Areole, Areola (pl. AREOLAE) Small space marked out on a surface, usually referring to the space bounded by small veins on the surface of a leaf; a vein-enclosed area.

Articulate (a) Jointed; joined; (b) with a visible discontinuity or place of separation.

Articulation A separation place; joint.

Ascending, Ascendent Rising at an oblique angle.

Asexual Without sex; reproducing without sex.

Asymmetrical Without symmetry.

Atropurpureous Dark purple; purple-black.

Attenuate Gradually tapering to a very slender tip, the taper more gradual than in acuminate.

Atypical Not typical; deviating from the norm.
Auricle Earlobe-like lobe or appendage; e.g., at the base of some leaves or other structures.
Auriculate With an auricle.
Autopolyploid An organism with multiple chromosome sets derived from a single parental species.
Awl-Shaped (= Subulate) Tapering from the base to a slender or stiff point; narrow and sharp-pointed.
Awn Terminal slender bristle or hair-like extension or projection.
Axil Angle between two organs; e.g., upper angle formed by a leaf and a stem.
Axis (pl. AXES) (a) the central stem from which organs arise; (b) a portion of a plant from which a series of organs arises radially.

B

Base The proximal portion of a structure, that part nearest to the point of attachment.
Basal (a) Located at the base of a plant or of an organ; (b) the lowest branch on a phylogenetic tree.
Basiscopic Located on the side facing or directed toward the leaf base.
Basionym The original epithet assigned to a species (or other taxon) by its author.
Bi-, Bis- Latin prefix signifying two, twice, or doubly.
Bicolored Two-colored.
Bilateral Arranged on two sides; two-sided.
Bilaterally Symmetrical With only one plane of symmetry; divisible into halves in one plane only.
Binomial The combination of a generic name and a specific epithet given to each species.
Binomial Nomenclature System of nomenclature where each species has a two-part name composed of a generic name and a specific epithet.
Biogeography Study of geographic distribution of plants and animals.
Bipinnate (=Twice-pinnate or 2-pinnate) Descriptive of a leaf with leaflets pinnately arranged on lateral axes that are themselves pinnately arranged on the main axis; with the primary divisions (pinnae) themselves pinnate.
Blade Flat, expanded portion, as the main part of a leaf or petal.
Blunt Not pointed.
Boss A protrusion or raised area.
Branch A shoot or secondary stem growing from the main stem.
Branchlet The ultimate division of a branch.
Bristle Stiff, strong but slender hair or trichome.
Bristly Bearing bristles.
Broad (= Wide) Distance across a structure (equal to diameter if tubular); sometimes restricted to signify the width or diameter of three-dimensional structures.
Bryophyta Group containing the mosses, liverworts, and hornworts, land plants that reproduce by spores but do not have vascular tissue. The Bryophyta is not treated in this flora.
Bud (a) undeveloped, much-condensed shoots, containing embryonic (meristematic or growing) tissue, usually covered by scales or bracts. Such buds are usually found at the tips of stems or in the axils of leaves; (b) in ferns, usually a lump of tissue that grows into a new fern plant.
Bulb Underground structure composed of a short, disc-like stem and one or more buds surrounded by layers of thickened fleshy leaf bases or scales.
Bulblet Small globular, asexual propagule borne along the rachis or costa that is capable of forming a new plant.
Bulbous, Bulbose (a) Having bulbs or bulb-like structures; (b) swollen or nearly spherical.

C

Ca. Latin, circa; abbreviation meaning about, around, approximately.
Calcareous Containing an excess of available calcium, usually in the form of the compound calcium carbonate; containing limestone or chalk.
Caliche A crust of calcium carbonate formed on stony soils in arid regions.
Cap specialized structure in *Azolla* attached to the megaspore, consisting of a collar (involved in attaching the floats), 3 (in TX species) floats, and a blue-green algal colony covered by remnants of the indusium.
Capillary Hair-like; very slender.
Capitate (a) In heads, head-like, or head-shaped; aggregated into a dense or compact cluster; (b) referring to capitate hairs, like a pin-head on a stalk.
Carcinogen A substance potentially inducing cancer or malignancy.
Caudate Having a tail or tail-like appendage.
Caudex (pl. CAUDICES) Woody or wood-like stem base.
Cell One of the living units of which a plant is composed.
Chlorophyll The light-capturing pigment giving the green color to plants. Because chlorophyll absorbs less green than other wavelengths of light (and thus reflects and transmits relatively more green), leaves appear green to the human eye.
Chromosomes Thread-like "colored bodies" occurring in the nuclei of cells and containing the genetic material.
Ciliate With a marginal fringe of hairs similar to eye lashes.
Cilium (pl. CILIA) Marginal hair or trichome.
Cinnamon In reference to color, indicating light reddish brown.
Circinate Coiled, with the apex innermost; most ferns have circinate vernation in which young unexpanded leaves are tightly coiled, with appearance of a fiddlehead.
Class The unit, category, or rank in classification made up of one or more orders; ending in -ae or -opsida; sometimes divided into subclasses which in turn are made of orders.
Clade A monophyletic lineage containing a common ancestor and all its descendants.
Clathrate Latticed or net-like in appearance; with a series of crossed members.
Clone A group of individuals of the same genotype, usually propagated vegetatively.

Clump A single plant with two to many, more or less crowded stems arising from a branched rootstock or short rhizome.

cm Centimeter; 10 mm; 1/100 of a meter; ca. 2/5 of an inch.

Cobwebby Cobweb-like, with entangled, slender, loose hairs; thinly pubescent with relatively long, usually appressed and interlaced hairs.

Colonial Forming colonies usually by means of underground rhizomes, stolons, etc. The term is commonly used to describe groups of plants with asexual reproduction.

Colony A stand, group, or population of plants of one species, spreading vegetatively, or from seeds, or both.

Compound (= Composite) Made up of several distinct parts.

Compound Leaf A leaf/frond that is cut completely to the base or midrib into segments (= leaflets/pinnae) resembling miniature leaves; a leaf with two or more leaflets.

Compressed Flattened.

Concave Hollow; with a depression on the surface.

Concolor, Concolorous Of a uniform color.

Conduplicate Folded together lengthwise.

Cone (= Strobilus) A usually globose or cylindrical structure involved in reproduction and composed of an axis with a spiral, usually dense aggregation of sporophylls, bracts, or scales (in pteridophytes these bearing spores, or in seed plants bearing pollen or seeds).

Confluent Blending of one part into another.

Congested Crowded together.

Conical, Conic Cone-shaped.

Coniferous Cone-bearing.

Connate United or fused, when the fusion involves two or more similar structures.

Conservation Concern Concern that a plant might become extinct locally because it is either so rare or of such limited distribution.

Conserved (a) Term applied to a scientific name whose use, even though illegitimate according to nomenclatural rules, is allowed by the International Code of Botanical Nomenclature; e.g., many family names long in use, have been conserved to prevent confusion; (b) unchanged over long periods of evolutionary time.

Conspecific Of the same species.

Constricted Tightened or drawn together or narrowed.

Continuous Not interrupted; not articulated; not jointed.

Contorted Twisted or distorted.

Contracted Narrowed or shortened; reduced in size.

Contrary In an opposite direction or at right angles to.

Convergent Coming together or approaching.

Convex Rounded or bulged on the surface.

Cordate, Cordiform (a) Heart-shaped; with a notch at the base and ovate in outline (the words apply specifically to flat surfaces and to solid shapes, respectively); (b) often referring only to the notched base of a structure; e.g., leaf base.

Coriaceous With texture like leather; tough; leathery.

Corm Bulb-like usually subterranean stem base, solid instead of with layers of modified leaves as in a true bulb.

Cormous Having a corm.

Cosmopolitan Occurring essentially worldwide.

Costa (pl. COSTAE) Central axis, midvein, or midrib of a pinna.

Costule Midvein or midrib of a pinnule.

Creeping Growing along the surface of the ground and emitting roots at intervals, usually from the nodes.

Crenate Scalloped with rounded teeth; shallowly round-toothed or with teeth obtuse.

Cross-Section A slice cut across an object; e.g., a slice of bread.

Crozier A young coiled leaf of some ferns.

Cryptogam An old term for plants that reproduce without flowers or seeds, instead typically reproducing by spores; contrasting with phanerogam.

Cultivar A variety or race of a cultivated plant, generally chosen for a particular desirable trait; abbreviated cv.

Cuneate, Cuneiform Wedge-shaped; triangular with tapering, straight-sided, narrow base.

Curvi- A prefix to denote curved or bent.

Cut A general term for any dissection of a leaf or petal deeper than a lobe.

Cuticle The waxy, more or less waterproof coating secreted by the cells of the epidermis. The cuticle prevents water loss.

Cylindrical, Cylindric Elongate, circular in cross-section; having the form of a cylinder.

D

Dc. De Candolle, name of a distinguished family of Swiss botanists; specifically Augustin Pyramus, who sponsored early botanical exploration in Texas by Berlandier, and named many Texas species; A. DC.:Alphonse, son of the preceding.

Deciduous Falling away; not persistent over a long period of time.

Decompound More than once compound.

Decumbent Lying flat or reclining with terminal shoots or stem tips ascending.

Decurrent Extending down the stem and united with it, as in the continuation of leaf bases down the stem as wings.

Dehisce, Dehiscent To open at maturity to discharge the contents; e.g.; sporangia releasing spores.

Dehiscence The process or act of opening.

Deltate, Deltoid Shaped like an equilateral triangle, like the Greek letter delta (Δ).

Dentate With sharp teeth not directed forward.

Denticulate Minutely dentate.

Depressed Low as if flattened.

Descending With a gradual downward tendency.

Di-, Dis- Greek prefix meaning two or double.

Dichotomous Forking regularly into two equal or nearly equal branches.

Dimorphic, Dimorphism Occurring in two forms.

Diploid Possessing two sets of chromosomes in each nucleus; twice the haploid number typical for gametes.

Discoid Shaped like a disc.

Discrete Separate.

Disjunct (a) Outside the main range of a species; (b) being divided into separate groups; disconnected.

Dissected Divided into numerous narrow or slender segments, the divisions usually deeper than lobes.

Distal Located at or toward the apex of a plant or organ; the terminal portion; the end opposite the attachment; contrasting with proximal.
Distinct (= Free) Separate, not united or fused.
Disturbed Referring to a habitat that has been altered or modified but not completely destroyed.
Divergent Spreading, drawing apart from a common point.
Divided (a) Cut 3/4–completely the distance from margin to midrib or from apex to base; (b) generally, cut deeply.
Division The highest rank, category, or taxon in the plant kingdom; made up of classes; ending in -phyta; equivalent to the rank of phylum in the animal kingdom.
Dolomite A sedimentary rock, and mineral of the same name, composed of calcium magnesium carbonate, which gives rise to alkaline/basic soils.
dm Decimeter; 10 cm.
Dormant Not active.
Dorsal Referring to the back or outer surface of an organ; the side away from the axis; the lower or abaxial surface; contrasting with ventral.
Downy Closely covered with short, weak, soft hairs.
Drooping More or less erect at base but with upper part bending downward.

E

E East.
E-, Ex- Latin prefixes denoting without, that parts are missing.
Echinate Prickly.
Eciliate Without cilia.
Ecological Indicator An organism that is sensitive to pollution or some other environmental problem and can therefore be used as an indicator or gauge of the condition of an ecosystem.
Edaphic Pertaining to soil conditions including such factors as texture, drainage, mineral composition, or pH.
Egg A female gamete or sex cell, in flowering plants contained in an ovule.
Eglandular Without glands.
Elater Strap- or ribbon-like appendages of *Equisetum* spores which are hygroscopic, coil and uncoil with changes in humidity, and function in spore dispersal.
Ellipsoid A solid that is elliptic in outline.
Elliptic Shaped like an ellipse, with widest part at the middle; in the form of a flattened circle usually more than twice as long as wide.
Elongate Lengthened; stretched out.
Embossed Apparently pushed into the surface on one side of an object resulting in a raised area on the surface above.
Embryo The very young multicellular sporophyte developing from the zygote; a baby sporophyte.
Emersed, Emergent Raised above and out of the water.
Enation Outgrowth on the surface; epidermal outgrowth; e.g., the small, veinless, scale-like outgrowth on the stems of *Psilotum*.
Endemic Confined geographically to or native to a single area.
Entire With smooth margins; without teeth.
Enzyme Proteins that catalyze (i.e., increase the rates of) chemical reactions in living organisms; some serve as plant defense chamicals; e.g., thiaminase can break down thiamine (Vitamin B1) and cause a fatal deficiency in livestock upon eating *Pteridium* foliage.
Ephemeral Lasting for a brief period; e.g., for only one day.
Epi- Greek prefix meaning upon or on.
Epidermal Relating to the epidermis.
Epidermis The cellular covering of plant tissue below the cuticle.
Epiphyte A plant growing on another plant for physical support only and not parasitic; e.g., a *Pleopeltis* species growing on a tree limb.
Epiphytic Having the character of an epiphyte.
Epipetric (= Lithophytic) Growing on rocks.
Epithet Second word in the scientific name or binomial of a species, e.g., *sensibilis* in *Onoclea sensibilis*.
Erect (a) Growing essentially in a vertical position (e.g., whole plant); (b) a structure perpendicular to the object to which it is attached.
Erose With ragged margin, as if nibbled or chewed.
Escape A cultivated plant not purposely planted but found growing as though wild.
Escarpment A steep slope.
Etiolate Lengthened and deprived of color by absence of light.
Eusporangiate the condition of having large, thick-walled sporangia, each containing from several hundred to thousands of spores, the sporangia developing from several initial cells; found in less than 5% of living ferns, those on the basal branches of the fern evolutionary tree.
Evergreen Remaining green through the winter.
Exotic Foreign; not native; from another geographic area
Exserted Projecting out or beyond; contrasting with included.
Extant Still existing; contrast with extinct.
Extinct No longer in existence; descriptive of a species for which living representatives no longer exist. Locally extinct refers to extinction in a given geographic region.
Exudate Material coming out slowly through small pores or openings.

F

F. (a) After an author's name: abbreviation of *filius*, the son, or "jr."; (b) abbreviation of forma or form.
Falcate Sickle-shaped, with the tip curved to one side.
False Indusium The underrolled, turned-under, or recurved edge of a frond/leaf that serves the same function as an indusium in protecting the sporangia; characteristic of many Pteridaceae.
Family The unit, category, or rank in classification made up of one or more genera; ending in -aceae; sometimes divided into subfamilies, which in turn are made up of genera.
Farina A wax-like or mealy appearing whitish or yellowish flavonoid material produced by short, capitate, glandular hairs.
Farinaceous Bearing farina.
Farinose Covered with farina.

Fern Allies Term used in the past for all vascular spore-bearing plants not considered to be ferns (equisetophytes, lycophytes, and psilophytes); no longer used since equisetophytes and psilophytes have been shown to be ferns with unusual morphologies).
Fertile Capable of normal reproductive functions.
Fertilization Union of two gametes (e.g., egg and a sperm) to form a zygote.
Fibrous Resembling or having fibers.
Fibrous Root System One with several roots about equal in size and arising from about the same place; contrasting with taproot.
-Fid A suffix meaning deeply cut.
Filament A thread or thread-like structure.
Filamentous, Filamentose Composed of filaments or threads; thread-like.
Filiform Slender; having the form of a thread; filamentous.
Flange A rim-like structure.
Fleshy Succulent, juicy, or pulpy.
Flexuous Zigzag; bending or curving alternately in opposite directions.
Flora (a) Collective term for the plants of an area; (b) a taxonomic work on the plants of an area.
Flower An axis bearing stamen(s), pistil(s), or both, and in addition, often floral envelopes (= calyx and corolla); the reproductive structure of an angiosperm.
Fluted With alternating ridges and grooves.
Foliage Collective term for the leaves of a plant.
Forked Dichotomous; divided into two equal or nearly equal branches.
Forma, Form (abbreviated f.) A taxon below the rank of variety used to refer to minor variations without distinctive geographic occurrence; e.g., occasional albinos or seasonal growth forms.
Free (= Distinct) Separate from one another.
Frond The leaf of a fern, often compound or decompound.
Furrowed With longitudinal channels or grooves.
Fused United by normal growth.

G

Gamete A sex cell; an egg or sperm.
Gametophyte The tiny, inconspicuous, gamete-producing, typically haploid (with one set of chromosomes) generation alternating with the sporophyte (= spore-producing, typically diploid); the stage in the life-history of a plant that produces male or female cells (= gametes).
Gemma A tiny asexual propagule sometimes appearing as, but not homologous with, a vegetative bud, found on some gametophytes.
Gene Unit of heredity, composed of DNA and located on a chromosome, that determines the inheritance of a particular character.
Genus (pl. GENERA) The unit, category, or rank in classification between family and species; composed of one or more closely related species; sometimes divided into subgenera, which in turn are made up of species.
Glabrate, Glabrescent Becoming hairless with age.
Glabrous Without hairs.
Gland A secreting part or appendage, often protruding or wart-like.
Glandular Having or bearing secreting organs, glands, or trichomes.
Glandular-Pubescent With gland-tipped, pinhead-like hairs.
Glaucescent Becoming glaucous.
Glaucous With waxy substances forming a whitish or gray-silvery covering or bloom.
Globose Nearly spherical or rounded.
Glycoside Complex, two-component chemical compound that can break down or hydrolyze under certain conditions, yielding a sugar plus another compound (= aglycone) that can be physiologically active including poisonous; a type of plant chemical defense mechanism.
Gymnosperms Literally, "naked seed"; plants with seeds but without flowers, the seeds "naked," (= not enclosed in a special structure), often on the surface of thick or thin, sometimes woody cone scales; including conifers, cycads, ginkgo, and joint-firs.
Gypsum A soft mineral composed of calcium sulfate.

H

Habit Style or arrangement of growth; general appearance.
Habitat Type of locality in which a plant grows; e.g., forest.
Hair An epidermal appendage that is usually slender, sometimes branched, not stiff enough to be called a spine, not flattened as a scale; often used synonymously with trichome.
Haploid Having the reduced number of chromosomes typical of gametes; usually with a single set of chromosomes in each nucleus.
Hemi- Greek prefix meaning half.
Hemidimorphic (= Partly dimorphic) With different sterile and fertile pinnae on the same leaf blade (e.g., as in *Osmunda regalis*)
Herb A vascular plant lacking a persistent woody stem and typically dying back to the ground each season.
Herbaceous (a) Referring to the aerial shoot of a plant that does not become woody; typically dying back to the ground each year; (b) of a soft texture, as green leaves.
Herbarium (pl. HERBARIA) A collection of dried pressed plants prepared for permanent preservation.
Hetero- Greek prefix meaning other, various, or having more than one kind.
Heterochrony An evolutionary change in the relative timing of events during development; e.g., in Marsileaceae the leaves complete development at a stage comparable to juvenile leaf stages of other ferns.
Heterogeneous Not uniform in kind.
Heterophyllous Having more than one form of leaf.
Heterosporous Having two spore types (male and female); e.g., *Selaginella*.
Hexagonal Six-angled.
Hexaploid Having six sets of chromosomes.
Hirsute With straight moderately stiff hairs.

Hispid Resembling hirsute but the hairs stiffer, ± bristly, feeling rough to the touch.
Hispidulose Minutely hispid.
Holotype The one specimen used or designated by the author of a species or other taxon as the nomenclatural type in the original publication. The holotype is the specimen to which the scientific name is permanently attached; it is not necessarily the most typical or representative element of a taxon.
Homo- Greek prefix meaning all alike, very similar, same, or of one sort.
Homogenous Of the same kind or nature; uniform; contrasting with heterogeneous.
Homosporous With spores all of one type.
Humus Decomposing organic matter in the soil.
Hyaline Thin, membranous, and transparent or translucent.
Hybrid (a) A cross between two unlike parents; (b) specifically, the offspring resulting from a cross between two species.
Hydathode A microscopic epidermal structure, usually marginal or terminal, that excretes water.
Hydrophyte A plant typically growing partially or wholly immersed in water; contrasting with mesophyte and xerophyte.
Hygroscopic Susceptible of expanding, shrinking, twisting, or untwisting on the application or removal of water or water vapor.

I

Illegitimate Name Name unacceptable as the accepted scientific name because it is not the earliest one given to the plant in question, or published without description, or violating some other specific requirement of the International Code of Botanical Nomenclature.
Imbricate Overlapping like shingles on a roof.
Incipient Beginning to be; coming into being.
Incised Cut rather deeply and sharply; intermediate between toothed and lobed.
Included Not exserted; contained within; not projecting beyond the surrounding organ, as a veinlet within an areola.
Incurved Curved inward.
Indeterminate Growth Growth that is not pre-determined in amount; e.g., the elongating leaf midribs (= rachises) of *Lygodium* have terminal meristems that are able to continue growing indefinitely.
Indigenous Native to an area; not introduced.
Indument, Indumentum Surface coating such as hairs, roughening, bloom, or glands.
Indusium (pl. INDUSIA) Usually thin, delicate epidermal outgrowth covering the sori or sporangia on fern fronds.
Infertile Incapable of normal reproductive functions.
Inflated With an internal air space; bladdery.
Inflexed Bent inward.
Infra- Latin prefix meaning below.
Infraspecific Within the species; referring to a unit of classification below the species; e.g., subspecies, variety, form.
Inrolled Rolled inward.
Inserted Attached to another part or organ.
Inter- Latin prefix meaning between.
Intercalary Medial in position.
Intercostal Located between the ribs or costae.
Intergametophytic Selfing A type of cross-fertilization in which a sperm from one gametophyte swims to and fertilizes an egg produced by a different gametophyte.
Internode Area of stem or other structure between two nodes.
Interrupted Not continuous or regular.
Interspecific Between different species.
Intra- Latin prefix used to denote within.
Intragametophytic Selfing Condition in which a single bisexual gametophyte can produce both sperm and eggs that can successfully join and produce a sporophyte; the result is a type of self-fertilzation.
Introduced Brought from another geographic region; not native.
Introgression, Introgressive Hybridization Successive crosses, first between plants of two species, then between the offspring of this cross and plants of one parent species, followed by further interbreeding between mongrels of varying percentage of impurity with purebreds of the parent line.This eventually leads to whole populations of one parent species being contaminated with genes derived from the other.
Intruded Projecting inward or forward.
Irregular Structures not similar in size or shape; asymmetrical.
-Ish Suffix meaning "slightly," often used with color terms.
Isotype A specimen of the type collection other than the holotype; an extra or duplicate specimen made at the same time and place as the holotype.

J

Jointed With or apparently with nodes or points of articulation; e.g., jointed stems of *Equisetum*.

K

Keel Prominent longitudinal ridge, shaped like the keel of a boat.
Keeled With a ridge or keel.

L

L. Linnaeus, Swedish naturalist who established the binomial system of nomenclature; L. f., his son.
Lacerate Irregularly cleft as if torn.
Lanceolate Lance-shaped; several times longer than wide, tapering at both ends, widest about a third above the base.
Lateral Belonging to or borne on the sides.
Lateral Bud Bud in a leaf axil; contrasting with terminal or apical bud.
Lax (a) Not rigid; (b) spread apart, loose, distant.
Leaf (= Frond in ferns) The primary photosynthetic organ of most plants, usually composed of a expanded blade and a stalk-like petiole.
Leaf Gap Break or gap in the vascular tissue of a stem above the point of attachment of a leaf trace; found in stems of plants with megaphylls, but not in those with lycophylls.
Leaflet A single, expanded segment or division of a compound leaf.

Leaf Trace A vascular bundle, one or more in number, extending from the stem into the leaf.

Leathery With a tough, leather-like texture; e.g., the leaves of many dry-adapted ferns.

Lectotype A specimen or other material selected by a later worker from the original material studied by the author of the species (or other taxon) to serve as the nomenclatural type when a holotype was not originally designated or was lost or destroyed.

Leptosporangiate the condition of having small delicate sporangia, each usually containing 128 spores or fewer, with each sporangium usually developing from a single cell and with sporangial walls only one cell thick; found in more than 95% of living ferns.

Ligule A minute, tongue-like or triangular, basal protuberance on a leaf; just distal to the sporangium on the adaxial surface of the leaf base in *Isoëtes*.

Limestone A sedimentary rock composed largely of the mineral calcite (calcium carbonate), which gives rise to alkaline/basic soils.

Linear Resembling a line, long and narrow, with margins parallel to one another.

Lithophytic (= Epipetric) Growing on rocks.

Lithophyte Plant that grows on rocks but derives its nourishment from the atmosphere and from accumulated humus.

Lobe A usually rounded segment or division of a leaf, petal, or other organ.

Lobed Having deep or coarse indentations of the margin, larger than mere teeth (However, there is no sharp distinction between large teeth and small lobes).

Lobulate Having small lobes.

Long-Creeping Growing horizontally for at least a few centimeters; usually in reference to rhizomes, bearing petiole bases attached some distance from one another.

Long-Distance Dispersal Dispersal of spores, seeds, or other propagules over long distances.

Longitudinal Lengthwise; along the long axis.

Lumper A taxonomist who in general has the tendency to lump segregates into larger groups; contrasting with splitter.

Lustrous Shining.

Lycophyll (= Microphyll) A relatively small leaf with a single unbranched vein, typical of the lycophytes.

Lycophyte A member of the Lycopodiophyta, the ancient group that among living vascular plants is the basal lineage; it includes club-mosses, spike-mosses, and quillworts.

M

m Meter; 10 decimeters; 39.37 inches.

Macro- Greek prefix denoting large or long.

Marcophyll (= Megaphyll) Leaf with branched veins, typical of ferns and seed plants.

Macrospore (= Megaspore) A large spore giving rise to the female gametophyte; the larger of two kinds of spores produced by heterosporous plants; a female spore.

Margin Edge; the outer portion of a blade or other structure.

Marginal Attached to the edge or pertaining to the edge.

Marsh Wet or periodically wet, treeless area.

Massula (pl. MASSULAE) Mucilaginous structure enclosing a cluster or mass of microspores in *Azolla* and usually covered with barbed hairs.

Medial, Median Central, middle.

Mega- Greek prefix meaning very large.

Megaphyll (= Marcophyll) Leaf with branched veins, typical of ferns and seed plants

Megasporangium Specialized sporangium in which megaspores are formed in heterosporous pteridophytes.

Megaspore (= Macrospore) A large spore giving rise to the female gametophyte in heterosporous pteridophytes; the larger of two kinds of spores produced by heterosporous plants; a female spore.

Megaspore Apparatus (also referred to a megaspore complex) Specialized term used to collectively describe the whole unit in *Azolla* consisting of the megaspore, covered by a perispore layer (which is sometimes distinctively marked, pitted, etc.) and topped with triangular cap consisting of a collar, floats, and a blue-green algal colony covered by remnants of the indusium.

Megasporocarp A specialized structure in *Azolla* containing the megasporangium; a specialized sorus enclosed by a modified indusium.

Megasporophyll A sporophyll (= spore-bearing leaf) bearing one or more megaspores.

Meiosis Type of cell division in which the number of chromosomes in the daughter cells is half that of the parental cells; in pteridophytes usually leading to the formation of spores; sometimes called reduction division.

Membranaceous, Membranous Having the nature of a membrane, thin, somewhat flexible, translucent.

Meristem Embryonic or undifferentiated tissue, capable of developing into various organs.

Meristematic Pertaining to or with the nature of a meristem.

Mesophyte Plant that grows under medium moisture conditions; contrasting with hydrophyte and xerophyte.

Micro- Greek prefix meaning small.

Microhabitat A small, localized, specialized habitat within a larger different habitat, sometimes capable of sustaining rare or unusual species; e.g., cave entrances, granite outcrops.

Microphyll (= Lycophyll) A relatively small leaf with a single unbranched vein, typical of the lycophytes; in contrast to megaphylls/macrophylls.

Microphyllous Having small leaves (note: this term is usually used differently from the microphyll/megaphyll distinction), typical of many dry-adapted plants, the reduction in surface area (and thus water loss) apparently being beneficial in dry climates.

Microphylly The condition of having small leaves.

Microsporangium Specialized sporangium in which microspores are produced in heterosporous pteridophytes.

Microspore A small spore giving rise to the male gametophyte in heterosporous pteridophytes; the smaller of two kinds of spores produced by heterosporous plants; a male spore.

Microsporocarp A specialized structure in *Azolla* containing the microsporangia; a specialized sorus enclosed by a modified indusium.

Microsporophyll The sporophyll (= spore-bearing leaf) upon which microspores are produced.

Midrib The central or main rib or vein of a leaf or other similar structure.

Mitosis Regular cell division where the number of chromosomes in the daughter cells is the same as that of the parent cell.

mm Millimeter; 1000 microns or 1/1000 of a meter.

Moniliform Like a string of beads.

Monilophyte Term referring to all species now known to be in the fern lineage including *Equisetum* and *Psilotum*.

Mono- Greek prefix meaning one or of one.

Monomorphic One form; contrasting with polymorphic.

Monophyletic A term previously used to describe a group of organisms with a common ancestor; more recently it has been used to describe a group consisting of a common ancestor and all of its descendants. Some authorities believe that a different term, holophyletic, should be used for a group consisting of a common ancestor and all of its descendants.

Monotypic Having a single type or representative; e.g., a genus with only one species.

Montane Pertaining to or living in mountains.

Morphology That aspect of biology concerned with form and structure of organisms.

Mostly A quantitative term meaning "most of them."

Mucro A short and small abrupt tip, as with the midrib extending as a short point.

Mucronate With a mucro.

Multi- Latin prefix for many.

Multifid Divided into many narrow segments or lobes.

Mycoheterotrophic Obtaining food from decaying organic material via a special relationship with a symbiotic fungus.

Mycorrhiza (pl. MYCHORRHIZAE) A mutually beneficial, symbiotic association of a fungus and the root of a plant. Mychorrhizal relationships are characteristic of most vascular plants, and are particularly important in the Ophioglossaceae and Psilotaceae (with subterranean mycoheterotropic gametophytes).

Mycorrhizal Pertaining to mycorrhiza.

N

N North.

Naked Lacking various coverings, organs, or appendages, almost always referring to organs or appendages present in other similar plants; e.g., a naked flower lacks perianth.

Naturalized Referring to an introduced foreign plant that has become part of the spontaneous, self-perpetuating flora of a region; a species intentionally or accidentally introduced to an area, but now growing wild without human help.

Nectar A sugar-rich solution secreted by plants, typically produced in nectaries. Nectar production has apparently evolved to attract insects or other animals for pollination or other purposes.

Nectary A specialized nectar-secreting structure or area; there can be floral nectaries (in the flowers) or extra-floral nectaries (not associated with the flowers).

Neotype A specimen selected by a later worker to serve as the nomenclatural type of a taxon when all material studied by the original author has been lost or destroyed.

Nerve A simple vein or slender rib of a leaf, bract, or other structure.

Nerved Having nerves.

Net-Veined With veinlets branching irregularly and not uniformly angular, forming a net-like pattern.

New World Term used synonymously with Western Hemisphere to refer to the Americas; the new signifying that its presence was not known by people inhabiting other parts of the world until relatively recently.

Nitrogen Fixation Capture and conversion of atmospheric nitrogen, which is unusable by plants, into nitrogen containing compounds that are of use by plants; the process is carried out by certain bacteria such as those that inhabit the leaves of *Azolla*.

Nodal Located at or pertaining to a node.

Node Area of stem or axis at which branches, leaves, bracts, or flower stalks are attached; joint of a stem.

Nodding Hanging down.

Nomenclature The scientific naming of organisms; the naming of plants is governed by the rules of the International Code of Botanical Nomenclature (McNeill et al. 2006).

O

Ob- Latin prefix indicating the reverse or upside-down, as obcordate, meaning cordate or ovate with wider end at top or away from point of attachment.

Obdeltoid Inversely deltoid; triangle-shaped with base pointed.

Oblanceolate Lanceolate with broadest part above the middle and tapering toward the base.

Oblique Slanting; unequal-sided.

Oblong Longer than wide with sides nearly parallel.

Obovate Egg-shaped with attachment at narrow end; inversely ovate.

Obovoid Inversely ovoid; a solid that is obovate in outline.

Obscure Hidden; not obvious.

Obsolete Not apparent or evident; rudimentary; vestigial; extinct.

Obtuse Forming a blunt or rounded angle of more than 90 degrees; not pointed.

Old World Term used to collectively refer to Europe, Asia, Africa, and Australia.

Ontogeny The developmental cycle of an individual organism.

Opaque Impervious to light.

Opposite Arranged two at each node, on opposite sides of the axis.

Orbicular, Orbiculate With round, approximately circular outline.

Order The unit, category, or rank in classification made up of one or more families; ending in -ales; sometimes divided into suborders or superfamilies, which in turn are made up of families.

Oval Broadly elliptic.

Ovate Egg-shaped with widest part at the base.

Ovoid Solid oval or solid ovate.

P

Palmate Attached or radiating from one point, as leaflets in a palmately compound leaf, veins in some leaf blades, or fingers of a hand.

Palmately Compound With the leaflets attached at one point at the apex of the petiole, like the fingers all attached to the palm of a hand.

Panicle-Like Resembling a panicle, an indeterminate, branched, usually elongate (not flat-topped) inflorescence with flowers on the branches of the primary axis; e.g., the arrangement of the sporangia in *Botrychium*.

Papilla (pl. PAPILLAE) Small pimple-like or nipple-like projection.

Papillate, Papilliform, Papillose Shaped like or bearing papillae.

Parallel-Veined With main veins parallel to each other. Such leaves are typical of many monocots.

Paraphyletic A term used to describe a taxonomic group consisting of an ancestral species and some but not all of its descendants. Paraphyletic groups can give an inaccurate view of phylogeny but are often useful in classification.

Parasite A plant that derives its food, mineral nutrition, and/or water wholly or chiefly from another plant (the host) to which it is attached; contrasting with epiphyte, saprophyte, or autophyte.

Pedate Palmately divided with the lateral segments again divided.

Peduncle Stem or stalk supporting reproductive structures.

Peltate Shield-shaped, with stalk attached on the undersurface away from the margin or base (sometimes attached at the middle like the axis of an umbrella).

Pendent, Pendulous Hanging down or suspended.

Pentagonal Shaped like a pentagon; five-sided or -angled.

Perennial Root system or plant living at least three growing seasons (years); lasting from year to year.

Peripheral On or near the margin.

Perispore The outer covering of a spore, sometimes distinctively marked, pitted, etc.

Persistent Remaining attached; not falling off; contrasting with deciduous.

Petiolar Pertaining to or located on the petiole.

Petiolate With a petiole.

Petiole (= Stipe) Stalk of a leaf/frond supporting the blade.

Petiolule Stalk supporting a leaflet/pinna.

Phanerogam A seed plant or spermatophyte; contrasting with cryptogam.

Phenology Study of the times at which various events occur in the life of a plant.

Phloem The food-conducting tissue in a plant.

Photosynthesis The process by which plants convert carbon dioxide and water into carbohydrates in the presence of light.

Phylogeny The evolutionary history of a group.

Phytoecdysones A class of hormone-like compounds that disrupt ecdysis (= molting) in insects and thus serve as a plant defense mechanism.

Pilose With long, soft, ± straight hairs; softer than hirsute, not flexuous or curved as in villous.

Pinna (pl. PINNAE) A primary division of a compound leaf. A pinna can be simple (and thus equivalent to a leaflet) or compound and divided into leaflets.

Pinnate, Pinnatley Compound Descriptive of a compound leaf with leaflets distributed along both sides of an elongate axis; feather-like. Bipinnate or 2-pinnate leaves have the leaflets distributed along a secondary axis; tripinnate or 3-pinnate leaves have the leaflets along a tertiary axis, etc.

Pinnatifid Pinnately divided into stalkless segments, but the segments not distinct leaflets (i.e., not divided all the way to the midrib).

Pinnule A secondary pinna; division of a pinna or a leaflet.

Pitted Having little depressions or cavities.

Plane With a flat even surface.

Plano- A suffix denoting flat.

-Ploid A suffix used in genetics, prefixed by a term indicating number, to denote the number of sets of chromosomes in the nucleus.

Poly- Greek prefix meaning many.

Polymorphic Of various forms; with three or more forms.

Polyphyletic A taxonomic group having species derived from more than one common ancestor, the species having been placed in the same group because of similarities due to convergent or parallel evolution. Polyphyletic taxa give an inaccurate view of phylogeny.

Polyploid A plant with three or more basic sets of chromosomes.

Pore A small aperature or opening.

Posterior Describing the position of an organ located on the side adjacent to the axis.

Procumbent Trailing or prostrate, not rooting.

Proliferous Bearing or developing offshoots or redundant parts; producing numerous new individuals, or parts, of the same kind; bearing offsets, bulbils, or other vegetative progeny.

Prostrate Lying flat.

Prothallus, Prothallium Gametophyte stage or generation in ferns and lycophytes, bearing the sexual organs.

Protuberance A protrusion, swelling, bump, or bulge.

Proximal Located nearest to the base or attachment point of a structure; contrasting with distal.

Pseudo- Greek prefix meaning false, not genuine.

Pteridology The study of ferns and lycophytes.

Pteridophyte General collective term for ferns and lycophytes; any member of the Lycopodiophyta and Polypodiophyta; seedless, spore-bearing, vascular plants.

Puberulent Minutely pubescent.

Pubescence The covering or indumentum of hairs on a plant without reference to specific type (e.g., pilose, hirsute).
Pubescent (a) General term for covering or indumentum of hairs; (b) sometimes used in a more restricted sense to refer to fine short hairs; downy.
Pulvinate With a pulvinus; cushion-shaped.
Pulvinus A joint-like thickening at the base of a leaf or leaflet involved in movement (e.g., in *Marsilea* they orient leaflets during the day and cause the leaflets to fold together at night).

Q

Quadr- Latin prefix meaning four.
Quadrangular Four-cornered or sided; square.
Quill-Like Resembling a quill, the hollow stem-like main shaft of a feather, formerly used as writing implements.

R

Rachis Main axis of a leaf blade; the midvein or midrib of the blade or the pinnae-bearing central axis of a pinnately compound leaf.
Radially Symmetrical Descriptive of a structure that can be cut into halves from any marginal point through the center.
-Ranked Suffix, when combined with a numerical prefix, indicating the number of longitudinal rows in which leaves or other structures are arranged along an axis or rachis.
Reclined, Reclinate Bent or turned downward.
Recurved Curved downward or backward.
Reduced Small but probably derived from larger forerunners.
Reflexed Abruptly bent downward.
Relic (a) A long-surviving species whose relatives have become extinct; (b) a plant persisting in a relatively small portion or portions of its former range.
Remote Widely or distantly spaced.
Reniform Kidney-shaped.
Resin, Resinoid A miscellaneous or catchall term for a variety of amorphous, sometimes semisolid, perhaps gummy substances from plants.
Resinous, Resiniferous Producing or bearing resin.
Reticulate Net-veined or with a net-like pattern.
Revolute With margins rolled down and inward; e.g., revolute leaves.
Rhizoid A filamentous root-like structure (without the anatomy of a root) on the gametophyte of ferns or other non-seed-producing plants.
Rhizomatous Possessing a rhizome.
Rhizome Stem (typically underground or at ground level) with nodes and scales and bearing the fronds and adventitious roots; the type of stem found in most ferns.
Rhizomorphs Root-like stems in *Isoëtes* (quillworts) that have the nutrient absorption and anchoring functions usually carried out by roots.
Rhizophore In lycophytes, a specialized leafless stem emitting roots.
Rhizotaxy A regular and precise spiral pattern of "rootlets" around the main "root" (rhizomorph) in *Isoëtes* unlike the irregular arrangement seen in other plants.
Rhombic Somewhat diamond-shaped; shaped like two adjacent equilateral triangles.
Rhomboid, Rhomboidal A three-dimensional rhombic figure.
Rib One of the principal longitudinal veins of a leaf or other organ.
Ribbed With prominent ribs or veins.
Root The portion of the main axis (or one of its subdivisions) of a plant usually found below ground and lacking nodes, internodes, or leaves.
Rootlet (a) A little root; (b) term often applied to the holdfast roots of certain climbing plants.
Rootstock According to Shinners (1958), a "weasel-word" indicating an elongate crown, rhizome, or rhizome-like structure; an old inaccurate term for rhizome.
Rosette A cluster or whorl of leaves or other organs closely arranged in a radial pattern.
Rotund Essentially circular in outline.
Rudiment A structure very imperfectly developed, nonfunctional, or represented only by a vestige; e.g., rudimentary nonfunctional florets in some grass spikelets.
Rudimentary Having the character of a rudiment.
Rugose Wrinkled.
Rugulose With slight ridges or wrinkles, a diminutive of rugose.

S

S South.
Saline Of or pertaining to salt.
Sap The juice of a plant.
Scabrous Rough to the touch due to short stiff hairs or the structure of the epidermis.
Scale A small, muticellular, usually flat and elongate epidermal outgrowth 2 to many cells wide (hairs are similar but only 1 cell wide).
Scarious Membranous, dry, papery, translucent or transparent, not green.
Scattered Distributed in an irregular manner; not clustered together.
Scrub Vegetation of stunted or densely crowded bushes.
Secondary Compounds Naturally occurring plant materials not essential to the primary (= life-sustaining) metabolism of the plant; examples of categories include alkaloids and glycosides. Many are significant because of their physiological activity when given to animals; they are probably important to plants in defense against herbivores or microbes.
Seed An embryo (baby plant) packaged with stored food and surrounded by a protective waterproof coating.
Seep A moist spot where underground water comes to or near the surface.
Segment One of the parts of a leaf that is divided but not truly compound.
Segregate Term used as a noun or adjective to refer to or describe a taxon that is sometimes recognized separately from a more inclusive group; e.g., depending on authority, the genus *Botrychium* is sometimes separated into a number of segregate genera.

Self-Fertile Capable of self-fertilization (= union of gametes from same plant).
Self-Incompatible Incapable of self-fertilization.
Semi- Latin prefix meaning half.
Sensu Lato "In a broad sense"; used to refer to the broad treatment of taxa; e.g., a genus sensu lato is one that has not been split into a number of segregates.
Sensu Stricto "In a narrow sense"; used to refer to a restricted or narrow treatment of a taxonomic group; e.g., a genus sensu stricto is viewed in a more restricted sense than previously as the result of segregating or splitting out various taxa.
Separate Not joined; of individual units.
Septate With partitions or divisions.
Serrate With pointed teeth sloping forward; saw-toothed.
Serrulate Finely serrate.
Sessile Without a pedicel, petiole, or stalk; inserted directly.
Seta (pl. SETAE) A bristle.
Sheath ± tubular structure surrounding an organ; portion that clasps or encloses.
Shoot (a) A young stem or branch; (b) the ascending axis of a plant.
Short-Creeping Growing horizontally for a few centimeters at most; usually in reference to rhizomes, bearing petiole bases all attached in close proximity.
Shrub A woody perennial usually branching from the base with several main stems.
Silica Silicon dioxide; a white or colorless, very hard, crystalline mineral substance.
Siliceous Containing or composed of silica (the principal component of glass and sand).
Silvery With a whitish, metallic, more or less shining luster.
Simple (a) Single, of one piece, not compound; (b) descriptive of an unbranched stem, inflorescence, or other structure; (c) descriptive of a leaf that is not compound.
Simple Leaf Single-bladed leaf, not divided into individual leaflets.
Sinuate, Sinuous Having the margin wavy with regular strong indentations.
Sinus The space or recess between two lobes, segments, or divisions of a leaf or other expanded organ.
Sister, Sister Group A monophyletic group more closely related to the group in question than any other group; e.g., the Osmundaceae is the sister group to the group composed of all other leptosporangiate ferns.
Smooth Not rough to the touch; without vestiture or other special covering.
Solitary Borne singly.
Sorus (pl. SORI) A cluster or heap of sporangia.The term is used mainly to refer to the sporangial clusters of ferns.
Species Unit of classification below the rank of genus; a group of individuals that are actually or potentially capable of interbreeding in natural populations and are reproductively isolated from other such groups. Generally there are morphological characteristics that distinguish and can thus be used to separate such groups; the definition is complicated by instances of asexual reproduction.
Specific Epithet The second half of the scientific name of a species, the scientific name or binomial being composed of the genus name and the specific epithet.
Spermatophyte A plant that produces seeds; all gymnosperms and angiosperms.
Spherical Globular; orbicular.
Spike-Like Resembling a spike, an indeterminate inflorescence with sessile flowers on an elongate floral axis; e.g., the arrangement of the sporangia in *Ophioglossum*.
Spine A sharp-pointed structure, usually vascularized and generally modified from part or all of a leaf.
Spiral Describing the arrangement of like organs, such as leaves, at regular angular intervals along an axis.
Splitter A taxonomist who in general has the tendency to split or divide larger taxa into a number of segregates; contrasting with lumper.
Sporadic Of irregular occurrence.
Sporangiophore An appendage holding a sporangium; in *Equisetum,* a highly modified peltate sporophyll bearing a ring of 5–10 sporangia.
Sporangium (pl. SPORANGIA) The structure within which spores are produced; a spore case or spore sac.
Spore An asexual, usually one-celled reproductive body; a cell resulting from meiotic cell division in a sporangium representing the first cell of the gametophyte generation.
Sporocarp A specialized structure containing sporangia; e.g., in *Marsilea*.
Sporophore The fertile sporangia-bearing portion of the leaves in the Ophioglossaceae, the leaves divided into a fertile sporangia-bearing portion and a sterile blade portion or trophophore.
Sporophyll Spore-bearing leaf.
Sporophyte The spore-producing, typically diploid generation that alternates with the gamatophyte (= gamete-producing, typically haploid); the dominant generation in most plants except mosses and liverworts.
Spreading Diverging to the side, almost to the horizontal.
Spreading Hairs Hairs that are ± erect, not at all appressed.
Stalk The supporting structure of an organ; e.g., petiole.
Stellate Star-shaped or star-like; when used in reference to hairs it means those branched hairs with a central stalk and branch hairs arising at the top of the stalk (like points of light coming out of a star).
Stem A major division of the plant-body in contrast to root and leaf, distinguished from both by certain anatomical features and commonly also by general aspect; the main axis or axes of a plant; the portion of the plant axis bearing nodes, leaves, and buds and usually found above ground.
Sterile Unproductive; without functional sex organs; without spores.
Stipe (= Petiole) Stalk of a leaf/frond supporting the blade of a fern.
Stolon A creeping horizontal stem that loops or runs along the surface of the ground and roots at the nodes.
Stoloniferous Producing stolons.

Stoma, Stomate (pl. STOMATA) A minute opening (= "breathing" pore) between the epidermal cells of a leaf or stem through which gases and water-vapor enter and leave the plant.
Strobilus (pl. STROBILI) (= Cone) A usually cone-like, globose or cylindrical structure involved in reproduction and composed of an axis with a spiral, usually dense aggregation of sporophylls, bracts, or scales (in pteridophytes these bearing spores, or in seed plants bearing pollen or seeds).
Sub- Latin prefix meaning almost, somewhat, of inferior rank, beneath.
Subapical Almost at the apex.
Subbasal Almost at the base.
Subclass The unit, category, or rank in classification between class and order, composed of one or more orders; e.g., the Liliidae is a subclass of class Monocotyledonae.
Subcoriaceous Somewhat leathery in texture.
Subfamily The unit, category, or rank in classification between family and genus, composed of one or more genera.
Subgenus A unit, category, or rank in classification between genus and species, composed of one or more species.
Suborder The unit, category, or rank in classification between order and family, composed of one or more families.
Submerged, Submersed (= Immersed) Growing under water.
Subsessile Almost sessile.
Subsp. Abbreviation for subspecies.
Subspecies A unit, category, or rank in classification below the level of species and between species and variety; a geographically distinct variant. The categories of subspecies and variety are not used consistently by taxonomists.
Subterranean Below ground.
Subulate (= Awl-shaped) Tapering from the base to a slender or stiff point; narrow and sharp-pointed.
Succulent Fleshy, thickened.
Supra-, Super- Latin prefix meaning above.
Supramedial Above the middle; when used to refer to the location of fern sori, it means somewhat beyond the middle of the distance between the leaf segment midvein and margin, but not so much so as to be called submarginal.
Swamp Wet or periodically wet area with some trees.
Sym- Greek prefix meaning with or together.
Symmetrical Possessing one or more planes of symmetry; regular in number and size of parts.
Sympatric Growing together with or having the same range as.
Syn- Greek prefix meaning united.
Synangium (pl. SYNANGIA) A cluster of sporangia that have become fused in development; in *Psilotum* the sporangia are fused in groups of 3 (rarely 2).
Synonym A currently unaccepted alternative scientific name for a taxon.
Synonymy Referring to the series of names no longer used for a taxon.
Systematics Scientific study of the kinds and diversity of living organisms and of the relationships between them. The term is often used synonymously with taxonomy.

T

Taxon (pl. TAXA) (a) General term referring to any unit of classification such as variety, subspecies, species, genus, family, etc.; (b) term used to refer to a specific variety, subspecies, etc.
Taxonomy The branch of science that deals with classification, identification, and nomenclature.
Teeth (pl. of tooth) Marginal projections, protuberances, serrations, or dentations, usually sharply pointed.
Terminal At the tip or apex; distal.
Terminal Bud (= Apical bud) Bud at the end (= apex) of a stem or branch.
Ternate, Ternary In threes.
Terrestrial Growing in the ground; supported by soil; contrasting with aquatic.
Tetra- Greek prefix referring to four.
Tetrad A group of four similar objects; e.g., in Ericaceae, the four pollen grains remaining together.
Tetragonal, Tetragonous Four-angled.
Tetrahedal Four-sided, as a three-sided pyramid and its base.
Tetraploid With four sets of chromosomes; twice the normal diploid level.
Tomentose Covered with short, soft, curly, densely matted or entangled hairs.
Tomentum Densely matted wool.
Tooth (pl. TEETH) Any marginal projection, protuberance, serration, or dentation, usually sharp pointed.
Toothed With minor projections and indentations alternating along the margin.
Topotype A specimen from the original or type locality of that species or other taxon.
Trailing Prostrate, but not rooting.
Translucent Allowing the passage of light rays, but not transparent.
Transverse Lying or being across or in a cross direction.
Tree A woody perennial with usually a solitary trunk or main stem.
Tri- Latin prefix indicating three, or three times.
Tribe The unit, category, or rank in classification between subfamily and genus, composed of one or more genera.
Trichome Any hair, hair-like projection, or bristle from the epidermal surface.
Trichotomous Forking into three equal parts.
Trigonous, Trigonal Three-angled.
Trilobed, Trilobate Three-lobed.
Trimerous Having the parts in threes.
Triploid Having three sets of chromosomes.
Trophophore The sterile blade portion of the leaves in the Ophioglossaceae, the leaves divided into a blade portion and a fertile sporangia-bearing portion or sporophore.
Trophopod Modified petiole bases that accumulate food reserves and persist after the leaf blade has withered.
Trowel-Shaped Shaped like a gardener's trowel, somewhat wider basally.
Truncate Ending abruptly as if cut off squarely at the end; appearing "chopped off."

Tubercle A small rounded protuberance or projection from a surface.

Tuberculate Covered with tubercles or warty or nipple-like protuberances.

Tubular With the shape of a hollow cylinder.

Tuft; Tufted A cluster or fascicle of trichomes, leaves, or other elongate structures.

Twice-Pinnately Compound (= Bipinnate) Descriptive of a leaf with leaflets pinnately arranged on lateral axes that are themselves pinnately arranged on the main axis; with the primary divisions (= pinnae) themselves pinnate.

Type A plant specimen to which the name of a taxon is permanently attached.When any new taxon (e.g., species, variety) is named, the name has to be associated with a particular "type" specimen.

U

Ultimate The smallest subdivision; of the lowest order.

Ultimate Segment The smallest subdivision of a leaf.

Umbo A rounded, sometimes knob-like elevation or protuberance at the end of or on the side of a solid organ.

Umbonate With an umbo or projection.

Undulate Gently wavy, less pronounced than sinuate.

Uni- Latin prefix meaning one.

Unilateral One-sided; developed or hanging on one side.

United Fused into one unit.

V

Var. Abbreviation of variety.

Variety A unit, category, or rank in classification below the level of species, sometimes treated as a subdivision of subspecies; group of plants with minor characters or differences separating them from other similar plants. The terms variety and subspecies are used inconsistently by taxonomists.

Vascular Pertaining to the conducting tissues (xylem and phloem), which transport water, minerals, and sugars.

Vascular Bundle Thread-like fiber of xylem and phloem in a stem or other organ.

Vegetative Organ Root, stem, leaf, or other non-reproductive organ of a plant.

Vegetative Reproduction A form of asexual reproduction by which new individuals arise without production of spores or seeds; e.g., via leaf borne bud in *Cystopteris bulbifera.*

Vein Strand or bundle of vascular tissue.

Veinlet A little or ultimate vein.

Velum In *Isoëtes*, a thin, membranous flap of tissue extending downward over the sporangium.

Venation The pattern or arrangement of veins.

Ventral Situated on or pertaining to the adaxial side (= side toward axis) of an organ; typically the upper or inner surface; contrasting with dorsal.

Vernal Appearing in spring.

Vernation The arrangement of leaves or other structures in an unopened bud; e.g., circinate vernation in most ferns.

Vestigial Reduced to a trace, rudiment, or vestige; degenerate; referring to a once more fully developed structure.

Vesture, Vestiture Any covering on a surface making it other than glabrous; e.g., hairs, scales.

Vicariance The fragmentation of a more continuous distribution, with survivors in only parts of the former range.

Villous, Villose With long, soft, spreading, or ascending, unmatted hairs; shaggy.

Vine A plant that climbs by tendrils or other means, or that trails or creeps along the ground.

W

W West.

Whorl, Whorled (= Verticillate) With three or more leaves or other parts attached at the same node; in a circle or ring; e.g., the whorled, scale-like leaves of *Equisetum*.

Wide (= Broad) Distance across a structure (equal to diameter if tubular).

Woolly (= Lanate) With long, soft, and more or less matted or entangled hairs; wool-like.

Wort An old word of Anglo-Saxon origin meaning the equivalent of herbaceous plant.

X

Xeric Characterized by or pertaining to conditions of scanty moisture supply; dry.

Xero- Greek prefix signifying dry.

Xerophyte A plant that can subsist with a small amount of moisture, such as a desert plant; contrasting with hydrophyte and mesophyte.

Xerophytic Dry-adapted; drought resistant; contrasting with mesophytic and hydrophytic.

Xylem Vascular tissue that conducts water and minerals.

Z

Zygote Cell produced from fertilization or the union of two gametes; a fertilized egg; the first cell of the sporophyte.

LITERATURE CITED

ADJIE, B., S. MASUYAMA, H. ISHIKAWA, and **H. WATANO.** 2007. Independent origins of tetraploid cryptic species in the fern *Ceratopteris thalictroides*. J. Pl. Res. 120:129–138.

ALEXANDER, P. 2006. Two ferns not occurring in New Mexico. New Mexico Botanist Newslett. 35:2.

ALLISON, J.R. 2001. Supplement to "Vascular flora of Ketona Dolomite outcrops in Bibb County, Alabama." http://jimbotany.com/10a-NotewSpp-StateRec-Astro_&_Bapt.htm. Accessed Sep 2010.

ALLISON, J.R. and **T.E. STEVENS.** 2001. Vascular flora of Ketona Dolomite outcrops in Bibb County, Alabama. Castanea 66:154–205.

ALLRED, K.W. 1990. Elmer Ottis Wooton and the botanizing of New Mexico. Syst. Bot. 15:700–719.

ALLRED, K. 2008. E.O. Wooton: New Mexico's pioneer botanist. Rangelands 30:11–17.

ALLSOP, A. 1952. Longevity of *Marsilea* sporocarps. Nature 169:79–80.

ALSTON, A.H.G. 1955. The heterophyllous Selaginellae of continental North America. Bull. Brit. Mus. (Nat. Hist.), Bot. 1:219–274.

ALPERT, P. 2000. The discovery, scope, and puzzle of desiccation tolerance in plants. Pl. Ecol. 151:5–17.

AMES, O. and **D.S. CORRELL.** 1952–1953. Orchids of Guatemala. Fieldiana 26:1–727. Field Museum of Natural History, Chicago, IL.

ANDERSEN, B.A. 1996. Desert plants of Utah. Utah State Univ. Extension. HG505. http://www.ut.nrcs.usda.gov/despub.pdf. Accessed Aug 2010.

ANDERSON, J.E. 1987. Development and structure of sagebrush steppe plant communities. In: P.J. Joss, P.W. Lynch, and O.B. Williams. Rangelands: A resource under siege. Proceedings of the Second International Rangeland Congress. Pp. 10–12. Cambridge Univ. Press, Cambridge, England, U.K.

ANDREWS, E.G. and **M.D. WINDHAM.** 1993. *Pleopeltis*. In: Flora of North America Editorial Committee, eds. Fl. North Amer. 2:324–327. Oxford Univ. Press, New York and Oxford.

ARCHANGELSKY, A., C.J. PHIPPS, T.N. TAYLOR, and **E.L. TAYLOR.** 1999. *Paleoazolla*, a new heterosporous fern from the Upper Cretaceous of Argentina. Amer. J. Bot. 86:1200–1206.

ARBER, A. 1912. Herbals, their origin and evolution: A chapter in the history of botany 1470–1670. Cambridge Univ. Press, Cambridge, England, U.K.

ARMSTRONG, J. and **W. ARMSTRONG.** 2009. Record rates of pressurized gas-flow in the great horsetail, *Equisetum telmateia*. Were Carboniferous Calamites similarly aerated? New Phytol. 184:202–215.

AXELROD, D.I. 1958. Evolution of the Madro-Tertiary Geoflora. Bot. Rev. (Lancaster) 24:433–509.

BAILEY, L.H. 1924. Manual of cultivated plants. Macmillan Co., New York.

BALICK, M.J. and **J.M BEITEL.** 1988. *Lycopodium* spores found in condom dusting agent. Nature 332:591.

BALICK, M.J. and **J.M BEITEL.** 1989. *Lycopodium* spores used in condom manufacture: Associated health hazards. Econ. Bot. 43:373–377.

BANKS, J.A., T. NISHIYAMA, M. HASABE, and **MANY OTHERS.** 2011. The *Selaginella* genome identifies genetic changes associated with the evolution of vascular plants. Science 332:960–963.

BARKER, M.S. and **W.D. HAUK.** 2003. An evaluation of *Sceptridium dissectum* (Ophioglossaceae) with ISSR markers: Implications for *Sceptridium* systematics. Amer. Fern J. 93:1–19.

BARKER, M.S. and **P.G. WOLF.** 2010. Unfurling fern biology in the genomics age. BioScience 60:177–185.

BARKER, W.R., R.M.BARKER, J.P. JESSOP, and **H.P. VONOW,** eds. 2005. Census of south Australian vascular plants, 5th ed. J. Adelaide Bot. Gard. Suppl. 1. http://www.flora.sa.gov.au/pdfs/Census_5.0_web.pdf. Accessed Jul 2011.

BARRINGTON, D.S. 1993. Ecological and historical factors in fern biogeography. J. Biogeogr. 20:275–279.

BASHALON. 2008. Bashalon-Hebrew Language Detective. http://www.balashon.com/2008/03/cinnamon.html. Accessed Jul 2010.

BASINGER, J.F. 1991. The fossil forests of the Buchanan Lake Formation (early Tertiary), Axel Heiberg Island, Canadian Arctic Archipelago: Preliminary floristics and paleoclimate. Bull. Geol. Surv. Canada 403:39–65.

BATEMAN, R.M. 1991. Palaeobiological and phylogenetic implications of anatomically-preserved Archaeocalamites from the Dinantian of Oxroad Bay and Loch Humphrey Burn, southern Scotland. Palaeontographica B 223:1–59.

BATEMAN, R.M. 1996. An overview of lycophytes phylogeny. In: J.M. Camus, M. Gibby, and R.J. Johns, eds. Pteridology in Perspective. Pp. 405–415. Royal Botanic Gardens, Kew, England, U.K.

BATES, V.M. and **E.T. BROWNE.** 1981. *Azolla filiculoides* new to the southeastern United States. Amer. Fern J. 71:33–34.

BECK, J.B., J.R. ALLISON, K.M. PRYER, and **M.D. WINDHAM.** 2012. Identifying multiple origins of polyploid taxa: A multilocus study of the hybrid cloak fern (*Astrolepis integerrima*; Pteridaceae). Amer. J. Bot. 99:1857–1865.

BECK, J.B., M.D. WINDHAM, and **K.M. PRYER.** 2011. Do asexual polyploid lineages lead short evolutionary lives? A case-study from the fern genus *Astrolepis*. Evolution 65:3217–3229.

BECK, J.B., M.D. WINDHAM, G. YATSKIEVYCH, and **K.M. PRYER.** 2010. A diploids-first approach to species delimitation and interpreting polyploid evolution in the fern genus *Astrolepis* (Pteridaceae). Syst. Bot. 35:223–234.

BECK, J.T. 2005. Noteworthy collections: Tennessee. Castanea 70:314–316.

BEITEL, J.M., W.H. WAGNER, JR., and **K.S. WALTER.** 1981. Unusual frond development in sensitive fern *Onoclea sensibilis* L. Amer. Midl. Naturalist 105:396–400.

Bell, P. and **C. Woodcock.** 1983. The diversity of green plants. Edward Arnold, London, England, U.K.

Benedict, R.C. 1909. The genus *Ceratopteris*: A preliminary revision. Bull. Torrey Bot. Club 36:463–476.

Benedict, R.C. 1916. The origin of new varieties of *Nephrolepis* by orthogenetic saltation. I. Progressive variations. Bull. Torrey Bot. Club 43:207–234.

Benedict, R.C. 1934. Can anyone readily distinguish the northern and southern lady fern species. Amer. Fern J. 24:117–119.

Benham, D.M. 1989. A biosystematic revision of the fern genus *Astrolepis* (Adiantaceae). Ph.D. dissertation, Northern Arizona Univ., Flagstaff.

Benham, D.M. 1992. Additional taxa in *Astrolepis*. Amer. Fern J. 82:59–62.

Benham, D.M. and **M.D. Windham.** 1992. Generic affinities of the star-scaled cloak ferns. Amer. Fern J. 82:47–58.

Benham, D.M. and **M.D. Windham.** 1993. *Astrolepis*. In: Flora of North America Editorial Committee, eds. Fl. North Amer. 2:140–143. Oxford Univ. Press, New York and Oxford.

Bennert, W., M. Lubienski, S. Korner, and **M. Steinberg.** 2005. Triploidy in *Equisetum* subgenus *Hippochaete* (Equisetaceae, Pteridophyta). Ann. Bot. (Oxford) 95:807–815. http://aob.oxfordjournals.org/cgi/reprint/95/5/807. Accessed May 2010.

Benson, L. 1940. The North American subdivision of *Ranunculus*. Amer. J. Bot. 27:799–807.

Bierhorst, D.W. 1977. The systematic position of *Psilotum* and *Tmesipteris*. Brittonia 29:3–13.

Blasdell, R.F. 1963. A monographic study of the fern genus *Cystopteris*. Mem. Torrey Bot. Club 21(4):1–102.

Blassingame, J. (Jim). Personal communication. Collector and specialist in ferns of western Texas; retired from career at South Plains College, Levelland, TX.

Bold, H.C., C.J. Alexopoulos, and **T. Delevoryas.** 1987. Morphology of plants and fungi. Harper & Row Publishers, New York.

Bomar, G.W. 1995. Texas weather, 2nd ed. Univ. of Texas Press, Austin.

Bondada, B.R. and **L.Q. Ma.** 2003. Tolerance of heavy metals in vascular plants: Arsenic hyperaccumulation by Chinese brake ferns (*Pteris vittata* L.). In: S. Chandra and M. Srivastava, eds. Pteridology in the New Millennium. Pp. 397–420. Kluwer, Boston, MA.

Boom, B.M. 1982. Synopsis of *Isoëtes* in the southeastern United States. Castanea 47:38–59.

Boufford, D.E. and **S.A. Spongberg.** 1983. Eastern Asian-eastern North American phytogeographical relationships–a history from the time of Linnaeus to the twentieth century. Ann. Missouri Bot. Gard. 70:423–439.

Brandt, L.A. and **D.W. Black.** 2001. Impacts of the introduced fern, *Lygodium microphyllum*, on the native vegetation of tree islands in the Arther R. Marshall Loxahatchee National Wildlife Refuge. Florida Sci. 64:191–196.

Bray, J.R. 1996. *Palhinhaea cernua* new to Arkansas. Amer. Fern J. 86:124.

Brinkhuis, H., S. Schouten, M.E. Collinson, and **numerous others.** 2006. Episodic fresh surface waters in the Eocene Arctic Ocean. Nature 441:606–609.

Brown, C.A. and **D.S. Correll.** 1942. Ferns and fern allies of Louisiana. Louisiana State Univ. Press, Baton Rouge.

Brown, L.E. 2008. A checklist of the vascular plants of the Houston area. Copy in the library of the Botanical Research Institute of Texas.

Bruce, J.G. 1976. Comparative studies in the biology of *Lycopodium carolinianum*. Amer. Fern J. 66:125–137.

Bruneton, J. 1999. Toxic plants dangerous to humans. Lavoisier Publishing, Paris, France.

Brunton, D.F. and **D.M. Britton.** 2006. *Isoetes melanopoda* spp. *silvatica* (subsp. nov.), a new quillwort (Isoetaceae) from eastern North America. Castanea 71:15–30.

Brune, T., M. Thiv, and **K. Haas.** 2008. *Equisetum* (Equisetaceae) species or hybrids? ISSR fingerprinting profiles help improve diagnoses based on morphology and anatomy. Pl. Syst. Evol. 274:67–81.

Bryant, V.M., Jr. 1977. A 16,000 year pollen record of vegetational change in central Texas. Palynology 1:143–156.

Bryant, V.M., Jr. and **R.G. Holloway.** 1985. A late-Quaternary paleoenvironmental record of Texas: An overview of the pollen evidence. In: V.M. Bryant, Jr. and R.G. Holloway, eds. Pollen records of late-Quaternary North American sediments. Pp. 39–70. Amer. Assoc. of Stratigraphic Palynologists Foundation, Dallas, TX.

Budke, J.M., R.J. Hickey, and **K.D. Heafner.** 2005. Analysis of morphological and anatomical characteristics of *Isoetes* using *Isoetes tennesseensis*. Brittonia 57:167–182.

Burlage, H.M. 1968. Index of plants of Texas with reputed medicinal and poisonous properties. Published by the author, Austin, TX.

Burrows, G.E. and **R.J. Tyrl.** 2001. Toxic plants of North America. Iowa State Univ. Press, Ames.

Bush, B.F. 1903. A list of the ferns of Texas. Bull. Torrey Bot. Club 30:343–358.

Bush, E.M. 1986. A second eastern North American occurrence for forked spleenwort, *Asplenium septentrionale*. Amer. Fern J. 76:26–28.

Bush, S.P., W.E. Kutz, and **J.M. Anderton.** 1999. RAPD variation in temperate populations of the epiphytic orchid *Epidendrum conopseum* and the epiphytic fern *Pleopeltis polypodioides*. Selbyana 20:120–124.

Butters, F.K. 1917a. Taxonomic and geographic studies in North American ferns. The genus *Athyrium* and the North American ferns allied to *A. filix-femina*. Rhodora 19:170–207.

Butters, F.K. 1917b. Taxonomic and geographic studies in North American ferns. II. *Botrychium virginianum* and its American allies. Rhodora 19:207–216.

Butterwick, M. 1979. A survey of the flora of Enchanted Rock and vicinity, Llano and Gillespie counties, Texas. In:

Enchanted Rock, a Natural Area Survey No. 14, LBJ School of Public Affairs, Univ. of Texas, Austin.

Calvert, H.E., S.K. Perkins, and **G.A. Peters.** 1983. Sporocarp structure in the heterosporus water fern *Azolla mexicana* Presl. Scan. Electron Microsc. 3:1499–1510.

Camus, J.M. 1990. Marattiaceae. In: K.U. Kramer and P.S. Green, eds. The families and genera of vascular plants, Vol. I. Pteridophytes and gymnosperms. Pp. 174–180. Springer-Verlag, Berlin, Germany.

Carlquist, S. and **E.L. Schneider.** 1997. SEM studies on vessels in ferns. 4. *Astrolepis*. Amer. Fern J. 87:43–50.

Carlquist, S. and E.L. Schneider. 1998. SEM studies on vessels in ferns. X. Selected Osmundaceae and Schizaeaceae. Int. J. Pl. Sci. 159:788–797.

Carlquist, S. and **E.L. Schneider.** 2000a. SEM studies on vessels in ferns. 18. Montane cheilanthoid ferns (Pteridaceae) of North America. Aliso 19:31–39.

Carlquist, S. and **E.L. Schneider.** 2000b. SEM studies of vessels in ferns: 14. *Ceratopteris*. Aquatic Bot. 66:1–8.

Carlson, M.C. 1954. Floral elements of the pine-oak-liquidambar forest of Montebello, Chiapas, Mexico. Bull. Torrey Bot. Club 81:387–399.

Carlson, T.J. and **W.H. Wagner, Jr.** 1982. The North American distribution of the genus *Dryopteris*. Contr. Univ. Michigan Herb. 15:141–162.

Carr, W. (William). Personal communication. Texas plant collector; previously with Texas Parks and Wildlife and The Nature Conservancy; currently a botanical consultant at Acme Botanical Services.

Carr, W. 1994. Preliminary checklist of vascular plants, Camp Maxey, Lamar County, Texas. Unpublished October 1994 draft, copy in the library of the Botanical Research Institute of Texas.

Carr, W. 2002a. An annotated list of the G3/T3 and rarer plant taxa of Texas. Texas Conservation Data Center, The Nature Conservancy of Texas. Unpublished Jul 2002 draft, copy in the library of the Botanical Research Institute of Texas.

Carr, W. 2002b. No place but Texas: An annotated list of plant taxa endemic to the lone star state. The Nature Conservancy of Texas. Unpublished Feb 2002 draft, copy in the library of the Botanical Research Institute of Texas.

Carr, W. 2002c. Plant taxa endemic to the state of Texas. The Nature Conservancy of Texas. Unpublished Mar 2002 draft, copy in the library of the Botanical Research Institute of Texas.

Carr, W. 2002d. An annotated checklist of the vascular plants of Travis County, Texas. Unpublished Feb 2002 draft, copy in the library of the Botanical Research Institute of Texas.

Chandler, A. 1941. *Pteris multifida* in Arkansas. Amer. Fern J. 31:112.

Chapman, A.W. 1883. Flora of the Southern United States. Ivison, Blakeman, Taylor and Co., New York.

Cheatham, S. and **M.C. Johnston.** 1995. The useful wild plants of Texas, the southeastern and southwestern United States, the southern Plains, and northern Mexico. Vol. 1. *Abronia–Arundo*. Useful Wild Plants, Inc., Austin, TX.

Chopra, N. 1960. Chromosome number of *Pilularia caroliniana* A. Br. Castanea 25:116–119.

Clausen, R.T. 1946. *Selaginella* subgenus *Euselaginella*, in the southeastern United States. Amer. Fern J. 36:65–82.

Christensen, C. 1930. The genus *Cyrtomium*. Amer. Fern J. 20:41–52.

Cole, K. 1985. Past rates of change, species richness, and a model of vegetational inertia in the Grand Canyon, Arizona. Amer. Naturalist 125:289–303.

Collinson, M.E. 1980. A new multiple-floated *Azolla* from the Eocene of Britain with a brief review of the genus. Palaeontology 23:213–229.

Collinson, M.E. 2001. Cainozoic ferns and their distribution. Brittonia 53:173–235.

Collinson, M.E. 2002. The ecology of Cainozoic ferns. Rev. Palaeobot. Palynol. 119:51–68.

Collinson, M.E., J. Barkeb, J. van der Burgh, J.H.A. van Konijnenburg-van Cittert, C. Heilmann-Clausen, L.E. Howard, and **H. Brinkhuib.** 2009. Did a single species of Eocene *Azolla* spread from the Arctic Basin to the southern North Sea? Rev. Paleobot. Palynol. 159:152–165.

Combs, S. 2012. Texas State Parks: Natural economic Assets. http://www.window.state.tx.us/specialrpt/parks/overview.html. Accessed Jul 2012.

Cooper-Driver, G. 1976. Chemotaxonomy and phytoecological ecology of bracken. Bot. J. Linn. Soc. 73:35–46.

Cooper-Driver, G. 1977. Chemical evidence for separating the Psilotaceae from the Filicales. Science 198:1260–1262.

Cooper-Driver, G.A. 1990. Defense strategies in bracken, *Pteridium aquilinum* (L.) Kuhn. Ann. Missouri Bot. Gard. 77:281–286.

Copeland, E.B. 1947. Genera filicum. Chronica Botanica Co., Waltham, MA.

Correll, D.S. 1949. A preliminary survey of the distribution of Texas Pteridophyta. Wrightia 1:247–278.

Correll, D.S. 1950. Native Orchids of North America, North of Mexico. Chronica Botanica Co., Waltham, MA.

Correll, D.S. 1951. *Dryopteris setigera* in Texas. Amer. Fern. J. 41:124.

Correll, D.S. 1956. Ferns and fern allies of Texas. Texas Research Foundation, Renner.

Correll, D.S. 1960a. A fern new to the United States. Wrightia 2:108.

Correll, D.S. 1960b. *Psilotum* in Texas. Wrightia 2:163–165.

Correll, D.S. 1961. Two Texas-Chihuahuan ferns. Wrightia 2:200–203.

Correll, D.S. 1965a. *Cystopteris bulbifera* (L.) Bernh. new to Texas. Amer. Fern J. 55:84–85.

Correll, D.S. 1966. Pteridophyta. In: C.L. Lundell, ed. Flora of Texas 1:3–121. Texas Research Foundation, Renner.

Correll, D.S. 1972a. Manual of the vascular plants of Texas 1. Additions and corrections. Amer. Midl. Naturalist 88:490–496.

Correll, D.S. 1978. The little runt: A botanical Christmas story. Castanea 43:185–187.

Correll, D.S. and **H.B. Correll.** 1972. Aquatic and wetland plants of southwestern United States. U.S. Environmental Protection Agency. U.S. Government Printing Office, Washington, D.C.

Correll, D.S. and **H.B. Correll.** 1982. Flora of the Bahama Archipelago, including the Turks and Caicos. J. Cramer in A.R.G. Gantner Verlag K.-G., Vaduz, Liechtenstein.

Correll, D.S. and **M.C. Johnston.** 1970. Manual of the vascular plants of Texas. Texas Research Foundation, Renner.

Correll, D.S. and **I.W. Knobloch.** 1962. Ferns and fern allies of Chihuahua, Mexico. Texas Research Foundation, Renner, TX.

Cory, V.L. and **H.B. Parks.** 1937. Catalogue of the flora of the state of Texas. Texas Agric. Exp. Sta. Bull. No. 550.

Croxdale, J.G. 1978. *Salvinia* leaves. I. Origin and early morphogenesis of floating and submerged leaves. Canad. J. Bot. 56:1982–1991.

Croxdale, J.G. 1979. *Salvinia* leaves. II. Morphogenesis of the floating leaf. Canad. J. Bot. 57:1951–1959.

Croxdale, J.G. 1981. *Salvinia* leaves. III. Morphogenesis of the submerged leaf. Canad. J. Bot. 59:2065–2072.

Coulter, J.M. 1890. Upon a collection of plants made by Mr. G.C. Nealley in the region of the Rio Grande, in Texas, from Brazos Santiago to El Paso County. Contr. U.S. Natl. Herb. 1:29–61.

Crane, P.R. and **P. Kenrick.** 1997. Problems in cladistic classification: Higher-level relationships in land plants. Aliso 15:87–104.

Cranfill, R. 1981. Bog clubmosses (*Lycopodiella*) in Kentucky. Amer. Fern J. 71:97–100.

Cranfill, R. 1983. The distribution of *Woodwardia areolata*. Amer. Fern J. 73:46–52.

Cranfill, R.B. 1993a. Dennstaedtiaceae. In: Flora of North America Editorial Committee, eds. Fl. North Amer. 2:198–199. Oxford Univ. Press, New York and Oxford.

Cranfill, R.B. 1993b. Blechnaceae. In: Flora of North America Editorial Committee, eds. Fl. North Amer. 2:223–227. Oxford Univ. Press, New York and Oxford.

Cranfill, R. and **M. Kato.** 2003. Phylogenetics, biogeography, and classification of the woodwardioid ferns (Blechnaceae). In: S. Chandra and M. Srivastava, eds. Pteridology in the New Millennium. Pp. 25–48. Kluwer, Boston, MA.

Creagh, C. 1991. A marauding weed in check. Ecos 70:26–29.

Crosthwaite, S. (Sally). 1916–2000. Personal communication. Botanist trained at Vanderbilt and Columbia, plant collector in North Central TX, and former instructor at Austin College, Sherman, TX.

Culwell, D.E. 1994. Large population of *Pilularia americana* in Arkansas. Amer. Fern J. 84:133–134.

Darwin Azolla Project. 2009. Introduction: Darwin Azolla Project. http://www3.bio.uu.nl/palaeo/Azolla/introduction.htm. Accessed Jul. 2009.

Davenport, G.E. 1883. Distribution of ferns in the United States. Proc. Amer. Phil. Soc. 20:605–612.

Degener, O. and **A.D. Hawkes.** 1951. The holly fern. Amer. Fern J. 41:117–120.

de la Sota, E.R. 1973. On the classification and phylogeny of the Polypodiaceae. In: A.C. Jermy, J.A. Crabbe, and B.A. Thomas, eds. The phylogeny and classification of the ferns. Bot. J. Linn. Soc. 67, Suppl. 1:229–244.

de la Sota, E.R. and **L.A. Cassa de Pazos.** 2001. Two cytotypes and a new hybrid in *Salvinia* Seguier. Acta Amazonica 31:557–564. http://acta.inpa.gov.br/fasciculos/31-4/PDF/v31n4a04.pdf. Accessed Jul 2011.

Dennis, W.M. and **D.H. Webb.** 1981. The distribution of *Pilularia americana* A. Br. (Marsileaceae) in North America, north of Mexico. Sida 9:19–24.

Der, J.P., J.A. Thomson, J.K. Stratford, and **P.G. Wolf.** 2009. Global chloroplast phylogeny and biogeography of bracken (*Pteridium*; Dennstaedtiaceae). Amer. J. Bot. 96:1041–1049.

Des Marais, D.L., A.R. Smith, D.M. Britton, and **K.M. Pryer.** 2003. Phylogenetic relationships and evolution of extant horsetails, *Equisetum*, based on chloroplast DNA sequence data (rbcl and trnL-F). Int. J. Pl. Sci. 164:737–751.

Dias, E., C. Mendes, C. Melo, D. Pereira, and **R. Elias.** 2005. Azores central islands vegetation and flora field guide. Quercetea 7:123–173.

Dickinson, M.B. and **T.E. Miller.** 1998. Competition among small, free-floating, aquatic plants. Amer. Midl. Naturalist 140:55–67.

Diggs, G.M., Jr., B.L. Lipscomb, and **R.J. O'Kennon.** 1999. Shinners and Mahler's illustrated flora of North Central Texas. Sida, Bot. Misc. 16.

Diggs, G.M., Jr., B.L. Lipscomb, M.D. Reed, and **R.J. O'Kennon.** 2006. Illustrated flora of East Texas, Vol. 1 (Introduction, Ferns, Gymnosperms, Monocots). Sida, Bot. Misc. 26.

Diggs, G.M., Jr. and **R.J. O'Kennon.** 2003. A rockhouse microhabitat in the West Cross Timbers of North Central Texas. Sida 20:1327–1333.

Dilcher, D. 2000. Geological history of the vegetation in southeast United States. In: B.L. Lipscomb, J.J. Pipoly, III, and R.W. Sanders. Floristics in the new millennium: Proceedings of the flora of the southeast US symposium. Sida, Bot. Misc. 18:1–21.

DiMichele, W.A. and **J.E. Skog.** 1992. The Lycopsida: A symposium. Ann. Missouri Bot. Gard. 79:447–449.

Dimmitt, M. 2009. Adaptations of desert plants. In: Encyclopedia of Earth. Environmental Information Coalition, National Council for Science and the Environment. http://www.eoearth.org/article/Adaptations_of_desert_plants. Accessed Aug 2010.

Dittmer, H.J., E.F. Castetter, and **O.M. Clark.** 1954. The ferns and fern allies of New Mexico. Univ. of New Mexico Press, Albuquerque.

Do, L.H., R.D. Gooch, J.R. Stevens, W.C. Holmes, and **J.R. Singhurst.** 1996. New county records of *Botrychium lunarioides* in Texas. Amer. Fern J. 86:28–31

Donoghue, M.J., C.D. Bell, and **J. Li.** 2001. Phylogenetic patterns in northern hemisphere plant geography. Int. J. Pl. Sci. 162(6 Suppl.):S41–S52.

Doyle, J.A. 1998. Phylogeny of vascular plants. In: D.G. Fautin, ed. Ann. Rev. Ecol. Syst. 29:567–599.

Driscoll, H.E., D.S. Barrington, and **A.V. Gilman.** 2004. A reexamination of the apogamous tetraploid *Phegopteris* (Thelypteridaceae) from northeastern North America. Rhodora 105:309–321.

Duek, J.J. 1978. A taxonomic revision of *Lygodium* (Filicinae) in America. Feddes Repert. 89:411–423.

Duek, J.J. 1980. A taxonomical monograph of *Anemia* subgenus *Anemorrhiza* (Filicinae). Feddes Repert. 91:69–87.

Duff, R.J. 2000. The chloroplast genome structure of the vascular plant *Isoetes* is similar to that of the liverwort *Marchantia*. Amer. Fern J. 90:51–59.

Durant, M. 1976. Who named the daisy? Who named the rose? A roving dictionary of North American wild flowers. Congdon and Weed, Inc., New York.

Dyer, A.F. and **S. Lindsay.** 1992. Soil spore banks of temperate ferns. Amer. Fern J. 82:89–123.

Earl, J.W. and **B.V. McCleary.** 1994. Mystery of the poisoned expedition. Nature 368:683–684.

Eaton, D.C. 1878. Ferns of the southwest: An account of the ferns which have been collected in so much of the territory of the United States of America as is west of the 105th degree of w. longitude, and south of the 40th degree of n. latitude. In: G.M. Wheeler. Report upon the United States geographical surveys west of the one hundredth meridian. Vol. VI—Botany. U.S. Government Printing Office, Washington, D.C.

Elder, H. (Howard). Personal communication. Aquatic habitat biologist at Texas Parks and Wildlife, Inland Fisheries Division, Jasper, TX.

Evrard, C. and **C. van Hove.** 2004. Taxonomy of the American *Azolla* species (Azollaceae): A critical review. Syst. Geogr. Pl. 74:301–318.

Emory, D.L. 1970. A major North American range extension for the Forked Spleenwort. Amer. Fern J. 60:132–133.

Eriksson, T. 2004. Evolutionary biology: Ferns reawakened. Nature 428:480–481.

Farrar, D.R. 1998. The tropical flora of rockhouse cliff formations in the eastern United States. Bull. Torrey Bot. Soc. 125:91–108.

Fernald, M.L. 1950a. Gray's manual of botany, 8th ed. Reprinted 1987. Dioscorides Press, Portland, OR.

Fernald, M.L. 1950b. *Adiantum capillus-veneris* in the United States. Rhodora 52:201–208.

Fernie, W.T. 2008. Herbal simples approved for modern uses of cure. Echo Library, Gloucestershire, England, U.K. Originally published 1895.

Filippini-de Giorgi, A., R. Holderegger, and **J.J. Schneller.** 1997. Aspects of spore dispersal in *Selaginella*. Amer. Fern J. 87:93–103.

Fleenor, S.B. and **S.W. Taber.** 2009. Plants of central Texas wetlands. Texas Tech Univ. Press, Lubbock.

FLEPPC. 2009. Florida Exotic Pest Plant Council. http://www.fleppc.org/list/List-WW-F09-final.pdf. Accessed Aug 2010.

Flora of North America Editorial Committee, eds. 1993. Flora of North America north of Mexico, Vol. 2, Pteridophytes and Gymnosperms. Oxford Univ. Press, New York and Oxford.

Flores, D. and **J.W. Carlson.** 2006. Biological control of giant *Salvinia* in east Texas waterways and the impact on dissolved oxygen levels. J. Aquat. Pl. Managem. 44:115–121. http://www.apms.org/japm/vol44/v44p115.pdf. Accessed Jul 2010.

Forno, I.W. 1983. Native distribution of the *Salvinia auriculata* complex and keys to species identification. Aquatic Bot. 17:71–83.

Forno, I.W. and **K.L.S. Harley.** 1979. The occurrence of *Salvinia molesta* in Brazil. Aquatic Bot. 6:185–187.

Foster, S. and **R. Caras.** 1994. Venomous amimals and poisonous plants of North America north of Mexico. Peterson Field Guide Series, Houghton Mifflin Co., Boston, MA and New York.

Fowler, K. 1975. An escape mechanism for spermatozoids in *Azolla* massulae. Amer. Fern J. 65:7–10.

Fraser-Jenkins, C.R. 1986. A classification of the genus *Dryopteris* (Pteridophyta: Dryopteridaceae). Bull. Brit. Mus. (Nat. Hist.), Bot. 14:183–218.

Frederiksen, N.O. 1995. Differing Eocene floral histories in southeastern North America and Western Europe: Influence of paleogeography. Historical Biol. 10:13–23.

Frohne, D. and **H.J. Pfänder.** 2005. Poisonous plants: A handbook for doctors, pharmacists, toxicologists, biologists and veterinarians, 2nd ed. Timber Press, Portland, OR.

Fuller, T.C. and **E. McClintock.** 1986. Poisonous plants of California. Univ. of California Press, Berkeley.

Garbari, F., A. Giovannini, and **D. Marchetti.** 2000. *Salvinia molesta* D.S. Mitchell (Salviniaceae) nuova per la flora d'Italia. Arch. Geobot. 6:73–78.

Gastrony, G.J. 1988. The *Pellaea glabella* complex: Electrophoretic evidence for the derivations of the agamosporous taxa and a revised taxonomy. Amer. Fern J. 78:44–67.

Gastrony, G.J. and **C.H. Haufler.** 1976. Chromosome numbers and apomixis in the fern genus *Bommeria* (Gymnogrammaceae). Biotropica 8:1–11.

Gastony, G.J. and **D.R. Rollo.** 1995. Phylogeny and generic circumscriptions of cheilanthoid ferns (Pteridaceae: Cheilanthoideae) inferred from rbcL nucleotide sequences. Amer. Fern J. 85:341–360.

Gastony, G.J. and **D.R. Rollo.** 1998. Cheilanthoid ferns (Pteridaceae: Cheilanthoideae) in the southwestern United States and adjacent Mexico—A molecular phylogenetic reassessment of generic lines. Aliso 17:131–144.

Gastony, G. and **M. Ungerer.** 1997. Molecular systematics and a revised taxonomy of the onocleoid ferns (Dryopteridaceae: Onocleeae). Amer. J. Bot. 84:840–849.

Gastony, G.J. and **M.D. Windham.** 1989. Species concepts in pteridophytes: The treatment and definition of agamosporous species. Amer. Fern J. 79:65–77.

Gastony, G.J., G. Yatskievych, and **C.K. Dixon.** 1992. Chloroplast DNA restriction site variation in the fern genus *Pellaea*: Phylogenetic relationships of the *Pellaea glabella* complex. Amer. J. Bot. 79:1072–1080.

Geiger, J.M.O., T.A. Ranker, J.M. Ramp Neale, and **S.T. Klimas.** 2007. Molecular biogeography and origins of the Hawaiian fern flora. Brittonia 59:142–158.

Gensel, P.G. 1977. Morphologic and taxonomic relationships of the Psilotaceae relative to evolutionary lines in early land vascular plants. Brittonia 29:14–29.

Gensel, P.G. 1992. Phylogenetic relationships of the zosterophylls and lycopsids: Evidence from morphology, paleoecology, and cladistic methods of inference. Ann. Missouri Bot. Gard. 79:450–473.

Gensel, P.G. and **C.M. Berry.** 2001. Early lycophyte evolution. Amer. Fern. J. 91:74–98.

Given, D.R. 1993. Changing aspects of endemism and endangerment in pteridophyta. J. Biogeogr. 20:293–302.

Gledhill, D. 2008. The names of plants. Cambridge Univ. Press, Cambridge, England, U.K.

Gliessman, S.R. and **C.H. Muller.** 1978. The allelopathic mechanisms of dominance in bracken (*Pteridium aquilinum*) in southern California. J. Chem. Ecol. 4:337–362.

Gorham, E. 1957. Development of peatlands. Q. Rev.Biol. 32:145–166.

Gorman, C.E., M.S. Bruton, and **L.D. Estes.** 2011. *Macrothelypteris torresiana* (Thelypteriaceae) new to Kentucky. J. Bot. Res. Inst. Texas 5:343–344.

Gould, S.J. 1977. Ontogeny and phylogeny. Harvard Univ. Press, Cambridge, MA.

Graham, A. 1993a. History of the vegetation: Cretaceous (Maastrichtian)–Tertiary. In: Flora of North America Editorial Committee, eds. Fl. North Amer. 1:57–70. Oxford Univ. Press, New York and Oxford.

Graham, A. 1993b. Historical factors and biological diversity in Mexico. In: T.P. Ramamoorthy, R. Bye, A. Lot, and J. Fa, eds. Biological diversity of Mexico: Origins and distribution. Pp. 109–127. Oxford Univ. Press, New York and Oxford.

Graham, A. 1999a. Late Cretaceous and Cenozoic history of North American vegetation north of Mexico. Oxford Univ. Press, New York and Oxford.

Graham, A. 1999b. The Tertiary History of the Northern Temperate Element in the Northern Latin American Biota. Amer. J. Bot. 86:32–38.

Graham, A. and **C. Heimsch.** 1960. Pollen studies of some Texas peat deposits. Ecology 41:751–763.

Grant, M. 1995. *Marsilea* poisoning in 19th century Australia. Pteridologist 2:255–256.

Greenwood, D.R. and **J.F. Basinger.** 1994. The paleoecology of high latitude Eocene swamp forests from Axel Heiberg Island, Canadian High Arctic. Rev. Palaeobot. Palynol. 81:83–97.

Greenwood, D.R. and **J.F. Basinger.** 1993. Stratigraphy and floristics of Eocene swamp forests from Axel Heiberg Island, Canadian Arctic Archipelago. Canadian J. Earth Sci. 30:1914–1923.

Greer, G.K. and **B.C. McCarthy.** 2000. Patterns of growth and reproduction in a natural population of the fern *Polystichum acrostichoides*. Amer. Fern J. 90:60–76.

Grusz, A.L., M.D. Windham, and **K.M. Pryer.** 2009. Deciphering the origins of apomictic polyploids in the *Cheilanthes yavapensis* complex (Pteridaceae). Amer. J. Bot. 96:1636–1645.

Guillon, J.M. 2004. Phylogeny of horsetails (*Equisetum*) based on the chloroplast rps4 gene and adjacent noncoding sequences. Syst. Bot. 29:251–259.

Guillon, J.M. 2007. Molecular phylogeny of horsetails (*Equisetum*) including chloroplast atpB sequences. J. Pl. Res. 120:569–574.

Gupta, K.M. 1957. Some American species of *Marsilea* with special reference to their epidermal and soral characters. Madroño 14:113–127.

Gutierrez, R., Jr. 2007. Vascular plants of Arizona: Psilotaceae. Canotia 3:32–34. http://www.canotia.org/volumes/CANOTIA_2007_Vol3_2_Guitierrez_Psilotaceae.pdf. Accessed Jun 2011.

Hall, C.C. 1950. *Notholaena copelandii*, a newly recognized species of the Texano-Mexican region. Amer. Fern J. 40:178–187.

Hannen, H.H. 1969. The introduction and establishment of *Ceratopteris* in Texas. Amer. Fern J. 59:122.

Hansen, B.F. and **R.P. Wunderlin.** 1998. Two new combinations in Florida selaginellas. Phytologia 85:300–302.

Harvey, W.D. 1998. Harmful or potentially harmful fish, shellfish, and aquatic plants. In: Texas Administrative Code. Title 31, Part 2 (Texas Parks and Wildlife Department regulations). Chapter 57, subchapter A, section 57.111–112.

Hasebe, M. et al. (multiple authors). 1995. Fern phylogeny based on rbcL nucleotide sequences. Amer. Fern J. 85:134–181.

Hatch, S.L. 1995. *Salvinia minima* (Salviniaceae), new to Texas. Sida 16:595.

Hatch, S.L., K.N. Gandhi, and **L.E. Brown.** 1990. Checklist of the vascular plants of Texas. Texas Agric. Exp. Sta. Misc. Publ. 1655:1–158.

Haufler, C.H. 1979. A biosystematic revision of *Bommeria*. J. Arnold Arbor. 60:445–476.

Haufler, C.H. 1985. Enzyme variability and modes of evolution in *Bommeria* (Pteridaceae). Syst. Bot. 10:92–104.

Haufler, C.H. 1987. Electrophoresis is modifying our concepts of evolution in homosporous pteridophytes. Amer. J. Bot. 74:953–966.

Haufler, C.H. 1993. *Bommeria*. In: Flora of North America Editorial Committee, eds. Fl. North Amer. 2:151–152. Oxford Univ. Press, New York and Oxford.

Haufler, C.H. 1996. Species concepts and speciation in pteridophytes. In: J.M. Camus, M. Gibby, and R.J. Johns, eds.

Pteridology in Perspective. Pp. 291–305. Royal Botanic Gardens, Kew, England, U.K.

Haufler, C.H. 2007. Genetics, phylogenetics, and biogeography: Considering how shifting paradigms and continents influence fern diversity. Brittonia 59:108–114.

Haufler, C.H. 2008. Species and speciation. In: T.A. Ranker and C.H. Haufler, eds. The biology and evolution of ferns and lycophytes. Ch. 12. Pp. 303–331. Cambridge Univ. Press, Cambridge, England, U.K.

Haufler, C.H. and **D.E. Giannasi.** 1982. A chemosystematic survey of the fern genus *Bommeria*. Biochem. Syst. Ecol. 10:107–110.

Haufler, C.H., R.C. Moran, and **M.D. Windham.** 1993. *Cystopteris*. In: Flora of North America Editorial Committee, eds. Fl. North Amer. 2:263–270. Oxford Univ. Press, New York and Oxford.

Haufler, C.H. and **D.E. Soltis.** 1984. Obligate outcrossing in a homosporous fern: Field confirmation of a laboratory prediction. Amer. J. Bot. 71:878–881.

Haufler, C.H. and **C.B. Welling.** 1994. Antheridiogen, dark spore germination, and outcrossing mechanisms in *Bommeria* (Adiantaceae). Amer. J. Bot. 81:616–621.

Haufler, C.H. and **M.D. Windham.** 1991. New species of North American *Cystopteris* and *Polypodium*, with comments on their reticulate relationships. Amer. Fern J. 81:7–23.

Haufler, C.H., M.D. Windham, D.M. Britton, and **S.J. Robinson.** 1985. Triploidy and its evolutionary significance in *Cystopteris protrusa*. Canad. J. Bot. 63:1855–1863.

Haufler, C.H., M.D. Windham, and **T.A. Ranker.** 1990. Biosystematic analysis of the *Cystopteris tennesseensis* (Dryopteridaceae) complex. Ann. Missouri Bot. Gard. 77:314–329.

Hauk, W.D. 1995. A molecular assessment of relationships among cryptic species of *Botrychium* subgenus *Botrychium* (Ophioglossaceae). Amer. Fern J. 85:375–394.

Hauk, W.D. and **C.H. Haufler.** 1999. Isozyme variability among cryptic species of *Botrychium* subgenus *Botrychium* (Ophioglossaceae). Amer. J. Bot. 86:614–633.

Hauk, W.D., C.R. Parks, C.H. Haufler, and **M.W. Chase.** 2003. Phylogenetic studies of Ophioglossaceae: Evidence from *rbcL* and *trnL-F* plastid DNA sequences and morphology. Molec. Phylogen. Evol. 28:131–151.

Hauke, R.L. 1960. The smooth scouring rush and its complexities. Amer. Fern. J. 5:185–193.

Hauke, R.L. 1963. A taxonomic monograph of the genus *Equisetum* subgenus *Hippochaete*. Beih. Nova Hedwigia 8:1–123.

Hauke, R.L. 1978. A taxonomic monograph of *Equisetum* subgenus *Equisetum*. Nova Hedwigia 30:385–455.

Hauke, R.L. 1990. Equisetaceae. In: K.U. Kramer and P.S. Green, eds. The families and genera of vascular plants, Vol. I. Pteridophytes and gymnosperms. Pp. 46–48. Springer-Verlag, Berlin, Germany.

Hauke, R.L. 1993. Equisetaceae. In: Flora of North America Editorial Committee, eds. Fl. North Amer. 2:76–84. Oxford Univ. Press, New York and Oxford.

Heafner, K.D. 1997. Taxonomic studies of southeastern quillworts (*Isoetes*). Chinquapin 5(3):27.

Heafner, K.D. 2001. *Pellaea wrightiana* Hooker (Pteridaceae) in North Carolina revisited with a new record for eastern North America and a key to *Pellaea* species in the Carolinas. Castanea 66:319–326.

Heafner, K.D. and **R.D. Bray.** 2005. Taxonomic reassessment of North American granite outcrop *Isoetes* species with emphasis on vegetative morphology and *I. piedmontana* (Pfeiffer) Reed sensu lato. Castanea 70:204–221.

Helton, R. (Randy). 1999. Personal communication. Conservation Scientist at Texas Parks and Wildlife, Inland Fisheries Division, Jasper, TX.

Helton, R.J. and **E. Chilton.** 2001. Giant salvinia (*Salvinia molesta* D.S. Mitchell) in Texas: A destructive menace. Aquatics (Tallahassee) 23:12–18.

Hennipman, E., P. Veldhoen, K.U. Kramer, and **M.G. Price.** 1990. Polypodiaceae. In: K.U. Kramer and P.S. Green, eds. The families and genera of vascular plants, Vol. I. Pteridophytes and gymnosperms. Pp. 203–230. Springer-Verlag, Berlin, Germany.

Hernandez-Castillo, G.R., R.A. Stockey, and **G.W. Rothwell.** 2006. *Anemia quatsinoensis* sp. nov. (Schizaeaceae), a permineralized fern from the Lower Cretaceous of Vancouver Island. Int. J. Pl. Sci. 167:665–674.

Herzog, T.K.J. 1935. Ein Beitrag zur Systematik der Gattung *Salvinia*. Hedwigia 74:257–284.

Hevly, R.H. 1963. Adaptations of cheilanthoid ferns to desert environments. J. Arizona Acad. Sci. 2:164–175.

Hewitson, W. 1962. Comparative morphology of the Osmundaceae. Ann. Missouri Bot. Gard. 49:57–93.

Hickey, R.J. 1986a. *Isoetes* megaspore surface morphology: Nomenclature, variation, and systematic importance. Amer. Fern J. 76:1–16.

Hickey, R.J. 1986b. The early evolutionary and morphological diversity of *Isoetes*, with description of two neotropical species. Sys. Bot. 11:309–321.

Hickey, R.J., C. Macluf, and **W.C. Taylor.** 2003. A reevaluation of *Isoetes savatieri* Franchet in Argentina and Chile. Amer. Fern J. 93:126–136.

Hickman, J.C., ed. 1993. The Jepson manual: Higher plants of California. Univ. of California Press, Berkeley.

Hietz, P. 2010. Fern adaptations to xeric environments. In: K. Mehltreter, L.R. Walker, and J.M. Sharpe, eds. 2010. Fern ecology. Ch. 3. Pp. 140–176. Cambridge Univ. Press, Cambridge, England, U.K.

Hill, S.R. 1980. A new country record for *Pilularia americana* in Texas. Amer. Fern. J. 70:28.

Hillebrand, H. 2004. On the generality of the latitudinal diversity gradient. Amer. Naturalist 163:192–211.

Hodge, W.H. 1973. Fern foods of Japan and the problem of toxicity. Amer. Fern J. 63:77–80.

Hodgdon, A.R. 1951. Is *Onoclea sensibilis* poisonous to horses? Amer. Fern J. 41:61–62.

Holderegger, R. and **J.J. Schneller.** 1994. Are small isolated populations of *Asplenium septentrionale* variable? Biol. J. Linn. Soc. 51:377–385.

Holmes, W.C., T.L. Morgan, J.R. Stevens, and **R.D. Gooch.** 1996. Comments on the distribution of *Botrychium lunarioides* (Ophioglossaceae) in Texas. Phytologia 80:280–283.

Holmes, W.C., A.E. Rushing, and **J.R. Singhurst.** 2005. Taxonomy and identification of *Isoetes* (Isoetaceae) in Texas based on megaspore features. Lundellia 8:1–6.

Holmes, W.C., J.R. Singhurst, and **J.N. Mink.** 2011. A commentary on *Cheilanthes lanosa* (Pteridaceae) in Texas. Phytoneuron 2011-34:1–5.

Holttum, R.E. 1969. Studies in the family Thelypteridaceae. The genera *Phegopteris*, *Pseudophegopteris*, and *Macrothelypteris*. Blumea 17:5–32.

Holub, J. 1983. Validation of generic names in Lycopodiaceae: With a description of a new genus *Pseudolycopodiella*. Folia Geobot. Phytotax. 18:439–442.

Hooker, W.J. 1860. *Llavea cordifolia* (Cordate-leaved Llavea). Curtis's Bot. Mag. 86:pl. 5159.

Hooper, E.A. 1995. New combinations in the *Pleopeltis macrocarpa* group (Polypodiaceae: Polypodieae). Amer. Fern. J. 85:75–82.

Hoot, S.B. and **W.C. Taylor.** 2001. The utility of nuclear ITS, a *LEAFY* homolog intron, and chloroplast *atpB-rbcL* spacer region data in phylogenetic analyses and species delimitation in *Isoëtes*. Amer. Fern J. 91:166–177.

Hoot, S.B., N.S. Napier, and **W.C. Taylor.** 2004. Revealing unknown or extinct lineages within *Isoetes* (Isoëtaceae) using DNA sequences from hybrids. Amer. J. Bot. 91:899–904.

Hoot, S.B., W.C. Taylor, and **N.S. Napier.** 2006. Phylogeny and biogeography of *Isoëtes* based on nuclear and chloroplast DNA sequence data. Syst. Bot. 31:449–460.

Hoshizaki, B.J. and **R.C. Moran.** 2001. Fern grower's manual (revised and expanded ed.). Timber Press, Portland, OR.

Hoshizaki, B.J. and **K.A. Wilson.** 1999. The cultivated species of the fern genus *Dryopteris* in the United States. Amer. Fern J. 89:1–98.

Howard, V. 2010. *Salvinia molesta*. USGS nonindigenous aquatic species database, Gainesville, FL. http://nas.er.usgs.gov/queries/factsheet.aspx?SpeciesID=298. Accessed May 2010.

Huang, T.-C. et al. (Editorial Committee of the flora of Taiwan). 1994. Flora of Taiwan, 2nd ed., Vol. 1 (Pteridophyta and Gymnosperms). Editorial Committee of the Flora of Taiwan, Taipei, Taiwan, ROC.

Hughes, J.S. and **S.P. Otto.** 1999. Ecology and evolution of biphasic life cycles. Amer. Naturalist 154:306–320.

Jacobs, C.A. and **J.H. Peck.** 1993. *Pteridium*. In: Flora of North America Editorial Committee, eds. Fl. North Amer. 2:201–204. Oxford Univ. Press, New York and Oxford.

Jacobsen, W.B.G. 1983. The ferns and fern allies of southern Africa. Butterworths, Durban/Pretoria, South Africa.

Jacono, C. 1999a. Have you seen this plant? Texas Parks and Wildlife information sheet. Texas Parks and Wildlife, Austin.

Jacono, C. 1999b. *Salvinia molesta* (Salviniaceae), new to Texas and Louisiana. Sida 18:927–928.

Jacono, C. 1999c. *Salvinia molesta* (D.S. Mitchell) invades the United States! Aquatics 21:4–9.

Jacono, C. 2001a. *Salvinia molesta*: Giant salvinia. http://nas.er.usgs.gov/ferns/. Accessed 2001.

Jacono, C. 2001b. Toledo Bend Bi-State Alliance. http://www.toledo-bend.com/alliance/docs/salvinia2.html. Accessed 2001. http://nas.er.usgs.gov/ferns/.

Jacono, C., T.R. Davern, and **T.D. Center.** 2001. The adventive status of *Salvinia minima* and *S. molesta* in the southern United States and the related distribution of the weevil *Cyrtobagous salviniae*. Castanea 66:214–226. http://fl.biology.usgs.gov/posters/Nonindigenous/Status_of_Salvinia/status_of_salvinia.html. Accessed Apr 2010.

Jacono, C.C. and **D.M. Johnson.** 2006. Water-cloverferns, *Marsilea*, in the southeastern United States. Castanea 71:1–14.

Jacono, C.C. and **M. M. Richerson.** 2005. Giant salvinia - *Salvinia molesta* [Distribution map]. USGS. http://salvinia.er.usgs.gov/html/distribution_map.html. Accessed Jun 2011.

Jahren, A.H. 2007. The Arctic forest of the Middle Eocene. Ann. Rev. Earth Planet. Sci. 35:509–540.

Jahren, A.H. and **L.S.L. Sternberg.** 2003. Humidity estimate for the middle Eocene Arctic rain forest. Geology 31:463–466.

Jansson, I.-M., S. McLoughlin, V. Vajda, and **M. Pole.** 2008. An early Jurassic flora from the Clarence-Moreton Basin, Australia. Rev. Palaebot. Palynol. 150:5–21.

Jaramillo, I.R., B. Pérez-García, and **A. Mendoza.** 2000. Fase gametofítica del helecho *Llavea cordifolia* (Pteridaceae). Rev. Biol. Trop. 48:19–23.

Jermy, A.C. 1990a. Isoetaceae. In: K.U. Kramer and P.S. Green, eds. The families and genera of vascular plants, Vol. I. Pteridophytes and gymnosperms. Pp. 26–31. Springer-Verlag, Berlin, Germany.

Jermy, A.C. 1990b. Selaginellaceae. In: K.U. Kramer and P.S. Green, eds. The families and genera of vascular plants, Vol. I. Pteridophytes and gymnosperms. Pp. 39–45. Springer-Verlag, Berlin, Germany.

Jermy, A.C., J.A. Crabbe, and **B.A. Thomas,** eds. 1973. The phylogeny and classification of the ferns. Bot. J. Linn. Soc. 67, Suppl. 1.

Johnson, D.M. 1985. New records for longevity of *Marsilea* sporocarps. Amer. Fern J. 75:30–31.

Johnson, D.M. 1986. Systematics of the New World species of *Marsilea* (Marsileaceae). Syst. Bot. Monogr. 11:1–87.

Johnson, D.M. 1988. Proposal to conserve *Marsilea* L. (Pteridophyta: Marsileaceae) with *Marsilea quadrifolia* as *typ. conserv*. Taxon 37:483–486.

Johnson, D.M. 1993a. Marsileaceae. In: Flora of North America Editorial Committee, eds. Fl. North Amer. 2:331–335. Oxford Univ. Press, New York and Oxford.

Johnson, D.M. 1993b. *Onoclea*. In: Flora of North America Editorial Committee, eds. Fl. North Amer. 2:251. Oxford Univ. Press, New York and Oxford.

Johnson, D.M. 1993c. *Matteuccia*. In: Flora of North America Editorial Committee, eds. Fl. North Amer. 2:249–250. Oxford Univ. Press, New York and Oxford.

Johnston, M.C. 1990. The vascular plants of Texas: A list, updating the manual of the vascular plants of Texas, 2nd ed. Published by the author, Austin, TX.

Jones, D.L. 1987. Encyclopaedia of ferns: An introduction to ferns, their structure, biology, economic importance, cultivation and propagation. Timber Press, Portland, OR.

Jones, G.N. 1955. On the occurrence of *Pteris multifida* in the United States. Amer. Fern J. 45:107–113.

Jones, R.L. 2005. Plant life of Kentucky: An illustrated guide to the vascular flora. Univ. of Kentucky Press, Lexington.

Jones, S.B. and **A.E. Luchsinger.** 1986. Plant systematics, 2nd ed. McGraw-Hill Book Co., New York.

Jones, S.D., J.K. Wipff, and **P.M. Montgomery.** 1997. Vascular plants of Texas: A comprehensive checklist including synonymy, bibliography, and index. Univ. of Texas Press, Austin.

Jud, N.A., G.W. Rothwell, and **R.A. Stockey.** 2008. *Todea* from the Lower Cretaceous of western North America: Implications for the phylogeny, systematics, and evolution of modern Osmundaceae. Amer. J. Bot. 95:330–339.

Julien, M.H., T.D. Center, and **P.W. Tipping.** 2002. Floating fern (*Salvinia*). http://dnr.state.il.us/stewardship/cd/biocontrol/2FloatingFern.html. Accessed Aug 2010.

Kaplan, D.R. 1977. Morphological status of the shoot systems of Psilotaceae. Brittonia 29:30–53.

Kartesz, J.T. 1994. A synonymized checklist of the vascular flora of the United States, Canada, and Greenland, 2nd ed. (2 vols.). Timber Press, Portland, OR.

Kartesz, J.T. 1999. A synonymized checklist and atlas with biological attributes for the vascular flora of the United States, Canada, and Greenland, 1st ed. In: J.T. Kartesz and C.A. Meacham. Synthesis of the North American Flora, CD format, Version 1.0. North Carolina Botanical Garden, Chapel Hill.

Kato, M. 1988. The phylogenetic relationship of Ophioglossaceae. Taxon 37:381–386.

Kato, M. 1993a. *Athyrium*. In: Flora of North America Editorial Committee, eds. Fl. North Amer. 2:255–258. Oxford Univ. Press, New York and Oxford.

Kato, M. 1993b. Biogeography of ferns: Dispersal and vicariance. J. Biogeogr. 20:265–274.

Kato, M. 2007. Distribution of Osmundaceae. Bull. Natl. Mus. Nat. Sci., Ser. B 33:81–90.

Kato, M. and **K. Iwatsuki.** 1983. Phytogeographic relationships of pteridophytes between temperate North America and Japan. Ann. Missouri Bot. Gard. 70:724–733.

Kaul, R.B., D.M. Sutherland, and **S.B. Rolfsmeier.** 2006. The flora of Nebraska: Keys, descriptions, and distributional maps of all native and introduces species that grow outside cultivation. Conservation and Survey Division, School of Natural Resources, Institute of Agriculture and Natural Resources, Univ. of Nebraska-Lincoln.

Kaur, S. 1974. The family Lomariopsidaceae (Filicopsida) and its probable ancestors. Bot. J. Linn. Soc. 68:153–162.

Kawai, H., T. Kanegae, S. Christensn, T. Kiyosue, Y. Sato, T. Imazumi, A. Kadota, and **M. Wada.** 2003. Responses of ferns to red light are mediated by an unconventional photoreceptor. Nature 421:287–290.

Keith, E.L. (Eric). Personal communication. Environmental consultant, East Texas plant collector, and expert on *Yucca* in East TX. Raven Environmental Services, Huntsville, TX.

Kelloff, C.L., J. Skog, L. Adamkewicz, and **C.R. Werth.** 2002. Differentiation of eastern North American *Athyrium filix-femina* taxa: Evidence from allozymes and spores. Amer. Fern J. 92:185–213.

Kendrick, P. and **P.R. Crane.** 1997. The origin and early diversification of land plants. Smithsonian Institution Press, Washington, D.C.

Kessler, M. 2010. Biogeography of ferns. In: K. Mehltreter, L.R. Walker, and J.M. Sharpe, eds. 2010. Fern ecology. Ch. 2. Pp. 21–60. Cambridge Univ. Press, Cambridge, England, U.K.

Khandelwal, S. 1990. Chromosome evolution in the genus *Ophioglossum* L. Bot. J. Linn. Soc. 102:205–217.

King, L.C. 1961. "Visible" monomolecular films: Estimation of molecular size and Avogadro's number. J. Chem. Educ. 38:A751.

King, L.C. and **E. K. Neilsen.** 1958. Estimation of Avogadro's number: An experiment for general chemistry laboratory. J. Chem. Educ. 35:198–200.

Kingsbury, J.M. 1964. Poisonous plants of the United States and Canada. Prentice-Hall, Englewood Cliffs, NJ.

Klingaman, G. 2011. Boston fern Latin: *Nephrolepis exaltata* 'Bostoniesis.' http://www.thegardengeeks.com/home/index.php?option=com_content&view=article&id=2348:boston-fern-latin-nephrolepis-exaltata-bostoniensis&catid=34:specific-plant-articles&Itemid=144. Accessed Jul 2011.

Kirkpatrick, R.E. 2007. Investigating the monophyly of *Pellaea* (Pteridaceae) in the context of a phylogenetic analysis of cheilanthoid ferns. Syst. Bot. 32:504–518.

Kitfield, A. 1974. Range extension of *Woodwardia areolata*. Rhodora 76:312.

Knobloch, I.W. 1967. Chromosome numbers in *Cheilanthes, Notholaena, Llavea,* and *Polypodium*. Amer. J. Bot. 54:461–464.

Knobloch, I.W. and **D.M. Britton.** 1963. The chromosome number and possible ancestry of *Pellaea wrightiana*. Amer. J. Bot. 50:52–55.

Knobloch, I.W. and **D.B. Lellinger.** 1969. *Cheilanthes castanea* and its allies in Virginia and West Virginia. Castanea 34:59–61.

Knobloch, I.W., W. Tai, and **T.A. Ninan.** 1973. The cytology of some species of the genus *Notholaena*. Amer. J. Bot. 60:92–95.

Korall, P. and **P. Kenrick.** 2002. Phylogenetic relationships in Selaginellaceae based on rbcL sequences. Amer. J. Bot. 89:506–517.

Korall, P., P. Kenrick, and **J.P. Therrien.** 1999. Phylogeny of Selaginellaceae: Evaluation of generic/subgeneric relationships based on *rbc*L gene sequences. Int. J. Pl. Sci. 160:585–594.

Korpelainen, H. and **M. Kolkkala.** 1996. Genetic diversity and population structure in the outcrossing populations of *Equisetum arvense* and *E. hyemale* (Equisetaceae). Amer. J. Bot. 83:58–62.

Kramer, K.U. 1990a. Psilotaceae. In: K.U. Kramer and P.S. Green, eds. The families and genera of vascular plants, Vol. I. Pteridophytes and gymnosperms. Pp. 22–25. Springer-Verlag, Berlin, Germany.

Kramer, K.U. 1990b. Dennstaedtiaceae. In: K.U. Kramer and P.S. Green, eds. The families and genera of vascular plants, Vol. I. Pteridophytes and gymnosperms. Pp. 81–94. Springer-Verlag, Berlin, Germany.

Kramer, K.U. 1990c. Marsileacae. In: K.U. Kramer and P.S. Green, eds. The families and genera of vascular plants, Vol. I. Pteridophytes and gymnosperms. Pp. 180–183. Springer-Verlag, Berlin, Germany.

Kramer, K.U. 1990d. Nephrolepidaceae. In: K.U. Kramer and P.S. Green, eds. The families and genera of vascular plants, Vol. I. Pteridophytes and gymnosperms. Pp. 188–190. Springer-Verlag, Berlin, Germany.

Kramer, K.U. 1990e. Osmundaceae. In: K.U. Kramer and P.S. Green, eds. The families and genera of vascular plants, Vol. I. Pteridophytes and gymnosperms. Pp. 197–200. Springer-Verlag, Berlin, Germany.

Kramer, K.U. 1990f. Schizaeaceae. In: K.U. Kramer and P.S. Green, eds. The families and genera of vascular plants, Vol. I. Pteridophytes and gymnosperms. Pp. 258–263. Springer-Verlag, Berlin, Germany.

Kramer, K.U. 1990g. Lomariopsidaceae. In: K.U. Kramer and P.S. Green, eds. The families and genera of vascular plants, Vol. I. Pteridophytes and gymnosperms. Pp. 164–170. Springer-Verlag, Berlin, Germany.

Kramer, K.U. 1993. Distribution patterns in major pteridophyte taxa relative to those of angiosperms. J. Biogeogr. 20:287–291.

Kramer, K.U., T.C. Chambers, and **E. Hennipman.** 1990a. Blechnaceae. In: K.U. Kramer and P.S. Green, eds. The families and genera of vascular plants, Vol. I. Pteridophytes and gymnosperms. Pp. 60–68. Springer-Verlag, Berlin, Germany.

Kramer, K.U. and **P.S. Green,** eds. 1990. The families and genera of vascular plants, Vol. I. Pteridophytes and gymnosperms. Springer-Verlag, Berlin, Germany.

Kramer, K.U., R.E. Holttum, R.C. Moran, and **A.R. Smith.** 1990b. Dryopteridaceae. In: K.U. Kramer and P.S. Green, eds. The families and genera of vascular plants, Vol. I. Pteridophytes and gymnosperms. Pp. 101–144. Springer-Verlag, Berlin, Germany.

Kramer, K.U. and **R. Viane.** 1990. Aspleniaceae. In: K.U. Kramer and P.S. Green, eds. The families and genera of vascular plants, Vol. I. Pteridophytes and gymnosperms. Pp. 52–57. Springer-Verlag, Berlin, Germany.

LaMotte, C. 1940. *Pilularia* in Texas. Amer. Fern J. 30:99–101.

Landry, G.P. and **S.W. Holder.** 1985. *Marsilea macropoda* new to Louisiana. Amer. Fern J. 75:133.

Landry, G.P., M. Israel, R. Schwarzwalder, Jr., and **R.D. Thomas.** 1979. *Cyrtomium fortunei* in Louisiana and Mississippi. Amer. Fern J. 69:85–86.

Landry, G. and **J.W. Thieret.** 1973. *Isoetes lousianensis* (Isoetaceae), a new species from Louisiana. Sida 5:129–130.

Lebkuecher, J.G. and **W.G. Eickmeier.** 1991. Reduced photoinhibition with stem curling in the resurrection plant *Selaginella lepidophylla*. Oecologia 88:597–604.

Lebkuecher, J.G. and **W.G. Eickmeier.** 1993. Physiological benefits of stem curling for resurrection plants in the field. Ecology 74:1073–1080.

Lellinger, D.B. 1981. Notes on North American Ferns. Amer. Fern J. 71:90–94.

Lellinger, D.B. 1985. A field manual of the ferns & fern allies of the United States & Canada. Smithsonian Institution Press, Washington, D.C.

Lellinger, D.B. 1989. The ferns and fern-allies of Costa Rica, Panama, and the Choco (Part I: Psilotaceae through Dicksoniaceae). Pteridologia 2A:1–364.

Lemke, D.E. (David). Personal communication. Specialist in the flora of Texas and aquatic plants; Professor of Biology at Texas State Univ.-San Marcos.

Lemke, D.E. 1994. New collection records for the aquatic macrophytes *Ceratopteris thalictroides* (Parkeriaceae) and *Limnophila sessiliflora* (Scrophulariaceae) in Texas. Sida 16:379–380.

Leonard, S.W. 1972. The distribution of *Thelypteris torresiana* in the southeastern United States. Amer. Fern J. 62:97–99.

Lesho, C.L. 1994. A summary of chromosome numbers in the Marsileaceae, with counts for additional species of *Marsilea*. Amer. Fern J. 84:121–125.

Li, C.-X. and **C.G. Lu.** 2006. Phylogenetic analysis of Dryopteridaceae based on chloroplast *rbcL* sequences. Acta Phytotax. Sinica 44:503–515.

Li, H.-L. 1952. Floristic relationships between eastern Asia and eastern North America. Trans. Amer. Philos. Soc. 42:371–409.

Li, J. and **C.H. Haufler.** 1994. Phylogeny, biogeography, and population biology of *Osmunda* species: Insights from isozymes. Amer. Fern J. 84:105–114.

Liew, F.S. 1972. Numerical taxonomic studies on North American lady ferns and their allies. Taiwaniana 17:190–221.

Little, D.P. and **D.S. Barrington.** 2003. Major evolutionary events in the origin and diversification of the fern genus *Polystichum* (Dryopteridaceae). Amer. J. Bot. 90:508–514.

Liu, H.-M., X.-C. Zhang, Z.-D. Chen, S.-Y. Dong, and **Y.-L. Qiu.** 2007a. Polyphyly of the fern family Tectariaceae sensu Ching: Insights from cpDNA sequence data. Sci. China C Life Sci. 50:789–798.

Liu, H.-M., X.-C. Zhang, W. Wang, Y.L. Qiu, and **Z.-D. Chen.** 2007b. Molecular phylogeny of the fern family Dryopteridaceae inferred from chloroplast rbcL and atpB genes. Int. J. Pl. Sci. 168:1311–1323.

Liu, H.-M., X.-C. Zhang, W. Wang, and **H. Zeng.** 2010. Molecular phylogeny of the endemic fern genera *Cyrtomidictyum* and *Cyrtogonellum* (Dryopteridaceae) from East Asia. http://www.springerlink.com/content/un26v2000n002438/. Accessed Apr 2010.

Lloyd, R.M. 1971. Systematics of the onocleoid ferns. Univ. Calif. Publ. Bot. 61:1–86.

Lloyd, R.M. 1974. Systematics of the genus *Ceratopteris* Brongn. (Parkeriaceae) II. Taxonomy. Brittonia 26:139–160.

Lloyd, R.M. 1993. Parkeriaceae. In: Flora of North America Editorial Committee, eds. Fl. North Amer. 2:119–121. Oxford Univ. Press, New York and Oxford.

Lloyd, F.E. and **L.M. Underwood.** 1900. A review of the species of *Lycopodium* of North America. Bull. Torrey Bot. Club 27:147–168.

Lodwick, L.N. 1975. A second collection of *Psilotum nudum* from Texas. Amer. Fern J. 65:62.

Lott, E.J., B.M. Boom, and **F. Chiang.** 1982. *Isoetes butleri* (Isoetaceae) in Texas. Sida 9:264–266.

Lott, M.S., J.C. Volin, R.W. Pemberton, and **D.F. Austin.** 2003. The reproductive biology of the invasive ferns *Lygodium microphyllum* and *L. japonicum* (Schizaeaceae): Implications for invasive potential. Amer. J. Bot. 90:1144–1152.

Lovis, D.J. 1977. Evolutionary patterns and processes in ferns. In: R.D. Preston and H.W. Woolhouse, eds. Advances in botanical research. Pp. 229–240. Academic Press, London, England, U.K. and NewYork.

Lu, J.M., D.S. Barrington, and **D.Z. Li.** 2007. Molecular phylogeny of the polystichoid ferns in Asia based on *rbcL* sequences. Syst. Bot. 32:26–33.

Lu, J. M. and **X. Cheng.** 2003. Distributional study of the genus *Cyrtomium* C. Presl (Dryopteridaceae). In: S. Chandra and M. Srivastava, eds. Pteridology in the new millennium. Pp.133–141. Kluwer Academic, Dordrecht, Netherlands.

Lu, J.M., X. Cheng, D. Wu, and **D.Z. Li.** 2006. Chromosome study of the fern genus *Cyrtomium* (Dryopteridaceae). Bot. J. Linn. Soc. 150:221–228.

Lu, J.M., D.Z. Li, L.M. Gao, and **X. Cheng.** 2005. Paraphyly of *Cyrtomium* (Dryopteridaceae): Evidence from rbcL and trnL-F sequence data. J. Pl. Res. 118:129–135.

Lumpkin, T.A. 1993. Azollaceae. In: Flora of North America Editorial Committee, eds. Fl. North Amer. 2:338–342. Oxford Univ. Press, New York and Oxford.

Lumpkin, T.A. and **D.L. Plucknett.** 1980. *Azolla*: Botany, physiology, and use as a green manure. Econ. Bot. 34:111–153.

Lupia, R., H. Schneider, G.M. Moeser, K.M. Pryer, and **P.R. Crane.** 2000. Marsileaceae sporocarps and spores from the Late Cretaceous of Georgia, U.S.A. Int. J. Pl. Sci. 161:975–988.

Ma, L.Q., K.M. Komar, C. Tu, W. Zhang, Y. Cai, and **E.D. Kennelley.** 2001. A fern that hyperaccumulates arsenic. Nature 409(6820):579.

Mabberley, D.J. 1987. The plant book, a portable dictionary of the higher plants. Reprinted with corrections in 1989. Cambridge Univ. Press, Cambridge, England, U.K.

Mabberley, D.J. 1997. The plant book, a portable dictionary of the higher plants, 2nd ed. Cambridge Univ. Press, Cambridge, England, U.K.

Mabberley, D.J. 2009. The plant book, a portable dictionary of the higher plants, 3nd ed. Cambridge Univ. Press, Cambridge, England, U.K.

MacDougal, J.M. 1976. Naturalization of *Cyrtomium fortunei* in North America. Amer. Fern J. 66:25–26.

MacRoberts, B.R. and **M.H. MacRoberts.** 1997a. Floristics of beech-hardwood forests in East Texas. Phytologia 82:20–29.

MacRoberts, B.R., M.H. MacRoberts, and **J.C. Cathey.** 2002b. Floristics of xeric sandylands in the post oak savanna region of East Texas. Sida 20:373–386.

MacRoberts, M.H. and **B.R. MacRoberts.** 1995a. Noteworthy vascular plant collections on the Kisatchie National Forest, Louisiana. Phytologia 78:291–313.

MacRoberts, M.H. and **B.R. MacRoberts.** 2001. Bog communities of the West Gulf Coastal Plain: A profile. Bog Research Papers in Botany & Ecology, No. 1. Bog Research, Shreveport, LA.

Madeira, P.T. C.C. Jacono, P. Tipping, T.K. Van, and **T.D. Center.** 2003. A genetic survey of *Salvinia minima* in southern United States. Aquatic Bot. 76:127–139.

Madeira, P.T., R.W. Pemberton, and **T.D. Center.** 2008. A molecular phylogeny of the genus *Lygodium* (Schizaeaceae) with special reference to the biological control and host range testing of *Lygodium microphyllum*. Biol. Control 45:308–318. http://ddr.nal.usda.gov/bitstream/10113/17096/1/IND44061222.pdf. Accessed Jul 2010.

Malone, C.R. and **V.W. Proctor.** 1965. Dispersal of *Marsilea mucronata* by water birds. Amer. Fern J. 55:167–170.

Manhart, J.R. 1995. Chloroplast 16S rDNA sequences and phylogenetic relationships of fern allies and ferns. Amer. Fern J. 85:182–192.

Manton, I. 1950. Problems of cytology and evolution in the pteridophyta. Cambridge Univ. Press, Cambridge, England, U.K.

Marcon, A.B., I.C.L. Barros, and **M. Guerra.** 2005. Variation in chromosome numbers, CMA bands and 45S rDNA sites in species of *Selaginella* (Pteridophyta). Ann. Bot. (Oxford) 95:271–276.

Marks, P.L. and **P.A. Harcombe.** 1975. Community diversity of coastal plain forests in southern East Texas. Ecology 56:1004–1008.

Mason, C.T., Jr. 1975. Walter Sargeant Phillips (1905–1975). Amer. Fern J. 65:70.

Masuyama, S. 2008. Cryptic species in the fern *Ceratopteris thalictroides* (Parkeriaceae). III. Referential diagnostic characters of three cryptic species. J. Pl. Res. 121:279–286.

Masuyama, S. and **Y. Watano.** 2005. Cryptic species in the fern *Ceratopteris thalictroides* (L.) Brongn. (Parkeriaceae). II. Cytological characteristics of three cryptic species. Acta Phytotax. Geobot. 56:231–240.

Masuyama, S., Y. Yataba, N. Murakami, and **Y. Watano.** 2002. Cryptic species in the fern *Ceratopteris thalictroides* (L.) Brongn. (Parkeriaceae). I. Molecular analyses and crossing tests. J. Pl. Res. 115:87–97.

Matsumoto, S. 2003. Species ecological study on reproductive systems and speciation of *Cyrtomium falcatum* complex (Dryopteridaceae) in Japanese Archipelago. Ann. Tsukuba Bot. Gard. 22:1–141.

Maxon, W.R. 1912. Notes on the North American species of *Phanerophlebia*. Bull. Torrey Bot. Club 39:23–28.

Maxon, W.R. 1913. Studies of tropical American ferns. No. 4. Contr. U.S. Natl. Herb. 17:125–179.

Maxon, W.R. 1916. Studies of tropical American ferns. Notes on *Notholaena*. Contr. U.S. Natl. Herb. 17:601–607.

Maxon, W.R. 1921. Notes on North American ferns—XVII. Amer. Fern J. 11:33–39.

May, L.W. 1978. The economic uses and associated folklore of ferns and fern allies. Bot. Rev. (Lancaster) 44:491–528.

McFarland, D.G., L.S. Nelson, M.J. Grodowitz, R.M. Smart, and **C.S. Owens.** 2004. *Salvinia molesta* D.S. Mitchell (giant salvinia) in the United States: A review of species ecology and approaches to management. U.S. Army Corps of Engineers, Engineeer Research and Development Center, Vicksburg, MS, U.S.A. Special Report 04-2. http://el.erdc.usace.army.mil/elpubs/pdf/srel04-2.pdf. Accessed Aug 2010.

McIver, E.E. and **J.F. Basinger.** 1999. Early Tertiary floral evolution in the Canadian high Arctic. Ann. Missouri Bot. Gard. 86:523–545.

McLaughlin, S.P. 2007. Tundra to tropics: The floristic plant geography of North America. Sida, Bot. Misc. 30.

McNeill, J., F.R. Barrie, H.M. Burdet, V. Demoulin, D.L. Hawksworth, K. Marhold, D.H. Nicolson, J. Prado, P.C. Silva, J.E. Skog, J.H. Wiersema, and **N.J. Turland,** eds. 2006. International code of botanical nomenclature (Vienna Code). Regnum Veg. 146. http://ibot.sav.sk/icbn/main.htm. Accessed Jun 2011.

McNeill, J. and **K.M. Pryer.** 1985. The status and typification of *Phegopteris* and *Gymnocarpium*. Taxon 34:136–143.

Meharg, A.A. 2003. Variation in arsenic accumulation: Hyperaccumulation in ferns and their allies. New Phytol. 157:25–31.

Mehltreter, K., L.R. Walker, and **J.M. Sharpe,** eds. 2010. Fern ecology. Cambridge Univ. Press, Cambridge, England, U.K.

Mendoza Ruiz, A. and **B. Pérez-García.** 2003. Comparative analysis of the sexual phase of *Phanerophlebia* (Dryopteridaceae) in Mexico. Canad. J. Bot. 81:501–516.

Mendoza, A., M. Windham, B. Pérez-García, and **G. Yatskievych.** 2001. Una nueva especie de *Pellaea* (Pteridaceae) del estado de San Luis Potosí, Mexico. Acta Bot. Mex. 57:15–21.

Metzgar, J.S., H. Schneider, and **K.M. Pryer.** 2007. Phylogeny and divergence time estimates for the fern genus *Azolla* (Salviniaceae). Int. J. Pl. Sci. 168:1045–1053.

Metzgar, J.S., J.E. Skog, E.A. Zimmer, and **K.M. Pryer.** 2008. The paraphyly of *Osmunda* is confirmed by phylogenetic analyses of seven plastid loci. Syst. Bot. 33:31–36.

Mickel, J.T. 1962. A monographic study of the fern genus *Anemia*, subgenus *Coptophyllum*. Iowa State Coll. J. Sci. 36:349–482.

Mickel, J.T. 1973. Position and classification within the Dennstaedtiaceae. In: A.C. Jermy, J.A. Crabbe, and B.A. Thomas, eds. The phylogeny and classification of the ferns. Bot. J. Linn. Soc. 67, Suppl. 1:135–144.

Mickel, J.T. 1979. The fern genus *Cheilanthes* in the continental United States. Phytologia 41:431–437.

Mickel, J.T. 1981. Revision of *Anemia* subgenus *Anemiorrhiza* (Schizaeaceae). Brittonia 33:413–429.

Mickel, J.T. 1982. The genus *Anemia* (Schizaeaceae) in Mexico. Brittonia 34:388–413.

Mickel, J.T. 1993. Anemiaceae. In: Flora of North America Editorial Committee, eds. Fl. North Amer. 2:117–118. Oxford Univ. Press, New York and Oxford.

Mickel, J.T. and **J.M. Beitel.** 1988. Pteridophyte flora of Oaxaca, Mexico. Mem. New York Bot. Gard. 46:1–568.

Mickel, J.T. and **A.R. Smith.** 2004. The pteridophytes of Mexico. New York Bot. Gard., Bronx.

Miller, C.N., Jr. 1967. Evolution of the fern genus *Osmunda*. Contrib. Mus. Paleontol. Univ. Mich. 21:139–203.

Mink, J.N. (Jeffrey). Personal communication. Graduate student and author of a number of recent papers on the distribution of Texas plants (e.g., Mink et al. 2010, 2011a, 2011b).

Mink, J.N., J.R. Singhurst, and **W.C. Holmes.** 2010. *Dryopteris marginalis* (Dryopteridaceae): New to Texas. Phytoneuron 2010-53:1–6. http://www.phytoneuron.net/PhytoN-Dryopterismarginalis.pdf. Accessed Dec 2010.

Mink, J.N., J.R. Singhurst, and **W.C. Holmes.** 2011a. *Dryopteris celsa* (Dryopteridaceae) and E.J. Palmer 29404: Solution of a Texas mystery. Phytoneuron 2011-11:1–8. http://www.phytoneuron.net/PhytoN-Dryopteriscelsa.pdf. Accessed Feb 2011.

Mink, J.N., J.R. Singhurst, and **W.C. Holmes.** 2011b. Remarks on *Equisetum arvense* (Equisetaceae) in Texas. Phytoneuron 2011-21:1–3. http://www.phytoneuron.net/PhytoN-Equisetumarvense.pdf. Accessed Jun 2011.

Mitchell, D.S. and **N.M. Tur.** 1975. The rate of growth of *Salvinia molesta (S. auriculata* Auct.*)* in laboratory and natural conditions. J. Applied Ecol. 2:213–225.

Montgomery, J.D. 1982. *Dryopteris* in North America. Part II. The hybrids. Fiddlehead Forum, Bull. Amer. Fern Soc. 9:23–30.

Montgomery, J.D. and **E.M. Paulton.** 1981. *Dryopteris* in North America. Fiddlehead Forum, Bull. Amer. Fern Soc. 8:25–31.

Montgomery, J.D. and **W.H. Wagner, Jr.** 1993. *Dryopteris*. In: Flora of North America Editorial Committee, eds. Fl.

North Amer. 2:280–288. Oxford Univ. Press, New York and Oxford.

Montz, G.N. 1989. Distribution of *Salvinia minima* in Louisiana. In: Proceedings 23rd Annual Meeting, Aquatic Pl. Control Res. Prog., 14–17 November 1988, West Palm Beach, FL, Misc. Paper A-89-1, USACOE, Waterways Exp. Sta., Vicksburg, MS. Pp. 312–316.

Moore, A.W. 1969. *Azolla*: Biology and agronomic significance. Bot. Rev. 35:17–34.

Moore, L.B. 2008. Arizona desert ferns. http://www.mineralarts.com/ferns/desertferns.html. Accessed Aug 2010.

Moran, K. and **numerous coauthors.** 2006. The Cenozoic palaeoenvironment of the Arctic Ocean. Nature 441:601–605.

Moran, R.C. (Robbin). Personal communication. Pteridologist at the New York Botanical Garden, editor of the pteridophyte volume for *Flora Mesoamericana*, and author of *A Natural History of Ferns*.

Moran, R.C. 1992. The story of the molesting *Salvinia*. Fiddlehead Forum, Bull. Amer. Fern Soc. 19:26–28.

Moran, R.C. 1993a. Bracken, the poisoner. Fiddlehead Forum, Bull. Amer. Fern Soc. 20:18–19, 22.

Moran, R.C. 1993b. *Tectaria*. In: Flora of North America Editorial Committee, eds. Fl. North Amer. 2:302–304. Oxford Univ. Press, New York and Oxford.

Moran, R.C. 1995. Nardoo. Fiddlehead Forum, Bull. Amer. Fern Soc. 22:20–22.

Moran, R.C. 1997. The little nitrogen factories. Fiddlehead Forum, Bull. Amer. Fern Soc. 24:9, 12–14.

Moran, R.C. 1998. Palea-botany. Fiddlehead Forum, Bull. Amer. Fern Soc. 25:19, 21–23.

Moran, R.C. 2000. Monograph of the neotropical species of *Lomariopsis* (Lomariopsidaceae). Brittonia 52:55–111.

Moran, R.C. 2004. A natural history of ferns. Timber Press, Portland, OR.

Moran, R.C. 2007. Introduction to papers from the fern biogeography symposium. International Botanical Congress, 2005. Brittonia 59:107.

Moran, R.C. 2008. Diversity, biogeography, and floristics. In: T.A. Ranker and C.H. Haufler, eds. Biology and evolution of ferns and lycophytes. Ch. 14. Pp. 367–394. Cambridge Univ. Press, Cambridge, England, U.K.

Moran, R.C. and **A.R. Smith.** 1999. *Salvinia adnata* Desv. versus *S. molesta* D.S. Mitch. Amer. Fern J. 89:268–269.

Moran, R.C. and **A.R. Smith.** 2001. Phytogeographic relationships between neotropical and African-Madagascan pteridophytes. Brittonia 53:304–351.

Mora-Olivo, A. and **G. Yatskievych.** 2009. *Salvinia molesta* in Mexico. Amer. Fern J. 99:56–58.

Morgan, V.H. 2010. *Salvinia minima*. USGS Nonindigenous aquatic species database, Gainesville, FL. http://nas.er.usgs.gov/queries/factsheet.asp?SpeciesID=297. Accessed May 2010.

Morrell, J.C. and **W.F. Faragher.** 1929. Cracking of tars from cannel coal. Ind. Eng. Chem. 21:1084–1086.

Morton, C.V. 1939. Our southwestern resurrection-plants. Amer. Fern J. 29:14–16.

Morton, C.V. 1957a. Observations on cultivated ferns. Amer. Fern J. 47:7–14.

Morton, C.V. 1957b. Observations on cultivated ferns. II. The proper generic name of the holly fern. Amer. Fern J. 47:52–55.

Morton, C.V. 1966. The Mexican species of *Tectaria*. Amer. Fern J. 56:120–137.

Morton, C.V. 1967a. *Ceratopteris thalictroides*, a fern new to Texas. Amer. Fern J. 57:13–14.

Morton, C.V. 1967c. Studies of fern types. I. Contr. U.S. Natl. Herb. 38:29–83.

Münster, T., W. Faigl, H. Saedler, and **G. Theiben.** 2002. Evolutionary aspects of MADS-box genes in the eusporangiate fern *Ophioglossum*. Pl. Biol. 4:474–483.

Murakami, N., S. Nogami, M. Watanabe, and **K. Iwatsuki.** 1999. Phylogeny of Aspleniaceae inferred from *rbcL* nucleotide sequences. Amer. Fern. J. 89:232–243.

Musselman, L.J. 2002. Ornamentation of *Isoetes* (Isoetaceae, Lycophyta) Microspores. Bot. Rev. (Lancaster) 68:474–487.

Nagalingum, N.S., M.D. Nowak, and **K.M. Pryer.** 2008. Assessing phylogenetic relationships in extant heterosporous ferns (Salviniales), with a focus on *Pilularia* and *Salvinia*. Bot. J. Linn. Soc. 157:673–685.

Nagalingum, N.S., H. Schneider, and **K.M. Pryer.** 2006. Comparative morphology of reproductive structures in heterosporous water ferns and a reevaluation of the sporocarp. Int. J. Pl. Sci. 167:805–815.

Nagalingum, N.S., H. Schneider, and **K.M. Pryer.** 2007. Molecular phylogenetic relationships and morphological evolution in the heterosporous fern genus *Marsilea*. Syst. Bot. 32:16–25.

Nakazato, T., M.-K. Jung, E.A. Housworth, L.H. Rieseberg, and **G.J. Gastony.** 2006. Genetic map-based analysis of genome structure in the homosporous fern *Ceratopteris richardii*. Genetics 173:1585–1597.

Nauman, C.E. 1981. The genus *Nephrolepis* in Florida. Amer. Fern J. 71:35–40.

Nauman, C.E. 1987. Schizaeaceae in Florida. Sida 12:69–74.

Nauman, C.E. 1993a. Lygodiaceae. In: Flora of North America Editorial Committee, eds. Fl. North Amer. 2:114–116. Oxford Univ. Press, New York and Oxford.

Nauman, C.E. 1993b. *Pteris*. In: Flora of North America Editorial Committee, eds. Fl. North Amer. 2:132–135. Oxford Univ. Press, New York and Oxford.

Nauman, C.E. 1993c. *Nephrolepis*. In: Flora of North America Editorial Committee, eds. Fl. North Amer. 2:305–308. Oxford Univ. Press, New York and Oxford.

Nauman, C.E. 1993d. Salviniaceae. In: Flora of North America Editorial Committee, eds. Fl. North Amer. 2:336–337. Oxford Univ. Press, New York and Oxford.

Nauman, C.E. 1993e. *Blechnum*. In: Flora of North America Editorial Committee, eds. Fl. North Amer. 2:224–225. Oxford Univ. Press, New York and Oxford.

Nauman, C.E. and **A.M. Evans.** 1993. *Dennstaedtia*. In: Flora of North America Editorial Committee, eds. Fl. North Amer. 2:199–201. Oxford Univ. Press, New York and Oxford.

Nauman C.E., R.P. Wunderlin, and **B.F. Hansen.** 2000. Pteridophytes (Pteridophyta). In: R.P. Wunderlin and B.F. Hansen. Flora of Florida Vol. 1(Pteridophytes and gymnosperms):101–300. Univ. Press of Florida, Gainesville.

Neill, A.K. and **H.D. Wilson.** 2001. The vascular flora of Madison County, Texas. Sida 19:1083–1121.

Nelson, G. 2000. The ferns of Florida: A reference and field guide. Pineapple Press, Sarasota, FL.

Noodén, L.D. and **W.H. Wagner, Jr.** 1997. Photosynthetic capacity and leaf reorientation in two wintergreen ferns, *Polystichum acrostichoides* and *Dryopteris intermedia*. Amer. Fern. J. 87:143–149.

Norton, D.A., J.M. Lord, D.R. Given, and **P.J. De Lange.** 1994. Over-collecting: An overlooked factor in the decline of plant taxa. Taxon 43:181–185.

Oka, T. 2008. How to identify *Cyrtomium* of the Kanto region. Fiddlehead Forum, Bull. Amer. Fern Soc. 35:4–7.

O'Kennon, R.J. (Bob). Personal communication. Texas plant collector, botanist, and co-author of floras on TX plants. Botanical Research Institute of Texas, Fort Worth.

O'Kennon, R.J., T.M. Barkley, G.M. Diggs, Jr., and **B. Lipscomb.** 1999. *Lapsana communis* (Asteraceae) new for Texas and notes on invasive exotics. Sida 18:1277–1283.

O'Kennon, R.J. and **G.M. Diggs, Jr.** 2011. *Llavea cordifolia* (Pteridaceae): New for Texas and the United States. J. Bot. Res. Inst. Texas 5:351–355.(In Press.)

Oklahoma Native Plant Database. 2009. http://www.biosurvey.ou.edu/atlasdesc.html. Accessed Jul 2009.

Øllgaard, B. 1987. A revised classification of the Lycopodiaceae s. lat. Opera Bot. 92:153–178.

Øllgaard, B. 1990. Lycopodiaceae. In: K.U. Kramer and P.S. Green, eds. The families and genera of vascular plants, Vol. I. Pteridophytes and gymnosperms. Pp. 31–39. Springer-Verlag, Berlin, Germany.

Øllgaard, B. 1992. Neotropical Lycopodiaceae—an overview. Ann. Missouri Bot. Gard. 79:687–717.

Pacheco, L. 1995. *Llavea*. In: G. Davidse, M. Sousa S., and A.O. Chater, eds. Flora Mesoamericana, Vol. 1. Psilotaceae to Salviniaceae. (R. Moran & R. Riba, eds for Pteridophyta). Pp. 133–134. Universidad Nacional Autónoma de México, Mexico City.

Page, C.N. 1976. The taxonomy and phytogeography of bracken—a review. Bot. J. Linn. Soc. 73:1–34.

Page, C.N. 1979. Experimental aspects of fern ecology. In: A.F. Dyer. The experimental biology of ferns. Pp. 552–589. Academic Press, London, England, U.K. and New York.

Pakeman, R.J., R.H. Marrs, D.C. Howard, C.J. Barr, and **R.M. Fuller.** 1996. The bracken problem in Great Britain; its present extent and future changes. Appl. Geogr. 16:65–86. http://aob.oxfordjournals.org/cgi/reprint/85/suppl_2/37.pdf. Accessed Aug 2010.

Palmer, E.J. 1919a. Texas Pteridophyta: I. Amer. Fern J. 9:17–22.

Palmer, E.J. 1919b. Texas Pteridophyta: I. Amer. Fern J. 9:50–56.

Palmer, E.J. 1919c. Texas Pteridophyta: III. Amer. Fern J. 9:81–85.

Palmer, E.J. 1927. Ferns of the Davis Mountains. Amer. Fern J. 17:77–85.

Palmer, E.J. 1930. Additional Notes on Texas Ferns. Amer. Fern J. 20:138–142.

Palmer, E.J. 1934. Adventures in fern collecting. Amer. Fern J. 24:104–109.

Panero, J.L. and **P. Dávila A.** 1998. The family Schisandraceae: A new record for the flora of Mexico. Brittonia 50:87–90.

Papaefthimiou, D., C. Van Hove, A. Lejeune, U. Rasmussen, and **A. Wilmotte.** 2008. Diversity and host specificity of *Azolla* cyanobionts. J. Phycology 44:60–70.

Paris, C.A. 1993. *Adiantum*. In: Flora of North America Editorial Committee, eds. Fl. North Amer. 2:125–130. Oxford Univ. Press, New York and Oxford.

Parks, H.B. 1940. Notes on Texas ferns. Amer. Fern J. 30:56–61.

Peck, J.H. (James). Personal communication. Pteridologist and expert on the ferns of Arkansas and the upper midwest. Univ. of Arkansas at Little Rock.

Peck, J.H. 1980. Life history and reproductive biology of the ferns of Woodman Hollow, Webster County, Iowa. Ph.D. dissertation, Iowa State Univ., Ames.

Peck, J.H. 1982. Ferns and fern allies of the Driftless Area of Illinois, Iowa, Minnesota and Wisconsin. Milwaukee Public Mus. Contr. Biol. Geol. 53. Milwaukee, WI.

Peck, J.H. 2000. Floristic status of log ferns (*Dryopteris*) in Arkansas. J. Arkansas Acad. Sci. 54:88–102.

Peck, J.H. 2002. *Nephrolepis exaltata* new to Arkansas. Sida 20:833–835.

Peck, J.H. 2003. Arkansas flora: Additions, reinstatements, exclusions, and reexclusions. Sida 20:1737–1757.

Peck, J.H. 2011a. New and noteworthy additions to the Arkansas fern flora. Phytoneuron 2011-30:1–33. http://www.phytoneuron.net/PhytoN-Arkansasferns.pdf. Accessed Jun 2011.

Peck, J.H. 2011b. History of Arkansas pteridophyte studies with a new annotated checklist and floristic analysis. Phytoneuron 2011-38:1–39. http://www.phytoneuron.net/PhytoN-Arkansasfernfloristics.pdf. Accessed Jun 2012.

Peck, J.H. and **C.J. Peck.** 1988. Distribution, abundance, status and phytogeography of log ferns (*Dryopteris*: Woodsiaceae) in Arkansas. Proc. Arkansas Acad. Sci. 42:74–78.

Peck, J.H., C.J. Peck, and **D.R. Farrar.** 1990. Influences of life history attributes on formation of local and distant fern populations. Amer. Fern J. 80:126–142.

Peck, J.H. and **W.C. Taylor.** 1995. Checklist and distribution of Arkansas pteridophytes. Proc. Arkansas Acad. Sci. 49:130–137.

Peck, J.H., C.T. Witsell, and **E. Hendrix.** 2000. *Dryopteris goldiana* and its hybrid with *D. celsa* new to Arkansas. Amer. Fern J. 90:110–111.

Pemberton, R.W. 1998. The potential of biological control to manage Old World climbing fern (*Lygodium microphyllum*), an invasive weed in Florida. Amer. Fern J. 88:176–182.

Pemberton, R.W. 2003. Northward range extension in Florida of the invasive fern *Lygodium microphyllum* (Lygodiaceae). Sida 20:1759–1761.

Pemberton, R.W. and **A.P. Ferriter.** 1998. Old World climbing fern (*Lygodium microphyllum*), a dangerous invasive weed in Florida. Amer. Fern J. 88:165–175.

Perkins, S.K. and **G.A. Peters.** 1993. The *Azolla-Anabaena* symbiosis: Endophyte continuity in the *Azolla* life-cycle is facilitated by epidermal trichomes. I. Partitioning of the endophytic *Anabaena* into developing sporocarps. New Phytol. 123:53–64.

Perkins, S.K., G.A. Peters, T.A. Lumpkin, and **H.E. Calvert.** 1985. Scanning electron microscopy of perine architecture as a taxonomic tool in the genus *Azolla* Lamarck. Scan. Electron Microsc. 4:1719–1734.

Perry, J.P., III and **L.J. Musselman.** 1994. *Psilotum nudum* new to North Carolina. Amer. Fern J. 84:102–104.

Petrik-Ott, A.J. and **F.D. Ott.** 1976. Two new sites for *Ceratopteris thalictroides* in Texas. Amer. Fern. J. 66:111.

Pfeiffer, N.E. 1922. Monograph of the Isoëtaceae. Ann. Missouri Bot. Gard. 9:79–233.

Pfeiffer, N.E. 1939. A new variety of *Isoetes virginica*. Bull. Torrey. Bot. Club 66:411–413.

Philips, W.S. 1946. A checklist of Arizona ferns. Amer. Fern J. 36:97–108.

Philips, W.S. 1947. A checklist of Arizona ferns. Amer. Fern J. 37:13–20, 39–51.

Philips, W.S. 1970. *Psilotum nudum* new to Arizona. Amer. Fern J. 60:30–32.

Phipps, C.J., T.N. Taylor, E.L. Taylor, N.R. Cúneo, L.D. Boucher, and **X. Yao.** 1998. *Osmunda* (Osmundaceae) from the Triassic of Antarctica: An example of evolutionary stasis. Amer. J. Bot. 85:888–895.

Pichi Sermolli, R.E.G. 1963. Adumbratio florae aethiopicae 9. Cryptogrammaceae. Webbia 17:299–315.

Pickering, I.J., L. Gumaelius, H.H. Harris, R.C. Prince, G. Hirsch, J.A. Banks, D.E. Salt, and **G.N. George.** 2006. Localizing the biochemical transformations of arsenate in a hyperaccumulating fern. Environ. Sci. Technol. 40:5010–5014.

Pickett, F.L. 1931. Notes on xerophytic ferns. Amer. Fern J. 21:49–57.

Pigg, K.B. 1992. Evolution of isoetalean lycopsids. Ann. Missouri Bot. Gard. 79:589–612.

Pigg, K.B. 2001. Isoetalean lycopsid evolution: From the Devonian to the present. Amer. Fern J. 91:99–214.

Pigg, K.B., M.L. DeVore, and **W.C. Wehr.** 2006. Filicalean ferns from the tertiary of western North America: *Osmunda* L. (Osmundaceae: Pteridophyta), *Woodwardia* Sm. (Blechnaceae: Pteridophyta) and onocleoid forms (Filicales: Pteridophyta). Fern Gaz. 17:279–286.

Pigg, K.B. and **G.W. Rothwell.** 2001. Anatomically preserved *Woodwardia virginica* (Blechnaceae) and a new filicalean fern from the Middle Miocene Yakima Canyon flora of central Washington, U.S.A. Amer. J. Bot. 88:777–787.

Plant Resources Center. 2009. Plant Resources Center (Univ. of Texas) Flora of Texas database. http://129.116.69.198:427/Texas.html. Accessed Aug 2009.

PlantSystematics.org. 2008. Equisetaceae: *Equisetum ferrissii*. http://www.plantsystematics.org/imgs/robbin/r/Equisetaceae_Equisetum_ferrissii_7443.html.Accessed Jul 2010.

Platt, S.G. and **J.F. Townsend.** 1996. Noteworthy collections: *Pellaea wrightiana* in Pickens County, South Carolina. Castanea 61:397–398.

Poole, J.M. (Jackie). Personal communication. Botanist and expert on rare plants of Texas (e.g., Poole et al. 2007a). Wildlife Diversity Program. Texas Parks and Wildlife Dept.

Poole, J.M., W.R. Carr, D.M. Price, and **J.R. Singhurst.** 2007a. Rare plants of Texas. Texas A&M Univ. Press, College. Station.

Poole, J.M., J.R. Singhurst, and **W.R. Carr.** 2010. A list of the rare plants of Texas (Dec 2010 ed.). Wildlife Diversity Program, Texas Parks and Wildlife Dept. and Texas Conservation Data Center, The Nature Conservancy of Texas. http://www.tpwd.state.tx.us/publications/pwdpubs/media/pwd_rp_w7000_1142.pdf. Accessed Jun 2011.

Poole, J.M., J.R. Singhurst, D.M. Price, and **W.R. Carr.** 2002. A list of the rare plants of Texas. Wildlife Diversity Program, Texas Parks and Wildlife Dept. and Texas Conservation Data Center, The Nature Conservancy of Texas, Unpublished January 2002 edition, copy in the library of the Botanical Research Institute of Texas.

Poole, J.M., J.R. Singhurst, D.M. Price, and **W.R. Carr.** 2007b. A list of the rare plants of Texas. Wildlife Diversity Program, Texas Parks and Wildlife Dept. and Texas Conservation Data Center, The Nature Conservancy of Texas. Unpublished January 2007 edition, copy in the library of the Botanical Research Institute of Texas.

Power, M.S. and **J.E. Skog.** 1987. Ultrastructure of the extrafloral nectaries of *Pteridium aquilinum*. Amer. Fern J. 77:1–15.

Prado, J., C.N. Rodrigues, A. Salatino, and **M.L.F. Salatino.** 2007. Phylogenetic relationships among Pteridaceae, including Brazilian species, infered from rbcL sequences. Taxon 56:355–368.

Pryer, K.M. 1999. Phylogeny of marsileaceous ferns and relationships of the fossil *Hydropteris pinnata* reconsidered. Int. J. Pl. Sci. 160:931–954.

Pryer, K.M. and **D.J. Hearn.** 2009. Evolution of leaf form in marsilaceous ferns: Evidence for heterochrony. Evolution 63:498–513.

Pryer, K.M., H. Schneider, and **S. Magallón.** 2004a. The radiation of vascular plants. In J. Cracraft and M. J. Donoghue, eds. Assembling the tree of life. Pp. 138–153. Oxford Univ. Press, New York.

Pryer, K.M., H. Schneider, A.R. Smith, R. Cranfill, P.G. Wolf, J.S. Hunt, and **S.D. Sipes.** 2001. Horsetails and ferns are a monophyletic group and the closest living relatives to seed plants. Nature 409:618–622.

Pryer, K.M. and **E. Schuettpelz.** 2009. Ferns. Chapter 14. In: S.B. Hedges and S. Kumar, eds. Timetree of life. Pp. 153–156. Oxford Univ. Press, New York. http://www.pryerlab.net/publication/fichier1177.pdf. Accessed Jul 2010.

Pryer, K.M., E. Schuettpelz, L. Huiet, A.L. Grusz, C.J. Rothfels, T. Avent, D. Schwartz and **M.D. Windham.** 2010. DNA barcoding exposes a case of mistaken identity in the fern horticultural trade. Molec. Ecol. Resources 10:979–985. http://www.pryerlab.net/publication/fichier1215.pdf. Accessed Mar 2011.

Pryer, K.M., E. Schuettpelz, P.G. Wolf, H. Schneider, A.R. Smith, and **R. Cranfill.** 2004b. Phylogeny and evolution of ferns (monilophytes) with a focus on the early leptosporangiate divergences. Amer. J. Bot. 91:1582–1598.

Pryer, K.M., A.R. Smith, and **J.E. Skog.** 1995. Phylogenetic relationships of extant ferns based on evidence from morphology and *rbcL* sequences. Amer. Fern J. 85:205–282.

Qian, H. 2002. Floristic relationships between eastern Asia and North America: Test of Gray's hypothesis. Amer. Nautralist 160:317–332.

Qiu, Y.-L., L. Li, B. Wang, Z. Chen, V. Knoop, M. Groth-Malonek, O. Dombrovska, J. Lee, L. Kent, J. Rest, G.F. Estabrook, T.A. Hendry, D.W. Taylor, C.M. Testa, M. Ambros, B. Crandall-Stotler, R.J. Duff, M. Stech, W. Frey, D. Quandt, and **C.C. Davis.** 2006. The deepest divergences in land plants inferred from phylogenomic evidence. Proc. Natl. Acad. Sci. U.S.A. 103:15511–15516.

Quattrocchi, U. 2000. CRC world dictionary of plant names: Common names, scientific names, eponyms, synonyms, and etymology. Four volumes. CRC Press, Boca Raton, FL.

Radford, A.E., H.E. Ahles, and **C.R. Bell.** 1968. Manual of the vascular flora of the Carolinas. Univ. of North Carolina Press, Chapel Hill.

Ranker, T.A. 1990. Phylogenetic systematics of neotropical *Hemionitis* and *Bommeria* (Adiantaceae) based on morphology, allozymes, and flavonoids. Syst. Bot. 15:442–453.

Ranker, T.A. and **J.M.O. Geiger.** 2008. Population genetics. In: T.A. Ranker and C.H. Haufler, eds. The biology and evolution of ferns and lycophytes. Ch. 4. Pp. 107–133. Cambridge Univ. Press, Cambridge, England, U.K.

Ranker, T.A. and **C.H. Haufler,** eds. 2008. Biology and evolution of ferns and lycophytes. Cambridge Univ Press, Cambridge, England, U.K.

Ranker, T.A., A.R. Smith, B.S. Parris, J.M.O. Geiger, C.H. Haufler, S.C.K. Straub, and **H. Schneider.** 2004. Phylogeny and evolution of grammitid ferns (Grammitidaceae): A case of rampant morphological homoplasy. Taxon 53:415–428.

Rashbrook, V.K. 1992; Ant-herbivore interactions: Reasons for the absence of benefits to a fern with foliar nectaries. Ecology 73:2167–2174.

Raubeson, L.A. and **R.K. Jansen.** 1992. Chloroplast DNA evidence on the ancient evolutionary split in vascular land plants. Science 255:1697–1699.

Raubeson, L.A. and **D.B. Stein.** 1995. Insights into fern evolution from mapping chloroplast genomes. Amer. Fern J. 85:193–204.

Raven, P.H., R.F. Evert, and **S.E. Eichhorn.** 1986. Biology of plants, 4th ed. Worth Publishers, Inc., New York.

Rehman, T. Franklin (Tiana). 2011. Personal communication. Collections Manager at BRIT who is doing research on *Isoëtes*. Botanical Research Institute of Texas, Fort Worth.

Reid, J.D. (Jill). Personal communication. Student of *Azolla* (e.g., Reid et al. 2006). Virginia Commonwealth Univ., Richmond, VA.

Reid, J.D., G.M. Plunkett, and **G.A. Peters.** 2006. Phylogenetic relationships in the heterosporous fern genus *Azolla* (Azollaceae) based on DNA sequence data from three noncoding regions. Int. J. Pl. Sci. 167:529–538.

Renzaglia, K.S., R.J. Duff, D.L. Nickrent, and **D.J. Garbary.** 2000. Vegetative and reproductive innovations of early land plants; implications for a unified phylogeny. Philos. Trans., Ser. B 355:769–793.

Renzaglia, K.S., T.H. Johnson, H.D. Gates, and **D.P. Whittier.** 2001. Architecture of the sperm cell of *Psilotum*. Amer. J. Bot. 88:1151–1163.

Renzaglia, K.S., S.B. Dengate, S.J. Schmitt, and **J.G. Duckett.** 2002. Novel features of *Equisetum arvense* spermatozoids: Insights into pteridophyte evolution. New Phytol. 154:159–174.

Renzaglia, K.S. and **D.J. Garbary.** 2001. Motile gametes of land plants: Diversity, development, and evolution. Crit. Rev. Pl. Sci. 20:107–213.

Retallack, G.J. 1997. Earliest Triassic origin of *Isoetes* and quillwort evolutionary radiation. J. Paleont. 71:500–521. http://pages.uoregon.edu/gregr/Papers/Isoetes%20final.pdf. Accessed Jun 2011.

Reudink, M.W., J.P. Snyder, B. Xu, A. Cunkelman, and **R.A. Balsamo.** 2005. A comparison of physiological and morphological properties of deciduous and wintergreen ferns in southeastern Pennsylvania. Amer. Fern J. 95:45–56.

Reverchon, J. 1903. The fern flora of Texas. Fern Bull. 11:33–38.

Richter, S.L. and **B.A. LePage.** 2005. A high-resolution palynological analysis, Axel Heiberg Island, Canadian High Arctic. In: B.A. LePage, C.J. Williams, and H. Yang, eds. The geobiology and ecology of *Metasequoia*. Pp. 137–158. Springer, Dordrecht, The Netherlands.

Riefner, R.E., Jr. and **A.R. Smith.** 2009. *Salvinia minima* and *S. oblongifolia* (Salviniaceae) new to California, with notes on the *S. auriculata* complex. J. Bot. Res. Inst. Texas 3:855–866.

Robbins, W.J. and **H.W. Rickett.** 1939. Botany: A textbook for college and university students, 3rd ed. D. Van Nostrand Co. Inc., New York.

Robinson, R.C., E. Sheffield, and **J.M. Sharpe.** 2010. Problem ferns: Their impact and management. In: K. Mehltreter, L.R. Walker, and J.M. Sharpe, eds. 2010. Fern ecology.

Ch. 8. Pp. 255–322. Cambridge Univ. Press, Cambridge, England, U.K.

Rolfsmeier, S. (Steven). Personal communication. Expert on Nebraska flora and Herbarium Director, Chadron State College, Chadron, NE.

Roos, M. 1996. Mapping the world's pteriophyte diversity—Systematics and floras. In: J.M. Camus, M. Gibby, and R.J. Johns, eds. Pteridology in perspective. Pp. 29–42. Royal Botanic Gardens, Kew, England, U.K.

Roper, J.M., S.K. Hansen, P.G. Wolf, K.G. Karol, D.F. Mandoli, K.D.E. Everett, J. Kuehl, and **J.L. Boore.** 2007. The complete plastid genome sequence of *Angiopteris evecta* (G. Forst.) Hoffm. (Marattiaceae). Amer. Fern J. 97:95–106.

Roth, D. 2001. Species account for *Cystopteris utahensis*. Navajo Natural Heritage Program, Window Rock, AZ. http://nnhp.nndfw.org/Plants/cyut.pdf. Accessed Aug 2012.

Rothfels, C. 2009. Tree of life web project. Cheilanthoids: Cheilanthoid ferns. http://tolweb.org/cheilanthoids/133070. Accessed Jul 2009.

Rothfels, C.J., A. Larsson, L.-Y. Kuo, P. Korall, W.-L. Chiou, and K.M. Pryer. 2012a. Overcoming deep roots, fast rates, and short internodes to resolve the ancient rapid radiation of eupolypods II ferns. Syst. Biol. 61:490–509.

Rothfels, C.J., M.A. Sundue, L.-Y. Kuo, A. Larsson, M. Kato, E. Schuettpelz, and K.M. Pryer. 2012b. A revised family-level classification for eupolypod II ferns (Polypodiidae: Polypodiales). Taxon 61:515–533.

Rothfels, C.J., M.D. Windham, A.L. Grusz, G.J. Gastony, and **K.M. Pryer.** 2008. Toward a monophyletic *Notholaena* (Pteridaceae): Resolving patterns of evolutionary convergence in xeric-adapted ferns. Taxon 57:712–724.

Rothwell, G.W. 1999. Fossils and ferns in the resolution of land plant phylogeny. Bot. Rev. (Lancaster) 65:188–218.

Rothwell, G.W. and **E. Karrfalt.** 2008. Growth, development, and systematics of ferns: Does *Botrychium* s. l. (Ophioglossales) really produce secondary xylem? Amer. J. Bot. 95:414–423.

Rothwell, G.W. and **K.C. Nixon.** 2006. How does the inclusion of fossil data change our conclusions about the phylogenetic history of euphyllophytes? Int. J. Pl. Sci. 167:737–749.

Rothwell, G.W. and **R.A. Stockey.** 1989. Fossil Ophioglossales in the Paleocene of western North America. Amer. J. Bot. 76, 637–644.

Rothwell, G.W. and **R.A. Stockey.** 1991. *Onoclea sensibilis* in the Paleocene of North America: A dramatic example of structural and ecological stasis. Rev. Palaeobot. Palynol. 70:113–124.

Rothwell, G.W. and **R.A. Stockey.** 1994. The role of *Hydropteris pinnata* gen. et sp. nov. in reconstructing the cladistics of heterosporous ferns. Amer. J. Bot. 81:479–492.

Rothwell, G.W. and **R.A. Stockey.** 2008. Phylogeny and evolution of ferns: A paleontological perspective. In: T.A. Ranker and C.H. Haufler, eds. Biology and evolution of ferns and lycophytes. Ch. 13. Pp. 332–366. Cambridge Univ. Press, Cambridge, England, U.K.

Rouhan, G., J.G. Hanks, D. McClelland, and **R.C. Moran.** 2007. Preliminary phylogenetic analysis of the fern genus *Lomariopsis* (Lomariopsidaceae). Brittonia 59:115–128.

Roux, J.P. 1995. Systematic studies in the genus *Mohria* (Pteriophyta: Anemiaceae): 6. Taxonomic review. Bothalia 25:1–12.

Rowe, N.P. 1988. A herbaceous lycophyte from the Lower Carboniferous Drybrook Sandstone of the Forest of Dean, Gloucestershire. Palaeontology 31:69–83.

Roy, S.K. and **I. Manton.** 2006. A new base number in the genus *Lygodium*. New Phytologist 64:286–292.

Runneson, U.T. 2010. Common fern and fern-ally species of the northwest forest, *Lycopodium annotinum*, stiff clubmoss. http://www.borealforest.org/ferns/fern8.htm. Accessed Sep 2010.

Rutz, L.M. and **D.R. Farrar.** 1984. The habitat characteristics and abundance of *Equisetum* ×*ferrissii* and its parent species, *Equisetum hyemale* and *Equisetum laevigatum*, in Iowa. Amer. Fern J. 74:65–76.

Rydin, C. and **N. Wikström.** 2002. Phylogeny of *Isoetes* (Lycopsida): Resolving basal relationships using rbcL sequences. Taxon 51:83–89.

Sano, R., M. Takamiya, M. Ito, S. Kurita, and **M. Hasebe.** 2000. Phylogeny of the lady fern group, tribe Physematieae (Dryopteridaceae), based on chloroplast *rbcL* gene sequences. Molec. Phylogen. Evol. 15:403–413.

Sauer, J.D. 1988. Plant migration: The dynamics of geographic pattern in seed plant species. Univ. of California Press, Berkley. http://ark.cdlib.org/ark:/13030/ft196n99v8/. Accessed Jul 2010.

Saunders, R.M.K. and **K. Fowler.** 1992. A morphological taxonomic revision of *Azolla* Lam. section *Rhizosperma* (Mey.) Mett. (Azollaceae). Bot. J. Linn. Soc. 109:329–357.

Saunders, R.M.K. and **K. Fowler.** 1993. The supraspecific taxonomy and evolution of the fern genus *Azolla* (Azollaceae). Pl. Syst. Evol. 184:175–193.

Sciarretta, K.L., E.P. Arbuckle, C.H. Haufler, and **C.R. Werth.** 2005. Patterns of genetic variation in southern Appalachian populations of *Athyrium filix-femina* var. *asplenioides* (Dryopteridaceae). Int. J. Pl. Sci. 166:761–780.

Schneider, E.L. and **S. Carlquist.** 1997. SEM studies on vessels in ferns. III. *Phlebodium* and *Polystichum*. Int. J. Pl. Sci. 158:343–349.

Schneider, E.L. and **S. Carlquist.** 1999a. SEM studies on vessels in ferns. 11. *Ophioglossum*. Bot. J. Linn. Soc. 129:105–114.

Schneider, E.L. and **S. Carlquist.** 1999b. SEM studies on vessels in ferns. 13. *Nephrolepis*. Amer. Fern J. 89:171–177.

Schneider, E.L. and **S. Carlquist.** 2000a. SEM studies on vessels in ferns. 17. Psilotaceae. Amer. J. Bot. 87:176–181.

Schneider, E.L. and **S. Carlquist.** 2000b. SEM studies on vessels in ferns. 19. *Marsilea*. Amer. Fern J. 90:32–41.

Schneider, H. 1996. Root anatomy of Aspleniaceae and the implications for systematics of this fern family. Fern. Gaz. 15:160–168.

Schneider, H. and **K.M. Pryer.** 2002. Structure and function of spores in the aquatic heterosporous fern family Marsileaceae. Int. J. Pl. Sci. 163:485–505.

Schneider, H., S.J. Russell, C.J. Cox, F. Bakker, S. Henderson, F. Rumsey, M. Gibby, and **J.C. Vogel.** 2004a. Chloroplast phylogeny of asplenioid ferns based on *rbcL* and *trnL-F* spacer sequences (Polypodiidae, Aspleniaceae) and its implications for biogeography. Syst. Bot. 29:260–274.

Schneider, H., E. Schuettpelz, K.M. Pryer, R. Cranfill, S. Magallón, and **R. Lupia.** 2004b. Ferns diversified in the shadow of angiosperms. Nature 428:553–557.

Schneider, H., E. Schuettpelz, K.M. Pryer, R. Cranfill, S. Magallón, and **R. Lupia.** 2004c. Supplementary information for: Schneider et al. Ferns diversified in the shadow of angiosperms (published on-line). http://www.nature.com/nature/journal/v428/n6982/suppinfo/nature02361.html. Accessed Aug 2009.

Schneider, H., A.R. Smith, R. Cranfill, T.E. Hildebrand, C.H. Haufler, and **T.A. Ranker.** 2004d. Unraveling the phylogeny of polygrammoid ferns (Polypodiaceae & Grammitidaceae): Exploring aspects of the diversification of epiphytic plants. Molec. Phylogen. Evol. 31:1041–1063.

Schneider, H., A.R. Smith, and **K.M. Pryer.** 2009. Is morphology really at odds with molecules in estimating fern phylogeny? Syst. Bot. 34:455–475.

Schneller, J.J. 1979. Biosystematic investigations on the lady fern (*Athyrium filix-femina*). Pl. Syst. Evol. 132:255–277.

Schneller, J.J. 1990a. Azollaceae. In: K.U. Kramer and P.S. Green, eds. The families and genera of vascular plants, Vol. I. Pteridophytes and gymnosperms. Pp. 57–60. Springer-Verlag, Berlin, Germany.

Schneller, J.J. 1990b. Salviniaceae. In: K.U. Kramer and P.S. Green, eds. The families and genera of vascular plants, Vol. I. Pteridophytes and gymnosperms. Pp. 256–258. Springer-Verlag, Berlin, Germany.

Schneller, J.J. 2008. Antheridiogens. In: T.A. Ranker and C.H. Haufler, eds. The biology and evolution of ferns and lycophytes. Ch. 5. Pp. 134–158. Cambridge Univ. Press, Cambridge, England, U.K.

Schölch, A. 2007. Relations between submarginal and marginal sori in ferns IV. The sori of *Osmunda* and selected Schizaeaceae and their morphological relations. Pl. Syst. Evol. 263:227–251.

Schubert, B.G. 1984. Donovan Stewart Correll, 1908–1983. Econ. Bot. 38:134–136.

Schuettpelz, E., A.L. Grusz, M.D. Windham, and **K.M. Pryer.** 2008. The utility of nuclear *gapCp* in resolving polyploid fern origins. Syst. Bot. 33:621–629.

Schuettpelz, E. and **S.B. Hoot.** 2006. Inferring the root of *Isoëtes*: Exploring the alternatives in the absence of an acceptable outgroup. Syst. Bot. 31:258–270.

Schuettpelz, E., P. Korall, and **K.M. Pryer.** 2006. Plastid atpA data provide improved support for deep relationships among ferns. Taxon 55:897–906.

Schuettpelz, E. and **K.M. Pryer.** 2007. Fern phylogeny inferred from 400 leptosporangiate species and three plastid genes. Taxon 56:1037–1050.

Schuettpelz, E. and **K.M. Pryer.** 2008. Fern phylogeny. In: T.A. Ranker, and C.H. Haufler, eds. The biology and evolution of ferns and lycophytes. Ch. 15. Pp. 395–416. Cambridge Univ. Press, Cambridge, England, U.K.

Schuettpelz, E. and **K.M. Pryer.** 2009. Evidence for a Cenozoic radiation of ferns in an angiosperm-dominated canopy. Proc. Natl. Acad. Sci. USA. 106:11200–11205.

Schuettpelz E., H. Schneider, L. Huiet, M.D. Windham, and **K.M. Pryer.** 2007. A molecular phylogeny of the fern family Pteridaceae: Assessing overall relationships and the affinities of previously unsampled genera. Molec. Phylogen. Evol. 44:1172–1185.

Schuettpelz, E. and **D.W. Trapnell.** 2006. Exceptional epiphyte diversity on a single tree in Costa Rica. Selbyana 27:65–71.

Seigler, D.S. 1977. Confirmation of several cheilanthoid fern records from the Edwards Plateau, Texas. Southw. Naturalist 22:137–139.

Seigler, D.S. and **T.E. Lockwood.** 1975. *Blechnum occidentale* new to Texas. Amer. Fern J. 65:96.

Seigler, D.S. and **E. Wollenweber.** 1983. Chemical Variation in *Notholaena standleyi*. Amer. J. Bot. 70:790–798.

Serbet, R. and **G.W. Rothwell.** 1999. *Osmunda cinnamomea* (Osmundaceae) in the Upper Cretaceous of western North America: Additional evidence for exceptional species longevity among filicalean ferns. Int. J. Pl. Sci. 160:425–433.

Shaver, J.M. 1970. Ferns of the eastern central states with special reference to Tennessee. Dover Publications, New York.

Sheffield, E. 2008. Alternation of generations. In: T.A. Ranker and C.H. Haufler, eds. The biology and evolution of ferns and lycophytes. Ch. 2. Pp. 49–74. Cambridge Univ. Press, Cambridge, England, U.K.

Sheffield, E. and **P.R. Bell.** 1987. Current studies of the pteridophyte life cycle. Bot. Rev. (Lancaster) 53:442–490.

Shi, D.-J. and **D.O. Hall.** 1988. The *Azolla-Anabaena* association: Historical perspective, symbiosis and energy metabolism. Bot. Rev. (Lancaster) 54:353–386.

Shmida, A. and **A. Liston.** 1987. *Ophioglossum polyphyllum* - An addition to the fern flora of Israel and its ecogeographical significance. Bot. Jahrb. Syst. 109:145–151. http://bio.huji.ac.il/upload/E048-C%20%20Ophioglossum%20polyphyllum.pdf. Accessed Jul 2011.

Sigel, E.M., M.D. Windham, L. Huiet, G. Yatskievych, and **K.M. Pryer.** 2011. Species relationships and farina evolution in the cheilanthoid fern genus *Argyrochosma* (Pteridaceae). Syst. Bot. 36:554–564.

Simpson, K. and **A.F. Dyer.** 1999. The survival of fern dormant spores. Pteridologist 3:98–103.

Singhurst, J.R. (Jason). Personal communication. Specialist in rare and endandered plants of Texas, plant colletor, and phytogeographer, Wildlife Diversity Program, Texas Parks and Wildlife Dept., Austin.

Singhurst, J.R., A.E. Rushing, C.K. Hanks, and **W.C. Holmes.** 2011. *Isoetes texana* (Isoetaceae): A new species from the Texas Coastal Bend. Phytoneuron 2011-22:1–6. http://www.phytoneuron.net/PhytoN-Isoetestexana.pdf. Accessed Jun 2011.

Skoda, B. 1997. Taxonomic comments on the "Flora of North America north of Mexico," vol. 2, with some new nomenclatural combinations for Pteridophyta. Preslia 68:341–359.

Skog, J.E. 1992. The Lower Cretaceous ferns in the genus *Anemia* (Schizaeaceae), Potomac Group of Virginia, and relationships within the genus. Rev. Palaeobot. Palynol. 70:279–295.

Skog, J.E. 2001. Biogeography of Mesozoic leptosporangiate ferns related to extant ferns. Brittonia 53:236–269.

Skog, J.E., E.A. Zimmer, and **J.T. Mickel.** 2002. Additional support for two subgenera of *Anemia* (Schizaeaceae) from data for the chloroplast intergenic spacer region trnL-F and morphology. Amer. Fern J. 92:119–130.

Small, J.K. 1931. Ferns of Florida: Being descriptions of and notes on the fern-plants growing naturally in Florida (illustrated). The Science Press, New York.

Small, J.K. 1938. Ferns of the southeastern United States. The Science Press Printing Co., Lancaster, PA.

Small, R.L., E.B. Lickey, J. Shaw, and **W.D. Hauk.** 2005. Amplification of noncoding chloroplast DNA for phylogenetic studies in lycophytes and monilophytes with a comparative example of relative phylogenetic utility from Ophioglossaceae. Molec. Phylogen. Evol. 36:509–522.

Smith, A.R. 1971a. Systematics of the neotropical species of *Thelpyteris* section *Cyclosorus*. Univ. Calif. Publ. Bot. 59:1–136.

Smith, A.R. 1971b. The *Thelypteris normalis* complex in the southeastern United States. Amer. Fern J. 61:21–33.

Smith, A.R. 1972. Comparison of fern and flowering plant distributions with some evolutionary interpretations for ferns. Biotropica 4:4–9.

Smith, A.R. 1990. Thelypteridaceae. In: K.U. Kramer and P.S. Green, eds. The families and genera of vascular plants, Vol. I. Pteridophytes and gymnosperms. Pp. 263–272. Springer-Verlag, Berlin, Germany.

Smith, A.R. 1993a. Phytogeographic principles and their use in understanding fern relationships. J. Biogeogr. 20:255–264.

Smith, A.R. 1993b. Dryopteridaceae [Family description and key to genera]. In: Flora of North America Editorial Committee, eds. Fl. North Amer. 2:246–249. Oxford Univ. Press, New York and Oxford.

Smith, A.R. 1993c. Polypodiaceae [Family description and key to genera]. In: Flora of North America Editorial Committee, eds. Fl. North Amer. 2:312–313. Oxford Univ. Press, New York and Oxford.

Smith, A.R. 1995. Non-molecular phylogenetic hypotheses for ferns. Amer. Fern J. 85:104–122.

Smith, A.R. and **R.B. Cranfill.** 2002. Intrafamilial relationships of the thelypteroid ferns (Thelypteridaceae). Amer. Fern. J. 92:131–149.

Smith, A.R., K.M. Pryer, E. Schuettpelz, P. Korall, H. Schneider, and **P.G. Wolf.** 2006. A classification for extant ferns. Taxon 55:705–731.

Smith, A.R., K.M. Pryer, E. Schuettpelz, P. Korall, H. Schneider, and **P.G. Wolf.** 2008. Fern classification. In: T.A. Ranker and C.H. Haufler, eds. The biology and evolution of ferns and lycophytes. Ch. 16. Pp. 417–467. Cambridge Univ. Press, Cambridge, England, U.K.

Smith, B.A. 2010. Four western cheilanthoid ferns in Oklahoma. Oklahoma Native Pl. Rec. 10:65–76. ojs.library.okstate.edu/osu/index.php/ONPR/article/download/421/401. Accessed Jun 2011.

Smith, B.L., D.R. Lauren, and **A.S. Prakash.** 2000. Bracken fern (*Pteridium*): Toixicity in animal and human health. In: J.A. Taylor and R. Thornton Smith, eds. Bracken fern: Toxicity, biology and control. Proceedings of the IV International Bracken Conference, Univ. of Manchester, 20–23 July 1999. International Bracken Group, Special publ. no. 4:76–85. Aberystwyth, Wales, UK.

Snyder, L.H., Jr. and **J.G. Bruce.** 1986. Field guide to the ferns and other pteridophytes of Georgia. Univ. of Georgia Press, Athens.

Soltis, D.E., P.S. Soltis, and **R.D. Noyes.** 1988. An electrophoretic investigationof intragametophytic selfing in *Equisetum arvense*. Amer. J. Bot. 75:231–237.

Somers, P. and **W.R. Buck.** 1975. *Selaginella ludoviciana*, *S. apoda*, and their hybrids in the southeastern United States. Amer. Fern J. 65:76–82.

Soxman, G.M. 1939. *Lygodium japonicum* in Texas. Amer. Fern J. 29:111.

Soxman, G.M. 1945. *Pteris multifida* in Texas. Amer. Fern J. 35:53.

Speer, W.D. 2000. A systematic assessment of British and North American *Pteridium* using cpDNA gene sequences. In: J.A. Taylor and R. Thornton Smith, eds. Bracken fern: Toxicity, biology and control. Proceedings of the IV International Bracken Conference, Univ. of Manchester, 20–23 July 1999. International Bracken Group, Spec. Publ. no. 4:37–42. Aberystwyth, Wales, UK.

Speer, W.D. 2008. Phylogenetic and biogeographic relationships among North American and Hawaiian *Pteridium aquilinum* (L.) Kuhn (Dennstaedtiaceae) based on chloroplast rps4 and rps4-renS intergenic spacer sequences. Amer. Fern J. 98:179–193.

Speer, W.D. and **K.W. Hilu.** 1998 [1999]. Relationships between two infraspecific taxa of *Pteridium aquilinum* (Dennstaedtiaceae). I. Morphological evidence. Syst. Bot. 23:305–312.

Speer, W.D., C.R. Werth, and **K.W. Hilu.** 1998 [1999]. Relationships between two infraspecific taxa of *Pteridium aquilinum* (Dennstaedtiaceae). II. Isozyme evidence. Syst. Bot. 23:313–325.

Stanford, J.W. (Jack). Personal communication. Specialist in plants of the Edwards Plateau, Fabaceae, especially

Mimosa, and ferns; retired from career at Howard Payne Univ., Brownwood, TX.

Stanford, J.W. 1971. Vascular plants of the three central Texas counties of Brown, Comanche, and Hamilton. Ph.D. dissertation, Oklahoma State Univ., Stillwater.

Stanford, J.W. and **G.M. Diggs, Jr.** 1998. *Pteris vittata* (Pteridaceae), a new fern for Texas. Sida 18:359–360.

Stanich, N.A., G.W. Rothwell, and **R.A. Stockey.** 2009. Phylogenetic diversification of *Equisetum* (Equisetales) as inferred from Lower Cretaceous species of British Columbia, Canada. Amer. J. Bot. 96:1289–1299.

Stein, D.B., G. Yatskievych, and **G.J. Gastony.** 1989. Chloroplast DNA evolution and phylogeny of some polystichoid ferns. Biochem. Syst. Ecol. 17:93–101.

Stensvold, M. 2010. Ferns of the National Forests in Alaska. USDA Forest Service, Alaska Region Botany program. http://www.fs.fed.us/r10/tongass/maps/Ferns_web.pdf. Accessed Jul 2012.

Stergianou, K.K. and **K. Fowler.** 1989. Preliminaty report of chromosome counts in the genus *Azolla* (Pteridophyta). Fern Gaz. 13:317–319.

Stergianou, K.K. and **K. Fowler.** 1990. Chromosome numbers and taxonomic implications in the fern genus *Azolla*. Pl. Syst. Evol. 173:223–239.

Stevens, C.E. 1977. An occurrence of *Pteris multifida* in Virginia. Amer. Fern J. 67:63.

Stewart, W.N. and **G.W. Rothwell.** 1993. Paleobotany and the evolution of plants. Cambridge Univ. Press, Cambridge, England, U.K.

Stolze, R.G. 1981. Ferns and fern allies of Guatemala. Part 2: (genus *Elaphoglossum* by J.T. Mickel, genus *Thelypteris* by A.R. Smith). Fieldiana Bot. N.S. 6 (*Llavea* pp. 293–295).

Summerfield, R.J. 1972. Biological inertia—an example. J. Ecol. 60:793–798.

Sun, B.-Y., M.H. Kim, C.H. Kim, and **C.-W. Park.** 2001. *Mankyua* (Ophioglossaceae): A new fern genus from Cheju Island, Korea. Taxon 50:1019–1024.

Sun, B.-Y., T.G. Baek, Y.D. Kim, and **C.S. Kim.** 2009. Phylogeny of the family Ophioglossaceae with special emphasis on genus *Mankyua*. Korean J. Pl. Tax. 39:135–142.

Svenson, H.K. 1944. The New World species of *Azolla*. Amer. Fern J. 34:69–84.

Swain, T. and **G. Cooper-Driver.** 1973. Biochemical systematics in the Filicopsida. In: A.C. Jermy, J.A. Crabbe and B.A. Thomas, eds. The phylogeny and classification of the ferns. Bot. J. Linn. Soc. 67, Suppl. 1:111–134.

Taber, S.W. 2008. Biogeographic characterization of a glacially relict pine forest in east central Texas as measured by invertebrate composition. Southw. Entomol. 33:91–109.

Taber, S.W. and **S.B. Fleenor.** 2005. Invertebrates of Central Texas Wetlands. Texas Tech Univ. Press, Lubbock, TX.

Taylor, W.C. 1984. Arkansas ferns and fern allies. Milwaukee Public Museum, Milwaukee, WI.

Taylor, W.C., R.H. Mohlenbrock, and **J.A. Murray.** 1975. The spores and taxonomy of *Isoëtes butleri* and *I. melanopoda*. Amer. Fern J. 65:33–38.

Taylor, R.J. and **C.E. Taylor.** 1981. Plants new to Arkansas, Oklahoma and Texas. Sida 9:25–28.

Taylor, R.J. and **C.E. Taylor.** 1994. An annotated list of the ferns, fern allies, gymnosperms and flowering plants of Oklahoma, 3rd ed. Southeastern Oklahoma State Univ. Herbarium, Durant.

Taylor, J.A. and **R. Thornton Smith,** eds. 2000. Bracken fern: Toxicity, biology and control. Proceedings of the IV International Bracken Conference, Univ. of Manchester, 20–23 July 1999. International Bracken Group, Special publ. no.4. Aberystwyth, Wales, UK.

Taylor, W.C. and **R.J. Hickey.** 1992. Habitat, evolution, and speciation in *Isoetes*. Ann. Missouri Bot. Gard. 79:613–622.

Taylor, W.C., N.T. Luebke, D.M. Britton, R.J. Hickey, and **D.F. Brunton.** 1993. Isoëtaceae. In: Flora of North America Editorial Committee, eds. Fl. North Amer. 2:64–75. Oxford Univ. Press, New York and Oxford.

Taylor, W.C., N.T. Luebke, and **M.B. Smith.** 1985. Speciation and hybridization in North American quillworts. Proc. Royal Soc. Edinburgh 86B:259–263.

Taylor, W.C., R.H. Mohlenbrock, and **F.J. Burton.** 1976. Variation in North American *Asplenium platyneuron*. Amer. Fern J. 66:63–68.

The Plant List. 2011. The plant list: A working list of all plant species. A collaboration between the Royal Botanic Gardens, Kew and Missouri Botanical Garden, St. Louis. http://www.theplantlist.org/. Accessed Jun 2011.

Thieret, J.W. 1977. Juvenile leaves in Oklahoma *Marsilea* (Marsileaceae). Sida 7:218.

Thieret, J.W. 1980. Louisiana ferns and fern allies. Lafayette Natural History Museum published in conjunction with the Univ. of Southwestern Louisiana, Lafayette.

Thieret, J.W. 1993. Psilotaceae. In: Flora of North America Editorial Committee, eds. Fl. North Amer. 2:16–17. Oxford Univ. Press, New York and Oxford.

Thomas, B.A. 1992. Paleozoic herbaceous lycopsids and the beginnings of extant *Lycopodium* sens. lat. and *Selaginella* sens. lat. Ann. Missouri Bot. Gard. 79:623–631.

Thomas, B.A. and **N. Quansah.** 1991. The palaeobotanical case for dividing *Selaginella* (Selaginellaceae: Pteridophyta). Fern Gaz. 14:59–64.

Thomas, P.A. and **P.M. Room.** 1986. Taxonomy and control of *Salvinia molesta*. Nature 320:581–584.

Thomas, R.D. 1979. First record of *Botrychium lunarioides* and *Ophioglossum nudicaule* var. *tenerum* (Ophioglossaceae) from Texas. Southw. Naturalist 24:395–396.

Thomas, R.D. 1980. Ophioglossaceae. In: J.W. Thieret. Louisiana ferns and fern allies. Pp. 47–54. Lafayette Natural History Museum published in conjunction with Univ. of Southwestern Louisiana, Lafayette.

Thomas, R.D., T. Briley, and **N. Carroll.** 1981. Additional collections of *Botrychium lunarioides* from Texas and Oklahoma and comments on its dormancy. Phytologia 48:276–278.

Thomas, R.D., P.B. Cox, and **G.P. Landry.** 1987. The distribution of *Ophioglossum petiolatum* Hooker in the United

States with a report of it from Virginia. Phytologia 62:1–4.

Thomas, R.D., W.H. Wagner, Jr., and **M.R. Mesler.** 1973. Log fern (*Dryopteris celsa*) and related species in Louisiana. Castanea 38:269–274.

Thomson, J.A. 2000a. Morphological and genomic diversity in the genus *Pteridium* (Dennstaedtiaceae). Ann. Bot. (Oxford) 85 (Suppl. B):77–99. http://aob.oxfordjournals.org/cgi/reprint/85/suppl_2/77.pdf. Accessed Jul 2010.

Thomson, J.A. 2000b. New perspectives on taxonomic relationships in *Pteridium.* In: J.A. Taylor and R. Thornton Smith, eds. Bracken fern: Toxicity, biology and control. Proceedings of the IV International Bracken Conference, Univ. of Manchester, 20–23 July 1999. International Bracken Group, Special publ. no. 4:15–34. Aberystwyth, Wales, UK.

Thomson, J.A. 2004a. Towards a taxonomic revision of *Pteridium* (Dennstaedtiaceae) in Central and South America. Bot. J. Linn. Soc. 140:237–248.

Thomson, J.A. 2004b. Towards a taxonomic revision of *Pteridium* (Dennstaedtiaceae). Telopea 10:793–803.

Thomson, J. A. 2008. Morphotype and conflicting taxonomies in *Pteridium* (Dennstaedtiaceae: Pteridophyta). Fern Gaz. 18:101–109.

Thomson, J.A., J.T. Mickel, and **K. Mehltreter.** 2008. Taxonomic status and relationships of bracken ferns (*Pteridium*: Dennstaedtiaceae) of Laurasian affinity in Central and North America. Bot. J. Linn. Soc. 157:1–17.

Thomson, J.A., P.H Weston, and **M.D. Tan.** 2000. A molecular approach to tracing major lineages in *Pteridium*: Update and amendment. In: J.A. Taylor and R. Thornton Smith, eds. Bracken fern: Toxicity, biology and control. Proceedings of the IV International Bracken Conference, Univ. of Manchester, 20–23 July 1999. International Bracken Group, Special publ. no. 4:35–36. Aberystwyth, Wales, UK.

Thorne, R.F. 1993. Phytogeography. In: Flora of North America Editorial Committee, eds. Fl. North Amer. 1:132–153. Oxford Univ. Press, New York and Oxford.

Tidwell, W.D. and **S.R. Ash.** 1994. A review of selected Triassic to early Cretaceous ferns. J. Pl. Res. 107:417–442.

Tiffney, B.H. 1985a. Perspectives on the origin of the floristic similarity between eastern Asia and eastern North America. J. Arnold Arbor. 66:73–94.

Tiffney, B.H. 1985b. The Eocene North Atlantic Land Bridge: Its importance in Tertiary and modern phytogeography of the Northern Hemisphere. J. Arnold Arbor. 66:243–273.

Tipping, P.W., L. Bauer, M.R. Martin, and **T.D. Center.** 2009. Competition between *Salvinia minima* and *Spirodela polyrhiza* mediated by nutrient levels and herbivory. Aquat. Bot. 90:231–234.

Tipping, P.W. and **T.D. Center.** 2005. Population dynamics of *Cyrtobagous salviniae* on common *Salvinia* in South Florida. J. Aquat. Pl. Managem. 43:47–50. http://www.apms.org/japm/vol43/v43p47.pdf. Accessed Jul 2010.

Tipping, P.W., M.R. Martin, T.D. Center, and **T.M. Davern.** 2008. Suppression of *Salvinia molesta* Mitchell in Texas and Louisiana by *Cyrtobagous salviniae* Calder and Sands. Aquatic Bot. 88:196–202. http://ddr.nal.usda.gov/bitstream/10113/32420/1/IND44244538.pdf. Accessed Jul 2010.

TOES (Texas Organization for Endangered Species). 1993. Endangered, threatened and watch lists of Texas plants. Publ. No. 9. Texas Organization for Endangered Species, Austin.

TPWD (Texas Parks & Wildlife Department). 2010. Texas launches new front in war on invasive species. http://www.tpwd.state.tx.us/newsmedia/releases/?req=20100401b. Accessed Aug 2010.

Tropicos. 2011. Tropicos botanical database, Missouri Botanical Garden. http://www.tropicos.org/. Accessed Jun 2011.

Tryon, A.F. 1957. A revision of the fern genus *Pellaea* section *Pellaea*. Ann. Missouri Bot. Gard. 44:125–193.

Tryon, A.F. 1968. Comparisons of sexual and apogamous races in the fern genus *Pellaea*. Rhodora 70:1–24.

Tryon, A.F. 1972. Spores, chromosomes, and relations of the fern *Pellaea atropurpurea*. Rhodora 74:220–241.

Tryon, R.M., Jr. 1941. A revision of the genus *Pteridium*. Rhodora 43:1–31, 37–67.

Tryon, R.M., Jr. 1955. *Selaginella rupestris* and its allies. Ann. Missouri Bot. Gard. 42:1–99.

Tryon, R.M., Jr. 1956. A revision of the American species of *Notholaena*. Contr. Gray Herb. 179:1–106.

Tryon, R.M., [Jr.]. 1960. A review of the genus *Dennstaedtia* in America. Contr. Gray Herb. 187:23–52.

Tryon, R.M., Jr. 1969. Taxonomic problems in the geography of North American ferns. BioScience 19:790–795.

Tryon, R.M., Jr. 1987. Some new names and combinations in Pteridaceae. Amer. Fern. J. 76:184–186.

Tryon, R.M., Jr. and **A.F. Tryon.** 1973. Geography, spores and evolutionary relationships in the cheilanthoid ferns. In: A.C. Jermy, J.A. Crabbe, and B.A. Thomas, eds. The phylogeny and classification of the ferns. Bot. J. Linn. Soc. 67, Suppl. 1:145–153.

Tryon, R.M. [Jr.] and **A.F. Tryon.** 1982. Ferns and allied plants with special reference to tropical America. Springer-Verlag, Berlin, Germany.

Tryon, R.M. [Jr.], A.F. Tryon, and **K.U. Kramer.** Pteridaceae. 1990. In: K.U. Kramer and P.S. Green, eds. The families and genera of vascular plants, Vol. I. Pteridophytes and gymnosperms. Pp. 230–256. Springer-Verlag, Berlin, Germany.

Turner, B.L., H. Nichols, G. Denny, and **O. Doron.** 2003. Atlas of the vascular plants of Texas, Vol. 1: Dicots, Vol. 2: Ferns, gymnosperms, monocots. Sida, Bot. Misc. 24.

Turner, N.W. and **A.F. Szczawinski.** 1991. Common poisonous plants and mushrooms of North America. Timber Press, Portland, OR.

Turner, N.A., W.C. Taylor, S. Masi , and **M.E. Stupen.** 2005. Confirming dioecy in *Isoetes butleri*. Amer. Fern J. 95:85–87.

Underwood, L.M. 1899. American ferns—II. The genus *Phanerophlebia*. Bull. Torrey Bot. Club 26:205–216.

USDA. 2009. United States Department of Agriculture. Germplasm Resources Information Network. Federal and state noxious weed links. http://www.ars-grin.gov/cgi-bin/npgs/html/taxwdlx.pl. Accessed Aug 2010.

USDA PLANTS. 2011. PLANTS Profile for *Astrolepis sinuata* (Lag. ex Sw.) Benham & Windham, wavy scaly cloakfern. USDA Natural Resources Conservation Service. PLANTS Database. http://plants.usda.gov/java/profile?symbol=ASSI9. Accessed Jun 2011.

USDA PLANTS. 2012. PLANTS Profile for *Cyrtomium falcatum* (L. f.) C. Presl. Japanese netvein hollyfern. USDA Natural Resources Conservation Service. PLANTS Database. http://plants.usda.gov/java/profile?symbol=CYFA2. Accessed Aug 2012.

USDA Natural Resources Conservation Service. 2010. United States Department of Agriculture. Natural Resources Conservation Service. Invasive and noxious weeds. http://plants.usda.gov/java/noxious?rptType=Federal. Accessed Aug 2010.

USGS (U.S. Geological Survey). 2000. *Salvinia molesta*, giant salvinia. http://nas.er.usgs.gov/ferns/. Accessed 2000.

Vajda, V., J.I. Raine, and **C.J. Hollis.** 2001. Indication of global deforestation at the Createaceous-Tertiary boundary by New Zealand fern spike. Science 294:1700–1702. http://www.geol.lu.se/personal/viv/Vajda%20et%20al.%202001.pdf. Accessed Jul 2010.

Valdespino, I.A. 1993. Selaginellaceae. In: Flora of North America Editorial Committee, eds. Fl. North Amer. 2:38–63. Oxford Univ. Press, New York and Oxford.

Van Cat, D., I. Watanabe, and **T. de Waha Baillonville.** 1989. Sexual hybridization among *Azolla* species.. Canad. J. Bot. 67:3482–3485.

Van Coppenolle, B., I. Watanabe, C. Van Hove, and **S.R. McCouch.** 1993. Genetic diversity and phylogeny analysis of *Azolla* based on DNA amplification by arbitrary primers. Genome 46:686–693.

Vangerow, S., T. Teerkorn, and **V. Knoop.** 1999. Phylogenetic information in the mitochondrial nad5 gene of pteridophytes: RNA editing and intro sequences. Pl. Biol. 1:235–243.

van Hove, C. and **A. Lejune.** 2002. The *Azolla-Anabaena* symbiosis. Biology and environment: Proc. Royal Irish Acad. 102B(1):23–26.

Volin, J. C., M.S. Lott, J.D. Muss, and **D. Owen.** 2004. Predicting rapid invasion of the Florida Everglades by Old World climbing fern *Lygodium microphyllum*). Diversity & Distrib. 10:439–446.

Von Holle, B., H.R. Delcourt, and **D. Simberloff.** 2003. The importance of biological inertia in plant community resistance to invasion. J. Veg. Sci. 14:425–432.

Wagner, D.H. 1993. *Polystichum*. In: Flora of North America Editorial Committee, eds. Fl. North Amer. 2:290–299. Oxford Univ. Press, New York and Oxford.

Wagner, G.M. 1997. *Azolla*: A review of its biology and utilization. Bot. Rev. (Lancaster) 63:1–26.

Wagner, W.H., Jr. (Herb), 1920–2000. Personal communication. Specialist in pteridophytes and evolution of vascular plants; Flora of North America Editorial Committee; retired from a career at the Univ. of Michigan, Ann Arbor (see Farrar 2000 for epitaph and selected bibliography).

Wagner, W.H., Jr. 1954. Reticulate evolution in the Appalachian Aspleniums. Evolution 8:103–118.

Wagner, W.H., Jr. 1960. Periodicity and pigmentation in *Botrychium* subg. *Sceptridium* in the northeastern United States. Bull. Torrey Bot. Club 87:303–325.

Wagner, W.H. Jr. 1965. *Pellaea wrightiana* in North Carolina and the question of its origin. J. Elisha Mitchell Sci. Soc. 81:95–103.

Wagner, H.W., Jr. 1970. Evolution of *Dryopteris* in relation to the Appalachians. In: P.C. Holt and R.A. Paterson, eds. The distributional history of the biota of the southern Appalachians, Part II: Flora. Pp. 147–192. Research Division Monogr. 2, Virginia Polytechnic Institute and State Univ., Blacksburg.

Wagner, W.H. Jr. 1972. Disjunctions in homosporous vascular plants. Ann. Missouri Bot. Gard. 59:203–217.

Wagner, W.H., Jr. 1977. Systematic implication of the Psilotaceae. Brittonia 29:54–63.

Wagner, W.H. Jr. 1979. Veins in the systematics of modern ferns. Taxon 28:87–95.

Wagner, W.H., Jr. 1990. Ophioglossaceae. In: K.U. Kramer and P.S. Green, eds. The families and genera of vascular plants, Vol. I. Pteridophytes and gymnosperms. Pp. 193–197. Springer-Verlag, Berlin, Germany.

Wagner, W.H., Jr. 1992. *Hiemobotrychium*, a new section of *Botrychium* subgenus *Sceptridium* from the southeastern United States. Novon 2:267–268.

Wagner, W.H., Jr., C.M. Allen, and **G.P. Landry.** 1984. *Ophioglossum ellipticum* Hook. & Grev. in Louisiana and the taxonomy of *O. nudicaule* L.f. Castanea 49:99–110.

Wagner, W.H., Jr. and **J.M. Beitel.** 1992. Generic classification of modern North American Lycopodiaceae. Ann. Missouri. Bot. Gard. 79:676–686.

Wagner, W.H., Jr. and **J.M. Beitel.** 1993. Lycopodiaceae. In: Flora of North America Editorial Committee, eds. Fl. North Amer. 2:18–37. Oxford Univ. Press, New York and Oxford.

Wagner, W.H., Jr., D.R. Farrar, and **B.W. McAlpin.** 1970. Pteridology of the Highlands Biological Station area, southern Appalachians. J. Elisha Mitchell Sci. Soc. 86:1–27.

Wagner, W.H., Jr. and **D.M. Johnson.** 1981. Natural history of the ebony spleenwort, *Asplenium platyneuron* (Aspleniaceae) in the Great Lakes area. Canad. Field-Naturalist 95:156–166.

Wagner, W.H., Jr. and **D.M. Johnson.** 1983. Trophopod, a commonly overlooked storage structure of potential systematic value in ferns. Taxon 32:268–269.

Wagner, W.H., Jr. and **A.R. Smith.** 1993. Pteridophytes. In: Flora of North America Editorial Committee, eds. Fl. North Amer. 1:247–266. Oxford Univ. Press, New York and Oxford.

Wagner, W.H., Jr. and **F.S. Wagner.** 1993. Ophioglossaceae. In: Flora of North America Editorial Committee, eds. Fl.

North Amer. 2:85–106. Oxford Univ. Press, New York and Oxford.

Walck, J.L., J.M. Baskin, and **C.C. Baskin.** 1996. Sandstone rockhouses of the eastern United States with particular reference to the ecology and evolution of the endemic plant taxa. Bot. Rev. 62:311–362.

Walker, L.R., K. Mehltreter, and **J.M. Sharpe.** 2010. Current and future directions in fern ecology. In: K. Mehltreter, L.R. Walker, and J.M. Sharpe, eds. 2010. Fern ecology. Ch. 10. Pp. 360–378. Cambridge Univ. Press, Cambridge, England, U.K.

Walker, S. 1962. Further studies in the genus *Dryopteris*: The origin of *D. clintoniana, D. celsa* and related taxa. Amer. J. Bot. 49:497–503.

Walker, T.G. 1984. Chromosomes and evolution in pteridophytes. In: A.K. Sharma and A. Sharma, eds. Chromosomes in evolution of eukaryotic groups, Vol. 2. CRC Press, Boca Raton, FL.

Walters, T.W. and **R. Wyatt.** 1982. The vascular flora of granite outcrops in the Central Mineral Region of Texas. Bull. Torrey Bot. Club 109:344–364.

Wang, M.-L., Z.-D. Chen, X.-C. Zhang, S.-G. Lu, and **G.-F. Zhao.** 2003. Phylogeny of the Athyriaceae: Evidence from chloroplast *trnL-F* region sequences. Acta Phytotax. Sin. 41:416–426.

Wang, M.-L., Y.-T. Hsieh, and **G.-F. Zhao.** 2004. A revised subdivision of the Athyriaceae. Acta Phytotax. Sin. 42:524–527.

Watano, Y. and **S. Masuyama.** 1991. Inbreeding in natural populations of the annual polyploid fern *Ceratopteris thalictroides* (Parkeriaceae). Syst. Bot. 16:705–714.

Watano, Y. and **N. Sahashi.** 1992. Predominant inbreeding and its genetic consequences in a homosporous fern genus, *Sceptridium* (Ophioglossaceae). Syst. Bot. 17:486–502.

Weakley, A. 2010. Flora of the southern and mid-Atlantic states. Univ. North Carolina Herbarium. http://www.herbarium.unc.edu/flora.htm. Accessed Apr 2010.

Weakley, A.S., J.C. Ludwig, and **J.F. Townsend.** 2012. Flora of Virginia. Foundation of the Flora of Virginia Project Inc., Richmond. Botanical Research Institute of Texas, Fort Worth.

Weatherby, C.A. 1937. A further note on *Salvinia*. Amer. Fern J. 27:98–102.

Weatherby, C.A. 1939. The group of *Polypodium polypodioides*. Contr. Gray Herb. 124:22–35.

Weatherby, C.A. 1944. The group of *Selaginella oregana* in North America. J. Arnold Arbor. 25:407–419.

Weathers, S.A. 1998. Field guide to plants poisonous to livestock: Western U.S. Rosebud Press, Fruitland, UT.

Weber, W.A. 2003. The middle Asian element in the southern Rocky Mountain flora of the western United States: A critical biogeographical review. J. Biogeogr. 30:649–685.

Webster, T.R. 1992. Developmental problems in *Selaginella* (Selaginellaceae) in an evolutionary context. Ann. Missouri Bot. Gard. 79:632–647.

Weed-Killer.net. 2010. Water fern. http://www.weed-killer.net/water-fern.html. Accessed 13 Sep 2010.

Wen, J. 1999. Evolution of eastern Asian and eastern North American disjunct distributions in flowering plants. Ann. Rev. Ecol. Syst. 30:421–455.

Wen, J. 2001. Evolution of eastern Asian-North American biogeographic disjunctions: A few additional items. Int. J. Pl. Sci. 162 (Suppl.):S117–S122.

Wendt, T. 1980. Notes on some *Pleopeltis* and *Polypodium* species of the Chihuahuan Desert region. Amer. Fern J. 70:5–11.

Werth, C.R. 1991. Isozyme studies on the *Dryopteris 'spinulosa'* complex, I: The origin of the log fern *Dryopteris celsa*. Syst. Bot. 16:446–461.

Werth, C.R., A.M. Evans, and **D.M.E. Ware.** 1988. The southern woodfern *Dryopteris* [X]*australis* in Tennessee and Virginia. Castanea 53:263–271.

Wherry, E.T. 1964. The southern fern guide (southeastern and midland United States; Doubleday Nature Guide Series). Illustrated by James C. W. Chen and Keith C.Y. Chen. First edition: Doubleday & Co., Garden City, New York.

Whetstone, R.D. and **T.A. Atkinson.** 1993. Osmundaceae. In: Flora of North America Editorial Committee, eds. Fl. North Amer. 2:107–109. Oxford Univ. Press, New York and Oxford.

White, R.A. 1977. Taxonomic and morphological relationships of the Psilotaceae. Brittonia 29:1–2.

Whitebread, C. 1941. Beware of "Lycopodium"! Amer. Fern J. 31:100–102.

Whitten, W.M and **C. Jacono.** 2009. Molecular identification of invasive water-clover ferns: Final report for a contract from the Florida Department of Environmental Protection. http://www.flmnh.ufl.edu/herbarium/research/barcoding/dep-marsilea1-report.htm. Accessed Jul 2010.

Whittier, D.P. 2003. Rapid gametophyte maturation on *Ophioglossum crotalaphoroides*. Amer. Fern J. 93:137–145.

Whittier, D.P. 2008. Forcing autumnal growth of *Ophioglossum*. Amer. Fern J. 98:47–48.

Wieboldt, T.F. and **S. Bentley.** 1982. *Cheilanthes feei* new to Virginia. Amer. Fern. J. 72:76–78.

Wikström, N. 2001. Diversification and relationships of extant homosporous lycopods. Amer. Fern J. 91:150–165.

Wikström, N. and **P. Kenrick.** 1997. Phylogeny of Lycopodiaceae (Lycopsida) and the relationships of *Phylloglossum drummondii* Kunze based on *rbc*L sequences. Int. J. Pl. Sci. 158:862–871.

Wikström, N. and **P. Kenrick.** 2000. Relationships of *Lycopodium* and *Lycopodiella* based on combined plastid *rbcL* gene and *trnL* intron sequence data. Syst. Bot. 25:495–510.

Wikström, N., P. Kenrick, and **J.C. Vogel.** 2002. Schizaeaceae: A phylogenetic approach. Rev. Palaeobot. Palynol. 119:35–50.

Wikström, N. and **K.M. Pryer.** 2005. Incongruence between primary sequence data and the distribution of a mitochondrial *atp1* group II intron among ferns and horsetails. Molec. Phylogen. Evol. 36:484–493. http://www.pryerlab.net/publication/fichier724.pdf. Accessed Jul 2010.

Wilbur, R.L. and **M.K. Whitson.** 2005. The identity of Riddell's seven validly published but over-looked pteridophytic binomials. Amer. Fern J. 95:160–169.

Williams, C.J., A.H. Johnson, B.A. LePage, D.R. Vann, and **K.D. Taylor.** 2003a. Reconstruction of Tertiary *Metasequoia* forests. I. Test of a method for biomass determination based on stem dimensions. Paleobiology 29:256–270.

Williams, C.J., A.H. Johnson, B.A. LePage, D.R. Vann, and **T. Sweda.** 2003b. Reconstruction of Tertiary *Metasequoia* forests. II. Structure, biomass, and productivity of Eocene floodplain forests in the Canadian Arctic. Paleobiology 29:271–292.

Willis, T.E. and **J.E. Nester-Hudson.** 2006. Characterization of a *Thelypteris* hybrid from Walker County, Texas. Amer. Fern J. 96:127–133.

Windham, M.D. 1986. Reassessment of the phylogenetic relationships of *Notholaena*. Amer. J. Bot. 73:742.

Windham, M.D. 1987a. Chromosomal and electrophoretic studies of the genus *Woodsia* in North America. Amer. J. Bot. 74:715.

Windham, M.D. 1987b. *Argyrochosma*, a new genus of cheilanthoid ferns. Amer. Fern J. 77:37–41.

Windham, M.D. 1993a. Pteridaceae [Family description and key to genera]. In: Flora of North America Editorial Committee, eds. Fl. North Amer. 2:122–124. Oxford Univ. Press, New York and Oxford.

Windham, M.D. 1993b. *Argyrochosma*. In: Flora of North America Editorial Committee, eds. Fl. North Amer. 2:171–175. Oxford Univ. Press, New York and Oxford.

Windham, M.D. 1993c. *Pellaea*. In: Flora of North America Editorial Committee, eds. Fl. North Amer. 2:175–186. Oxford Univ. Press, New York and Oxford.

Windham, M.D. 1993d. *Woodsia*. In: Flora of North America Editorial Committee, eds. Fl. North Amer. 2:270–280. Oxford Univ. Press, New York and Oxford.

Windham, M.D. 1993e. New taxa and nomenclatural changes in the North American fern flora. Contr. Univ. Michigan Herb. 19:31–61.

Windham, M.D. 1993f. *Notholaena*. In: Flora of North America Editorial Committee, eds. Fl. North Amer. 2:143–149. Oxford Univ. Press, New York and Oxford.

Windham, M.D., L. Huiet, E. Schuettpelz, A.L. Grusz, C. Rothfels, J. Beck, G. Yatskievych, and **K.M. Pryer.** 2009. Using plastid and nuclear DNA sequences to redraw generic boundaries and demystify species complexes in cheilanthoid ferns. Amer. Fern J. 99:128–132.

Windham, M.D. and **E.W. Rabe.** 1993. *Cheilanthes*. In: Flora of North America Editorial Committee, eds. Fl. North Amer. 2:152–169. Oxford Univ. Press, New York and Oxford.

Windham, M.D. and **G. Yatskievych.** 2003a. Revised nomenclature and a new North American record for the villose cliff brake (*Pellaea*, Pteridophyta). Novon 13:358–362.

Windham, M.D. and **G. Yatskievych.** 2003b. Chromosome studies of cheilanthoid ferns (Pteridaceae: Cheilanthoideae) from the western United States and Mexico Amer. J. Bot. 90:1788–1800.

Winther, J.L. and **W.F. Friedman.** 2007. Arbuscular mycorrhizal symbionts in *Botrychium* (Ophioglossaceae). Amer. J. Bot. 94:1248–1255.

Witsell, C.T. and **W.M. Shepherd.** 1998. Rediscovery of *Marsilea vestita* subsp. *vestita* in Pulaski County, Arkansas after 162 years. J. Arkansas Acad. Sci. 52:142–143. http://www.naturalheritage.com/%21userfiles/Marsilea.pdf. Accessed Jun 2011.

Wolf, P.G. 1995. Phylogenetic analyses of *rbcL* and nuclear ribosomal RNA gene sequences in Dennstaeditiaceae. Amer. Fern. J. 85:306–327.

Wolf, P.G. 1997. Evaluation of *atpB* nucleotide sequences for phylogenetic studies of ferns and other pteridophytes. Amer. J. Bot. 84:1429–1440.

Wolf, P.G., A.M. Duffy, and **J.M. Roper.** 2009. Phylogenetic use of inversions in fern chloroplast genomes. Amer. Fern J. 99:132–134.

Wolf, P.G. and **J.M. Roper.** 2008. Structure and evolution of fern plastid genomes. In: T.A. Ranker and C.H. Haufler, eds. The biology and evolution of ferns and lycophytes. Ch. 6. Pp. 159–174. Cambridge Univ. Press, Cambridge, England, U.K.

Wolf, P.G., C.A. Rowe, R.B. Sinclair, and **M. Hasebe.** 2003. Complete nucleotide sequence of the chloroplast genome from a leptosporangiate fern, *Adiantum capillus-veneris* L. DNA Res. 10:59–65.

Wolf, P.G., H. Schneider, and **T.A. Ranker.** 2001. Geographic distributions of homosporous ferns: Does dispersal obscure evidence of vicariance? J. Biogeogr. 28:263–270.

Wollenweber, E. and **H. Schneider.** 2000. Lipophilic exudates of Pteridaceae—chemistry and chemotaxonomy. Biochem. Syst. Ecol. 28:751–777.

Wood, M., B. Hardin, and **J. García.** 2001. Attack on giant salvinia. U.S.D.A. Agric. Res. 49:4–6. http://www.ars.usda.gov/is/AR/archive/nov01/giant1101.htm. Accessed Aug 2004.

Wood, T.E., N. Takebayashi, M.S. Barker, I. Mayrose, P.B. Greenspoon, and **L.H. Rieseberg.** 2009. The frequency of polyploid speciation in vascular plants. Proc. Natl. Acad. Sci. USA. 106:13875–13879.

Woodland, D.W. 1997. Contemporary plant systematics, 2nd ed. Andrews Univ. Press, Berrien Springs, MI.

Wu, T.-C. and W-Y. Kao. 2009. The function of trichomdes of an amphibious fern, *Marsilea quadrifolia*. Amer. Fern J. 99:323–332.

Wu, Z. 1983. On the significance of Pacific intercontinental discontinuity. Ann. Missouri Bot. Gard. 70:577–590.

Wunderlin, R.P. and **B.F. Hansen.** 2000. Flora of Florida. Vol. 1. Pteridophytes and gymnosperms. Univ. Press of Florida, Gainesville.

Xiang, Q.-Y. and **D.E. Soltis.** 2001. Dispersal-vicariance analysis on intercontinental disjuncts: Historical biogeographic implications for angiosperms in the Northern Hemisphere. Int. J. Pl. Sci. 162(Suppl.):S29–S39.

Xiang, Q.-Y., D.E. Soltis, and **P.S. Soltis.** 1998. The eastern Asian and eastern and western North American floristic disjunction: Congruent phylogenetic patterns in seven diverse genera. Molec. Phylogen. Evol. 10:178–190.

Yamada, T. and **M. Kato.** 2002. *Regnellites nagashimae* gen. et sp. nov., the oldest macrofossil of Marsileaceae, from the Upper Jurassic to Lower Cretaceous of western Japan. Int. J. Pl. Sci. 163:715–722.

Xiang, Q.Y., W.H. Zhang, R.E. Ricklefs, H. Qian, Z.D. Chen, J. Wen, and **J.H. Li.** 2004. Regional differences in speciation and ITS evolution: A comparison between eastern Asia and eastern North America. Evolution 58:2175–2184.

Yarborough, S.C. and **A.M. Powell.** 2002. Ferns and fern allies of the Trans-Pecos and adjacent areas. Texas Tech Univ. Press, Lubbock.

Yatabe, Y., N. Murakami, and **K. Iwatsuki.** 2005. *Claytosmunda*; a new subgenus of *Osmunda* (Osmundaceae). Acta Phytotax. Geobot. 56:127–128.

Yatabe, Y., H. Nishida, and **N. Murakami.** 1999. Phylogeny of Osmundaceae inferred from *rbcL* nucleotide sequences and comparison to the fossil evidences. J. Pl. Res. 112:397–404.

Yatskievych, G. (George). Personal communication. Director of the Flora of Missouri Project and expert in Pteridophytes. Missouri Bot. Gard., St. Louis.

Yatskievych, G. 1990. Systematics of the fern genus *Cyrtomium* in the New World. Ph.D. Dissertation, Indiana Univ., Bloomington.

Yatskievych G. 1992. Innovations in the fern genus *Phanerophlebia*. Novon 2:445–446.

Yatskievych, G. 1993a. *Cyrtomium*. In: Flora of North America Editorial Committee, eds. Fl. North Amer. 2:299–300. Oxford Univ. Press, New York and Oxford.

Yatskievych, G. 1993b. *Phanerophlebia*. In: Flora of North America Editorial Committee, eds. Fl. North Amer. 2:200–301. Oxford Univ. Press, New York and Oxford.

Yatskievych, G. 1993c. Antheridiogen response in *Phanerophlebia* and related fern genera. Amer. Fern J. 83:30–36.

Yatskievych, G. 1996. A revision of the fern genus *Phanerophlebia* (Dryopteridaceae). Ann. Missouri Bot. Gard. 83:168–199.

Yatskievych, G. 1999. Steyermark's flora of Missouri. Vol. 1 (Pteridophytes, gymnosperms, and monocots). Missouri Dept. of Conservation, Jefferson City, in cooperation with the Missouri Botanical Garden Press, St. Louis.

Yatskievych, G.A. 2003. Conservation and endangerment of North American pteridophytes. Fiddlehead Forum, Bull. Amer. Fern Soc. 3:25–26.

Yatskievych, G.A. and **A.L. Arbeláez A.** 2008. A new species and three generic transfers in the fern genus *Notholaena* (Pteridaceae). Novon 18:120–124.

Yatskievych, G., D.B. Stein, and **G.J. Gastony.** 1988. Chloroplast DNA evolution and systematics of *Phanerophlebia* (Dryopteridaceae) and related fern genera. Proc. Natl. Acad. Sci. USA. 85:2589–2593.

Yatskievych, G. and **M.D. Windham.** 2008a. Vascular plants of Arizona: Azollaceae. Canotia 4:31–34. http://www.canotia.org/volumes/CANOTIA_2008_Vol4_2_Yatskievych_et_Windham_Azollaceae.pdf. Accessed Jun 2010.

Yatskievych, G. and **M.D. Windham.** 2008b. Vascular plants of Arizona: Dennstaedtiaceae. Canotia 4:38–40. http://www.canotia.org/volumes/CANOTIA_2008_Vol4_2_Yatskievych_et_Windham_Dennstaedtiaceae.pdf. Accessed Jul 2010.

Yatskievych, G. and **M.D. Windham.** 2008c. Vascular plants of Arizona: Equisetaceae. Canotia 4:41–45. http://nhc.asu.edu/vpherbarium/canotia/Yatskievych_Windham_CANOTIA_Dec08_Vol4_2_Equisetaceae.pdf. Accessed Jun 2010.

Yatskievych, G. and **M.D. Windham.** 2008d. Vascular plants of Arizona: Salviniaceae. Canotia 4:46–49. http://www.canotia.org/volumes/CANOTIA_2008_Vol4_2_Yatskievych_et_Windham_Salviniaceae.pdf. Accessed Jul 2010.

Yatskievych, G. and **M.D. Windham.** 2009a. Vascular plants of Arizona: Polypodiaceae. Canotia 5:34–38. http://www.canotia.org/volumes/CANOTIA_2009_Vol5_1_Yatskievych_et_Windham_Polypodiaceae.pdf. Accessed Jul 2010.

Yatskievych, G. and **M.D. Windham.** 2009b. Vascular plants of Arizona: Selaginellaceae. Canotia 5:39–48. http://www.canotia.org/volumes/CANOTIA_2009_Vol5_1_Yatskievych_et_Windham_Selaginellaceae.pdf. Accessed Jul 2010.

Yatskievych, G. and **M.D. Windham.** 2009c. Vascular plants of Arizona: Thelypteridaceae. Canotia 5:49–52. http://www.canotia.org/volumes/CANOTIA_2009_Vol5_1_Yatskievych_et_Windham_Thelypteridaceae.pdf. Accessed Jun 2011.

Yatskievych, G. and **M.D. Windham.** 2009d. Vascular plants of Arizona: Marsileaceae. Canotia 5:30–33. http://www.canotia.org/volumes/CANOTIA_2009_Vol5_1_Yatskievych_et_Windham_Marsileaceae.pdf. Accessed Jun 2011.

Zech, J.C. and **P.R. Manning.** 1996. *Ophioglossum engelmannii* in Trans-Pecos, Texas. Amer. Fern J. 86:102.

Zech, J.C., P.R. Manning, and **W.H. Wagner, Jr.** 1998. A new adder's-tongue for North America. Sida 18:307–313.

Zhao K., F. Hai, and **I.A. Ungar.** 2002. Survey of halophyte species in China. Pl. Sci. 163:491–498.

Zhao, K., J. Song, G. Feng, M. Zhao, and **J. Liu.** 2011. Species, types, distribution, and economic potential of halophytes in China. Pl. Soil 342:495–509.

Zimmerman, W.J., T.A. Lumpkin, and **I. Watanabe.** 1989. Classification of *Azolla* spp., section *Azolla*. Euphytica 43:223–232.

Zimmerman, W.J., I. Watanabe, T. Ventura, P. Payawal, and **T.A. Lumpkin.** 1991. Aspects of the genetic and botanical status of neotropical *Azolla* species. New Phytol. 119:561–566.

INDEX

The accepted scientific names of native or naturalized members of the Texas flora (and other plants discussed in detailed notes) are given in [Roman type]. In addition, accepted generic and family names of ferns and lycophytes treated in the flora are in [**bold**]. Taxonomic synonyms and names of plants casually mentioned are in [*italics*]. Common names are in [SMALL CAPS]. In order to help locate specific and infraspecific epithets, they are given in [Roman type] as separate entries followed on the next line by the genus or genera in which they are named.

Acer, 37
acrostichoides
Polystichum, 152
Acrostichum, 226
bonariense, 236
septentrionale, 106
sinuatum, 225
thalictroides, 228
ADDER'S-TONGUE, 5, 184, 194
BULBOUS, 58, 185, 190
DWARF, 185
ENGELMANN'S, 187
FAMILY, 177
LARGE, 190
LEAST, 187
LIMESTONE, 18, 187
LONGSTEM, 190
SLENDER, 58, 187
SOUTHERN, 7, 194
STALKED, 190
UPRIGHT, 190
Adiantum, 210
capillus-veneris, 11, 18, **210**
var. *modestum*, 213
var. *rimicola*, 213
cordifolium, 272
tricholepis, 30, 40, 41, 42, **213**
aemula
Cheilanthes, 232
affine
Equisetum hyemale subsp., 158
AFRICAN PYLE, 291
agamospory, 22
ALABAMA LIP FERN, 232
alabamensis
Cheilanthes, 232
aliena
Notholaena, 259
allelochemicals, 30
allelopathy, 30
Allosorus karwinskii, 258
alopecuroides
Lycopodiella, 65
AMERICAN
BASSWOOD, 37
BEECH, 37
CLIMBING FERN, 164
PILLWORT, 2, 173
SCOURING-RUSH, 158
americana
Pilularia, 173
Anabaena azollae, 29, 110, 111
Anchistea, 119
ANDEAN HORSETAIL, 153
Anemia, 6, **97,** 177
FAMILY, 97
mexicana, 10, 41, **99**
Anemiaceae, 6, **97**
annulus, 6, 8, 13
ANTLER FERN, 226
Apios americana, 37
apoda
Selaginella, 76
apogamy, 22, 29
apomixis, 22
Appalachian disjuncts, 31
appendiculatum
Blechnum, 117
appressa
Lycopodiella, 69
APPRESSED
BOG CLUB-MOSS, 69
CLUB-MOSS, 69
SOUTHERN, 69
AQUARIUM WATER-MOSS, 291
aquilinum
Pteridium, 132
Arcto-Tertiary, 12
flora, 33
areolata
Woodwardia, 119
Argyrochosma, 213
dealbata, 27, **214**
limitanea
subsp. *limitanea*, 214
subsp. **mexicana, 214**
microphylla, 218
ARIZONA
CLIFF-BRAKE, 32
SPIKE-MOSS, 79
arizonica
Pellaea ternifolia subsp., 280
Selaginella, 79
arvense
Equisetum, 155
aschenborniana
Notholaena, 261
asexual reproduction, 22, 23
Asian connection, 32
ASIAN NET-VEIN HOLLYFERN, 139
ASIATIC HOLLY FERN, 137
Aspidium ludovicianum, 147
Aspleniaceae, 99
asplenioides
Athyrium filix-femina subsp., 107
Asplenium, 99
×*alternifolium*, 106
melanocaulon, 107
nidus, 99
palmeri, 2, 22, 23, 29, **102**
platyneuron, 2, 32, **102**
var. *bacculum-rubrum*, 102
var. *incisum*, 102
resiliens, 23, **106**
septentrionale, 2, 32, **106**
trichomanes
subsp. *quadrivalens*, 107
subsp. **trichomanes, 106**
Astrolepis, 13, 23, 25, 26, **218,** 271
cochisensis, 2, 23, 30, 42, **219**
subsp. **chihuahuensis, 219**
subsp. **cochisensis, 219**
crassifolia, 225
integerrima, 32, **219**
mexicana, 41
sinuata, 28, 31, 32, 41, **221**
subsp. **mexicana, 221**
subsp. **sinuata, 225**
windhamii, 25, 27, **225**
Athyriaceae, 9, **107**
Athyrium, 9, **107**, 122, 135, 307
angustum, 109
asplenioides, 109
subsp. *angustum*, 109
filix-femina
subsp. **asplenioides,** 12, 142, **107**
var. *asplenioides*, 109
subsp. *filix-femina*, 109
atropurpurea
Pellaea, 272
auriculata
Phanerophlebia, 149
AUSTRALIAN WATER-CLOVER, 172
Azolla, 8, 13, 16, 22, 29, 33, 42, **110,** 166
caroliniana, 2, 18, 22, **112**
filiculoides, 112
mexicana, 115
microphylla, 8, 11, **115**
pinnata, 115
Azollaceae, 8, **109,** 166

B

BANDED NARDOO, 172
BASKET SELAGINELLA, 76
BASSWOOD, AMERICAN, 37
BEAD FERN, 175
BEADED
CUPLET FERN, 2, 41, 42, 130
LIP FERN, 242

BEECH FERN, 296
  BROAD, 12, 18, 296
  SOUTHERN, 296
BEECH, AMERICAN, 37
BEECHDROPS, 37
BERRY, PARTRIDGE, 37
Betula, 33
BIG-FOOT WATER-CLOVER, 168, 168
Bignonia capreolata, 37
biogeography, 30
BIRCH, 33
BIRD'S-NEST FERN, 99
biternatum
  Botrychium, 178
BLACK-GUM, 37
BLACK-STEM SPLEENWORT, 106
BLADDER FERN, 123
  BULBLET, 22, 23, 123
  FAMILY, 122
  LOWLAND, 125
  SOUTHERN, 37, 125, 310
  UTAH, 129
Blechnaceae, 117
Blechnum, 117
  appendiculatum, 30, 42, 213, **117**
  occidentale
    var. *minor,* 119, 213
    forma *pubirachis,* 119
    var. *pubirachis,* 119
BLUE FERN, 272
BLUNT SPIKE-MOSS, 82
BLUNT-LOBE
  CLIFF FERN, 310
  WOODSIA, 125, 310
BOG CLUB-MOSS, 65, 72
  APPRESSED, 69
  FEATHER-STEM, 69
  PROSTRATE, 69
  SLENDER, 18, 72
  SOUTHERN, 69
BOMMER, HAIRY, 226
Bommeria, 225
  hispida, 28, **226,** 271
  schaffneri, 226
BOMMERIA, 225
  DANCING, 226
BONAIRE LIP FERN, 232
bonariensis
  Cheilanthes, 232
Boreotropical, 12
  flora, 33
BOSTON FERN, 163
  WILD, 163
BOSTON SWORD FERN, 163
Botrychium, 5, **178**
  biternatum, 5, 7, **178**
  dissectum, 180
    forma *obliquum,* 180
    var. *obliquum,* 180
    var. *oblongifolium,* 180
    var. *tenuifolium,* 180
  lunaria, 180
  lunarioides, 180
  obliquum, 180
    var. *elongatum,* 180
  tenuifolium, 180
  virginianum, 184
Botryogramme karwinskii, 258
Botrypus, 178
  virginianus, 184
BOTTLE-BRUSH, 155
BRAKE, 285
  BUCKHORN, 199
  CHINESE, 285, 288
  CHINESE LADDER, 288
  CLIFF-, 271, 272
  CREEPING CLIFF-, 276
  FAMILY, 208
  HEART-LEAF CLIFF-, 272
  INTERMEDIATE CLIFF-, 276
  LADDER, 42, 288
  OVATE-LEAF CLIFF-, 276
  PASTURE, 132
  PURPLE CLIFF-, 272
  PURPLE-STEM CLIFF-, 272
  SMOOTH CLIFF-, 276
  SPIDER, 285
  SPINY CLIFF-, 280
  TERNATE CLIFF-, 276
  TRANS-PECOS CLIFF-, 276
  VILLOSE CLIFF-, 282
  ZIGZAG CLIFF-, 276
BRACKEN
  FERN, 2, 30, 130, 132, 134
    DOWNY, 134
    EASTERN, 135
    FAMILY, 129
    HAIRY, 134
    TAILED, 134
    WESTERN, 134
  PASTURE, 134
  SAW-LEAF, 285
BRAUN'S SCOURING-RUSH, 161
BRITTLE FERN, 123, 129
  LOWLAND, 125
  SOUTHWESTERN, 129
BROAD
  BEECH FERN, 12, 18, 296
  HALBERD FERN, 2, 42, 293
BROWN-STEM SPLEENWORT, 102
BRUSH, BOTTLE-, 155
Bryodesma
  arizonica, 79
  muticum, 82
  peruviana, 89
  rupincola, 89
  underwoodii, 92
  viridissimum, 95
  wrightii, 95
BUCKHORN
  BRAKE, 199
  FERN, 199
BUCKLER FERN FAMILY, 293
BULB LIP FERN, 221
bulbifera
  Cystopteris, 123
BULBLET BLADDER FERN, 22, 23, 123
BULBOUS ADDER'S-TONGUE, 58, 185, 190
butleri
  Isoëtes, 56
BUTLER'S QUILLWORT, 5, 190

C

CABALLO, COLA DE, 153, 161
Calamites, 152
CAÑUELA, 158, 161
capillus-veneris
  Adiantum, 210
CAROLINA
  CLUB-MOSS, 72
  MOSQUITO FERN, 2, 112
caroliniana
  Azolla, 112
  Pseudolycopodiella, 72
Carpinus, 37
Carya, 36
celsa
  Dryopteris, 140
Central Mineral Region, 11, 38, 40
Ceratodactylis osmundoides, 258
Ceratopteris, 208, **226**
  cornuta, 228
  thalictroides, 228
Cercis, 37
cernua
  Palhinhaea, 71
CHAIN FERN, 13, 117, 119
  NARROW-LEAF, 119
  NETTED, 119
  VIRGINIA, 122
CHAPMAN'S CLUB-MOSS, 5, 69
Cheilanthes, 12, 13, 25, 26, **228**
  aemula, 41, **232,** 253
  alabamensis, 232, 253
  aschenborniana, 261
  bonariensis, 232
  candida var. *copelandii,* 261
  castanea, 236
  cochisensis, 219
  dealbata, 214
  eatonii, 32, **236**
    forma *castanea,* 236
  feei, 32, **238**
  fendleri, 238
  grayi, 265
  greggii, 265
  horridula, 238
  integerrima, 221
  kaulfussii, 41, **242**
  lanosa, 242
  lendigera, 242, 282
  leucopoda, 40, 41, 42, **246**, 258
  limitanea var. *mexicana,* 214

Cheilanthes *(cont.)*
lindheimeri, 28, **246**
microphylla, 232
nealleyi, 265
neglecta, 268
parvifolia, 218
sinuata, 225
var. *cochisensis*, 219
standleyi, 268
tomentosa, 249
var. *eatonii*, 236
villosa, 249
wootonii, 249
wrightii, 21, **253**
yavapensis, 40, **253**
CHIHUAHUA SCALY CLOAK FERN, 219
Chihuahuan Desert, 10, 13, 40
chihuahuensis
Astrolepis cochisensis subsp., 219
CHINESE
BRAKE, 285, 288
LADDER BRAKE, 288
CHISOS HOLLY FERN, 41, 149
Christella
dentata, 300
hispidula, 300
normalis, 304
CHRISTMAS FERN, 149, 152
Chrysochosma
aliena, 261
aschenborniana, 261
candida var. *copelandii*, 261
grayii, 265
greggii, 265
hookeri, 268
neglecta, 268
schaffneri var. *nealleyi*, 265
cinnamomea
Dryopteris, 142
cinnamomeum
Osmundastrum, 122, 197
Cinnamomum verum, 197
CINNAMON
FERN, 6, 9, 36, 37, 122, 197, 199, 307
FERN FAMILY, 194
WOOD FERN, 41, 42, 142
Class Polypodiopsida, 4
classification
fern, 5, 7
lycophyte, 7
pteridophyte, 7
CLIFF FERN, 308
BLUND-LOBE, 310
NEW MEXICO, 308
PHILLIPS'S, 310
PLUMMER'S, 314
CLIFF-BRAKE, 271, 272
ARIZONA, 32
CREEPING, 276
HEART-LEAF, 16, 41, 272
INTERMEDIATE, 276
OVATE-LEAF, 276
PURPLE, 24, 272
-STEM, 272
SMOOTH, 276
SPINY, 280
TERNATE, 276
TRANS-PECOS, 33, 41, 276
VILLOSE, 282
WRIGHT'S, 31, 32
ZIGZAG, 41, 276
CLIMBING FERN, 6, 164
AMERICAN, 164
FAMILY, 163
HOLLY FERN, 163
JAPANESE, 2, 164
OLD WORLD, 166
CLOAK FERN, 26, 258
CHIHUAHUA SCALY, 219
COCHISE SCALY, 219
COPELAND'S, 41, 261
FALSE, 213, 214
FOREIGN, 259
GRAY'S, 261
GREGG'S, 41, 265
HYBRID, 221
SCALY, 221
LONG, 221
MAXON'S, 268
MEXICAN, 41, 259
SCALY, 41, 221
NEALLEY'S, 41, 265
POWDERY, 214
FALSE, 214
SCALED, 261
SCALY, 218
SMALL-LEAF FALSE, 218
SOUTHWESTERN, 32, 221
SOUTHWESTERN FALSE, 214
STAR, 28, 268
-SCALED, 13, 23, 25, 218
WAVY, 28, 221
SCALE, 32, 221, 225
WHOLE-LEAF, 221
WINDHAM'S, 225
SCALY, 27, 225
CLOVER
AUSTRALIAN WATER-, 172
BIG-FOOT WATER-, 168
EUROPEAN WATER-, 167
GOLDEN WATER-, 168
HAIRY WATER-, 172
LARGE-FOOT WATER, 168
MEXICAN WATER-, 168
VARIEGATED WATER-, 172
WATER-, 8, 167, 168, 172
CLUB-MOSS
APPRESSED BOG, 69
BOG, 65, 72
CAROLINA, 72
CHAPMAN'S, 5, 69
CREEPING, 69
FAMILY, 63
FEATHER-STEM, 69
FEATHER-STEM BOG, 69
FOXTAIL, 65
BOG, 65
NODDING, 71
PROSTRATE BOG, 69
SLENDER, 69, 72
BOG, 18, 72
SOUTHERN, 69
APPRESSED, 69
BOG, 69
STAG-HORN, 71
TIGHT-LEAF, 69
CLUB-MOSSES, 3, 4, 42, 53, 54, 64
COCHISE SCALY CLOAK FERN, 219
cochisensis
Astrolepis, 219
COLA DE CABALLO, 153, 161
COMMON
GRAPE FERN, 184
HORSETAIL, 18, 155
SALVINIA, 289
SCOURING-RUSH, 158
WOODSIA, 310
Coniogramme, 256
Conservation, 43
COPELAND'S CLOAK FERN, 41, 261
copelandii
Notholaena, 261
COPPER FERN, 225, 226
corallina
Selaginella, 79
CORDATE-LEAF LLAVEA, 256
cordifolia
Llavea, 256
Pellaea, 272
Correll, Donovan Stuart, 1, 9, 44
CRAPE FERN, 194
CREEPING
CLIFF-BRAKE, 276
CLUB-MOSS, 69
CRESTED WOOD FERN, 140
CROSS-VINE, 37
crotalophoroides
Ophioglossum, 185
crosiers, 16
cryptogams, vascular, 17
Cryptogramma, 256
Cryptogrammaceae, 256
Ctenitis, 135
CULANTRILLO, 210
CUP FERN, 130
CUPLET FERN
BEADED, 2, 41, 42, 130
FAMILY, 129
CUT-LEAF GRAPE FERN, 180
Cyclosorus, 299
Cyrtobagous, 293

Cyrtomium, 147, **137**
auriculatum, 149
falcatum, 137
fortunei, 139
CYRTOMIUM, FORTUNE'S, 139
Cystopteridaceae, 9, **122**
Cystopteris, 9, 135, 122, **123,** 135, 307
bulbifera, 22, 23, 29, 36, 37, 40, **123**
fragilis
var. *protrusa*, 125
subsp. *tenuifolia*, 129
protrusa, 36, 37, 125, **125**, 310
reevesiana, 129
tennesseensis, 125
utahensis, 129

D
DAGGER FERN, 152
DALLAS
FERN, 163
JEWEL FERN™, 163
DANCING BOMMERIA, 226
DAWN REDWOOD, 33
dealbata
Argyrochosma, 214
DEER FERN, 117
FAMILY, 117
Dennstaedtia, 130
globulifera, 2, 30, 31, 41, 42, 213, **130**
punctilobula, 130
Dennstaedtiaceae, 129
dentata
Thelypteris, 300
Desert, Chihuahuan, 10, 13, 40
Dicksonia globulifera, 130
Diplazium, 107, 307
Diplostachyum, 74
apodum, 76
disjunct distribution, 30
disjuncts, Appalachian, 31
dispersal, long-distance, 30
DISSECTED GRAPE FERN, 180
dissectum
Botrychium, 180
diversity, 9
Division Polypodiophyta, 95
DORADILLA, 89
DOWNY
BRACKEN FERN, 134
MAIDEN FERN, 300
SHIELD FERN, 300
Dryopteridaceae, 9, 107, 122, **135**
Dryopteris, 31, 135, **139**
celsa, 40, 44, **140**
cinnamomea, 31, 41, 42, **142**
Dryopteris
cristata, 140
dentata, 300
filix-mas, 41, 107, **142**
goldiana, 140
subsp. *celsa*, 140
hexagonoptera, 299
knoblochii, 142
ludoviciana, 12, 40, **147**
marginalis, 44, **147**
normalis, 11, 304
var. *lindheimeri*, 304
setigera, 296
thelypteris var. *pubescens*, 304
versicolor, 300
DWARF
ADDER'S-TONGUE FERN, 185
PALMETTO, 36

E
EARED HOLLY FERN, 149
East Texas, 9, 10, 11, 12, 13, 14, 35, 37, 40
EASTERN
BRACKEN FERN, 135
MARSH FERN, 304
MOSQUITO FERN, 112
EATON'S LIP FERN, 32, 236
eatonii
Cheilanthes, 236
EBONY SPLEENWORT, 2, 102
LITTLE, 106
Edwards Plateau, 10, 11, 13, 14, 36, 37, 40
Elaphoglossum, 135
Enchanted Rock, 11, 15, 38
endemic, 38, 40
ENGELMANN'S ADDER'S-TONGUE, 187
engelmannii
Ophioglossum, 187
Epifagus virginiana, 37
Equisetaceae, 152
Equisetales, 6
Equisetites, 154
Equisetophyta, 3, 5
Equisetopsida, 5
Equisetum, 4, 33, 42, 152, **153**
arvense, 18, **155**
bogotense, 153
×ferrissii, 9, **155**
fluviatile, 154
funstonii, 161
giganteum, 154
subgenus *Hippochaete*, 153
hyemale, 8, 13, 34, 35
subsp. **affine, 158**
var. *elatum*, 158
var. *intermedium*, 158
var. *robustum*, 158
kansanum, 161
laevigatum, 161
subsp. *funstonii*, 161
prealtum, 158
robustum var. *affine*, 158
erythrolepis
Pleopeltis polylepis var., 201
EUPHYLLOPHYTES, 95
EUROPEAN WATER-CLOVER, 167
exaltata
Nephrolepis, 163

F
Fagus grandifolia, 37
FAIRY
-MOSS, 110
-SWORDS, 246
falcatum
Cyrtomium, 137
FALSE
CLOAK FERN, 213, 214,
POWDERY, 214
SMALL-LEAF, 218
SOUTHWESTERN, 214
MAIDEN FERN, 296
farina, 26
FEATHER-STEM
BOG CLUB-MOSS, 69
CLUB-MOSS, 69
FEE'S LIP FERN, 238
feei
Cheilanthes, 238
FEMALE FERN, 299
FENDLER'S LIP FERN, 238
fendleri
Cheilanthes, 238
fern
allies, 1
classification, 5, 7
phylogeny, 6
FERN
ALABAMA LIP, 232
AMERICAN CLIMBING, 164
ANTLER, 226
ASIATIC HOLLY, 137
BEAD, 175
BEADED
CUPLET, 130
LIP, 242
BEECH, 296
BIRD'S-NEST, 99
BLADDER, 123
BLUE, 272
BLUNT-LOBE CLIFF, 310
BONAIRE LIP, 232
BOSTON, 163
BOSTON SWORD, 163
BRACKEN, 130, 132, 134
BRITTLE, 123, 129
BROAD
BEECH, 296
HALBERD, 293
BUCKHORN, 199
BULB LIP, 221
BULBLET BLADDER, 123
CAROLINA MOSQUITO, 112
CHAIN, 119
CHIHUAHUA SCALY CLOAK, 219

FERN *(cont.)*
CHISOS HOLLY, 149
CHRISTMAS, 149, 152
CINNAMON, 122, 197, 199, 307
WOOD, 142
CLIFF, 308
CLIMBING, 164
HOLLY, 163
CLOAK, 258
COCHISE SCALY CLOAK, 219
COMMON GRAPE, 184
COPELAND'S CLOAK, 261
COPPER, 225, 226
CRAPE, 194
CRESTED WOOD, 140
CUP, 130
CUT-LEAF GRAPE, 180
DAGGER, 152
DALLAS JEWEL, 163
DEER, 117
DISSECTED GRAPE, 180
DOWNY
BRACKEN, 134
MAIDEN, 300
SHIELD, 300
DWARF ADDER'S-TONGUE, 185
EARED HOLLY, 149
EASTERN
BRACKEN, 135
MARSH, 304
MOSQUITO, 112
EATON'S LIP, 236
FALSE
CLOAK, 213, 214
MAIDEN, 296
FEE'S LIP, 238
FEMALE, 299
FENDLER'S LIP, 238
FLOATING, 226, 289
FLOWERCUP, 314
FLOWERING, 97, 195, 199
FOREIGN CLOAK, 259
FORTUNE'S NET-VEIN HOLLY, 139
FRANKINCENSE, 97
FRINGED, 163
FUZZY MAIDENHAIR, 213
GLANDULAR LIP, 242
GOLDEN LIP, 232
GRACEFUL LIP, 253
GRAPE, 178
GRAY'S CLOAK, 261
GREGG'S CLOAK, 265
HAIRY
BRACKEN, 134
LIP, 242
MAIDEN, 300
MAIDENHAIR, 213
HALBERD, 293
HAMMOCK, 117
HARDY JAPANESE HOLLY, 139
HAY-SCENTED, 130, 130
HOLLY, 137, 147, 149
HOUSE HOLLY, 137
HUGUENOT, 285
HYBRID
CLOAK, 221
SCALY CLOAK, 221
INDIAN-HAIR, 102
JAPANESE
CLIMBING, 164
HOLLY, 137
NET-VEIN HOLLY, 137
JIMMY, 219, 221
KAULFUSS' LIP, 242
KING, 194
KUNTH'S MAIDEN, 300
LADY, 107, 142
LEATHER WOOD, 147
LINDHEIMER'S
LIP, 246
SHIELD, 304
LIP, 228
LLAVE´S, 256
LOG, 140
LONG CLOAK, 221
LOUISIANA WOOD, 147
LOWLAND
BLADDER, 125
BRITTLE, 125
LADY, 107
MAIDEN, 210, 299
MALE, 107, 142
MARGINAL
SHIELD, 147
WOOD, 147
MARIANA MAIDEN, 296
MARSH, 299, 304
MATTRESS, 210
MAXON'S CLOAK, 268
MEXICAN, 99
CLOAK, 259
FLOWERING, 99
HOLLY, 147
MOSQUITO, 115
SCALY CLOAK, 221
SILVER, 214
MOSQUITO, 110, 112
NARDOO, 168
NARROW-LEAF CHAIN, 119
NEALLEY'S CLOAK, 265
NETTED CHAIN, 119
NET-VEIN HOLLY, 137
NEW MEXICO CLIFF, 308
NEW WORLD MIDSORUS, 117
NIT-BEARING LIP, 242
OBLIQUE GRAPE, 180
OLD WORLD CLIMBING, 166
OSTRICH, 173
OVATE MARSH, 304
PHILLIPS'S CLIFF, 310
PINELAND, 97
PLUMMER'S CLIFF, 314
POWDERY
CLOAK, 214
FALSE CLOAK, 214
PRICKLY LIP, 238
PROSTRATE GRAPE, 180
RATTLESNAKE, 184
RESURRECTION, 201
RIVAL LIP, 232
ROUGH
LIP, 238
-HAIRY MAIDEN, 300
ROYAL, 195, 258
SAN SABA, 304
SCALED CLOAK, 261
SCALY CLOAK, 218
SENSITIVE, 58, 175, 190
SHIELD, 139
SHIELD-SORUS, 200
SILVER, 213
SLENDER LIP, 238
SMALL-LEAF
FALSE CLOAK, 218
SILVER, 218
SMOOTH LIP, 232
SOUTHERN
BEECH, 296
BLADDER, 125, 310
GRAPE, 178
HAY-SCENTED, 130
LADY, 107
MARSH, 304
SHIELD, 300
WOOD, 147
SOUTHWESTERN
BRITTLE, 129
CLOAK, 221
FALSE CLOAK, 214
SPARSE-LOBE GRAPE, 178
SPRING, 107
ST. JOHN'S SHIELD, 300
STAR, 268
CLOAK, 268
-SCALED CLOAK, 218
SWORD, 149, 161, 163
TAILED BRACKEN, 134
TEXAS LIP, 232
TORRES', 296
TREE, 201
UTAH BLADDER, 129
VARIABLE MAIDEN, 300
VENUS'-HAIR, 210
VILLOUS LIP, 249
VIRGINIA
CHAIN, 122
GRAPE, 184
WATER, 110, 112, 226, 228
WAVY
CLOAK, 221
SCALY CLOAK, 221, 225
WESTERN
BRACKEN, 134

FERN *(cont.)*
    MOSQUITO, 112
  WHISK, 206
  WHITE-FOOT LIP, 246
  WHOLE-LEAF CLOAK, 221
  WIDESPREAD MAIDEN, 300
  WILD BOSTON, 163
  WINDHAM'S
    CLOAK, 225
    SCALY CLOAK, 225
  WINTER GRAPE, 178, 180
  WOOD, 139
  WOOLLY LIP, 238, 242, 249
  WRIGHT'S LIP, 253
  YANKEE, 163
  YAVAPAI LIP, 253
FERRISS'
  HORSETAIL, 155
  SCOURING-RUSH, 155
×ferrissii
  Equisetum, 155
fiddleheads, 16
FIELD HORSETAIL, 155
filiculoides
  Azolla, 112
filix-femina
  Athyrium, 107
filix-mas
  Dryopteris, 142
FLOATING FERN, 8, 226, 289
  FAMILY, 288
FLOWER OF STONE, 79
FLOWERCUP FERN, 314
FLOWERING
  FERN, 6, 97, 195, 199
  FERN FAMILY, 97
FOOT LIP FERN, WHITE-, 246
FOREIGN CLOAK FERN, 259
FORKED SPLEENWORT, 2, 32, 106
FORK-FERN, 206
FORTUNE'S
  CYRTOMIUM, 139
  NET-VEIN HOLLY FERN, 139
FOXTAIL
  BOG CLUB-MOSS, 65
  CLUB-MOSS, 65
FRANKINCENSE FERN, 97
Fraxinus, 37
FRINGED FERN, 163
FUZZY MAIDENHAIR FERN, 213

G
gametophyte, 17
GIANT SALVINIA, 2, 30, 41, 291
glabella
  Pellaea, 276
GLANDULAR LIP FERN, 242
globulifera
  Dennstaedtia, 130
GOLDEN
  LIP FERN, 232
  WATER-CLOVER, 168
GOLDIE'S WOOD FERN, 140
GRACEFUL LIP FERN, 253
Grammitidaceae, 200
GRAPE FERN, 5, 178
  COMMON, 184
  CUT-LEAF, 180
  DISSECTED, 180
  OBLIQUE, 180
  PROSTRATE, 180
  SOUTHERN, 7, 178
  SPARSE-LOBE, 178
  VIRGINIA, 184
  WINTER, 178, 180
GRAY POLYPODY, 201
GRAY'S CLOAK FERN, 261
grayi
  Notholaena, 261
  Notholaena grayi subsp., 261
GREAT SCOURING-RUSH, 158
GREEN SPIKE-MOSS, 41, 92
GREGG'S CLOAK FERN, 41, 265
greggii
  Notholaena, 265
GROUNDNUT, 37
Gymnogramma
  ehrenbergiana var. *muralis,* 226
  hispida, 226
  schaffneri, 226
Gymnopteris hispida, 226

H
HAIR FERN
  INDIAN-, 102
  VENUS'-, 18, 210
HAIRY
  BOMMER, 226
  BRACKEN FERN, 134
  LIP FERN, 242
  MAIDEN FERN, 41, 42, 300
  MAIDEN FERN, ROUGH-, 300
  MAIDENHAIR FERN, 213
  PEPPERWORT, 172
  RESURRECTION PLANT, 89
  WATER-CLOVER, 172
HALBERD FERN, 293
  BROAD, 2, 42, 293
  FAMILY, 293
Hamamelis virginiana, 37
HAMMOCK FERN, 42, 117
HARDY JAPANESE HOLLY FERN, 139
HAY-SCENTED FERN, 130
  SOUTHERN, 130
HAZEL, WITCH-, 37
HEART-LEAF CLIFF-BRAKE, 16, 41, 272
HELECHO DE LA LLAVE, 256
heracleifolia
  Tectaria, 293
Heracleum, 293
heterospory, 22
hexagonoptera
  Phegopteris, 296
HICKORIES, 36
hispida
  Bommeria, 226
hispidula
  Thelypteris, 300
HOLLY FERN, 137, 147, 149
  ASIAN NET-VEIN, 139
  ASIATIC, 137
  CHISOS, 41, 149
  CLIMBING, 163
  EARED, 149
  FORTUNE'S NET-VEIN, 139
  HARDY JAPANESE, 139
  HOUSE, 137
  JAPANESE, 137
    NET-VEIN, 137
  MEXICAN, 147
  NET-VEIN, 137
Holubiella lunarioides, 184
HOOKED PEPPERWORT, 172
HORN CLUB-MOSS, STAG-, 71
horridula
  Cheilanthes, 238
HORSETAIL, 3, 4, 6, 53, 95, 153
  ANDEAN, 153
  COMMON, 18, 155
  FAMILY, 152
  FERRISS', 155
  FIELD, 155
  INTERMEDIATE, 155
  KANSAS, 161
  ROUGH, 158
  SCOURING-RUSH, 35, 158
  SMOOTH, 161
HOUSE HOLLY FERN, 137
HUGUENOT FERN, 285
HYBRID
  CLOAK FERN, 221
  SCALY CLOAK FERN, 221
Hydrilla verticillata, 293
hyemale subsp. affine
  Equisetum, 158
Hymenoasplenium, 99

I
ice age, 36, 44
INDIAN-HAIR FERN, 102
indusium, 17, 19
integerrima
  Astrolepis, 219
intermedia
  Pellaea, 276
INTERMEDIATE
  CLIFF-BRAKE, 276
  HORSETAIL, 155
intragametophytic selfing, 21
Isoëtaceae, 4, **54**
Isoëtes, 22, 40, **55**
  butleri, 5, **56,** 190
  flaccid, 63
  lithophila, 2, 38, 41, 42, 43, **58**

Isoëtes *(cont.)*
  melanopoda, 58
    subsp. *melanopoda,* 58
    subsp. *silvatica,* 58
  piedmontana, 31, 42, 44, **63**
  texana, 41, 43, 44, **63**
  virginica var. *piedmontana,* 63

J

JAPANESE
  CLIMBING FERN, 2, 164
  HOLLY FERN, 137
    HARDY, 139
  NET-VEIN HOLLY FERN, 137
japonicum
  Lygodium, 164
JEWEL FERN™, DALLAS, 163
JIMMY FERN, 2, 42, 219, 221
Juglans, 37
KANSAS
  HORSETAIL, 161
  SCOURING-RUSH, 161
KARIBA WEED, 291
KAULFUSS' LIP FERN, 41, 242
kaulfussii
  Cheilanthes, 242
KING FERN, 194
KUNTH'S MAIDEN FERN, 300
kunthii
  Thelypteris, 300

L

LADDER BRAKE, 42, 288
  CHINESE, 288
LADY FERN, 107, 142
  FAMILY, 107, 307
  LOWLAND, 107
  SOUTHERN, 12, 107
laevigatum
  Equisetum, 161
lanosa
  Cheilanthes, 242
LARGE
  ADDER'S-TONGUE, 190
  -FOOT PEPPERWORT, 11, 168
  -FOOT WATER CLOVER, 168
  WOODSIA, 310
Lauraceae, 197
LEAF
  BRACKEN, SAW-, 285
  CHAIN FERN, NARROW-, 119
  CLIFF-BRAKE, HEART-, 272
  CLIFF-BRAKE, OVATE-, 276
  CLUB-MOSS, TIGHT-, 69
  LLAVEA, CORDATE-, 256
LEAST
  ADDER'S-TONGUE, 187
  WATER-SPANGLES, 289
LEATHER WOOD FERN, 147
LEDGE SPIKE-MOSS, 89
lendigera
  Cheilanthes, 242
lepidophylla
  Selaginella, 79
Leptopteris, 194
leptosporangiate ferns, 6
LIMESTONE ADDER'S-TONGUE, 18, 187
limitanea
  Argyrochosma, 214
  Selaginella mutica var., 82
Lindera benzoin, 37
LINDHEIMER'S
  LIP FERN, 246
  MAIDEN FERN, 41, 304
lindheimeri
  Cheilanthes, 246
  Thelypteris ovata var., 304
LIP FERN, 13, 26, 228
  ALABAMA, 232
  BEADED, 242
  BONAIRE, 232
  BULB, 221
  EATON'S, 236
  FEE'S, 238
  FENDLER'S, 238
  GLANDULAR, 242
  GOLDEN, 232
  GRACEFUL, 253
  HAIRY, 242
  KAULFUSS', 41, 242
  LINDHEIMER'S, 246
  NIT-BEARING, 242
  PRICKLY, 238
  RIVAL, 232
  ROUGH, 238
  SLENDER, 32, 238
  SMOOTH, 232
  TEXAS, 41, 232
  VILLOUS, 249
  WHITE-FOOT, 41, 42, 246
  WOOLLY, 238, 242, 249
  WRIGHT'S, 21, 253
  YAVAPAI, 253
Liquidambar styraciflua, 37
lithophila
  Isoëtes, 58
LITTLE EBONY SPLEENWORT, 106
Llano Uplift, 38
LLAVE, HELECHO DE LA, 256
LLAVE'S FERN, 41, 256
Llavea, 256
  cordifolia, 40, 41, 44, **256**
LLAVEA, CORDATE-LEAF, 256
LOBE CLIFF FERN, BLUNT-, 310
LOBE WOODSIA, BLUNT-, 125
LOG FERN, 140
Lomariopsidaceae, 9, 135, **161**
Lomariopsis, 163
LONG CLOAK FERN, 221
long-distance dispersal, 30
LONGSTEM ADDER'S-TONGUE, 190
Lorinseria, 122
  areolata, 122
LOUISIANA WOOD FERN, 147
LOVING SPIKE-MOSS, ROCK-, 89
LOWLAND
  BLADDER FERN, 125
  BRITTLE FERN, 125
  LADY FERN, 107
ludoviciana
  Dryopteris, 147
lunarioides
  Botrychium, 180
Lycophylls, 4
lycophyte, 3, 4
  classification, 7
Lycopodiaceae, 4, 42, **63**
Lycopodiella, 65
  alopecuroides, 37, **65**
  × *L. appressa,* 69
  appressa, 5, **69**
  cernua, 71
  ×*copelandii,* 69
  prostrata, 69
Lycopodiophyta, 3, 4, **54**
Lycopodium, 42
  alopecuroides, 69
  var. *pinnatum,* 69
  apodum, 76
  appressum, 69
  carolinianum, 72
  cernuum, 71
  corallinum, 79
  inundatum
  var. *adpressum,* 69
  var. *appressum,* 69
  var. *pinnatum,* 71
  lepidophyllum, 82
  nudum, 206
  prostratum, 71
Lygodiaceae, 6, **163**
Lygodium, 6, **164**
  japonicum, 2, 40, 42, 43, **164**
  microphyllum, 166
  palmatum, 43, 164

M

macropoda
  Marsilea, 168
Macrothelypteris, 296
  torresiana, 296
Madrean sky islands, 43
Madro-Tertiary, 13
Magnolia, 37
MAIDEN FERN, 299
  DOWNY, 300
  FALSE, 296
  FAMILY, 295
  HAIRY, 300
  KUNTH'S, 300
  LINDHEIMER'S, 41
  MARIANA, 296

MAIDEN FERN *(cont.)*
ROUGH-HAIRY, 300
VARIABLE, 300
WIDESPREAD, 12, 300
MAIDENHAIR FERN, 201
FAMILY, 208
FUZZY, 213
HAIRY, 41, 42, 213
SPLEENWORT, 106
SOUTHERN, 210
MALE FERN, 107, 142
Mankyua, 177
Marattiopsida, 5
MARGINAL
SHIELD FERN, 147
WOOD FERN, 147
marginalis
Dryopteris, 147
MARIANA MAIDEN FERN, 296
MARSH FERN, 299, 304
EASTERN, 304
FAMILY, 295
OVATE, 304
SOUTHERN, 35, 304
Marsilea, 8, 22, 166, **167**
ancylopoda, 172
drummondii, 168
macropoda, 8, 11, **168**
mexicana, 172
mollis, 25, 29, 167, **168**
mucronata, 172
mutica, 172
quadrifolia, 167
tenuifolia, 172
uncinata, 172
vestita, 172
subsp. *tenuifolia*, 172
Marsileaceae, 8, 29, **166**
Matteuccia struthiopteris, 173
MATTRESS FERN, 210
MAXON'S CLOAK FERN, 268
MEADOW
SPIKE-MOSS, 6, 39, 76
-RUE, 228
melanopoda
Isoëtes, 58
Metasequoia, 33
MEXICAN
CLOAK FERN, 41, 259
FERN, 10, 41, 99
FLOWERING FERN, 99
HOLLY FERN, 147
MOSQUITO FERN, 11, 115
SCALY CLOAK FERN, 41, 221
SILVER FERN, 214
WATER-CLOVER, 168
Mexican connection, 37
mexicana
Anemia, 99
Argyrochosma limitanea subsp., 214
Astrolepis sinuata subsp., 221
Mexico, 13, 37, 40
michauxiana
Pleopeltis polypodioides var., 201
microphylla
Argyrochosma, 218
Azolla, 115
microphylls, 4
microphylly, 27
MIDSORUS FERN
FAMILY, 117
NEW WORLD, 117
minima
Salvinia, 289
Mitchella repens, 37
Mohria caffrorum, 97
molesta
Salvinia, 291
mollis
Marsilea, 168
Monilophyta, 4
Monilophytes, 4
MOONWORT, 178
MOSQUITO FERN, 8, 13, 18, 29, 42, 110, 112
CAROLINA, 2, 112
EASTERN, 112
FAMILY, 109
MEXICAN, 11, 115
WESTERN, 112
MOSS
APPRESSED BOG CLUB-, 69
AQUARIUM WATER-, 291
ARIZONA SPIKE-, 79
BLUNT SPIKE-, 82
BOG CLUB-, 65, 72
CAROLINA CLUB-, 72
CHAPMAN'S CLUB-, 69
CREEPING CLUB-, 69
FAIRY-, 110
FEATHER-STEM
CLUB-, 69
BOG CLUB-, 69
FOXTAIL
CLUB-, 65
BOG CLUB-, 65
GREEN SPIKE-, 92
LEDGE SPIKE-, 89
MEADOW SPIKE-, 76
NEW MEXICO SPIKE-, 86
NODDING CLUB-, 71
PERUVIAN SPIKE-, 86
PROSTRATE BOG CLUB-, 69
RIDDELL'S SPIKE-, 79
ROCK-LOVING SPIKE-, 89
ROCKY MOUNTAIN SPIKE-, 92
SAND SPIKE-, 79
SLENDER
BOG CLUB-, 72
CLUB-, 69, 72
SPIKE-, 92
SOUTHERN
APPRESSED CLUB-, 69
BOG CLUB-, 69
CLUB-, 69
SPIKE-, 72
STAG-HORN CLUB-, 71
TIGHT-LEAF CLUB-, 69
UNDERWOOD'S SPIKE-, 92
WATER-, 289
WRIGHT'S SPIKE-, 95
MOSSES
CLUB-, 54
SPIKE-, 54
MOUNTAIN SPIKE-MOSS, ROCKY, 92
Muhlenbergia reverchonii, 58
multifida
Pteris, 285
mutica
Selaginella, 82, 86
Myriopteris
fendleri, 238
lendigera, 246
lindheimeri, 246
tomentosa, 249

N

NARDOO
FERN, 168
BANDED, 172
NARROW-LEAF
CHAIN FERN, 119
PEPPERWORT, 172
naturalized species, 40
NEALLEY'S CLOAK FERN, 41, 265
nealleyi
Notholaena, 265
neglecta
Notholaena, 268
neomexicana
Selaginella, 86
Woodsia, 308
Nephrodium
acrostichoides, 152
marginale, 147
Nephrolepis, 9, 135, **163**
exaltata, 161, **163**
cv. 'Bostoniensis', 163
NEST FERN, BIRD'S-, 99
NETTED CHAIN FERN, 119
NET-VEIN HOLLY FERN
ASIAN, 139
FORTUNE'S, 139
JAPANESE, 137
Neurogramma hispida, 226
NEW MEXICO
CLIFF FERN, 308
SPIKE-MOSS, 86
NEW WORLD MIDSORUS FERN, 117
NIT-BEARING LIP FERN, 242
NODDING CLUB-MOSS, 71
NORTHERN SPLEENWORT, 106
Notholaena, 12, 25, 26, 27, 28, **258**
aliena, 23, 41, **259**

Notholaena *(cont.)*
aschenborniana, 261, 282
aurea, 236
bipinnata, 261
candida var. *copelandii*, 261
cochisensis, 219
copelandii, 28, 41, **261**
dealbata, 214
ferruginea, 236
grayi
subsp. **grayi, 261**
subsp. *sonorensis*, 265
greggii, 40, 41, **265**
hookeri, 268
integerrima, 221
limitanea
subsp. *mexicana*, 214
var. *mexicana*, 214
nealleyi, 41, **265**
neglecta, 268, 282
parvifolia, 218
schaffneri var. *nealleyi*, 265
sinuata, 225
var. *cochisensis*, 219
var. *integerrima*, 221
standleyi, 27, 28, **268**
nudicaule
Ophioglossum, 187
nudum
Psilotum, 206
Nyssa sylvatica, 37

O
OBLIQUE GRAPE FERN, 180
obtusa
Woodsia, 310
occidentalis
Woodsia obtusa subsp., 310
OLD WORLD CLIMBING FERN, 166
Onoclea, 9, 33, 135, **175**
sensibilis, 2, 13, 32, 34, 42, 58, **175,** 190
var. *interrupta*, 175
var. *sensibilis*, 175
Onocleaceae, 9, 135, **173**
Ophioglossaceae, 5, **177**
Ophioglossales, 5
Ophioglossum, 5, **184**
californicum, 185, 190
crotalophoroides, 58, **185,** 190
dendroneuron, 187
ellipticum, 187
engelmannii, 18, **187**
japonicum, 164
mononeuron, 187
nudicaule, 58, **187**
var. *minus*, 187
var. *tenerum*, 187
petiolatum, 5, 41, **190**
polyphyllum, 40, 44, **190**
pycnostichum, 194
tenerum, 187
vulgatum, 7, 37, **194**
var. *pycnostichum*, 194
Ornithopteris mexicana, 99
Osmunda, 33, **195**
cinnamomea, 197, 199
claytoniana, 195
claytoniites, 195
japonica, 197
lancea, 197
regalis, 6, 13, 37, 42, **195,** 258
var. *regalis*, 197
var. **spectabilis, 195**
Osmundaceae, 6, **194**
Osmundales, 6
Osmundastrum, 33, 122, **197**
cinnamomeum, 6, 9, 13, 36, 37, 42, 122, **197,** 307
OSTRICH FERN, 173
ovata
Pellaea, 276
Thelypteris, 304
OVATE MARSH FERN, 304
OVATE-LEAF CLIFF-BRAKE, 276

P
Paleocene/Eocene thermal maximum (PETM) 33
Palhinhaea, 71
cernua, 40, 43, 44, **71**
PALMER'S SPLEENWORT, 2, 102
palmeri
Asplenium, 102
PALMETTO, DWARF, 36
palustris
Thelypteris, 304
Paragymnopteris, 271
PARTRIDGE BERRY, 37
PASTURE BRAKE, 132, 134
PECAN, 36
PECOS CLIFF-BRAKE, TRANS-, 276
PEIDMONT QUILLWORT, 31
Pellaea, 12, 25, **271**
alabamensis, 232
arizonica, 32
atropurpurea, 22, 23, 24, 25, **272**
var. *bushii*, 276
var. *cristata*, 272
cardiomorpha, 272
cordifolia, 16, 41, **272**
dealbata, 214
flexuosa, 276
glabella, 40, **276**
subsp. **glabella, 276**
greggii, 265
intermedia, 276
longimucronata, 282
microphylla, 218
ovata, 41, **276**
sagittata var. *cordata*, 272
ternifolia, 32, 33, 41, **276**
subsp. **arizonica, 280,** 282
subsp. *brandegeei*, 280
subsp. **ternifolia, 280**
subsp. *villosa*, 282
var. *wrightiana*, 285
truncata, 25, **280**
villosa, 44, **282**
wrightiana, 31, 32, **282**
var. *longimucronata*, 282
PEPPER, WATER-, 173
PEPPERWORT, 167
FAMILY, 166
HAIRY, 172
HOOKED, 172
LARGE-FOOT, 11, 168
NARROW-LEAF, 172
PERUVIAN SPIKE-MOSS, 86
peruviana
Selaginella, 86
petiolatum
Ophioglossum, 190
PETM (Paleocene/Eocene thermal maximum) 33
PEWTERWORT, 154
Phanerophlebia, 147
auriculata, 149
falcata, 139
fortunei, 139
umbonata, 41, **149**
Phegopteris, 296
hexagonoptera, 12, 18, 40, **296**
PHILLIPS'S CLIFF FERN, 310
phillipsii
Woodsia, 310
Phlebodium erythrolepis, 201
Photoinhibition, 28
Photoprotection, 28
phylogeny, 6
PIEDMONT QUILLWORT, 42
piedmontana
Isoëtes, 63
pilifera
Selaginella, 89
PILLWORT, 172
AMERICAN, 2, 173
Pilularia, 8, 22, 166, **172**
americana, 2, **173**
PINELAND FERN, 97
Pityrogramma, 226
Plantago, 184
PLANTAINS, 184
Platyloma, 271
platyneuron
Asplenium, 102
Pleopeltis, 25, **200**
erythrolepis, 201
polylepis, 40, 41, **201**
var. **erythrolepis, 201**
polypodioides, 29, **201**
var. **michauxiana, 201**
riograndensis, 205

PLUMMER'S CLIFF FERN, 314
plummerae
  Woodsia, 314
polylepis
  Pleopeltis, 201
polyphyllum
  Ophioglossum, 190
Polyploidy, 23
Polypodiaceae, 199
Polypodiales, 5, 8, 9
polypodioides
  Pleopeltis, 201
Polypodiophyta, 3, 4, 5
Polypodium, 200
  bulbiferum, 125
  dentatum, 300
  erythrolepis, 201
  exaltatum, 163
  falcatum, 139
  filix-mas, 142
  globuliferum, 130
  hexagonopterum, 299
  marginale, 147
  polypodioides
    var. *knoblochianum*, 205
    var. *michauxianum*, 201
  riograndense, 205
  thyssanolepis, 205
    var. *riograndense*, 205
POLYPODY, 200
  FERN FAMILY, 199
  GRAY, 201
  RED-SCALE, 41, 201
    SCALY-, 201
  RIO GRANDE, 205
    SCALY-, 205
  SCALY-, 200
Polystichum, 135, **149**
  acrostichoides, 152
  falcatum, 139
    var. *fortunei*, 139
  fortunei, 139
  torresianum, 296
POWDERY
  CLOAK FERN, 214
  FALSE CLOAK FERN, 214
precipitation, 15
PRICKLY LIP FERN, 238
prize-winners, 1, 2
prostrata
  Lycopodiella, 69
PROSTRATE
  BOG CLUB-MOSS, 69
  GRAPE FERN, 180
protrusa
  Cystopteris, 125
Prunus, 37
pseudocaudatum
  Pteridium aquilinum subsp., 134
Pseudolycopodiella caroliniana, 18
Pseudolycopodiella, 72
  caroliniana, 72
Psilophyta, 3, 5
PSILOPHYTES, 3, 4, 53, 95
Psilotaceae, 5, 177, **205**
Psilotales, 5
Psilotopsida, 5
Psilotum, 4, 5, **206**
  nudum, 2, 5, 8, 18, 40, **206**
Pteridaceae, 23, **208**
Pteridium, 130
  aquilinum, 2, 22, 30, 41, 42, **132**
    subsp. *latiusculum*, 135
    var. *latiusculum*, 135
    subsp. **pseudocaudatum, 134**
    subsp. **pubescens, 134**
  caudatum, 134
  latiusculum, 135
pteridology, 1
Pteridophyta, 3
pteridophytes, 1, 3
  classification, 7
Pteris, 285
  alabamensis, 232
  aquilina
    var. *pseudocaudata*, 134
    var. *pubescens*, 134
  atropurpurea, 272
  latiuscula, 135
  lendigera, 246
  multifida, 285
  ovata, 276
  ternifolia, 280
  vittata, 42, 44, **288**
pubescens
  Pteridium aquilinum subsp., 134
  Thelypteris palustris var., 304
PURPLE
  CLIFF-BRAKE, 24, 272
  -STEM CLIFF-BRAKE, 272
Pycnodoria
  multifida, 288
  vittata, 288
PYLE, AFRICAN, 291

Q
QUILLWORT, 3, 4, 53, 54, 55
  BUTLER'S, 5, 190
  FAMILY, 54
  PEIDMONT, 31, 42
  ROCK, 2, 38, 41, 42
  TEJAS, 41

R
Ranunculaceae, 228
RATTLESNAKE FERN, 184
record-holders, 1, 2
RED-SCALE
  POLYPODY, 41, 201
  SCALY-POLYPODY, 201
REDWOOD, DAWN, 33
reevesiana
  Cystopteris, 129
regalis
  Osmunda, 195
Regnellidium, 22, 166
reproduction, asexual, 22, 23
resiliens
  Asplenium, 106
RESURRECTION
  FERN, 29, 201
  HAIRY, 89
  PLANT, 2, 25, 27, 79, 89
Rhynia, 205
RIDDELL'S
  SELAGINELLA, 79
  SPIKE-MOSS, 79
RIO GRANDE
  POLYPODY, 205
  SCALY-POLYPODY, 205
riograndensis
  Pleopeltis, 205
RIVAL LIP FERN, 232
ROCK QUILLWORT, 2, 38, 41, 42
rockhouses, 37
ROCK-LOVING SPIKE-MOSS, 89
ROCKY MOUNTAIN SPIKE-MOSS, 92
ROUGH
  -HAIRY MAIDEN FERN, 300
  HORSETAIL, 158
  LIP FERN, 238
ROYAL FERN, 6, 195, 258
rupincola
  Selaginella, 89
RUSH
  AMERICAN SCOURING-, 155
  BRAUN'S SCOURING-, 161
  FERRISS' SCOURING-, 155
  GREAT SCOURING-, 155
  HORSETAIL, SCOURING-, 155
  KANSAS SCOURING-, 161
  SCOURING-, 6, 42, 153
  SMOOTH SCOURING-, 161
  SUMMER SCOURING-, 161
  TALL SCOURING-, 155
  WINTER SCOURING-, 155

S
Sabal minor, 36
Salvinia, 8, 22, 166, **289**
  adnata, 293
  minima, 30, **289**
  molesta, 2, 22, 29, 30, 40, 42, 44, **291**
  rotundifolia, 291
SALVINIA
  COMMON, 289
  GIANT, 2, 30, 41, 291
Salviniaceae, 8, 166, **288**
Salviniales, 8
SAN SABA FERN, 304
SAND SPIKE-MOSS, 79

Sassafras, 36
SASSAFRAS, 36
SAW-LEAF BRACKEN, 285
SCALED CLOAK FERN, 261
CHIHUAHUA, 219
COCHISE, 219
HYBRID, 221
MEXICAN, 221
STAR-, 218
SCALY CLOAK FERN, 218
MEXICAN, 41
WAVY, 221, 225
WINDHAM'S, 27, 225
SCALY-POLYPODY, 200
RED-SCALE, 201
RIO GRANDE, 205
SCENTED FERN
HAY-, 130
SOUTHERN HAY-, 130
Sceptridium, 178
biternatum, 180
dissectum, 180
lunarioides, 184
Schizaeales, 6
scopulorum
Selaginella, 89
SCOURING-RUSH, 6, 42, 153
AMERICAN, 158
BRAUN'S, 161
FERRISS', 155
HORSETAIL, 35, 158
GREAT, 158
KANSAS, 161
SMOOTH, 161
SUMMER, 161
TALL, 158
WINTER, 158
SEEP MUHLY, 58
Selaginella, 12, 22, 25, 29, **74**
apoda, 6, 37, 39, **76**
var. *ludoviciana*, 76
arenicola
subsp. *riddellii*, 79
var. *riddellii*, 79
arizonica, 79
corallina, 79
coryi, 95
densa var. *scopulorum*, 92
engelmannii var. *scopulorum*, 92
fendleri, 92
lepidophylla, 2, 25, 27, 29, **79**
ludoviciana, 76
mutica, 82
var. **limitanea, 82**
var. **mutica, 86**
var. *texana*, 86
×**neomexicana,** 9, **86**
peruviana, 79, **86**
pilifera, 89
var. *pringlei*, 89
pringlei, 89
riddellii, 79
rupestris
var. *rupincola*, 89
forma *wrightii*, 95
rupincola, 89
× *S. underwoodii*, 86
scopulorum, 89
underwoodii, 86, **92**
viridissima, 41, 43, **92**
wrightii, 95
SELAGINELLA
BASKET, 76
RIDDELL'S, 79
Selaginellaceae, 4, **72**
selfing, intragametophytic, 21
sensibilis
Onoclea, 175
SENSITIVE FERN, 2, 32, 34, 36, 58, 175, 190
FAMILY, 173
septentrionale
Asplenium, 106
sexual reproduction, 19
SHIELD FERN, 130
DOWNY, 300
LINDHEIMER'S, 304
MARGINAL, 147
SOUTHERN, 300
ST. JOHN'S, 300
SHIELD-SORUS FERN, 200
SIEMPRE VIVA, 79
SILVER FERN, 213
MEXICAN, 214
SMALL-LEAF, 218
sinuata
Astrolepis, 221
sky islands, Madrean, 43
SLENDER
ADDER'S-TONGUE, 58, 187
BOG CLUB-MOSS, 18, 69, 72
LIP FERN, 32, 238
SPIKE-MOSS, 92
SMALL-LEAF
CLOAK FERN, 218
FALSE CLOAK FERN, 218
SMOOTH
CLIFF-BRAKE, 276
HORSETAIL, 161
LIP FERN, 232
SCOURING-RUSH, 161
sori, 17, 19
SORUS FERN, SHIELD-, 200
SOUTHERN
ADDER'S-TONGUE, 7, 194
APPRESSED CLUB-MOSS, 69
BEECH FERN, 296
BLADDER FERN, 37, 125, 310
BOG CLUB-MOSS, 69
CLUB-MOSS, 69
GRAPE FERN, 7, 178
HAY-SCENTED FERN, 130
LADY FERN, 12, 107
MAIDENHAIR, 210
MARSH FERN, 35, 304
SHIELD FERN, 300
WOOD FERN, 12, 147
SOUTHWESTERN
BRITTLE FERN, 129
CLOAK FERN, 32, 221
FALSE CLOAK FERN, 214
SPANGLES
LEAST WATER-, 289
WATER-, 289
SPARSE-LOBE GRAPE FERN, 178
spectabilis
Osmunda regalis var., 195
SPICEBUSH, 37
SPIDER BRAKE, SPIDER, 285
SPIKE-MOSS, 3, 4, 53, 54, 72
ARIZONA, 79
BLUNT, 82
GREEN, 41, 92
LEDGE, 89
MEADOW, 6, 39, 76
NEW MEXICO, 86
PERUVIAN, 86
RIDDELL'S, 79
ROCK-LOVING, 89
ROCKY MOUNTAIN, 92
SAND, 79
SLENDER, 92
UNDERWOOD'S, 92
WRIGHT'S, 95
SPINY CLIFF-BRAKE, 280
SPLEENWORT, 99
BLACK-STEM, 106
BROWN-STEM, 102
EBONY, 102
FAMILY, 99
FORKED, 2, 32, 106
LITTLE EBONY, 106
MAIDENHAIR, 106
NORTHERN, 106
PALMER'S, 2, 102
sporangium, 4
spores, 2, 3, 17
sporophyte, 17
SPRING FERN, 107
SPRITE, WATER, 228
ST. JOHN'S SHIELD FERN, 300
STAG-HORN CLUB-MOSS, 71
STALKED ADDER'S-TONGUE, 190
standleyi
Notholaena, 268
STAR
CLOAK FERN, 28, 268
FERN, 268
-SCALED CLOAK FERN, 13, 23, 25, 218
STEM
BOG CLUB-MOSS, FEATHER-, 69
CLUB-MOSS, FEATHER-, 69
SPLEENWORT
BLACK-, 106
BROWN-, 102
Stenopelmus rufinasus, 115
STONE, FLOWER OF, 79

SUMMER SCOURING-RUSH, 161
SWEETGUM, 37
SWORD FERN, 149, 161, 163
  BOSTON, 163
  FAMLY, 161
SWORDS, FAIRY-, 246

T

TAILED BRACKEN FERN, 134
TALL SCOURING-RUSH, 158
Tectaria, 9, 135, **293**
  cinnamomea, 142
  heracleifolia, 2, 30, 40, 42, 44, 213, **293**
Tectariaceae, 9, 135, **293**
TEJAS QUILLWORT, 41
TERNATE CLIFF-BRAKE, 276
ternifolia
  Pellaea, 280
Tertiary refugia, 34
texana
  Isoëtes, 63
TEXAS LIP FERN, 41, 232
Texas, East, 9, 10, 11, 12, 13, 14, 35, 37, 40
thalictroides
  Ceratopteris, 228
Thalictrum, 228
Thelypteridaceae, 122, **295**
Thelypteris, 33, **299**
  dentata, 300
  hexagonoptera, 299
  hispidula, 40, **300**
    var. **versicolor, 300**
  kunthii, 11, 12, **300**
    × *Thelypteris ovata* var. *lindheimeri*, 304
  ludoviciana, 147
  normalis, 304
  ovata, 41, **304**
    var. **lindheimeri, 304**
  palustris, 13, 34, 35, **304**
    var. *haleana*, 304
    var. **pubescens, 304**
  quadrangularis var. *versicolor,* 300
  torresiana, 296
  versicolor, 300
THYME, WATER-, 293
TIGHT-LEAF CLUB-MOSS, 69
Tilia Americana, 37
Tmesipteris, 205
Todea, 194
  barbara, 194
tomentosa
  Cheilanthes, 249
TONGUE
  ADDER'S-, 184, 194
  BULBOUS ADDER'S-, 58, 185, 190
  ENGELMANN'S ADDER'S-, 187
  FERN DWARF ADDER'S-, 185
  LARGE ADDER'S-, 190
  LEAST ADDER'S-, 187
  LIMESTONE ADDER'S-, 187
  LONGSTEM ADDER'S-, 190
  SLENDER ADDER'S-, 58, 187
  SOUTHERN ADDER'S-, 194
  STALKED ADDER'S-, 190
  UPRIGHT ADDER'S-, 190
TORRES' FERN, 296
torresiana
  Macrothelypteris, 296
Trans-Pecos, 10, 11, 13, 14, 29, 31, 40
  CLIFF-BRAKE, 33, 41, 276
TREE FERN, 201
tricholepis
  Adiantum, 213
trichomanes
  Asplenium
true ferns, 3
truncata
  Pellaea, 280

U

Ulmus, 37
umbonata
  Phanerophlebia, 149
UNDERWOOD'S SPIKE-MOSS, 92
underwoodii
  Selaginella, 92
UPRIGHT ADDER'S-TONGUE, 190
UTAH BLADDER FERN, 129
utahensis
  Cystopteris, 129

V

VARIABLE MAIDEN FERN, 300
VARIEGATED WATER-CLOVER, 172
vascular cryptogams, 17
VEIN HOLLY FERN, NET-, 137
VENUS'-HAIR FERN, 18, 210
versicolor
  Thelypteris hispidula var., 300
vestita
  Marsilea, 172
vicariance, 30, 32
villosa
  Cheilanthes, 249
  Pellaea, 282
VILLOSE
  CLIFF-BRAKE, 282
  LIP FERN, 249
VIRGINIA
  CHAIN FERN, 122
  GRAPE FERN, 184
virginianum
  Botrychium, 184
virginica
  Woodwardia, 122
viridissima
  Selaginella, 92
vittata
  Pteris, 288
VIVA, SIEMPRE, 79
vulgatum
  Ophioglossum, 194

W

WATER
  FERN, 110, 112, 226, 228
    FAMILY, 109
  SPRITE, 228
  -CLOVER, 8, 167, 168, 172
    AUSTRALIAN, 172
    BIG-FOOT, 168
    EUROPEAN, 167
    FAMILY, 166
    GOLDEN, 168
    HAIRY, 172
    LARGE-FOOT, 168
    MEXICAN, 168
    VARIEGATED, 172
  -MOSS, 289
  AQUARIUM, 291
  -PEPPER, 173
  -SPANGLE FAMILY, 288
  -SPANGLES, 289
  LEAST, 289
  -THYME, 293
WAVY
  CLOAK FERN, 28, 221
  SCALY CLOAK FERN, 32, 221, 225
WESTERN
  BRACKEN FERN, 134
  MOSQUITO FERN, 112
WHISK FERN, 2, 3, 4, 5, 8, 18, 177, 206
  FAMILY, 205
WHITE-FOOT LIP FERN, 41, 42, 246
WHOLE-LEAF CLOAK FERN, 221
WIDESPREAD MAIDEN FERN, 12, 300
WILD BOSTON FERN, 163
WINDHAM'S
  CLOAK FERN, 225
  SCALY CLOAK FERN, 27, 225
windhamii
  Astrolepis, 225
WINTER
  GRAPE FERN, 178, 180
  SCOURING-RUSH, 158
Wisteria, 36
WISTERIA, 36
WITCH-HAZEL, 37
WOOD FERN, 31, 139
  CINNAMON, 41, 42, 142
  CRESTED, 140
  FAMILY, 135
  GOLDIE'S, 140
  LEATHER, 147
  LOUISIANA, 147
  MARGINAL, 147
  SOUTHERN, 12, 147
Woodsia, 9, 107, 122, 135, 307, **308**
  mexicana, 310
  neomexicana, 308
  obtusa, 36, 125, **310**

Woodsia *(cont.)*
var. *glandulosa*, 314
subsp. **obtusa, 310**
subsp. **occidentalis, 310**
var. *plummerae*, 314
phillipsii, 310
plummerae, 314
pusilla var. *glandulosa*, 314
WOODSIA
BLUNT-LOBE, 125, 310
COMMON, 310
LARGE, 310
Woodsiaceae, 9, 107, 122, 135, **307**
Woodwardia, 13, 33, **119**
areolata, 36, **119,** 175
virginica, 36, **122**
WOOLLY LIP FERN, 238, 242, 249
WOOTON'S LIP FERN, 249
WRIGHT'S
CLIFF-BRAKE, 31, 32, 282
LIP FERN, 253
SPIKE-MOSS, 95
wrightiana
Pellaea, 282
wrightii
Cheilanthes, 253
Selaginella, 95
WRIGHT'S LIP FERN, 21

Y

YANKEE FERN, 163
YAVAPAI LIP FERN, 253
yavapensis
Cheilanthes, 253

Z

ZIGZAG CLIFF-BRAKE, 41, 276

COLOPHON

One thousand five hundred copies were printed using soy-based inks via offset lithography.

All digitization of color photos, and most other b/w images throughout the book, and all prepress work up to actual printing of the book were done in Dallas, Texas by *MILLET THE PRINTER, INC.*—craftsmen for four generations.
This book was printed and bound in Korea by *SUNG IN PRINTING AMERICA, INC.*
The binding is a flexbound using 300gsm MGB, C2S/4c + Gloss film lamination.
Smythe sewn, round back with heads & tail bands.
Stock for the endpaper binding, 150gsm WOODFREE/ACID-FREE, white NAMTLAN PAPER, manufactured by *NAMHAN PAPER COMPANY, LTD.*, South Korea.
Text pages are printed on 100 GSM MATTE ART (564 PPI), manufactured by *HANSOL PAPER MILL*, South Korea.
Archival permanence was kept in mind throughout the project.
Layout and production of the entire book was done by Rebecca Horn, *RHORNGRAPHICS*, Plano, Texas.
Robert George, M.S., Project Research Assistant: Texas Floras Projects, was responsible for the selection, screening, scanning, and proofing of all botanical illustrations as well as managed the layout and production of the botanical illustration/map pages.
The introduction and treatments are set in *BERKELEY OLDSTYLE MEDIUM* and *ITALIC*, *BERKELEY BLACK*, *UNIVERSAL-GREEKWITHMATHPI*, and *SYMBOL*.
The keys, along with the book's captions and the majority of the appendices, glossary, literature cited, and index, are set in *MYRIAD* (a multiple master typeface) and *UNIVERSAL-GREEKWITHMATHPI*.
Visual icons (called dingbats) are used for ease of reference throughout the book.
Some of these icons were found in existing commercial typefaces: *BOTANICAL MT* and *WARNING PI*.
Special designs were created to represent other specific "at-a-glance" icons.
These designs were digitized into a typeface which we call *Texicons*.
The majority of the book was electronically produced on the Macintosh platform with Adobe Illustrator®, Adobe InDesign®, and Adobe Photoshop®.
The PC platform was used to produce the botanical illustration color figures, and map pages in the InDesign and Photoshop programs.

These specifications are given in the spirit of sharing within the botanical community.
We hope our efforts are well-received and will contribute to furthering an interest in the "art & science" found throughout the botanical world.

AUTHORS & INSTITUTIONS

George M. Diggs, Jr. and Barney L. Lipscomb

George M. Diggs, Jr.

George M. Diggs, Jr. was born 4 February 1952 in Charlottesville, Virginia. He attended the College of William and Mary (B.S. 1974, M.A. 1976) and the University of Wisconsin-Madison (Ph.D. 1981). He has been a faculty member in the Biology Department of Austin College, Sherman, Texas, since 1981 and is currently the Donald MacGregor Chair of Natural Sciences. In 1994, he was made a Research Associate at the Botanical Research Institute of Texas. His research interests include the flora of Texas, the systematics of the Ericaceae (the blueberry family), evolution as it relates to human health, and biogeography. He has done field work in Africa, Australia, Mexico, Central and South America, Canada, and the U.S. In 1999, Diggs, together with co-authors Lipscomb and Robert O'Kennon, published *Shinners & Mahler's Illustrated Flora of North Central Texas*, the first fully illustrated flora for any part of Texas. In 2000, with Lipscomb, O'Kennon, and Monique Dubrule Reed, he co-authored the *Illustrated Flora of East Texas (Vol. 1)*. Diggs has taken numerous groups of undergraduates on tropical natural history field courses to Latin America and Africa in an effort to increase knowledge of tropical ecosystems and to raise awareness of current ecological problems including the accelerating destruction of tropical forests. He is active in Austin College's Center for Environmental Studies.

Barney L. Lipscomb

Barney L. Lipscomb was born 24 October 1950 in Temple, Oklahoma. He attended Cameron University, Lawton, Oklahoma (B.S. 1973) and the University of Arkansas, Fayetteville (M.S. 1975). He began his career as herbarium botanist at Southern Methodist University in 1975. Two years later he became the assistant editor of *Sida, Contributions to Botany*. In 1983 he became editor of *Sida*, and in 1987, with Dr. William F. Mahler, founded *Sida, Botanical Miscellany*. He continues as the editor of the *Journal of the Botanical Research Institute of Texas*, the successor to *Sida*. Also in 1987, Lipscomb, Mahler, and Andrea McFadden, were instrumental in the establishment of a freestanding research institution, the Botanical Research Institute of Texas (BRIT), located in Fort Worth. Lipscomb has served at BRIT since its inception in a variety of capacities including as its Assistant Director and in charge of the library. Currently he is the Dorothea Leonhardt Chair of Texas Botany and the Head of BRIT Press. His research specialties include the flora of Texas and the genus *Cyperus* (Cyperaceae). He has done field work in various parts of the U.S. as well as Mexico and Central America. In an effort to increase public awareness, Lipscomb frequently gives talks on plants and conservation to preschoolers, garden clubs, plant-oriented societies, high school and college groups, and civic organizations.

Austin College

Austin College is a private, residential, coeducational college dedicated primarily to educating undergraduate students in the liberal arts and sciences. The Austin College education emphasizes academic excellence, high achievement, intellectual and personal integrity, and participation in community life. Founded in 1849 by the Presbyterian Church, Austin College is the oldest college in Texas operating under its original charter.

The mission of Austin College is to educate individuals to prepare them for productive and meaningful lives in an increasingly complex world, providing diverse educational opportunities both on campus and in other countries, as well as select pre-professional programs and a graduate teacher program. Approximately 70 percent of Austin College students spend some time abroad during their four years at the school, either during the special January Term semester or through a study abroad program.

Austin College has a long history of excellence in the sciences, being particularly well known for its pre-medical program. The College also has a tradition of strength in the area of field biology and the environment, and has an active Center for Environmental Studies. This Center promotes multidisciplinary inquiry of environmental issues and problems through education, research, and outreach programs. These programs are designed to increase scientific knowledge, expand community awareness, and foster greater appreciation for the interdependence of humans and other species.

Botanical Research Institute of Texas (BRIT)

The Botanical Research Institute of Texas is a private, nonprofit, international botanical resource center. Its mission is to conserve our natural heritage by deepening our knowledge of the plant world and achieving public understanding of the value plants bring to life. This mission is realized by discovering new information through research and interpreting this research so the public may understand its meaning and importance, leading to greater stewardship of these vital resources. BRIT is comprised of two main collections: an herbarium of about 1,000,000 dried plant specimens and a research library containing over 100,000 items. The core of the collections is the Lloyd H. Shinners Collection in Systematic Botany, originally at Southern Methodist University, and the Vanderbilt University Herbarium, acquired in 1997. The BRIT library has one of the finest collections of botanical literature in the United States with books dating to 1549. A special part of the collection is the Oliver G. Burk Memorial Library of children's botanical literature—a rare collection of over 4,000 children's books. BRIT is also well known for its scientific press. Two scientific journals, *Journal of the Botanical Research Institute of Texas* and *Sida, Botanical Miscellany*, and other botanical book publications are published by the institute as well as an online newsletter, *The Leaflet*. Annually BRIT gives its Award of Excellence in Conservation to an individual whose life and work epitomize the ideals set forth in the BRIT mission. Along the same lines of excellence in conservation, BRIT also gives an International Award of Excellence in Sustainable Winegrowing. The institute has special programs in research, education, and public outreach. Several of the main activities of the research program include biodiversity exploration of the Andes to Amazon Botany Program in Peru, LBJ National Grasslands, Cider apple agrobiodiversity, the Illustrated Texas Floras Project, and Biodiversity Informatics. The education and public outreach programs are actively engaged in providing lectures, workshops, and teaching materials to local schools and other educational organizations in order to bring botany alive to students and teachers. The new BRIT SEED (Science-based Experiential Education Design) School for educators, located in the Suzanne Rall Peacock Education Center, inspires and supports educators in connecting children to nature through the use of outdoor classrooms, nature, and urban environments.

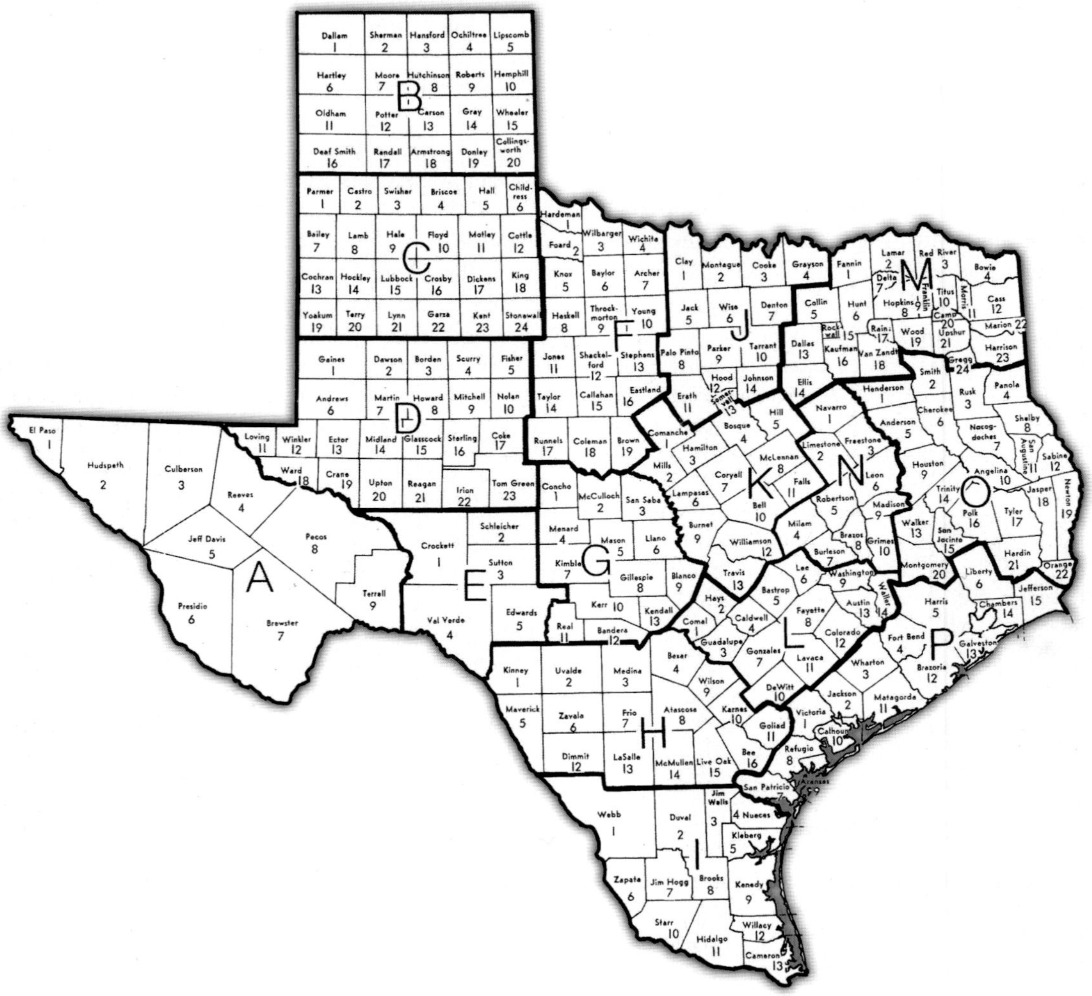

From Turner et al. 2003.

Counties of the State of Texas

Anderson	O-5
Andrews	D-6
Angelina	O-10
Aransas	P-9
Archer	F-7
Armstrong	B-18
Atascosa	H-8
Austin	L-13
Bailey	C-7
Bandera	G-12
Bastrop	L-5
Baylor	F-6
Bee	H-16
Bell	K-10
Bexar	H-4
Blanco	G-9
Borden	D-3
Bosque	K-4
Bowie	M-4
Brazoria	P-12
Brazos	N-8
Brewster	A-7
Briscoe	C-4
Brooks	I-8
Brown	F-19
Burleson	N-7
Burnet	K-9
Caldwell	L-4
Calhoun	P-10
Callahan	F-15
Cameron	I-13
Camp	M-20
Carson	B-13
Cass	M-12
Castro	C-2
Chambers	P-14
Cherokee	O-6
Childress	C-6
Clay	J-1
Cochran	C-13
Coke	D-17
Coleman	F-18
Collin	M-5
Collingsworth	B-20
Colorado	L-12
Comal	L-1
Comanche	K-1
Concho	G-1
Cooke	J-3
Coryell	K-7
Cottle	C-12
Crane	D-19
Crockett	E-1
Crosby	C-16
Culberson	A-3
Dallam	B-1
Dallas	M-13
Dawson	D-2
Deaf Smith	B-1-6
Delta	M-7
Denton	J-7
DeWitt	L-10
Dickens	C-17
Dimmit	H-12
Donley	B-19
Duval	I-2
Eastland	F-16
Ector	D-13
Edwards	E-5
Ellis	M-14
El Paso	A-1
Erath	J-11
Falls	L-11
Fannin	M-1
Fayette	L-8
Fisher	D-5
Floyd	C-10
Foard	F-2
Fort Bend	P-4
Franklin	M-9
Freestone	N-3
Frio	H-7
Gaines	D-1
Galveston	P-13
Garza	C-22
Gillespie	G-8
Glasscock	D-15
Goliad	H-11
Gonzales	L-7
Gray	B-14
Grayson	J-4
Gregg	M-24
Grimes	N-10
Guadalupe	L-3
Hale	C-9
Hall	C-5
Hamilton	K-3
Hansford	B-3
Hardeman	F-1
Hardin	O-21
Harris	P-5
Harrison	M-23
Hartley	B-6
Haskell	F-8
Hays	L-2
Hemphill	B-10
Henderson	O-1
Hidalgo	I-11
Hill	K-5
Hockley	C-14
Hood	J-12
Hopkins	M-8
Houston	O-9
Howard	D-8
Hudspeth	A-2
Hunt	M-6
Hutchinson	B-8
Irion	D-22
Jack	J-5
Jackson	P-2
Jasper	O-18
Jeff Davis	A-5
Jefferson	P-1-5
Jim Hogg	I-7
Jim Wells	I-3
Johnson	J-14
Jones	F-11
Karnes	H-10
Kaufman	M-16
Kendall	G-13
Kenedy	I-9
Kent	C-23
Kerr	G-10
Kimble	G-7
King	C-18
Kinney	H-1
Kleberg	I-5
Knox	F-5
Lamar	M-2
Lamb	C-8
Lampasas	K-6
La Salle	H-13
Lavaca	L-11
Lee	L-6
Leon	N-6
Liberty	P-6
Limestone	N-2
Lipscomb	B-5
Live Oak	H-15
Llano	G-6
Loving	D-11
Lubbock	C-15
Lynn	C-21
Madison	N-9
Marion	M-22
Martin	D-7
Mason	G-5
Matagorda	P-11
Maverick	H-5
McCulloch	G-2
McLennan	K-8
McMullen	H-14
Medina	H-3
Menard	G-4
Midland	D-14
Milam	N-4
Mills	K-2
Mitchell	D-9
Montague	J-2
Montgomery	O-20
Moore	B-7
Morris	M-11
Motley	C-11
Nacogdoches	O-7
Navarro	N-1
Newton	O-19
Nolan	D-10
Nueces	I-4
Ochiltree	B-4
Oldham	B-11
Orange	O-22
Palo Pinto	J-8
Panola	O-4
Parker	J-9
Parmer	C-1
Pecos	A-8
Polk	O-16
Potter	B-12
Presidio	A-6
Rains	M-17
Randall	B-17
Reagan	D-21
Real	G-11
Red River	M-3
Reeves	A-4
Refugio	P-8
Roberts	B-9
Robertson	N-5
Rockwall	M-15
Runnels	F-17
Rusk	O-3
Sabine	O-12
San Augustine	O-11
San Jacinto	O-15
San Patricio	P-7
San Saba	G-3
Schleicher	E-2
Scurry	D-4
Shackelford	F-12
Shelby	O-3
Sherman	B-2
Smith	O-2
Somervell	J-13
Starr	I-10
Stephens	F-13
Sterling	D-16
Stonewall	C-24
Sutton	E-3
Swisher	C-3
Tarrant	J-10
Taylor	F-14
Terrell	A-9
Terry	C-20
Throckmorton	F-9
Titus	M-10
Tom Green	D-2-3
Travis	K-13
Trinity	O-14
Tyler	O-17
Upshur	M-21
Upton	D-20
Uvalde	H-2
Val Verde	E-4
Van Zandt	M-18
Victoria	P-1
Walker	O-13
Waller	L-14
Ward	D-18
Washington	L-9
Webb	L-1
Wharton	P-3
Wheeler	B-15
Wichita	F-4
Wilbarger	F-3
Willacy	I-12
Williamson	K-12
Wilson	H-9
Winkler	D-12
Wise	J-6
Wood	M-19
Yoakum	C-19
Young	F-10
Zapata	I-6
Zavala	H-6

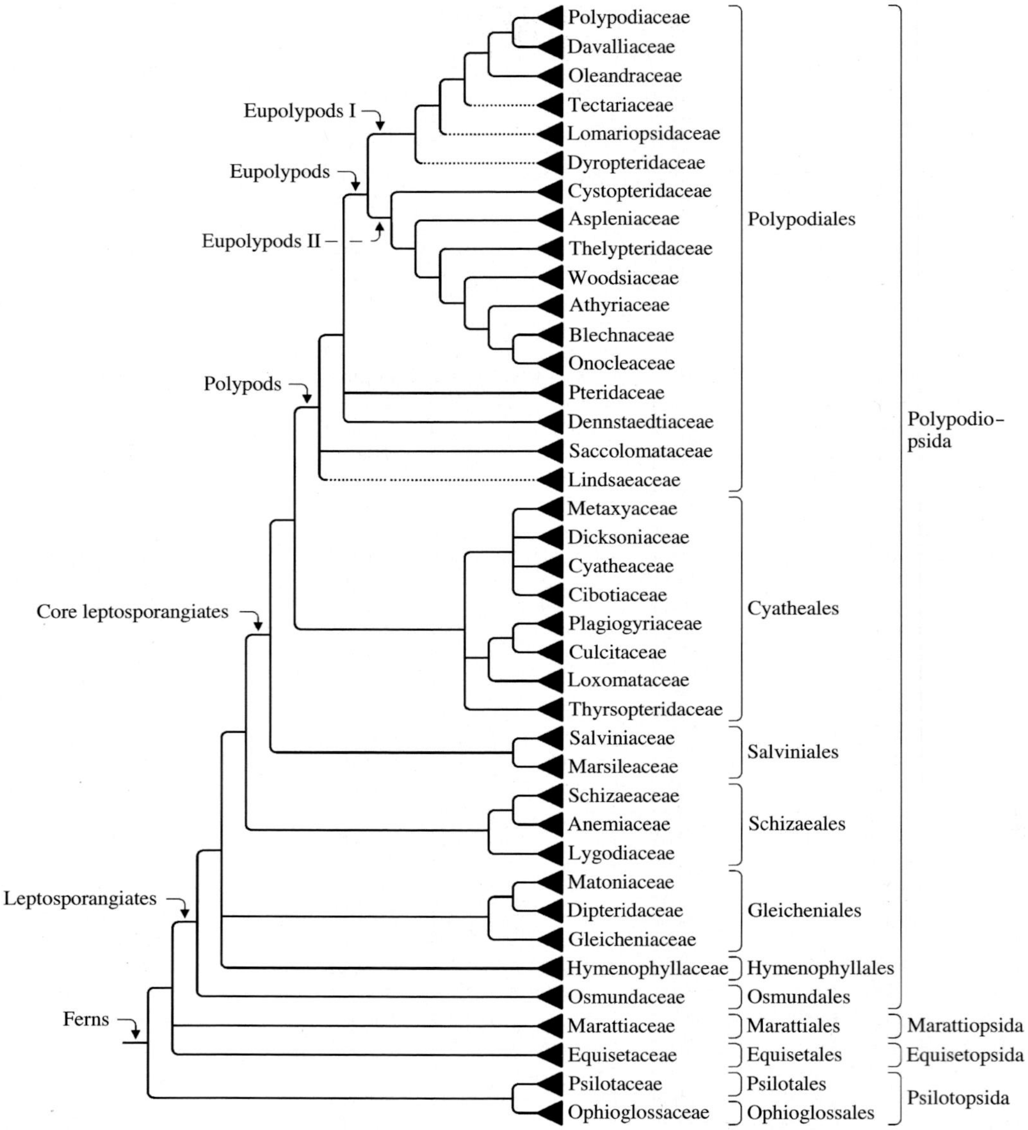

Phylogeny showing fern relationships from Smith et al. 2006 with modifications from Rothfels et al. 2012b.

FAMILIES AND GENERA OF TEXAS FERNS AND LYCOPHYTES (WITH PAGE #S)

Families

Anemiaceae, 97 (FLOWERING FERN FAMILY)
Aspleniaceae, 99 (SPLEENWORT FAMILY)
Athyriaceae, 107 (LADY FERN FAMILY)
Azollaceae, 109 (MOSQUITO FERN FAMILY)
Blechnaceae, 117 (CHAIN FERN FAMILY)
Cystopteridaceae, 122 (BLADDER FERN FAMILY)
Dennstaedtiaceae, 129 (BRACKEN OR CUPLET FERN FAMILY)
Dryopteridaceae, 135 (WOOD FERN FAMILY)
Equisetaceae, 152 (HORSETAIL FAMILY)
Isoëtaceae, 54 (QUILLWORT FAMILY)
Lomariopsidaceae, 161 (SWORD FERN FAMILY)
Lycopodiaceae, 63 (CLUB-MOSS FAMILY)
Lygodiaceae, 163 (CLIMBING FERN FAMILY)
Marsileaceae, 166 (WATER-CLOVER FAMILY)
Onocleaceae, 173 (SENSITIVE FERN FAMILY)
Ophioglossaceae, 177 (ADDER'S-TONGUE FAMILY)
Osmundaceae, 194 (ROYAL FERN FAMILY)
Polypodiaceae, 199 (POLYPODY FERN FAMILY)
Psilotaceae, 205 (WHISK FERN FAMILY)
Pteridaceae, 208 (MAIDENHAIR FERN FAMILY)
Salviniaceae, 288 (FLOATING FERN FAMILY)
Selaginellaceae, 72 (SPIKE-MOSS FAMILY)
Tectariaceae, 293 (HALBERD FERN FAMILY)
Thelypteridaceae, 295 (MARSH FERN FAMILY)
Woodsiaceae, 307 (CLIFF FERN FAMILY)

Genera

Adiantum, 210
Anemia, 97
Argyrochosma, 213
Asplenium, 99
Astrolepis, 218
Athyrium, 107
Azolla, 110
Blechnum, 117
Bommeria, 225
Botrychium, 178
Ceratopteris, 226
Cheilanthes, 228
Cyrtomium, 137
Cystopteris, 123
Dennstaedtia, 130
Dryopteris, 139
Equisetum, 153
Isoëtes, 55
Llavea, 256
Lycopodiella, 65
Lygodium, 164
Macrothelypteris, 296
Marsilea, 167
Nephrolepis, 163
Notholaena, 258
Onoclea, 175
Ophioglossum, 184
Osmunda, 195
Osmundastrum, 197
Palhinhaea, 71
Pellaea, 271
Phanerophlebia, 147
Phegopteris, 296
Pilularia, 172
Pleopeltis, 200
Polystichum, 149
Pseudolycopodiella, 72
Psilotum, 206
Pteridium, 130
Pteris, 285
Salvinia, 289
Selaginella, 74
Tectaria, 293
Thelypteris, 299
Woodsia, 308
Woodwardia, 119